Heterogeneous Catalysis for Energy Applications

Energy and Environment Series

Editor-in-chief:
Heinz Frei, *Lawrence Berkeley National Laboratory, USA*

Series editors:
Nigel Brandon OBE FREng, *Imperial College London, UK*
Roberto Rinaldi, *Imperial College London, UK*
Vivian Wing-Wah Yam, *University of Hong Kong, Hong Kong*

Titles in the series:
1: Thermochemical Conversion of Biomass to Liquid Fuels and Chemicals
2: Innovations in Fuel Cell Technologies
3: Energy Crops
4: Chemical and Biochemical Catalysis for Next Generation Biofuels
5: Molecular Solar Fuels
6: Catalysts for Alcohol-Fuelled Direct Oxidation Fuel Cells
7: Solid Oxide Fuel Cells: From Materials to System Modeling
8: Solar Energy Conversion: Dynamics of Interfacial Electron and Excitation Transfer
9: Photoelectrochemical Water Splitting: Materials, Processes and Architectures
10: Biological Conversion of Biomass for Fuels and Chemicals: Explorations from Natural Utilization Systems
11: Advanced Concepts in Photovoltaics
12: Materials Challenges: Inorganic Photovoltaic Solar Energy
13: Catalytic Hydrogenation for Biomass Valorization
14: Photocatalysis: Fundamentals and Perspectives
15: Photocatalysis: Applications
16: Unconventional Thin Film Photovoltaics
17: Thermoelectric Materials and Devices
18: X-Ray Free Electron Lasers: Applications in Materials, Chemistry and Biology
19: Lignin Valorization: Emerging Approaches
20: Advances in Photoelectrochemical Water Splitting: Theory, Experiment and Systems Analysis
21: Electrochemical Reduction of Carbon Dioxide: Overcoming the Limitations of Photosynthesis
22: Integrated Solar Fuel Generators
23: Magnesium Batteries: Research and Applications
24: Organic Thermoelectric Materials
25: Electrochemical Methods for Hydrogen Production
26: Carbon Capture and Storage
27: Heterogeneous Catalysis for Energy Applications

How to obtain future titles on publication:
A standing order plan is available for this series. A standing order will bring delivery of each new volume immediately on publication.

For further information please contact:
Book Sales Department, Royal Society of Chemistry, Thomas Graham House, Science Park, Milton Road, Cambridge, CB4 0WF, UK
Telephone: +44 (0)1223 420066, Fax: +44 (0)1223 420247
Email: booksales@rsc.org

Heterogeneous Catalysis for Energy Applications

Edited by

Tomas R. Reina
University of Surrey, UK
Email: t.ramirezreina@surrey.ac.uk

and

Jose A. Odriozola
Universidad de Sevilla, Spain
Email: odrio@us.es

ROYAL SOCIETY OF CHEMISTRY

Energy and Environment Series No. 27

Print ISBN: 978-1-78801-718-3
PDF ISBN: 978-1-78801-957-6
EPUB ISBN: 978-1-78801-958-3
Print ISSN: 2044-0774
Electronic ISSN: 2044-0782

A catalogue record for this book is available from the British Library

The Royal Society of Chemistry is a charity, registered in England and Wales, Number 207890, and a company incorporated in England by Royal Charter (Registered No. RC000524), registered office: Burlington House, Piccadilly, London W1J 0BA, UK, Telephone: +44 (0) 20 7437 8656.

For further information see our web site at www.rsc.org

Printed in the United Kingdom by CPI Group (UK) Ltd, Croydon, CR0 4YY, UK

Preface

Heterogeneous catalysis is a cornerstone in the global energy paradigm with practically all energy-related process relying on a catalyst at a certain point. Traditional processes using fossil fuels and new sustainable processes based on green alternatives depend on carefully designed catalytic materials to deliver an energy vector or to achieve the transformation of such an energy vector on a particular source of energy (*i.e.* electrical, chemical, *etc.*).

The central role played by heterogeneous catalysis in traditional and modern energy applications constitutes the genesis of this book. The spirit of this work is to illustrate the progress on heterogeneous catalysis for energy-related applications. This constitutes a hot topic with an increasing number of publications in the open literature and also an increasing effort in terms of funding for catalysis-related projects worldwide. Catalytic reactions constitute the heart of most chemical processes and the application of heterogeneous catalysts will be of paramount importance in achieving the transition to low-carbon, sustainable societies.

This book provides an overview of the design, limitations and challenges of heterogeneous catalysts for energy applications. In an attempt to cover a broad range of scenarios, the book considers traditional processes linked to fossil fuels such as reforming, Fischer–Tropsch and hydrocracking, but it also deals with catalysis for sustainable energy applications such as hydrogen production, photocatalysis, biomass upgrading, biorefineries and conversion of CO_2 to clean fuels. Novel approaches in catalyst design are addressed, including microchannel reactors and structured catalysts, ionic liquids and the catalytic aspects of fuel cells.

The editors of this book have been working on catalysis for energy and environmental applications for more than 10 and 35 years, respectively.

Heterogeneous Catalysis for Energy Applications
Energy and Environment Series No. 27
Edited by Tomas R. Reina and Jose A. Odriozola

Published by the Royal Society of Chemistry, www.rsc.org

At this point of their careers, they have a panoramic perspective of the trends in catalysis, the knowledge gaps and the challenges to be tackled in the near future. We are committed to excellence in academic research and scientific education and we are convinced that we should maintain heterogeneous catalysis at the forefront of the chemical sciences to foster economic prosperity and pursue a more sustainable world.

Overall, this book is an essential toolkit and a relevant contribution to the fascinating area of catalysis. Since catalysis is an interdisciplinary science *per se*, the book is appealing for chemists, physicists, chemical engineers and industrialists working on energy. This multidisciplinary approach makes the book a unique piece of work that was conceived to ignite inspiration among the scientific community in order to address our current global challenges by applying catalytic solutions. The international relevance of this work is also reflected in the multiple nationalities and affiliations within the book chapters' contributors. These include Spain, the UK, Mexico, China, France, Russia, the USA, Colombia, Romania and Italy. The editors would like to express their sincere gratitude to all of the authors for their outstanding contributions. Personally, it has been a pleasure to work with all of them during this journey.

In addition, the editors would like to express their special acknowledgement of Professor Roberto Rinaldi (Imperial College London) who, in his role of Royal Society of Chemistry series editor, encouraged us to take this project forward. Roberto has always been an excellent academic reference and a source of inspiration. We would also like to extend our gratitude to the Royal Society of Chemistry team who have provided wonderful support at every step of the book's production – Connor Sheppard, Robin Driscoll and Sylvia Pegg.

We hope the readers enjoy this inspiring book, which we have edited with great excitement. Progress in catalysis science and energy applications seems to share a common destiny that brings hope for a greener future using renewable sources for clean energy and chemicals production. We invite the readers to reflect on this matter and navigate through the book while sharing our passion for applied catalysis.

Tomas R. Reina

Jose A. Odriozola

Contents

Heterogeneous Catalysis for Energy Applications
Energy and Environment Series No. 27
Edited by Tomas R. Reina and Jose A. Odriozola

Published by the Royal Society of Chemistry, www.rsc.org

CHAPTER 1

Design of Advanced Catalysts for Natural Gas Reforming Reactions

A. C. ROGER* AND K. PARKHOMENKO

University of Strasbourg, ICPEES UMR CNRS 7515, 25 rue Becquerel, Strasbourg, 67087, France
*Corresponding contributor. Email: annececile.roger@unistra.fr

1.1 Introduction

Natural gas is one of the main energy resources that occupies a pivotal role in the global energy paradigm. Natural gas reforming is among the most investigated reactions in heterogeneous catalysis due its direct connection with key industrial processes such as methanol or ammonia synthesis. This chapter will focus on the design of robust reforming catalysts able to overcome the traditional drawbacks of commercial catalysts such as sintering or coking.

1.2 Natural Gas as an Energy Resource

The global consumption of natural gas has very much increased over recent decades.[1] This is mainly due to the industrialization of developing economies in regions with large gas resources (Africa and the Middle East) and to the progressive coal-to-gas switching,[2] especially in the USA and China,

Heterogeneous Catalysis for Energy Applications
Energy and Environment Series No. 27
Edited by Tomas R. Reina and Jose A. Odriozola

Published by the Royal Society of Chemistry, www.rsc.org

which has been initiated to decarbonize energy and decrease energy-related CO_2 emissions.

Gas consumption is still supposed to grow by 50% until 2040, reaching a total volume of 5370 billion cubic metres (Bcm) (see Table 1.1). Although the direct use of gas as a transportation fuel is growing rapidly, this sector remains small, at less than 2% of total consumption in 2017, compared to industry (32%), buildings (21%) and power generation (39%). These four uses imply the combustion of natural gas to generate heat or power. In 2017, only 206 Bcm of natural gas (around 6% of total consumption) was not combusted but chemically converted.

The flow diagram in Figure 1.1 depicts the use or natural gas in 2017. Among the 206 Bcm that has been converted, around 90% has been converted by reforming reactions to hydrogen for ammonia synthesis or petrochemistry. Around 10%, after a step of reforming to produce synthesis gas (syngas), was further converted into methanol through the Fischer–Tropsch (FT) process to produce chemicals or fuels. Note that despite the strong industrial interest in direct routes of methane conversion to fuels and chemicals,[3] this direct use, avoiding the energy-consuming reforming step, is almost equal to 0%.

Table 1.1 Global gas consumption (Bcm) by sector in 1990 and 2017 and forecast for 2040.

Natural gas consumption	*1990*	*2017*	*2040*
Transport	2	55	192
Industry	771	1182	1749
Buildings	530	784	971
Power	543	1443	2109
Non-combusted	103	206	349
Total	1949	3670	5370

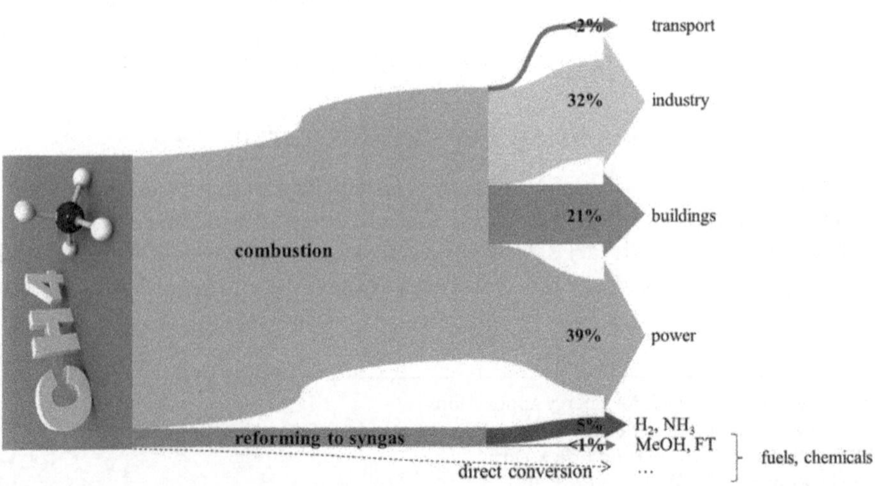

Figure 1.1 Natural gas use by sector in 2017.

Only one oxidative coupling of methane (OCM) industrial unit producing ethylene has been identified to date.[4]

1.3 Thermodynamic Aspects of Natural Gas Reforming

The steam reforming of methane (SRM) (eqn (1.1)) is a mature technology that is used industrially to produce hydrogen from natural gas.[5,6] The syngas thus obtained is compatible with methanol synthesis and the FT process.[7] Syngas can also be generated by dry reforming of methane (DRM), using CO_2 as an oxidizing agent according to eqn (1.2). This process would be of particular interest in the context of the decarbonization of fuels, starting from biogas. The recent considerations around reducing greenhouse gas emissions clearly drove the research for efficient and stable DRM processes.[8]

$$CH_4 + H_2O \rightarrow 3H_2 + CO \qquad \Delta_r H^\circ_{298K} = 206 \text{ kJ mol}^{-1} \tag{1.1}$$

$$CH_4 + CO_2 \rightarrow 2H_2 + 2CO \qquad \Delta_r H^\circ_{298K} = 247 \text{ kJ mol}^{-1} \tag{1.2}$$

$$CO_2 + H_2 \rightarrow H_2O + CO \qquad \Delta_r H^\circ_{298K} = 41 \text{ kJ mol}^{-1} \tag{1.3}$$

$$CH_4 + 0.5\,O_2 \rightarrow 2H_2 + CO \qquad \Delta_r H^\circ_{298K} = -36 \text{ kJ mol}^{-1} \tag{1.4}$$

The steam reforming and the dry reforming reactions are linked by the reverse water gas shift reaction (RWGS) (eqn (1.3)). Both reforming reactions are strongly endothermic and can be combined with the exothermic partial oxidation of methane (POX) (eqn (1.4)) to decrease energy demand[9] and to allow for better thermal control of the catalytic reactor.[10]

The thermodynamics of the reforming reactions imply that they have to be operated at high temperature.[11] Depending on temperature, pressure and inlet composition, the formation of solid carbon, through methane decomposition (eqn (1.5)) or the Boudouard reaction (eqn (1.6)), may be favoured, which would be detrimental to the stability of the catalytic materials.[12]

$$CH_4 \rightarrow C(s) + 2H_2 \qquad \Delta_r H^\circ_{298K} = 75 \text{ kJ mol}^{-1} \tag{1.5}$$

$$2CO \rightarrow C(s) + CO_2 \qquad \Delta_r H^\circ_{298K} = -171 \text{ kJ mol}^{-1} \tag{1.6}$$

For methane steam reforming under a stoichiometric mixture, the maximal formation of carbon is expected at around 600 °C (Figure 1.2). The formation of carbon decreases at higher temperatures, in the range where high conversions are reached. At temperatures higher than 850 °C, almost total conversion can be achieved with close to 100% selectivity.[11] The use of steam to methane (S : C) ratios higher than 1 allows for the total suppression of the formation of solid carbon.[13]

The management of carbon formation is much trickier in the DRM reaction, which is one of the reasons why it has not been operated at an

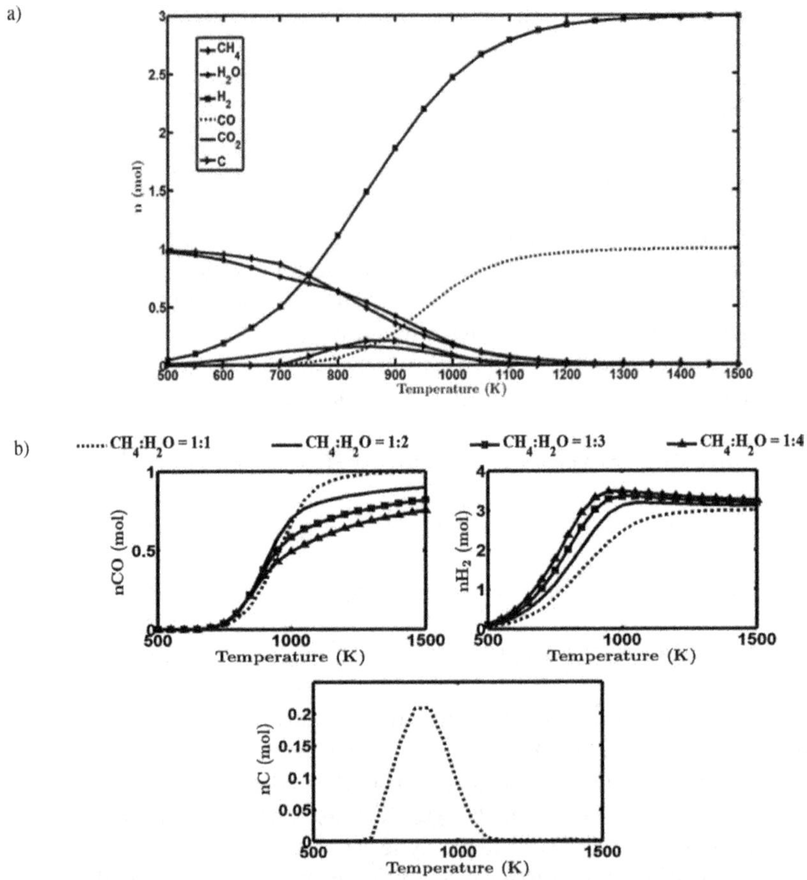

Figure 1.2 Thermodynamic equilibrium for methane steam reforming at atmospheric pressure. (a) Equilibrium composition *versus* temperature for a stoichiometric mixture $CH_4:H_2O = 1:1$. (b) Effect of the inlet ratio on the formation of CO, H_2 and C. Reproduced from Ref. 11 with permission from American Chemical Society, Copyright 2014.

industrial scale. As is shown in Figure 1.3a, the formation of solid carbon is favoured at low temperatures and tends to decrease at higher temperatures. For a stoichiometric inlet mixture $CH_4:CO_2$, the formation of solid carbon is negligible at temperatures above 950 °C. These high temperatures are favourable for high conversions and total selectivities to syngas, but the energy required to operate the reactor under such conditions is prohibitive. Note that high temperatures would, in addition, promote catalyst deactivation by metal phase sintering. An excess of CO_2 in the mixture tends to decrease the temperature above which the solid carbon is no longer favoured (Figure 1.3b).

The zones of thermodynamic stability of solid carbon are illustrated in Figure 1.4.[14]

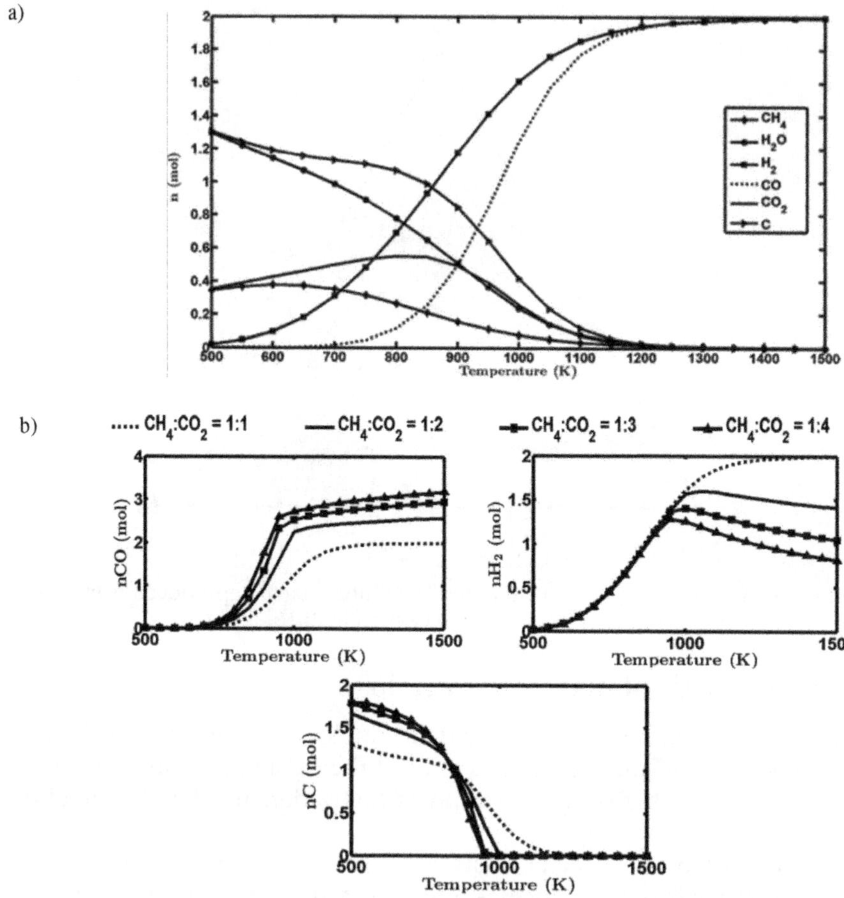

Figure 1.3 Thermodynamic equilibrium for methane dry reforming at atmospheric pressure. (a) Equilibrium composition *versus* temperature for a stoichiometric mixture $CH_4 : CO_2 = 1:1$. (b) Effect of the inlet ratio on the formation of CO, H_2 and C. Reproduced from Ref. 11 with permission from American Chemical Society, Copyright 2014.

1.4 Challenges for Reforming Catalysts: Overcoming Coking and Sintering

The two main problems for the efficient and stable activity of the catalytic materials in methane reforming reactions are the intensive carbon deposition and the active metal sintering, which are promoted by high reaction temperatures needed for thermodynamic and kinetic reasons.

Many researchers all over the world are trying to overcome these issues. Here, a general view on the currently developed strategies is presented combining the approaches for the improvement of catalytic materials, as well as the approaches for the modification of the reforming process itself.

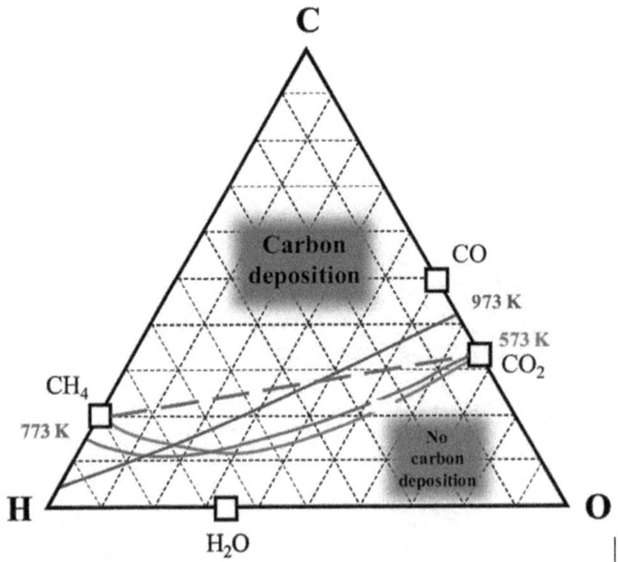

Figure 1.4 C–H–O phase ternary diagram (graphite, 1 bar). Reproduced from Ref. 14 with permission from Elsevier, Copyright 2018.

1.4.1 Catalytic Material Modifications

The essential requirements of catalytic technologies for methane reforming reactions are high conversion rates to react the tremendous amounts of the feed gas, as well as high tolerance against deactivation due to active metal sintering and due to coke deposition.

The supported noble metal catalysts[15,16] are those that present the most attractive catalytic performances in methane reforming, with no strong tendency to deactivation by carbon deposition. Nevertheless, due to their high cost, they are not used at an industrial scale. Ni-based catalytic materials are generally preferred.

As discussed in the previous section, the catalysts for DRM are much more difficult to stabilize against the formation of carbon than are the catalysts for SRM, which can be efficiently operated under an excess of steam. In that latter case, they have to be resistant to sintering, accentuated by the steam and high temperatures.[17]

Based on the mechanism of coke formation, numerous attempts to reduce or to prevent coke deposition have been carried out by many researchers. Many of them compare the effect of using different kinds of catalyst components and supports. Taking into account the catalytic performance and the cost of catalysts, Ni-supported catalysts on alkaline-earth oxides and rare-earth oxides, especially magnesia, have been intensively studied. The intelligent design of Ni-based catalysts for methane reforming reactions should take into consideration the synthesis method, allowing different Ni structures to be obtained, as the geometric structure of the Ni catalysts plays an important role in the carbon deposition rate and carbon type and morphology. For example, simple thermal decomposition of precursor salts for the preparation of Ni reforming catalysts leads to the Ni (1 0 0) and Ni (1 1 0) orientation with many defects, thus leading to intensive carbon deposition during reforming reactions. The recently developed dielectric barrier discharge (DBD) plasma decomposition method mainly forms the Ni (1 1 1) structure with fewer defects, which suppressed carbon deposition.[18] It was shown that carbon formed on the catalysts obtained by the plasma decomposition method was easy to react with O_2, H_2 and CO_2, indicating strong coke resistance.

The new synthetic methods, such as atomic layer deposition of active metal or the plasma decomposition method, still need further development for the design of coke-resistant catalysts. New characterization technologies should be developed first for convenient monitoring of Ni catalysts in real environments (*e.g. in situ* Transmission Electron Microscopy (TEM)). It was proposed[18] to combine the DBD synthesis method with reduction by hydrogen for catalyst preparation as an alternative route.

The structure of the active metal is not the only factor that plays an important role. If the metal–support interface is expanded, the metallic Ni dispersion will be better and the metallic nanoparticle size will be smaller, ensuring superior and stable performance of the reforming catalysts. For Ni/La_2O_3 catalysts, the beneficial metallic Ni particles have a size of 4–6 nm with guarantied stability in DRM due to the suppression of coke deposition.

Figure 1.5 shows schematically coke precursor formation over larger metallic particles but no coke formation on smaller Ni particles.[19] According to theoretical and experimental results, the DRM reaction mechanism on large and small Ni particles is proposed as follows: CH_4 is activated on the Ni surface with the formation of H_2 and activated forms of carbon; CO_2 adsorption forms bidentate carbonates at the interface between Ni and the support; the bidentate carbonates react with activated forms of carbon to form CO. Thus, only Ni in close contact with the support at the interface can avoid carbon deposition. If the Ni metal particles are too large, the activated forms of carbon will accumulate over the metallic Ni surface without close contact with the support and thus form coke

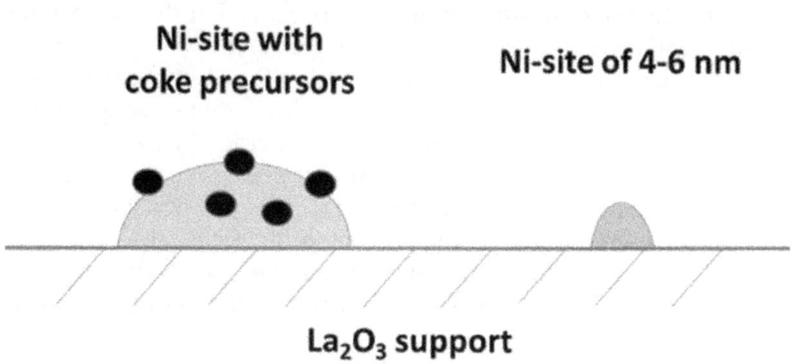

Figure 1.5 Ni/La_2O_3 model catalyst in the DRM reaction.

(carbon fibres, graphite or other forms of coke). Another study confirmed that carbon deposition could be suppressed over the Ni metallic nanoparticles ≤8 nm in size.[20] However, small Ni particles are prone to aggregation at high methane reforming temperatures (≥700 °C) due to its thermodynamic instability. This low-stability issue can be addressed by increasing metal–support interaction or by immobilizing Ni nanoparticles inside porous supports. It would be beneficial to investigate the synergetic effect of small particle sizes coupled with the "confinement effect" using porous supports, which is a promising strategy for developing highly active and stable Ni catalysts with good economic viability.

The method of preparation/activation can strongly affect the stability of the Ni-based catalysts. For Ni/silica systems, it is reported[21] that reduction–oxidation–reduction pre-treatment decreased the average Ni particle size and significantly enhanced catalytic performance in DRM. Accordingly, the initial reduction of Ni precursors after incipient wetness impregnation of silica was efficient at stabilizing catalytic performances, leading to negligible sintering and no carbon deposition.[22]

Stabilization of small Ni metallic nanoparticles could be also achieved by promoting the effect of noble metals, as well as through the modification of the support itself. A recent study[23] showed a comparison of the effects of different noble metals and also the effects of different supports for Ni reforming catalysts. Consequently, an economically promising catalyst for methane reforming reactions is expected to be based on an inexpensive non-noble metal (*e.g.* Ni) while adding a small yet influential amount of noble and/or non-noble metal, noble metals to enhance the catalytic properties and bimetallic materials showing better performance. Additionally, trimetallic catalysts can also produce similar outcomes. Researchers should investigate the alloying and synergetic effect of nickel and cobalt along with the excellent promoting effect of rhodium in Ni-Co-Rh trimetallic catalysts. From the number of different supports, perovskite-type catalysts could stand out from the crowd. Alternatively, a cheaper catalyst without Rh but with strong potential would confine Ni and Co nanoparticles in mesoporous structures, with catalytic properties enhanced by the "confinement effect". The recommendations inspired by this study[23] for methane reforming catalyst design are presented schematically in Figure 1.6.

Regarding the choice of support for the active metal, it should be noted that anti-coking abilities can be enhanced by the presence of basic sites on the surface of the catalyst, promoting CO_2 dissociation for DRM of super-reforming reactions, for example. Metal oxide supports with higher basicity can alleviate the deactivation of the Ni surface due to coke formation by promoting CO_2 dissociation at basic sites. They can provide the oxygen atoms to neighbouring Ni sites covered with carbon species and thus facilitate gasification of the carbon species from the metal surface, thus regenerating the catalytically active site for reforming reaction.[20] Moreover, it has been reported that the promotion of alumina by alkali or alkaline earth was beneficial to protecting the Ni particles from sintering.[24]

One more promising strategy for the support choice in the design of highly active and stable Ni-based catalysts is the use of mesoporous materials, as was mentioned above. A porous support represents a synergism of the

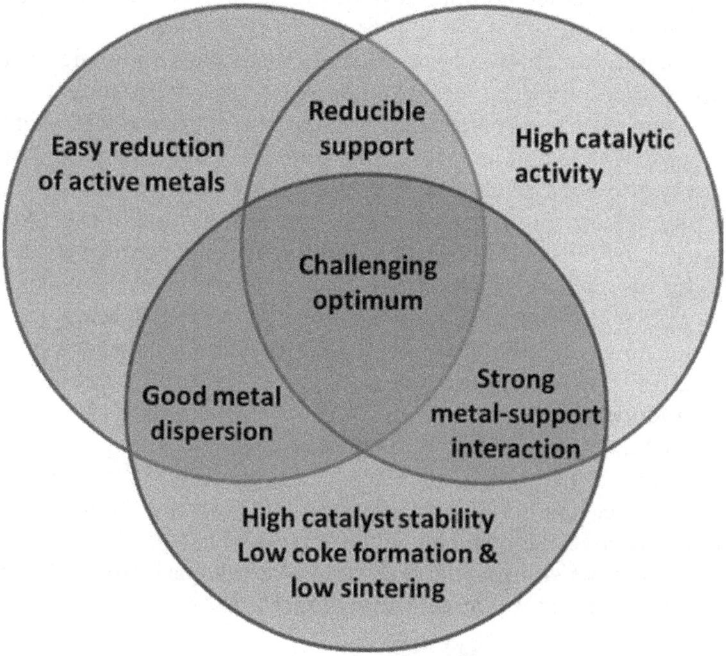

Figure 1.6 Summary of methane reforming catalyst design.[23]

"confinement effects" characteristics: control of metallic particles' small size and suppression of coke due to the restricted pore volume of porous materials, and more precisely of mesoporous oxides,[25] SBA-15 or different core–shell or yolk–shell composite catalytic materials. Comparing catalysts where Ni is encapsulated in inner cavities of different diameters ranging from 5 to 50 nm, a larger inner cavity space is of vital importance to provide sufficient inner room for the survival of small, independent and highly active Ni core particles, as well as to the elimination of absorbed carbon intermediates over the Ni site under very high space velocity conditions. Most importantly, the unique confined structure of a shell with a small radius dimension ($\leq$7 nm) can therefore act as a hard anti-coking framework. This important feature facilitates high coking resistance for the reforming processes with high space velocity. However, a cautionary point is that despite the advantages of inner space, an overly spacious interior cavity could reduce any anti-coking ability. In other studies, it was confirmed that the cavity space in the yolk–shell structures plays an important role in the elimination of coke precursors from the Ni metallic yolk (core), while at the same time preserving the metal yolk particles from sintering.[26] Various shell materials, including SiO_2, zeolite, CeO_2 and Al_2O_3, have been studied, and various synthetic methods were used for active metal incorporation, such as sol–gel, micro-emulsion, self-assembly, precipitation, atomic layer deposition, physical mixing, *etc.* In general, the core/yolk–shell catalysts exhibited better hydrocarbon (not only methane) reforming performance compared with the supported counterparts due to the "confinement effect".

Even though the core–shell and yolk–shell catalysts show high selectivity to certain desired products, the conversions may be greatly affected.[27] A common and known problem of these structures is the high mass transfer barrier, especially when the shell has worm-like pores of small pore size. Therefore, the porosity and shell thickness should be tuned to achieve both high conversion and selectivity. In terms of the practical application of core/yolk–shell heterogeneous catalysts, they are unfortunately regarded as being less effective for the mass-scale production of catalysts.

The rapid deactivation and high coke formation rate associated with non-noble-based monometallic Ni catalysts, coupled with the expensive price of the more coke-resistant and more stable noble metal-containing catalysts, dictate that new and innovative approaches for catalyst design must be investigated to overcome these issues. In general, there is a need for a catalyst that is active with small particle size, well dispersed over a support with basic character and porous structure, has good metal–support interface but nonetheless good reducibility of active metal and has a certain tolerance to carbon formation.

The following is a summary of recommendations for further efficient catalyst development for methane reforming reactions:

- Innovative synthetic method preferentially allowing Ni (1 1 1) to be obtained.
- Control metallic Ni particle size to $\leq$8 nm.
- Ni-Co bimetallic and Ni-Co-Rh trimetallic catalysts seem to be most promising.
- Perovskite-type supports stand out from the crowd.
- Prefer supports with high basicity.
- Porous supports are desired for enhancing the "confinement effect" for Ni particles.

1.4.2 Engineering Solutions

Methane transformation processes vary by their energetic character (exo- or endo-thermicity). Thus, the reaction could be directly affected by modification of such operating parameters as reaction temperature, gas hour space velocity (GHSV), reactor shape, packing loading and others. To obtain higher CH_4 conversions and increase the catalyst's lifespan, it is mandatory to have an optimum set of the conditions. Temperature range and GHSV will not be discussed here; instead, this section will focus on recent developments in natural gas transformation technologies.

On the example of DRM as a promising technology for syngas production, a new *reactor packing* method was developed.[28] Different packing strategies allow the maintenance of the reaction temperature along the reactor within the specified range, as well as the conversion in an acceptable range for $\pm$20% change in the feed flow rate. The proposed design can maintain the temperature in the region where coke is rarely generated.

Another efficient strategy for suppressing coke formation could be *admixture* of the feed with other different hydrocarbons. In thermo-catalytic decomposition of methane for the production of pure hydrogen, the main drawback limiting its industrial applications is catalyst deactivation due to carbon formation. Thus, co-feeding of CH_4 with other molecules such as C_6H_6, H_2S, CO_2, alkanes and alcohols can be utilized as an efficient method for significantly improving H_2 production yields and for switching the coke formed towards more active forms of carbon.[29] The co-feeding of methane with other feedstock molecules increases the production of more active carbon on the catalyst surface as compared to carbon eliminated from methane, and it is an effective solution to overcome rapid catalyst deactivation. The activity originating from different molecules is in the order $C_{benzene} > C_{acetylene} > C_{ethylene} > C_{propane}$ and $C_{methane}$.[30] Limited work has been done on co-feeding of methane with other hydrocarbons to stabilize catalyst activity and deactivation time.

Sorption-enhanced reforming of methane is the combination of the reforming process and the CO_2 abstraction reaction to yield high-purity hydrogen. For implementation of this strategy, a bi-functional, highly active and stable material is always demanded, usually based on CaO in the mixture with $Ca_3Al_2O_6$ or other efficient CO_2 sorbents. The sorption–desorption cycling of the process in this case provides greater H_2 production together with little coke formation and almost completely suppressed active metal sintering.[31]

An emerging alternative to conventional technologies is CH_4 reforming based on *chemical looping*, which is a process designed mostly for syngas production. This is an option that is applicable in a diverse range of processes and could be divided into sub-reactions.[32] The link between the sub-reactions is a solid mediator, usually in the form of a metal oxide (*e.g.* Fe_3O_4, CeO_2, NiO, CuO), which is reduced and re-oxidized in a cyclic progression of the sub-reactions.[33] Figure 1.7 shows the scheme of the chemical looping for DRM, where the process globally corresponds to DRM, although in practice it consists of two independent gas–solid reactions:

$$CH_4 + \text{O-vector} \rightarrow 2\,H_2 + CO + \text{R-vector},$$

$$CO_2 + \text{R-vector} \rightarrow CO + \text{O-vector}.$$

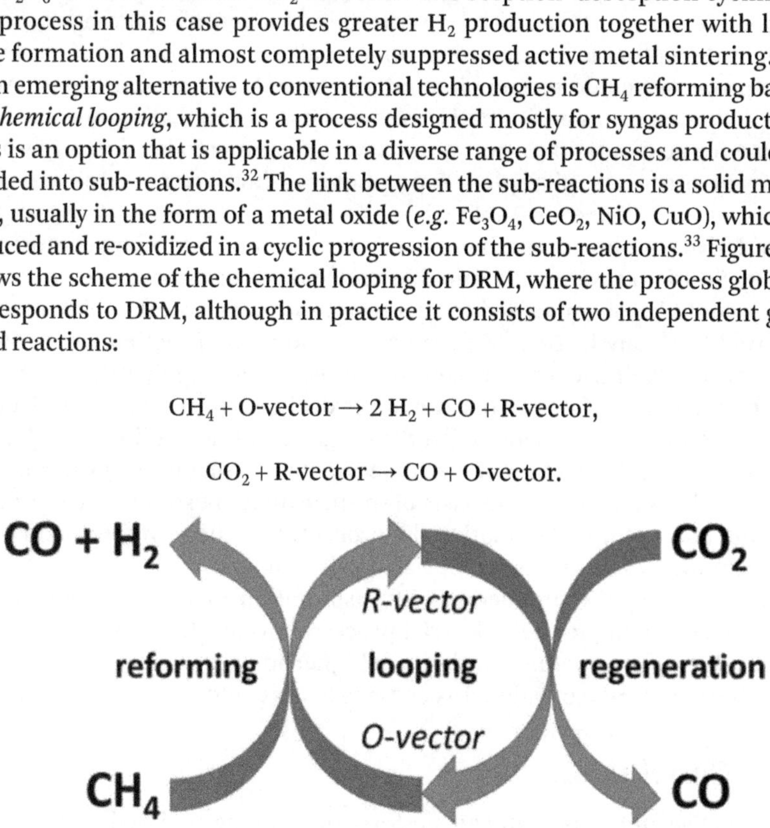

Figure 1.7 Schematic illustration of the chemical looping in the example of DRM where the O-vector is oxidized catalyst and the R-vector is reduced catalyst.

The results of DRM chemical looping showed that carbon formed during exposure to methane can be fully removed during the re-oxidation by CO_2.[34] This technology allows for the avoidance of carbon accumulation and catalyst deactivation.

To equilibrate the energetic balance of methane reforming reactions for syngas or for hydrogen production, a combination of endothermic and exothermic reactions is used – so-called bi-reforming, co-reforming, tri-reforming or super-reforming.[35] Even a low concentration of O_2 in the feed gas dramatically suppresses carbon deposition. The simultaneous use of O_2 and H_2O may not only decrease the formation of carbon, but also gives an appropriate H_2:CO molar ratio to match the downstream use of the produced syngas. Additional combination with a membrane reactor has the advantages that the reaction can be operated at much lower temperatures and CO_x-free hydrogen can be obtained. This is favourable for use in fuel cells as an ideal small-scale energy system. The catalytic results using the membrane approach are still problematic because they are limited by the intrinsic performance of the catalysts and by diffusion across the membrane.[36]

From an industrial point of view, the natural gas transformations in syngas and in pure hydrogen are characterized by severe economies of scale: plants must be very large. Thus, direct approaches for converting methane to valuable products that are economical on a small scale are of extreme interest. Such processes could rescue bio-methane or so-called stranded natural gas (flared, wasted or unused) that is produced in too small amounts and is too remote for economic transportation. Several commercialization attempts of emerging processes, pilot plant experiments or both for methane activation processes are collected in Table 1.2.[37]

Oxidative coupling of methane to ethylene is currently being commercialized by Siluria. Ethylene is a valuable molecule of interest; as an alternative use, it could be oligomerized to fuel-type molecules. Commercial attempts reported by Catalytic, Inc., of Pt-catalysed oxidation of methane with SO_3 in concentrated sulfuric acid or oleum resulted in the highly selective formation of methylbisulfate.[38] This can be hydrolysed to methanol and under well-chosen conditions can be produced at rates that are in the industrial processes range.[39] Reaction of methane with SO_3 in oleum with substantially different chemistry is also the basis of the new methanesulfonic acid process; Methion has attempted to market the process.[40] Methanesulfonic acid could replace mineral acids in many special applications.

Currently, only the production of transportation fuels has a demand at a level corresponding to the industrial processes scale. However, other chemicals produced from methane with much higher added value compared to fuels and other chemicals considered as energy vectors could be attractive options.[41]

1.5 Conclusion

Even if active and stable catalytic performance can be achieved from the process of steam reforming of methane to produce syngas, the limitations of the

Table 1.2 Making more from methane: different pathways of methane transformation.

	Reaction	*Initial product*	*End product*	*Companies*
Classical Large-scale	Steam reforming (H_2O, Ni/alumina) Partial oxidation (O_2)	Syngas	H_2 NH_3 Methanol Higher hydrocarbons	Several worldwide
Laboratory-scale	Dry reforming (CO_2, catalyst)	Syngas	Methanol Higher hydrocarbons	
Laboratory-scale	Partial reforming (O_2, catalyst)	Syngas	Methanol Higher hydrocarbons	
Commercialization phase	Oxidative coupling (O_2, catalyst)	Ethylene	Higher hydrocarbons	Siluria
Commercialization attempts	Sulfonation (SO_3, catalyst)	Methylbisulfate Methanesulfonic acid	Methanol Replacement of mineral acids	Catalytica Methion
Laboratory-scale	Membrane reaction Chemical looping	H_2 free from CO_x		
Laboratory-scale	Thermo-catalytic decomposition	Pure hydrogen		

development of dry reforming of methane are mainly due to rapid deactivation by carbon deposition.

Despite extensive work on the design and stabilization of DRM catalysts, coupled with engineering strategies to overcome deactivation, the industrial development of DRM seems hardly achievable.

Direct conversion of methane to chemicals and fuels remains the best strategy in the future to avoid the intermediate stage of reforming.

References

1. BP, Natural gas, https://www.bp.com/en/global/corporate/energy-economics/energy-outlook/demand-by-fuel/natural-gas.html (accessed Jan 2020).
2. *The Role of Gas in Today's Energy Transitions*, World Energy Outlook special report, IEA 2019.
3. V. Swarup, ExxonMobil, Plenary lecture "Energy Outlook: Role of Natural Gas", presented at NGCS12, San Antonio, TX, USA, June 2019.
4. Silura, Oxidative Coupling of Methane, http://siluria.com/Technology/Oxidative_Coupling_of_Methane (accessed Jan 2020).
5. J. R. Rostrup-Nielsen, *J. Catal.*, 1973, **31**, 173.
6. J. R. Rostrup-Nielsen, *Catal. Sci. Technol.*, 1984, **5**, 1.
7. G. Prieto, *ChemSusChem*, 2017, **10**, 1056.

8. W. J. Jang, J. O. Shim, H. K. Kim, S. Y. Yoo and H. S. Roh, *Catal. Today*, 2019, **324**, 15.
9. M. S. Challiwala, M. M. Ghouri, P. Linke, M. M. El-Halwagi and N. O. Elbashir, *J. CO2 Util.*, 2017, **17**, 99.
10. M. M. V. M Souza, O. R. Macedo Neto and M. Schmal, *J. Nat. Gas Chem.*, 2006, **15**(1), 21.
11. M. M. B. Noureldin, N. O. Elbashir and M. M. El-Hawagi, *Ind. Eng. Chem. Res.*, 2014, **53**, 1841.
12. C. H. Bartholomew, *App. Catal. A Gen.*, 2001, **212**, 17.
13. C. H. Bartholomew, *Catal. Rev. Sci. Eng.*, 1982, **24**(1), 67.
14. T. Yabe and Y. Sekine, *Fuel Process Technol.*, 2018, **181**, 187.
15. J. R. Rostrup Nielsen and J. H. B. Hansen, *J. Catal*, 1993, **144**, 38.
16. J. Wei and E. Iglesia, *J. Phys. Chem. B*, 2004, **108**, 4094.
17. J. Sehested, *J. Catal.*, 2004, **223**, 432.
18. X. Yan, B. Zhao, Y. Liu and Y. Li, *Catal. Today*, 2015, **256**, 29.
19. K. Li, X. Chang, C. Pei, X. Li, S. Chen, X. Zhang, S. Assabumrungrat, Z. J. Zhao, L. Zeng and J. Gong, *App. Catal. B Env.*, 2019, **259**, 118092.
20. H. O. Seo, *Catalysts*, 2018, **8**, 110.
21. E. C. Lovell, J. Scott and A. Fuller, *App. Catal. B Env.*, 2016, **199**, 155.
22. Z. Bian and S. Kawi, *J. CO2 Util.*, 2018, **27**, 238.
23. N. A. K. Aramouni, J. G. Touma, B. A. Tarboush, J. Zeatiter and M. N. Ahmad, *Renewable Sustainable Energy Rev.*, 2018, **82**, 2570.
24. I. Chen and F. L. Chen, *Ind. Eng. Chem. Res.*, 1990, **29**(4), 534.
25. S. Y. Kim, B. J. Cha, Sh; Saqlain, H. O. Seo and Y. D. Kim, *Catalysts*, 2019, **9**, 266.
26. C. Wang, X. Jie, Y. Qiu, Y. Zhao, H. A. Al-Megren, S. Alshihri, P. P. Edwards and T. Xiao, *App. Catal. B Env.*, 2019, **259**, 118019.
27. Z. Li, Z. Wang and S. Kawi, *ChemCatChem*, 2019, **11**, 202.
28. J. Lee, B. Kim and M. Han, *Ind. Eng. Chem. Res.*, 2019, **58**, 17433.
29. A. F. S. Muhammad, A. Awad, R. Saidur, N. Masiran, A. Salam and B. Abdullah, *Int. J. Hydrogen Energy*, 2018, **43**, 18713.
30. A. Malaika and M. Kozlowski, *Int. J. Hydrogen Energy*, 2010, **35**, 10302.
31. S. M. Kim, P. M. Abdala, D. Hosseini, A. Armutlulu, T. Margossian, Ch. Coperet and Ch. Muller, *Catal. Sci. Technol.*, 2019, **9**, 5745.
32. J. Adanez, A. Abad, F. Garcia-Labiano, P. Gayan and L. F. de Diego, *Prog. Energy Combust. Sci.*, 2012, **38**(2), 215.
33. J. Hu, L. Buelens, S.-A. Theofanidis, V. V. Galvita, H. Poelman and G. B. Marin, *J. CO2 Util.*, 2016, **16**, 8.
34. A. Lofberg, J. Guerrero-Caballero, T. Kane, A. Rubens and L. Jalowiecki-Duhamel, *App. Catal. B Env.*, 2017, **212**, 159.
35. L. C. Buelens, V. V. Galvita, H. Poelman, Ch. Detavernier and G. B. Marin, *Science*, 2016, **354**, 449.
36. J. J. Spivey and T. Inui, *Catalysis*, 2002, **16**, 133.
37. F. Schuth, *Science*, 2019, **363**, 1282.
38. R. A. Periana, D. J. Taube, S. Gamble, H. Taube, T. Satoh and H. Fujii, *Science*, 1998, **280**, 560.

39. T. Zimmermann, M. Soorholtz, M. Bilke and F. Schuth, *J. Am. Chem. Soc.*, 2016, **138**, 12395.
40. A. K. Richards, Patent WO 2005069751A2, 2003.
41. C. D. Elvidge, M. D. Bazilian, M. Zhizhin, T. Ghosh, K. Baugh and F.-C. Hsu, *Energy Strateg. Rev.*, 2018, **20**, 156.

CHAPTER 2

Natural Clay Minerals for Hydrocracking Reactions

JAHAZIEL AMAYA, SONIA MORENO AND RAFAEL MOLINA*

Estado Sólido y Catálisis Ambiental (ESCA), Departamento de Química, Facultad de Ciencias, Universidad Nacional de Colombia, Kra 30 No. 45-03, Bogotá, Colombia
*Corresponding contributor. Email: ramolinag@unal.edu.co

2.1 Introduction

The sources to sustain humanity's energy consumption are currently focused on three primary energy resources: fossil fuels, nuclear energy and renewable energy. However, the current global energy basket clearly indicates that the dependence on fossil sources (oil, coal and natural gas) is by far the most important, at about 87% (Figure. 2.1)[1] and, on the other hand, that the rate growth of alternative energy is still marginal.[2] This scenario suggests that the oil refinery industry will continue to grow and an abrupt decline in its production could be the cause of an energy crisis in the near future. Indeed, by 2025, the US Department of Energy projects a demand of roughly 125 million barrels per day (MBD), while the Organization of the Petroleum Exporting Countries (OPEC) projects an offer of less than half (61 MBD).[3] Consequently, greater efforts are required to implement the use of new energy sources in the near future, but at the same time it is urgent and

Heterogeneous Catalysis for Energy Applications
Energy and Environment Series No. 27
Edited by Tomas R. Reina and Jose A. Odriozola

Published by the Royal Society of Chemistry, www.rsc.org

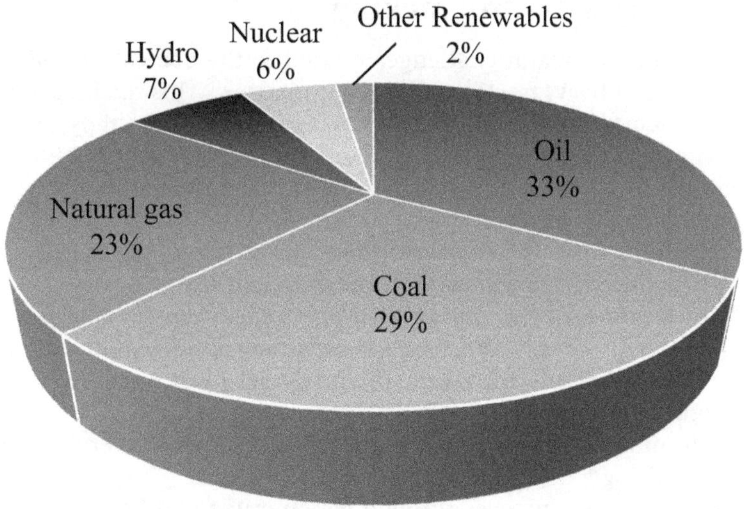

Figure 2.1 World energy panorama.[1]

indispensable to improve the reasonable use of current fossil fuels with an emphasis on achieving environmentally appropriate processes for the transformation of heavy crudes.

In this scenario, the specific and well-known hydrocracking (HCK) processes are of particular interest, where the design and optimization of catalysts that respond in the best way to current requirements should occupy an urgent and central place in scientific research and technological development.

The use of catalysts for HCK has been the subject of extensive studies, among which mainly different types of zeolites are reported and, to a lesser extent, silica-alumina with different variables in its composition, used primarily as a support that provides acidic function and additional characteristics for the control of molecular traffic. Many of these catalysts have been brought to industrial use and recorded in broad reports over the past two decades.

On the other hand, natural clay minerals have attracted attention, particularly in recent years, for various catalytic reactions, including hydroconversion, not only because of their abundance, low cost,[4] modular catalytic properties[5] and because they are environmentally friendly, but also mainly because these qualities are united by great advances in synthesis methods that overcome the obstacles of the past and today offer viable alternatives for stacking at the industrial level. However, systematic and rigorous studies are still needed to evaluate the possible positioning of these natural materials in industrial HCK processes. To contribute in this regard, this chapter addresses the main aspects of the current theoretical framework of HCK processes and the respective reported catalytic systems, paralleling the beneficial properties of natural clay minerals as potential catalytic supports for this process.

2.2 Catalytic Hydrocracking

The refineries face major challenges related to the decrease in oil reserves and the waste of heavy fractions for low-profit uses.[6] This requires evaluating and permanently improving processes and treating increasingly heavy loads, which require the reconfiguration or adaptation of their treatments to achieve a greater benefit in generating cleaner fuels within the framework of current environmental regulations.

The challenge of a better use of the units that process the heavy loads implies a knowledge and control of the acid and textural properties of the solid, which allow for innovating with new generations of catalysts for the corresponding process.[6,7] There are a wide variety of conventional treatments that are adjusted regularly to make the processes involved in the conversion of light crudes even better and thus achieve the most appropriate conditions to take advantage of heavy crudes. These processes include thermal cracking,[8] catalytic cracking[9,10] and HCK.[11]

Catalytic HCK is a process through which crude and heavy fractions are transformed into liquid fractions with molecules of lower molecular weight, significantly preventing coke formation. The process includes cracking, hydrogenation reactions and some hydrotreatment reactions (HDS hydrodesulfurization [HDS], hydrodenitrogenation [HDN] and hydrodemetalation [HDM]) to remove contaminants (such as sulfur, nitrogen and metals)[12–15] present in the feed.

The type of feed for HCK covers a range that can go from vacuum funds or waste to atmospheric waste, and the products can range from heavy diesel to naphtha. However, the main industrial use of HCK is the conversion of vacuum distillates whose boiling point range is between 330 and 550 °C. The process is also characterized by requiring the presence of hydrogen (H_2) at high pressures, high temperatures and the use of bifunctional catalysts comprising a hydrogenating/dehydrogenating (H/DH) function and an acid function.

The conditions under which HCK is carried out at the industrial level depend on factors such as the type of oil fed or loaded, the design of the reactor used (usually fixed-bed down-flow reactors) and the desired products. These conditions generally correspond to hydrogen pressures between 1250 and 2900 psi (85 and 200 bar) and temperatures between 300 and 450 °C (570 and 840 °F).[12,16]

2.2.1 Catalytic Hydrocracking Chemistry of Materials

Due to the physicochemical complexity and impurities of the feeds, the catalytic HCK process usually comprises two reaction stages: (1) treatment (also called pre-treatment) and (2) HCK as such.[12,16,17] Chemically, there are no boundaries between hydrotreatment and HCK, since they use similar temperatures and pressures. Once the hydrotreatment process is carried out, the next stage is HCK, where smaller molecules are formed from larger ones that can be grouped into different categories,[18] usually paraffins, naphthenes and aromatics.

2.2.1.1 *Pre-treatment or Treatment Reactions*

In hydrotreatment, the conversion focuses on the removal of sulfur (HDS), nitrogen (HDN), oxygen (HDO) and metals (HDM),[19–30] which are eliminated in the form of H_2S, NH_3 and H_2O.[12,16] HDS removes the sulfur contained in the hydrocarbons from the load without prior hydrogenation (direct route) or with it (indirect route).[31] In HDN, the organic nitrogen is removed from the charge, which is almost entirely based on aromatic and non-aromatic rings.[31,32] This is undertaken by means of a fast mechanism, which needs seven hydrogen molecules per nitrogen atom, or a slow one, which uses four hydrogen molecules per nitrogen atom. Finally, the HDO reaction, which is carried out before the cracking of unsaturated cyclic hydrocarbons, is necessary because the aromatic rings are hardly cracked directly under the reaction conditions, and the naphthenes resulting from their hydrogenation are easily cracked.[16]

2.2.1.2 *Hydrocracking Reaction*

HCK comprises a set of reactions whose main objective is the breaking of C–C bonds, which is carried out following a bifunctional mechanism that requires a catalyst with two active sites of different natures to catalyse separate steps in the reaction sequence. Among these are hydrodecyclization, isomerization and dealkylation of aromatic and naphthenic rings, among others.[12,16]

The active sites of the bifunctional catalyst comprise an H/DH function, which usually corresponds to transition metals responsible for the hydrogenation (H) and dehydrogenation (DH) of the hydrocarbons involved in the reaction, and an acid function, usually provided by the support, where cracking occurs.[33–38]

2.2.1.2.1 Paraffin Hydrocracking. The understanding of the HCK mechanism and the influence of bifunctional catalysts has been studied more extensively in the presence of repeating model compounds of petroleum fractions, where paraffins are the most used group of molecules.[39,40]

The reaction mechanism of the HCK of paraffins on a bifunctional catalyst begins with the generation of an olefin or a cyclo-olefin product of its dehydrogenation on the H/DH site. Next, the olefin migrates to the acid site, where it is converted into a carbocation (carbenium ion) by adding a proton (step 1). The carbocation may undergo a rearrangement of its structure to produce a more stable branched isomer, or it may undergo a β cleavage to form a smaller carbenium ion and a smaller olefin (step 2). The branched carbocation resulting from the rearrangement of the structure can migrate to an H/DH site to form a branched alkene (step 3) or it can be cracked by a β cleavage. Finally, cracking and isomerization products are desorbed from acid sites and hydrogenated to form the corresponding alkenes.[11,31,36,40]

2.2.1.2.2 Hydrocracking of Aromatic Molecules. The mechanism described for the HCK of paraffins can be applied to that of aromatic molecules. However, the rupture of the C–C bonds that are part of the naphthenic rings have a greater resistance to β cleavage, which makes them less reactive compared to alkanes.[41] For aromatics with less than 10 carbon atoms, it is very common to find triramified compounds associated with consecutive isomerization reactions in the HCK products. As for aromatics with 10 carbon atoms or more, in addition to isomerization reactions in which contractions or ring expansions can occur, after obtaining triramified isomers, β cleavages of exocyclic C–C bonds can occur.[42] This type of reaction leads to the selective formation of methyl-cyclopentane (C_6) and isobutane (C_4).[41]

Monoaromatic Molecules: These correspond to unsubstituted benzenes and those substituted with alkyl chains. In these molecules, the alkyl chains that replace the benzene ring undergo the paraffin reaction mechanism. Some reactions such as dealkylation are favoured, where branching occurs on the α-carbon with respect to the aromatic ring, because the carbocation formed is secondary or tertiary (more stable).[12] The reactions are followed by HCK on the naphthene formed after saturation of the aromatic ring.[12,17,31]

Polyaromatic Molecules: These correspond to polyaromatic nuclei that contain at least two benzene rings. Studies related to the HCK of polyaromatic compounds include both unsubstituted and substituted molecules, the latter being important for the understanding of what happens in aliphatic chains, which link polyaromatic fragments into molecules such as asphaltenes. Feeds as vacuum bottoms may contain compounds with at least seven benzene rings in a molecule. HCK of this type of complex ring in smaller molecules can follow the following sequence of events:[43] (1) saturation of an aromatic ring; (2) removal of lateral paraffinic chains (dealkylation); (3) isomerization of paraffins and naphthalene aromatic rings; and, finally, (4) HCK of paraffins including large side chains. However, HCK of polyaromatic compounds follows complex reaction paths in which hydrogenation reactions occur and, in a parallel or consecutive manner, isomerization, ring opening and dealkylation reactions.[44]

In the case of aromatic nuclei, phenanthrene is one of the polyaromatic compounds that has been used for the understanding of HCK reactions in heavy petroleum fractions.[45,46] This molecule has been reported in the presence of catalysts based on noble metals, obtaining hydrogenation of the three aromatic rings. This reaction is so rapid that the reagent should be considered as the naphthenic molecule (perhydrofenanthrene) and, therefore, the isomerization, ring opening and cracking reactions can be present to a greater extent compared to products obtained in the presence of sulfides.[47] In the case of the use of transition metal sulfides as catalysts, the complete hydrogenation product of the phenanthrene (perhydrofenantrene) is not obtained.[44,45]

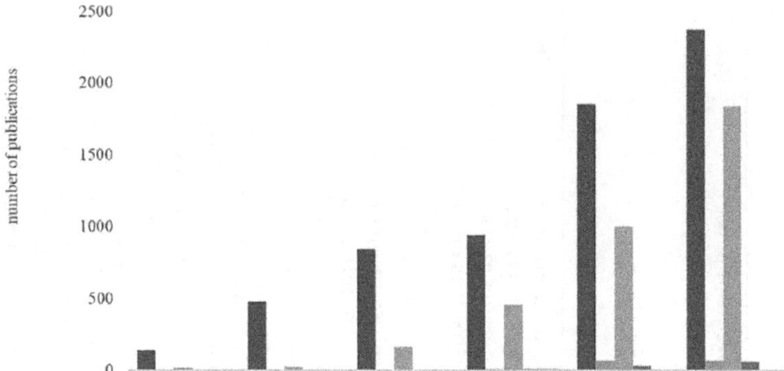

	1960-1969	1970-1979	1980-1989	1990-1999	2000-2009	2010-2019
"hydrocracking all documents"	143	484	848	944	1856	2373
"hydrocracking-reviews"	1	0	2	9	66	66
"clays" "as" "catalysts all documents"	20	22	166	460	1004	1842
"clays" "as" "catalysts"-"reviews"	0	0	2	9	31	62
"clays" "in" "hydroconversion"	0	0	2	10	6	6

Figure 2.2 Trends in the number of publications with reference to "hydrocracking", hydrocracking reviews, "clay" "as" "catalysts", "clays" "as" catalysts" – "reviews" and "clays" "in" "hydroconversion" in the last six decades. Note: no reviews were found in the bibliographic sources corresponding to "clays" "in" "hydroconversion". Adapted from Scopus.

2.3 Catalysts for Hydrocracking

Figure 2.2 illustrates the evolution of published scientific documents on HCK, showing a notable increase particularly in the last two decades, where more than 4200 documents and 130 reviews were published corresponding to approximately 63% and 91%, respectively, of all publications are registered since the 1960s. This marked interest in the subject could be attributed, among other aspects, to the energy panorama described above and, consequently, to the increasingly indispensable increase in transforming heavy crudes, in which HCK constitutes a fundamental stage.

Table 2.1 records a selection, in our opinion, that is representative of the catalysts and catalytic HCK processes reported in recent years. Ultra-dispersed systems generate a very interesting route, particularly using MoS_2 with Ni or Co as promoters at temperatures between 200 and 500 °C and hydrogen pressures between 6 and 25 MPa for catalytic cracking of heavy fractions (vacuum residue, heavy oil and bitumen, among others). The most commonly used supports to generate suitable acidic and textural properties are generally zeolites (mainly HZSM-5, HY, HUSY and MCM-41) and silica-alumina. The behaviour of these same supports with noble metals (Pt or Pd) or transition metals (mainly of the Co-Ni-MoS_2 type) at hydrogen pressures between 1 and 40 MPa and temperatures between 250 and 500 °C demonstrates potentially good performances for the catalytic cracking of heavy fractions of the same type as those already mentioned, or allows for milder conditions for model molecules (such as benzene or toluene, H of 1-methylnaphthalene and naphthalene, among others).

Table 2.1 Catalytic hydrocracking.

System	Catalyst (metal-support)	Type of reaction (HC = hydroconversion; HK = hydrocracking; HI = hydroisomerization; H = hydrogenation; HT = hydrotreatment)	Reaction conditions P (MPa); T (°C)	Conversion (C)	Interpretations	Ref.
Ultra-disperse	M = MoS_2 or MoS_2-ASA	HK (phenanthrene)	P_{H2} = 5 MPa T = 300 °C t = 4 h	C = 60–70% to 300 °C	An opposite effect of temperature, degree of dispersion and sulfuration of MoS_2 was demonstrated. In addition, the presence of the catalysts affected by the addition of ASA and promoted cracking, isomerization and ring opening products	46
	M = MoS_2 S = ASA	HK (phenanthrene and decalin)	P_{H2} = 2.8 MPa T = 400 °C t = 4 h	C = 40–50% to 300 °C of phenanthrene and decalin	The catalyst retained the advantages of the unsupported MoS_2 with the additional functionality of the acid component of the bifunctional catalyst; hence, conversion was promoted due to both hydrogenation and hydrocracking reactions	167
	M = NiMo, MoS_2, FeMo, CoMo	HK (PFO, vacuum residue, heavy oil, bitumen of Athabasca, oil of Wyoming)	P_{H2} = 6–17 MPa T = 200–500 °C	C = 80–90% $Y_{Liq. prod.}$ = 70–80% (250–400 °C)	Significant advantages have been achieved in obtaining dispersed catalysts for the hydrocracking reaction. Improved catalyst performance is the top priority	168

	M = MoS_2	HK of organic raw materials (coal, bitumen, biomass or plastic polymers) or heavy oil	P_{H2} = 10 MPa T = 425 °C	$Y_{Distill.\ prod.}$ = 90% to 425 °C	MoS_2 is the catalytically active form regardless of the method of introduction and composition of precursors used	169
	M = MoS_2 and promoters Ni or Co	HK of vacuum residues, tar, carbon, bitumen	P_{H2} = 6 y 25 MPa t = 400 y T = 500 °C	C = 85–97% to 350–500 °C	A synergistic effect is evident between the Ni or Co promoters with the MoS_2 species. Synthesis methods of ultra-dispersed catalysts are increasingly striking where MoS_2-xCx catalysts can be promising	170
Acid supports	S = HZSM-5 HY, HUSY, MCM-41 or SiO_2-Al_2O_3	HK of plastic materials (PET, PVC, PS, PUR, PE-HD, PE-MD, PE-LL, PE-LLD, PP)	P_{H2} = 6 MPa T = 450 °C	$Y_{Liq.\ prod.}$ = 64%, to 450 °C S = 55% gasoline range	The hydrocracking process for plastic waste can be an interesting and feasible alternative for its recovery in refinery units, generating a wide distribution of products	62
	S = HZSM-5, HZSM-5/Al_2O_3	HK of vacuum residue	P_{H2} = 0.1 MPa T = 125–360 °C t = 4–20 h	HK = 52–77% to 300–360 °C	HZSM-5 improves its catalytic activity by loading it in pseudo-bohemite, the activity and catalytic stability of HZSM-5 improves and benefits the dewaxing reaction in the vacuum residue	63

(*continued*)

Table 2.1 (*continued*)

System	Catalyst (metal-support)	Type of reaction (HC = hydroconversion; HK = hydrocracking; HI = hydroisomerization; H = hydrogenation; HT = hydrotreatment)	Reaction conditions P (MPa); T (°C)	Conversion (C)	Interpretations	Ref.
Bifunctionals	M = Pt S = HY-30, HY-A	HK of n-hexadecane	T = 210–280 °C	C = 100% to 260 °C	The good catalytic activity of zeolite-based catalysts is demonstrated by following the reaction products in the conversion of n-hexadecane	171
	M = CoMo, NiMo S = Al_2O_3, C	HC of vacuum residue	P_{H2} = 3–10 MPa T = 430–470 °C	C = 90.6% S = 53.9% of light fraction to 470 °C	The performance and selectivity of a catalyst used for hydrocracking is subject to the type of support, composition, preparation method and process conditions; therefore, attention should be paid to the details of the development of a catalyst to obtain high activity, considering as well the impact of impurities on them	75
	M = MoS_2, NiMoS S = Al_2O_3, TiO_2, SiO_2, MgO, USY	HK of benzene or toluene, H of 1-metilnaftalene	P_{H2} = 7 MPa T = 300–400 °C	C = 80% to 400 °C	The formation of layered layers of MoS_2 on alumina is ideal for hydrogenation reactions, obtaining high catalytic activity	172

M = Pd, Pt, Cr, Mn, Fe, Co, Ni, Cu, Zn, Ru, Rh S = ZSM-5, ZSM-12, USYZ, SAPO-11, Meso-Y, HZSM-5, SBA-15, Al-MCM-41, β-zeolite	HK of bio-oil	P_{H2} = 1–40 MPa T = 250–550 °C t = 6–360 min	$Y_{Jet\ fuel}$ = 69% (Pd/ β-zeolite) to 270 °C	The zeolites presented the best catalytic activity in the hydrocracking of oils of vegetable origin compared to the other supports, attributed mainly to their structure and textural properties	173
M = NiS or NiMoS S = silica or zeolites	HK of plastic materials (PET, PC, PA66, PA6, PVC, PP, HDPE, LDPE, PS)	P_{H2} = 7 MPa T = 375–450 °C	$S_{Parafines}$ = 32% for gasolines and 43% for diesel 29% of solids residues to 375 and 425 °C	Different technological aspects of classification and reprocessing of plastics are presented, highlighting hydrocracking as a process with good industrial potential	64, 65
M = NiW, CoMo, NiMoP, Mo_2P, Rh, Pt S = USY, Al_2O_3, HSMS	HC of naphthalene	P_{H2} = 2–4 MPa T = 350–450 °C t = 1–3 h	C = 80–90% to 450 °C	The reaction leads mostly to tetralins and decalins. The best catalyst is $PtSiAl_2O_3$, which has a 100% conversion rate with a large amount of decalins	174
M = Mo, NiK, Fe, Ru S = $yCe_xZr1\text{-}_xO_2$-$MaAl_2O_3$	HK of bitumen of Athabasca and heavy oils	P_{H2O} = 0.2–20 MPa H_2O/ HOF = 45 : 150 T = 300–450 °C t = 1–3 h	C = 68–80% to 450 °C	Catalysts containing Fe, Mo and bifunctional systems based on NiK have good catalytic activity, comparable to catalysts with noble metals such as Ru	175
M = Ni, Co or Fe 5 wt%, Mo or W 15 wt% S = delaminated vermiculite	HK of bitumen of Athabasca	P_{H2} = 10 MPa T = 380 °C t = 6 h	C = 32–40% a 325 °C HK = 35–39% to 350 °C	The best catalyst is $V_5Ni_{15}Mo$, which exceeds the conversion of solid $5Ni_{15}Mo/Al_2O_3$ taken as a reference	166

(continued)

Table 2.1 (*continued*)

System	Catalyst (metal-support)	Type of reaction (HC = hydroconversion; HK = hydrocracking; HI = hydroisomerization; H = hydrogenation; HT = hydrotreatment)	Reaction conditions P (MPa); T (°C)	Conversion (C)	Interpretations	Ref.
	M = Pt or Pd or group VI or VIII S = amorphous silica-alumina, zeolites or mixtures of both	HK of raw material, from light gasoline to heavy crudes	P_{H2} =16 MPa T = 385 °C	$Y_{Liq. prod.}$ = 15–24% to 385 °C	Hydrocracking is the most versatile process in the oil industry allowing for the transformation of different raw materials, as well as the mixing among them, allowing for industries to have a greater margin for the treatment of different sources obtaining different products	49
	M = CoMo, WNi S = NaY, Al_2O_3, ZrO_2, TiO_2	HK of vacuum residues of bitumen of Atabasca, HVGO and LVGO	P_{H2} = 7–13 MPa Cat/feed = 1 : 100 T = 340–430 °C t = 0.5–24 h	C = 60–90% to 430 °C	The sulfurized phase of CoMo is essential to increase liquid products and to obtain a good dispersion to decrease coke formation	176
	M = $CoNiMoS_2$, Ru, Pd, Pt, Ni, Co, Cu, Fe, MoS_2 S = Al_2O_3, SiO_2, USY, TiO_2, ZrO_2, TiO_2-ZrO_2, C	HK of bio-oil	P_{H2} = 4 MPa T = 250–360 °C	C = 30–100% to 250 °C	MoS_2 exhibits a complete conversion of triglycerides mainly through the hydrogenation pathway and allows for the formation of carbon monoxide during co-hydrotreatment	177, 178

2.3.1 Hydrogenating/dehydrogenating Function

The H/DH function is responsible for activating the hydrogen present in the hydrocracker and for carrying out the H/DH of the hydrocarbon. The elements used for this function include noble metals such as palladium or platinum[48,49] or transition metal sulfides of group 6 (VIA group) (molybdenum or tungsten) and of groups 9 or 10 (VIIIA; cobalt or nickel).[50,51] Transition metal sulfides (Mo, W, Co and Ni) are widely used in the industry because of their resistance to poisoning due to sulfur and nitrogen content in the feed.[52–55] In this sense, Co and Ni are used as promoters of Mo and W, which generate species that increase activity.[12,50,51]

2.3.2 Acid Function

The acid support is responsible for catalysing cracking and isomerization reactions and usually consists of oxides such as amorphous silica-alumina,[56,57] zeolites,[58,59] a binder such as γ-alumina,[60] or a mixture of zeolites and amorphous oxides (Table 2.1).[61] As already mentioned, zeolites[59,62,63] and amorphous silica-alumina[64,65] are the most commonly used materials as supports for this type of catalyst because in addition to providing acid function, they also offer some textural properties (meso- and macro-pores) that allow for the access of high-molecular-weight molecules (present in the charge) to acid sites.[12,16]

2.3.3 Relationship between Hydrogenating/dehydrogenating and Acid Functions

One of the key parameters for the design of a HCK catalyst is the balance between the H/HD and acid functions. In an ideal bifunctional catalyst, the functions are almost in equilibrium and the limiting step occurs in the acid function.[27,66,67] Achieving an ideal bifunctional catalyst depends on the concentration and strength of the Brønsted acid sites and the average distance between the H/DH function and the acid function.[68,69] For an ideal bifunctional catalytic reaction to occur, it is necessary that the two functions be in a balance,[70] which is determined by the number and strength of each type of site. When the acid function is weak and the hydrogenation function is strong, low activity for HCK occurs, so the reaction conditions must be modified, either by increasing the temperature or decreasing the space velocity to increase the selectivity towards the products of isomerization. When the acid function is strong and the hydrogenating function is weak, high activity for HCK and low selectivity towards the isomers is present, because there is a greater number of side reactions.[71]

2.3.4 Catalyst Activation

Catalysts based on non-noble metals (which comprise transition metals) must be activated by transforming their precursor oxidic forms to their active forms (sulfides). This sulfurization process is usually carried out at temperatures between 300 and 400 °C, hydrogen pressures between 1.2 and 5.0 MPa and in the presence of compounds that are a source of sulfur such as H_2S, dimethyl disulfide (DMDS), dimethyl sulfide (DMS) and carbon disulfide (CS_2), among others. The process can be carried out "*in situ*" or "*ex situ*", and these can occur in the gas or liquid phase.[12,16]

Sulfurization "*in situ*" is the industrially preferred method in refineries, where the sulfurizing agent is usually added to the load, avoiding the use of H_2S (to avoid environmental problems). On the laboratory scale, the gas phase is preferred, using a mixture of H_2S/H_2 or H_2S/He containing 5–10% H_2S.[12,16]

On the other hand, "*ex situ*" sulfurization occurs outside the reactor, where sulfur compounds are added to the catalyst and activation occurs when the catalyst is charged into the reactor and heated in the presence of hydrogen. Some refineries prefer to receive the pre-sulfurized catalyst in order to accelerate the start-up of the HCK unit; however, this activation is more expensive because it requires an additional sulfurization step.[72,73] For three decades, the preferred catalysts at the industrial level for this reaction comprised a metal phase consisting of Mo or W promoted with Co or Ni and a support such as Al_2O_3, silica-alumina or zeolites.[74–77] However, the catalysts used have increased in their price by 300% in the last 10 years, thus reducing the profit margins in the process.[78]

2.4 Clays as Acid Catalyst Supports

The importance of acid catalysts in oil refining has stimulated the search for solids with these surface properties to be used in different catalytic processes. In general terms, the catalysts used in the treatment of different feeds in the oil industry present three fundamental problems:[18,79] (1) deactivation due to the formation of coke on the catalyst surface; (2) the deposition of metals such as vanadium and nickel present in the charge on the active phase of the catalyst; and (3) diffusional limitations due to the pore size of the support and the large size of the molecules mainly present in the vacuum bottoms.

Among the challenges at the level of the catalyst is the ability to modulate the acidic properties of the same depending on the strength, type and number of acids sites[82] to obtain an adequate balance with the metal function. If the parameters that direct the deactivation processes together with the dynamics inherent in the catalytic process are clear, solid-state chemistry can provide the necessary tools to design a catalyst that has the best properties to lead to a marked decrease in the formation of coke, a significant increase in acidity and a good structure–activity relationship.

On an industrial level, the first systems for HCK reactions were based on acid-treated montmorillonite and later, in the middle of the last century, these systems were replaced by amorphous silica-aluminas.[81] Currently, the use of zeolites and silica-alumina predominates in the catalyst market for this type of reaction. Despite their recognized important qualities, materials such as zeolites still carry relatively high costs and do not always have simple methodologies for their synthesis and purification[82] due to the low purity of the zeolitic phase[82] and the long crystallization times.[83] These as a whole make it more complicated to meet environmental standards, which fortunately have become more and more rigorous, both for obtaining the catalyst and for the utilization of the material after use.

Natural clay minerals offer very interesting advantages for handling, modification and adaptation to different catalytic processes, with the additional advantage of being natural, economical and abundant materials.[4] One of the most striking properties is the acid nature that, although it depends on the type of mineral, can be modulated by physical, chemical and mechanical processes such as pillarization, delamination and the incorporation of metals,[84–90] which gives them great versatility in a wide range of applications.[91] However, these processes of modification of the natural material, developed during the last two decades of the last century, require different unit stages, long reaction times and excessive volumes of water, generating little interest in industrial application.[84] However, these studies provided fundamental knowledge that, along with innovative methods such as the use of microwaves, ultrasound and other modern synthesis strategies, have again boosted interest in the use of natural clay minerals[92,93] in the search for new active catalysts that are selective and stable in a great diversity of reactions. This new trend is seen in Figure 2.2, which reveals a strong growth in the number of reports, particularly in the last decade, concerning the use of clays as catalysts.

The processes today for the synthesis of metal catalysts or metal oxides supported on clays are achieved in process times of a few hours and with minimal or non-existent amounts of water.[94] These methods allow for modulation of both acidity (type, quantity and strength) and texture and, as reflected in Table 2.2, lead to active materials in hydroconversion reactions. In conjunction with the use of microwaves, ultrasound has been used primarily in the interleaving stage. The heating that is generated when using ultrasound in synthesis results from the formation, growth and implosion of microbubbles due to the effect of soundwaves of 1–15 MHz in frequency.[79] The reactions can be carried out easily in the presence of the high pressures and temperatures generated by the process of implosion of the bubbles, and thanks to these effects, the reactivity of the solid surfaces is favoured and the syntheses are verified in less time.

In parallel, progress has been made in the synthesis of solid mixed polymers (used as a template to insert active sites in the mineral) that, in addition to allowing for the modification of clays with minimal amounts of water, thanks to their recognized stability, can be stored for several months and even years without altering their structures and properties.

2.4.1 Structure of Clay Minerals

The starting point for designing and synthesizing catalysts from clay minerals is, of course, the deep knowledge of their structure, as well as their physical and chemical properties.[81,89,90,95–98] The structure of clays, which is widely known, is based on the stacking of planes formed by tetrahedral (T) and octahedral

(O) units. The way in which these planes are joined governs a large part of the properties of each type of clay and allows for generating the principle of the classification of these materials, which can be type T:O (one-dimensional), giving rise to 1:1 clays, or T:O:T (two-dimensional), which gives rise to clays of the 2:1 type. The latter have the ability to increase their original interplanar spacing $d_{(001)}$, which is generally between 14 and 15 Å, which makes them susceptible to modification by chemical manipulation in the interlayer.

Among the T:O:T minerals is the group of vermiculites and smectites, whose fundamental difference is in the interlaminar charge, which is higher in vermiculite clay due to the greater number of isomorphic substitutions of silicon for aluminium in the tetrahedral layers compared with those present in smectites. This charge decompensation generated by the described isomorphic substitution promotes the formation of acidic sites, either because the deprotection of positive charges allows for the acceptance of electronic pairs (Lewis acidity) or because they can capture protons and then yield them (Brønsted acidity).[99]

Thus, the acidity of the clays is associated with the presence of different functional groups on the surface of the solid, which can affect the chemical environment of the elements that make it up and depends on both covalent environments and coordinated states, including interaction with adsorbed species.[100]

In this way, the clays could intrinsically be used as acid catalysts, but unfortunately, when they are at temperatures above 200 °C the loss of water molecules causes the adhesion of the sheets, which inevitably leads to the destruction of the structure and an almost total decrease in catalytic activity. However, their acid nature and the way in which said property can be modulated based on variables such as the nature of the material, the process and the different chemical species with which they can be modified make them very useful and increasingly promising materials for industrial catalytic processes.

Clay minerals are considered nanostructured materials because they have a well-defined structure of nanometric size, in which at least one of its phases has one or more dimensions on the nanometric scale typically between 1 and 100 nm[101,102] with systems of regular pores, being similar to other nanostructured systems such as zeolites,[103] metal phosphates, mesoporous periodic materials, nanoporous organic materials, organometallic materials and graphite oxides,[104] among others. These minerals have a lateral dimension of 30 nm, and the thickness of the individual platelets is in the order of 1 nm,[105] presenting dimensions in the plane that reach the micrometric order (Figure 2.3). These nanodimensions give the material unique physicochemical properties such as tensile strength, modulation, gas permeability, a low coefficient of thermal expansion and optical homogeneity,[105] which facilitate the modulation of its acidic, textural and structural properties.[106]

2.4.2 Methods of Modifying and Modulating the Acidity of Clay Minerals

Although natural clay minerals can be used directly as adsorbents or catalysts, there are modifications or activation procedures to increase the surface area and acidity of these solids.[107] The great variety of clay minerals found in nature

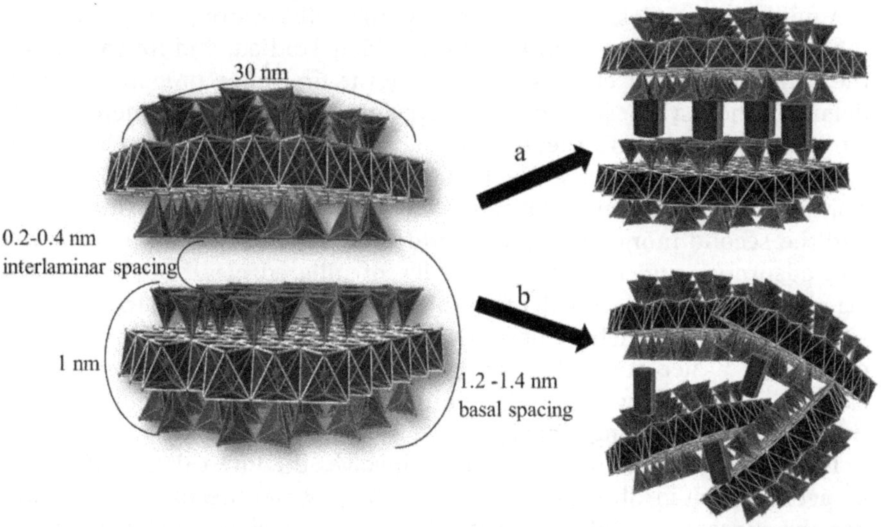

Figure 2.3 Average dimensions of a clay mineral and method of modification *via* (a) pillarization and (b) delamination.

provides a wide range of natural materials with different physicochemical characteristics, thus, different minerals can fulfill requirements in different processes. Additionally, the various modification methods allow the variation of the intrinsic characteristics of minerals, generating materials which may satisfy future requirements.

The modification methodologies include microwave or ultrasound, cold drying,[108–111], treatment with peroxide,[112] acid[113,114] and incorporation of large polyhydroxycations (organic molecules and surfactants such as polyvinyl alcohol [PVA]),[115,116] methods that allow for modulating the porosity of the clay generating highly micro- and meso-porous materials, with large volumes of pores[116] and high surface areas, thereby improving access to the intrinsic acid sites of the mineral[117] and obtaining through this modification pillared or delaminated solids, as illustrated in Figure 2.3.

2.4.2.1 Pillared Clays

Pillarization was proposed in the 1950s, but its extensive development began in the late 1970s.[118,119] The preparation of pillared clays basically uses the concept of exchanging the ions present in the interlaminar space of natural clay for other larger ones. The introduction of pillars, in addition to increasing the strength and stability of the material, provides micro- and/or meso-porosity, increases the surface area, gives greater access to the acidic sites present in natural clay and generates the presence of potentially active species for a specific reaction.[118,120] In the conventionally used process, diluted solutions of inorganic polyhydroxycation species are used, which are exchanged within the interlaminar spaces and subsequently transformed into oxides by calcination.[118,121–123]

Among the species used to carry out the pillarization, the products of the hydrolysis of aluminium have been widely studied, and for this reason the aluminium-pillar clays (Al-PILC) were the first to be prepared, and the details of their synthesis and characterization are widely documented.[124,125] However, the chemical structure of the abutment can be modified by the synthesis of mixed abutments, which are obtained from the hydrolysis of solutions containing various metal cations. The main pillar cation is aluminium, and the second inorganic cation is added to improve the thermal stability and adsorptive and catalytic properties of pillared clay.[96,98] The catalytic activity of modified clays with mixed pillars has been evidenced in a variety of reactions, among which we have the alkylation of benzene with ethylene,[126] cracking,[127,128] heptane hydroconversion,[129,131] propane dehydrocyclodimerization,[131] HDS,[132,133] wet oxidation of phenol[136] and selective catalytic reduction of NO_x,[135–137] among others.

The classical method of pillarization involves the intercalation of the pillar agent, which involves the synthesis of the polymer precursor in solution, and the subsequent addition on the diluted suspension of the clay mineral (typically 2%). Parameters such as ageing time, stirring conditions, washing method (filtration, centrifugation, dialysis) and drying method influence the final characteristics of the solid. Variants of this method have been investigated in order to reduce water volume and synthesis times. With the information available to date, it is possible to perform the synthesis of pillared clays using concentrated suspensions of both clay and pillarizing agent. As for the pillar agent, the tendency is towards the use of mixed systems in concentrated medium and, as far as possible, in solid form. It has also been established that ultrasound-assisted intercalation considerably reduces the exchange time of species.[94,138]

2.4.2.2 Delaminated Clays

The delamination process allows for nanostructures to be achieved through the generation of edge–edge or edge–face aggregates of clay materials using water-soluble polymeric surfactants such as PVA and Keggin-type polyhydroxycation such as with Al_{13}. The materials obtained are generally of the mesoporous type with metallic species widely distributed on the surface.[139,140] These porosity characteristics determine their adsorption properties and their catalytic behaviour.[139,141] The wider distribution of pore size that characterizes delaminated structures favours the arrangement of well-dispersed active phases and the passage of large molecules through their pores. Additionally, the increase in mesoporosity can be an advantage in processes where the porous structure is blocked with coke, since the plugging of the material could be less severe compared to that of microporous structures, converting these clay minerals into interesting and economical materials in the search for catalytic supports with the characteristics of texture and acidity suitable for the hydroprocessing of crudes and heavy fractions.[141]

2.4.2.3 Clay Acidity Modulation

The clays are intrinsically acid catalysts, but unfortunately, when they are at temperatures above 200 °C, the loss of water molecules causes the adhesion of the sheets, which leads to the destruction of the structure and an almost total decrease of the catalytic activity. The strategy of the insertion of species with acid characteristics is part of the processes of potentiation of said chemical properties in solids. Thus, acidity modulation by means of modifying a clay requires the careful selection of the precursor cations, for which you can turn to the basic tools of inorganic chemistry.

A comparison between pillared clays with Zr and Al demonstrates a considerable increase in the content and acid strength achieved in Zr-PILC compared to Al-PILC.[142,143] On the other hand, if the insertion of mixed AlZr species in delaminated and pillared minerals is compared, there is a notable increase in acidity in the delaminated solid in contrast with the pillarized solid.[143,144] This difference is attributed to the fact that delamination allows for greater access to the intrinsic acid sites of the natural mineral since a larger mesoporous area is generated compared to pillared minerals,[99,143] which would guarantee a better distribution of AlZr species. AlZr species generate Lewis-type acidity,[145] as well as proton production caused by dehydration of the polyhydroxycation introduced into the interlaminar space[146] as a result of the protonation of the Si–O(H)–Al bond induced by a decrease in the characteristic pH of solutions of Zr.[130,134] These systems exhibit greater selectivity towards cracking products in hydrocarbon hydroconversion reactions compared to Al[143,147] due to the increase in acid strength and the number of Brønsted acid sites.[147,148]

On the other hand, it has been reported that Ce acts as a promoter of catalytic activity,[149,150] and in the case of acid catalysts, it increases the strength of acidity observed by means of the increase in selectivity towards cracking products in hydroconversion reactions.[149] Some studies indicate that the AlCe polyhydroxycation, formed by the addition of small amounts of Ce^{3+} to an Al^{3+} solution, is probably made up of four Al_{13}^{7+} polymer units joined by a tetrahedrally coordinated Ce atom.[98,151,152] The greater catalytic activity and thermal stability caused by the addition of Ce have been related to the stabilization of Brønsted-type acid centres.[150]

In this way, when comparing delaminated minerals with AlCe against pillars with the same species, a similar trend is observed to that discussed for AlZr. In fact, with delaminated material, greater areas are achieved that favour the greater distribution of AlCe species on the surface and improve the acid properties of the material.[153]

Clay minerals of the 2 : 1 type with high interlaminar load manage to significantly increase their acid properties by being modified with AlHf pillars, which become even more acidic and active in the hydroconversion of decane than pillars with AlCe.[99,154]

In addition, there are reports of pillared montmorillonite with ZrAl and SiAl species for the hydroconversion reaction of n-heptane[155,156] and, more recently, the use of a pillared montmorillonite with Al species for the HCK of C_{30}.[157]

Similarly, a modified vermiculite with Zr and Ti species has been reported using a sulfating method, generating greater acidity in the material.[158]

2.5 Clays as Catalysts in Hydrocracking Reactions

The study of clay minerals at a structural, physical and chemical level, together with the knowledge of solid-state chemistry, has shown the enormous potential of this type of material to be used as catalysts in various chemical processes, mainly in reactions that need acidic sites.[122,159,160] Table 2.2 records a selection of reports on the use of natural and modified clay minerals as catalysts in hydroconversion reactions. The main use of these materials is reserved for the hydroconversion of linear paraffins, with some exceptions that include the successful use of these minerals in HCK reactions of heavier feeds as a bitumen. In the first case, the similarity and even superiority in catalytic performance that can be achieved with modified natural clay minerals compared to recognized zeolites is evidenced in some reports.[99,154,157,161]

Clay minerals as catalyst precursors for catalytic HCK in the treatment of vacuum bottoms have also been reported.[128,162] In particular, montmorillonite- and laponite-type clays with Sn, Cr and Al species have been evaluated in this reaction, registering good catalytic performance compared to commercial catalysts ($NiMo/Al_2O_3$, $Mo[CO]_6$).[128,162] This behaviour is attributed to the good adsorption of large molecules in the porosity of the support, the presence of strong acid sites suitable for catalytic HCK, the high thermal stability and the high redox efficiency of the oxides of the metals present in the catalyst.[128,162]

On the other hand, reports of delaminated clays for HCK processes have been presented since the 1980s.[141,163–165] However, currently there are fewer reports showing the use of these materials for HCK reactions. The use of 2 : 1 clays delaminated and impregnated with NiMo results in excellent catalysts for the hydroconversion of decane[117] and, more recently, the use of delaminated vermiculite for the HCK of Athabasca bitumen using Mo or W promoted with Ni, Co or Fe as the metal phase reached conversions that were very close to those of commercial catalysts.[166]

2.6 Conclusion

In recent years, HCK research has been directed towards the development of catalytic systems capable of treating increasingly complex oil fractions for processing at the refinery. Among the different strategies that have been developed to deal with this type of feed, it is very important to consider the design of bifunctional catalysts that have high thermal stability, resistance to poisoning, good molecular traffic of bulky compounds, adequate acid/metal site balance and very low or no coking levels. Likewise, the use of environmentally friendly materials that are economical should be considered.

In this way, clays come forth as promising candidates for obtaining supports with suitable and modular acid and textural properties capable of providing active catalysts in HCK reactions. Additionally, they constitute a very

Table 2.2 Natural clay minerals in hydroconversion processes.

Catalyst (metal-support)	Synthesis method	Feed	Reaction conditions P (MPa); T (°C)	Conversion (C)[a], hydrogenation (H), hydrocracking (HK) and hydroconversion (HI)	Interpretations	Ref.
M = Pt 0.5 wt% S = pillared beidellites with Si, Si/Al and Al	Impregnation in beidellites pillared with Si, Si/Al and Al	n-Decane	P_{H2} = 0.4 MPa T = 100–300 °C	C = 100% to 200 °C HI = 70% to 200 °C	Catalysts obtained from pillared clays record greater activity than zeolites such as CBV-720 and CBV-760	161
M = Pt 1 wt% S = pillared beidellites	Impregnation in pillared beidellites (pillars of Si)	n-Heptane	P_{H2} = 0.1 MPa T = 190–320 °C	C = 87% to 260 °C HI = 76% to 265 °C	Pillared clay reaches higher isomerization values compared to HY, HB and EU-1 zeolites	179
M = Pd 0.4 wt% S = pillared montmorillonite	Impregnation in pillared montmorillonite with aluminium using ultrasound	n-Heptane	P_{H2} = 3 atm. T = 400 °C t = 2 h	C = 40% to 359 °C	It was determined that the optimal ultrasound time to obtain the most active catalysts is 20 min	180
M = Pt 1 wt% S = pillared vermiculite	Impregnation in vermiculite pillared with Al, influence of the conditions of the mineral load reduction method	n-Heptane	P_{H2} = 0.1 MPa T = 150–350 °C	C = 80% to 300 °C HI = 57% to 300 °C	The method of reducing the load of the mineral was evaluated, establishing optimal conditions to reduce the high load density, allowing for subsequent modification, obtaining active solids in the hydroconversion reaction of heptane	181
M = Pt 1 wt% S = pillared vermiculite	Impregnation in pillared vermiculite (pillars of AlZr, AlCe and AlHf)	n-Decane	P_{H2} = 0.4 MPa T = 150–300 °C	C = 100% to 230 °C HK = 100% to 230 °C HI = 60–80% to 200 °C	Pillared clays with AlHf showed greater activity, and most of the solids obtained were more active than a USY zeolite	154, 99

(*continued*)

Table 2.2 (*continued*)

Catalyst (metal-support)	Synthesis method	Feed	Reaction conditions *P* (MPa); *T* (°C)	Conversion (C)[a], hydrogenation (H), hydrocracking (HK) and hydroconversion (HI)	Interpretations	Ref.
M = Pt 1 wt% S = pillared bentonite	Impregnation in bentonite pillared with Al in concentrated medium using microwave radiation	n-Heptane	P_{H2} = 0.1 MPa T = 150–350 °C	C = 55% to 350 °C HK = 10% to 350 °C HI = 43% to 300 °C	It was demonstrated that the pillared solids under this new methodology presented better catalytic activity than those obtained by conventional methods in heptane hydroconversion	182
M = Pd 0.4 wt% S = saponite	Saponite modified with Mg, Zn, Co, Ni or the combination of these	n-Heptane	P_{H2} = 3 MPa T = 200–400 °C	C = 40% to 280 °C	A linear relationship between the acidic properties and the catalytic activity of the different solids is shown	183
M = NiMo S = metakaolinite	Synthesis of ZSM-5 with gel metakaolinite – hydrothermal	Jatropha oil	P_{H2} = 5–8 MPa T = 350–425 °C	C = 100% to 375 °C HK = 100% to 425 °C HI = 90% to 375 °C	The conversion to kerosene is comparable to commercial catalysts, but the clay material exceeds the stability of mesoporous zeolites	184
M = Ni 5 wt%, Mo 15 wt% S = delaminated bentonite	Incorporation of the metal phase by wet impregnation and self-combustion in delaminated bentonite	n-Decane	P_{H2} = 0.1 MPa T = 150–400 °C	C = 100% to 325 °C HK = 100% to 325 °C HI = 50% to 250 °C	The self-combustion method showed better metal dispersion, which was reflected in greater catalytic activity	117

M = Ni 5 wt%, Mo 15 wt% S = halloysite	Impregnation in halloysite previously treated with acid	n-Decane	P_{H2} = 0.1 MPa T = 150–400 °C	C = 100% to 250 °C HK = 100% to 275 °C HI = 56% to 205 °C		A good correlation was established between the acidic properties of the catalyst, as well as the textural properties, with its catalytic performance and selectivity to isomerization products	185
M = Ni 5 wt% S = pillared bentonite	Impregnation in pillared bentonite with Keggin (Al or AlZr) using ultrasound	n-Decane	P_{H2} = 0.1 MPa T = 150–400 °C	C = 100% to 325 °C HK = 100% to 350 °C HI = 50% to 225 °C	The catalysts pillared with AlZr presented greater activity and selectivity to cracking products compared to solids modified only with Al, indicating that the incorporation of the second cation in the pillars improves the catalytic performance	143
M = Ni, Co or Fe 5 wt%, Mo or W 15 wt% S = delaminated vermiculite	Wet impregnation in delaminated clay with aluminium Keggin (Al_{13})	Bitumen of Athabasca	P_{H2} = 10 MPa T = 380 °C t = 6 h	C = 32–40% to 325 °C HK = 35–39% to 350 °C	The best catalyst is $V_5Ni_{15}Mo$, which exceeds the conversion of solid $5Ni_{15}Mo/Al_2O_3$ taken as a reference	166
M = Ni 3 wt% and Mo 15 wt% S = pillared montmorillonite	Montmorillonite pillared with aluminium	Squalene (C_{30})	P = 4 atm. T = 300 °C t = 6 h	C = 100% to 300 °C H = 91.2% to 300 °C HK = 35–53% to 300 °C	The selectivity of clay minerals to products such as diesel or aeroplane gasoline was higher compared to the values obtained for the zeolites HSZM-5 or HY, associated with moderate acidity on the surface of clay minerals	157

[a]temperature at which the maximum conversion is obtained

economical and environmentally friendly alternative, since, due to their chemical structure, they do not become a soil contaminant once they must be disposed of.

Acknowledgements

Thanks are given to Cesar Rodriguez (chemist, MSc) of the ESCA group for his valuable logistical support.

References

1. M. A. Salam, K. Ahmed, N. Akter, T. Hossain and B. Abdullah, *Int. J. Hydrogen Energy*, 2018, **43**(32), 14944–14973.
2. V. Kumaravel and A. Abdel-Wahab, *Energy Fuels*, 2018, **32**(6), 6423–6437.
3. World Energy Council, "World Energy Issues Monitor 2018, Perspectives on the Grand Energy Transition", https://www.worldenergy.org/publications/entry/world-energy-issues-monitor-2018-perspectives-on-the-grand-energy-transition (accessed 27/05/2020).
4. C. G. Fonseca, V. S. Vaiss, F. Wypych, R. Diniz and A. A. Leitão, *Appl. Clay Sci.*, 2018, **165**, 170–178.
5. G. D. Yadav, *Catal. Surv. Asia*, 2005, **9**(2), 117–137.
6. D. A. Castelo Branco, G. L. Gomes and A. S. Szklo, *Energy Policy*, 2010, **38**(6), 3098–3105.
7. P. Wang, Y. Zhao and J. Liu, *Sci. Bull.*, 2018, **63**(4), 252–266.
8. A. Corma, L. Sauvanaud, Y. Mathieu, S. Al-Bogami, A. Bourane and M. Al-Ghrami, *Fuel*, 2018, **211**, 726–736.
9. E. Furimsky, *Appl. Catal., A*, 1998, **171**(2), 177–206.
10. A. T. Jarullah, N. A. Awad and I. M. Mujtaba, *Fuel*, 2017, **206**, 657–674.
11. J. Weitkamp, *ChemCatChem*, 2012, **4**(3), 292–306.
12. P. Raybaud and H. Toulout, *Catalysis by Transition Metal Sulphides: From Molecular Theory to Industrial Application*, Editions Technip, Paris, 2013.
13. J. G. Speight, Enhanced Recovery Methods for Heavy Oil and Tar Sands, Gulf Publishing Company, Houston, TX, 2013.
14. E. Furimsky, Hydroprocessing Reactions, in *Catalyst for Upgrading Heavy Petroleum Feeds*, ed. E. Furimsky, Studies in Surface Science and Catalysis, Elsevier, 2007, ch. 6, pp. 95–139.
15. H. Martinez-Grimaldo, H. Ortiz-Moreno, F. Sanchez-Minero, J. Ramírez, R. Cuevas-Garcia and J. Ancheyta-Juarez, *Catal. Today*, 2014, **220–222**, 295–300.
16. *Handbook of Petroleum Processing*, ed. D. S. J. Jones and P. P. Pujadó, Springer Netherlands, Dordrecht, 2006.
17. *Practical Advances in Petroleum Processing*, ed. C. S. Hsu and P. R. Robinson, Springer New York, New York, NY, 2006.
18. T. M. Nguyen, J. Jung, C. W. Lee and J. Cho, *Fuel* 2018, **214**, 174–186.
19. J. Ancheyta, M. S. Rana and E. Furimsky, *Catal. Today*, 2005, **109**(1–4), 3–15.

20. S. Y. Gómez-Orozco, R. Huirache-Acuña, B. Pawelec, J. L. G. Fierro, E. M. Rivera-Muñoz, J. Lara-Romero and G. Alonso-Nuñez, *Catal. Today*, 2018, **305**, 152–161.
21. L. Ding, Y. Zheng, Z. Zhang, Z. Ring and J. Chen, *Appl. Catal., A*, 2007, **319**, 25–37.
22. S. Harris and R. R. Chianelli, *J. Catal.*, 1986, **98**(1), 17–31.
23. M. J. Ledoux and B. Djellouli, *J. Catal.*, 1989, **115**(2), 580–590.
24. J. Weitkamp, in *New Horizons in Catalysis*, ed. T. Seiyama and K. Tanabe, Studies in Surface Science and Catalysis, Elsevier, Amsterdam, 1981, pp. 1404–1405.
25. K. D. Oliveira, R. C. Santana, C. N. Ávila-Neto and D. Cardoso, *Appl. Catal., A*, 2015, **495**, 173–183.
26. P. M. Lima, T. Garetto, C. L. Cavalcante Jr. and D. Cardoso, *Catal. Today*, 2011, **172**(1), 195–202.
27. M. Guisnet, *Catal. Today*, 2013, **218–219**, 123–134.
28. O. D. Mante, F. A. Agblevor and R. McClung, *Fuel*, 2013, **108**, 451–464.
29. P. O'Connor, J. P. J. Verlaan and S. J. Yanik, *Catal. Today*, 1998, **43**(3–4), 305–313.
30. A. Duan, T. Li, H. Niu, X. Yang, Z. Wang, Z. Zhao, G. Jiang, J. Liu, Y. Wei and H. Pan, *Catal. Today*, 2015, **245**, 163–171.
31. J. A. Martens and P. A. Jacobs, in *Introduction to Zeolite Science and Practice*, ed. P. A. J. H. van Bekkum, E. M. Flanigen and J. C. Jansen, Studies in Surface Science and Catalysis, Elsevier B.V., 2001, vol. 137, pp. 633–671.
32. J. G. Speight, in *Petroleum Chemistry and Refining*, CRC Press USA, 1998, vol. 121, p. 122.
33. E. Verheyen, C. Jo, M. Kurttepeli, G. Vanbutsele, E. Gobechiya, T. I. Korányi, *et al.*, *J. Catal.*, 2013, **300**, 70–80.
34. J. E. Samad, J. Blanchard, C. Sayag, C. Louis and J. R. Regalbuto, *J. Catal.*, 2016, **342**, 213–225.
35. T. Pinto, P. Arquilliere, G. P. Niccolai, F. Lefebvre and V. Dufaud, *New J. Chem.* 2015, **39**(7), 5300–5308.
36. J. Zecevic, G. Vanbutsele, K. P. De Jong and J. A. Martens, *Nature*, 2015, **528**(7581), 245–254.
37. X. Su, L. Yang, Z. Feng, Y. Wang, W. Wu, X. Bai and W. Wang, *Microporous Mesoporous Mater.*, 2018, **274**, 1–8.
38. B. C. Gagea, Y. Lorgouilloux, Y. Altintas, P. A. Jacobs and J. A. Martens, *J. Catal.*, 2009, **265**(1), 99–108.
39. J. Weitkamp, *Ind. Eng. Chem. Prod. Res. Dev.*, 1982, **21**(4), 550–558.
40. J. A. Martens, P. A. Jacobs and J. Weitkamp, *Appl. Catal.*, 1986, **20**(1–2), 283–303.
41. D. M. Brouwer and H. Hogeveen, *Recl. Trav. Chim. Pays Bas*, 1970, **89**(2), 211–224.
42. C. J. Egan, G. E. Langlois and R. J. White, *J. Am. Chem. Soc.*, 1962, **84**(7), 1204–1212.

43. A. T. Lapinas, M. T. Klein, B. C. Gates, A. Macris and J. E. Lyons, *Ind. Eng. Chem. Res.*, 1991, **30**(1), 42–50.
44. S. C. Korre, M. T. Klein and R. J. Quann, *Ind. Eng. Chem. Res.*, 1997, **36**(6), 2041–2050.
45. H. Du, C. Fairbridge, H. Yang and Z. Ring, *Appl. Catal., A*, 2005, **294**(1), 1–21.
46. J. Sánchez, A. Moreno, F. Mondragón and K. J. Smith, *Energy Fuels*, 2018, **32**(6), 7066–7077.
47. E. Benazzi, L. Leite, N. Marchal-George, H. Toulhoat and P. Raybaud, *J. Catal.*, 2003, **217**(2), 376–387.
48. N. D. Hesse and R. L. White, *J. Appl. Polym. Sci.*, 2004, **92**(2), 1293–1301.
49. C. Peng, Y. Du, X. Feng, Y. Hu and X. Fang, *Front. Chem. Sci. Eng.*, 2018, **12**(4), 867–877.
50. M. Marafi, A. Stanislaus and E. Furimsky, *Handbook of Spent Hydroprocessing Catalysts*, Elsevier, 2017, pp. 405–411.
51. J. W. Thybaut and G. B. Marin, in *Advances in Catalysis*, ed. C. Song, Elsevier Inc., 2016, pp. 109–238.
52. C. Li, B. Shi, M. Cui, H. Shang and G. Que, *J. Fuel Chem. Technol.*, 2007, **35**(4), 407–411.
53. M. R. Rad, A. Rashidi, L. Vafajoo and M. Rashtchi, *J. Ind. Eng. Chem.*, 2014, **20**(6), 4298–4303.
54. S. Mohanty, D. Kunzru and D. N. Saraf, *Fuel*, 1990, **69**(12), 1467–1473.
55. A. Ishihara, T. Itoh, H. Nasu, T. Hashimoto and T. Doi, *Fuel Process. Technol.*, 2013, **116**, 222–227.
56. I. H. Metecan, A. R. Ozkan, R. Isler, J. Yanik, M. Saglam and M. Yuksel, *Fuel*, 2005, **84**(5), 619–628.
57. S. Karagöz, T. Karayildirim, S. Uçar, M. Yuksel and J. Yanik, *Fuel*, 2003, **82**(4), 415–423.
58. J. M. Escola, J. Aguado, D. P. Serrano, L. Briones, J. L. Díaz De Tuesta, R. Calvo and E. Fernandez, *Energy Fuels*, 2012, **26**(6), 3187–3195.
59. J. Francis, E. Guillon, N. Bats, C. Pichon, A. Corma and L. Simon, *Appl. Catal., A*, 2011, **409–410**, 140–147.
60. A. Dhar, R. L. Vekariya and P. Sharma, *Petroleum*, 2017, **3**(4), 489–495.
61. L. C. Gomes, D. de Oliveira Rosas, R. C. Chistone, F. M. Z. Zotin, L. R. R. de Araujo and J. L. Zotin, *Fuel*, 2017, **209**, 521–528.
62. G. Lopez, M. Artetxe, M. Amutio, J. Bilbao and M. Olazar, *Renewable Sustainable Energy Rev.*, 2017, **73**, 346–368.
63. J. Zhu and W. T. Wu, *Pet. Sci. Technol.*, 2017, **35**(5), 465–470.
64. K. Ragaert, L. Delva and K. Van Geem, *Waste Manag.*, 2017, **69**, 24–58.
65. D. Munir, M. F. Irfan and M. R. Usman, *Renewable Sustainable Energy Rev.*, 2018, **90**, 490–515.
66. J. Weitkamp and S. Ernst, in *Guidelines for Mastering the Properties of Molecular Sieves*, ed. D. Barthomeuf, E. G. Derouane and W. Hölderich, Springer US, Boston, 1990, vol. 221, pp. 343–354.
67. F. Alvarez, F. Ribeiro, G. Perot, C. Thomazeau and M. Guisnet, *J. Catal.*, 1996, **162**(2), 179–189.

68. P. B. Weisz, in *Advances in Catalysis*, Academic Press, New York/London, 1962, vol. 13, pp. 137–190.
69. W. Wang, C.-J. Liu and W. Wu, *Catal. Sci. Technol.*, 2019, **9**, 4162–4187.
70. J. Weitkamp, P. A. Jacobs and J. A. Martens, *Appl. Catal.*, 1983, **8**(1), 123–141.
71. E. Verheyen, C. Jo, M. Kurttepeli, G. Vanbutsele, E. Gobechiya, T. I. Korányi, *et al.*, *J. Catal.*, 2013, **300**, 70–80.
72. M. S. Rana, V. Sámano, J. Ancheyta and J. A. I. Diaz, *Fuel*, 2007, **86**(9), 1216–1231.
73. J. Speight, *The Refinery of the Future*, William Andrew, Norwich, NY, 2010.
74. J. Li, H. Xia, Q. Wu, Z. Hu, Z. Hao and Z. Zhu, *Catal. Today*, 2016, **271**, 172–178.
75. R. Sahu, B. J. Song, J. S. Im, Y.-P. Jeon,and C. W. Lee, *J. Ind. Eng. Chem.*, 2015, **27**, 12–24.
76. M. O. Kazakov, K. A. Nadeina, I. G. Danilova, P. P. Dik, O. V. Klimov, V. Y. Pereyma, *et al.*, *Catal. Today*, 2018, **305**, 117–125.
77. J. G. Speight, in *The Refinery of the Future*, Elsevier, Oxford, 2011, pp. 275–313.
78. Y. G. Hur, M.-S. Kim, D.-W. Lee, S. Kim, H.-J. Eom, G. Jeong *et al.*, *Fuel*, 2014, **137**, 237–244.
79. B. M. Vogelaar, S. Eijsbouts, J. A. Bergwerff and J. J. Heiszwolf, *Catal. Today*, 2010, **154**(3–4), 256–263.
80. E. L. Moreno and K. Rajagopal, *Quim. Nova*, 2009, **32**, 538–542.
81. C. Yue, C. Liu, B. Mezari, A. Brückner, E. A. Pidko, M. S. Rigutto and E. J. M. Hensen, *Appl. Catal., B*, 2016, **191**, 62–75.
82. J. Behin, H. Kazemian and S. Rohani, *Ultrason. Sonochem.*, 2016, **28**, 400–408.
83. J. Zhang, M. Li, Y. Lin, C. Liu, X. Liu, L. Bai, *et al.*, *Microporous Mesoporous Mater.*, 2016, **219**, 103–111.
84. T. J. Pinnavaia, M.-S. Tzou, S. D. Landau and R. H. Raythatha, *J. Mol. Catal.*, 1984, **27**(1–2), 195–212.
85. H. Pérez-Vidal, E. Custodio-García, E. López-Alejandro, J. Morales-Hidalgo and D. M. Frías-Márquez, *Sol. Energy Mater. Sol. Cells*, 2006, **90**(6), 841–846.
86. A. Elmchaouri and R. Mahboub, *Colloids Surfaces A Physicochem. Eng. Asp.*, 2005, **259**(1–3), 135–141.
87. J. G. Carriazo, M. A. Centeno, J. A. Odriozola, S. Moreno and R. Molina, *Appl. Catal., A*, 2007, **317**(1), 120–128.
88. R. W. McCabe and J. M. Adams, in *Handbook of Clay Science*, ed. F. Bergaya and G. Lagaly, Elsevier Ltd, Amsterdam, 2013, vol. 5, pp. 491–538.
89. D. K. Dutta, in *Developments in Clay Science*, ed. R. Schoonheydt, C. Johnston and F. Bergaya, Elsevier Ltd, Amsterdam, 2018, vol. 9, pp. 289–329.
90. J. M. Adams and R. W. McCabe, in *Developments in Clay Science*, ed. F. Bergaya, K. G. B. Theng and G. Lagaly, Elsevier B.V., Amsterdam, 2006, vol. 1, pp. 541–581.

91. E. Gutierrez-Acebo, C. Leroux, C. Chizallet, Y. Schuurman and C. Bouchy, *ACS Catal.*, 2018, **8**(7), 6035–6046.
92. X. Wenyang, Y. Yizhao, X. Xianmei, L. Shizheng and Z. Taoying, *Appl. Catal.*, 1991, **75**(1), 33–40.
93. L. D. Volkova, N. A. Zakarina and A. K. Akurpekova, *Pet. Chem.*, 2014, **54**(1), 37–41.
94. N. Sanabria, A. Álvarez, R. Molina and S. Moreno, *Catal. Today*, 2008, **133–135**(1–4), 530–533.
95. J. M. Adams, J. J. Hooper and A. Gil, in *Reference Module in Materials Science and Materials Engineering*, Elsevier Inc., 2016, pp. 1–8.
96. M. A. Vicente, A. Gil and F. Bergaya, in *Handbook of Clay Science*, ed. F. Bergaya and G. Lagaly, Elsevier Ltd, Amsterdam, 2013, vol. 5, pp. 523–557.
97. R. A. Schoonheydt, C. T. Johnston and F. Bergaya, *Dev. Clay Sci.*, 2018, **9**, 1–21.
98. A. Gil, L. M. Gandía and M. A. Vicente, *Catal. Rev. Sci. Eng.*, 2000, **42**(1–2), 145–212.
99. A. Campos, B. Gagea, S. Moreno, P. Jacobs and R. Molina, *Appl. Catal.*,, 2008, **345**(1), 112–118.
100. C. Sun and J. C. Berg, *Adv. Colloid Interface Sci.*, 2003, **105**(1), 151–175.
101. M. Oyharçabal, T. Olinga, M.-P. Foulc and V. Vigneras, *Synth. Met.*, 2012, **162**(7–8), 555–562.
102. R. Luque, *An. R. Soc. Esp. Quim.*, 2010, **106**, 296–303.
103. B. S. Mojarad, A. Nourbakhsh, R. E. Kahrizsangi, M. Masoud and K. J. D. MacKenzie, *Ceram. Int.*, 2015, **41**(4), 5287–5293.
104. I. Sheet, A. Kabbani and H. Holail, *Energy Procedia*, 2014, **50**, 130–138.
105. M. Kotal and A. K. Bhowmick, *Prog. Polym. Sci.*, 2015, **51**, 127–187.
106. V. Singh, V. Sapehiyia, V. Srivastava and S. Kaur, *Catal. Commun.*, 2006, **7**(8), 571–578.
107. B. Sarkar, R. Rusmin, U. C. Ugochukwu, R. Mukhopadhyay and K. M. Manjaiah, *Modified Clay Minerals for Environmental Applications*, Elsevier, Amserdam, 2019.
108. F. B. Gebretsadik, D. Mance, M. Baldus, P. Salagre and Y. Cesteros, *Appl. Clay Sci.*, 2015, **114**, 20–30.
109. J. P. Chen, M. C. Hausladen and R. T. Yang, *J. Catal.*, 1995, **151**(1), 135–146.
110. P. Yuan, H. He, F. Bergaya, D. Wu, Q. Zhou and J. Zhu, *Microporous Mesoporous Mater.*, 2006, **88**(1–3), 8–15.
111. F. Franco, L. A. Pérez-Maqueda and J. L. Pérez-Rodríguez, *J. Colloid Interface Sci.*, 2004, **274**(1), 107–117.
112. M. Zhang, P. Li, M. Zhu, Z. Tian, J. Dan, J. Li, *et al.*, *Chin. J. Chem. Eng.*, 2018, **26**(9), 1873–1878.
113. L. Chmielarz, M. Rutkowska, M. Jabłońska, A. Wegrzyn, A. Kowalczyk, P. Boroń, *et al.*, *Appl. Clay Sci.*, 2014, **101**, 237–245.
114. S. S. G. Santos, H. R. M. Silva, A. G. De Souza, A. P. M. Alves, E. C. Silva and M. G. Fonseca, *Appl. Clay Sci.*, 2015, **104**, 286–294.

115. C. M. Urruchurto, J. G. Carriazo, C. Osorio, S. Moreno and R. A. Molina, *Powder Technol.*, 2013, **239**, 451–457.
116. O. Gamba, S. Moreno and R. Molina, *Int. J. Hydrogen Energy*, 2011, **36**(2), 1540–1550.
117. J. C. Cortés, M. Muñoz, L. Macías, R. Molina and S. Moreno, *Catal. Today*, 2017, **296**, 205–213.
118. R. A. Schoonheydt, T. O. M. Pinnavaia, G. Lagaly, N. Gangas, È. Kiel and È. A. Chemie, *Pure Appl. Chem.*, 1999, **71**(12), 2367–2371.
119. *Handbook of Clay Science*, ed. F. Bergaya, B. K. G. Theng and G. Lagaly, Elsevier Science, 2006, vol. 1.
120. D. E. W. Vaughan, *Catal. Today*, 1988, **2**(2–3), 187–198.
121. A. Vaccari, *Appl. Clay Sci.*, 1999, **14**(4), 161–198.
122. F. Belkhadem, J.-M. Clacens, A. Bengueddach and F. Figueras, *Appl. Catal., A*, 2006, **298**, 188–193.
123. A. Gil, M. A. Vicente and S. A. Korili, *J. Catal.*, 2005, **229**(1), 119–126.
124. F. Figueras, Z. Klapyta, P. Massiani, Z. Mountassir, D. Tichit, F. Fajula, *et al.*, *Clays Clay Miner.*, 1990, **38**(3), 257–264.
125. J. Pires, M. Brotas de Carvalho and A. P. Carvalho, *Zeolites*, 1997, **19**(2), 107–113.
126. S. V. Awate, S. B. Waghmode and M. S. Agashe, *Catal. Commun.*, 2004, **5**(8), 407–411.
127. M. E. Gyftopoulou, M. Millan, A. V. Bridgwater, D. Dugwell, R. Kandiyoti and J. A. Hriljac, *Appl. Catal., A*, 2005, **282**(1), 205–214.
128. S. D. Bodman, W. R. McWhinnie, V. Begon, I. Suelves, M.-J. Lazaro, T. J. Morgan, *et al.*, *Fuel*, 2002, **81**(4), 449–459.
129. J. T. Kloprogge, E. Booy, J. B. H. Jansen and J. W. Geus, *Catal. Lett.*, 1994, **29**(3), 293–297.
130. C. Pesquera, F. González, M. J. Hernando, C. Blanco and I. Benito, *React. Kinet. Catal. Lett.*, 1995, **55**(2), 267–274.
131. S. M. Bradley and R. A. Kydd, *J. Catal.*, 1993, **142**(2), 448–454.
132. J. T. Kloprogge, W. J. J. Welters, E. Booy, V. H. J. de Beer, R. A. van Santen, J. W. Geus and J. B. H. Jansen, *Appl. Catal., A*, 1993, **97**(1), 77–85.
133. M. Sugioka and T. Kurosaka, *Sekiyu Gakkaishi (J. Japan Pet. Inst.)*, 2002, **45**(6), 342–354.
134. C. Catrinescu, C. Teodosiu, M. Macoveanu, J. Miehe-Brendlé and R. Le Dred, *Water Res.*, 2003, **37**(5), 1154–1160.
135. B.-S. Kim, S.-H. Lee, Y.-T. Park, S.-W. Ham, H.-J. Chae and I.-S. Nam, *Korean J. Chem. Eng.*, 2001, **18**(5), 704–710.
136. S. Mendioroz, A. B. Martín-Rojo, F. Rivera, J. C. Martín, A. Bahamonde and M. Yates, *Appl. Catal., B*, 2006, **64**(3), 161–170.
137. D. Olszewska, *Catal. Today*, 2006, **114**(2), 326–332.
138. N. R. Sanabria, M. A. Centeno, R. Molina and S. Moreno, *Appl. Catal., A*, 2009, **356**(2), 243–249.
139. L. C. A. Oliveira, R. M. Lago, J. D. Fabris and K. Sapag, *Appl. Clay Sci.*, 2008, **39**(3–4), 218–222.

140. P. Yuan, X. Yin, H. He, D. Yang, L. Wang and J. Zhu, *Microporous Mesoporous Mater.*, 2006, **93**(1–3), 240–247.
141. M. L. Occelli, *Catal. Today*, 1988, **2**(2–3), 339–355.
142. F. Kooli and W. Jones, *Chem. Mater.*, 1997, **9**(12), 2913–2920.
143. A. Castro, J. Amaya, R. Molina and S. Moreno, *Catal. Today*, 2019, DOI: 10.1016/j.cattod.2019.04.006.
144. D. V. Cristiano, A. M. Campos and R. Molina, *J. Phys. Chem. B*, 2005, **109**(40), 19026–19033.
145. H. Ming-Yuan, L. Zhonghui and M. Enze, *Catal. Today*, 1988, **2**(2–3), 321–338.
146. F. Figueras, *Catal. Rev.*, 1988, **30**(3), 457–499.
147. S. Moreno, R. S. Kou, R. Molina and G. Poncelet, *J. Catal.*, 1999, **182**(1), 174–185.
148. A. P. Carvalho, A. Martins, J. M. Silva, J. Pires, H. Vasques and M. Brotas de Carvalho, *Clays Clay Miner.*, 2003, **51**(3), 340–349.
149. L. Heller-Kallai, *Appl. Clay Sci.*, 2001, **20**(1–2), 27–38.
150. G. Fetter, P. Salas, L. A. Velazquez and P. Bosch, *Ind. Eng. Chem. Res.*, 2000, **39**(6), 1944–1949.
151. E. Booij, J. T. Kloprogge and J. A. R. Van Veen, *Appl. Clay Sci.*, 1996, **11**(2–4), 155–162.
152. E. Booij, J. T. Kloprogge and J. A. R. van Veen, *Clays Clay Miner.*, 1996, **44**(6), 774–782.
153. S. Moreno, N. Sanabria and R. Molina, Recent tendencies in the synthesis of pillared clays for phenol oxidation, in *Focus on Water Resource Research*, ed. E. Heikkine, Nova Science Publisher, New York, NY, USA, 2008, pp. 185–209.
154. A. Campos, B. C. Gagea, S. Moreno, P. Jacobs and R. Molina, in *From Zeolites to Porous MOF Materials - The 40th Anniversary of International Zeolite Conference*, Studies in Surface Science and Catalysis, Elsevier B.V., vol. 170, 2007, pp. 1405–1410.
155. M. F. Molina, R. Molina and S. Moreno, *Catal. Today*, 2005, **107–108**, 426–430.
156. N. Bouchenafa-Saïb, R. Issaadi and P. Grange, *Appl. Catal., A*, 2004, **259**(1), 9–15.
157. Y. Liu, K. Murata and M. Inaba, *Catal. Today*, 2019, **332**, 115–121.
158. W. Y. Hernández, M. A. Centeno, J. A. Odriozola, S. Moreno and R. Molina, *Mater. Res. Bull.*, 2008, **43**(7), 1630–1640.
159. P. Pushpaletha, S. Rugmini and M. Lalithambika, *Appl. Clay Sci.*, 2005, **30**(3), 141–153.
160. B. Caglar, O. Cubuk, E. Demir, F. Coldur, M. Catir, C. Topcu and A. Tabak, *J. Mol. Struct.*, 2015, **1089**, 59–65.
161. J. A. Martens, E. Benazzi, J. Brendlé, S. Lacombe and R. Le Dred, *Stud. Surf. Sci. Catal.*, 2000, **130A**, 293–298.
162. S. D. Bodman, W. R. McWhinnie, V. Begon, M. Millan, I. Suelves, M.-J. Lazaro, *et al.*, *Fuel*, 2003, **82**(18), 2309–2321.

163. M. L. Occelli, S. D. Landau and T. J. Pinnavaia, *J. Catal.*, 1984, **90**(2), 256–260.
164. O. Mario, *US Pat.*, US4844790 (A), 1989.
165. O. Mario, *US Pat.*, US4761391 (A), 1988.
166. J. C. Cortés, C. E. Scott, P. Pereira-Almao, R. Molina and S. Moreno, *Energy Fuels*, 2019, **33**(6), 5153–5161.
167. J. Sánchez, A. Moreno, F. Mondragón and K. J. Smith, *Energy Fuels*, 2018, **32**(10), 10910–10922.
168. M. T. Nguyen, N. T. Nguyen, J. Cho, C. Park, S. Park, J. Jung and C. W. Lee, *J. Ind. Eng. Chem.*, 2016, **43**, 1–12.
169. M. Y. Visaliev, K. M. Kadiev, L. A. Zekel' and M. Y. Shpirt, *Solid Fuel Chem.*, 2018, **52**(5), 313–319.
170. K. H. Kang, G. T. Kim, S. Park, P. W. Seo, H. Seo and C. W. Lee, *J. Ind. Eng. Chem.*, 2019, **76**, 1–16.
171. K. Li, J. Valla and J. Garcia-Martinez, *ChemCatChem*, 2014, **6**(1), 46–66.
172. H. Shimada, *J. Japan Pet. Inst.*, 2016, **59**(2), 46–58.
173. M. Shahinuzzaman, Z. Yaakob and Y. Ahmed, *Renewable Sustainable Energy Rev.*, 2017, **77**, 1375–1384.
174. G. C. Laredo, P. Pérez-Romo, J. Escobar, J. L. Garcia-Gutierrez and P. M. Vega-Merino, *Ind. Eng. Chem. Res.*, 2017, **56**(39), 10939–10948.
175. P. M. Eletskii, O. O. Mironenko, R. G. Kukushkin, G. A. Sosnin and V. A. Yakovlev, *Catal. Ind.*, 2018, **10**(3), 185–201.
176. T. A. Al-Attas, S. A. Ali, M. H. Zahir, Q. Xiong, S. A. Al-Bogami, Z. O. Malaibari, *et al.*, *Energy Fuels*, 2019, **33**(9), 7917–7949.
177. A. N. Varakin, A. V Mozhaev, A. A. Pimerzin and P. A. Nikulshin, *Catal. Today*, 2019, DOI: 10.1016/j.cattod.2019.06.005.
178. Y. Han, M. Gholizadeh, C.-C. Tran, S. Kaliaguine, C.-Z. Li, M. Olarte and M. Garcia-Perez, *Fuel Process. Technol.*, 2019, **195**, 106140.
179. A. Patrigeon, E. Benazzi, C. Travers and J. Y. Bernhard, *Catal. Today*, 2001, **65**(2–4), 149–155.
180. L. V. Duong, J. T. Kloprogge, R. L. Frost and J. A. R. Van Veen, *J. Porous Mater.*, 2007, **14**(1), 71–79.
181. A. M. Campos, S. Moreno and R. Molina, *Catal. Today*, 2008, **133–135**, 351–356.
182. A. Olaya, S. Moreno and R. Molina, *Catal. Commun.*, 2009, **10**(5), 697–701.
183. E. J. M. Hensen, D. G. Poduval, D. A. J. M. Ligthart, J. A. R. Van Veen and M. S. Rigutto, *J. Phys. Chem. C*, 2010, **114**(18), 8363–8374.
184. R. Kumar, B. S. Rana, D. Verma, S. Rayaroth, V. S. Prasad and A. K. Sinha, *RSC Adv.*, 2015, **5**(49), 39342–39349.
185. J. A. Torres-Luna, S. Moreno, R. Molina and J. G. Carriazo, *Energy Fuels*, 2018, **32**(9), 9782–9792.

CHAPTER 3

Catalytic Conversion of Fossil and Renewable Fuel Resources: Approaches Using Sub and Supercritical Water as a Reaction Medium

P. ARCELUS-ARRILLAGA,[a*] J. L. PINILLA[b] AND M. MILLAN[c]

[a]Universidad Iberoamericana Ciudad de México, Department of Chemical, Industrial and Food Engineering, Prolongación Paseo de la Reforma 880, Col. Lomas de Santa Fe, Álvaro Obregón, Mexico City 01219, Mexico
[b]Instituto de Carboquímica, CSIC, Calle Miguel Luesma Castán 4, Zaragoza 50018, Spain
[c]Imperial College London, Department of Chemical Engineering, London SW7 2AZ, UK
*Corresponding contributor. Email: pedro.arcelus@ibero.mx

3.1 Introduction

The world's population reached 7.7 billion people in 2019, which represents an increase of 35% over a period of 25 years. Although the population growth rate has slowed down in the past few years, it is expected that the global

Heterogeneous Catalysis for Energy Applications
Energy and Environment Series No. 27
Edited by Tomas R. Reina and Jose A. Odriozola

Published by the Royal Society of Chemistry, www.rsc.org

population will be approximately 26% higher by the year 2050.[1] This will present great challenges to society such as global warming, poverty, energy shortages and waste generation, among others. The impact is expected to be especially hard in undeveloped countries where population growth will be considerably higher.[1] It is forecasted that the world's energy consumption will continue to increase in the coming decades and is expected to reach around 22,950 million tonnes of oil equivalent (Mtoe) by the year 2050, which is a 45% increase from current consumption.[2] Most of this energy comes from fossil fuel sources, especially from liquid fuels that account for approximately 32% of the energy used worldwide. Although the share of fossil fuels in the energy market is expected to drop in the coming decades, liquid fuels will remain among the main sources of energy. Therefore, in order to meet the world's future energy demand, it is necessary to improve and develop cleaner and more effective technologies to transform unconventional energy sources such as heavy crude oil and biomass into fuels. Crude oil production will have to increase from 83.2 to approximately 107 million barrels per day (mbpd) by the year 2050,[2] which means that an increase of around 34% in the production of heavy and extra-heavy oil has to be reached by 2035.[3] These oils have a high propensity to cause processing problems due to their chemical nature, resulting in catalyst deactivation, coke deposition and low liquid fuel yields.[4,5] It is also necessary to increase the share of renewable sources of liquid fuels such as biomass and organic waste in the energy mix to meet future energy targets. There is a lack of complete agreement regarding the real potential of biomass as an energy source, with studies estimating it will contribute 12–47% of the world's energy consumption in 2050.[6] The reason for such a wide range may be the lack of certainty in terms of biomass supply, land availability, environmental impacts and competition with other socioeconomic sectors such as food and material industries.[7] Another important challenge is ensuring that adequate technology is in place to transform, in an efficient way, unconventional liquid fuel sources to reduce by as much as possible any negative impacts on society and the environment. The use of water in hydrothermal conditions near or above its critical point can provide an efficient route to obtain fuel from unconventional sources due to its particular physicochemical properties that make it an excellent reaction medium for organic chemical reactions. The use of catalysts for hydrothermal processing plays an important role in ensuring that the quality of the fuel is adequate and that high yields and selectivity are obtained.

3.2 Overview of Water at Sub and Supercritical Conditions

Hydrothermal treatment is an effective route to transform unconventional fuel sources like heavy oil, biomass and organic waste into fuel and other value-added products. A hydrothermal process is one that is carried out in water at high temperatures as a reaction medium. Water is very attractive as a reaction medium due to its abundance and because it is a green and non-toxic solvent that presents no health risks. In addition, it is a very versatile solvent that changes its physicochemical properties and reactivity with small variations in process conditions (temperature and pressure) in the presence or absence of a catalyst.[8] This allows for fine-tuning of the process to obtain a wide range of products of different natures by simply modifying process conditions. It is also to be noted that the implementation of water at sub or supercritical conditions has an advantage over other upgrading technologies in that it allows for the treatment of feedstocks with high humidity without the need for an energy-intensive pre-drying stage.[9,10]

Water at ambient conditions is a strong solvent for dissolving polar compounds such as salt and a weak solvent for organic compounds and gases.[10] However, the characteristic properties of water change with increasing temperature, resulting in greater movement of molecules and lower density, leading to fewer and weaker hydrogen bonds.[10,11] Water turns into a supercritical fluid when it reaches a temperature and a pressure of 374 °C and 221 bar or above, respectively.[12] A phase diagram of pure water at sub and supercritical conditions as a function of pressure and temperature is shown in Figure 3.1.

Although it has been found that some degree of hydrogen bonding takes place in supercritical water (SCW), this is almost completely disrupted, changing water solvent properties from a good solvent for ionic compounds to a good one for gases and non-ionic organic compounds of an oily nature.[10–12]

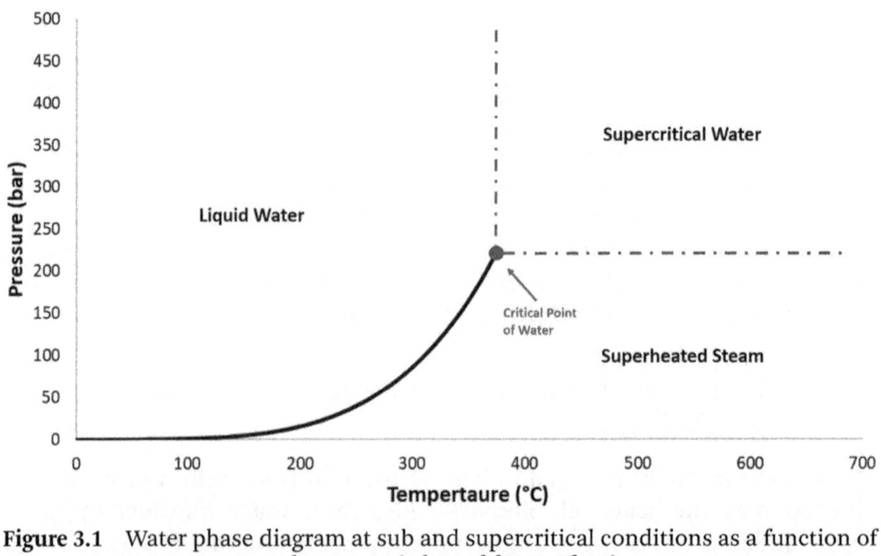

Figure 3.1 Water phase diagram at sub and supercritical conditions as a function of temperature and pressure (adapted from ref. 13).

3.2.1 Properties of Sub and Supercritical Water

In the vicinity or above a critical point, the physicochemical and thermal properties of water, such as density, viscosity, heat capacity and thermal conductivity, have large variations and differ from those of water at ambient conditions, as is shown in Figure 3.2. These variations are related to other changes occurring in water at a molecular level that can affect chemical reactions, such as the dielectric constant, ion product, molecular diffusivity and hydrogen bonding.[14]

Figure 3.2 shows that SCW has liquid-like density and gas-like viscosity, which make it a good reaction medium with favourable transport properties. In addition, water is highly compressible in the vicinity of this critical point. It demonstrates large changes in density with minor changes in temperature or pressure, which enable the fine-tuning of the reaction medium properties. Solvation power in water is greatly influenced by density and shows increasingly non-polar solvent behaviour when approaching supercritical conditions.[14,15] This is reflected in an important decrease in the dielectric constant that influences the solubility of chemical species. SCW is of great technological interest as it can provide an opportunity to find new reaction mechanisms. The dependence of water density and its solvation properties on temperature and pressure allows for thermodynamic control of reactivity, which can determine the dominant reaction mechanisms at supercritical conditions.[16]

The ion product or autoionization constant of water is dependent on temperature, showing an initial increase and reaching a maximum at near-critical

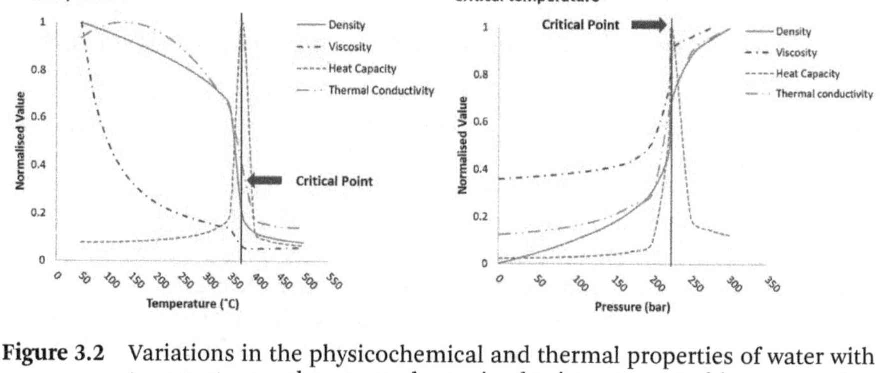

Figure 3.2 Variations in the physicochemical and thermal properties of water with temperature and pressure determined using a Peng–Robinson equation of state.

water (NCW) conditions, then decreasing sharply at temperatures above the critical point.[17] This behaviour is thought to be the result of two competing factors. The first is related to the dissociation of water, which is an endothermic process favoured by higher temperatures. The second is the decrease in density and the dielectric constant with temperature, which disfavours ion formation. Above the critical point, ion products decrease by about five orders of magnitude due to the breakdown of the solvent shell at lower densities.[10,14,17] Therefore, subcritical water is a stronger acid and base than water at ambient conditions due to its considerably higher concentrations of H_3O^+ and OH^-, which suggests that it is a good medium for acid- and base-catalysed reactions.[18] By contrast, the lower density of SCW promotes free radical-based reaction mechanisms.[19,20]

Water near or above the critical point shows transport properties similar to those of vapour. An increment in temperature results in a decrease in viscosity that reaches values of around 10^{-2} cP at sub and supercritical conditions. Low vapour-like viscosity values result in high diffusion coefficients adequate for increasing reaction rates in mass transfer-limited processes.[19]

The thermal properties of water show large variations with small changes in temperature and pressure, especially near the critical point, as is seen in Figure 3.2. Heat transfer at these conditions is considerably enhanced and differs greatly from that observed at subcritical conditions.[21] For instance, the heat capacity of water finds a maximum near the critical point, and similar behaviour is seen for the heat transfer coefficient. Large variations in the physicochemical properties of water at these conditions can lead to heat transfer deterioration. This is potentially caused by an increase in heat flux that results in a decrease in the peak of the heat transfer coefficient.[21,22]

The thermal conductivity of high-pressure water shows interesting behaviour with increments in temperature, as is shown in Figure 3.2. Initially, it shows an increase in thermal conductivity due to the decomposition of hydrogen bonds.[23] Then it sharply decreases with further temperature increments when approaching and surpassing the critical temperature.

3.2.2 Near-critical and Supercritical Water as Reaction Media

Hydrothermal upgrading is a thermochemical process that uses hot compressed water to transform fossil and renewable feedstocks, such as heavy oil, biomass and organic residues, into hydrocarbon fuels and other high-value products. As previously mentioned, at near-critical and supercritical conditions, water possesses particular physicochemical and thermal properties that can be finely tuned with small changes in operating parameters. In this way, it can provide adequate conditions to upgrade different raw materials. Additionally, water is regarded as a green, non-toxic substance that at near-critical and supercritical conditions has multiple roles in organic chemical reactions. It can act as a reactant, catalyst and solvent in the same process.[24] Hydrothermal processes at these conditions have been studied for the upgrading of unconventional fossil resources such as heavy and extra-heavy oil,[4,25] bitumen[26,27] and residue fractions,[28] as well as to produce fuel and value-added products from renewable feedstocks such as microalgae,[29,30] macroalgae,[31] terrestrial biomass[9,14,15] and organic residues.[32,33]

The properties of water at near or above the critical point are responsible for the solubilization and favourable dispersion of heavy organic components. It also helps to suppress polymerization, preventing recombination reactions by surrounding active organic intermediates (radicals) with water molecules, thereby reducing reaction rates in what is known as the cage effect.[34] This results in a higher collision frequency between water molecules and heavy organic radicals than that observed between two heavy organic radicals,[34] preventing their recombination. By contrast, heavy organic radicals can continue to react with gases and other small molecules, as well as further crack to produce lighter organic products, as is shown in Figure 3.3.

Therefore, the cage effect suppresses the condensation reactions that commonly occur in transformation and upgrading water-free processes. For example, during heavy oil upgrading, coke is formed from the polymerization reaction between heavy organic radicals formed during asphaltene thermal cracking.[36] By contrast, when water in hydrothermal conditions is used as a reaction medium, it acts as a physical barrier that prevents coke formation. In addition, coke suppression in hydrothermal conditions is aided by the capacity of the reaction medium to donate hydrogen (quench heavy organic radicals) and extract the aqueous-phase coke precursors resulting from cracking reactions, which prevents the occurrence of recombination reactions leading to coke formation. On the one hand, a high extraction capacity is observed in

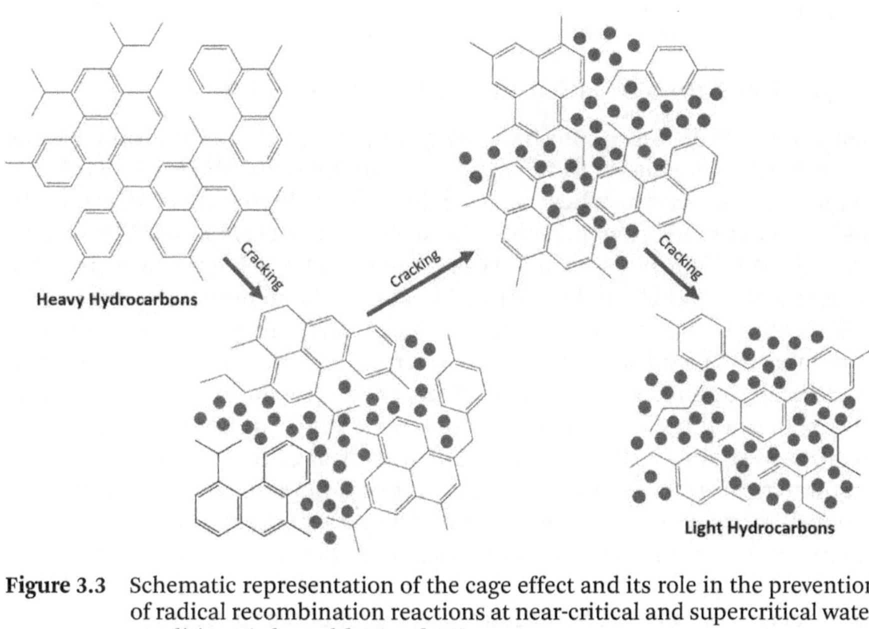

Figure 3.3 Schematic representation of the cage effect and its role in the prevention of radical recombination reactions at near-critical and supercritical water conditions (adapted from ref. 35).

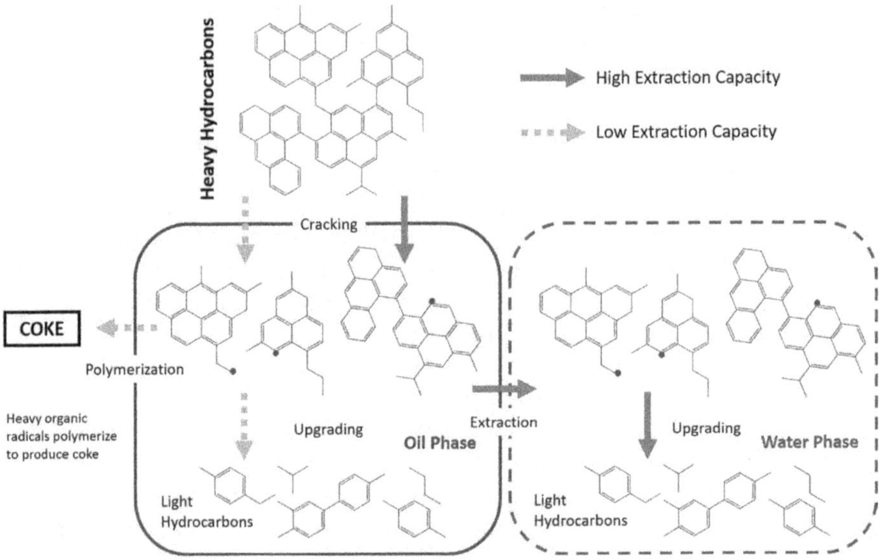

Figure 3.4 Coke suppression in hydrothermal conditions: high- and low-extraction capacity scenarios (adapted from ref. 4, 35 and 37).

high-density water at near or above the critical point where coke precursors are removed to the water-rich phase, thereby avoiding coke formation. On the other hand, at conditions far from the critical point, a low organic molecule extraction capacity is observed, leading to re-polymerization reactions that form coke, as is shown in Figure 3.4.[4,37]

3.2.3 Reactions in Sub and Supercritical Water

Reactions in SCW conditions have many advantages over conventional liquid or gas reaction media. In SCW, higher reaction rates are obtained mainly due to the reduced mass and heat transfer limitations (high diffusivities) and the high solvent capacity that prevents coke deposition over catalysts.[19,38,39] In addition, the solubility of gases in SCW is greater, enabling higher gas concentrations than those that can be achieved in conventional liquid–gas systems, as well as eliminating any mass transfer limitations through the interphase of both fluids.[40] The role of water in hydrothermal processes is still under debate. It has been reported that SCW participates as a reactant in oil upgrading processes, promoting the cleavage of heterocyclic sulfur and nitrogen structures in oil. Early studies on the influence of water at near-critical conditions showed that water can promote hydrolytic decomposition reactions and retard solid product formation as it almost completely suppresses polycondensation reactions.[41] Near the critical point, water becomes a favourable medium for hydrolysis reactions where water can simultaneously play three roles, namely as a reactant, as an acid homogeneous catalyst and as a reaction medium.[24]

It has also been reported that salt in hydrothermal conditions can produce an acidic or alkaline solution that can participate in bond scission reactions.[19]

The generation of *in situ* hydrogen is of great relevance for the production and upgrading of fuel in hydrothermal conditions. As has been mentioned, at near or above the critical point of water, hydrogen bonds characteristic of water at ambient conditions are almost completely disrupted. This enables water to become a hydrogen donor for chemical reactions.[19] Hydrogen donation was demonstrated in experiments using deuterated water in hydrothermal conditions to track hydrogen fate in reactions. It was observed that hydrogen exchange from water into the products took place.[42,43]

Another important route for the production of *in situ* hydrogen is the water gas shift reaction. This considers the reaction between CO and water to generate H_2 and CO_2.

$$CO + H_2O \leftrightarrow CO_2 + H_2$$

CO is formed from the partial oxidation of organic compounds, and it reacts with water in the presence of a catalyst that can provide oxygen vacancies, has a high CO conversion rate, has a low CO adsorption strength and is capable of participating in the dissociation of water. Different catalytic materials have been proposed for water gas shift reactions, such as Cu-, Al-, Zn-, Ni-, Ce- and Fe-based catalysts, supported metal nanoparticles, supported noble metals over metal oxides, *etc.*[44,45] The presence of CO and high-density water may contribute to favouring hydrogen production and accelerate reaction rates. It is believed that water participates in the formation of a transition-state complex that modifies the activation energy and enables homogeneous water gas shift reactions in the absence of a catalyst. This proceeds in a two-step mechanism in which formic acid as an intermediate product is formed from the reaction between CO and H_2O and then decomposes into CO_2 and H_2.[46]

Oxidation reactions also take advantage of the physicochemical properties of NCW and SCW, in which oxygen has very high solubility and diffusivity, enabling reactions to occur in a pseudo-homogeneous phase. In SCW, oxidation reactions proceed at very high rates, resulting in high organic compound conversions to other intermediate molecules, water and CO_2.[47,48] SCW oxidation is thought to proceed through two main reaction pathways, the first being the direct oxidation of organic compounds into H_2O and CO_2 and the second being a potential two-step process in which a stable intermediate product is formed from the partial oxidation of an organic compound followed by further oxidation reactions leading to the formation of H_2O and CO_2. The formation of these stable intermediates is considered to be the rate-determining step in the process.[49] Hydrothermal oxidation reactions with limited oxygen supply (partial oxidation) can also lead to valuable intermediate products such as quinones, alcohols, anhydrides, esthers, *etc.*[50–53] These can also be the starting reactions to form CO, which can be transformed into H_2 and CO_2 through water gas shift reactions, as explained earlier.

3.2.4 Catalysis in Sub and Supercritical Water

Research into catalytic NCW and SCW processes has received great attention in the past two decades. The majority of investigations were focused on the improvement of chemical synthesis, fuel production from biomass, coal and oil and the destruction of waste materials.[54] The addition of a catalyst to hydrothermal upgrading processes has the purpose of increasing reaction rates and product yields, suppressing the formation of solid residues, reducing the rate of repolymerization of intermediate species and accelerating cracking reactions.[30,33,40] The key to the successful introduction of catalytic hydrothermal conditions for commercial organic chemical processes is the ability of the material to catalyse C–H, C–C and C–heteroatom bond formation or cleavage.[40] It is also desirable to find a catalyst that is sufficiently active and able to promote reactions such as partial oxidation coupled with water gas shift and hydrolysis to obtain high-value products in NCW and SCW.[4] The nature of this reaction medium presents great challenges for researchers to find a catalyst that shows high activity and is sufficiently stable to endure the reaction conditions without deactivation.[4,55]

Homogeneous (soluble in the medium) and heterogeneous (having two or more phases) catalysts have been studied in hydrothermal conditions. Traditional homogeneous catalysts consist of various acidic and alkaline species as well as transition metal salts.[54] It has been reported that SCW reactions for chemical synthesis such as hydration, dehydration, hydrogenation, dehydrogenation, oxidation, decarboxylation and isomerization occur in the presence of catalysts of this nature, while no reactions are observed in their absence.[56,57] The desired reaction dictates the catalyst that will be implemented. In a study on the hydrothermal liquefaction (HTL) of microalgae, a comparison of the bio-oil obtained with acidic or alkaline homogeneous catalysts showed that yields of bio-oil are higher in the presence of organic acids. It was also reported that acid-catalysed processes resulted in a bio-oil with better flow properties and a lower boiling point. However, the heating value (HHV) of the bio-oil was higher with alkaline catalysts. Overall, the presence of a homogeneous catalyst was not enough to reduce the nitrogen content in bio-oils.[58] The use of metal salts has been considered for SCW gasification reactions. In a study on the gasification of glucose as a biomass model compound, the introduction of Ni, Co and Fe acetylacetonate salts did not produce much enhancement compared to the yields obtained in the absence of a catalyst. However, in the same work, experiments performed with Ni-Raney catalysts showed great activity, increasing gas yields considerably compared to those obtained with metal salts.[59]

Transition metal heterogeneous catalysts are classified into supported or unsupported materials. Unsupported metal catalysts are subdivided into powders, wires or skeletal structures. The drawback of using unsupported metal catalysts, especially powders or wires, is the small surface area and the limited number of active sites available for reaction. Unsupported metal catalysts with a skeletal structure (Raney) are spongy metal structures obtained from an aluminium–metal alloy through an aluminium leaching process.[60] Different Raney catalysts (Ni-Raney, Co-Raney and Cu-Raney) were tested under SCW gasification conditions. Ni-Raney catalysts showed greater catalytic activity than Co-Raney or Cu-Raney catalysts, as well as Ru supported on carbon and Al_2O_3. This is thought to be the result of the higher active metal concentration per weight of catalyst found in Ni-Raney catalysts compared to the rest, which results in more active sites being available for reaction.[61] Ni-Raney catalysts may contain greater amounts of hydrogen in their structures than can be chemisorbed onto the surfaces of other catalysts. This hydrogen can be released under hydrothermal conditions in significant amounts.[60] The released hydrogen can then participate in the *in situ* upgrading of products under hydrothermal conditions.

One of the most important properties in supported metal catalysts, which is normally overlooked, is the nature and stability of the support. Different supports result in activity variations in the supported catalysts. This depends on the nature of the support and the pore size of the catalyst. If pores are small, only the metal supported on the external surface contributes to reaction.[62] Another important catalyst design variable to consider is the acidity of the support, which can promote specific reactions under hydrothermal conditions.[60] Materials that have been studied as supports for heterogeneous catalysts in hydrothermal applications are metal oxides such as SiO_2,[63,64] ZrO_2 and ZrO_2–Al_2O_3,[65,66] Al_2O_3[67,68] and TiO_2,[69,70] as well as carbon materials such as activated carbon,[71,72] carbon nanofibres[72,73] and carbon nanotubes.[74] Metal-supported catalysts are normally synthetized with nickel[60,74] or noble metals[75] as active catalytic species. Nickel-based catalysts are cheaper than noble metal catalysts and have been shown to be active under high-pressure hydrothermal conditions.[76] However, noble metal catalysts present better metal dispersion on the catalytic surface and also higher resistance to deactivation due to sintering.[60] Heterogeneous catalysts to be applied in hydrothermal processes near or above the critical point need to be resistant to reaction conditions (catalytic support degradation or oxidation) and to the deactivation (catalyst coking and poisoning from inorganic deposition) that is common to hydrocarbon upgrading processes.[14] One good option is to use materials that have been synthetized in similar conditions to those of hydrothermal processing, which can increase resistance to deactivation as they have already been exposed to sintering, water adsorption or removal of catalytic components through dissolution.[74]

The nature of the process will dictate the catalyst to be used. Homogeneous catalysts have the advantage of being dissolved in the same phase as the reactants, reducing limitations due to mass transfer and not presenting problems derived from deactivation or coking. However, heterogeneous catalysts are easily recovered after the process, which makes them highly desirable from an economic point of view.[14] Application of catalysts in the hydrothermal processing of unconventional fuel sources in near-critical and supercritical conditions is extensively discussed in the following sections.

3.3 Catalytic Upgrading of Oil in Sub and Supercritical Water

Oil conversion research involving SCW systems has focused on the upgrading of oil residues and unconventional oils, such as tar sands. A number of studies were carried out on vacuum residues (VRs), which are low-value products of crude oil distillation and can represent a significant proportion of heavy crude oils. This fraction is rich in asphaltenes, sulfur, nitrogen and metals such as V and Ni. Competing technologies are fluid catalytic cracking, visbreaking and, arguably to a lesser extent, hydrocracking, as the industrial application of the latter is typically confined to lighter fractions. A summary of non-catalytic reactions in SCW will be presented before focusing on catalytic systems.

Most of the literature on heavy fractions deals with processes where the objective is a decrease in molecular weight, which in turn improves other oil properties, most notably by reducing viscosity and heteroatom content, resulting in a particularly promising technology for desulfurization. In the several studies conducted with VR and bitumen as feedstocks, assessment of the extent of upgrading is typically based on changes in saturates, aromatics, resins and asphaltenes (SARA)[77] analysis, variations in boiling point distributions and the production of a liquid phase. These experiments showed that, in SCW, VRs can be converted into lighter fractions as well as coke. Rather than presenting a comprehensive list of feedstocks and process conditions, this section aims to summarize the main effects of SCW on these upgrading reactions.

Research on the role of SCW in oil upgrading processes has led to two potential ways in which SCW may affect the reactions being postulated and assessed. First, SCW could act as a solvent, stabilizing some cracking products and inhibiting them from taking part in the condensation reactions that would produce coke. Second, SCW could act as hydrogen donor, where H from water is incorporated into the organic species generated thermally by a radical mechanism. These hypotheses have been examined in work with model compounds as well as oil-derived fractions.

Addressing the solvent effect of SCW, studies with hexadecane (C16) and polyethylene[78] separately reacting in SCW resulted in similar reaction rates, product yields and distributions as the reaction taking place in an inert gas (Ar). This suggests little influence of SCW, and it was proposed that the reactions are the same in SCW as in gas-phase pyrolysis, with water acting as an inert solvent being no more advantageous than traditional pyrolysis. On the other hand, it has been reported that coke yields decreased in water compared to what would be expected in systems purely consisting of pyrolysis. Experiments with a VR in the absence of a catalyst showed a decrease in the concentration of aromatics, resins and asphaltenes and an increase in saturates, with coke yields staying at low levels (gas was unreported).[77] A comparison between heavy oil visbreaking in N_2 and SCW[79] led to the conclusion that although the same reaction mechanism was followed in both media, the reaction time for dealkylation of aromatics, a key reaction leading to decreases in oil viscosity, was shorter in SCW. This had the positive effect of suppressing the extent of condensation reactions for aromatics, which is a pathway leading to coke. The ease with which bonds between aromatic cores and alkyl substituents can be cleaved was confirmed in studies with heptyl-benzene and heptyl-naphthalene.[80]

Whether water can act a hydrogen donor was also addressed in research involving model compounds. The use of a free radical precursor such as tert-butylbenzene[81] as a model compound led to the determination that no hydrogen donation from deuterated water appeared in the product. Supporting this observation, mechanisms assessed by quantum mechanic techniques showed that direct hydrogen donation from water to hydrocarbons is unfavourable.[82] Although this seemingly contradicts evidence from the steam cracking of $C_{32}H_{66}$ using deuterated water, which showed deuterium exchange between hydrocarbons and water, the incorporation of H from water is not ascribed to direct radical exchange.[83] Instead, a mechanism involving olefin isomerization reactions, by which the position of a C=C double bond is shifted in an alkene molecule, was identified as the route of hydrogen exchange between water and hydrocarbons. It is worth highlighting that this mechanism was observed in subcritical conditions, where ionic pathways are favoured over radical ones, and the reactions were acid-catalysed.[84]

It has been reported that the introduction of a hydrogen precursor was more effective than the use of molecular H_2. Upgrading of a bitumen derived from oil sand by steam-assisted gravity drainage was carried out in SCW with the addition of formic acid.[85] This technology led to significant conversion of asphaltenes, suppression of coke formation and an overall conversion to compounds with lower molecular weights. On the other hand, it was reported that asphaltenes in SCW formed a relatively small amount of gas (4.3%) but a large coke residue amounting to nearly 48.6% of the yield.[86] Despite the relatively low yields of the hexane-soluble fraction of interest, analysis carried out on these products revealed that it was composed of lighter aliphatic and aromatic species.

The use of heterogeneous catalysts has been investigated as a way of activating hydrocarbons and promoting hydrogen generation. Research has been carried out on a range of catalysts, which has included: metal oxides and sulfides; typical hydroprocessing catalysts based on $NiMo/Al_2O_3$ and $CoMo/Al_2O_3$; other metals supported on Al_2O_3; carbon materials, mainly activated carbons; a phosphomolybdic acid hydrate; and zeolites. Some of these reports are summarized below.

Metal oxides originally attracted interest as catalysts for the SCW oxidation of organic-containing waste. These catalysts have included MnO_2, CeO_2, TiO_2, CuO, CoO, ZnO and some of their mixed oxides. These oxides are typically inexpensive, and in some cases the possibility of synthesizing them in hydrothermal conditions may become an additional advantage.[87] Work by Ahmadi and co-workers addressed the activity, selectivity and potential mechanisms of various metal oxides in a series of articles.[88–90] These oxides have been observed to increase maltene yields at the expense of asphaltenes over what was achieved in non-catalytic reactions in SCW, with some concurrent coke formation. It has been suggested that redox cycles between the different oxidation states of a metal oxide were instrumental in these reactions, a mechanism found in the chemical looping oxidation of hydrocarbons.[91] The metal is reduced by oxidizing the hydrocarbon feed with lattice oxygen. Hydrogen is produced by the water gas shift reaction and the regeneration of the oxidation state of the catalyst by water, and this hydrogen is available to react with the feed.

The redox cycle described above was observed in experiments on upgrading of a VR in SCW with α-Fe_2O_3 nanoparticles, which were partially reduced to Fe_3O_4 (magnetite).[88] Whether this partial reduction reflects a stage in a redox cycle or a net conversion of the metal oxide has not been elucidated. Experiments without catalyst led to the coke yield increasing in greater proportion than asphaltene conversion, suggesting that asphaltenes and a fraction of maltenes formed coke. The addition of the catalyst led to a decrease in the coke and asphaltene yields combined, but the relative maltene and gas yields were not provided.[88] Supporting Fe_2O_3 in SiO_2 improved the stability of this phase, although some degree of particle agglomeration took place.[92] Other oxides, such as ZnO and MoO_3, have been investigated in work with model compounds.[93] MoO_3 was also found to react with the sulfur present in the feed and to undergo structural changes, forming a mixture of Mo sulfide and oxide characterized as $MoS_{0.12}O_{1.88}$. The tendency of oxides to undergo reduction in SCW has also been observed for MnO_2 and Co_3O_4, which were converted to Mn_2O_3 and CoO, respectively.[90] However, other oxides such as CeO_2 did not show changes in oxidation state.[90] The latter catalyst also showed higher conversion of asphaltenes to maltenes.

Iron(III) oxide catalysts have also been supported over Zr and Zr–Al_2O_3 and their catalytic effect on upgrading heavy oils with steam was assessed. Active oxygen species were detected, leading to high light oil yields and nearly no coke formation, which supports the redox mechanism postulated above.[66,94–97] The conditions studied were bitumen in water, iron or ruthenium water solutions at different temperatures. Fe(II) was also used as a catalyst for bitumen upgrading and it was found to minimize the formation of insoluble products.[98] Among other active oxides, PtO_2 was observed to catalyse SCW dehydrogenation of model compounds such as cyclohexanol.[56,99]

The use of zinc and aluminium shavings led to greater bitumen conversion and hydrogenation in SCW, leading to a higher H : C ratio. There were clear variations in the composition of products compared to the use of non-catalytic reactions in SCW alone. Yields to volatiles increased by 15.3% and 38.2% with zinc and aluminium, respectively.[100] Zinc generated a product yield of up to 62.3% with a high yield of resins of up to 33.5% and a high oxygen content. On the other hand, yields of resins were 7.5% with aluminium and there was an absence of oxygen in the liquid products. In both cases, the viscosity of bitumen was reduced as the molecular weight of asphaltenes decreased.[100] Similar results were obtained when heavy oil was upgraded *via in situ* hydrogen formation with a range of oil- and water-soluble metallic catalyst precursors in a water–synthesis gas atmosphere.[101]

A phosphomolybdic acid hydrate as catalyst showed an improvement over a non-catalytic SCW processes, mainly in decreasing asphaltene and resin contents and coke formation. Experiments were carried out with hydrogen sources such as CO added to the water to form hydrogen *via* water-gas shift reaction (WGSR). This caused a further reduction in yield of a toluene-insoluble fraction and resulted in the highest yields of cracking products between at 300 and 500 °C.[101]

Catalytic cracking of hydrocarbon model compounds including, among others, dodecane and hexadecane was performed using zeolites as a catalyst.[102] The stability of the zeolite was found to depend on its structure. Three zeolites, ZSM-5, HY and β-zeolite, were tested in cracking dodecane in a fully miscible mixture of supercritical dodecane and SCW. HY and β-zeolite lost their crystallinity, but ZSM-5 managed to retain it despite the loss of porosity and some of the acid sites.

Other model compounds employed in studying reactions of oil fractions in SCW are polycyclic aromatics, which are component structures found in asphaltenes. As mentioned above, these molecules are stable in SCW, but the presence of a catalyst triggers cracking.[63] Yeletski *et al.* showed that stable aromatic structures can be converted in the presence of a catalyst in SCW. Their experiments tested $NiMo/SiO_2$ catalysts for phenanthrene cracking. Conversions after 1 h were between 12.3% and 24.0%, increasing with the Ni content of the catalyst. Both oxidation (benzophenone and xanthene) and hydrogenation (dihydrophenanthrene) products were detected.[63]

A large share of the attention and detailed studies on upgrading process in SCW have been devoted to heteroatom removal, mainly desulfurization, where SCW can have a clearly positive effect. Desulfurization of aliphatic species can be achieved in SCW without the addition of catalysts. However, aromatic sulfur needs a catalyst in order to be removed. Deoxygenation is more relevant to biomass (and coal) feedstocks and is covered in a separate section. Denitrogenation has been studied with model compounds such as pyridine. Although in general little work has been carried out on catalytic denitrogenation in SCW, Duan and Savage investigated the conversion of pyridine with the use of a range of catalysts, including noble metals such as Pt and Rh, and a variety of supports, with C and Al_2O_3 among them. Pyridine conversion was complete within up to 4 h with noble metal catalysts. The most active catalysts, PtO_2 and $Pt/\gamma\text{-}Al_2O_3$, yielded total conversion in 1.5 h.[103]

Model compounds have also played a key role in developing an understanding of the type of desulfurization reactions taking place in SCW. In the absence of a catalyst, dibenzylsulfide led to the formation of toluene, xylene, diphenyl and diphenyl methane as well as benzaldehyde, which is evidence that oxygen from water was incorporated into the organic species.[104] It was also observed that H atoms derived from water were incorporated into the H_2S formed in the process. It was found that a Ru-based catalyst promoted C–S bond cleavage, and this led to higher amounts of insoluble products due to recombination of the reactive fragments generated.

A study by Ates *et al.* tested ZnO, MoO_3 and MoS_2 as catalysts for upgrading Arabian heavy crude oil in SCW. Particular emphasis was placed on assessing the desulfurization of this crude with 3 wt% S content. It was found that the desulfurization in SCW in the absence of a catalyst was limited to below 10%. The use of catalysts increased it to nearly 12%, but it was much lower than in work with model compounds such as DBT and hexyl sulfide in hexadecane, where sulfur removal reached greater than 85%. MoS_2 was resistant to change under these the reaction conditions, but ZnO and MoO_3 suffered changes due to sulfidation reactions. Desulfurization in the absence of hydrogen was considered a promising result of treatment in SCW.[93]

CoMo and NiMo hydrotreating catalysts were also used in these conditions. They work in the presence of H_2. CO was an effective precursor for H_2 production through the water gas shift reaction. Typical hydrotreating catalysts such as NiMo/Al_2O_3 have worked in SCW for the desulfurization of DBT as a model compound in the presence of H_2 and H_2 precursors such as CO and formic acid. The latter led to higher S conversions, showing that *in situ*-produced H_2 (through WGSR) is more reactive than molecular H_2 itself.[105]

Catalyst deactivation is key for processes in SCW. In addition to poisoning and sintering, which are common deactivation processes in heavy hydrocarbon processing, there is the issue of the stability of the supports in SCW. Common supports such as γ-Al_2O_3 degrade within 1 h, with a large decrease in surface area. On the other hand, α-Al_2O_3 is stable under hydrothermal conditions. Similarly, supports based on carbon, ZrO_2 and rutile titania (but not anatase) seem to be stable.[106]

3.4 Catalytic Upgrading of Biomass and Organic Waste in Sub and Supercritical Water

Biomass is a general term that includes by definition any organic matter available on a recurring basis, such as agricultural and forestry residues, herbaceous crops, algae and aquatic plants and wastes from different origins (animal, food and feed processing, municipal and industrial).[107–109] Biomass compositions vary significantly depending on the biomass origin, and carbohydrates, lignin, protein and lipids are the main components commonly found in various biomass feedstocks.[108]

Biomass, given its availability, is considered an alternative carbon-based resource that can be used for the production of valuable products, including biofuels, bio-based materials and chemicals.[110,111] Among the different strategies that can be applied in order to achieve this goal, thermochemical conversion is the most widely studied, and mainly includes combustion, pyrolysis, gasification and liquefaction. The ultimate objective of thermochemical conversion technologies is the degradation of biomass into small-molecule fragments by heat application. These processes can be classified according to their reaction temperature, reaction medium and target products. Liquefaction, in contrast to gasification and pyrolysis, involves the liquid-phase thermal deconstruction of biomass under low temperature (200–400 °C) and high pressure (5–20 MPa) in the presence of solvents. During this thermochemical process, biomass is converted into three products: a bio-oil fraction, a gas fraction and a solid residue fraction. The type of solvent used classifies the liquefaction technologies into solvent liquefaction, which employs organic solvents such as methanol and ethanol, and HTL, in which water is used. The use of water as a solvent in biomass liquefaction technologies is probably the most commonly used mainly because it is an environmentally benign and cheap solvent. Another advantage of the HTL of biomass is the potential for direct processing of wet biomass, making it possible to eliminate the drying process of wet feedstock, with the associated improvement to the economics of the process as compared to other conventional thermochemical biomass conversion technologies.[108,110] Liquefaction of biomass includes different reactions pathways, including: (1) depolymerization of the biomass into biomass monomers; (2) decomposition of biomass monomers by cleavage, dehydration, decarboxylation and deamination, forming light fragments of small molecules, which are unstable and active; and (3) rearrangement of light fragments through condensation, cyclization and polymerization, leading to new compounds.[9,112,113]

In the case of HTL, and depending on the operating conditions used, the target product obtained can be either a bio-oil, also called biocrude, or a solid product called hydro-char. There are several recent reviews that have compiled the yields of the different products obtained in the HTL of different types of real biomass and its different constituents, such as cellulose, lignin, lipids and so on,[9,110,114] and it is clear that the large variation reported is mainly due to the large differences in the biomass origin and process conditions. On average, bio-oil yields obtained from HTL range between 15 and 40 wt% and are composed of a complex mixture of hydrocarbons that result in HHVs of around 20–40 kJ kg^{-1}.

3.4.1 Catalytic Hydrothermal Liquefaction of Biomass and Solid Waste

It is generally accepted that the use of a catalyst in the HTL of biomass helps to reduce char and tar formation, as well as simultaneously to improve the bio-oil product distribution and physicochemical properties in order to obtain an upgraded bio-oil with more desirable combustion properties. In this section, the main findings regarding the use of catalysts in the HTL of different types of biomass will be briefly reviewed. From a literature survey, it can be observed that HTL for upgrading biomass in the presence or absence of a catalyst has been increasingly studied, as is shown in Figure 3.5. The use of catalysts for HTL is spreading, with homogeneous catalysts, mainly alkali salts, being the most commonly used, although heterogeneous catalysts have also attracted the attention of the scientific community during recent years.

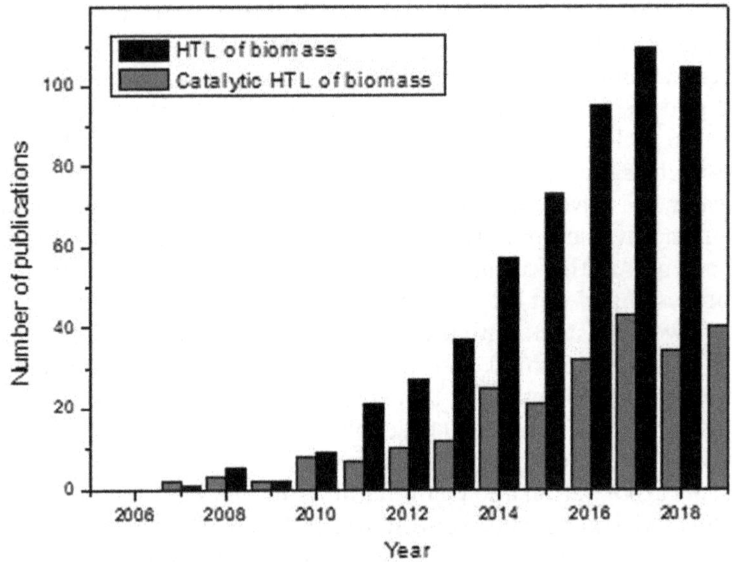

Figure 3.5 Evolution of the number of papers published per year from 2007 to 2018 on HTL. The graph was constructed using the Scopus platform with the topics HTL of biomass and catalyst or catalytic HTL of biomass.

3.4.1.1 Homogeneous Hydrothermal Liquefaction

The use of homogeneous catalysts to enhance the process of the HTL of biomass has been extensively considered. In this section, the main studies regarding homogeneous catalysis in HTL will be reviewed based on the type of biomass (aquatic or terrestrial) or solid waste to upgrade.

3.4.1.1.1 Algae/microalgae The HTL of *Cyanidioschyzon merolae* algal species was studied using different homogeneous catalysts at 300 °C for 30 min and at 10% solid loading in a bench-scale autoclave micro-reactor. Bio-oil yields increased from 16.98% without catalyst to 21.23% with CH_3COOH, to 21.78% with NaOH and to 22.67% with KOH, whereas the use of H_2SO_4 did not result in any noticeable increase in the yield of bio-oil. The HHV of bio-oil obtained with a catalyst was not improved as compared to that obtained without catalysts, although the use of a catalyst enhanced carbohydrate extraction.[115]

As well as the catalyst, temperature has an important effect on bio-oil yields and physicochemical properties. The effect of temperature (250, 300 and 350 °C) and the addition of Na_2CO_3 as a catalyst in the HTL of three different algae (*Nannochloropsis*, *Pavlova* and *Isochrysis*) was studied in a 100 mL batch reactor with a holding time of 60 min. The impact of the catalyst's presence was highly dependent on the type of algae studied and the process temperature; thus, for the case of high protein-containing algae (*Nannochloropsis*), the highest bio-oil yield was obtained at 250 °C, whereas for high carbohydrate-containing algae (*Pavlova* and *Isochrysis*), the bio-oil yield was maximized at higher temperatures (300 and 350 °C). The bio-oil produced had a HHV in the range of 32–37 MJ kg^{-1}. It was reported that the addition of Na_2CO_3 had no significant impact on the physicochemical properties of the bio-oil when compared to those obtained in the absence of catalyst.[116]

3.4.1.1.2 Agriculture/forestry Residues The transformation of biomass produced as a residue from the agriculture and forestry sectors is receiving increasing attention as an interesting way of obtaining value from waste. HTL presents an interesting alternative for upgrading this waste into bio-oil without the need for pre-treatment. Catalytic HTL of furniture waste sawdust was studied using different alkalis (NaOH, Na_2CO_3, KOH and K_2CO_3) as catalysts at 280 °C in a batch reactor for 15 min. The catalyst reactivity order was $K_2CO_3 > KOH > Na_2CO_3 > NaOH$. The total bio-oil yield obtained in the absence of catalyst (12.7%) was substantially improved as the K_2CO_3 concentration increased, obtaining the best results at 1.0 M K_2CO_3 concentration (34.9%). This was accompanied by a reduction in the solid residue, from 40% when no catalyst was used down to 6.8% with the catalyst.[117] In a different study, the liquefaction of pinewood sawdust using K_2CO_3 was studied with different solvents (carbon dioxide, water, acetone and ethanol). Interestingly, it was reported that when water was used as a solvent, the bio-oil yield was 17.3%, a value lower than those obtained with organic solvents (27.9–30.8%). This result was explained based on the fact that a fraction of the bio-oil was dissolved in the aqueous phase, which can account for the relatively low bio-oil yield as compared to those of other organic solvents.[118] In another work on the HTL of birchwood sawdust, KOH and K_2CO_3 were used as catalysts in reactions carried out in a benchtop autoclave reactor at 300 °C for 30 min. Similar results were observed, as the impact of the use of homogeneous catalysts was clearly demonstrated in terms of both bio-oil yields and the properties of the bio-oil. The use of KOH increased the yield up to around 40 wt%, more than double the yield of the uncatalysed reaction (18 wt%), and also significantly improved the occurrence of phenol derivatives (mainly 2-methoxy-phenol) and aliphatic compounds. It was also reported that the use of a catalyst had a profound effect on solid residue formation, which reduced from approximately 33 wt% to 12 wt%.[119] Similarly, the use of K_2CO_3 as a catalyst in the HTL of woody biomass (pine sawdust) had a positive impact on the upgrading process. Experiments with different catalyst concentrations were carried out in a 200 mL autoclave at 280 °C for 15 min. Best results in terms of biomass conversion and oil product formation were obtained when the highest K_2CO_3 concentration was used (1 M), simultaneously reducing the solid residue yield down to 4%.[120]

The effect of the addition of a co-solvent to enhance catalyst activity was studied. Glycerol was used as a co-solvent in the catalytic HTL of rice straw using Na_2CO_3 as a catalyst. Temperature (220–300 °C) and glycerol : water ratio (0 : 1 to 1 : 0) were the parameters studied. The best results were obtained with the addition of the co-solvent in a water : glycerol volume ratio of 1 : 1 at 265 °C for 1 h, resulting in a bio-yield of 50.31%. The catalytic effect on the bio-oil and solid residue yields was more pronounced for a water : glycerol volume ratio higher than 1 : 5, whereas lower ratios did not have much impact on the product yields.[121]

Semi-continuous operation of the HTL of corn stalk in the presence of Na_2CO_3 was studied with a non-isothermal fluid extraction technique at sub and supercritical water conditions (from 100 to 410 °C) and 25 MPa. The results showed that the bio-oil yield not only increased from 33.4 to 47.2 wt% upon catalyst addition, but also that the quality of the liquid product was considerably improved.[122]

Overall, the use of homogeneous catalysts in the HTL of terrestrial biomass has shown good results in terms of biomass conversion, bio-oil yields and product distribution. Different alkali catalysts (NaOH, K_2CO_3 and Na_2CO_3) were used in the catalytic HTL of oil palm shell in a high-pressure autoclave at temperatures ranging between 210 and 330 °C. The reactivity order in terms of liquid yield (including bio-oil and water-soluble products) was NaOH > $Na_2CO_3 \approx K_2CO_3$ > no catalyst. Highest biomass conversions (83.4%) and liquid yields (53.4%) were obtained with 10% NaOH at 290 °C, a two-fold increase as compared to using no catalyst.[123]

Temperature has an important effect on the yields and product distributions during the HTL of biomass. This was shown in a study on the catalytic HTL of barley straw using K_2CO_3 in a temperature range between 280 and 400 °C, thus covering both sub and supercritical conditions. High temperature was detrimental to the bio-oil yield, and the highest bio-oil yield (35.24 wt%) was obtained at 300 °C.[124] In the absence of K_2CO_3, the bio-oil yield obtained during the HTL of barley straw at 300 °C was 17.88 wt%. The positive effect of catalyst addition was evident not only in the improvement of bio-oil yields, but also in its properties. The bio-oil obtained with the addition of K_2CO_3 was of better quality with a higher carbon to oxygen ratio and hence higher HHV than that obtained from HTL in the absence of a catalyst.[125]

3.4.1.1.3 Organic Wastes Catalytic HTL has also been considered for treating urban or agro-industrial organic wastes and transforming them into fuel or other value-added products. Three homogeneous catalysts were studied for the HTL of pre-treated and washed sweet sorghum (solid left over after starch and juice extraction), namely K_2CO_3, KOH and formic acid, in a 25 mL tubular reactor made from stainless steel at 300 and 350 °C. In the study, the highest bio-oil yield and HHV of 61.8% and 33.1 MJ kg^{-1}, respectively, were obtained with K_2CO_3 at a temperature of 300 °C.[126]

A study evaluating the effect of process conditions on the HTL of blackcurrant pomace (*Ribes nigrum* L.), a fruit residue obtained after berry pressing, in the presence of NaOH as a catalyst was optimized in a 0.6 L stainless steel stirred batch reactor. Several process parameters were evaluated, namely the reaction temperature (290–335 °C), holding time (0–240 min), mass fraction of dry biomass in the slurry (0.05–0.29) and initial pH of the feed by adding NaOH (3.1–12.8). An optimum bio-oil yield (30%) was obtained at 310 °C, while holding time did not have any considerable impact on the results. Increasing biomass concentrations decreased bio-oil yields from 29% to 24%. The addition of NaOH decreased char yields from 35% to 24%, although the effect on the bio-oil yield was small (from 26% to 31%). The main impact of the catalyst addition an increase in the carbon recovery in the aqueous phase (from 12% to 25%). This was again attributed to the presence of larger amounts of water-soluble molecules when NaOH was used.[127]

As in studies described in the previous section, K_2CO_3 was tested as a potential homogeneous catalyst for the upgrading of waste bamboo chopsticks into bio-oil *via* catalytic HTL. It was reported that operating at a range of temperatures between 290 and 380 °C for 30 min, bio-oil yields were very low (3.8 wt%) in the absence of a catalyst. However, the addition of K_2CO_3 significantly improved this value, reaching a 21.2 wt% bio-oil yield. Nonetheless, catalyst addition did not have much impact on the HHV of the resulting bio-oil (30.9 *vs.* 31.6 MJ kg^{-1}).[128] K_2CO_3 as a catalyst was also used to study the effect of recycling the process water from the HTL of dried distiller grains with soluble compounds in a bomb-type reactor at 340 °C and a reaction time of 20 min. This strategy was beneficial to the process, increasing the yield of bio-oil, with this effect being more marked when K_2CO_3 was used. However, it is interesting to note that the yield of bio-oil in freshwater (without recycling) was not improved upon the incorporation of K_2CO_3 into the reaction system. The high protein and lipid content of this type of biomass is thought to be the main cause of such worse catalytic HTL performance.[129]

The use of different alkaline and acidic homogeneous catalysts for the HTL of urban waste has also been studied. Different additives (alkalis: K_2CO_3 and Na_2CO_3; organic acids: HCOOH and CH_3COOH) were used as catalysts in the HTL of sewage sludge at relative high temperatures (400 and 500 °C), with little positive effect on the bio-oil yield regardless of the heating mode (isothermal *vs.* fast heating conditions). In fact, upon catalyst addition, in all cases the bio-oil yield decreased at the expense of an increase in the gas yield, most likely due to the occurrence of the water gas shift reaction at such relatively high temperatures.[130] This result is in good agreement with previous results regarding the HTL of wet sewage sludge using Na_2CO_3 as a catalyst, even at temperatures around 300 °C,[131,132] which had very little impact or even negative effects on bio-oil yield for this particular biomass.

The HTL of lignin at NCW conditions (350 °C, 25 MPa) was studied using phenol as a co-solvent (~4.1 wt%) and an aqueous solution containing K_2CO_3 (from 0.4% to 2.2%) as a catalyst in a continuous fixed-bed catalytic reactor filled with ZrO_2 pellets. Regardless of the catalyst concentration, bio-oil yields were approximately 70% with respect to the dry lignin fed into the system. The bio-oil obtained had a HHV of 32 MJ kg^{-1}, which represents an increase of 15% with respect to the original value. The total yield of one-ring aromatic compounds (mainly anisoles, alkylphenols, guaiacols and catechols) increased from 17% to 27% (dry lignin basis) as K_2CO_3 concentration increased from 0.4% to 2.2%.[133] The HTL of lignin using K_2CO_3 (1.7 wt%) as a catalyst and phenol as a co-solvent (0.0–9.7 wt%) as a function of temperature (280–350 °C) was studied. In this case, a specially designed reactor was used that allowed for the injection of lignin slurry, enabling fast heating of the biomass, effective control of pressure and withdrawal/quenching of the product. It was found that the yield of one-ring aromatic components was significantly improved with the addition of small amounts of phenol into the reacting system. For a given phenol content (3.4 wt%), the maximum total yield of monomeric aromatics was obtained at 320 °C.[134]

3.4.1.2 Heterogeneous Hydrothermal Liquefaction

The use of heterogeneous catalysts is highly desirable for the larger-scale implementation of HTL technology as it facilitates separation processes downstream. However, mass transfer limitations and developing catalytic materials that maintain high activity for long operation times remain big challenges for researchers. There has been much work done in the design and implementation of heterogeneous catalysts for the HTL of different biomass and organic residues, as will be reviewed in the following sections.

3.4.1.2.1 Algae/microalgae The catalytic HTL of a microalgae (*Dunaliella tertiolecta*) was studied in an autoclave reactor at 320 °C for 30 min. A set of heterogeneous catalysts based on Co, Ni and Pt (5 wt%) supported on carbon nanotubes (CNTs) were prepared by incipient wetness impregnation. The presence of the catalysts slightly improved the conversion and bio-oil yield. Thus, for the experiment carried out without catalyst, ~85% conversion and 37% bio-oil yield were obtained. The best catalytic results in terms of conversion were obtained with the Co/CNT catalyst (95.78%), although the Ni/CNT and Pt/CNT results only differed slightly (92.36% and 93.26%, respectively). On the other hand, no differences were observed in the resulting bio-oil yields among the three catalysts tested. However, the bio-oil obtained with the Co/CNT catalyst had a higher percentage of hydrocarbons, a lower content of fatty acid and a lower content of N-containing compounds and therefore Co/CNT was considered the best option among the catalysts tested.[135]

The performance of two heterogeneous catalysts (Ni/SiO_2 and analcime zeolite) in the HTL of a microalga (*Nannochloropsis* sp.) was compared with that of Na_2CO_3 as a homogeneous catalyst at relatively low temperatures (210–250 °C) in a 500 mL batch reactor. Interestingly, bio-oil yields obtained with heterogeneous catalysts surpassed that obtained in the absence of a catalyst (20.2 wt%) and that with Na_2CO_3 (24.2 wt%), obtaining the highest bio-oil yield (30 wt%) with the Ni/SiO_2 catalyst at 250 °C. In addition, the resulting bio-oil had a lower oxygen and nitrogen content, and hence a higher HHV as compared to the bio-oil obtained in the absence of a catalyst. However, it was observed that the catalyst suffered from a progressive deactivation after successive reutilization tests.[136]

Another Ni-based heterogeneous catalyst (Ni/TiO_2) was studied for the HTL of *Spirulina* microalgae using a 100 mL batch reactor. The bio-oil yield increased upon the addition of catalyst from 34.15 wt% using water alone up to 43.05 wt% using a catalyst dosage of 5%. Optimum reaction conditions were found at 260 °C and a 30 min reaction time, with an algae : water ratio of 1 : 4. The use of Ni/TiO_2 as a catalyst not only promoted the formation of desirable products, including hydrocarbons and esters, but also the formation of value-added lighter compounds.[137]

Traditionally, zeolite-based catalysts have been widely used for fuel upgrading processes with good results. A series of acidic microporous zeolite materials (HZSM-22, HZSM-5, H-β, MCM-22 and SAPO-11) were used as HTL catalysts for the upgrading of *Euglena* sp. microalgae. Despite the fact that bio-oil yields slightly decreased in the presence of zeolite catalysts, it was observed that H-β zeolite exhibited the best catalytic activity, as it reduced N-containing compounds and carboxylic acids while increasing hydrocarbon content in bio-oil. No data regarding the reusability nor the stability of the catalysts under hydrothermal conditions were reported.[138] A general reaction network for the HTL of *Euglena* sp. microalgae with the principal products obtained is shown in Figure 3.6.

Noble metal heterogeneous catalysts have been extensively used for fuel upgrading applications with good results. In a study on the catalytic HTL of *Nannochloropsis* sp. microalgae, a series of noble metal catalysts supported on carbon (Pd/C, Pt/C, Ru/C) and traditional hydrotreating catalysts (Ni/SiO_2–Al_2O_3 and sulfided $CoMo/\gamma\text{-}Al_2O_3$) were tested at 350 °C in a batch reactor with

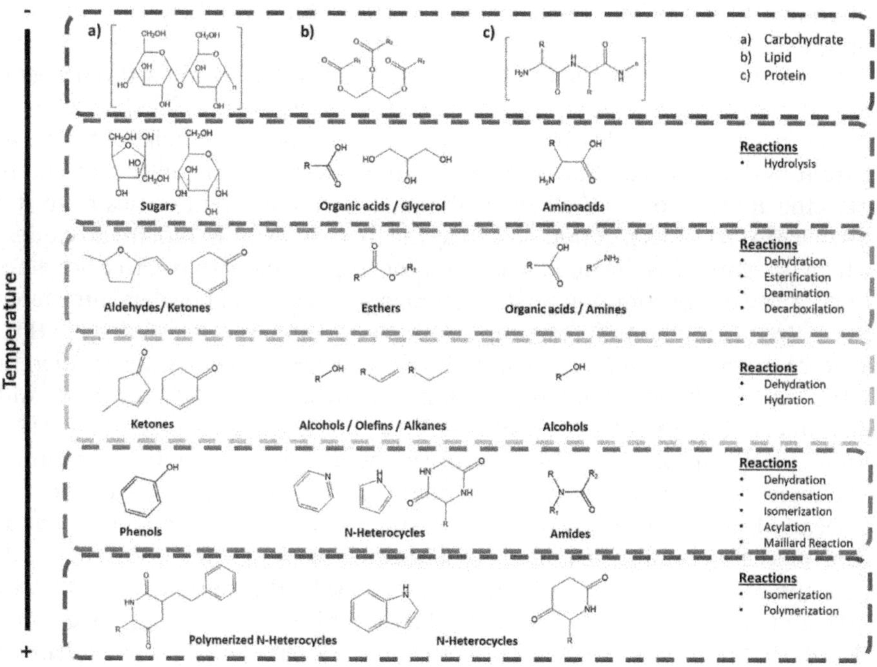

Figure 3.6 Principal products obtained and main reactions occurring in the HTL of *Euglena* sp. microalgae (adapted from ref. 138).

or without H_2 addition. In the absence of H_2, bio-oil yields were higher compared to those obtained in tests without a catalyst. However, bio-oil compositions and HHVs (around 38 MJ kg^{-1}) were unaffected. Bio-oil obtained with Pd/C was analysed in greater detail. The main differences observed between the bio-oil obtained in the presence of a catalyst compared to that obtained without one was a lower proportion of fatty acids and a higher content of alkanes.[139]

3.4.1.2.2 Agriculture/forestry Residues The HTL of empty fruit bunch derived from oil palm residues was studied at supercritical conditions (390 °C, 25 MPa and 1 h holding time) with several metal oxide catalysts (CaO, MgO, MnO, ZnO, NiO, SnO, CeO_2, Al_2O_3 and La_2O_3) in an 8.8 mL Inconel batch reactor. Best results in terms of bio-oil yield were obtained with those metal oxides that have lower electronegativity such as CaO, MnO, La_2O_3 and CeO_2. An increase of 40% in the yield of bio-oil was observed with catalysts when compared to the HTL test without a catalyst. The main components found in the bio-oil obtained included phenol and its derivatives, followed by ketones, other aromatic compounds and carboxylic acids.[140] Similarly, the HTL of bagasse was studied over solid alkali catalysts of metal oxide composites of the form $MgMO_x$ (M = Mn, Ni, Fe, Cr, Zn and Al) in a 500 mL stainless steel autoclave. All catalysts improved the degree of liquefaction and the yield of water-soluble products. The best results were obtained with $MgMnO_2$ at 270 °C, 5 min reaction time and a water to solid bagasse ratio of 10, where the degree of liquefaction was 93.7% and a yield of water-soluble products of 59.5% was obtained.

Based on results from the comparative study of HTL with and without a catalyst, it was observed that carbohydrate self-catalyses in subcritical water, and in the absence of a catalyst lignin remains as a residue. Nonetheless, this recalcitrant lignin is successfully converted into value-added phenolic monomers with the addition of solid alkali catalysts. The enhancement comes from the increase in OH concentration caused by the thermal hydrolysis of metal oxides, which is regarded as the main reason for the improvement in bagasse liquefaction.[141] The use of common waste materials such as metal powders and sieves as catalysts could be an interesting alternative to reduce costs in catalytic HTL processes. The HTL of *Paulownia* in the presence of Fe powder was studied in a 1000 mL batch reactor at temperatures ranging between 260 and 320 °C and 10 min reaction time. The yield of bio-oil was maximized in the presence of Fe catalysts (36% at 340 °C), and an increase in the content of phenolic compounds and their derivatives was observed compared to the non-catalysed process.[142]

Depending on the nature of the biomass, it has been reported that the addition of a catalyst failed to enhance HTL or to improve the quality of the products obtained. A study on the HTL of native grassland perennials at 300 °C, 10 MPa and 10 min reaction time showed that neither the use of homogeneous (NaOH, H_2SO_4) nor heterogeneous (superacid SO_4/ZrO_2–Al_2O_3 and solid alkali CaO–ZrO_2) catalysts improved the yield of bio-oil when compared to that obtained in just water. Therefore, it was concluded that the non-catalysed

HTL of this particular biomass was more economically viable considering the associated costs of adding, removing and regenerating catalysts.[143]

3.4.1.2.3 Organic Wastes Various studies have considered the implementation of heterogeneous catalysts to upgrade organic wastes through HTL. As an example, three heterogeneous catalysts, namely nickel phosphide (Ni_2P), nickel on silica alumina (Ni/Si-Al) and zeolite (ZSM-5), were studied in the HTL of pre-treated and washed sweet sorghum in a 25 mL tubular reactor made from stainless steel. No improvement on the yield of bio-oil was observed with the use of catalyst at 300 °C. The only catalyst that gave a significantly higher bio-oil yield (45.5 wt%) than the experiment without a catalyst (23.3 wt%) was Ni/Si-Al. However, the catalytic performance shown by Ni/Si-Al is not as good as that of K_2CO_3 as a homogeneous catalyst that, under the same experimental conditions, resulted in a considerably higher bio-oil yield (61.8%).[126]

Metal oxide catalysts (La_2O_3 and Dy_2O_3) were also studied in the HTL of rice husks in a 10 mL stainless steel microreactor at varying temperatures, reaction times and water : rice husk mass ratios. Bio-oil yields ranged from 11.8 to 23.8 wt%, depending on reaction conditions. The use of metal oxides clearly improved the yield of bio-oil, with the highest yield of 32.5 wt% obtained at 300 °C, 10 min and a water : rice husk mass ratio of 5 using La_2O_3 as a catalyst. This catalyst also improved the HHV of bio-oil, reaching 31.78 MJ kg^{-1}, which is a great improvement when compared to the 16.19 MJ kg^{-1} obtained without a catalyst. This result is thought to be due to a reduction in oxygen content and the modified chemical composition of bio-oil that showed a decrease in the content of phenols and organic acids, while the formation of hydrocarbons (alkenes and alkynes) and esters was promoted. It was concluded that the use of these metal oxides as catalysts plays an important role in the deoxygenation of bio-oil, improving the HHV of the product.[144]

A different series of HTL heterogeneous catalysts based on transition metal chlorides ($ZnCl_2$, $CuCl_2$ and $NiCl_2$) were tested on coconut shells at different temperatures (from 270 to 330 °C) in a batch-type reactor. A significant decrease in the yield of bio-oil was observed in the presence of these catalysts. This is thought to be the result of the high reaction rates promoted by the acidic nature of the metal chloride in water, which may accelerate undesired chemical reactions turning bio-oil into gas or chars.[145]

Similar observations to those reported in the previous section on the catalytic HTL of native grassland perennials were made in a study on the catalytic HTL of dried distiller grains using a nanocrystalline zirconia catalyst in a stop-flow reaction system. The addition of ZrO_2 had no significant effects on the yields and properties of the bio-oil obtained at 350 °C, 250 bar and 15 min reaction time. In addition, the potential catalytic effect of the steel reactor walls was also disproven by comparing these results with those obtained using an inert gold liner.[146]

The use of structured carbon materials as catalyst supports has caught the attention of researchers. In a study on the HTL of Kraft lignin, the catalytic effect of two series of Ni and Ni-CeO_2 catalysts supported on different carbon materials (activated carbon and carbon nanofibres) under SCW conditions was tested in a microbomb type reactor. Best results were observed when Ni-CeO_2 supported on carbon nanofibres was used as a catalyst, with a 56 wt% yield of the light bio-oil fraction at 400 °C and 230 bar. It was concluded that operating at short reaction times was key to obtaining high yields of bio-oil and preventing secondary reactions leading to the formation of gas and char.[72]

3.4.2 Catalytic Bio-oil Deoxygenation

Bio-oil produced from biomass *via* either HTL or pyrolysis typically contains large amounts of oxygenated compounds that hinder its direct use as a fuel. The typical upgrading processes aimed at overcoming the undesirable properties of bio-oils (low heating value, low volatility and high viscosity) are catalytic cracking and catalytic hydrodeoxygenation (HDO). HDO is an exothermic process by which oxygen present in bio-oil components reacts with hydrogen over a catalyst, resulting in water and a pool of hydrocarbons chemically similar to crude oil. Recently, HDO in the presence of either sub or supercritical water has attracted the attention of the scientific community, mainly regarding algae-derived bio-oil obtained from HTL processes.

Catalytic upgrading of bio-oil from the HTL of *Chlorella* sp. microalgae was performed in an autoclave reactor at 400 °C for 240 min with the addition of 6 MPa of H_2 and 10 wt% of nine different zeolite-based catalysts in SCW. Yields of upgraded bio-oil ranged from 42.4 to 54.5 wt%. All catalysts used promoted denitrogenation, deoxygenation and desulfurization of the pre-treated bio-oil to a greater extent than the reaction carried out without catalysts, obtaining the best results with HY (5% Na_2O), HY (0.8% Na_2O) and HZSM-5 (SiO_2 : Al_2O_3 = 25:1) zeolites. The behaviour observed was attributed to the catalyst acid sites. The use of these catalysts reduced oxygen content down to values lower than 2%, significantly lower than the original bio-oil (22.73%). However, no data regarding the reusability of the catalyst nor the zeolite stability at supercritical conditions were provided.[147]

The SCW upgrading (with and without catalysts) of the bio-oils obtained from the HTL of *Nannochloropsis* microalgae was studied at 2.5 MPa of H_2 pressure. Among the catalysts studied (Ru/C, Mo_2C, Pt/C), Pt/C proved to be the most effective for the upgrading of water-insoluble bio-oil from algae liquefaction. The resulting bio-oil had a HHV of 41.6 MJ kg^{-1} and halved the heteroatom content of the original bio-oil.[148] The same authors had previously reported the positive effect of using a Pt/C catalyst for the upgrading of a *Nannochloropsis* microalga bio-oil in SCW. The upgraded bio-oil had a HHV of 43 MJ kg^{-1} and a very high content of hydrocarbon molecules, including alkanes and aromatic compounds.[149]

Ni-Ru/CeO_2 and Ni/CeO_2 were also tested as catalysts for the upgrading of a bio-oil obtained by the HTL of *Nannochloropsis* sp. microalgae. Reaction conditions were 450 °C, 60 min and 2 MPa of H_2 pressure (when used). The combination of Ni-Ru/CeO_2 in the presence of added H_2 resulted in the best catalytic performance, namely the best catalytic desulfurization, highest upgraded bio-oil yield and a higher HHV (39.94 MJ kg^{-1}). The chemical analysis of the upgraded bio-oil showed that it was mainly composed of a series of aliphatic saturated hydrocarbons such as pentadecane, tridecane, tetradecane, heptadecane and dodecane.[150]

The upgrading of water-insoluble algal (*Chlorella*) bio-oil in hydrothermal conditions was studied using with a series of Al_2O_3-supported catalysts (NiMoW, NiCoMo, CoMoW and CoNiMoW) at 400 °C with added H_2 for 1 or 4 h. The improved bio-oil obtained with the CoMoW/γ-Al_2O_3 catalysts after 1 h showed very interesting properties, such as a low oxygen content (6.31 wt% as compared to 8.36 wt% obtained without catalyst and 15.53 wt% for the original bio-oil) and a HHV of 38.7 MJ kg^{-1}. Overall, this catalyst significantly reduced the boiling point of the upgraded oil.[151]

A series of activated carbons were used as catalysts for the upgrading in SCW (400 °C, 1 h and 6 MPa of H_2) of bio-oil produced from the HTL of duckweed. Interestingly, all carbon-based materials showed activity similar to Ru/C catalysts towards the denitrogenation and deoxygenation of the bio-oil. The best catalyst (an AC derived from bamboo) resulted in an upgraded bio-oil with a HHV of 44 MJ kg^{-1} and 2.2 wt% O content. The ions and weak acids leached from ACs are believed to promote denitrogenation and desulfurization reactions, whereas the mesoporosity favoured the conversion of heavy compounds and restricted coke formation.[152]

NiCo catalysts supported on carbon nanofibres were used for the supercritical upgrading of bio-oil obtained from the pyrolysis of pine sawdust. Reactions were carried out in a microbomb-type reactor at 380 °C and a pressure of 230 bar with a catalyst : dry bio-oil ratio of 0.25, in the absence of any added hydrogen. The surface chemistry of the carbon support had a profound influence on the resulting bio-oil properties. Higher liquid yields and less gas formation were obtained in the presence of the catalyst when compared to those obtained in the absence of a catalyst. However, coke formation increased. The upgraded bio-oil contained higher amounts of phenolic and cyclic compounds, together with lower amounts of carboxylic acids than the original bio-oil.[72] The SCW upgrading of wood-derived bio-oil was performed under similar reaction conditions using a NiCo/AlMg catalyst in the presence of crude glycerol as a hydrogen donor. A synergistic interaction between crude glycerol and bio-oil was reported. Optimum conditions in terms of the bio-oil : glycerol ratio and catalyst loading were determined, resulting in an upgraded bio-oil yield of 37% with a HHV of 32 MJ kg^{-1}.[153]

3.5 Conclusion

As the world's energy demand increases, conventional energy resources are being depleted at a rapid pace. It is therefore necessary to develop cleaner and more effective technologies to convert unconventional renewable and fossil resources into fuel in order to meet future energy targets. The nature of these feedstocks causes severe processing problems resulting in low fuel yields and catalyst deactivation due to coking. NCW and SCW have characteristic physicochemical properties and solvent properties such as reduced mass and heat transfer limitations, a high capacity to donate hydrogen and good solubility of organic compounds, which are favourable for fuel conversion and upgrading processes. The addition of a catalyst to hydrothermal conversion and upgrading processes increases reaction rates, suppresses the formation of solid residues, prevents repolymerization of intermediate species and promotes cracking reactions. The key to its implementation is finding catalysts that are active and stable at these reaction conditions and are resistant to deactivation, mainly due to sintering, water adsorption or lixiviation of catalytic components.

Upgrading of heavy oil with NCW and SCW leads to the formation of lighter oil with improved properties such as lower viscosity and reduced S, N, O and metal contents. The high extraction capacity of NCW and SCW aid in the suppression of condensation reactions of the organic radicals formed, which is a pathway leading to undesired coke formation and deposition on the catalyst surface. The presence of hydrogen is an important factor, as hydrogenation aids in the upgrading of heavy oil molecules. Several reports state that *in situ*-generated hydrogen has a higher reactivity than supplied molecular H_2. *In situ*-generated hydrogenating species and hydrocarbon activation are promoted by the presence of catalysts. Common catalysts tested for heavy oil upgrading include acids, metal oxides and traditional hydrotreating catalysts. The main challenges encountered in their implementation are their low activity levels and lack of stability. It has been suggested that a way of designing catalysts that show greater stability is to synthetize them in hydrothermal or similarly severe conditions. The presence of a catalyst improved the quality of the oil recovered, as well as increased S and N conversions.

The production of bio-oils from the HTL of high-moisture biomass is a promising technology that requires research in order to become economically and commercially viable. The use of homogeneous or heterogeneous catalysts has great potential in the HTL of different types of biomass. They enhance bio-oil yields and improve their physicochemical properties. However, the performance of a catalyst is highly dependent on the feedstock and reaction conditions. In general, alkali salts such as Na_2CO_3 or K_2CO_3 have shown positive impacts on both bio-oil yields and HHV. The mechanism involved in the catalytic HTL of biomass with alkali salts is believed to proceed through decarboxylation reactions between hydroxyl groups in biomass and formate ions in alkali carbonates that result in the formation of esters, H_2 and CO_2. H_2 may act as a reducing agent, increasing the HHV and quality of bio-oils. Alkali salts raise the pH, which inhibits dehydration of biomass monomers and so limits char and tar formation. The complex reaction mechanisms of the catalytic HTL of biomass include cyclization, polymerization and condensation of smaller molecules formed from different biomass compounds *via* dehydration, deoxygenation and decarboxylation.[9,154]

However, the use of homogeneous catalysts has some drawbacks due to the expensive recovery of the catalyst stemming from the cost-intensive separation process. Even though the use of heterogeneous catalysts is desirable from the industrial point of view, their impact on HTL remains an open question and must be studied for each particular biomass. In addition to this, catalyst deactivation by char/coke formation adsorbed on the catalyst surface is often overlooked. Since most of the studies on catalytic HTL are performed in batch-type reactors, reutilization of the catalyst in successive reaction–regeneration cycles would be highly desirable in order to gather more information about the performance of different types of catalysts, with the prospect of the development of continuous processes.

Acknowledgements

José Luis Pinilla is grateful for the financial support of FEDER and the Spanish Economy and Competitiveness Ministry (MINECO) (ENE2017-83854-R). Pedro Arcelus acknowledges the funding for research provided by Universidad Iberoamericana through the grants Proyecto de Ciencia Básica 0055 and the Capital Semilla funds.

References

1. United Nations, Department of Economic and Social Affairs, Population Division, World Population Prospects 2019: Highlights, Report No.: ST/ESA/SER.A/423, United Nations, Department of Economic and Social Affairs, Population Division, New York, 2019.
2. Energy Information Administration, International Energy Outlook 2019, Report No.: DOE/EIA-0484(2019), US Department of Energy, Washington, DC, 2019.
3. Hart Energy Research Group, Heavy Crude Oil: Global Perspective, Analysis & Outlook to 2035, Hart Energy, Houston, TX, 2011.
4. P. Arcelus-Arrillaga, J. L. Pinilla, K. Hellgardt and M. Millan, *Energy Fuel*, 2017, **31**(5), 4571–4587.
5. F. Trejo, G. Centeno and J. Ancheyta, *Fuel*, 2004, **83**(16), 2169–2175.
6. J. Heinimö and M. Junginger, *Biomass Bioenergy*, 2009, **33**(9), 1310–1320.
7. G. Berndes, M. Hoogwijk and R. van den Broek, *Biomass Bioenergy*, 2003, **25**(1), 1–28.
8. G. Brunner, *J. Supercrit. Fluid.*, 2009, **47**(3), 373–381.
9. S. S. Toor, L. Rosendahl and A. Rudolf, *Energy*, 2011, **36**(5), 2328–2342.
10. A. Kruse and N. Dahmen, *J. Supercrit. Fluid.*, 2015, **96**, 36–45.
11. N. Matubayasi, C. Wakai and M. Nakahara, *J Chem. Phys.*, 1997, **107**(21), 9133–9140.
12. H. Weingärtner and E. U. Franck, *Angew. Chem., Int. Ed.*, 2005, **44**(18), 2672–2692.
13. E. W. Lemmon, M. O. McLinden and D. G. Friend, Thermophysical Properties of Fluid Systems in NIST Chemistry WebBook, in *NIST Standard Reference Database Number 69*, ed. P. J. Linstrom and W. G. Mallard, National Institute of Standards and Technology, Gaithersburg, MD, 2019.

14. A. A. Peterson, F. Vogel, R. P. Lachance, M. Fröling, J. Michael, J. Antal and J. W. Tester, *Energy Environ. Sci.*, 2008, **1**(1), 32–65.
15. P. C. Torres-Mayanga, D. Lachos-Perez, A. Mudhoo, S. Kumar, A. B. Brown, M. Tyufekchiev, G. Dragone, S. I. Mussatto, M.A. Rostagno, M. Timko and T. Forster-Carneiro., *Biomass Bioenergy*, 2019, **130**, 105397.
16. H. Luo and S. C. Tucker, *J. Am. Chem. Soc.*, 1995, **117**(45), 11359–11360.
17. S. J. Halstead and A. J. Masters, *Mol. Phys.* 2010, **108**(2), 193–203.
18. S. E. Hunter and P. E. Savage, *Chem. Eng. Sci.*, 2004, **59**(22), 4903–4909.
19. Y. Guo, S. Z. Wang, D. H. Xu, Y. M. Gong, H. H. Ma and X. Y. Tang, *Renewable Sustainable Energy Rev.* 2010, **14**(1), 334–343.
20. J. T. Henrikson, Z. Chen and P. E. Savage, *Ind. Eng. Chem. Res.*, 2003, **42**(25), 6303–6309.
21. Q. L. Wen and H. Y. Gu, *Prog. Nucl. Energy*, 2011, **53**(5), 480–486.
22. K. Yamagata, K. Nishikawa, S. Hasegawa, T. Fujii and S. Yoshida, *Int. J. Heat. Mass Transfer*, 1972, **15**(12), 2575–2593.
23. E. U. Franck, *Ber. Bunsen-Ges.*, 1984, **88**(9), 820–825.
24. J. Fraga-Dubreuil and M. Poliakoff, *Pure. Appl. Chem.*, 2006, **78**(11), 1971–1982.
25. A. A. Al-Muntaser, M. A. Varfolomeev, M. A. Suwaid, C. Yuan, A. E. Chemodanov, D. A. Feoktistov, I. Z. Rakhmatullin, M. Abbas, E. Domínguez-Álvarez, A. A. Akhmadiyarov, V. V. Klochkov and M. I. Amerkhanov, *J. Petrol. Sci. Eng.*, 2019, 106592.
26. N. Berkowitz and J. Calderon, *Fuel Process. Technol.*, 1990, **25**(1), 33–44.
27. J. H. Park and S. H. Son, *Korean. J. Chem. Eng.*, 2010, **28**(2), 455–460.
28. L.-Q. Zhao, Z.-M. Cheng, Y. Ding, P.-Q. Yuan, S.-X. Lu and W.-K. Yuan, *Energy Fuel*, 2006, **20**(5), 2067–2071.
29. E. P. Bennion, D. M. Ginosar, J. Moses, F. Agblevor and J. C. Quinn, *Appl. Energy*, 2015, **154**, 1062–1071.
30. D. López Barreiro, W. Prins, F. Ronsse and W. Brilman, *Biomass Bioenergy*, 2013, **53**, 113–127.
31. M. Plaza, M. Amigo-Benavent, M. D. del Castillo, E. Ibáñez and M. Herrero, *Food. Res. Int.*, 2010, **43**(10), 2341–2348.
32. S. Xiu, A. Shahbazi, V. Shirley and D. Cheng, *J. Anal. Appl. Pyrol.*, 2010, **88**(1), 73–79.
33. A. Dimitriadis and S. Bezergianni, *Renewable Sustainable Energy Rev.*, 2017, **68**, 113–125.
34. A. Kruse and E. Dinjus, *J. Supercrit. Fluid.*, 2007, **39**(3), 362–380.
35. P. Arcelus-Arrillaga, PhD thesis, Heavy Oil Upgrading through Oxidative Cracking in Near-critical and Supercritical Water, PhD Thesis, Imperial College London, 2016.
36. F. Trejo, M. S. Rana and J. Ancheyta, *Catal. Today*, 2010, **150**(3–4), 272–278.
37. J. Vilcáez, M. Watanabe, N. Watanabe, A. Kishita and T. Adschiri, *Fuel*, 2012, **102**, 379–385.
38. T. Adschiri, T. Suzuki and K. Arai, *Fuel*, 1991, **70**(12), 1483–1484.
39. A. Loppinet-Serani, C. Aymonier and F. Cansell, *Chem. Sus. Chem.*, 2008, **1**(6), 486–503.

40. P. E. Savage, *Catal. Today*, 2000, **62**(2–3), 167–173.
41. J. M. L. Penninger, *Fuel*, 1988, **67**(4), 490–496.
42. K. C. Park and H. Tomiyasu, *Chem. Commun.*, 2003, (6), 694–695.
43. B. Kuhlmann, E. M. Arnett and M. Siskin, *J. Org. Chem.*, 1994, **59**(18), 5377–5380.
44. T. L. LeValley, A. R. Richard and M. Fan, *Int. J. Hydrogen Energy*, 2014, **39**(30), 16983–17000.
45. S. C. Ammal and A. Heyden, *J. Catal.*, 2013, **306**, 78–90.
46. S. F. Rice, R. R. Steeper and J. D. Aiken, *J. Phys. Chem. A*, 1998, **102**(16), 2673–2678.
47. M. D. Bermejo and M. J. Cocero, *AIChE J.*, 2006, **52**(11), 3933–3951.
48. P. E. Savage, *Chem. Rev.*, 1999, **99**(2), 603–622.
49. L. Li, P. Chen and E. F. Gloyna, *AIChE J.*, 1991, **37**(11), 1687–1697.
50. P. Arcelus-Arrillaga, K. Hellgardt and M. Millan, *Fuel*, 2017, **209**, 434–41.
51. J. A. Onwudili and P. T. Williams, *J. Supercrit. Fluid.*, 2007, **43**(1), 81–90.
52. J. A. Onwudili and P. T. Williams, *Int. J. Energy Res.*, 2006, **30**(7), 523–533.
53. J. A. Onwudili and P. T. Williams, *J. Supercrit. Fluid.*, 2007, **39**(3), 399–408.
54. P. E. Savage, *J. Supercrit. Fluid.*, 2009, **47**(3), 407–414.
55. A. Kruse and H. Vogel, *Chem. Eng. Technol.*, 2008, **31**(9), 1241–1245.
56. R. C. Crittendon and E. J. Parsons, *Organometallics*, 1994, **13**(7), 2587–2591.
57. P. G. Jessop, *J. Supercrit. Fluid.*, 2006, **38**(2), 211–231.
58. A. B. Ross, P. Biller, M. L. Kubacki, H. Li, A. Lea-Langtonand and J. M. Jones, *Fuel*, 2010, **89**(9), 2234–2243.
59. P. Azadi, A. A. Khodadadi, Y. Mortazavi and R. Farnood, *Fuel Process. Technol.*, 2009, **90**(1), 145–151.
60. P. Azadi and R. Farnood, *Int. J. Hydrogen Energy*, 2011, **36**(16), 9529–9541.
61. P. Azadi, K. M. Syed and R. Farnood, *Appl. Catal., A*, 2009, **358**(1), 65–72.
62. T. Minowa and T. Ogi, *Catal. Today*, 1998, **45**(1), 411–416.
63. P. M. Yeletsky, T. R. Reina, O. A. Bulavchenko, A. A. Saraev, E. Yu. Gerasimov, O. O. Zaikina, J. M. Bermúdez, P. Arcelus-Arrillaga, V. A. Yakovlev and M. Millan, *Catal. Today*, 2019, **329**, 197–205.
64. T. R. Reina, P. Yeletsky, J. M. Bermúdez, P. Arcelus-Arrillaga, V. A. Yakovlev and M. Millan, *Fuel*, 2016, **182**, 740–748.
65. E. Fumoto, T. Tago and T. Masuda, *Energy Fuel*, 2006, **20**(1), 1–6.
66. E. Fumoto, A. Matsumura, S. Sato and T. Takanohashi, *Energy Fuel*, 2009, **23**(3), 1338–1341.
67. J. L. Pinilla, P. Arcelus-Arrillaga, H. Puron and M. Millan, *Fuel*, 2013, **109**, 303–308.
68. J. L. Pinilla, P. Arcelus-Arrillaga, H. Puron and M. Millan, *Appl. Catal., A*, 2013, **459**, 17–25.
69. M. Akizuki and Y. Oshima, *Ind. Eng. Chem. Res.*, 2012, **51**(38), 12253–12257.
70. J. Yu and P. E. Savage, *Environ. Sci. Technol.*, 2000, **34**(15), 3191–3198.
71. I.-G. Lee and S.-K. Ihm, *Ind. Eng. Chem. Res.*, 2009, **48**(3), 1435–1442.
72. A. Cardoso, T. R. Reina, I. Suelves, J. L. Pinilla, M. Millan and K. Hellgardt, *Green. Chem.*, 2018, **20**(18), 4308–4318.

73. J. Remón, J. Arauzo, L. García, P. Arcelus-Arrillaga, M. Millan, I. Suelves and J.L. Pinilla, *Fuel Process. Technol.*, 2016, **154**, 178–187.
74. A. D. Taylor, G. J. DiLeo and K. Sun, *Appl. Catal., B*, 2009, **93**(1), 126–133.
75. E. Furimsky, *Ind. Eng. Chem. Res.*, 2013, **52**(50), 17695–17713.
76. D. C. Elliott, L. J. Sealock and E. G. Baker, *Ind. Eng. Chem. Res.*, 1993, **32**(8), 1542–1548.
77. Z.-M. Cheng, Y. Ding, L.-Q. Zhao, P.-Q. Yuan and W.-K. Yuan, *Energy Fuel*, 2009, **23**(6), 3178–3183.
78. M. Watanabe, H. Hirakoso, S. Sawamoto, A. Tadafumi and K. Arai, *J. Supercrit. Fluid.*, 1998, **13**(1), 247–252.
79. J. Liu, Y. Xing, Y.-X. Chen, P.-Q. Yuan, Z.-M. Cheng and W.-K. Yuan, *Ind. Eng. Chem. Res.*, 2018, **57**(3), 867–875.
80. C. Ma, R. Zhang and J. Bi, *J. Fuel Chem. Tech.*, 2003, **31**(2), 103–110.
81. H. J. Ederer, A. Kruse, C. Mas and K. H. Ebert, *J. Supercrit. Fluid.*, 1999, **15**(3), 191–204.
82. C.-C. Zhu, C. Ren, X.-C. Tan, G. Chen, P.-Q. Yuan, Z.M Cheng and W.-K Yuan, *Fuel Process. Technol.*, 2013, **111**, 111–117.
83. R. N. Leif and B. R. T. Simoneit, *Org. Geochem.*, 2000, **31**(11), 1189–1208.
84. R. P. Dutta, W. C. McCaffrey, M. R. Gray and K. Muehlenbachs, *Energy Fuel*, 2000, **14**(3), 671–676.
85. T. Sato, S. Mori, M. Watanabe, M. Sasaki and N. Itoh, *J. Supercrit. Fluid.*, 2010, **55**(1), 232–240.
86. I. V. Kozhevnikov, A. L. Nuzhdin and O. N. Martyanov, *J. Supercrit. Fluid.*, 2010, **55**(1), 217–222.
87. A. Cabañas, J. A. Darr, E. Lester and M. Poliakoff, *J. Mater. Chem.*, 2001, **11**(2), 561–568.
88. M. Hosseinpour, S. Fatemi and S. J. Ahmadi, *Fuel*, 2015, **159**, 538–549.
89. M. Kosari, M. Golmohammadi, S. J. Ahmadi J. Towfighi and A. H. Chenari, *J. Supercrit. Fluid.*, 2017, **126**, 14–24.
90. M. Golmohammadi, S. J. Ahmadi and J. Towfighi, *J. Supercrit. Fluid.*, 2016 **113**, 136–143.
91. D. Wang, Y. Li, L. Jin, K. Hao, B. Wei, D. Yao and H. Hu, *Appl. Catal., B Environ.*, 2019, **258**, 117944.
92. M. Hosseinpour, S. J. Ahmadi and S. Fatemi, *J. Supercrit. Fluid.*, 2015, **100**, 70–78.
93. A. Ates, G. Azimi, K.-H. Choi, W. H. Green and M. T. Timko, *Appl. Catal., B Environ.*, 2014, **147**, 144–155.
94. E. Fumoto, T. Tago, T. Tsuji and T. Masuda, *Energy Fuel*, 2004, **18**(6), 1770–1774.
95. E. Fumoto, A. Matsumura, S. Sato and T. Takanohashi, *Energy Fuel*, 2009, **23**(11), 5308–5311.
96. S. Funai, E. Fumoto, T. Tago and T. Masuda, *Chem. Eng. Sci.*, 2010, **65**(1), 60–65.
97. E. Fumoto, S. Sato and T. Takanohashi, *Energy Fuel*, 2011, **25**(2), 524–527.
98. P. D. Clark and M. J. Kirk, *Energy Fuel*, 1994, **8**(2), 380–387.

99. M. L. Myrick, J. Kolis, E. Parsons, K. Chike, M. Lovelace, W. Scrivens, R. Holliday and M. Williams, *J. Raman Spectrosc.*, 1994, **25**(1), 59–65.
100. O. N. Fedyaeva and A. A. Vostrikov, *J. Supercrit. Fluid.*, 2012, **72**, 100–110.
101. C. Jian, L. Yihong, L. Yunhua and Q. Guohe, *Petrol. Sci. Technol.*, 2006, **24**(11), 1339–1346.
102. A. Zaker, P. Guerra, Y. Wang, G. A. Tompsett, X. Huang, J. Q. Bond and M. T. Timko, *Catal. Today.*, 2018, **317**, 2–11.
103. P. Duan and P. E. Savage, *Appl. Catal., B Environ.*, 2011, **108–109**, 54–60.
104. T. Yan, J. Xu, L. Wang, Y. Liu, C. Yang and T. Fang, *RSC. Adv.*, 2015, **5**(92), 75129–75140.
105. T. Adschiri, R. Shibata, T. Sato, M. Watanabe and K. Arai, *Ind. Eng. Chem. Res.*, 1998, **37**(7), 2634–2638.
106. T. M. Yeh, J. G. Dickinson, A. Franck, S. Linic, L. T. Thompson and P. E. Savage, *J. Chem. Technol. Biot.*, 2013, **88**(1), 13–24.
107. J. Sadhukan, K. S. Ng and E. Martinez Hernandez, Introduction, in *Biorefineries and Chemical Processes: Design, Integration and Sustainability Analysis*, John Wiley & Sons, Ltd, Hoboken, NJ, 2014, pp. 1–41.
108. H. Huang and X. Yuan, *Prog. Energy Combust.*, 2015, **49**, 59–80.
109. J. J. Cheng and G. R. Timilsina, *Renewable Energy*, 2011, **36**(12), 3541–3549.
110. J.-Y. Kim, H. W. Lee, S. M. Lee, J. Jae and Y.-K. Park, *Bioresour. Technol.*, 2019, **279**, 373–384.
111. M.J. Climent, A. Corma and S. Iborra, *Green. Chem.*, 2014, **16**(2), 516–547.
112. A. Demirbaş, *Energy Convers. Manage.*, 2000, **41**(6), 633–646.
113. A. Demirbaş, *Energy Convers. Manage.*, 2001, **42**(11), 1357–1378.
114. G. Perkins, N. Batalha, A. Kumar, T. Bhaskar and M. Konarova, *Renewable Sustainable Energy Rev.*, 2019, **115**, 109400.
115. T. Muppaneni, H. K. Reddy, T. Selvaratnam, K. P. R. Dandamudi, B. Dungan, N. Nirmalakhandan, T. Schaub, F. Omar Olguin, W. Voorhies, P. Lammers and S. Deng. *Bioresour. Technol.*, 2017, **223**, 91–97.
116. R. Shakya, J. Whelen, S. Adhikari, R. Mahadevan and S. Neupane, *Algal. Res.*, 2015, **12**, 80–90.
117. M. K. Jindal and M. K. Jha, *Indian Chem. Eng.*, 2016, **58**(2), 157–171.
118. Y. Wang, H. Wang, H. Lin, Y. Zheng, J. Zhao, A. Pelletier and K. Li, *Biomass. Bioenergy*, 2013, **59**, 158–167.
119. L. Nazari, Z. Yuan, S. Souzanchi, M. B. Ray and C. C. Xu, *Fuel*, 2015, **162**, 74–83.
120. S. Karagöz, T. Bhaskar, A. Muto and Y. Sakata, *Bioresour. Technol.*, 2006, **97**(1), 90–98.
121. L. Cao, C. Zhang, S. Hao, G. Luo, S. Zhang and J. Chen, *Bioresour. Technol.*, 2016, **220**, 471–478.
122. C. Song, H. Hu, S. Zhu, G. Wang and G. Chen, *Energy Fuel*, 2004, **18**(1), 90–96.
123. H. Mazaheri, K. T. Lee and A. R. Mohamed, *Fuel Process. Technol.*, 2013, **110**, 197–205.

124. Z. Zhu, S. S. Toor, L. Rosendahl and G. Chen, *Environ. Prog, Sustain.*, 2014, **33**(3), 737–743.
125. Z. Zhu, S. S. Toor, L. Rosendahl, D. Yu and G. Chen, *Energy*, 2015, **80**, 284–292.
126. Z. Bi, J. Zhang, E. Peterson, Z. Zhu, C. Xia, Y. Liang and T. Wiltowski, *Fuel*, 2017, **188**, 112–120.
127. M. Déniel, G. Haarlemmer, A. Roubaud, E. Weiss-Hortala and J. Fages, *Biomass Bioenergy*, 2016, **95**, 273–285.
128. C.-C. Chang, C.-P. Chen, C.-S. Yang, Y.-H. Chen, M. Huang, C.-Y. Chang, J.-L. Shie, M.-H. Yuan, Y.-H. Chen, C. Ho, K. Li and M.-T. Yang, *Sustain. Environ. Res.*, 2016, **26**(6), 262–267.
129. P. Biller, R. B. Madsen, M. Klemmer, J. Becker, B. B. Iversen and M. Glasius, *Bioresour. Technol.*, 2016, **220**, 190–199.
130. L. Qian, S. Wang and P. E. Savage, *Bioresour. Technol.*, 2017, **232**, 27–34.
131. K. Malins, V. Kampars, J. Brinks, I. Neibolte, R. Murnieks and R. Kampare, *Bioresour. Technol.*, 2015, **187**, 23–29.
132. A. Suzuki, T. Nakamura and S. Yokoyama, *J. Chem. Eng. Jpn*, 1990, **23**(1), 6–11.
133. K. R. Arturi, M. Strandgaard, R. P. Nielsen, E. G. Søgaard and M. Maschietti, *J. Supercrit. Fluid*, 2017, **123**, 28–39.
134. T. D. H. Nguyen, M. Maschietti, T. Belkheiri, L.-E. Åmand, H. Theliander, L. Vamling, L. Olausson and S.-I. Andersson, *J. Supercrit. Fluid.*, 2014, **86**, 67–75.
135. Y. Chen, R. Mu, M. Yang, L. Fang, Y. Wu, K. Wu, Y. Liu and J. Gong, *Chem. Eng. Sci.*, 2017, **161**, 299–307.
136. M. Saber, A. Golzary, M. Hosseinpour, F. Takahashi and K. Yoshikawa, *Appl. Energy*, 2016, **183**, 566–576.
137. W. Tian, R. Liu, W. Wang, Z. Yin and X. Yi, *Bioresour. Technol.*, 2018, **263**, 569–575.
138. B. Zhang, Q. Lin, Q. Zhang Q, K. Wu, W. Pu, M. Yang and Y. Wu, *RSC Adv.*, 2017, 7, 8944–8951.
139. P. Duan and P. E. Savage, *Ind. Eng. Chem. Res.*, 2011, **50**(1), 52–61.
140. S. C. Yim, A. T. Quitain, S. Yusup, M. Sasaki, Y. Uemura and T. Kida, *J. Supercrit. Fluid.*, 2017, **120**, 384–394.
141. J. Long, Y. Li, X. Zhang, L. Tang, C. Song and F. Wang, *Fuel*, 2016, **186**, 685–693.
142. P. Sun, M. Heng, S. Sun and J. Chen, *Energy*, 2010, **35**(12), 5421–5429.
143. B. Zhang, M. von Keitz and K. Valentas, *J. Anal. Appl. Pyrol.*, 2009, **84**(1), 18–24.
144. W. Shi, S. Li, H. Jin, Y. Zhao and W. Yu, *Energy Sources, Part A*, 2013, **35**(22), 2149–2155.
145. J. H. Lee, H. Hwang, J. Moon and J. W. Choi, *J. Anal. Appl. Pyrol.*, 2016, **122**, 415–421.
146. P. R. Christensen, A. J. Mørup, A. Mamakhel, M. Glasius, J. Becker and B. B. Iversen, *Fuel*, 2014, **123**, 158–166.

147. P. Duan, Y. Xu, F. Wang, B. Wang and W. Yan, *Biochem. Eng. J.*, 2016, **116**, 105–112.
148. D. Xu, and P. E. Savage, *J. Supercrit. Fluid.*, 2018, **133**, 683–689.
149. P. Duan and P. E. Savage, *Bioresour. Technol.*, 2011, **102**(2), 1899–1906.
150. D. Xu, S. Guo, L. Liu, H. Hua, Y. Guo, S. Wang and Z. Jing, *Biomed. Res. Int.*, 2018, **2018**, 8376127.
151. D. Xu, L. Liu, Z. He, J. Yang, Z. Wu and Z. Jing, *J. Anal. Appl. Pyrol.*, 2019, **140**, 188–194.
152. P. Duan, C. Zhang, F. Wang, J. Fu, X. Lü, Y. Xu and X. Shi, *Catal. Today*, 2016, **274**, 73–81.
153. R. Remón, P. Arcelus-Arrillaga, L. García and J. Arauzo, *Appl. Energy*, 2018, **228**, 2275–2287.
154. P. Kumar, D. M. Barrett, M.J. Delwiche and P. Stroeve, *Ind. Eng. Chem. Res.*, 2009, **48**(8), 3713–3729.

CHAPTER 4

Recent Advances in Photocatalytic Materials for Solar Fuel Production from Water and Carbon Dioxide

MIRIAM REGUE, SANTOSH KUMAR, AND SALVADOR ESLAVA*

Department of Chemical Engineering, Imperial College London, London, UK
*Corresponding contributor. Email: s.eslava@imperial.ac.uk

4.1 Introduction

Rising atmospheric levels of carbon dioxide and depletion of fossil fuel reserves raise serious concerns about the ensuing effects on the global climate and future energy supply.[1] Therefore, the searches for renewable energy sources and sustainable environmental technologies are some of the foremost challenges for maintaining our quality of life.[2] Honda and Fujishima discovered in 1972 the use of TiO_2 electrodes for the photoelectrochemical (PEC) splitting of water,[3] opening up a great opportunity for the use of photocatalysis in the utilization of solar energy and environmental remediation.[4,5] Utilizing abundant solar energy to convert water and carbon dioxide into chemical fuels, so-called "artificial photosynthesis", could address the energy crisis and global climate change simultaneously.[4] Artificial photosynthesis is a chemical process, as its name suggests, which mimics nature's

Heterogeneous Catalysis for Energy Applications
Energy and Environment Series No. 27
Edited by Tomas R. Reina and Jose A. Odriozola

Published by the Royal Society of Chemistry, www.rsc.org

photosynthesis to produce chemical fuels from sunlight and Earth-abundant materials such as water, carbon dioxide and many more, as shown in Figure 4.1a. The most important advantages of this process are the use of solar energy and its ecological purity, which offer the possibility of accomplishing energy cycles without environmental pollution or additional warming of Earth.[6] Since Earth receives more solar energy than what is required to meet the human energy demands, the production of so-called "solar fuels" using photocatalysis is extremely relevant for a future portfolio of sustainable energy sources.[7] For example, if the solar fuel produced is hydrogen, this is a very useful energy carrier that is convertible to electrical power without generating by-products that are harmful to the environment.[8] Other solar fuels such as synthesis gas (syngas; $CO + H_2$) and light hydrocarbons would also be extremely useful as sustainable feedstocks in the chemical and energy industries.[9]

Photocatalysis or artificial photosynthesis is catalysed by a photoactive semiconductor material – a so-called "photocatalyst" – that harnesses solar energy and converts Earth-abundant materials and/or waste (such as water, carbon dioxide and biomass) into solar fuels.[10] Photocatalysis can be considered as an integrated system consisting of a light-harvesting part and a catalytic conversion part. Hence, it is important to maximize the performance of each part and design an efficient system based on a thorough understanding of each component and the interface chemistry between them. Although practical applications of photocatalysis are still limited due to a number of challenges, in recent years, tremendous efforts have been made for developing efficient photocatalytic systems for clean energy applications.[11,12] This chapter focuses on recent progress in sustainable materials and strategy developments for maximizing the use of the visible light fraction of the solar spectrum and improving the photoconversion efficiency and selectivity in water splitting and CO_2 reduction into solar fuels for greener chemical and energy industries.

4.2 Semiconductor Photocatalysis and Basic Mechanism

Photocatalysis is initiated by an electron–hole pair after band gap excitation in a semiconductor material called a "photocatalyst".[13] When a photocatalyst is illuminated by light with an energy equal to or greater than its band gap energy, the valence band (VB) electrons can be excited to the conduction band (CB), leaving a positive hole in the VB (step 1), as shown in Figure 4.1b. The excited electrons in the CB and holes in the VB can recombine, releasing the energy input as heat, with no chemical effect (step 2), which is a major drawback of the process. Alternatively, electrons and holes can reach the surface of the photocatalyst without recombination. These surface electrons and holes are potentially capable of participating in oxidation and reduction reactions with surface-adsorbed species such as H_2O, O_2 and many other inorganic and organic species (steps 3 and 4). The surface electrons and holes can also undergo surface recombination (step 2), which again reduces the photoconversion efficiency. The process described above could be briefly expressed by the following steps:

Light absorption	$SC + h\vartheta \rightarrow SC^{*}\left(e_{CB}^{-} + h_{VB}^{+}\right)$	Step 1
Recombination	$e_{CB}^{-} + h_{VB}^{+} \rightarrow heat$	Step 2
Reduction reaction	$A + e_{CB}^{-} \rightarrow A^{-}$	Step 3
Oxidation reaction	$B + h_{VB}^{+} \rightarrow B^{+}$	Step 4

The above oxidation and reduction reactions are the basic mechanisms of photocatalytic environmental remediation and solar fuel production.[14,15] The reduction potential of photo-excited electrons at the bottom of the CB and holes at the top of the VB determines the ability of a semiconductor to undergo oxidation and reduction reactions. Therefore, the redox potential along with the band gap energy of a semiconductor determine the likelihood and rate of the charge transfer process of electrons and holes for selective photocatalytic applications.[15]

4.3 Solar Fuel Production

Although semiconductor photocatalysis has been extensively studied for a decade in various applications,[14] it became especially attractive in recent years because of its potential to resolve the growing environmental pollution and the renewable energy demands of future generations, utilizing inexhaustible sunlight as an energy source and avoiding pollution.[6] In recent years, several materials and strategies have shown remarkable performance in renewable energy production, including hydrogen fuel from photocatalytic water splitting and CO_2 photoreduction.

4.3.1 Photocatalytic H_2 Production

Hydrogen has been considered as an alternative energy source and as the most feasible energy vector for the development of a more sustainable energy industry. Dissociation of water to produce hydrogen has garnered significant scientific attention. However, the unfavourably large energy barrier of this simple process makes it difficult from a scientific viewpoint:[16]

$$H_2O\,(l) \rightarrow H_2\,(g) + \tfrac{1}{2}\,O_2\,(g),\ \Delta G = +237\ \text{kJ mol}^{-1} \qquad (4.1)$$

It is worth noting that the positive Gibbs free energy change clearly indicates that the water splitting is thermodynamically an uphill reaction, which is opposite to the photocatalytic degradation of organic pollutants, for which the Gibbs free energy change is negative. The electrons in the CB and holes in the VB of the photocatalyst are crucial carriers involved in the process, since they are the responsible for reducing and oxidizing water into H_2 and O_2, respectively. In order to initiate hydrogen production, the bottom of the CB must be more negative than the reduction potential of H^+

to H_2 (0 V_{NHE} at pH = 0) and the top of the VB must be more positive than the oxidation potential of water to O_2 (1.23 V_{NHE} at pH = 0, where V_{NHE} stands for volts *vs.* the normal hydrogen electrode). The water splitting will proceed differently depending on whether the environment is acidic or basic. For an alkaline environment, the reduction and oxidation reactions are as follows:[17]

$$4H_2O + 4e^- \rightarrow 2H_2 + 4OH^- \quad (4.2)$$

$$4OH^- + 4H^+ \rightarrow 2H_2O + O_2 \quad (4.3)$$

For an acidic environment, the reduction and oxidation reactions are as follows:

$$4H^+ + 4e^- \rightarrow 2H_2 \quad (4.4)$$

$$2H_2O + 4h^+ \rightarrow 4H^+ + O_2 \quad (4.5)$$

In general, photocatalytic water-splitting experiments are conducted at around pH 7, at which the reduction and oxidation reactions are as follows:[18]

$$2H_2O + 2e^- \rightarrow H_2 + 2OH^- \quad (4.6)$$

$$2H_2O \rightarrow O_2 + 4H^+ + 4e^- \quad (4.7)$$

The redox potential for the overall reaction is $E_H = -1.23$ V, with the corresponding reduction and oxidation potentials of −0.41 and 0.82 V_{NHE} at pH 7, respectively, which gives a $\Delta G_o = +237$ kJ mol^{-1}.

To date, a large number of research articles on photocatalysis for water splitting have been published.[12,19] The design of a sustainable photocatalyst to efficiently split water to produce H_2 fuel using solar light is a challenge that captures the attention of chemists, chemical engineers and material scientists.[15] This chapter will highlight the key developments in recent years in terms of materials and strategies for improving photocatalytic H_2 production.[15]

4.3.2 Photocatalytic CO_2 Conversion

Photocatalytic CO_2 reduction is a multi-electron transfer process for converting CO_2 into low-carbon feedstocks such as formic acid, carbon monoxide, methanol and methane with appropriate reduction potential requirements.[20,21] Similarly to photocatalytic H_2 production, this process is also catalysed by a photocatalyst that harnesses solar energy and produces charges (electrons and holes). The photocatalytic CO_2 reduction process starts with the adsorption of a CO_2 molecule on the photocatalyst surface, where the molecule undergoes reduction by photogenerated electrons. These electrons result from light absorption, excitation and charge separation from the counterpart holes, which can drive a charge-balancing oxidation reaction. These oxidation and reduction reactions are the basic mechanisms of photocatalytic CO_2 reduction. The reduction potentials of both photo-excited electrons at the bottom of the CB and holes at the top of the VB determine the ability of a semiconductor to undergo oxidation and reduction reactions; the reduction potentials for CO_2 photoreduction with H_2O to various products with reference to NHE at pH 7 are as follows:[22]

$$CO_2 + e^- \rightarrow CO_2^{\cdot -} \quad E^o = -1.90\ V \tag{4.8}$$

$$CO_2 + 2H^+ + 2e^- \rightarrow HCOOH \quad E^o = -0.61\ V \tag{4.9}$$

$$CO_2 + 2H^+ + 2e^- \rightarrow CO + H_2O \quad E^o = -0.53\ V \tag{4.10}$$

$$CO_2 + 4H^+ + 4e^- \rightarrow HCHO + H_2O \quad E^o = -0.48\ V \tag{4.11}$$

$$CO_2 + 6H^+ + 6e^- \rightarrow CH_3OH + H_2O \quad E^o = -0.38\ V \tag{4.12}$$

$$CO_2 + 8H^+ + 8e^- \rightarrow CH_4 + 2H_2O \quad E^o = -0.24\ V \tag{4.13}$$

$$2CO_2 + 12H^+ + 12e^- \rightarrow C_2H_4 + 4H_2O \quad E^o = -0.22\ V \tag{4.14}$$

$$2CO_2 + 12H^+ + 12e^- \rightarrow C_2H_5OH + 3H_2O \quad E^o = -0.33\ V \tag{4.15}$$

Although the physics behind the separation of charge carriers varies with different applications and the surface electronic structure of a photocatalyst, it is unambiguously accepted that the primary reactions responsible for a positive photocatalytic effect are interfacial redox reactions of electrons and holes that are generated upon band gap excitation (Figure 4.1b). Diverse photocatalytic materials have been intensively studied for photocatalytic CO_2 reduction.[23,24] There are many insightful review articles that have reported the recent advances achieved in the field of CO_2 reduction.[25,26] The key factors of an ideal photocatalyst for optimal activity and selectivity for CO_2 reduction include efficient CO_2 adsorption, charge separation, CO_2 activation on the catalyst surface and suitable positions in the energy diagram at the bottom of the CB and at the top of the VB with respect to the redox potential of CO_2 conversion reactions. This chapter will also highlight the key developments in photocatalytic CO_2 reduction to low-carbon chemical fuels.

4.4 Photocatalytic Materials

So far, many insightful review articles have described the field of sustainable photocatalytic materials for various renewable energy applications.[12,15,27] In this chapter, we will focus on recent developments in novel photocatalytic

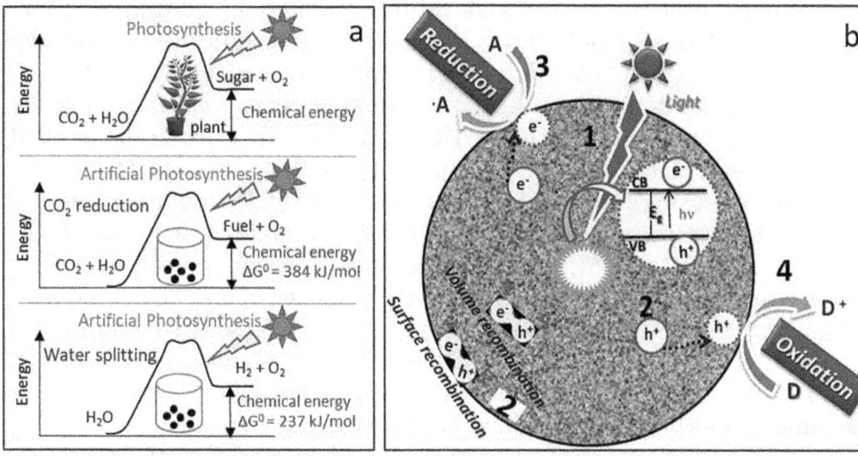

Figure 4.1 (a) Photosynthesis and artificial photosynthesis. (b) Schematic illustration of the major processes occurring on a semiconductor particle under light irradiation. Reproduced from Ref. 21, https://doi.org/10.3390/catal8020074, under the terms of the CC BY 4.0 licence, https://creativecommons.org/licenses/by/4.0/.

materials, including metal oxides, perovskites, plasmonic materials, metal-free materials, metal–organic frameworks (MOFs), metal sulfides and layered double hydroxides, and their renewable energy applications, including photocatalytic hydrogen production from water splitting and CO_2 reduction.

4.4.1 Metal Oxides

Metal oxide semiconductors are the most studied materials for photocatalytic applications, especially for water splitting and CO_2 photoreduction. These materials contain metal cations at the highest oxidation state having either a d^0 or d^{10} electronic configuration.[28,29] Metal oxides usually have a VB consisting of occupied O 2p orbitals, whereas its CB depends on the electronic configuration of the metal cation employed (d^0 or d^{10}). For instance, metal oxides with a d^0 electronic configuration such as TiO_2 or ZrO_2 have a CB consisting of empty d orbitals of the corresponding metal cation employed, whereas those with a d^{10} electronic configuration such as Ga_2O_3 have a CB consisting of hybridized orbitals of empty s and p orbitals of the metal cation.[29] In order to achieve overall water splitting *via* a one-step excitation process that involves both oxidation (to O_2) and reduction of water (to H_2), the CB and VB potentials must straddle the water oxidation and reduction potentials. The VB of most metal oxides is typically located at around 3.0 V_{NHE} and the CB must be more negative than 0 V_{NHE} (at pH 0) for the water reduction reaction, implying that the band gap required to achieve overall water splitting must be above 3.0 eV, which corresponds to the UV region of the solar spectrum. Therefore, common metal oxides with visible-light response such as Cu_2O, WO_3, Fe_2O_3

and $BiVO_4$ are discarded as materials for the overall water-splitting reaction. Examples of materials that have achieved overall water splitting *via* a one-step excitation process using UV irradiation are TiO_2, Ga_2O_3, ZrO_2 and Ta_2O_5.

TiO_2 has been extensively studied over recent decades, especially since the first report on water splitting using TiO_2 as a photoelectrode under applied bias.[3] A few years later, in 1977, Schrauzer and Guth reported photocatalytic water splitting using powdered TiO_2 loaded with small amounts of Pt and Rh.[30] TiO_2 is an n-type semiconductor with a d^0 electronic configuration and a band gap of 3.0–3.2 eV.[31] Since the early years of the water-splitting discovery, a lot of effort has been devoted to solar water splitting using TiO_2-based photocatalysts.[32] Unfortunately, true water splitting in the absence of hole sacrificial agents is still limited and very inefficient due to various challenges, such as it involving four electrons for the water oxidation reaction.[33] However, several publications based on TiO_2 have been reported for overall water splitting under UV light.[34,35] Very recently, Hainer *et al.*[36] investigated photocatalytic hydrogen production over metal-decorated TiO_2 in order to understand the role of hole sacrificial agents *vs.* true water splitting. As shown in Figure 4.2A, the production rates of H_2 for the seven TiO_2-based photocatalysts in the absence (Figure 4.2A-a) or presence of hole sacrificial agents (Figure 4.2A-b and c) showed almost two-fold greater performances than true water-splitting conditions. The photocatalytic activity order for true water splitting (Figure 4.2A) is quite different from the activity in the presence of hole sacrificial agents. Therefore, it would be impossible to predict the photocatalytic performance of true water splitting based on the performance in the presence of a hole sacrificial agent. Thus, knowledge gathered on photocatalytic performance in true water splitting and hole sacrificial agent-assisted systems is simply not transferable.[36]

Various materials other than TiO_2 have been investigated for water splitting. For example, Ga_2O_3, an n-type semiconductor with a d^{10} electronic configuration and a band gap of 4.6 eV, has been widely studied for photocatalytic applications, especially since the first demonstration of overall water splitting under UV light by Takashi *et al.* in 2004.[37] In this publication, the authors reported H_2 and O_2 activities of 46 and 23 μmol h^{-1}, respectively, when loading NiO onto Ga_2O_3 particles. Along the same lines, the effect of adding different metal ions on Ni-supported Ga_2O_3 in photocatalytic water splitting was also investigated by Sakata *et al.*[38] Among the different metal ions studied, incorporation of Ca, Zn, Cr, Sr, Ba and Ta resulted in an overall increase of the photocatalytic water-splitting reaction. The highest performance was achieved when Zn ions were added, giving rise to H_2 and O_2 activities of 4100 and 2200 μmol h^{-1}, respectively. The photocatalytic water-splitting performance of Ga_2O_3 has also been improved by designing a junction of the α and β phase of Ga_2O_3 along with NiO_x as a co-catalyst.[39] The photocatalytic performance was only improved when both phases were exposed on the surface of the photocatalyst.

ZrO_2, a wide band gap (5.0 eV), n-type semiconductor material, has also been successful in overall water splitting under UV light irradiation.[40–42] Over recent years, several parameters such as additives, hydrolysing agents used during ZrO_2 synthesis or pre-treatment methods to control the co-catalyst interface have been investigated to improve overall water splitting. For instance, Sayama and Arakawa reported that incorporation of $NaHCO_3$ and Na_2CO_3 in the reaction system can increase significantly the rate of H_2 and O_2 production.[40] Similarly, Reddy *et al.* demonstrated that incorporation of hydrolysing agents during the synthesis of ZrO_2 could modify the crystallinity and surface area of the photocatalysts.[41] The use of different hydrolysing agents resulted in ZrO_2 with different crystal structures and surface properties. Overall, it was found that variations in the crystal structure (monoclinic and tetragonal) had no correlations with the photocatalytic activity of ZrO_2, but variations in the surface properties did. In fact, ZrO_2 prepared using KOH had the highest surface area among all samples prepared, resulting in the highest photocatalytic H_2 and O_2 rates produced.

The design of the ZrO_2 co-catalyst interface is also key to improving photocatalytic performance. For instance, Zou *et al.* demonstrated that, when loading NiO on ZrO_2 using cold plasma treatment, higher surface areas with half-ellipsoid-shaped metal particles were obtained, significantly outperforming the photocatalytic water-splitting performance of conventional methods of co-catalyst treatments.[42] In 2016, H_2 production from a methanol–water solution using visible light was reported for a ZrO_2-based photocatalyst.[43] In this report, oxygen-deficient black zirconia (ZrO_{2-x}) having a band gap of 1.52 eV was prepared.[41] This photocatalyst showed improved photocatalytic activity for H_2 production under simulated solar light over ZrO_2, which showed no activity under these conditions. Therefore, even though overall water splitting is not reported, these results are promising for using ZrO_2-based materials under visible light.

Examples for overall water splitting using Ta_2O_5 have also been reported in the literature. Ta_2O_5 is an n-type semiconductor with a d^0 electronic configuration and a band gap of ~3.9–4.0 eV, implying that photocatalyst excitation can only be performed under UV light. The bottom of its CB and the top of its VB are well located for the reduction and oxidation of water, being at −0.47 V_{NHE} and 3.43 V_{NHE}, respectively.[44] A widely cited publication in the field was authored by Noda *et al.*[45] They synthesized a crystalline mesoporous Ta_2O_5 photocatalyst with a two-dimensional hexagonal structure for overall water splitting under UV light. Crystallization of Ta_2O_5 requires high annealing temperatures that often result in a collapse of the mesoporous structure and a significant loss in surface area, which is detrimental to photocatalytic performance. However, in this publication, a SiO_2 pre-treatment method allowed for the successful synthesis of crystalline mesoporous Ta_2O_5 photocatalysts that kept its mesoporous structure even at annealing temperatures of ~860 °C. Loading of NiO_x (3.0 wt%) in the mesoporous crystalline Ta_2O_5 sample for overall water splitting resulted in H_2 and O_2 production rates of 3.4 and 1.6 mmol h^{-1}, respectively, outperforming by factors of almost 7 and 8 the H_2 and O_2 production rates of a mesoporous (non-crystallized) and crystallized nonporous Ta_2O_5 samples, respectively. This indicated that both the crystallinity and the mesoporous structure were responsible for the increase in photocatalytic performance. Both features allowed for an efficient transport of photogenerated electrons and holes from the bulk of the photocatalyst to the surface through the thin-walled crystalline phase. This was the first report that demonstrated the application of crystalline mesoporous transition metal oxides in the field of photocatalysis.

More recent publications for Ta_2O_5 were reported by Gonçalves *et al.*[46] and Cherevan *et al.*[47] Although only H_2 production rates were reported for both publications, they still represent a breakthrough in the field. For instance, Gonçalves synthesized Ta_2O_5 nanotubes (NTs) from an anodization method of Ta sheets and performed a systematic study of the photocatalytic properties at different annealing temperatures.[46] The authors successfully synthesized free-standing Ta_2O_5 NTs by changing the electrolyte temperature during the anodization process. Increasing the temperature (to 50 °C) resulted in a decrease of Ta_2O_5 adherence on the Ta substrate that formed amorphous, powdered Ta_2O_5 NTs. Crystalline (orthorhombic) Ta_2O_5 NTs were obtained at annealing temperatures of 750 °C. Temperatures above 800 °C destroyed the NT morphology and were not considered in the study. Photocatalytic H_2 measurements in a water/ethanol mixture under UV light showed increases in H_2 production rates as annealing temperatures increased, from 2900 μmol h^{-1} g^{-1} (at 550 °C) to 4900 μmol h^{-1} g^{-1} (at 800 °C). Interestingly, other products, such as CO, CO_2, CH_4, C_2H_4 and C_2H_6, were also detected in the experiments, coming from ethanol photoreforming. In contrast to H_2 evolution rates, higher amounts were observed for amorphous and slightly sulfur-contaminated samples coming from the synthesis method employed. To the best knowledge of the authors, this was the first time that ethanol photoreforming side products were detected using Ta_2O_5. Overall, these results suggested that an increase in annealing temperature resulted in improved crystallinity and less sulfur content, which improved the selectivity towards H_2 evolution.

Similarly, Cherevan *et al.* reported photocatalytic H_2 evolution from hybrids consisting of carbon nanotubes (CNTs) and graphene oxide (GO) coupled to Ta_2O_5 from a water/methanol mixture under UV light irradiation and loading of Pt as a co-catalyst.[47] The authors found that optimized samples consisting of a hybrid of crystalline CNT–Ta_2O_5 exhibited significantly higher H_2 production rates than samples consisting of both pure Ta_2O_5 and GO–Ta_2O_5, producing rates of 1600, 140 and 45 μmol h^{-1}, respectively. Furthermore, optimized CNT–Ta_2O_5 demonstrated constant H_2 evolution activity even after 20 h of UV light irradiation. The optimized CNT–Ta_2O_5 hybrid was mainly composed of a single thin, crystalline layer of Ta_2O_5 (6–8 nm) and occasionally a second layer with platelet-like particles surrounding the CNTs. The high performance of this hybrid was attributed to reduced electron–hole recombination due to the transferring of photoelectrons from the Ta_2O_5 to the thin CNT layer and due to the small particle size that yielded a slightly larger band gap with a higher CB band level that favoured the H^+ reduction to H_2. Additionally, the enhanced photocatalytic performance of optimized CNT–Ta_2O_5 over CNT–Ta_2O_5 with a thicker coating of Ta_2O_5 surrounding the CNTs was due to fewer grain boundaries that reduced electron–hole recombination and formation of a Schottky-type junction (between Ta_2O_5 and CNTs). This reduced the charge transfer barrier and enhanced charge separation due to transferring of photogenerated electrons to CNTs.

Very few publications have reported overall water splitting under visible light *via* a one-step excitation process due to the difficulty in finding oxide materials able to straddle the water oxidation/reduction potentials. An early report in 1998 by Domen's group reported overall photocatalytic water splitting under visible light illumination from pure water when using Cu_2O.[48] They achieved high activities for more than 1900 h and reported quantum efficiencies of 0.3% at between 550 and 600 nm. In 2013, Liao *et al.* demonstrated overall water splitting (without co-catalysts or sacrificial agents) from nanocrystalline CoO (<10 nm) under AM 1.5G solar illumination with a solar-to-hydrogen efficiency of 5%.[49] The calculated band positions of these nanoparticles indicated a suitable band alignment for nanocrystalline CoO particles, which explained the overall water-splitting activity. Although this photocatalyst deactivated after 1 h, these results are promising for visible light overall water-splitting research. Very recently, Li *et al.* reported overall water splitting under visible light with a GaN : ZnO photocatalyst loaded with small amounts of Pd with an apparent quantum efficiency (QE) of 1.62%.[50] Tanaka *et al.*[51] reported overall water splitting under visible light from a suspension of plasmonic Au/TiO_2 photocatalysts. This was the first report showing that surface plasmon resonance effects could induce overall water splitting under visible light. Incorporation of NiO_x onto Au/TiO_2 photocatalysts enhanced the H_2 and O_2 production rates, reaching values of 5.5 and 2.7 $\mu mol\ h^{-1}$, respectively.

The Z-scheme approach, where H_2 evolution and O_2 evolution photocatalysts are combined along with a shuttle redox mediator, has also been studied for overall water splitting under visible light in metal oxide semiconductors. In this approach, two different metal oxides with narrow band gaps can be combined as long as one of them satisfies the CB requirement for the water reduction reaction and the other the VB requirement for the water oxidation reaction.[3] This implies that narrow band gap materials such as $BiVO_4$ or WO_3 can be used as the oxygen evolution photocatalyst for photocatalytic overall water splitting. For instance, $BiVO_4$ was successfully coupled with Pt/$SrTiO_3$: Rh, demonstrating apparent quantum efficiencies of 2.1% at 420 nm and a solar energy conversion efficiency of 0.06% using a solar simulator (AM 1.5, 100 mW cm^{-2}).[52] As shown in Figure 4.2B, a Z-scheme system consisting of photocatalytic sheets based on $SrTiO_3$: La, Rh and Mo-doped $BiVO_4$ powders embedded in a gold layer achieved a solar-to-hydrogen efficiency of 1.1% and apparent quantum yields of over 30% at 419 nm.[53] No noticeable decrease in photocatalytic activity was observed after more than 10 h of irradiation (Figure 4.2C). Similarly, WO_3 loaded with Pt was successfully coupled

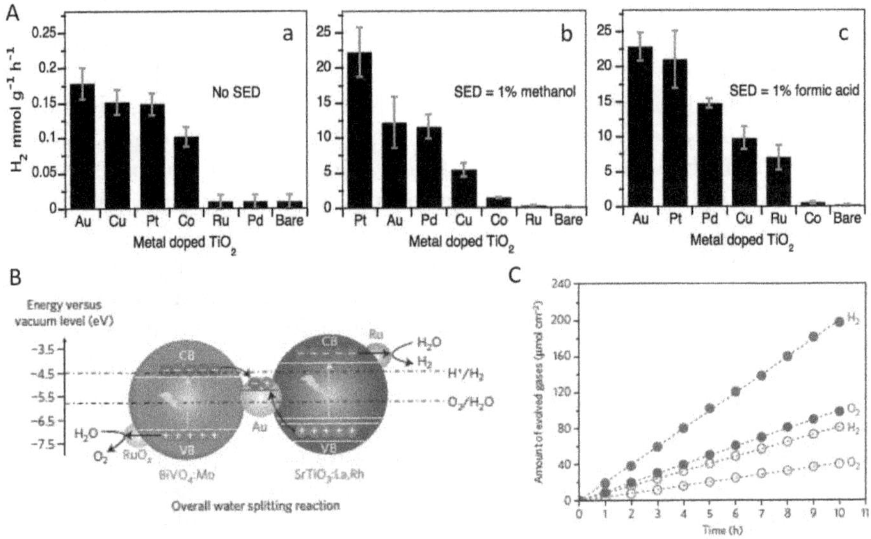

Figure 4.2 (A) H_2 production rates over different photocatalysts using 368 nm light for (a) true water-splitting conditions, (b) in the presence of 1% methanol and (c) in the presence of 1% formic acid (pH ≈ 2.2). Reproduced from Ref. 36 with permission from American Chemical Society, Copyright 2018. (B) Schematic of overall water splitting on the Ru-modified $SrTiO_3$:La,Rh/Au/$BiVO_4$:Mo sheet. (C) The water-splitting profile on a Cr_2O_3/Ru-modified $SrTiO_3$:La,Rh/Au/$BiVO_4$:Mo sheet under simulated sunlight (AM 1.5G) at 288 K and 5 kPa (open symbols) and 331 K and 10 kPa (closed symbols). Reproduced from Ref. 53 with permission from Springer Nature, Copyright 2016. CB = conduction band; VB = valence band.

with Pt/ZrO_2/TaON, demonstrating apparent quantum efficiencies of 6.3% at 420.5 nm for overall water splitting.[54]

Over recent years, metal oxides have also been explored for CO_2 reduction, among which TiO_2 is the most extensively studied metal oxide owing to its low cost, non-toxicity and exceptional stability. In 1987, Thampi *et al.*[55] reported TiO_2 as the first material used for CO_2 photoreduction. Xu *et al.*[56] prepared ultrathin anatase TiO_2 nanosheets (2 nm in thickness) with 95% of exposed {1 0 0} facet, which displayed about five times higher photocatalytic activity for CO_2 reduction to CH_4 over the reference TiO_2 cuboids with 53% of exposed {1 0 0} facets.[56] The enhancement in CO_2 reduction performance was due to the larger surface area and the higher percentage of exposed {1 0 0} facets of ultrathin anatase TiO_2 nanosheets. Furthermore, Tu *et al.*[57] fabricated a hollow sphere heterostructure assembly of $Ti_{0.91}O_2$ nanosheets and graphene nanosheets *via* a combination of a layer-by-layer assembly technique and a microwave irradiation technique for CO_2 photoreduction into CO and CH_4. The molecular-scale 2D contact of $Ti_{0.91}O_2$ nanosheets and graphene nanosheets in the hollow spheres displayed distinctive electronic and optical properties, such as rapid charge mobility, an enhanced lifetime of charge carriers and multi-scattering of incident light in comparison with simple integration. As a result, the hollow spheres showed nine times higher activity for CO_2 reduction compared to commercial P25 TiO_2. Although TiO_2-based materials have been thoroughly investigated for photocatalytic water splitting and environmental remediation applications, the studies on photocatalytic CO_2 reduction are still limited.[58] Many other oxides, such as Ga_2O_3, ZnO, NiO and WO_3, have been also investigated along with TiO_2.[59,60] For example, in 2018, Sun *et al.* reported a promising study on ultrathin $WO_3{\cdot}0.33H_2O$ nanotubes for CO_2 photoreduction to acetate under solar light.[61] These WO_3 nanotubes had a large amount of exposed surface oxygen vacancies, which were responsible for the high photocatalytic activity and selectivity (up to 85%) towards acetate generation. Yahaya *et al.*[62] successfully investigated photocatalytic reduction of CO_2 into methanol using TiO_2, NiO and ZnO under 355 nm UV light irradiation. Among the different metal oxides studied, NiO achieved the highest yield of the photoconversion of CO_2, whereas a poorer performance was reported for TiO_2.

In most cases, photocatalysts typically comprise two different components for CO_2 reduction : a semiconductor that can absorb light and generate excited charges to drive the reaction; and a co-catalyst that facilitates the charge transfer and conversion of the reactants to the products.[63] The nature of the co-catalyst can influence the catalytic activity and the product selectivity; understanding exactly how the co-catalyst's properties affect the catalytic performance is therefore important for designing optimal catalytic systems. Xing *et al.*[64] investigated the effect of the size of Pt co-catalysts in a photocatalytic CO_2 conversion system, and they demonstrated a significant role of Pt particle size in both photocatalytic activity and product selectivity (Figure 4.3). A set of photocatalysts were prepared by depositing Pt nanoparticles with an average particle size ranging from 1.8 to 7.2 nm and a constant loading of 2 wt% on hierarchically ordered TiO_2–SiO_2 materials as the light absorber (Figure 4.3a). The photocatalysts containing the smallest Pt nanoparticles showed the highest photocatalytic activity for the production of CH_4, as well as H_2, which was generated in the competing water reduction reaction (Figure 4.3b). Selectivity for CH_4, the desired product, increased from 39.1% to 79.1% on going from the smallest to the largest particle size (Figure 4.3c). The experimental results were in agreement with theoretical calculations, which suggested that it is the terrace sites on the Pt nanoparticles that are the most active sites for CH_4 production, while low-coordinate sites, such as those found at the corners and edges of the nanoparticles, are the most active for H_2 production. Li *et al.*[65] described the design and fabrication of a structure with reductive outer surfaces decorated with spatially separated co-catalysts, where oxidation co-catalysts were uniformly dispersed on inner surfaces, which were efficient for photocatalytic CO_2 reduction. Moreover, the resulting CO : H_2 ratio in the products could be adjusted to a large range of values, including the desired 1 : 2 for the chemical feedstocks industry, by changing the components and properties of the reduction co-catalysts. For example, TiO_2 hollow spheres, MnO_x and a CuPt alloy

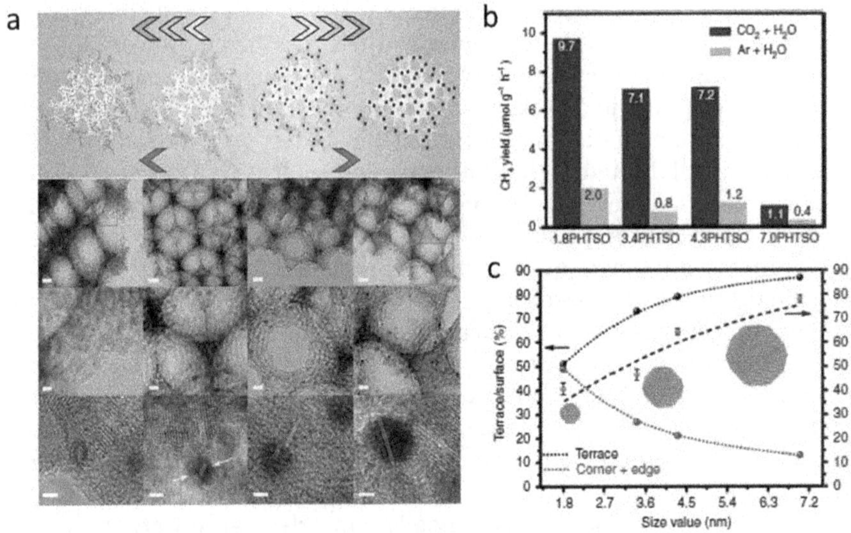

Figure 4.3 (a) Schematic illustration of the acid–base-mediated alcohol reduction method for manipulation of the size of Pt nanoparticles and transmission electron microscopy (TEM) and corresponding high-resolution transmission electron microscopy (HR-TEM) images; the squares indicate the stepwise magnifications of the local sites. (b) CH_4 yield comparison of Pt-loaded hierarchically ordered TiO_2–SiO_2 porous materials (xPHTSO) (x = 1.8, 3.4, 4.3 or 7.0) in different atmospheres, where dark bars denote the yield of CH_4 generated from a CO_2/H_2O atmosphere and light bars denote the yield of CH_4 generated from an Ar/H_2O atmosphere. (c) Size-dependent activity and selectivity of Pt nanoparticles in photocatalytic CO_2 reduction. Adapted from Ref. 64, https://doi.org/10.1038/s41467-018-03666-2, under the terms of the CC BY 4.0 licence, https://creativecommons.org/licenses/by/4.0/.

were prepared where TiO_2 acted as the main catalyst, MnO_x as the oxidation co-catalyst and CuPt as the reduction co-catalyst. In the MnO_x@TiO_2@CuPt mesoporous composite, driven by the spatially separated MnO_x particles and CuPt alloys, holes and electrons flew in opposite directions, enhancing the charge separation efficiency. When these well-separated electrons reached the surfaces of reduction co-catalysts (CuPt alloys) with optimized Cu contents, high photocatalytic activity with a desired $CO:H_2$ 1 : 2 product ratio was achieved owing to an adjusted energy of the CO_2 activation and CO* desorption, which they explained by density functional theory calculations and kinetic experiments. Colloidal nanoparticle systems seem to be promising photocatalysts for the control of the activity and selectivity for CO_2 reduction in aqueous water systems. For example, zinc oxide–copper(I) oxide hybrid nanoparticles[66] were reported as colloidal forms bearing copper(I) oxide nanocubes bound to zinc oxide spherical cores for the direct conversion of CO_2 to methane in an aqueous medium under ambient pressure and temperature. The catalysts produced methane with an activity of

1080 μmol g_{cat}^{-1} h^{-1}, an apparent quantum yield of 1.5% and a selectivity for methane of >99%.

Yu *et al.*[67] reported a Z-scheme system consisting of g-C_3N_4 and ZnO for CO_2 photoreduction into methanol under simulated solar light. The resultant material outperformed significantly pristine g-C_3N_4. Interestingly, the improved performance was due to the proposed Z-scheme mechanism instead of a conventional heterojunction-type mechanism in which, upon light excitation, both phases generated electron and hole pairs, and excited electrons from the CB of ZnO transferred to combine with the photogenerated holes of g-C_3N_4. Li *et al.*[68] carried out a detailed study of β-Ga_2O_3 doped with different metal ions (V, Cr, Mn, Fe, Co, Ni and Cu) for CO_2 photoreduction under visible light irradiation. Among the different metals studied, Co^{2+}, Fe^{3+} and Ni^{2+} enhanced significantly the photocatalytic activity, giving rise to CO and H_2 as reduction products, with Ni-doped β-Ga_2O_3 showing the best performance. The activity was further improved after Ag incorporation.

4.4.2 Perovskite Materials

Perovskites are a specific type of materials described as having the ABX_3 crystal structure (Figure 4.4a), where A and B are cations of different sizes (generally A ions are larger than B) and X is an anion. These structures are generally cubic, although less symmetric geometries such as tetragonal and orthorhombic also exist. Depending on the chemical nature of these ions, three main types of perovskite materials exist:[69,70] firstly, oxide perovskites, such as $SrTiO_3$, having the general formula of ABO_3. In this structure, the A site is typically an alkaline earth or rare-earth ion and the B site is a transition metal.[71] Similarly to the ABO_3 structure, double perovskites where two different cations are occupying the A or B sites also exist. These perovskites have either the structure A'AA''B_2O_6 or $A_2B'B''O_6$, such as Bi_2WO_6.[72] Secondly, all inorganic halide perovskites, such as $CsPbBr_3$, where both A and B are inorganic cations and X is a halide anion ($X = Br^-, I^-, Cl^-$). And thirdly, hybrid organic–inorganic halide perovskites, such as methylammonium lead triiodide ($CH_3NH_3PbI_3$ = $MAPbI_3$), where A is the organic cation, B is the metal cation and X is the halide anion.

One of the most studied oxide perovskites is $SrTiO_3$, with a band gap of 3.2 eV. This material finds application both in CO_2 photoreduction and in photocatalytic water splitting. For instance, a promising approach for the design of a 3D structure based on $SrTiO_3$ for photocatalytic CO_2 applications was reported by Zhou *et al.*[73] In this publication, the authors mimicked the natural photosynthetic system. They synthesized a 3D structure using fresh green leaves and $SrTiO_3$ and tested it for CO_2 photoreduction, giving rise to CO and CH_4 as the main photoreduction products. The 3D structured catalyst exhibited a 1.5-fold enhancement in photocatalytic CO_2 reduction when compared to powdered $SrTiO_3$, most likely due to enhanced light absorption. When Au was loaded onto the structure, both CO and CH_4 production rates increased, retaining its photocatalytic

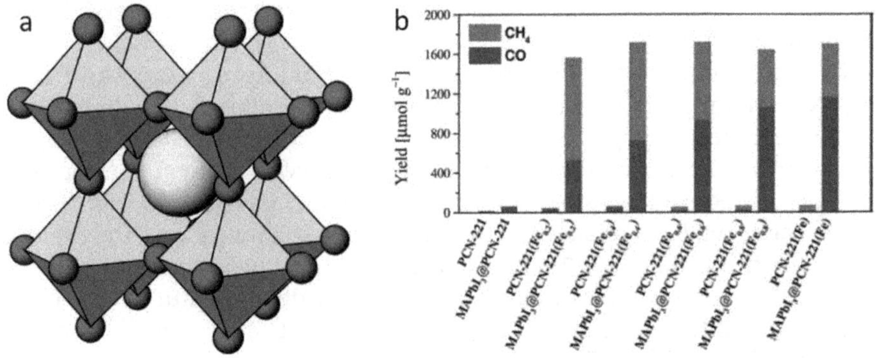

Figure 4.4 (a) ABX_3-type perovskite crystal structure. Reproduced from Ref. 89 with permission from American Chemical Society, Copyright 2012. (b) The production for CH_4 and CO from photocatalytic CO_2 reduction with perovskite quantum dots in iron-based metal–organic frameworks as photocatalysts in the CO_2-saturated ethyl acetate/water solution after 80 h of irradiation under a 300 W Xe lamp at a light intensity of 100 mW cm^{-2}. Reproduced from Ref. 88 with permission from John Wiley and Sons, © 2019 Wiley-VCH Verlag GmbH & Co. KGaA, Weinheim.

activity up to 80 h. Even though $SrTiO_3$ is a good material for photocatalytic applications, it is only active under UV light due to its large band gap (3.2 eV), which limits the use of solar light. In this regard, doping of $SrTiO_3$ has been investigated in order to tune the electronic properties of the material and reduce the band gap. In 2011, Xie *et al.* reported a self-doped $SrTiO_3$ strategy to enhance visible light absorption.[74] Self-doped $SrTiO_3$ photocatalysts were prepared using a carbon-free combustion method from TiN and $Sr(NO_3)_2$ followed by annealing in argon at a temperature range of 1200–1400 °C for 20 h. This annealing treatment induced the formation of Ti^{3+} and oxygen vacancies, resulting in a yellow powder with visible light absorption and enhanced chemical adsorption of CO_2 owing to the presence of these oxygen vacancies. Photocatalytic CO_2 experiments under visible light irradiation gave rise to CH_4 as the main photocatalytic product, achieving a maximum production rate of 0.25 µmol m^{-2} h^{-1}. Recently, Cr-doped $SrTiO_3$ has also been successful in CO_2 photoreduction under visible light, with CH_4 being the only CO_2 photoreduction product, with a yield of 8.8 µmol g^{-1} after 10 h of irradiation.[75] No photocatalytic activity was detected for pristine $SrTiO_3$.

As stated earlier, $SrTiO_3$ has also been tested for photocatalytic water splitting. Recently, Han *et al.* reported H_2 production rates of 8.8 µmol g^{-1} min^{-1} at peak efficiency and steady-state values of 1.2 µmol g^{-1} min^{-1} under solar light irradiation when incorporating Mg into the structure and using Ni/NiO as a co-catalyst.[76] Earlier works had shown H_2 production rates of 28 µmol g^{-1} h^{-1} for pristine $SrTiO_3$ and NiO as a co-catalyst when using UV light irradiation.[77] Other oxide perovskites, such as $NaTaO_3$, $LiTaO_3$, $KTaO_3$, $KNbO_3$ and $NaNbO_3$, have also shown promising photocatalytic activities.[71,72]

The simplest double-type oxide perovskite is Bi_2WO_6, with a band gap of 2.75 eV.[3] Several publications have demonstrated the CO_2 photoreduction activity of Bi_2WO_6 over recent years. For example, the synthesis of Bi_2WO_6 hollow microspheres,[78] which possess large surface areas and high CO_2 adsorption capacities, displayed efficient visible light CO_2 photocatalytic conversion into methanol without any co-catalyst. Along the same lines, a prototype single-unit-cell Bi_2WO_6 layer allowed for a three-fold larger CO_2 adsorption capacity and higher photoabsorption than bulk Bi_2WO_6, which resulted in a methanol formation rate of 75 μmol g^{-1} h^{-1}, 12.5 times higher than that of bulk Bi_2WO_6. The carrier lifetime increased from 14.7 to 83.2 ns, revealed by time-resolved fluorescence spectroscopy, demonstrating an improved electron–hole separation efficacy for the single-unit-cell structure. Very interestingly, the catalytic activity of the single-unit-cell layers proceeded without deactivation even after 2 days. Bi_2WO_6 in the form of square nanoplates of ~9.5 nm in thickness with {0 0 1} facets exposed also showed promising CO_2 photoreduction activities.[79] Only CH_4 as a photoreduction product was obtained with an approximate 1.1 μmol g^{-1} h^{-1} production rate. The high photocatalytic activity of these nanoplates over Bi_2WO_6 particles was due to their large surface area, fast charge carrier mobility and exposure of {0 0 1} facets.

Fewer publications are reported for photocatalytic water splitting when using Bi_2WO_6 as a photocatalyst due to its VB and CB edge positions. Its VB edge is more positive (~3.0 V) than the water oxidation reaction potential (1.23 V), indicating its suitability as an oxygen evolution photocatalyst. Nevertheless, its CB edge is slightly more positive than the reduction potential of water (~0.16 V).[80] This implies that modifications in the electronic structure of the material are required to promote the H_2 evolution reaction. In this regard, Sun *et al.* reported a reduced GO (RGO) and Bi_2WO_6 composite in which both O_2 and H_2 evolution performances were measured in the presence of scavengers.[80] In the preparation of the composite, GO was used as a support for the *in situ* growth of Bi_2WO_6 nanoparticles. Photocatalytic water splitting experiments under visible light revealed O_2 production values of 164.81 μmol after 8 h of the experiment in the presence of an electron scavenger. This performance was 2.17 times higher than a sample prepared by simply mixing GO and Bi_2WO_6 and 4.18 times higher than Bi_2WO_6 nanoparticles. This improved performance was ascribed to the better light absorption of the composite and its improved charge separation efficiency, as demonstrated by photoluminescence studies. H_2 evolution experiments under visible light and with the presence of a sacrificial reagent (a hole scavenger) resulted in no hydrogen being produced when only Bi_2WO_6 nanoparticles were used due to the inaccurate CB position in relation to the reduction potential of water. Nevertheless, the RGO–Bi_2WO_6 composite resulted in H_2 production values of 1273.60 μmol after 8 h of the experiment, being considerably higher than the sample prepared by simply mixing RGO and Bi_2WO_6 and higher than pure RGO, which alone also produced hydrogen. Therefore, the enhancement in the photocatalytic performance was attributed to the unique interaction of Bi_2WO_6 with RGO in the composite, in which RGO acted as an acceptor of photogenerated electrons and suppressed the electron and hole recombination.

Very recently, halide perovskite materials have also been tested in the field of photocatalysis, mainly for CO_2 photoreduction. Due to the instability of these materials in water, photocatalytic water splitting applications in pure water still remain challenging. However, Wang *et al.*[81] reported $MAPbI_3$ organic–inorganic hybrid perovskite hybridized with TiO_2/Pt nanoparticles, which showed visible light photocatalytic hydrogen evolution from hydrogen iodide (HI) splitting with an apparent QE of 70% for H_2 evolution at 420 nm and a solar-to-chemical conversion efficiency of 0.86%. Furthermore, Wu *et al.*[82] studied a mixed-halide perovskite $CH_3NH_3PbBr_{3-x}I_x$ that exhibited enhanced photocatalytic activity for H_2 evolution from HBr/HI solution under visible light at a rate of 2604 μmol g^{-1} h^{-1} and a solar-to-chemical conversion efficiency of 1.05%. More interestingly, the inorganic-type $CsPbBr_3$ perovskite has proven to be successful for photocatalytic CO_2 reduction, in the form of both nanocomposites and as $CsPbBr_3$ itself. For instance, Z. C. Kong *et al.* reported a composite of $CsPbBr_3$ quantum dots and a cobalt-based zeolitic imidazolate framework (ZIF-67) for photocatalytic CO_2 reduction. The photocatalytic performance for the production of CO and CH_4 was greatly enhanced in comparison with pristine $CsPbBr_3$, with production rates of CH_4 of 3.43 μmol g^{-1} h^{-1} in comparison with 0.5 μmol g^{-1} h^{-1}.[83] This excellent performance was attributed to improved moisture stability, CO_2 capture ability and charge separation. Similar approaches using other types of MOFs such as UiO-66(NH_2) and composite materials such as g-C_3N_4 and GO have also demonstrated enhancement in photocatalytic CO_2 conversion in comparison with pristine $CsPbBr_3$.[84–86] In particular, the $CsPbBr_3$–g-C_3N_4 composite showed improved charge separation efficiency owing to a good band alignment and bonding between the $CsPbBr_3$ and g-C_3N_4 (N–Br), with CO being the main photoreduction product.[85] Tuning the structure of $CsPbBr_3$ by amending the ratio of Br and Cl up to a certain amount ($CsPb[Br_{0.5}Cl_{0.5}]_3$) also enhanced significantly the photoreduction of CO_2 to CO and CH_4 in comparison to pristine $CsPbBr_3$ and $CsPbCl_3$.[87] The use of Fe-porphyrin-based MOFs (PCN-221[Fex]) was also successful for the encapsulation of $MAPbI_3$ quantum dots. This configuration achieved an optimum production rate of 19.4 μmol g^{-1} h^{-1} for CO_2 photoreduction (34% CO and 66% CH_4), as shown in Figure 4.4b.[88]

4.4.3 Plasmonic Materials

Plasmonic metallic materials represent another group of materials that have been widely used in the field of photocatalysis. These materials consist of metallic nanostructures such as gold, silver and copper that strongly interact with resonant photons through an excitation of surface plasmon resonance (SPR).[90] Surface plasmons are defined as the collective oscillation of valence electrons confined in the surface of a conducting material that strongly interacts with light. When these conducting materials are irradiated with light at their plasmon frequency (when the real part of the dielectric function goes to zero), SPR takes place.[90,91] Interestingly, plasmon frequency varies depending on the type of material, nanoparticle shape and size and proximity to other nanoparticles.

Plasmonic materials have been explored in the field of photocatalysis,[91,92] either as plasmonic nanostructures without the presence of a semiconductor material or as plasmonic composites of metals coupled to a semiconductor. The enhancement in photocatalytic performance of a plasmonic material/semiconductor over the pristine semiconductor is known to arise from three different possible mechanisms: firstly, charge injection from metal to semiconductor, where, upon light excitation, charge carriers are transferred from the plasmonic metal nanostructure to the semiconductor surface. Secondly, near-field electromagnetic mechanisms, where the semiconductor interacts with the SPR electric fields generated in the vicinity of the metallic nanoparticle. Typically, this mechanism takes place when the metallic nanoparticle and semiconductor are not directly in contact (*i.e.* both materials are separated by non-conductive specimens). And thirdly, scattering mechanisms that take place when the metallic particles of the composite are larger than 50 nm. Upon light irradiation, scattering of photons in the large nanoparticles occurs, increasing the photon path length in the composite and thus the number of electron–hole pairs generated.

Recently, Yu *et al.*[93] designed plasmonic Au nanoparticles for reduction of CO_2 to yield methane and ethane hydrocarbons. The light excitation attributes influenced the ethane and methane selectivity. The observed trends in photocatalytic activity and selectivity followed Poisson statistics of electron harvesting. Higher photon energy and flux favoured simultaneous harvesting of more than one electron from the photocharged Au nanoparticle catalyst, inducing the C–C coupling required for ethane production. These findings elucidate the nature of plasmonic photocatalysis, which involves strong light–matter coupling, and they set the stage for controlled chemical bond formation by light excitation. An unconventional photocatalysis process based on intrinsic plasmons, called intrinsic plasmonic photocatalysis, seems to be promising for photocatalytic hydrogen evolution under visible light irradiation.[94] In particular, a family of intrinsic plasmonic photocatalysts based on alkaline earth niobates ($MNbO_3$, where M = Ca, Sr or Ba) showed significant photocatalytic hydrogen evolution with an activity order of $CaNbO_3 > SrNbO_3 > BaNbO_3$. All of these intrinsic plasmonic photocatalysts have a high carrier density of ~1022 cm^{-3}, which resulted in strong plasmonic resonance in the visible range. As the plasmonic resonance provides the only available state within the visible range, it is concluded that the hot carriers solely generated through the decay of photon-induced plasmon resonances are responsible for the photocatalytic H_2 evolution over $MNbO_3$ intrinsic plasmonic photocatalysts. The intrinsic plasmonic photocatalyst showed good photocatalytic stability over conventional semiconductor photocatalysts.

Systems of plasmonic metal coupled to a semiconductor have also been extensively studied. For instance, Collado *et al.* reported a detailed mechanistic study on the photocatalytic CO_2 photoreduction pathway of Ag/TiO_2.[95] In the first stage, photocatalytic CO_2 experiments under UV light revealed that Ag/TiO_2 systems led to a 15-fold enhancement in CH_4 production rates compared with bare TiO_2, in which the main photoreduction product was CO. VB X-ray photoelectron spectra indicated that incorporation of Ag on the TiO_2 induced a modification of the interface band structure of TiO_2 by forming surface states near the TiO_2 Fermi level (E_F), which could lead to enhanced charged separation upon light excitation. To further investigate this, transient absorption spectroscopy (TAS) studies were also performed. Interestingly, the TAS experiments revealed that in Ag/TiO_2 systems under UV irradiation photogenerated holes accumulate at the VB of TiO_2, while electrons transfer to Ag nanoparticles, enhancing photocarrier separation. This improved charge separation mechanism promoted selectivity of higher-electron-demand products during the CO_2 photoreduction process, such as methane. Furthermore, the long lifetime of photogenerated holes ($t_{50\%} \approx 0.5$ s) further proved that water was the main electron source for the CO_2 reduction process, since prior studies showed that the water oxidation reaction at the surface of TiO_2 occurred over a 100 ms timescale. Although TiO_2 has been the main material investigated for SPR effects, other materials with large band gaps, such as Ta_2O_5, have also shown improved photocatalytic performance in the presence of a plasmonic metal. For instance, Zhou *et al.*[96] investigated SPR on mesoporous Ta_2O_5 loaded with Au nanoparticles for photocatalytic hydrogen production. Due to the large band gap of Ta_2O_5 (4.0 eV), the authors aimed to promote visible light-driven H_2 production by means of the plasmonic effect. Photocatalytic H_2 production from aqueous methanol suspensions under visible light showed no photocatalytic response for pristine Ta_2O_5, whereas maximum production rates of 29 μmol g^{-1} h^{-1} were achieved for Au/Ta_2O_5. In this regard, the H_2 production was attributed to the SPR effect. Upon visible light excitation, photoinduced electrons from Au nanoparticles were transferred to the CB of Ta_2O_5, leaving holes in the Au nanoparticles, which also resulted in an improvement in charge separation efficiencies.

Very recently, a metal–semiconductor system was constructed by exploiting the combination of whispering gallery-mode resonances and SPR.[97] This study showed that the photocatalytic activity of TiO_2–Au plasmonic photocatalysts in the visible range (*i.e.* where TiO_2 does not absorb) strongly depends on the size of the TiO_2 particle. Small TiO_2 particles with diameters of 60 nm (Figure 4.5a) do not express whispering gallery-mode resonances in the visible range. Hence, they do not significantly affect the absorption of the plasmonic Au nanoparticles in Au–TiO_2 composites (Figure 4.5b). By stark contrast, plasmonic absorption is greatly enhanced by the presence of whispering gallery-mode resonances for TiO_2 particles with diameters of 200–600 nm (Figure 4.5d–f). As is shown in Figure 4.5d–f, whispering gallery-mode resonances increased with the size of the TiO_2 particles. This plasmonic absorption enhancement is confirmed experimentally (Figure 4.5c). Furthermore, various Au–TiO_2 systems were investigated as photocatalysts for water splitting under visible to near-infrared (NIR) irradiation (Figure 4.5g and h). Importantly, the activity of Au-mono-TiO_2 was severely decreased (~40%) (Figure 4.5h) when TiO_2 particle size was reduced from 440 to 60 nm, which clearly highlights that whispering gallery-mode resonances increase light absorption in Au nanoparticles, where hot electrons with energy greater than the Schottky barrier are expected (Figure 4.5i), resulting in enhanced photocatalytic activity for water splitting. In addition, the simulation results (Figure 4.5g) proved that the presence of Au multimers further extends the absorption spectrum towards the NIR region. Therefore, the photocatalytic activity of the Janus Au multimer–TiO_2 system is much higher (~2–4 times higher) than that of the Au–mono-TiO_2 (Figure 4.5h), despite having the same TiO_2 support and Au content. Thus, by exploiting the combination of whispering gallery-mode resonances and SPR, one can easily construct photocatalysts that are active over a broad spectral range.

4.4.4 Metal-free Photocatalysts

Metal-free photocatalysts composed of non-metal elements such as Si, Se, S, P, Te, B and C-based materials such as graphitic carbon nitride (g-C_3N_4), GO, silicon carbide (SiC) and carbon quantum dots (CQDs) have attracted great scientific interest due to their unique advantages over metal photocatalysts.[98] They possess good thermal and electrical conductivity, excellent stability and strong bonding effects to create active sites, and they are environmentally friendly and abundant, making them feasible for large-scale applications. Among the different metal-free photocatalysts, the most common ones are g-C_3N_4 and SiC.

g-C_3N_4 is a layered material in which the layers, made of triazine and tri-s-triazine rings, are held together by van der Waals forces (Figure 4.6A). This specific structure and composition makes g-C_3N_4 highly thermochemically stable in both strongly acidic and basic environments. Its band gap ranges from 2.5 to 2.9 eV, depending on the kind of precursor, condensation temperature and reaction parameters employed, enabling it to harvest the visible light of the solar spectrum. In addition, its CB and VB are located at approximately −1.1 and 1.6 V_{NHE}, respectively, which are suitable both for the photocatalytic water-splitting reaction and CO_2 photoreduction (Figure 4.6B, right).[98–100] Despite its numerous advantages, it also suffers from several disadvantages such as fast recombination of electrons and holes and low electronic conductivity, resulting in poor photocatalytic performance and low quantum yield. Since the first publication on photocatalytic H_2 production from water using g-C_3N_4,[101] several approaches have been attempted to tackle the abovementioned limitations, including creating nanostructures and heterostructures.

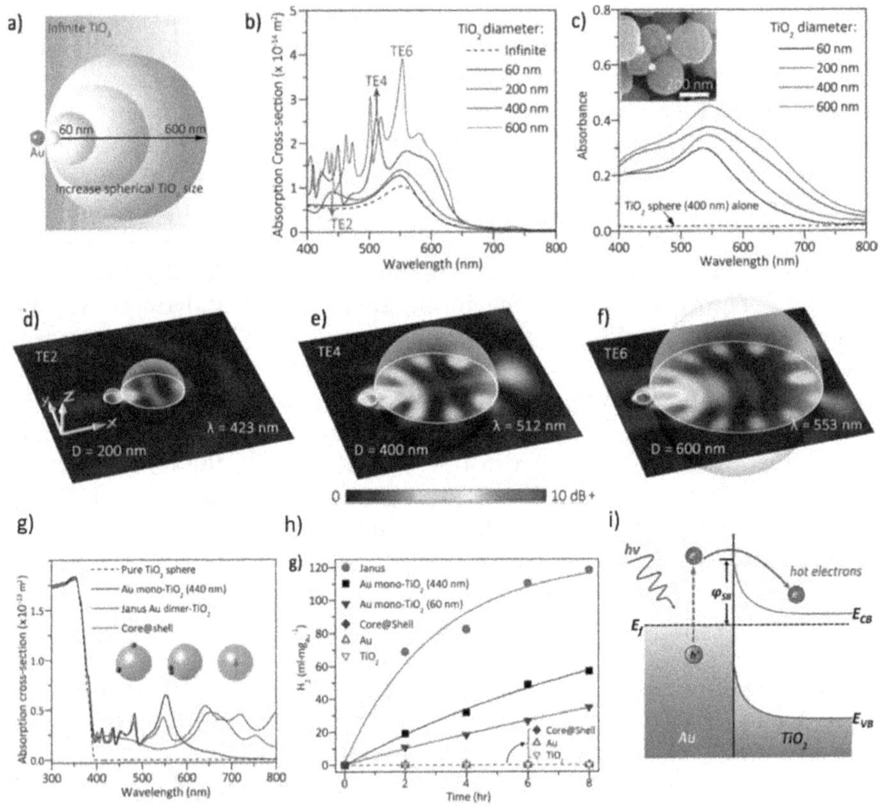

Figure 4.5 Plasmonic absorption engineering *via* whispering gallery-mode resonances: exploring the role of the TiO_2 particle size. (a) Schematic of the model system of a 60 nm Au nanoparticle supported on different TiO_2 spheres. (b) Averaged absorption cross-section calculated using finite-element method (FEM)-based simulations. (c) Experimentally measured absorption spectra of isolated 60 nm Au nanoparticles supported on different TiO_2 spheres. Inset shows a scanning electron micrograph of the Au on 200 nm TiO_2. (d–f) Simulated electric field distributions of Au-TiO_2 for TiO_2 diameters of 200, 400 and 600 nm, respectively. (g) UV–visible absorption characteristics of the four different structures obtained by FEM simulation. The inset in (g) shows the corresponding structural models used in the simulations. (h) Hydrogen production under visible–near-infrared light (λ >420 nm) irradiation using different photocatalysts. (i) Schematic illustration of hot electron generation and transfer from Au to TiO_2. Reproduced from Ref. 97 with permission from American Chemical Society, Copyright 2016.

g-C_3N_4-based heterostructured photocatalysts have resulted in improved photocatalytic performance due to increased light absorption, greater surface area and enhancement in the separation of photogenerated electrons and holes.[102,103] However, most of these g-C_3N_4-based heterostructures include an unwanted source of metal in the secondary component of the heterostructure, such as Pt in Pt/g-C_3N_4, Ti in TiO_2/ g-C_3N_4, Zn in ZnO/g-C_3N_4, and W in WO_3/g-C_3N_4.[102] For example, very recently, g-C_3N_4 nanospheres composed of nanosheets showed significant improvement in H_2 evolution with the use of Pt and reached an apparent quantum yield of 9.6% at 420 nm, compared with 3.8% for simple g-C_3N_4 nanosheets.[104] Yu *et al.* reported that the deposition of Pt on g-C_3N_4 using a post-impregnation method could also enhance the activity and selectivity of g-C_3N_4 for photocatalytic CO_2 reduction.[105] The production of CH_3OH, HCHO and CH_4 increased with an optimal addition of Pt, up to 0.75 wt% for HCHO and 1.0 wt% for CH_3OH and CH_4. The products HCHO and CH_4 formed more quickly in the presence of Pt. The enhanced photocatalytic activity was attributed to the role of Pt as a co-catalyst, facilitating electron transfer to the interface with the CO_2.

Recently, metal-free heterostructures consisting of CQDs, zeolitic imidazole-derived carbon frameworks (C-ZIFs), carbon nanotubes (CNTs), carbon nanofibres and carbon black coupled with g-C_3N_4 have demonstrated enhanced photocatalytic activity. For instance, Feng *et al.*[106] reported that when coupling non-polar CQDs onto g-C_3N_4, CO_2 photoreduction pathways could be modified and improved significantly to optimum production rates of 20.78 and 23.38 μmol g^{-1} h^{-1} of CH_4 and CO, respectively, whereas pristine g-C_3N_4 only formed small amounts of CO and H_2. A junction of zeolitic imidazole framework (ZIF-8) with g-C_3N_4 also resulted in improved photocatalytic H_2 performance of up to 36.2 times higher than pristine g-C_3N_4.[107] This improvement was not only due to a better charge separation and transfer efficiency, but also due to the ZIF acting as an efficient hydrogen evolution co-catalyst. Coupling g-C_3N_4 with carbon black, CNTs and carbon nanofibres also improved the charge separation efficiency of photocarriers, resulting in an overall improvement in photocatalytic hydrogen production.[108–110]

Nanostructuring of g-C_3N_4 is another approach to enhancing the photocatalytic performance of these materials since it allows for the formation of controllable morphologies with greater surface area that can contribute to the better separation and collection of photogenerated charges.[21] Different kinds of morphologies, including 0D, 1D, 2D and 3D g-C_3N_4, have been designed, showing improved performances over bulk g-C_3N_4 (Figure 4.6C). For instance, 2D g-C_3N_4 (Figure 4.6C-c) showed improved photocatalytic H_2 production due to its greater surface area and increased lifetime of photogenerated charge carriers.[111,112] Gu *et al.* synthesized 3D g-C_3N_4 microspheres (Figure 4.6C-d) *via* a template-free solvothermal method with a post-annealing treatment at 550 °C under argon flow.[113] These microspheres had a narrower band gap than bulk g-C_3N_4 (2.42 *vs.* 2.78 eV) and showed a higher rate of photocatalyitc H_2 production, attributed to improved visible light harvesting and more efficient charge transport and separation. Similarly, 1D oxygen-doped g-C_3N_4 nanorods (Figure 4.6C-b) were easily prepared by Zeng *et al.* As before, this unique architecture also presented improved photocatalytic performance.[114] Finally, 0D g-C_3N_4 in the form of quantum dots has also been prepared (Figure 4.6C-a).

In this particular case, these quantum dots have been coupled with other semiconductors such as TiO_2 to improve photocatalytic performance.[115]

SiC, another common metal-free photocatalyst, has a tuneable band gap, ranging from 2.4 to 3.3 eV depending on the morphology and structural features. Its CB potential is suitable for the reduction reaction of water, and it possesses high thermal conductivity, high chemical stability and mechanical strength, which make it an excellent candidate material for photocatalytic applications.[98,116] The first evidence of photocatalytic H_2 production using UV light irradiation was reported in 1990 by Nariki *et al.* Since then, many research efforts have aimed to overcome its main limitation of the fast recombination of photogenerated electrons and holes and to improve its photocatalytic performance.[117] As for g-C_3N_4, strategies such as nanostructures or the formation of heterostructures have been investigated.

Hao reported on the formation of SiC nanowires prepared *via* a carbothermal reduction route. These SiC nanowires produced H_2 from pure water using visible light irradiation. The authors also reported that by making the surface of SiC

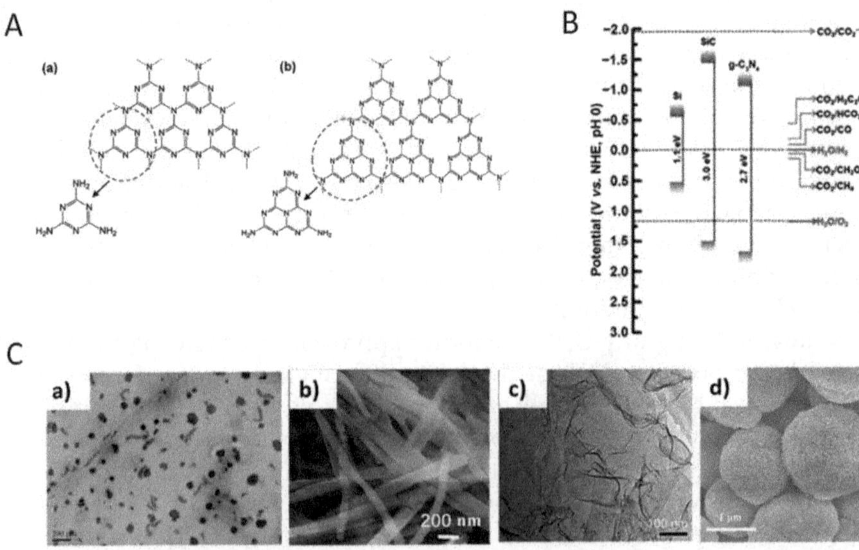

Figure 4.6 (A) (a) Triazine and (b) tri-s-triazine structures of g-C_3N_4. Reproduced from Ref. 100 with permission from American Chemical Society, Copyright 2016. (B) Band gap positions of some metal-free photocatalysts. Reproduced from Ref. 98 with permission from the Royal Society of Chemistry. (C) (a) 0D, (b) 1D, (c) 2D and (d) 3D g-C_3N_4 nanostructures. (C-a) Reproduced from Ref. 115 with permission from Elsevier, Copyright 2016. (C-b) Reproduced from Ref. 114 with permission from Elsevier, Copyright 2018. (C-c) Reproduced from Ref. 111 with permission from John Wiley and Sons, Copyright © 2012 Wiley-VCH Verlag GmbH & Co. KGaA, Weinheim. (C-d) Reproduced from Ref. 113 with permission from Elsevier, Copyright 2014.

more hydrophilic using an acid oxidation treatment, the photocatalytic H_2 performance could be enhanced even further due to an increase in oxygen-containing groups on the surface and a lower band gap.[116] Designing heterostructures of SiC with RGO and g-C_3N_4 has also resulted in improved photocatalytic performance of SiC. For instance, Han *et al.*[118] prepared a 2D SiC/RGO heterojunction for photoreduction of CO_2. This heterojunction allowed for fast transfer of electrons from the SiC to the RGO, resulting in an electron transfer rate (rate of photogenerated electrons that participate in CO_2 reduction) of 58.17μmol g^{-1} h^{-1} for CO_2 reduction, which is 2.7 times higher than pristine SiC. CO and CH_4 production rates were of 2.22 and 6.72 μmol g^{-1} h^{-1}, respectively. Heterostructures of the two main metal-free photocatalysts, g-C_3N_4 and SiC, were also prepared by Wang *et al.*[119] In this study, the authors prepared g-C_3N_4/SiC *via in situ* heating of thiourea on SiC particles. The optimized sample reached an H_2 production rate of 182 μmol g^{-1} h^{-1} under visible light illumination, which is 3.4 times higher than pristine g-C_3N_4. As in the other heterostructures, the improved performance of this junction was due to enhanced separation of photogenerated electrons and holes and increased hydrophilicity of the material. Furthermore, the authors also reported that the preparation method is key for the final photocatalytic performance; the same g-C_3N_4/SiC heterojunction prepared using a physical mixture of g-C_3N_4 and SiC resulted in only 34 μmol g^{-1} h^{-1} of H_2 produced. This can be explained due to a weaker interaction among these two materials when a physical mixture was employed as a synthesis method.

Other metal-free photocatalysts that have been reported in literature to a much lesser extent are ternary boron carbon nitrides (BCNs). These materials have also shown promising photocatalytic H_2 and O_2 evolution performances, as well as CO_2 photoreduction under visible light, as shown by Huang *et al.*[120] The samples were prepared by mixing glucose with boron oxide and urea followed by annealing in an ammonia atmosphere. The optimized sample resulted in an H_2 production rate of ~4 μmol h^{-1}. For O_2 evolution measurements, BCN was loaded with different co-catalysts, from which Ni–Co layered double hydroxides (Ni–Co LDHs) achieved the highest O_2 production of 11 μmol in 20 h. Photocatalytic CO_2 reduction to CO resulted in 9.3 and 2.9 μmol of CO and H_2, respectively, after 2 h of continuous visible light irradiation. Another report also showed that combining BCN with CdS resulted in a 10-fold enhancement in photocatalytic CO_2 reduction to CO under visible light over pristine BCN, mainly due to improved visible light harvesting and better separation of photogenerated charges.[121] In particular, pristine BCN resulted in a CO production rate of 24 μmol g^{-1} h^{-1}, whereas the optimized BCN/CdS sample presented a CO production rate of 250 μmol g^{-1} h^{-1}.

4.4.5 Metal–Organic Frameworks

In recent years, MOFs have emerged as a sustainable photocatalytic material for photocatalytic applications, including H_2 production and CO_2 reduction.[122] MOFs are a class of metal–organic hybrid structure where rigid bi- or multi-modal organic units are arranged in a 3D porous network with metal centres. These MOFs not only show the combined strengths of molecular and heterogeneous catalytic materials, but also offer a unique platform for catalysis within the confined dimensions of their reaction environment. Moreover, some of these MOFs possess high structural stability, combined with substantial flexibility in terms of not only design and synthesis, but also post-synthetic functionalization. Other advantages are high pore volume and surface area and very low framework density. MOFs can be photoresponsive *via* the organic linkers or metal centre nodes. Upon photoexcitation, the light-absorbing units in MOFs often produce a ligand-to-metal charge separation that induces photochemical reaction. More interestingly, it is possible to design MOFs with a visible light photoresponse at long wavelengths by tuning aromatic linkers or metal centres.[123] For example, unlike traditional inorganic metal oxide semiconductors such as TiO_2, visible light-driven photocatalytic activity can be finely tuned by direct modification of the organic linker to enhance light absorption in the visible region.[124–126] Horiuch *et al.*[125] reported a MOF (TiMOF-NH_2) photocatalyst containing a 2-amino-benzenedicarboxylic (bdc-NH_2) acid organic linker for hydrogen production from an aqueous medium under visible light irradiation. The observed visible light absorption at up to ~500 nm of Ti-MOF-NH_2 and the H_2 production under visible light were associated with the chromophore in the organic linker. Furthermore, the specific role of the –NH_2 group of the monoaminated bdc-NH_2 linker in lowering the optical band gap of a titanium-containing MIL-125-NH_2 showed that the functionalization of the linker with substituents could shift light absorption.[126] On the other hand, Castells-Gil *et al.*[127] reported a new family of titanium–organic frameworks, coined MUV-10(Ca or Mn) (MUV = Materials of Universidad de Valencia), by metal doping, and these are chemically robust and can be prepared as single crystals at multi-gram scale from multiple precursors. MUV-10(Ca) is a 3D porous solid built up of the interlinking of fully deprotonated trimesate anions and tetranuclear $Ti^{IV}{}_2Ca^{II}{}_2$–$(\mu_3\text{-}O)_2(H_2O)_4(CO_2)_8$ clusters. The advantages of these new MOFs are that the heterometallic structure can be engineered by metal doping rather than by organic linker functionalization because the incorporation of Mn^{II} units into the structure of MUV-10 affects its electronic structure, reducing the band gap from 3.1 to 2.6 eV. Basically, this approach allows for more control over metal dopant positioning at an atomic level for engineering the band gap and electronic properties of MOFs when compared to the post-synthetic metal functionalization of MOFs. The Mn phase doubled the photocatalytic H_2 production rate compared to the Ca phase (6500 mmol g^{-1}) after 24 h of irradiation when using methanol as a hole sacrificial agent.

There have also been examples of using other metal-based MOFs for photocatalysis, such as Fe, Al, Cd, Gd and Y-based MOFs.[123,128] For example, Ni^{II} coordination to aluminium-based MOFs containing a 2-aminoterephthalate linker showed efficient overall water splitting.[129] This MOF was obtained using 2-aminoterephthalic acid (H_2ATA) and consisted of $AlO_4(OH)_2$ octahedra linked by the carboxylate group of ATA^{2-}. The amine group present in the Al-ATA MOF allowed for coordination to Ni^{2+} cations. During photocatalysis, the benzene ring of the ATA^{2-} linker acted as the active site for O_2 evolution and Ni as the active sites for H_2 evolution. H_2 and O_2 were reported to be produced in a stoichiometric ratio of 2 : 1, indicating overall water splitting.

By combining the high adsorption capacity and photocatalytic activity of MOFs, photocatalytic CO_2 reduction over MOFs seems to be a promising route to overcome the low efficiency and product selectivity of the multi-electron-step photochemical reaction leading to various products, including CO, CH_3OH, CH_4, CH_2O, HCOOH and higher-order hydrocarbons.[122,128] Additional challenges include the low solubility of CO_2 if the reaction is performed in aqueous solution, as well as the competition of hydrogen generation with CO_2 reduction. One of the early examples of photocatalytic CO_2 reduction involved modified UiO-67 (Zr_6O_4 $[OH]_4$[BPYDC], where BPYDC = 2,2′-bipyridine-5,5′-dicarboxylate), which was obtained by replacing the para-biphenyldicarboxylate (BPDC) linker with 4.2 wt% BPYDC linker that had –$Re(CO)_3Cl$ attached to the nitrogen atoms.[130,131] This photoactive MOF was used to reduce CO_2 to CO with a turnover number of as high as 5 (6 h) and CO_2 and H_2 selectivity with a ratio of 10 : 1. The drawback of this MOF is that it is only active under UV light. Furthermore, the activity was enhanced seven-fold by depositing the MOF on Ag nanocubes to extend the light absorption into the visible region and improve the charge separation between the MOF and the Ag metal due to surface plasmonic effects.[131] Most commonly, these MOFs were used along with metal complex photosensitizers for CO_2 reduction. For example, photocatalytic CO_2 reduction to formate with UiO-67 modified with a Ru(ii) photosensitizer and Mn(i-) catalyst grafted onto the linker showed a high turnover number of ~110 in 18 h.[132] To extend the visible light region absorption, the linkers in UiO-66 and -67 were further modified with amino functional groups on the phenyl ring within the linker.[133] Another Zr-based MOF, NNU-28 ($Zr_6O_4[OH]_4[L]_6$, where L = 4,4′-[anthracene-9,10-diylbis{ethyne-2,1diyl}]dibenzoate), which contains an anthracene linker, showed high performance for photocatalytic CO_2 reduction to formate with a turnover number of 18 (10 h) under visible light without an external homogeneous photosensitizer, as shown in Figure 4.7.[134] This high performance was due to both the anthracene linker and the metal oxide cluster active centres in the framework. CO_2 reduction over Al-based MOFs containing a Cu porphyrin active centre, –CuTCPP-(AlOH–)$_2$, where TCPP-H_2 = 4,4′,4′′,4′′′-(porphyrin-5,10,15,20tetrayl)-tetrabenzoate, produced methanol as the major product.[135] Selective CH_4 production was seen over Ti-based MIL-125-NH_2 ($Ti_8O_8[OH]_4$[BDC-$NH_2]_6$) in composites with Au nanoparticles.[136] Although most studies were performed in the liquid phase, this photocatalysis was carried out in the gas phase, taking advantage of the high CO_2 uptake of MOF materials.

Considering that the metal nodes of MOFs provide high-density catalytic sites, the reasonable design of metal nodes may provide an efficient strategy for photocatalysis. For example, Ni MOF monolayers have been prepared using a traditional ultrasonication method with abundant coordinatively unsaturated Ni sites for metal node-dependent performance and selective photocatalytic reduction of diluted CO_2 to CO.[137] In diluted CO_2 (10%), Ni MOF monolayers exhibited the highest apparent quantum yield of 1.96% with a CO selectivity of 96.8%, which is not only much higher than what has been reported for systems in the photocatalytic reduction of diluted CO_2, but also is superior to most catalysts in the photocatalytic reduction of pure CO_2.

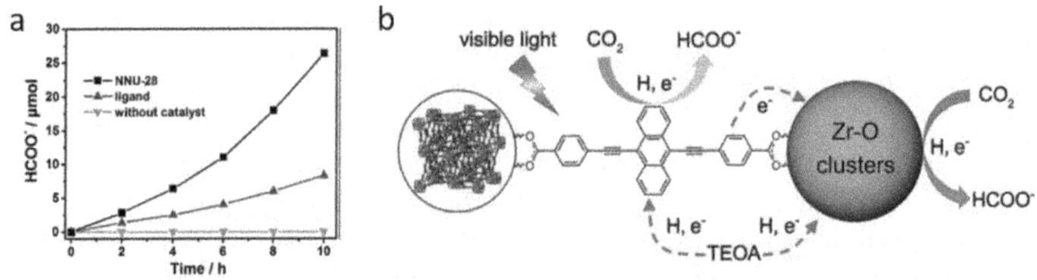

Figure 4.7 (a) Photocatalytic CO_2 reduction to formate as a function of irradiation time under visible light irradiation using MeCN/TEOA (30/1) solution as a hole sacrificial agent. (b) A scheme of dual photocatalytic routes for visible light-driven CO_2 reduction in NNU-28. Reproduced from Ref. 134 with permission from the Royal Society of Chemistry.

In summary, the various MOFs used and the products obtained, particularly for CO_2 reduction, show the versatility and future potential of MOFs for artificial photosynthesis. Although MOFs show high performance along with a co-catalyst in terms of turnover numbers, it seems to be more feasible to design MOFs containing a photosensitizer as part of the framework with high product selectivity and turnover numbers, thus fully utilizing the possibilities MOFs offer as sustainable photocatalytic systems for energy applications.

4.4.6 Miscellaneous: Metal Sulfides and Layered Double Hydroxides

In recent years, metal sulfides have also attracted extensive attention for visible light-driven photocatalytic H_2 evolution and CO_2 reduction due to their suitable band gap borders for conversion of the solar spectrum.[138,139] For example, porous CdS nanosheet-assembled flowers synthesized by a simple ion-exchange strategy using morphology-analogous $Cd(OH)_2$ and Na_2S as precursors exhibited high visible light photocatalytic H_2 production activity at a rate of 0.468 mmol h^{-1} and a corresponding apparent QE of 24.7% at 420 nm, which exceeded that obtained on CdS nanoparticles more than three-fold. Some other metal sulfides have also been proven to possess good photocatalytic activity for H_2 evolution and CO_2 reduction. For example, 3D hierarchical Cu_3SnS_4 flower-like microspheres exhibited good photocatalytic activity for H_2 evolution without any co-catalyst, and the rate of H_2 evolution was determined to be 1.1 mmol g^{-1} h^{-1} under simulated solar irradiation. Due to its unique electronic properties, MoS_2 has been extensively studied as a non-precious metal catalyst or co-catalyst combined with other photocatalysts to enhance the photocatalytic activity for H_2 evolution and CO_2 reduction. For instance, Tu *et al.* constructed a 2D MoS_2@TiO_2 hybrid material for selective photocatalytic reduction of CO_2 to CH_3OH under UV–visible light irradiation.[140] In another work, a hybrid TiO_2@graphene loaded with few-layered MoS_2 was studied for photocatalytic CO_2 reduction with an outstanding CO yield of up to 92.33 μmol g^{-1} h^{-1}, 14.5 times higher than the pristine TiO_2, with a higher selectivity of 97% for CO_2 reduction.[141] The use of electrostatic assemblies of negatively charged colloidal $CuInS_2/ZnS$ quantum dot sensitizers and positively charged trimethylamine-functionalized iron tetraphenylporphyrin (FeTMA) photocatalysts for the reduction of CO_2 to CO in water under irradiation of 450 nm light was also studied.[142] This system achieved a turnover number to CO of 450 after 30 h of irradiation, with a selectivity of 99%.

In recent years, layered double hydroxides (LDHs) also emerged as a promising photocatalyst for photocatalytic CO_2 reduction. LDHs are a class of 2D anionic clays made up of positively charged brucite-like host layers and exchangeable charge-balancing interlayer anions, usually expressed as $(M^{2+}{}_{1-x}M^{3+}{}_x[OH]_2]^{x+}(A^{n-})_{x/n}{\cdot}yH_2O$ (M = metal and A = interlayer anion).[143] On account of their 2D layered structure, abundant surface OH groups can accommodate CO_2 molecules. Furthermore, they offer controllable particle sizes and layer thicknesses from several micrometres to 2–3 nm, and their band gap energies can be tuned by changing/adjusting the metal cation composition.[144]

Tanaka's group synthesized various M^{2+}–M^{3+} LDHs (M^{2+} = Mg^{2+}, Ni^{2+} or Zn^{2+}; M^{3+} = Al^{3+}, Ga^{3+} or In^{3+}) for photocatalytic CO_2 photoreduction in water for the first time.[145] Among these, the Ni–In-LDH showed the highest activity for photocatalytic CO_2 reduction into CO, with a rate of 3.6 μmol g^{-1} h^{-1}. Zhao *et al.* reported useful insights into the effects of LDH morphology on photocatalytic activity for CO_2 reduction.[146] The defect-rich, ultrathin LDH nanosheets with an exposed {110} basal plane, composed of two repeated stacking layers with a lateral size of ~40 ± 20 nm and a platelet thickness of ~2.7 nm possessed abundant oxygen vacancies and showed high photocatalytic activity for selective CO_2 reduction to CO with a production rate of 7.6 μmol g^{-1} h^{-1} with water vapour.[147] Conversely, bulk ZnAl-LDHs showed barely any activity. A Zn/Ti-LDH photocatalyst showed very good photocatalytic activity for selective CO_2 reduction to CH_4, with a rate of 77 μmol g^{-1} h^{-1}. However, in this case, Zn/Ti-LDH served only as a UV-active photocatalyst. Very recently, Tonda *et al.* constructed a 2D/2D interface heterostructure of g-C_3N_4 and NiAl-LDH utilizing strong electrostatic interactions between positively charged 2D NiAl-LDH sheets and negatively charged 2D g-C_3N_4 nanosheets.[148] This new 2D/2D interface heterojunction showed significant improvement in terms of photocatalytic CO_2 reduction to CO and H_2 under visible light irradiation over the single-counterpart g-C_3N_4 or NiAl-LDH nanosheets. This high performance was mainly due to the excellent interfacial contact at the heterojunction of g-C_3N_4 and NiAl-LDH, which subsequently resulted in suppressed recombination and improved transfer and separation of photogenerated charge carriers. Furthermore, a hierarchical nanocomposite, comprising P25 (TiO_2) nanoparticles encapsulated within microporous CoAl-LDHs (Figure 4.8A-a and A-b), exhibited good activity and selectivity (>90%) for aqueous CO_2 photoreduction to CO without using a sacrificial hole acceptor, as shown in Figure 4.8A-c.[149] An investigation based on both the CoAl-LDH thickness and titania morphology showed a strong influence of both parameters regarding photocatalytic CO_2 reduction. More interestingly, when combining exfoliated CoAl-LDH nanosheets with TiO_2 nanotubes (Figure 4.8B-a), a rate

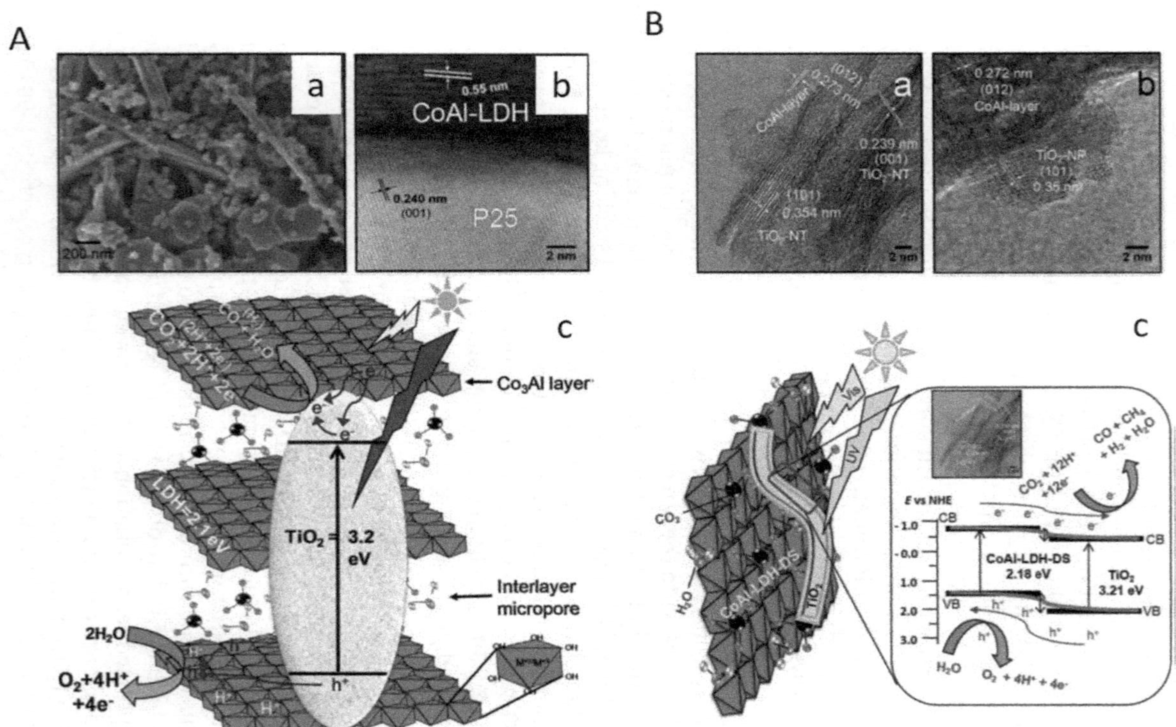

Figure 4.8 (A) (a) FE-SEM and (b) HR-TEM micrographs and (c) a proposed mechanism of CO_2 photoreduction due to heterojunction formation of a P25 (TiO_2)@CoAl-LDH nanocomposite. Reproduced from Ref. 149, https://doi.org/10.1016/j.apcatb.2017.03.006, under the terms of the CC BY 4.0 licence, https://creativecommons.org/licenses/by/4.0/. (B) (a) HR-TEM micrographs of TiO_2 nanotube@exfoliated CoAl-LDH nanosheets and (b) TiO_2 nanoparticles@exfoliated CoAl-LDH nanosheets and (c) a proposed mechanism of CO_2 photoreduction over TiO_2 nanotube@exfoliated CoAl-LDH nanosheet heterojunctions. Reproduced from Ref. 50 with permission from John Wiley and Sons, © 2017 Wiley-VCH Verlag GmbH & Co. KGaA, Weinheim.[149,150]

of 4.57 μmol g_{cat}^{-1} h^{-1} of CO and 0.41 μmol g_{cat}^{-1} h^{-1} of CH_4 under UV–visible irradiation was achieved through a stoichiometric redox process and in the absence of sacrificial agents (Figure 4.8B-c).[150] This CO_2 reduction performance was 5.0 and 7.5 times higher than that those pure TiO_2 nanotubes and delaminated CoAl-LDH nanosheets, respectively. The performance was two-fold better than composites containing either bulk CoAl-LDH or anatase nanoparticles (Figure 4.8B-b). These results represent a facile and cost-effective strategy for the design and development of exfoliated 2D LDH-based nanomaterials for efficient and selective photocatalytic CO_2 reduction to value-added chemical fuels.

4.5 Conclusions

A wide range of materials, namely metal oxides, perovskites, plasmonic materials, metal-free materials, MOFs, metal sulfides and LDHs, offer tuneable textural, electronic and optical properties that can be tailored for harvesting solar energy for photocatalytic clean energy applications such as solar fuel production, obtaining H_2 from water and carbon-based fuels from CO_2 reduction. Different synthetic strategies for tuning their nanostructures and surfaces with desired physical and chemical properties have been reviewed in this chapter. Furthermore, we have demonstrated that these materials can be easily integrated with other co-catalysts in order to extend their solar spectral utilization and/or improve their apparent quantum yield, charge separation and transport efficiencies and to achieve an enhancement in their photocatalytic activity and stability for H_2 production as well as CO_2 reduction. Photocatalytic CO_2 reduction is at a more preliminary stage compared to water splitting, although improvements in both activities and selectivity for products such as CO, CH_4, methanol, formic acid, C_2 and olefins or alkanes have been achieved. Despite extensive progress being made in discovering new materials and improving known ones, further progress is still needed in avoiding the use of sacrificial agents in photocatalytic tests. Moreover, scalable production for a solar-powered clean energy industry for a sustainable future remains a challenge. In addition, a lack of standardization methods in both reactor design and experimental protocols makes quantitative comparison in the photocatalysis research field very difficult. In any case, the potential of reducing CO_2 with water under solar light irradiation *via* artificial photosynthesis merits future research into photocatalysts and scalable technologies to promote carbon recycling, solar utilization and the production of renewable feedstocks that contribute to a more sustainable and environmentally friendly industry.

References

1. J. Jiang, B. Ye and J. Liu, *Appl. Energy*, 2019, **235**, 186–203.
2. T. D. Searchinger, T. Beringer, B. Holtsmark, D. M. Kammen, E. F. Lambin, W. Lucht, P. Raven and J.-P. van Ypersele, *Nat. Commun.*, 2018, **9**, 3741.
3. A. Fujishima and K. Honda, *Nature*, 1972, **238**, 37–38.

4. T. Faunce, S. Styring, M. R. Wasielewski, G. W. Brudvig, A. W. Rutherford, J. Messinger, A. F. Lee, C. L. Hill, H. deGroot, M. Fontecave, D. R. MacFarlane, B. Hankamer, D. G. Nocera, D. M. Tiede, H. Dau, W. Hillier, L. Wang and R. Amal, *Energy Environ. Sci.*, 2013, **6**, 1074–1076.
5. A. Mills, R. H. Davies and D. Worsley, *Chem. Soc. Rev.*, 1993, **22**, 417–425.
6. E. Kabir, P. Kumar, S. Kumar, A. A. Adelodun and K.-H. Kim, *Renewable Sustainable Energy Rev.*, 2018, **82**, 894–900.
7. T. E. Mallouk, *J. Phys. Chem. Lett.*, 2010, **1**, 2738–2739.
8. S. E. Hosseini and M. A. Wahid, *Renewable Sustainable Energy Rev.*, 2016, **57**, 850–866.
9. J. W. Ager and A. A. Lapkin, *Science*, 2018, **360**, 707–708.
10. S. Bensaid, G. Centi, E. Garrone, S. Perathoner and G. Saracco, *ChemSusChem*, 2012, **5**, 500–521.
11. B. Kumar, M. Llorente, J. Froehlich, T. Dang, A. Sathrum and C. P. Kubiak, *Ann. Rev. Phys. Chem.*, 2012, **63**, 541–569.
12. F. E. Osterloh, *J. Mater. Chem. A*, 2018, **6**, 9763–9764.
13. H. Kisch, *Angew. Chem., Int. Ed.*, 2013, **52**, 812–847.
14. M. R. Hoffmann, S. T. Martin, W. Choi and D. W. Bahnemann, *Chem. Rev.*, 1995, **95**, 69–96.
15. Y. Qu and X. Duan, *Chem. Soc. Rev.*, 2013, **42**, 2568–2580.
16. M. D. Hernández-Alonso, F. Fresno, S. Suárez and J. M. Coronado, *Energy Environ. Sci.*, 2009, **2**, 1231–1257.
17. R. van de Krol, in *Photoelectrochemical Hydrogen Production*, ed. R. van de Krol and M. Grätzel, Springer US, Boston, MA, 2012, pp. 13–67.
18. C. A. Grimes, O. K. Varghese and S. Ranjan, in *Light, Water, Hydrogen: The Solar Generation of Hydrogen by Water Photoelectrolysis*, ed. C. A. Grimes, O. K. Varghese and S. Ranjan, Springer US, Boston, MA, 2008, pp. 371–426.
19. K. Maeda, *J. Photochem. Photobiol. C*, 2011, **12**, 237–268.
20. D. Chen, X. Zhang and A. F. Lee, *J. Mater. Chem. A*, 2015, **3**, 14487–14516.
21. S. Kumar, S. Karthikeyan and A. F. Lee, *Catalysts*, 2018, **8**, 74.
22. W. Zhang, Y. Hu, L. Ma, G. Zhu, Y. Wang, X. Xue, R. Chen, S. Yang and Z. Jin, *Adv. Sci.*, 2018, **5**, 1700275.
23. J. K. Stolarczyk, S. Bhattacharyya, L. Polavarapu and J. Feldmann, *ACS Catal.*, 2018, **8**, 3602–3635.
24. Y. Y. Lee, H. S. Jung and Y. T. Kang, *J. CO2 Util.*, 2017, **20**, 163–177.
25. Y. Ma, X. Wang, Y. Jia, X. Chen, H. Han and C. Li, *Chem. Rev.*, 2014, **114**, 9987–10043.
26. H. S. N., S.-M. Lukas and S. J. K., *Angew. Chem., Int. Ed.*, 2013, **52**, 7372–7408.
27. N. Serpone and A. V. Emeline, *J. Phys. Chem. Lett.*, 2012, **3**, 673–677.
28. T. Hisatomi, J. Kubota and K. Domen, *Chem. Soc. Rev.*, 2014, **43**, 7520–7535.
29. A. Kudo and Y. Miseki, *Chem. Soc. Rev.*, 2009, **38**, 253–278.
30. G. N. Schrauzer and T. D. Guth, *J. Am. Chem. Soc.*, 1977, **99**, 7189–7193.
31. A. Fujishima, X. Zhang and D. A. Tryk, *Surf. Sci. Rep.*, 2008, **63**, 515–582.
32. S. Y. Tee, K. Y. Win, W. S. Teo, L.-D. Koh, S. Liu, C. P. Teng and M.-Y. Han, *Adv. Sci.*, 2017, **4**, 1600337.

33. H. Pan, *Renewable Sustainable Energy Rev.*, 2016, **57**, 584–601.
34. R. Li, Y. Weng, X. Zhou, X. Wang, Y. Mi, R. Chong, H. Han and C. Li, *Energy Environ. Sci.*, 2015, **8**, 2377–2382.
35. Z. Wang, C. Li and K. Domen, *Chem. Soc. Rev.*, 2019, **48**, 2109–2125.
36. A. S. Hainer, J. S. Hodgins, V. Sandre, M. Vallieres, A. E. Lanterna and J. C. Scaiano, *ACS Energy Lett.*, 2018, **3**, 542–545.
37. Y. Takashi, S. Yoshihisa and I. Hayao, *Chem. Lett.*, 2004, **33**, 726–727.
38. Y. Sakata, Y. Matsuda, T. Yanagida, K. Hirata, H. Imamura and K. Teramura, *Catal. Lett.*, 2008, **125**, 22–26.
39. X. Wang, Q. Xu, M. Li, S. Shen, X. Wang, Y. Wang, Z. Feng, J. Shi, H. Han and C. Li, *Angew. Chem., Int. Ed.*, 2012, **51**, 13089–13092.
40. K. Sayama and H. Arakawa, *J. Photochem. Photobiol. A Chem.*, 1996, **94**, 67–76.
41. V. R. Reddy, D. W. Hwang and J. S. Lee, *Korean J. Chem. Eng.*, 2003, **20**, 1026–1029.
42. J.-J. Zou, C.-J. Liu and Y.-P. Zhang, *Langmuir*, 2006, **22**, 2334–2339.
43. A. Sinhamahapatra, J.-P. Jeon, J. Kang, B. Han and J.-S. Yu, *Sci. Rep.*, 2016, **6**, 27218.
44. W.-J. Chun, A. Ishikawa, H. Fujisawa, T. Takata, J. N. Kondo, M. Hara, M. Kawai, Y. Matsumoto and K. Domen, *J. Phys. Chem. B*, 2003, **107**, 1798–1803.
45. Y. Noda, B. Lee, K. Domen and J. N. Kondo, *Chem. Mater.*, 2008, **20**, 5361–5367.
46. R. V. Gonçalves, P. Migowski, H. Wender, D. Eberhardt, D. E. Weibel, F. C. Sonaglio, M. J. M. Zapata, J. Dupont, A. F. Feil and S. R. Teixeira, *J. Phys. Chem. C*, 2012, **116**, 14022–14030.
47. A. S. Cherevan, P. Gebhardt, C. J. Shearer, M. Matsukawa, K. Domen and D. Eder, *Energy Environ. Sci.*, 2014, 7, 791–796.
48. M. Hara, T. Kondo, M. Komoda, S. Ikeda, J. N. Kondo, K. Domen, M. Hara, K. Shinohara and A. Tanaka, *Chem. Commun.*, 1998, **3**, 357–358.
49. L. Liao, Q. Zhang, Z. Su, Z. Zhao, Y. Wang, Y. Li, X. Lu, D. Wei, G. Feng, Q. Yu, X. Cai, J. Zhao, Z. Ren, H. Fang, F. Robles-Hernandez, S. Baldelli and J. Bao, *Nat. Nanotechnol.*, 2013, **9**, 69.
50. Z. Li, F. Zhang, J. Han, J. Zhu, M. Li, B. Zhang, W. Fan, J. Lu and C. Li, *Catal. Lett.*, 2018, **148**, 933–939.
51. A. Tanaka, K. Teramura, S. Hosokawa, H. Kominami and T. Tanaka, *Chem. Sci.*, 2017, **8**, 2574–2580.
52. Y. Sasaki, H. Kato and A. Kudo, *J. Am. Chem. Soc.*, 2013, **135**, 5441–5449.
53. Q. Wang, T. Hisatomi, Q. Jia, H. Tokudome, M. Zhong, C. Wang, Z. Pan, T. Takata, M. Nakabayashi, N. Shibata, Y. Li, I. D. Sharp, A. Kudo, T. Yamada and K. Domen, *Nat. Mater.*, 2016, **15**, 611.
54. K. Maeda, M. Higashi, D. Lu, R. Abe and K. Domen, *J. Am. Chem. Soc.*, 2010, **132**, 5858–5868.
55. K. R. Thampi, J. Kiwi and M. Grätzel, *Nature*, 1987, **327**, 506–508.
56. H. Xu, S. Ouyang, P. Li, T. Kako and J. Ye, *ACS Appl. Mater. Interfaces*, 2013, **5**, 1348–1354.

57. W. Tu, Y. Zhou, Q. Liu, Z. Tian, J. Gao, X. Chen, H. Zhang, J. Liu and Z. Zou, *Adv. Funct. Mater.*, 2012, **22**, 1215–1221.
58. L. Wang and T. Sasaki, *Chem. Rev.*, 2014, **114**, 9455–9486.
59. S. N. Habisreutinger, L. Schmidt-Mende and J. K. Stolarczyk, *Angew. Chem., Int. Ed.*, 2013, **52**, 7372–7408.
60. J. Z. Y. Tan and M. M. Maroto-Valer, *J. Mater. Chem. A*, 2019, **7**, 9368–9385.
61. S. Sun, M. Watanabe, J. Wu, Q. An and T. Ishihara, *J. Am. Chem. Soc.*, 2018, **140**, 6474–6482.
62. A. H. Yahaya, M. A. Gondal and A. Hameed, *Chem. Phys. Lett.*, 2004, **400**, 206–212.
63. R. Jingrun, J. Mietek and Q. Shi-Zhang, *Adv. Mater.*, 2018, **30**, 1704649.
64. C. Dong, C. Lian, S. Hu, Z. Deng, J. Gong, M. Li, H. Liu, M. Xing and J. Zhang, *Nat. Commun.*, 2018, **9**, 1252.
65. A. Li, T. Wang, X. Chang, Z.-J. Zhao, C. Li, Z. Huang, P. Yang, G. Zhou and J. Gong, *Chem. Sci.*, 2018, **9**, 5334–5340.
66. K.-L. Bae, J. Kim, C. K. Lim, K. M. Nam and H. Song, *Nat. Commun.*, 2017, **8**, 1156.
67. W. Yu, D. Xu and T. Peng, *J. Mater. Chem. A*, 2015, **3**, 19936–19947.
68. Q. Li, Y. Zhang, L. Zhang, J. Xia, X. Liu, L. Hu, F. Wang, X. Chu, P. Zhao, J. Yin and D. Yang, *Catal. Commun.*, 2019, **120**, 23–27.
69. M. A. Green, A. Ho-Baillie and H. J. Snaith, *Nat. Photonics*, 2014, **8**, 506.
70. S. D. Stranks and H. J. Snaith, *Nat. Nanotechnol.*, 2015, **10**, 391.
71. S. Zeng, P. Kar, U. K. Thakur and K. Shankar, *Nanotechnology*, 2018, **29**, 052001.
72. W.-J. Yin, B. Weng, J. Ge, Q. Sun, Z. Li and Y. Yan, *Energy Environ. Sci.*, 2019, **12**, 442–462.
73. H. Zhou, J. Guo, P. Li, T. Fan, D. Zhang and J. Ye, *Sci. Rep.*, 2013, **3**, 1667.
74. K. Xie, N. Umezawa, N. Zhang, P. Reunchan, Y. Zhang and J. Ye, *Energy Environ. Sci.*, 2011, **4**, 4211–4219.
75. Y. Bi, M. F. Ehsan, Y. Huang, J. Jin and T. He, *J. CO2 Util.*, 2015, **12**, 43–48.
76. K. Han, Y.-C. Lin, C.-M. Yang, R. Jong, G. Mul and B. Mei, *ChemSusChem*, 2017, **10**, 4510–4516.
77. T. K. Townsend, N. D. Browning and F. E. Osterloh, *ACS Nano*, 2012, **6**, 7420–7426.
78. H. Cheng, B. Huang, Y. Liu, Z. Wang, X. Qin, X. Zhang and Y. Dai, *Chem. Commun.*, 2012, **48**, 9729–9731.
79. Y. Zhou, Z. Tian, Z. Zhao, Q. Liu, J. Kou, X. Chen, J. Gao, S. Yan and Z. Zou, *ACS Appl. Mater. Interfaces*, 2011, **3**, 3594–3601.
80. Z. Sun, J. Guo, S. Zhu, L. Mao, J. Ma and D. Zhang, *Nanoscale*, 2014, **6**, 2186–2193.
81. X. Wang, H. Wang, H. Zhang, W. Yu, X. Wang, Y. Zhao, X. Zong and C. Li, *ACS Energy Lett.*, 2018, **3**, 1159–1164.
82. Y. Wu, P. Wang, Z. Guan, J. Liu, Z. Wang, Z. Zheng, S. Jin, Y. Dai, M.-H. Whangbo and B. Huang, *ACS Catal.*, 2018, **8**, 10349–10357.
83. Z.-C. Kong, J.-F. Liao, Y.-J. Dong, Y.-F. Xu, H.-Y. Chen, D.-B. Kuang and C.-Y. Su, *ACS Energy Lett.*, 2018, **3**, 2656–2662.

84. S. Wan, M. Ou, Q. Zhong and X. Wang, *Chem. Eng. J.*, 2019, **358**, 1287–1295.
85. M. Ou, W. Tu, S. Yin, W. Xing, S. Wu, H. Wang, S. Wan, Q. Zhong and R. Xu, *Angew. Chem., Int. Ed.*, 2018, **57**, 13570–13574.
86. Y.-F. Xu, M.-Z. Yang, B.-X. Chen, X.-D. Wang, H.-Y. Chen, D.-B. Kuang and C.-Y. Su, *J. Am. Chem. Soc.*, 2017, **139**, 5660–5663.
87. S.-H. Guo, J. Zhou, X. Zhao, C.-Y. Sun, S.-Q. You, X.-L. Wang and Z.-M. Su, *J. Catal.*, 2019, **369**, 201–208.
88. L.-Y. Wu, Y.-F. Mu, X.-X. Guo, W. Zhang, Z.-M. Zhang, M. Zhang and T.-B. Lu, *Angew. Chem., Int. Ed.*, 2019, **58**, 9491–9495.
89. A. Kubacka, M. Fernández-García and G. Colón, *Chem. Rev.*, 2012, **112**, 1555–1614.
90. S. Linic, P. Christopher and D. B. Ingram, *Nat. Mater.*, 2011, **10**, 911.
91. W. Hou and S. B. Cronin, *Adv. Funct. Mater.*, 2013, **23**, 1612–1619.
92. L. Zhou, C. Zhang, M. J. McClain, A. Manjavacas, C. M. Krauter, S. Tian, F. Berg, H. O. Everitt, E. A. Carter, P. Nordlander and N. J. Halas, *Nano Lett.*, 2016, **16**, 1478–1484.
93. X. Zhang, X. Li, D. Zhang, N. Q. Su, W. Yang, H. O. Everitt and J. Liu, *Nat. Commun.*, 2017, **8**, 14542.
94. D. Wan, B. Yan, J. Chen, S. Wu, J. Hong, D. Song, X. Zhao, X. Chi, S. Zeng, Z. Huang, C. Li, K. Han, W. Zhou, Y. Cao, A. Rusydi, S. J. Pennycook, P. Yang, Ariando, R. Xu, Q.-H. Xu, X. R. Wang and T. Venkatesan, *Chem. Mater.*, 2019, **31**, 2320–2327.
95. L. Collado, A. Reynal, F. Fresno, M. Barawi, C. Escudero, V. Perez-Dieste, J. M. Coronado, D. P. Serrano, J. R. Durrant and V. A. de la Peña O'Shea, *Nat. Commun.*, 2018, **9**, 4986.
96. C. Zhou, L. Shang, H. Yu, T. Bian, L.-Z. Wu, C.-H. Tung and T. Zhang, *Catal. Today*, 2014, **225**, 158–163.
97. J. Zhang, X. Jin, P. I. Morales-Guzman, X. Yu, H. Liu, H. Zhang, L. Razzari and J. P. Claverie, *ACS Nano*, 2016, **10**, 4496–4503.
98. C. Li, Y. Xu, W. Tu, G. Chen and R. Xu, *Green Chem.*, 2017, **19**, 882–899.
99. W. Iqbal, B. Yang, X. Zhao, M. Rauf, M. Waqas, Y. Gong, J. Zhang and Y. Mao, *Catal. Sci. Technol.*, 2018, **8**, 4576–4599.
100. W.-J. Ong, L.-L. Tan, Y. H. Ng, S.-T. Yong and S.-P. Chai, *Chem. Rev.*, 2016, **116**, 7159–7329.
101. X. Wang, K. Maeda, A. Thomas, K. Takanabe, G. Xin, J. M. Carlsson, K. Domen and M. Antonietti, *Nat. Mater.*, 2008, **8**, 76.
102. J. Fu, J. Yu, C. Jiang and B. Cheng, *Adv. Energy Mater.*, 2018, **8**, 1701503.
103. J. Wen, J. Xie, X. Chen and X. Li, *Appl. Surf. Sci.*, 2017, **391**, 72–123.
104. J. Zhang, M. Zhang, C. Yang and X. Wang, *Adv. Mater.*, 2014, **26**, 4121–4126.
105. J. Yu, K. Wang, W. Xiao and B. Cheng, *Phys. Chem. Chem. Phys.*, 2014, **16**, 11492–11501.
106. H. Feng, Q. Guo, Y. Xu, T. Chen, Y. Zhou, Y. Wang, M. Wang and D. Shen, *ChemSusChem*, 2018, **11**, 4256–4261.
107. F. He, G. Chen, Y. Zhou, Y. Yu, L. Li, S. Hao and B. Liu, *J. Mater. Chem. A*, 2016, **4**, 3822–3827.

108. Z. Wu, H. Gao, S. Yan and Z. Zou, *Dalton Trans.*, 2014, **43**, 12013–12017.
109. S. Ye, R. Wang, M.-Z. Wu and Y.-P. Yuan, *Appl. Surf. Sci.*, 2015, **358**, 15–27.
110. Y. Chen, J. Li, Z. Hong, B. Shen, B. Lin and B. Gao, *Phys. Chem. Chem. Phys.*, 2014, **16**, 8106–8113.
111. P. Niu, L. Zhang, G. Liu and H.-M. Cheng, *Adv. Funct. Mater.*, 2012, **22**, 4763–4770.
112. S. Yang, Y. Gong, J. Zhang, L. Zhan, L. Ma, Z. Fang, R. Vajtai, X. Wang and P. M. Ajayan, *Adv. Mater.*, 2013, **25**, 2452–2456.
113. Q. Gu, Y. Liao, L. Yin, J. Long, X. Wang and C. Xue, *Appl. Catal., B*, 2015, **165**, 503–510.
114. Y. Zeng, X. Liu, C. Liu, L. Wang, Y. Xia, S. Zhang, S. Luo and Y. Pei, *Appl. Catal., B*, 2018, **224**, 1–9.
115. G. Li, Z. Lian, W. Wang, D. Zhang and H. Li, *Nano Energy*, 2016, **19**, 446–454.
116. J.-Y. Hao, Y.-Y. Wang, X.-L. Tong, G.-Q. Jin and X.-Y. Guo, *Int. J. Hydrogen Energy*, 2012, **37**, 15038–15044.
117. Y. Nariki, Y. Inoue and K. Tanaka, *J. Mater. Sci.*, 1990, **25**, 3101–3104.
118. C. Han, Y. Lei, B. Wang and Y. Wang, *ChemSusChem*, 2018, **11**, 4237–4245.
119. B. Wang, J. Zhang and F. Huang, *Appl. Surf. Sci.*, 2017, **391**, 449–456.
120. C. Huang, C. Chen, M. Zhang, L. Lin, X. Ye, S. Lin, M. Antonietti and X. Wang, *Nat. Commun.*, 2015, **6**, 7698.
121. M. Zhou, S. Wang, P. Yang, C. Huang and X. Wang, *ACS Catal.*, 2018, **8**, 4928–4936.
122. A. Dhakshinamoorthy, A. M. Asiri and H. García, *Angew. Chem., Int. Ed.*, 2016, **55**, 5414–5445.
123. T. Zhang and W. Lin, *Chem. Soc. Rev.*, 2014, **43**, 5982–5993.
124. M. B. Chambers, X. Wang, L. Ellezam, O. Ersen, M. Fontecave, C. Sanchez, L. Rozes and C. Mellot-Draznieks, *J. Am. Chem. Soc.*, 2017, **139**, 8222–8228.
125. Y. Horiuchi, T. Toyao, M. Saito, K. Mochizuki, M. Iwata, H. Higashimura, M. Anpo and M. Matsuoka, *J. Phys. Chem. C*, 2012, **116**, 20848–20853.
126. C. H. Hendon, D. Tiana, M. Fontecave, C. Sanchez, L. D'arras, C. Sassoye, L. Rozes, C. Mellot-Draznieks and A. Walsh, *J. Am. Chem. Soc.*, 2013, **135**, 10942–10945.
127. J. Castells-Gil, N. M. Padial, N. Almora-Barrios, J. Albero, A. R. Ruiz-Salvador, J. González-Platas, H. García and C. Martí-Gastaldo, *Angew. Chem., Int. Ed.*, 2018, 57, 8453–8457.
128. C. A. Trickett, A. Helal, B. A. Al-Maythalony, Z. H. Yamani, K. E. Cordova and O. M. Yaghi, *Nat. Rev. Mater.*, 2017, **2**, 17045.
129. Y. An, Y. Liu, P. An, J. Dong, B. Xu, Y. Dai, X. Qin, X. Zhang, M.-H. Whangbo and B. Huang, *Angew. Chem., Int. Ed.*, 2017, **56**, 3036–3040.
130. C. Wang, Z. Xie, K. E. deKrafft and W. Lin, *J. Am. Chem. Soc.*, 2011, **133**, 13445–13454.
131. K. M. Choi, D. Kim, B. Rungtaweevoranit, C. A. Trickett, J. T. D. Barmanbek, A. S. Alshammari, P. Yang and O. M. Yaghi, *J. Am. Chem. Soc.*, 2017, **139**, 356–362.

132. H. Fei, M. D. Sampson, Y. Lee, C. P. Kubiak and S. M. Cohen, *Inorg. Chem.*, 2015, **54**, 6821–6828.
133. D. Sun, Y. Fu, W. Liu, L. Ye, D. Wang, L. Yang, X. Fu and Z. Li, *Chem. Eur. J.*, 2013, **19**, 14279–14285.
134. D. Chen, H. Xing, C. Wang and Z. Su, *J. Mater. Chem. A*, 2016, **4**, 2657–2662.
135. Y. Liu, Y. Yang, Q. Sun, Z. Wang, B. Huang, Y. Dai, X. Qin and X. Zhang, *ACS Appl. Mater. Interfaces*, 2013, **5**, 7654–7658.
136. K. Khaletskaya, A. Pougin, R. Medishetty, C. Rösler, C. Wiktor, J. Strunk and R. A. Fischer, *Chem. Mater.*, 2015, **27**, 7248–7257.
137. B. Han, X. Ou, Z. Deng, Y. Song, C. Tian, H. Deng, Y.-J. Xu and Z. Lin, *Angew. Chem., Int. Ed.*, 2018, **57**, 16811–16815.
138. K. Zhang and L. Guo, *Catal. Sci. Technol.*, 2013, **3**, 1672–1690.
139. S. Chandrasekaran, L. Yao, L. Deng, C. Bowen, Y. Zhang, S. Chen, Z. Lin, F. Peng and P. Zhang, *Chem. Soc. Rev.*, 2019, **48**, 4178–4280.
140. W. Tu, Y. Li, L. Kuai, Y. Zhou, Q. Xu, H. Li, X. Wang, M. Xiao and Z. Zou, *Nanoscale*, 2017, **9**, 9065–9070.
141. H. Jung, K. M. Cho, K. H. Kim, H.-W. Yoo, A. Al-Saggaf, I. Gereige and H.-T. Jung, *ACS Sust. Chem. Eng.*, 2018, **6**, 5718–5724.
142. S. Lian, M. S. Kodaimati and E. A. Weiss, *ACS Nano*, 2018, **12**, 568–575.
143. X. Long, Z. Wang, S. Xiao, Y. An and S. Yang, *Mater. Today*, 2016, **19**, 213–226.
144. Y. Zhao, X. Jia, G. I. N. Waterhouse, L.-Z. Wu, C.-H. Tung, D. O'Hare and T. Zhang, *Adv. Energy Mater.*, 2016, **6**, 1501974.
145. K. Teramura, S. Iguchi, Y. Mizuno, T. Shishido and T. Tanaka, *Angew. Chem., Int. Ed.*, 2012, **51**, 8008–8011.
146. Y. Zhao, G. Chen, T. Bian, C. Zhou, G. I. N. Waterhouse, L.-Z. Wu, C.-H. Tung, L. J. Smith, D. O'Hare and T. Zhang, *Adv. Mater.*, 2015, **27**, 7824–7831.
147. F. Sastre, A. Corma and H. García, *J. Am. Chem. Soc.*, 2012, **134**, 14137–14141.
148. S. Tonda, S. Kumar, M. Bhardwaj, P. Yadav and S. Ogale, *ACS Appl. Mater. Interfaces*, 2018, **10**, 2667–2678.
149. S. Kumar, M. A. Isaacs, R. Trofimovaite, L. Durndell, C. M. A. Parlett, R. E. Douthwaite, B. Coulson, M. C. R. Cockett, K. Wilson and A. F. Lee, *Appl. Catal., B*, 2017, **209**, 394–404.
150. S. Kumar, L. J. Durndell, J. C. Manayil, M. A. Isaacs, C. M. A. Parlett, S. Karthikeyan, R. E. Douthwaite, B. Coulson, K. Wilson and A. F. Lee, *Part. Part. Syst. Char.*, 2018, **35**, 1700317.

CHAPTER 5

Catalytic Technologies for Clean Hydrogen Production

SARA NAVARRO JAEN, CRISTINA MEGIAS SAYAGO, SVETLANA IVANOVA* AND JOSÉ ANTONIO ODRIOZOLA

Dpto. Química Inorgánica, Universidad de Sevilla, Instituto de Ciencia de Materiales de Sevilla, Centro mixto CSIC – Universidad de Sevilla, Av. Américo Vespucio 49, 41092 Sevilla, Spain
*Corresponding contributor. Email: sivanova@us.es

5.1 Hydrogen Economy: General Aspects

The current worldwide economy is based on the use of fossil fuels as the primary power and energy source for several sectors, such as industry, transport, electricity and heat.[1] However, the availability of carbon-based fuels seems uncertain in the long term and their use has had a negative impact on the environment, increasing CO_2 emissions and consequently contributing to climate change.[2]

Therefore, the search for alternative low-carbon energy sources has become a major concern for the international community in recent decades.[3] The use of hydrogen as an energy carrier has been positioned as one of the most promising options in this area due to its high energy content per unit mass and environmental advantages, since no greenhouse gas emissions are derived and the only product of its combustion is water.[4] Furthermore, hydrogen can be produced from a great variety of both renewable and carbon-based energy sources, although if produced by the latter, a CO_2 capture stage is necessary

Heterogeneous Catalysis for Energy Applications
Energy and Environment Series No. 27
Edited by Tomas R. Reina and Jose A. Odriozola

Published by the Royal Society of Chemistry, www.rsc.org

in the process. Consequently, one of the current socio-economic objectives consists of developing new energy strategies that allow the transition from the current "fossil fuel economy" to a "hydrogen economy" focused on the use of hydrogen as the primary energy carrier.[5] For the successful development of a hydrogen economy, the key factor lies in the implementation of such a technology, with all interest focused on the production, distribution and final uses of hydrogen.[6] Considerable techno-economic challenges remain to be solved.[7]

As mentioned above, hydrogen can be produced from fossil and renewable sources, with the ideal scenario including the latter, resulting in a minimized carbon footprint.[8] However, according to the International Energy agency (IEA), most of the hydrogen produced today originates from fossil fuels, with 76% of the global hydrogen obtained from natural gas and almost all the rest from coal. In terms of CO_2 emissions, hydrogen production is responsible for 830 Mt of CO_2 per year.[9]

5.2 Current Scenario: Prevalence of Fossil Fuel Sources

Current large-scale hydrogen production is based on hydrocarbon reforming. In this process, hydrocarbons are oxidized catalytically to produce hydrogen, water, CO and CO_2. Three chemical processes are distinguished as a function of the employed oxidant: steam reforming (SR), partial oxidation (POX) and autothermal reforming (ATR).

5.2.1 Steam Reforming

Today, SR of hydrocarbons (mainly natural gas and some light hydrocarbons) constitutes the most widespread large-scale hydrogen production method due to its high hydrogen production efficiency (about 74%) and its lower price compared to other production methods.[10] This process involves several steps: (1) hydrocarbon purification to eliminate the impurities; (2) catalytic reforming using water steam as an oxidant to produce synthesis gas (syngas) (eqn (5.1)); (3) gas purification through water gas shift (WGS) (eqn (5.2)) and methanation reactions (eqn (5.3)); and (4) CO_2 washing and other physical purification steps increasing the purity of the final product gas.[11,12]

$$C_nH_m + nH_2O \rightarrow nCO + (n + m/2)H_2 \tag{5.1}$$

$$CO + H_2O \rightarrow CO_2 + H_2 \tag{5.2}$$

$$CO + 3\,H_2 \rightarrow CH_4 + H_2O \tag{5.3}$$

5.2.2 Partial Oxidation

POX consists of oxygen-assisted exothermic conversion of heavy hydrocarbons (mainly heavy fuel oils and coal) into hydrogen (eqn (5.4)).

$$C_nH_m + (n/2)O_2 \rightarrow nCO + (m/2)H_2 \quad (5.4)$$

As in the case of SR, syngas is produced and afterwards enriched in hydrogen *via* the WGS reaction and subsequent gas purification. In contrast to SR, the POX process can convert a wider range of raw materials into hydrogen; nevertheless, the hydrogen yield decreases as the chain of the hydrocarbon increases, and it is energetically less efficient than SR.

5.2.3 Autothermal Reforming

ATR is the combination of both SR and POX processes. POX is used to produce heat (exothermic reaction) and SR (endothermic) to increase hydrogen yield. Even though no external heat supply is necessary for this process, the costs of air separation units (to produce oxygen) and the difficulties in gas purification make this process less favourable than the previously mentioned ones.

5.2.4 Coal Gasification

Due to the availability of coal deposits worldwide, coal gasification is also used to produce hydrogen and electricity.[13] Gasification refers to the reaction of coal with air or an oxygen/steam mixture to produce syngas (eqn (5.5)). As with the other processes, WGS is used to increase the hydrogen yield.

$$C + H_2O \rightarrow CO + H_2 \quad (5.5)$$

The major drawback of this method is the high carbon content and, consequently, higher CO_2 emissions compared to all of the other options. Hydrogen production from coal also differs in economic terms: whereas the costs of the raw materials are lower, the unit capital costs are higher for coal gasification plants.[14]

5.3 Hydrogen Production from Renewable Sources

Although the economic viability of hydrogen technology demands the use of fossil fuel sources in the near future, some technologies based on renewable sources open up new environmentally friendly horizons. Such methods are based on water electrolysis and thermochemical and biochemical biomass valorizations.

5.3.1 Electrolysis

Electrolysis currently accounts for 2% of the global hydrogen production. This process consists of splitting water into oxygen and hydrogen using electricity. An advantage of this technology lies in the possibility of scaling the size of the electrolyser, allowing for the production of hydrogen from small- to large-scale facilities. The efficiency of this process is determined by the amount of electricity used to produce hydrogen, with it currently being possible to obtain efficiencies between 60% and 80%.

However, its application still requires improvements to several aspects, such increasing the process efficiency and the lifespan of the electrolyser, as well as decreasing the costs and developing more flexible systems adapted to a fluctuating power supply. Furthermore, the economic advantages of this method directly depend on electricity prices, which are currently much higher than the price of SR-produced hydrogen. An alternative to this is the use of solar or wind energy for electrolysis.

5.3.2 Hydrogen Production from Biomass

This option constitutes one of the most promising technologies regarding hydrogen production in the future, considering the abundance of biomass, especially agricultural and domestic waste, and the low CO_2 emissions that are normally produced by this process. Nevertheless, reduction of the capital costs still constitutes an important challenge for the application of this technology. Two methods are used to produce hydrogen from biomass: thermochemical and biochemical methods.

Thermochemical methods generally refer to the gasification or pyrolysis of solid or liquid biomass and waste to produce syngas. As with the case of production from fossil fuels, a subsequent purification step is required. Regarding the gasification of solid biomass, untreated wood is the most suitable source due to the lower quantity of impurities such wood presents in comparison to stalky materials such as switchgrass and straw. Regarding liquid biomass, the reforming or POX of secondary energy sources such as biogas, bioethanol, vegetable oils, biodiesel or bioglycerol have been the most reported.[15,16]

In the case of *biochemical* methods, the conversion of biomass is carried out by microorganisms. The biomass is fermented and processed, or alternatively water splitting is carried out by biophotolysis using green algae or cyanobateria. Many methods in this field have been developed, with the most relevant being dark fermentation using heterotropic bacteria and photofermentation using photosynthetic bacteria. Nevertheless, these methods have only recently been applied at the lab scale, thus intense development of these techniques is still necessary.

It is clear that intense development of hydrogen production techniques from renewable sources is needed in order to achieve a sustainable energy vector. The non-existent large-scale production and less competitive costs of such renewable technologies hamper the implementation of this process as an environmentally friendly and sustainable hydrogen production route.

Thus, in the near future, the production of hydrogen from renewable sources remains infeasible and hydrogen production will be guaranteed by fossil fuel-based technologies. However, a new corner has been turned in the last decade, involving hydrogen production from fossil fuels coupled with the *carbon capture, utilization and storage (CCUS) technology*.

Fossil-based hydrogen production consumes 6% and 2% of global natural gas and coal, respectively, with more than 800 million tonnes of CO_2 ejected into the atmosphere per year. Since the production of hydrogen will continue in its *status quo*, a strategy for decreasing CO_2 emissions is mandatory. In this context, CCUS technologies play a fundamental role. This strategy consists of CO_2 being permanently captured in geological formations or temporarily for its use in the production of valuable products. Today, 19 large-scale CCUS facilities operate worldwide and are mainly incorporated in industries where relatively pure CO_2 is produced. The employed technology in this sector is already mature, uniquely requiring the purification or dehydration of the CO_2 feed for its subsequent transportation and organization. The use of these technologies could provide more time for the sustainable design of new hydrogen technologies.

Today, hydrogen is used in chemical product synthesis. Around 55% of the produced hydrogen is used for ammonia synthesis, 25% for refineries and for oil processing and around 10% for methanol production. The remaining 10% is used as a base material to produce other chemicals (nylon, polyurethane, paints, synthetic fibres and plastics). The use of hydrogen for energy applications as a feed for fuel cells constitutes the most promising aspect regarding the future scenario of using H_2 as an energy vector, but it is conditional on hydrogen storage technologies in both stationary and mobile applications. Concerning stationary applications, the main challenge is the selection of the tank materials, whereas in mobile applications, the requirements are more challenging, with the reduction of tank size being particularly important.[17] Hydrogen and fuel cells are envisaged as some of the most promising technologies to decarbonize the transportation sector. In order to avoid the safety concerns associated with the use of bottled hydrogen, the production of hydrogen on board is the best option for portable applications. The success of such *on-board* technology depends on the development of a suitable catalytic technology that makes possible the production of clean hydrogen to effectively feed low-temperature fuel cells. The state of the art of this technology is addressed in the following section.

5.4 Hydrogen Fuel Processors: Catalysis and Reaction Engineering

Let us imagine a hydrogen-based society, a world where power plant and car pollution emissions are reduced significantly by more than half. Let us imagine a fuel that only generates water as waste, originating from renewable sources and leading to a less fossil fuel-dependent community. Certainly, this positive future scenario is what motivates today's hydrogen fuel processor research.

The generation of electricity in a hydrogen fuel cell relies on pure electrochemistry, as does the basic battery, with the main difference being hydrogen's virtual inexhaustibility. Both devices produce electricity, but hydrogen fuel cells produce electricity as long as hydrogen is available, in contrast to a basic battery. On the other hand, the fuel cell's operating conditions are *a priori* very close to those of a hydrogen-based combustion engine, which is why it is worth briefly describing the key differentiating factor. In a conventional combustion engine, fuel (hydrogen) is burned, releasing heat:

$$H_2 + \frac{1}{2}O_2 \rightleftharpoons H_2O \tag{5.6}$$

The bond formation involves a reconfiguration of electrons, moving from one bonding state to another, a process that occurs in picoseconds at an intimate, subatomic scale. This explains why the energy difference between states is only recoverable as heat. To produce electricity, this heat must be converted first into mechanical energy and afterwards into electrical energy. It is evident, therefore, that undergoing all of these steps to produce electrical energy from hydrogen is complex and economically unviable. By contrast, in a fuel cell, hydrogen *combustion* proceeds through two spatially separated electrochemical reactions in such a way that the electrons coming from the fuel are forced to flow through an external circuit, directly generating an electric current:

$$H_2 \rightleftharpoons 2H^+ + 2e^- \tag{5.7}$$

$$\frac{1}{2}O_2 + 2H^+ + 2e^- \rightleftharpoons H_2O \tag{5.8}$$

The physical separation of these two electrochemical processes is accomplished by using two separated electrodes along with an electrolyte. In such a way, hydrogen is bubbled across one electrode, usually consisting of Pt, splitting into protons and electrons (according to eqn (5.7)). The protons migrate though the electrolyte (see Figure 5.1), while the electrons flow through an external circuit connecting both electrodes. Once the second Pt piece is reached, the electrons recombine with protons and bubbling oxygen to form water (eqn (5.8)). Although simple and apparently efficient enough, cost represents the main barrier to fuel cell implementation, along with the low power density (the factor that expresses how much power a fuel cell can produce per volume or weight), which implies that further improvements are needed to ensure successful fuel cell applications in portable devices.

As schematized in Figure 5.1, there are five major fuel cells types: polymer electrolyte membrane fuel cells (PEMFCs), phosphoric acid fuel cells (PAFCs), alkaline fuel cells (AFCs), molten carbonate fuel cells (MCFCs) and solid-oxide fuel cells (SOFCs). They mostly differ from one another by the type of electrolyte, electrode material and operating temperature range (from 60 to 1 000 °C).[18]

PEMFCs are usually the most attractive for many applications due to their acceptable operating temperature, high power density, reliability and fast

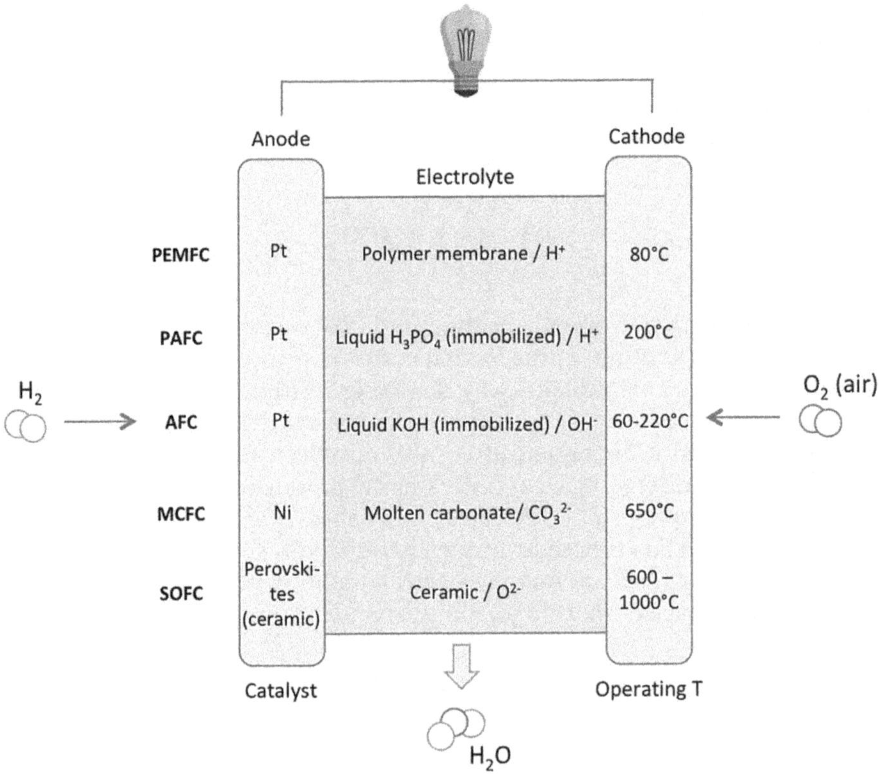

Figure 5.1 Major H_2 fuel cell types.

start up/shut down.[19] All of these features make PEMFCs the most promising alternatives for use in the transportation sector.[20]

The exchange current density (*i.e.* the fuel cell efficiency) mostly depends on the catalyst electrode design. For PEMFCs, highly porous, nanostructured Pt-based electrodes are the most used and so are the most commercialized for fuel cell electric vehicles.[21,22] However, the high cost of platinum group metals (PGMs) makes PEMFCs unviable for large-scale applications.[23] In order to overcome the economic barrier, an important approach is to replace the PGM-based electrocatalysts with costless homologues, and there have been a number of related investigations in the last 5 years.[24–29] Several attempts have been made to reduce the platinum amount, starting from bimetallic alloy preparation[30,31] and followed by ternary,[32,33] quaternary[33] and even quinary[34,35] alloys, in which Pt has been combined with different metals (Co, Fe, Ni, Ti, Cu, Ir, Ym, Sn, Cr, Mn, Mo, Ag, *etc.*). In addition, many PGM-free catalysts have been proposed and studied, mainly transition metal–nitrogen–carbon electrocatalysts (M–N–C, where M = Fe, Co, Ni, Cu, Mn, *etc.*), heteroatom-doped carbon, defective carbon, transition metal oxides, nitrides and/or carbides, among others.[35–37]

It is important to highlight that most studies dealing with a new catalyst electrode design date from the last two years (2018–2019), orientating the fuel cell state of the art towards costless materials for future hydrogen-fed fuel cells. Today, the only possibility of fuel cells for *on-board* application is a Pt cathode PEMFC. Although highly performing, a major drawback appears when fossil fuel-based H_2 production is employed on board. Pt cathodes are very sensitive to irreversible CO adsorption, with tolerance limits of less than 50 ppm.[38] This is why the development of a new generation of fuel cell cathodes must be closely linked to hydrogen clean-up technologies.

5.5 Successful Stories of Hydrogen Clean-up Heterogeneous Catalysis

Hydrogen clean-up in the classical stationary Haber–Bosch process includes two successive WGS reactions and methanation. No possible changes are envisaged in this kind of technology, but for mobile applications, new actors enter the scene. The WGS reaction could be combined with preferential CO oxidation (PrOx) instead of methanation. Both reactions should operate at low temperature using active, selective and stable catalysts able to operate in the presence of water and CO_2 and able to withstand continuous start-up/shut-down cycles.

5.5.1 WGS Catalysts

The WGS reaction in eqn (5.9) is a well-known process for hydrogen production at an industrial level, being a fundamental part of processes such as ammonia and methanol synthesis, Fischer–Tropsch synthesis and refining operations.[39] The importance of this process lies in its capacity to decrease the CO content while producing additional hydrogen.[40]

$$CO + H_2O \leftrightarrow CO_2 + H_2 \quad (5.9)$$

The WGS reaction is slightly exothermic ($\Delta H = -41.1$ kJ mol^{-1}), being limited thermodynamically at high temperatures and kinetically at low temperatures. Due to the equilibrium constraints, the industrial process is generally carried out in two adiabatic reactors: a high-temperature shift reactor, operating at 350–450 °C over Fe/Cr oxide catalysts; and a low-temperature shift reactor, operating at 200–250 °C using Cu-based catalysts.[41] Both catalytic systems are unsuitable for application in PEMFCs, where the operation conditions change substantially compared to the industrial ones. For fuel cell applications, the catalyst bed must have a reduced volume due to the restriction of space and weight, and the employed catalyst must be stable in start-up/shut-down cycles without any conditioning procedures. For example, while Fe catalysts in the presence of excess fuel from the reformer induces coke formation and subsequent deactivation, Cu catalysts are pyrophoric in their reduced state and experience deactivation in the presence of condensed water.[42–44] Consequently, the development of alternative catalyst formulations becomes essential. Numerous studies have already been devoted to the search for catalysts able to meet the previously mentioned requirements, with the general conclusion being the suitability of noble metals (Rh, Ru, Pd, Au and Pt) for this purpose.[41,45] Nevertheless, Rh- and Ru-based catalysts produce methane by means of the methanation reaction under WGS conditions,[46] which makes them less appropriate. On the other hand, Pd has been demonstrated to be less active than Pt, thus redirecting the focus of studies in recent years to the use of Au and Pt as WGS catalysts. Au has been demonstrated to be an exceptional catalyst in the low-temperature range (180–230 °C), whereas Pt presents higher rates and better tolerance to the stream composition, that is, to the presence of the reaction products (CO_2 and H_2) in the feed at 250 °C.[47] Indeed, the intrinsic reaction rate over Pt has been demonstrated to be higher than over other noble metals,[48] positioning Pt as the most suitable metal for WGS reactions. According to the aforementioned aspects, this section presents an overview of the most relevant results obtained using Au-based catalysts in the low-temperature WGS reaction and a more detailed view of the research focused on Pt-based catalysts.

5.5.2 Recent Advances in Gold WGS Catalysis

The use of gold catalysts for the WGS reaction has been extensively studied. Ramírez Reina *et al.*[49] presented a detailed review addressing the numerous studies performed until 2014, and Carter and Hutchings[50] and Tabakova[51] summarized the latest advances in the state of the art. Consequently, this section will be devoted to an overview of the most studied gold-based catalytic systems in order to understand the current trends in gold WGS catalysis.

The suitability of Au catalysts for the WGS reaction was first reported by Andreeva *et al.*[52] through the use of an Au/α-Fe_2O_3 catalyst in the low-temperature range. The exceptional catalytic performance of this solid was attributed to the metal–support interaction. Following this work, other authors studied several aspects of iron oxide-supported gold catalysts, such as the influence of the calcination temperature[53] or the reduction treatment prior to the reaction,[54] finding that parameters such as gold dispersion and support structure (OH group population) played key roles during the catalytic process.

The successful results derived from these studies paved the way for the exploration of other oxide-supported gold catalysts. In this sense, CeO_2 became the support of choice. Au/CeO_2 catalysts' excellent catalytic performance was attributed to the presence of oxygen vacancies in the oxide structure and their relation to the electronic properties of CeO_2. CeO_2 combines an elevated oxygen storage capacity along with redox behaviour through the Ce^{3+}/Ce^{4+} redox pair cycle.[55] The group of Flytzani-Stephanopoulos[56] studied a series of Au/CeO_2 catalysts and evaluated the influence of the preparation method on WGS catalytic performance. They found an increase in the reducibility of the CeO_2 surface oxygen in the presence of gold and related the amount of available oxygen to the CeO_2 crystal size. They also reported that gold particle size was not the most determining aspect of the reaction. However, gold particle size influences oxygen transfer, and gold is necessary for the adsorption of CO, which is not possible over hydroxylated CeO_2. Andreeva *et al.*[57] studied the influence of the gold content in a series of Au/CeO_2 catalysts, concluding that CeO_2 reducibility was improved in the presence of gold, with gold particle size decreasing under reaction conditions.

One of the first reports regarding the reaction mechanism over ceria-based gold catalysts was published by Tabakova *et al.*,[58] based on Fourier-transform infrared (FTIR) spectroscopic studies. This study revealed that the reduction of the catalyst exposures negatively charged gold atoms and oxygen vacancies on the CeO_2 surface, with a combination of these being necessary for WGS activity. The authors claimed that the WGS reaction was taking place at the boundary between the gold particles and ceria, with the CO being adsorbed on the Au particles and the H_2O being dissociated on CeO_2 defects.

After this work, many studies have been devoted to the elucidation of the WGS reaction mechanism on Au/CeO_2 catalysts, and controversial results have been obtained. However, there are some common conclusions derived from such works. For instance, the CeO_2 support is directly involved in the WGS reaction, lowering the energy required for the dissociation of water molecules, with this ability being related to the presence of oxygen vacancies on the CeO_2 surface. Nevertheless, the WGS catalytic performance is a result of the synergy between the ceria support and gold species, with the metal dispersion and the CeO_2 particle size being the crucial factors for higher interface perimeters.[59,61]

Despite the leading role of CeO_2, other oxide supports have also been studied. Idakiev *et al.*[61] reported the use of a high-surface area mesoporous TiO_2 as a promising support for gold-based WGS catalysts. The authors evaluated different metal loadings (1–5 wt%), which allowed for the correlation of gold loading, mean particle size and dispersion with catalytic performance and metal–support interactions. A sample with a 2.5 wt% Au loading presented unexpected catalytic activity, which was attributed to a greater number of gold active sites located at the Au–mesoporous TiO_2 interface, whereas increasing metal loading gave rise to a decrease in catalytic activity due to the increase in particle size. Based on these results, the same authors employed TiO_x nanotubes as supports for gold nanoparticles.[62] The average gold particle size and TiO_x nanotube diameter were similar, with some of the gold particles being inside the hollow titania tubes, thus blocking them and decreasing the WGS catalytic performance of these systems with respect to those obtained when using high-surface area mesoporous TiO_2.

Although TiO_2-supported gold catalysts presented good WGS catalytic performance, their deactivation at high temperatures due to gold sintering represents a significant problem. As a consequence, the modification of the titania with different metal oxide additives (M_xO_y, where M = Al, Ca, Fe, Zn, Ga, Y, Zr, La, Ce, Pr, Nd, Sm, Eu, Gd, Dy, Ho, Er or Yb) was proposed as a way to improve catalyst stability.[63] The use of CeO_2 as an additive to mesoporous TiO_2 improved the reducibility of the support and increased the number of active sites located at the Au/CeO_2–TiO_2 interface.[64]

Another interesting support for gold-based WGS catalysts is ZrO_2, whose crystal structure, either monoclinic (m) or tetragonal (t), was demonstrated to play an important role in WGS catalytic performance.[65] The Au/m-ZrO_2 demonstrated superior catalytic performance to the Au/t-ZrO_2 catalyst due to its higher CO adsorption capacity. As for Au/CeO_2 catalysts, the particle size of both gold and ZrO_2 negatively affected WGS activity, which decreases as particle size increases. The works developed from the 1990s allowed for the establishment of the most important aspects to be considered for the design of WGS gold-based catalysts. Thus, gold dispersion, the nature of the support and the metal–support interaction were demonstrated to be fundamental aspects for obtaining active and stable WGS catalysts. Consequently, in the last few years, several research works have focused on improving the existing catalytic formulations and evaluating new solids that are active in this reaction, being particularly focused on the use of CeO_2 as a support.

Leveraging the potential of Au/CeO_2 catalysts in the WGS reaction, Reina *et al.* proposed the use of a ZnO-promoted Au/CeO_2/Al_2O_3 catalyst.[66] The authors dispersed the CeO_2 support on a high-surface area carrier, γ-Al_2O_3, in order to obtain higher surface : bulk ratios and to improve oxygen mobility, and they also used ZnO as a ceria promoter due to its capacity to create structural defects on the ceria lattice. In every case, the solids containing ZnO (1 and 2 wt%) presented superior catalytic performance to those without promoters. The catalytic system containing 2 wt% ZnO was particularly promising, showing high catalytic activity and tolerance to start-up/shut-down cycles. The same group[67] used Fe_2O_3 as a CeO_2 promoter in the same Au/CeO_2/Al_2O_3 catalyst, relating the important activity improvement to Ce–Fe synergic effect favouring water activation, either by the redox or associative mechanism.

The advantages of dispersing CeO_2 on Al_2O_3 and the importance of the synergy between the catalytic components motivated a study of a series of Au/$Ce_{1-x}Cu_xO_2$/Al_2O_3 catalysts in which the copper content was varied. In this case, the authors added Cu to the catalytic formulations due to the well-known WGS activity of this metal in the low-temperature range.[68] The obtained results elucidated many important aspects: first, the exceptional WGS catalytic activity was the result of a cooperative effect between Au, Cu and CeO_2. The increase of the Cu content enhanced the WGS catalytic activity (Figure 5.2), with this result being explained by the substitution of Ce^{4+} ions by Cu^{2+} ions in the ceria lattice, which induces the formation of oxygen vacancies. Furthermore, a close interaction between CeO_x and Cu_xO (with x ranging between 0 and 2) species was observed by Raman spectroscopy. In this case, the role of ceria is not only to provide oxygen vacancies, but also to promote the oxygen mobility in the CeO_{2-x}/Cu interface sites, favouring the oxidation of CO. On the other hand, the elevated catalytic activity is explained by the presence of Au and the Au–Cu interaction, which facilitates the redox cycles of CeO_2, thus improving catalytic performance.

Tabakova *et al.*[69] also proposed the modification of the CeO_2 support with Y_2O_3. Despite the popularity of using CeO_2, other authors preferred TiO_2-supported catalysts. For example, Hakeem and co-workers prepared a series of Au/TiO_2 catalysts by deposition precipitation, studying the influence of pH during synthesis on the WGS catalytic performance in semi-industrial conditions.[70] The authors compared their catalyst with a commercial Cu-ZnO/Al_2O_3

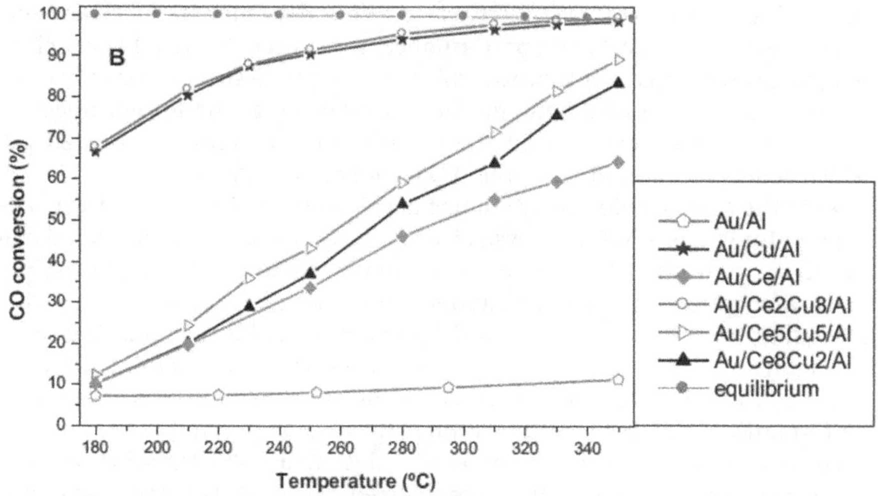

Figure 5.2 WGS activity of Au/CuO_x-CeO_{2-x}/Al_2O_3. Reproduced from Ref. 68 with permission from Elsevier, Copyright 2016.

and an AUROlite Au/TiO_2 catalyst used as reference. Both catalysts were more active than the studied catalyst in the evaluated conditions.

Considering the advantages of using CeO_2 and TiO_2 for the WGS reaction, some authors evaluated the influence of mixing both supports for generating gold-based WGS catalysts. For instance, Grinter and co-workers[71] studied an $Au/CeO_x/TiO_2$ model catalyst, evaluating the effect of a CeO_2 overlayer in contact with a TiO_2 (110) surface. The results showed that the prepared model catalyst presented significantly higher WGS catalytic activity than the individual Au/CeO_2 (111) and Au/TiO_2 (110) catalysts. However, the increase in CeO_2 coverage led to a decrease in catalytic activity related to a reduction in Ce^{3+} concentration during the WGS reaction. Other oxide supports such as magnesium[72] and uranium oxides[73] were also evaluated.

5.5.3 Recent Advances in Platinum WGS Catalysis

As mentioned in the introduction to this section, platinum is generally more active than gold for the WGS reaction, although the catalytic performance of gold in the low-temperature range is superior due to the higher activation energy of platinum.[74] Consequently, platinum catalysts are suitable systems for the medium-temperature WGS reaction. As both metals are extensively studied in this section, the most recent advances in the design of Pt-based WGS catalysts will be presented and sometimes compared to the corresponding gold catalysts.

Despite the differences regarding the temperature range of application, the considerations for the design of Au- and Pt-based WGS catalysts are similar. The importance of the support seems clear in both cases: the use of non-reducible supports such as Al_2O_3 gives rise to poor WGS catalytic performance for both Au and Pt catalysts.[75] In contrast, the use of reducible supports containing oxygen vacancies, such as CeO_2, has demonstrated a clear beneficial effect on WGS catalytic performance, no matter the metal used, since oxygen vacancies play a substantial role in the water dissociation step.[48] A significant difference exists between Au and Pt: whereas the former does not adsorb water molecules,[76] the latter adsorbs and in addition dissociates it, which could possibly influence the WGS reaction mechanism.[77]

As in the case of gold, many studies have focused on the use of CeO_2 as a support for Pt-based WGS catalysts. However, results have demonstrated that the activity of fresh Pt/CeO_2 catalysts was not high enough to meet the weight and cost requirements in mobile applications and suffered from deactivation upon start-up/shut-down cycles, which are necessary for the modification of the catalytic formulation in order to increase its activity and stability.[78] In this sense, most of the works have focused on the enhancement of WGS catalytic performance through the modification of the CeO_2 support.

For instance, González-Castaño *et al.*[79] designed a $Pt/CeO_2/Al_2O_3$ catalyst doped with ZrO_x and FeO_x. ZrO_x was selected based on its thermal stability, redox properties and ability to decrease CeO_2 sintering. On the other hand, the selection of FeO_x was based on its capacity to enhance the electronic properties and oxygen mobility of CeO_2.

With the aim of evaluating dopant influence, the authors prepared 4 wt% Pt catalysts on four different supports: CeO_2/Al_2O_3 (CeAl), $Ce_{0.9}Zr_{0.1}O_2/Al_2O_3$ (CeZrAl), $Ce_{0.9}Fe_{0.1}O_2/Al_2O_3$ (CeFeAl) and $Ce_{0.9}Zr_{0.05}Fe_{0.05}O_2/Al_2O_3$ (CeZrFeAl).

In order to discern the effect of the support, the WGS catalytic activity was normalized by the number of exposed Pt atoms. The results showed that the Pt/CeZrFeAl catalyst presented superior catalytic performance at higher space velocities, as shown in Figure 5.3.

In order to understand the influence of the promoters on the catalysts' reducibility, Temperature-Programmed Reduction (H_2-TPR) experiments were performed, showing that the reduction temperature of the CeO_2 support was shifted towards lower temperatures in the doped samples, with this effect being particularly evident in the samples containing Fe. The enhanced ceria reducibility was related to easier redox cycles and thus higher OH/H populations on the catalyst surface. Regarding the metal particles, the presence of Zr decreased their reduction temperature, whereas the presence of Fe led to higher H_2 consumption, demonstrating that the electronic changes on the support influenced the metal–support interaction. CO adsorption followed by FTIR spectroscopy demonstrated that the enhanced redox properties of the Pt/CeZrFeAl sample gave rise to enhanced transfer and reaction of the CO species to form carbonate and bicarbonate species on the support. However, these species could significantly deactivate the catalyst during the WGS reaction. Thus, the desorption of such species was evaluated, with it being concluded that the introduction of the doping agents facilitated the desorption of the carbonaceous species present on the catalyst surface, which could extend the lifespan of the catalyst. Whereas Fe contributed to accelerating the decomposition rate of these species, their stability was ruled out by the presence of Zr. Consequently, the enhanced catalytic activity of the

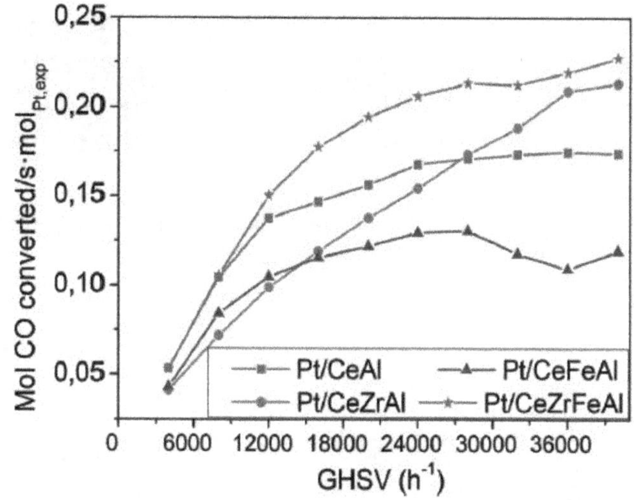

Figure 5.3 Normalized WGS catalytic activity of the promoted and unpromoted samples as a function of the gas hourly space velocity (GHSV) under 9% CO, 11% CO_2, 30% H_2O and 50% H_2 reaction flow. Reproduced from Ref. 79 with permission from the Royal Society of Chemistry.

Pt/CeZrFeAl was ascribed to both the enhanced reducibility and the ease of carbonate species formation and subsequent conversion into reaction products, which at the same time contributed to decreasing the deactivation degree of the catalyst.

The same group extrapolated this doping strategy to a structured 2 wt% Pt/CeO_2/Al_2O_3 catalyst.[80] The authors used metallic micromonoliths in order to obtain compact catalytic beds, thus increasing the efficiency of the process and bringing the application of WGS catalysts closer to their application in combination with fuel cells. Regarding the design of the catalyst, a careful multilayer approach was applied by adding to the Pt-based catalyst layer an additional inner layer consisting of a material able to absorb water and to continuously release the water-related species to the catalyst layer, where the WGS reaction would take place. Three inner (buffer) layers were evaluated: Al_2O_3 (Al), due to the absence of oxygen vacancies in its structure; CeO_2/Al_2O_3 (CeAl), with similar textural properties to Al_2O_3 but an increased number of oxygen vacancies; and $Ce_{0.9}Eu_{0.1}O_2$/Al_2O_3 (CeEuAl), due to the expected increase of oxygen vacancies through the substitution of Ce^{4+} by Eu^{3+} and solid solution formation. A scheme of the structured catalysts' configurations and their nomenclature is shown in Figure 5.4.

The supports (Al, CeAl and CeEuAl), powders, suspensions and micromonoliths containing only the buffer layer were not active in the WGS reaction. By contrast, the monoliths containing both of the catalysts and the buffer layer were active, following the order: M_Al < M_CeAl < M_CeEuAl, as shown in Figure 5.5.

The reaction mixture composition was also varied in order to evaluate the influence of every component on the catalytic activity. The presence of the

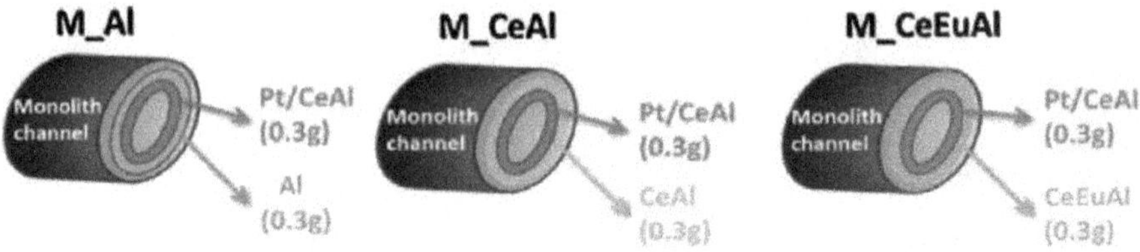

Figure 5.4 Scheme of the studied structured catalysts. Reproduced from Ref. 80 with permission from Elsevier, Copyright 2018.

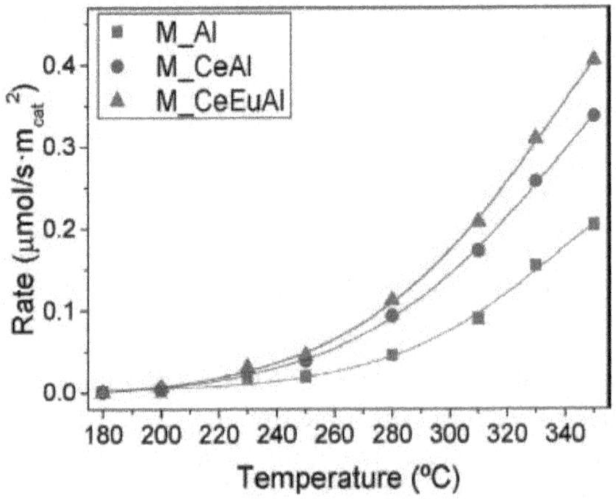

Figure 5.5 CO conversion as a function of the temperature of the structured catalysts under 7% CO, 9% CO_2, 30% H_2O and 50% H_2. Reproduced from Ref. 80 with permission from Elsevier, Copyright 2018.

reaction products (*i.e.* CO_2 and H_2) should provoke a decrease in CO conversion according to Le Chatelier's principle. However, no significant changes were observed when increasing the CO_2 concentration, indicating that higher concentrations of the latter would be necessary to inhibit the WGS catalytic activity. On the other hand, the increasing in the CO content had a negative effect on CO conversion, as expected from the increase in the number of CO molecules for the same number of active sites. At higher CO coverages, competition for CO and H_2O adsorption on the active sites takes place, and the OH population decreases, thus decreasing the extent of the reaction. The most significant changes were observed when the H_2O concentration was varied. As the H_2O concentration increased, a superior WGS catalytic performance was observed following the same order: M_CeEuAl > M_CeAl > M_Al. By means of water adsorption experiments followed by FTIR spectroscopy, it was possible to conclude that the oxygen vacancy concentration in the buffer layer had a significant influence on the water adsorption and dissociation capacity, directly reflected in the WGS catalytic performance.

The positive role of Eu doping was also evaluated by García-Moncada *et al.*[81] They prepared a series of Eu-doped ZrO_2 solid solutions in order to generate proton conductors, which were mixed with a $Pt/CeO_2/Al_2O_3$ catalyst. The novel idea of this work was to enhance the mobility of the OH/H species during the WGS reaction thanks to the contribution of rare earth-doped ZrO_2, a material with good ionic conductivity at low temperatures and high structural stability.

To this end, ZrO_2 doping was carried out with different Eu_2O_3 contents: 2, 5, 7, 9, 10, 12 and 15 mol%, named as ZrEu_*x*, where *x* refers to the dopant content. The evaluation of the crystal structure demonstrated that the samples with Eu_2O_3 loadings below 9% presented a cubic fluorite-type structure, whereas at higher loadings, segregation of the Eu_2O_3 phase or short-range defects appeared, indicating that 9 mol% is the maximum europium content allowed in the zirconia lattice. The most representative samples were evaluated in the WGS reaction and compared to the bare $Pt/CeO_2/Al_2O_3$ catalyst. The results showed that the addition of the ionic conductor improved the WGS catalytic activity in every case, as observed in Figure 5.6, although the samples with greater Eu contents presented the less improvement. It is also noteworthy that the differences in catalytic performance are observed between 270 and 290 °C, whereas from 300 °C, all of the conductor-containing samples presented similar behaviour.

With the aim of understanding the effects of the ionic conductors on WGS catalytic activity, ionic conductivity measurements under Ar, O_2/Ar and H_2O/Ar were performed. An increase in ionic conductivity with temperature was observed whatever the atmosphere, a typical behaviour observed for *p*-type

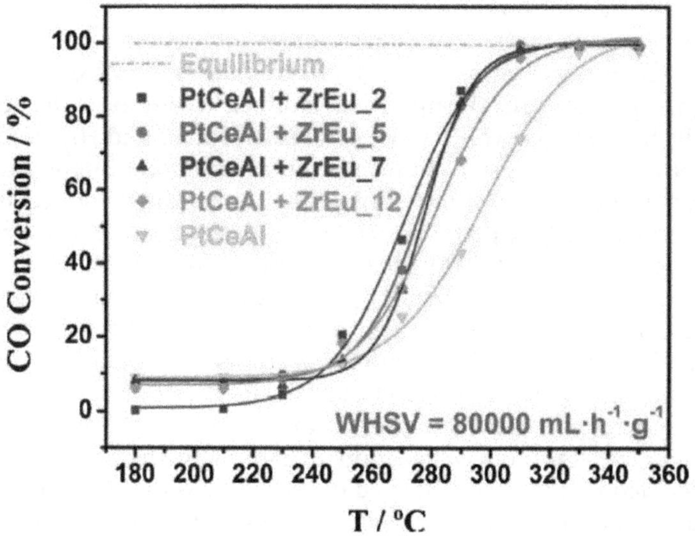

Figure 5.6 CO conversion of PtCeAl and PtCeAl + ZrEu_x (1 : 5) as a function of the temperature under 4.5% CO, 30% H_2O and N_2 as balance. Reproduced from Ref. 81 with permission from Elsevier, Copyright 2018.

semiconductors created in the presence of oxygen vacancies, which play a fundamental role during the WGS reaction. Furthermore, the presence of water in the atmosphere increased the conductivity at temperatures lower than 300 °C, which is in agreement with the improvement in the WGS catalytic activity in that temperature range. The highest conductivity value was that of the ZrEu_5 sample in all of the evaluated atmospheres. Although an increase of the dopant content should provoke an increase in the oxygen vacancy concentration, it has been reported that the association of the dopant with the oxygen vacancies would block the movement of the holes, explaining the conductivity decrease at higher Eu concentrations.

The aforementioned works determined the strategies to be performed in order to enhance the inherent properties of the well-known CeO_2 supports, but other research has focused on the application of new materials as supports for WGS Pt-based catalysts. For instance, phosphate-type supports for the WGS reaction have been reported. Miao *et al.*[82] developed several 1 wt% Pt/apatite catalysts. The authors synthesized hydroxyapatite (HAP) with different Ca : P ratios by varying the pH during synthesis (9.0, 9.5, 10.0 and 10.5), with the samples being named HAP-*x*, where *x* is the pH at which the synthesis was performed. They also increased the Ca content in the samples prepared at pH 9 (Ca-HAP-9) and 10 (Ca-HAP-10) and prepared a fluoride-substituted apatite (FAP).

The characterization results showed that the apatite's crystallinity and the Ca:P ratio increased with pH, with the HAP structure (Ca:P = 1.67) being obtained at pH 9.5. In every case, the maximum Ca:P ratio obtained during synthesis was 1.75, which was increased up to 2.00 by subsequent impregnation of the obtained apatites with $Ca(NO_3)_2$. As for the metal, well-dispersed Pt particles of about 1 nm were obtained.

The catalytic activity of the apatite-supported samples was compared with a reference 1 wt% Pt/CeO_2 catalyst, and the results demonstrated a superior WGS performance of the samples with greater Ca : P ratio (HAP-10 and HAP-10.5) with respect to the CeO_2-supported catalysts, in contrast to the behaviour showed by the samples with lower Ca:P ratios (HAP-9 and HAP-9.5), which presented lower WGS activity. These results pointed out the importance of the Ca:P ratio in the WGS reaction. On the other hand, the Pt/FAP catalyst presented a similar catalytic performance to the Pt/CeO_2 sample. The stability of the most active sample (Pt/HAP-10.5) was also studied. Although it suffered from deactivation during the first 24 h, the CO conversion was then stabilized, which was ascribed to a low degree of sintering of the Pt particles due to the structural characteristics of the apatite. The authors also evaluated the apparent activation energy value (E_a), and they obtained a value of 82 kJ mol^{-1} for the apatite-supported catalysts, which is in the upper limit of the previously reported energies for the WGS reaction. Regarding the reaction mechanism, they concluded that CO was activated on Pt, whereas the H_2O molecule was activated in the apatite structure, which adsorbs molecular water on the (001) surface and dissociates it on the (010) surface. Consequently, water dissociation seems to be the rate-limiting step in these solids, and the dissociation degree appears to be directly related to the Ca : P ratio in the apatite structure, thus being a fundamental factor for obtaining highly active catalysts for the WGS reaction.

The same authors later studied the addition of Sr^{2+} in the apatite structures, with the purpose of evaluating the cation and the Sr : P ratio influence on the WGS reaction.[83] As with the apatites, increasing the pH during synthesis give rise to an increase in the Sr : P ratio, but the incorporation of Sr led to a decrease in the specific surface area of the supports.

Regarding the catalytic tests, an increase in the Sr : P ratio gave rise to an increase in WGS catalytic performance, and the reaction rates were higher than when using the Pt/Ca apatite catalysts. Consequently, the specific surface area seemed not to be determining in this type of catalyst. In addition, the reducibility of the Pt species was enhanced when using apatite supports, which also contributed to an enhanced WGS catalytic performance.

Due to the suitability of phosphate supports for the WGS reaction, Navarro-Jaén *et al.*[84] studied $CePO_4$ as a support for Pt-based catalysts, based on the protonic conduction properties and exceptional water uptake capacity previously demonstrated by these solids. They studied the two crystalline structures of $CePO_4$: the hexagonal rhabdophane-type ($CePO_4 \cdot 0.53H_2O$) phase, named CeP400; and the monoclinic or monazite-type ($CePO_4$) phase, named CeP600. The characterization results showed that the atom arrangement and hydration degree were different in both supports. Whereas the rhabdophane-type phase presented a series of parallel channels where water molecules were accommodated, in the monazite-type phase, the channels vanished due to the dehydration of the $CePO_4$ structure.

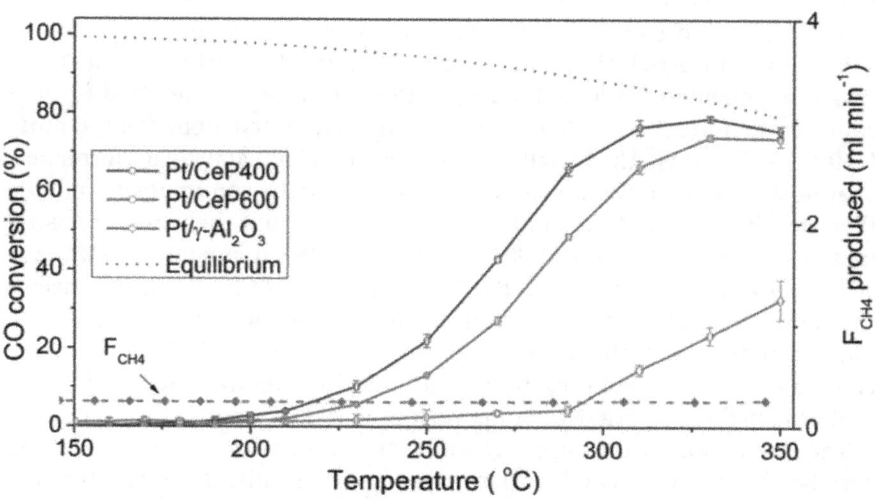

Figure 5.7 CO conversion as a function of the temperature under 7% CO, 9% CO_2, 30% H_2O, 50% H_2 and 4% N_2 reaction flow. Reproduced from Ref. 84 with permission from the Royal Society of Chemistry.

Catalytic activity measurements (Figure 5.7) demonstrated that both catalysts were highly active for the WGS reaction in the absence or presence of the reaction products. However, the catalytic activity of Pt/CeP400 was higher than that of Pt/CeP600, whatever the reaction conditions. The apparent activation energy values were similar to those reported for the dissociation of water on Pt (111) surfaces, suggesting that the activation of the water molecules was taking place not on the support, but on the platinum.

In order to explain the behaviour of the water molecules acting with the supports and their relationship with WGS catalytic activity, water adsorption experiments followed by Diffuse Reflectance Infrared Fourier Transform Spectroscopy (DRIFTS) were performed. The results exposed the greater water affinity of the CeP400 support, which is able to adsorb a greater quantity of water, even at 350 °C. Nevertheless, CeP600 also allowed for the inclusion of water in its structure, which was explained by the formation of a continuous layer or even multilayers of adsorbed water in the CeP600 structure. In every case, the water molecules seemed to be interacting with the phosphate groups of the $CePO_4$. No matter the support, no water dissociation was observed, indicating that $CePO_4$ was not responsible of water activation during the WGS reaction, thus acting as a water supplier, thereby increasing the number of available species to be dissociated on the Pt particles.

The successful application of the $CePO_4$ supports motivated the study of other phosphate-type supports for the WGS reaction. Thus, the same group[85] compared the catalytic performance of the $CePO_4$-supported Pt catalysts to Ca phosphates (CaP) and Ti phosphates (TiP). Additionally, the catalytic activity of these solids was compared to those of low- and medium-temperature Cu-based catalysts (HiFuel W220 and HiFuel W230, Alfa Aesar™). The characterization results demonstrated that the CaP support presented a HAP channel structure, similar to rhabdophane, whereas the TiP support resulted in a mixture of TiP_2O_7 and TiO_2 due to phosphorous loss during the preparation.

WGS catalytic activity followed the trend Pt/CaP > Pt/CeP400 > Pt/CeP600 > Pt/TiP, as shown in Figure 5.8. Whereas the Pt/CaP catalyst reached equilibrium conversion at 310 °C, both Pt/CeP catalysts required higher temperature (350 °C). By contrast, the Pt/TiP catalyst presented low catalytic performance, at around 10% CO conversion at 350 °C. Regarding the commercial catalysts, their activity was superior to the Pt/TiP catalyst across the whole temperature range, and to Pt/CaP and Pt/CeP catalysts at low temperatures (150–290 °C). However, Pt/CaP and Pt/CeP400 exceeded the catalytic performance of the commercial catalysts at a medium temperature range starting at 310 °C. Furthermore, none of the commercial catalysts reached the WGS equilibrium conversion in the studied reaction conditions, highlighting the suitability of well-designed phosphate-type supports for the WGS reaction.

The interaction of the supports with water molecules was evaluated and correlated to the WGS catalytic performance. The results demonstrated that supports containing structural channels, such as CaP and CeP400, favour the retention of water, although two different mechanisms of interaction were observed. In the case of CaP, the water molecule interactions take place

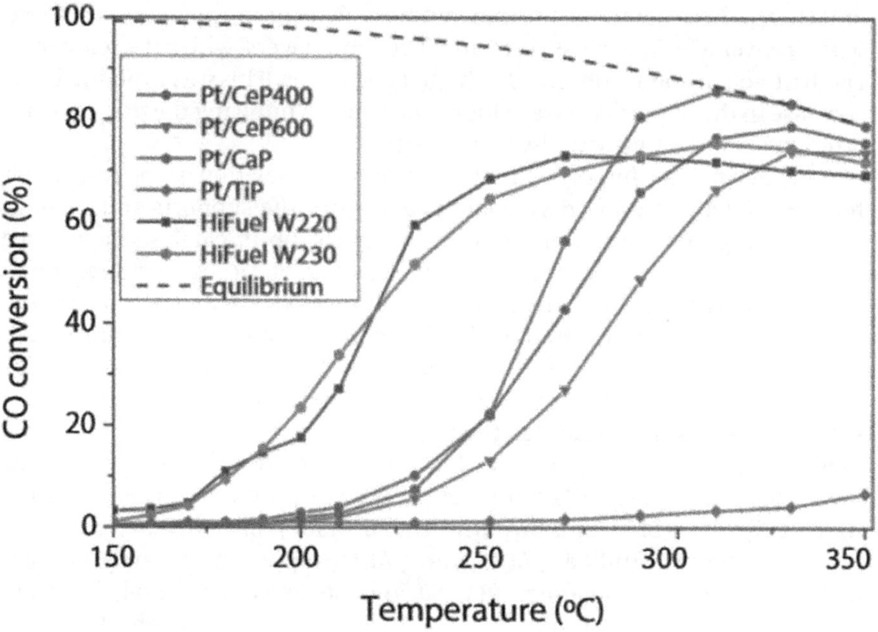

Figure 5.8 CO conversion as a function of the temperature of the prepared catalysts under a feed-stream containing 7% CO, 9% CO_2, 30% H_2O, 50% H_2 and 4% N_2. Reproduced from Ref. 85 with permission from Elsevier, Copyright 2019.

through hydrogen bonding to the phosphate groups and *via* strong electrostatic interactions with the Ca^{2+} ions. Furthermore, CaP presented the ability to dissociate water molecules due to the presence of oxygen vacancies in its structure. By contrast, CeP400 did not present the ability to dissociate water molecules, and interactions only took place by hydrogen bonding with the phosphate groups of the support without Ce^{3+} involvement, explaining the decrease in catalytic activity. Regarding CeP600, the lack of channels able to accommodate water molecules provoked a decrease in water availability during the reaction, thus decreasing the CO conversion. Finally, the loss of phosphorous in the TiP support limited the interaction of water with the phosphate groups, leading to poor water availability during the WGS reaction and thus limiting the catalytic activity of the Pt/TiP catalyst.

Continuing with the evaluation of new WGS catalytic supports, some authors focused on the implementation of carbide-supported Pt catalysts. For instance, Rodríguez *et al.*[86] studied for the first time MoC and TiC(001) as supports for Pt-based catalysts and compared them to a Pt/Mo_2C catalyst, evaluating the metal : carbon ratio and Pt coverage relationship with WGS catalytic activity. They found that both Pt/MoC and Pt/TiC(001) catalysts were active for the WGS reaction at temperatures below 200 °C, although the activity of the former was slightly superior. The activity of these catalytic systems was explained by an electronic perturbation of the Pt particles, since they receive electrons from the carbides, favouring the interaction of the metal with CO and H_2O. In both catalysts, the maximum catalytic performance was achieved with Pt coverages of between 0.15 and 0.20 monolayers, with a decrease in the catalytic activity being observed at higher coverages. This was explained by an increase in the Pt particle size at higher coverages, implying the presence of Pt atoms not interacting with the carbide surface.

Considering the importance of the water dissociation step, researchers have evaluated the water dissociation capacity of the supports and catalysts, with the superior formation of –OH groups in the Pt catalysts being clear compared to the bare supports. Nevertheless, the Pt/TiC catalyst presented a lower dissociation ability, which explains its lower WGS catalytic performance with respect to Pt/Mo_2C. On the other hand, the Pt/Mo_2C catalyst presented a higher water dissociation ability, which is problematic for the WGS reaction, since –OH and –O coming from water dissociation binds strongly on the Mo_2C surface. This result highlighted the importance of the metal : carbon ratio in these systems. Regarding stability, both systems were stable at temperatures lower than 200 °C, since a temperature increase promoted the reaction of water with the carbide, thus favouring the formation of an oxycarbide.

Other authors[87] studied η-Al_2O_3 and γ-Al_2O_3-supported Pt catalysts mixed with different Mo_2C loadings (33, 50 and 66 wt%). The catalytic activity results demonstrated an increase in WGS catalytic activity as the Mo_2C content increased, which was attributed to the dissociation capacity of the latter, and is in good agreement with the results obtained in the aforementioned work. However, the catalysts supported on η-Al_2O_3 presented higher catalytic activity, which was attributed to their higher pore volume compared to γ-Al_2O_3. Furthermore, it was proven that the use of different alumina phases with varied acidity gives rise to different Pt/Mo_2C interactions, with the synergistic effect of Pt and the carbide being superior in the catalysts supported on η-Al_2O_3.

Regarding the future use of WGS units for mobile applications, a few considerations have to be made. Firstly, it must be highlighted that WGS units are obligatory for preliminary purification of the reforming feed. The biggest problem to be solved is the necessity for them to run at high space velocities (small reactor volumes for mobile applications), as the reaction prefers high contact times. The use of combined high- and low-temperature shift units is then unviable, and thus medium shift catalysts are the best candidates for replacing the existing technology. Two metals are envisaged (Pt and Au), but only one (Pt) allows for activity at high space velocities,[80,81,84] although its temperature operation window does not permit purification in CO levels as low as Au does. On the other hand, Au better withstands the start-up/shut-down cycles, especially when some copper is added to the catalyst formulation.[69] However, in the battle between both metals, one very important fact plays in favour of platinum. Pt catalysts can be structured on monoliths (and hence offer a greater range of space velocities) without any problems, while Au catalysts suffer sintering and loss of activity.[88] To that fact we can also add the possibility of improving catalysts' work by acting directly on the reaction rate-limiting step (water activation) using different approaches, all directed at increasing the active sites for water dissociation and water transport to the active sites, as well as mobility and availability.[81,84,85,87]

Another point to be considered, which may also contribute to reducing the reactor volume, is the use of only single catalysts for the whole H_2 clean-up stream (*i.e.* WGS and CO-PrOx). This could be achieved with additional injection of oxygen, which also improves the performance of structured gold catalysts.[88] The recent developments in CO-PrOx catalysts will be summarized in the next section.

5.5.4 Recent Advances in CO-PrOx Catalysis

CO-PrOx with air is a cheap and efficient method to purify hydrogen-rich WGS outlets mainly due to the mild conditions required, including atmospheric pressure and working temperatures in the same range as PEMFCs.[89] The combination of WGS and CO-PrOx units is one of the most preferred technologies for portable and *on-board* applications and auxiliary power systems.[90] Despite the associated advantages, CO-PrOx is far from simple in practice due to the secondary processes that can occur based on the WGS feed-stream chemical composition. Besides H_2 and CO, significant amounts of H_2O and CO_2 are present, thus increasing the occurrence of side reactions such as the H_2 oxidation (eqn (5.10)), WGS (eqn (5.11)) and CO and CO_2 hydrogenation (methanation) reactions (eqn (5.12) and (5.13), respectively) that can affect the CO oxidation process (eqn (5.14)) and the final composition of the effluent.

$$H_2 + \frac{1}{2}O_2 \rightarrow H_2O \qquad \Delta H^\circ = -242\,\text{kJ mol}^{-1} \tag{5.10}$$

$$O + H_2O \rightleftharpoons CO_2 + H_2 \qquad \Delta H^\circ = -41.4\,\text{kJ mol}^{-1} \tag{5.11}$$

$$O + 3H_2 \rightleftharpoons CH_4 + H_2O \qquad \Delta H^\circ = -207\,\text{kJ mol}^{-1} \tag{5.12}$$

$$CO_2 + 4H_2 \rightleftharpoons CH_4 + 2H_2O \qquad \Delta H^\circ = -165.5\,\text{kJ mol}^{-1} \tag{5.13}$$

$$O + \frac{1}{2}O_2 \rightleftharpoons CO_2 \qquad \Delta H^\circ = -283\,\text{kJ mol}^{-1} \tag{5.14}$$

Amid these processes, H_2 oxidation and WGS and its reverse reaction (R-WGS) are the most problematic due to the low temperatures required for the CO-PrOx reaction (<300 °C),[91] a temperature range that matches the aforementioned processes.

The number of catalytic systems reported for the CO-PrOx reaction is extensive, although they can be classified into three main groups according to the metal present:[92] (1) the most studied group, copper–cerium oxide-based catalysts;[93,94] (2) gold-supported catalysts;[95,96] and (3) Pt-, Ru- and Rh-supported catalysts.[97,99] Regarding the first group of catalysts, they have been demonstrated to exhibit considerably higher selectivity to CO oxidation when compared to

Pt-based systems at similar conversion levels, which is very interesting from an economic viewpoint. Gold-supported catalysts have been demonstrated to be very active even at sub-ambient temperatures, which can be a potential advantage for mobile and on-board applications where low temperatures are preferred. In this section, the successful application of different systems in CO-PrOx will be described in detail, focusing in the two more interesting groups from a catalytic perspective: copper–cerium oxide catalysts and gold-supported catalysts. Copper–cerium oxide in combination with gold will be also discussed as an effective approach to developing efficient systems for CO-PrOx.

Recent studies have showcased the combination of copper and gold in several catalytic supports,[99–101] demonstrating a synergetic effect between gold and copper, mostly related to electronic interactions between both metals. The crystallographic similarities between them (*i.e.* the same crystal structure and similar lattice spacing) allows for their miscibility at all proportions, forming stable intermetallic alloys with 3 : 1, 1 : 1 and 1 : 3 molar Au : Cu ratios. The interaction between copper and gold determines the catalytic performance, which allows for precise control of various parameters. In fact, the enhancement of the catalytic activity is strongly related to the particle size, which is directly linked to the amount of Cu and Au and so to the Au : Cu molar ratio. The latter may be varied on the basis of the chosen synthesis procedure. Generally, alloying gold with copper gives rise to more stable metallic particles, preventing gold sintering during the oxidation process. This arises because copper is segregated on the surface, forming copper oxides, typically when large Au : Cu ratios (above 0.33) are used.[102]

Additionally, it is well known that the surface chemistry and the reactivity of the metal oxides are directly or indirectly related to the presence of defects, among which the presence of oxygen vacancies seems to be the most important.[103] Their concentration in a solid may be increased or reduced in several ways, lead to a negative charge imbalance when created[104] due to the pair of electrons that remains in the vacancy. Their natural occurrence in reducible supports, such as CeO_2, has been widely reported[105,106] and ascribed to the redox behaviour of the Ce^{4+}/Ce^{3+} cationic pair. In this sense, the creation of new oxygen vacancies in the ceria lattice is successfully achieved by means of transition metal doping, as is the case with Cu.[95,105,107] The formation of Ce–Cu solid solutions leads to a charge imbalance due to the different states of Ce/Cu, and this is compensated by the formation of oxygen vacancies. The role of these oxygen vacancies is far from negligible in catalysis, as well as in the synthesis of supported catalysts. In particular, the activation of oxygen species in the oxygen vacancies seems to be one of the most relevant contributions to the mechanism in oxidation reactions. Since oxygen activation is considered to be the rate-limiting step of the CO oxidation reaction,[108] the creation and presence of oxygen vacancies in different supports open the way to new oxygen activation routes, thus resulting in catalytic activity enhancement.

Concerning the synthesis of gold-supported catalysts, the most relevant feature is the direct interaction of gold species with oxygen vacancies during impregnation, which transforms the mentioned defects in preferential nucleation sites, thus increasing gold nanoparticle dispersion.[95,109] Experimental evidence of this was reported by Laguna *et al.*[102] on the basis of a Raman spectroscopic study of CeO_2, CuO_x/CeO_2 mixed oxide and $Au/CuO_x/CeO_2$. Three main signals have been observed, with the most intense one centred at 463 cm^{-1}, corresponding to the F_{2g} ceria band; the signal at 612 cm^{-1} is attributed to the presence of oxygen vacancies (O_v), and a small signal centred at 279 cm^{-1} is related to the presence of Ce^{3+}. The considerable increase in the oxygen vacancy signal observed for the CuO_x/CeO_2 sample in comparison to bare CeO_2 contrasts with the previously mentioned effect when doping with transition metal oxides. The preferential nucleation of gold nanoparticles in the oxygen vacancies is likewise evident when the Raman spectrum of AuCeCu is considered: the band attributed to the presence of oxygen vacancies disappears. Concerning the CO-PrOx reaction, both catalysts exhibit similar behaviour, with the CO conversion increasing with temperature, reaching a maximum value at about 98% and decreasing afterwards as the temperature increases over 140 °C. The maximum value of CO conversion was achieved first by CeCu (120 °C *vs.* 140 °C for the gold system). The observed differences in both reactions highlight the importance of the feed-stream chemical composition. The authors concluded that H_2 presence led to a decrease in CO conversion at higher temperatures due most probably to its oxidation and/or participation in the R-WGS. These reactions are promoted at higher temperatures, noticeably reducing the selectivity for CO oxidation.[110]

In line with the process intensification concept, several attempts have been made to find an optimal catalyst that is able to work well in the WGS and CO-PrOx reactions.[111] From an economic and practical point of view, it would be interesting to design a reactor in which both reactions are successfully carried out through the careful control of the temperature profile of the reactor as well as the reactant insertion zones, which would allow for the CO clean-up process to be performed in one reactor (Figure 5.9). Although the required experimental conditions (space velocity, temperature, equilibrium limitations) of both processes make this idea very challenging, the reduced costs along with the small volumes needed would entail wider applicability, especially in small portable devices.

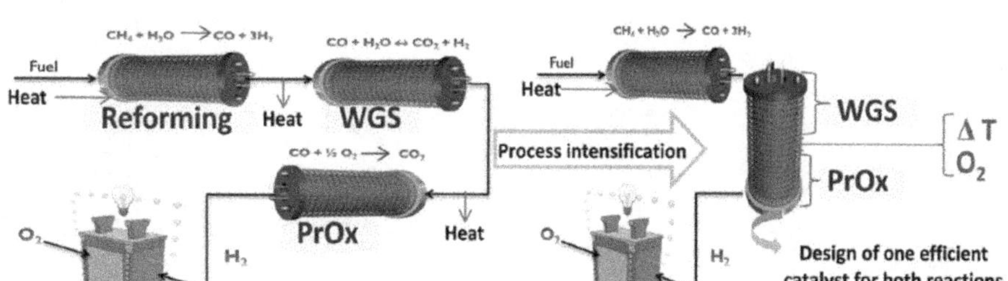

Figure 5.9 H_2 clean-up process *via* WGS–PrOx reactions: process intensification. Reproduced from Ref. 111 with permission from Elsevier, Copyright 2014.

The viability of CuO_x/CeO_2 mixed oxides and $Au/CuO_x/CeO_2$ systems has also been evaluated in the WGS reaction. Indeed, one of the main advantages of using CuCe for the WGS reaction is its ability to be used without further activation, in contrast to the traditional $Cu/Zn/Al_2O_3$ catalysts used in industrial applications.[112,113] Nevertheless, as the catalytic performance in both WGS and CO-PrOx reactions can be greatly improved with the inclusion of small amounts of gold in the catalysts' composition,[102,114] the use of gold-supported catalysts on reducible oxides is usually preferred. The main drawback associated with the use of gold is the cost increment of the final catalyst, which has led many authors to look for alternatives to reduce expenses. In an attempt to partially compensate for these costs, Reina *et al.*[115] have proposed reducing the amount of copper and ceria by dispersing them on a high-surface area carrier as alumina, thus increasing the surface to volume ratio of ceria particles and, as a consequence, the oxygen mobility.[116] In this study, the dispersion of both CuO_x/CeO_2 mixed oxides and $Au/CuO_x/CeO_2$ onto alumina carriers ($CuO_x/CeO_2/Al_2O_3$ and $Au/CuO_x/CeO_2/Al_2O_3$, respectively) was successfully achieved, and the systems were evaluated in both WGS and CO-PrOx processes. In order to optimize the catalyst, the authors have evaluated different Cu/Ce weight ratios, providing relevant correlations between the observed catalytic performances and the physicochemical properties of the designed materials.

Going through some intrinsic and general aspects of the abovementioned materials, the chemical composition of the catalysts using alumina represents a much lower amount of active phase (*i.e.* the CuCe solid solution), with this being demonstrated by the oxygen vacancy population being boosted when dispersed on the alumina carrier. Consequently, the surface/bulk dynamics (oxygen mobility) and the support surface/bulk reducibility change with respect to the non-dispersed systems. The observed catalytic behaviour in the WGS reaction over the proposed materials is presented in Figure 5.10.

Attending to the CuCe samples (Figure 5.10a), there are three very active samples in the shift process, namely CuCe-0.6, CuCe-0.8 and CuCe bulk, while

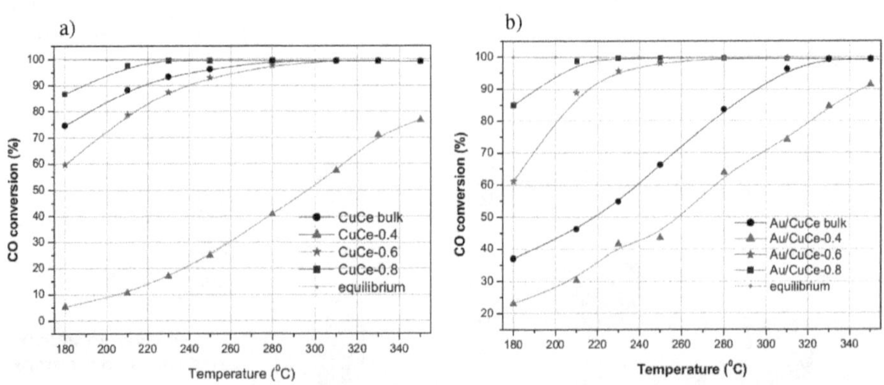

Figure 5.10 Catalytic behaviour in the WGS reaction using a model mixture (30% H_2O + 4.5% CO balanced in N_2). (a) CuCe supports, (b) Au catalysts. Reproduced from Ref. 115 with permission from Elsevier, Copyright 2016.

CuCe-0.4 shows medium to low levels of activity. The nomenclature refers to the constituent weight ratios; thus, for example, CuCe-0.6 possesses a CuO/CeO_2 weight ratio of 0.6, with the total amount of mixed oxide being less than 25 wt% in all cases (at least 75 wt% Al_2O_3 is present in each sample). CuCe bulk refers to the non-dispersed sample having 15 wt% CuO and 85 wt% CeO_2. The main reason behind the poor activity of the CuCe-0.4 solid might be the very low loading of active phases (around 10 wt% CuO + CeO_2). Within the active samples, CuCe-0.8 is the most active, exhibiting almost 90% CO conversion from the very beginning of the reaction, an outstanding result for a WGS catalyst at low temperature. The authors ascribe the excellent CuCe-0.8 catalytic performance to its superior concentration of oxygen vacancies that certainly play a pivotal role in the WGS reaction, as mentioned in Section 5.5.1. With water activation being the generally accepted rate-limiting step of the WGS reaction,[41,117] it is assumed that an enhancement in the water-splitting process will improve catalytic performance. In this regard, oxygen vacancies are viewed as the activation points for initiating water splitting, facilitating the reaction that afterwards proceeds at the Cu–CeO_2 interface.[118] The superior activity shown by the CuCe-0.8 sample in comparison with CuCe bulk demonstrates that advanced WGS catalysts with low active-phase loading can be obtained, which is advantageous from an economic and environmental point of view. The addition of gold (Figure 5.10b) variously affects the process; the activity of Au/CeCu bulk remarkably drops, while Al_2O_3-supported solids show activity improvement. The most active sample (Au/CuCe-0.8) reaches equilibrium conversion at 210 °C, demonstrating a gold promotion effect that is mostly based on enhancements in surface oxygen mobility.

It is clear that the dispersion of the classical CO-PrOx CuCe and Au/CuCe catalysts has positive aspects in terms of WGS catalytic performance, but could these systems also be more efficient in CO-PrOx? The results reported by Reina *et al.* are displayed in Figure 5.11. In the low-temperature window (130–200 °C), the CeCu bulk catalyst is clearly the best catalyst (Figure 5.11a), achieving practically complete conversion at all temperatures. The selectivity for CO_2 is unfortunately only complete (or almost so) at low temperatures (Figure 5.11b), as H_2 oxidation is promoted at higher temperatures. In fact, CO conversion sequentially drops when the temperature surpasses 200 °C, as the rate of hydrogen oxidation exceeds that of CO oxidation at higher temperatures.[119] Nevertheless, in the high-temperature range, CuCe-0.4 becomes the most efficient sample, with it also being superior to CuCe bulk in terms of selectivity at 200 °C. The addition of gold in this case does not produce much impact on activity (Figure 5.11c) and selectivity (Figure 5.11d), as only CuCe-0.4 is positively influenced, showing better activity in the low-temperature range. In general, gold benefits the alumina-based materials at low temperatures in both reactions, but is detrimental for the bulk configuration, which has been demonstrated to be the most efficient catalyst for CO-PrOx reactions under the studied conditions, while Au/CuCe-0.8 is the best for the WGS process.

With the aim of process intensification, the same authors have proposed several multicomponent catalysts, mostly based on gold nanoparticles,

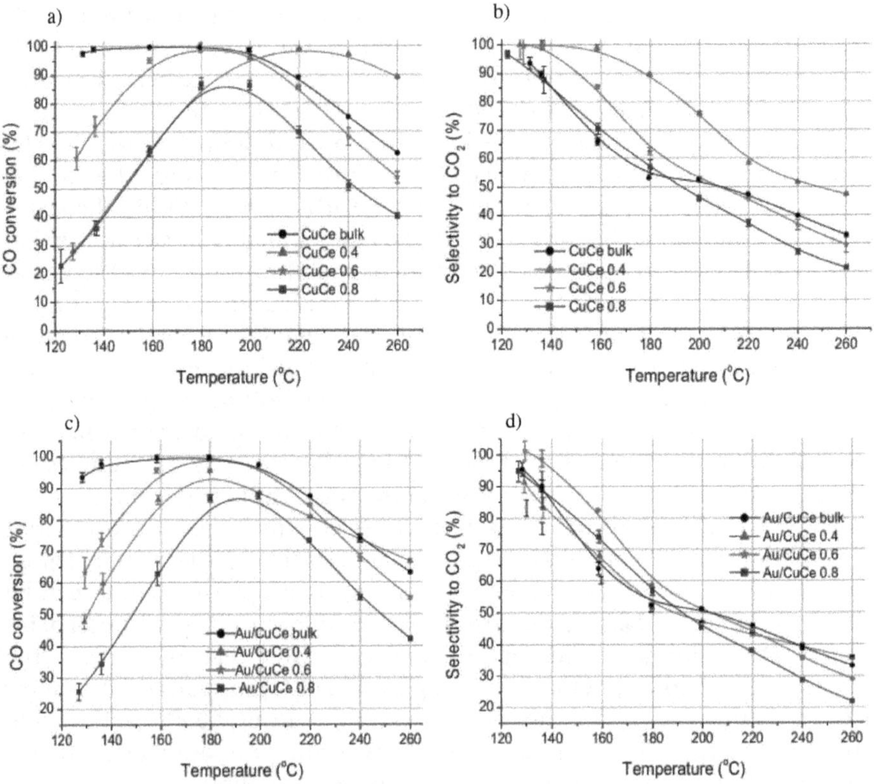

Figure 5.11 Catalytic screening in the CO-PrOx reaction using a hydrogen-rich mixture (1% CO + 1% O_2 + 50% H_2 balanced in N_2). (a & c) CO conversion obtained over CuCe- and Au/CuCe-based materials, respectively. (b & d) Selectivity towards CO_2 for CuCe- and Au/CuCe-based materials, respectively. Reproduced from Ref. 115 with permission from Elsevier, Copyright 2016.

supported on doped ceria and usually dispersed in alumina, which have been proven to be very efficient in the WGS reaction. One outstanding example comprises the iron-based family. Related to this, an interesting study dealing with several catalyst configurations has been reported elsewhere.[111] In this work, Au/Al_2O_3, Au/CeO_2/Al_2O_3, Au/FeO/Al_2O_3 and several Au/CeO_2-FeO_x/Al_2O_3 with different CeO_2-FeO_x weight ratios were evaluated in CO_2/H_2O-free CO-PrOx. The most efficient catalyst was a Fe-modified one in which the CeO_2-FeO_x ratio was 13 wt% for CeO_2 and 2 wt% for FeO_x. In order to understand this catalyst behaviour in a real post-WGS feed-stream, the authors evaluated the effects of CO_2, H_2O and CO_2/H_2O on the most efficient solid, namely Au/$CeFe_2$/Al (Figure 5.12). As deduced from the results, the addition of water led to decreases in activity (Figure 5.12a) and selectivity (Figure 5.12b), although the observed effect was not as detrimental as that with CO_2. The decreases in both activity and selectivity of the CO-PrOx catalysts due to the competitive

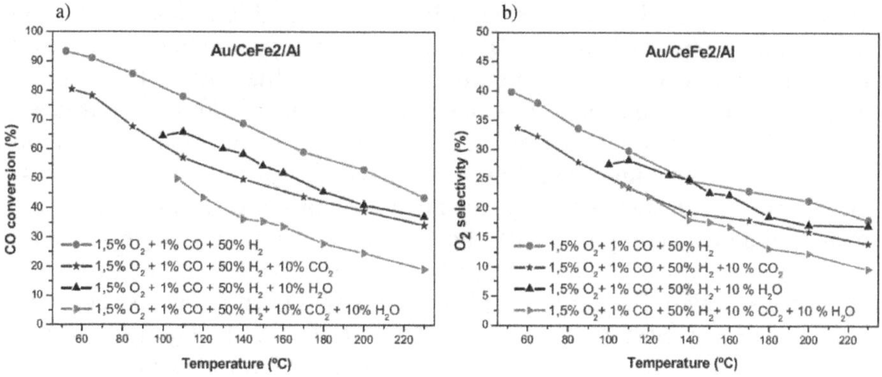

Figure 5.12 Influence of CO_2, H_2O and CO_2/H_2O on PrOx behaviour over Au/CeFe$_2$/Al catalyst. (a) CO conversion, (b) O_2 selectivity. Reproduced from Ref. 111 with permission from Elsevier, Copyright 2014.

adsorption of CO, CO_2 and H_2O on the catalyst surface have been broadly reported.[120–123] Additionally, the possible competition of water with oxygen for the electron-rich sites in the support (oxygen vacancies) where oxygen activation takes place should not be disregarded. This effect could also account for the decrease in the catalytic activity when water is included in the reaction mixture. As mentioned before, as a general trend, the activity decrease with temperature is related to a secondary non-desirable process (*i.e.* the oxidation of hydrogen). The ability of the sample to oxidize H_2 limits the selectivity of the process, but at the same time, the authors demonstrated by means of CO/H_2 pulses that the presence of H_2 increases the CO oxidation activity, most probably due to the formation of additional reactive peroxo-like intermediates. The authors conclude that although an excellent WGS catalyst such as Au/CeFe$_2$/Al could be also promising for the CO-PrOx reaction, it has to be taken into account that the conditions in both processes (temperature, space velocity and amount of CO, CO_2, H_2O and H_2) are very different. Indeed, a careful modulation of the reaction parameters must be performed if one reactor is to be used for both processes. For this particular case, and even after modulating the reaction parameters, Au/CeFe$_2$/Al is not able to satisfy the requirements of fuel cells, and further investigations are required in order to achieve this target with the same catalyst *via* WGS and CO-PrOx reactions.

The doping approach has been widely applied for the preparation of CO-PrOx catalysts. The viability of several systems based on Au/CeO_2-MO_x/Al_2O_3 (M = Co, Ni, Cr or Y) has been evaluated.[124,125] The conclusion of these studies is that the inclusion of small amounts of cobalt and yttrium oxides in the catalyst formula significantly improves the CO-PrOx activity in a real post-reforming stream (in the presence of H_2O and CO_2). Besides CeO_2, the rare earth oxides from the lanthanide series are seldom considered as supports for gold nanoparticles; they are usually considered efficient ceria promoters due to their similarities in electronegativity and ionic radius.[126] Regarding the similarity of all lanthanide metals within the lanthanide series and the good performance observed over CeO_2-based catalysts, promising results could be expected. In this sense, numerous systems with the formulation Au/MO_x/Al_2O_3 (M = Ce, La, Lu or Sm) have been further evaluated and compared with the Au/Al_2O_3 system.[127] The observed catalytic activity for total CO oxidation followed the order Au/CeO_2/Al_2O_3 > Au/Lu_2O_3/Al_2O_3 ≈ Au/La_2O_3/Al_2O_3 > Au/Sm_2O_3/Al_2O_3 > Au/Al_2O_3, showing that all of the lanthanide oxides promote CO oxidation activity compared with the Au/Al_2O_3 system. Au/CeO_2/Al_2O_3 shows the best activity, which is again attributed to its excellent redox properties, in contrast to all of the other samples that possess irreducible character. The similar catalytic performances of Au/Lu_2O_3/Al_2O_3, Au/La_2O_3/Al_2O_3 and Au/Sm_2O_3/Al_2O_3 were associated with their comparable physicochemical properties and with the presence of positively charged gold clusters in all of the samples, which is indirectly related to the existence of oxygen defects in all lanthanide sesquioxides. As for CO-PrOx, again the addition of a rare earth oxide can efficiently promote the activity of gold/alumina catalysts. The best system was alumina doped with lutetium oxide, a system combining high CO conversion and low hydrogen oxidation at high temperatures.

Regarding the future use of CO-PrOx units for mobile applications, we can conclude that is possible to purify the post-WGS effluent with this reaction; however, several more challenges must be confronted. It is highly desirable to use a single unique catalyst in WGS and CO-PrOx units, but a catalyst formulation that works at high space velocities in both reactions is necessary. We can imagine a combination of Pt and Cu as useful catalysts, taking into account that Pt performs well at high space velocities in WGS reactions and with Cu in CO-PrOx reactions. Besides both metals have demonstrated to be catalytically active in both the WGS and CO-PrOx processes. The input of sub-stoichiometric oxygen and the use of gold structured catalysts represent other possibilities, as the oxygen-assisted WGS reaction allows for good performance of the gold systems at the high space velocities needed for the CO-PrOx reaction as well. In any case, there is still room for innovation in catalyst configurations and reactor engineering in order to find the right catalyst formulations and to intensify the hydrogen purification reaction so as to facilitate the transition towards a green hydrogen economy.

References

1. I. Staffell, D. Scamman, A. Velazquez Abad, P. Balcombe, P. E. Dodds, P. Ekins, N. Shah and K. R. Ward, *Energy Environ. Sci.*, 2018, **12**, 463.
2. A. Sartbaeva, V. L. Kuznetsov, S. A. Wells and P. P. Edwards, *Energy Environ. Sci.*, 2008, **1**, 79.
3. R. Moradi and K. M. Groth, *Int. J. Hydrogen Energy*, 2019, **44**, 12254.
4. J. Ren, N. M. Musyoka, H. W. Langmi, M. Mathe and S. Liao, *Int. J. Hydrogen Energy*, 2016, **42**, 289.
5. Tracking Energy Integration, Hydrogen, https://www.iea.org/tcep/energyintegration/hydrogen/, accessed 9 March 2020.

6. R. Moliner, M. J. Lázaro and I. Suelves, *Int. J. Hydrogen Energy*, 2016, **41**, 19500.
7. M. S. Dresselhaus, G. W. Crabtree and M. V. Buchanan, *MRS Bull.*, 2005, **30**, 518.
8. J. O. Robles, S. De-León Almaraz and C. Azzaro-Pantel, Hydrogen as a Pillar of the Energy Transition, in *Hydrogen Supply Chains: Design, Deployment and Operation*, Academic Press, Cambridge, MA, 2018, ch. 1, p. 3.
9. https://www.iea.org/topics/hydrogen/production/
10. A. M. Abdalla, S. Hossain, O. B. Nisfindy, A. T. Azad, M. Dawood and A. K. Azad, *Energy Convers. Manag.*, 2018, **165**, 602.
11. C. Acar and I. Dincer, *Int. J. Hydrogen Energy*, 2014, **39**, 1.
12. Deutscher Wasserstoff und Brennstoffzellenverband (DWV), *Wasserstoff. Der neue Energieträger*, 5th edn, Hydrogeit Verlag, Oberkrämer, Berlin, 2015.
13. J. A. Turner. *Science*, 2004, **305**, 972.
14. C. Acar and I. Dincer, *Int. J. Hydrogen Energy*, 2014, **39**, 1.
15. P. Parthasarathy and K. SheebaNarayanan, *Renewable Energy*, 2014, **66**, 570–579.
16. V. Kirubakaran, V. Sivaramakrishnan, R. Nalini, T. Sekar, M. Premalatha and P. Subramanian, *Renewable Sustainable Energy Rev.*, 2009, **13**, 179–186.
17. P. P. Edwards, V. L. Kuznetsov and W. I. F. David, *Philos. Trans. R. Soc., A*, 2007, **365**, 1043–1056.
18. R. O'Hayre, S.-W. Cha, W. G. Colella and F. B. Prinz, *Fuel Cell Fundamentals*, John Wiley & Sons, Hoboken, NJ, 3rd edn, 2016.
19. G. Ercolino, M. A. Ashraf, V. Specchia and S. Specchia, *Appl. Energy*, 2015, **143**, 138–153.
20. S. Specchia, C. Francia and P. Spinelli, Polymer electrolyte membrane fuel cells, in *Electrochemical Technologies for Energy Storage and Conversion*, ed. R.-S. Liu, L. Zhang, X. Sun, H. Liu and J. Zhang, Wiley-VHC Verlag GmbH & Co, Weinheim, 2011, pp. 601–670.
21. M. Shao, Q. Chang, J. P. Dodelet and R. Chenitz, *Chem. Rev.*, 2016, **116**, 3594–3657.
22. O. Saritas, D. Meissner and A. Sokolov, *J. Knowl. Econ.*, 2019, **10**, 1183–1203.
23. S. Guo, S. Zhang and S. Sun, *Angew. Chem., Int. Ed.*, 2013, 52, 8526–8544.
24. P. C. Pei and H. C. Chen, *Appl. Energy*, 2014, **125**, 60–75.
25. L. P. Wang, W. S. Jia, X. F. Liu, J. Z. Li and M. M. Titirici, *J. Energy Chem.*, 2016, **25**, 566–570.
26. V. M. Bau, X. J. Bo and L. P. Guo, *J. Energy Chem.*, 2017, **26**, 63–71.
27. J. Liu, E. L. Li, M. B. Ruan, P. Song and W. L. Xu, *Catalysts*, 2015, **5**, 1167–1192.
28. G. Ren, L. Gao, C. Teng, Y. Li, H. Yang, J. Shui, X. Lu, Y. Zhu and L. Dai, *ACS Appl. Mater. Inter.*, 2018, **10**, 10778–10785.
29. W. Wei, X. Shi, P. Gao, S. Wang, W. Hu, X. Zhao, Y. Ni, X. Xu, Y. Xu, W. Yan, H. Ji and M. Cao, *Nano Energy*, 2018, **52**, 29–37.

30. B. A. Kakade, H. Wang, T. Tamaki, H. Ohashi and T. Yamaguchi, *RSC Adv.*, 2013, 3(26), 10487–10496.
31. K. Jayasayee, J. R. Van Veen, T. G. Manivasagam, S. Celebi, E. J. Hensen and F. A. de Brujin, *Appl. Catal., B*, 2012, **111**, 515–526.
32. Y. Zhou and D. J. Zhang, *Power Sources*, 2015, **278**, 396–403.
33. M. Ammam and E. B. Easton, *J. Power Sources*, 2013, **236**, 311–320.
34. X. Chen, C. Si, Y. Gao, J. Frenzel, J. Sun, G. Eggeler and Z. Zhang, *J. Power Sources*, 2015, **273**, 324–332.
35. H. Fei, J. Dong, Y. Feng, C.S. Allen, C. Wan, B. Volosskiy, M. Li, Z. Zhao, Y. Wang, H. Sun, P. An, W. Chen, Z. Guo, C. Lee, D. Chen, I. Shakir, M. Liu, T. Hu, Y. Li, A.I. Kirkland, X. Duan and Y. Huang, *Nat. Catal.*, 2018, **1**, 63–72.
36. L. F. Xue, Y. C. Li, X. F. Liu, Q. T. Liu, J. X. Shang, H. P. Duan, L. M. Dai and J. L. Shui, *Nat. Commun.*, 2018, **9**, 3819.
37. J. Z. Li, M. J. Chen, D. A. Cullen, S. Hwang, M. Y. Wang, B. Y. Li, K. X. Liu, S. Karakalos, M. Lucero, H. G. Zhang, C. Lei, H. Xu, G. E. Sterbinsky, Z. X. Feng, D. Su, K. L. More, G. F. Wang, Z. B. Wang and G. Wu, *Nat. Catal.*, 2018, **1**, 935–945.
38. C. Y. Chen and K. P. Huang, *Int. J. Hydrogen Energy*, 2017, **42**, 22250–22258.
39. P. Smirniotis and K. Gunugunuri, *Water Gas Shift Reaction: Research Developments and Applications*, Elesvier, Amsterdam, 2015.
40. T. L. LeValley, A. R. Richard and M. Fan, *Int. J. Hydrogen Energy*, 2014, **39**, 16983–17000.
41. C. Ratsanamy and J. P. Wagner, *Catal. Rev.*, 2009, **51**, 325–440.
42. D. B. Pal, R. Chand, S. N. Upadhyay and P. K. Mishra, *Renewable Sustainable Energy Rev.*, 2018, **93**, 549–565.
43. R. J. Byron Smith, M. Loganathan and M. S. Shantha, *Int. J. Chem. React. Eng.*, 2010, **8**, Review R4.
44. H. Kusar, S. Hocevar and J. Levec, *Appl. Catal., B*, 2006, **63**, 194–200.
45. P. Panagiotopoulou and D. I. Kondarides, *Catal. Today*, 2006, **112**, 49–52.
46. O. Thinon, F. Diehl, P. Avenier and Y. Schuurman, *Catal. Today*, 2008, **137**, 29–35.
47. M. González-Castaño, T. Ramirez Reina, S. Ivanova, M. A. Centeno and J. A. Odriozola, *J. Catal.*, 2014, **314**, 1–9.
48. J. P. Clay, J. P. Greeley, F. H. Ribeiro, W. N. Delgass and W. F. Schneider, *J. Catal.*, 2014, **320**, 106–117.
49. T. Ramírez Reina, M. González Castaño, S. Palma, S. Ivanova and J. A. Odriozola, *Twenty Years of Golden Future in the Water Gas Shift Reaction in Heterogeneous Gold Catalysts and Catalysis*, RSC Catalysis Series, Royal Society of Chemistry, Cambridge, 2014.
50. J. H. Carter and G. J. Hutchings, *Catalysts*, 2018, **8**, 627.
51. T. Tabakova, *Front. Chem.*, 2019, **7**, 517.
52. D. Andreeva, V. Idakiev, T. Tabakova and A. Andreev, *J. Catal.*, 1996, **158**, 354–355.
53. J. Hua, K. Wei, Q. Zheng and X. Lin, *Appl. Catal., A*, 2004, **259**, 121–130.

54. A. Venugopal and M. S. Scurrell, *Appl. Catal., A*, 2004, **258**, 241–249.
55. A. Trovarelli, *Catal. Rev.*, 1996, **38**, 439–520.
56. Q. Fu, A. Weber and M. Flytzani-Stephanopoulos, *Catal. Lett.*, 2001, **77**, 87–95.
57. D. Andreeva, V. Idakiev, T. Tabakova, L. Ilieva, P. Falaras, A. Bourlinos and A. Travlos, *Catal. Today*, 2002, **72**, 51–57.
58. T. Tabakova, F. Boccuzzi, M. Manzoli and D. Andreeva, *Appl. Catal., A*, 2003, **252**, 385–397.
59. T. V. W. Janssens, B. S. Clausen, B. Hvolbæk, H. Falsig, C. H. Christensen, T. Bligaard and J. K. Nørskov, *Top. Catal.*, 2007, **44**, 15–26.
60. W. Deng, A. I. Frenkel, R. Si and M. Flytzani-Stepahnopoulos, *J. Phys. Chem. C*, 2008, **112**, 12834–12840.
61. V. Idakiev, T. Tabakova, Z.-Y. Yuan and B.-L. Su, *Appl. Catal., A*, 2004, **270**, 135–141.
62. V. Idakiev, Z.-Y. Yuan, T. Tabakova and B.-L. Su, *Appl. Catal., A*, 2005, **281**, 149–155.
63. Z. Ma, H. Yin and S. Dai, *Catal. Lett.*, 2010, **136**, 83–91.
64. V. Idakiev, T. Tabakova, K. Tenchev, Z. Y. Yuan, T. Z. Ren and B. L. Su, *Catal. Today*, 2006, **128**, 223–229.
65. J. Li, J. Chen, W. Song, J. Liu and W. Shen, *Appl. Catal., A*, 2008, **334**, 321–329.
66. T. Ramírez Reina, S. Ivanova, J. J. Delgado, I. Ivanov, V. Idakiev, T. Tabakova, M. A. Centeno and J. A. Odriozola, *ChemCatChem*, 2014, **6**, 1401–1409.
67. T. Ramírez Reina, S. Ivanova, V. Idakiev, J. J. Delgado, I. Ivanov, T. Tabakova, M. A. Centeno and J. A. Odriozola, *Catal. Sci. Technol.*, 2013, **3**, 779–787.
68. T. R. Reina, S. Ivanova, M. A. Centeno and J. A. Odriozola, *Appl. Catal., B*, 2016, **187**, 98–107.
69. T. Tabakova, L. Ilieva, I. Ivanov, M. Manzoli, R. Zanella, P. Petrova and Z. Kaszkur, *J. Rare Earths*, 2019, **37**, 383–392.
70. A. A. Hakeem, Z. Zhao, F. Kapteijn and M. Makkee, *Catal. Today*, 2015, **244**, 19–28.
71. D. C. Grinter, J. B. Park, S. Agnoli, J. Evans, J. Hrbek, D. J. Stacchiola, S. D. Senanayake and J. A. Rodriguez, *Surf. Sci.*, 2016, **650**, 34–39.
72. Z.-J. Zhao, Z. Li, Y. Cui, H. Zhu, W. F. Schneider, W. Nicholas Delgass, F. Ribeiro and J. Greeley, *J. Catal.*, 2017, **345**, 157–169.
73. Y.-Y. Dong, W.-P. Liao and Z.-H. Suo, *Fuel Process. Technol.*, 2015, **137**, 164–169.
74. S. Liang and G. Veser, *Nanomater. Energy*, 2011, **1**, 117–135.
75. Y. Zhai, D. Pierre, R. Si, W. Deng, P. Ferrin, A. U. Nilekar, G. Peng, J. A. Herron, D. C. Bell, H. Saltsburg, M. Mavrikakis and M. Flytzani-Stephanopoulos, *Science*, 2010, **329**, 1633–1636.
76. H. Valdés, L. M. Molina and J. A. Alonso, *Appl. Surf. Sci.*, 2019, **487**, 244–252
77. P. Panagiotopoulou, J. Papavasiliou, G. Avgouropoulos, T. Ioannides and D. I. Kondarides, *Chem. Eng. J.*, 2007, **134**, 16–22.
78. W. Ruettinger, X. Liu and R. J. Farrauto, *Appl. Catal., B*, 2006, **65**, 135–141.

79. M. González-Castaño, S. Ivanova, T. Ioannides, M. A. Centeno and J. A. Odriozola, *Catal. Sci. Technol.*, 2017, 7, 1556.
80. M. González-Castaño, E. Le Saché, S. Ivanova, F. Romero-Sarria, M. A. Centeno and J. A. Odriozola, *Appl. Catal., B*, 2018, **222**, 124.
81. N. García-Moncada, L. F. Bobadilla, R. Poyato, C. López-Cartes, F. Romero-Sarria, M. A. Centeno and J. A. Odriozola, *Appl. Catal., B*, 2018, **231**, 343.
82. D. Miao, G. Cavusoglu, H. Lichtenberg, J. Yu, H. Xu, J.-D. Grunwaldt and A. Goldbach, *Appl. Catal., B*, 2017, **202**, 587.
83. D. Miao, G. Cavusoglu, H. Lichtenberg, J. Yu, H. Xu, J.-D. Grunwaldt and A. Goldbach, *Appl. Catal., B*, 2017, **202**, 587.
84. S. Navarro-Jaén, M. A. Centeno, O. H. Laguna and J. A. Odriozola, *J. Mater. Chem. A*, 2018, **6**, 17001.
85. S. Navarro-Jaén, F. Romero-Sarria, M. A. Centeno, O. H. Laguna and J. A. Odriozola, *Appl. Catal., B*, 2019, **244**, 853.
86. J. A. Rodríguez, P. J. Ramírez and R. A. Gutiérrez, *Catal. Today*, 2017, **289**, 47.
87. A. I. Osman, J. K. Abu-Dahrieh, N. Cherkasov, J. Fernandez-Garcia, D. Walker, R. I. Walton, D. W. Rooney and E. Rebrov, *Mol. Catal.*, 2018, **455**, 38.
88. M. González-Castaño, T. R. Reina, S. Ivanova, L. M. Martínez Tejada, M. A. Centeno and J. A. Odriozola, *Appl. Catal., B*, 2016, **185**, 337–343.
89. F. Mario, G. Baronetti, M. Laborde, N. Bion, A. Le Valant, F. Epron *et al.*, *Int. J. Hydrogen Energy*, 2008, **33**, 1345–1353.
90. J. D. Holladay, J. Hu, D. L. King and Y. Wang, *Catal. Today*, 2009, **139**, 244–260.
91. O. Korotkikh and R. Farrauto. *Catal. Today*, 2000, **62**, 249–254.
92. B. M. Abu-Zied and S. A. Soliman, *Thermochim. Acta*, 2008, **470**, 91–97.
93. G. Avgouropoulos, T. Ioannides, H. K. Matralis, J. Batista and S. Hocevar, *Catal. Lett.*, 2001, **73**, 33–40.
94. D. Gamarra, A. Horns, Z. Koppny, Z. Schay, G. Munuera, J. Soria *et al.* *J. Power Sources*, 2007, **169**, 110–116.
95. O. H. Laguna, F. Romero-Sarria, M. A. Centeno and J. A. Odriozola, *J. Catal.*, 2010, **276**, 360.
96. T. Tabakova, G. Avgouropoulos, J. Papavasiliou, M. Manzoli, F. Boccuzzi, K. Tenchev *et al.*, *Appl. Catal., B Environ.*, 2011, **101**, 256–265.
97. L. Chen, D. Ma, Z. Zhang, Y. Guo, D. Ye and B. Huang, *Catal Lett.*, 2012, **142**, 1–9.
98. A. Wörner, C. Friedrich and R. Tamme, *Appl. Catal., A*, 2003, **245**, 1–14.
99. X. Liu, A. Wang, L. Li, T. Zhang, C.-Y. Mou and J.-F. Lee, *J. Catal.*, 2011, **278**, 288–289.
100. J. Llorca, M. Domínguez, C. Ledesma, R. J. Chimentão, F. Medina, J. Sueiras *et al.*, *J. Catal.*, 2008, **258**, 187–198.
101. T. S. Mozer, D. A. Dziuba, C. T. P. Vieira and F. B. Passos, *J. Power Sources*, 2009, **187**, 209–215.
102. O. H. Laguna, W. Y. Hernández, G. Arzamendi, L. M. Gandía, M. A. Centeno and J. A. Odriozola, *Fuel*, 2014, **118**, 176–185.

103. G. Pacchioni, *Solid State Sci.*, 2000, **2**, 161.
104. S. Laursen and S. Linic, *Phys. Rev. Lett.*, 2006, **97**, 026101.
105. M. Nolan, V. Soto and H. Metiu, *Surf. Sci.*, 2008, **602**, 2734.
106. O. H. Laguna, M. I. Domínguez, F. Romero-Sarria, J. A. Odriozola and M. A. Centeno, Role of oxygen vacancies in gold oxidation catalysis, in *Heterogeneous Gold Catalysts and Catalysis*, ed. M. Zhen and D. Sheng, Royal Society of Chemistry, Cambridge, ch. 13, 2014.
107. M. Boronat and A. Corma, *Catal. Today*, 2011, **169**, 52.
108. G. C. Bond and D. T. Thompson, *Gold Bull.*, 2000, **33**, 41.
109. W. Y. Hernández, F. Romero-Sarria, M. A. Centeno and J. A. Odriozola, *J. Phys. Chem. C*, 2010, **114**, 10857–10865.
110. G. Arzamendi, I. Uriz, P. M. Dieguez, O. H. Laguna, W. Y. Hernández, A. Álvarez *et al.*, *Chem. Eng. J.*, 2011, **167**, 588–596.
111. T. R. Reina, E. Papadopoulou, S. Palma, S. Ivanova, M. A. Centeno, T. Ioannides and J. A. Odriozola, *Appl. Catal., B*, 2014, **150–151**, 554–563.
112. M. Boaro, M. Vicario, J. Llorca, C. Leitenburg, G. Dolcetti and A. Trovarelli, *Appl. Catal., B*, 2009, **88**, 272–282.
113. I. D. González, R. M. Navarro, M. C. Álvarez-Galván, F. Rosa and J. L. G. Fierro, *Catal. Commun.*, 2008, **9**, 1759–1765.
114. G. Avgouropoulos, M. Manzoli, F. Boccuzzi, T. Tabakova, J. Papavasiliou, T. Ioannides and V. Idakiev, *J. Catal.*, 2008, **256**, 237–247.
115. T. R. Reina, S. Ivanova, O. H. Laguna, M. A. Centeno and J. A. Odriozola, *Appl. Catal., B*, 2016, **197**, 62–72.
116. T. R. Reina, S. Ivanova, M. A. Centeno and J. A. Odriozola, *Catal. Today*, 2015, **253**, 149–154.
117. J. A. Rodriguez, *Catal. Today*, 2011, **160**, 3–10.
118. J. A. Rodriguez, J. Graciani, J. Evans, J. B. Park, F. Yang, D. Stacchiola, S.Senanayake, S. Ma, M. Pérez, P. Lui, J. Fdez Sanz and J. Hrbek, *Angew. Chem., Int. Ed.*, 2009, **48**, 8047–8050.
119. D. H. Lee and D. H. Kim, *Catal. Today*, 2008, **132**, 109–116.
120. G. Avgouropoulos and T. Ioannides, *Appl. Catal., A*, 2003, **244**, 155–167.
121. P. Snytnikov, V. Sobyanin, V. Belyaev, P. Tsyrulnikov, N. Shitova and D. Shlyapin, *Appl. Catal., A*, 2003, **239**, 149–156.
122. A. Luengnaruemitchai, S. Osuwan and E. Gulari, *Int. J. Hydrogen Energy*, 2004, **29**, 429–435.
123. A. Luengnaruemitchai, D. T. Kim Thoa, S. Osuwan and E. Gulari, *Int. J. Hydrogen Energy*, 2005, **30**, 981–987.
124. T. R. Reina, C. Megías-Sayago, A. Pérez Florez, S. Ivanova, M. A. Centeno and J. A. Odriozola, *J. Catal.*, 2015, **326**, 161–171.
125. T. R. Reina, S. Ivanova, M. A. Centeno and J. A. Odriozola, *Int. J. Hydrogen Energy*, 2015, **40**, 1782–1788.
126. W. Y. Hernández, O. H. Laguna, M. A. Centeno and J. A. Odriozola, *J. Solid State Chem.*, 2011, **184**, 3014.
127. T. R. Reina, A. Álvarez, S. Palma del Valle, S. Ivanova, F. Romero-Sarria, M. A. Centeno and J. A. Odriozola, *Adv. Chem. Lett.*, 2013, **1**, 237–246.

CHAPTER 6

Application of Heterogeneous Catalysts for the Conversion of Biomass-derived Feedstocks into Fuel Components and Eco-additives

P. M. YELETSKY,[a*] R. G. KUKUSHKIN,[a,b] M. V. ALEKSEEVA[a,b] AND A. A. SMIRNOV[a,b]

[a]Boreskov Institute of Catalysis SB RAS, Lavrentiev Ave. 5, Novosibirsk, 630090, Russian Federation
[b]Novosibirsk State University, Pirogova Str. 1, Novosibirsk, 630090, Russian Federation
*Corresponding contributor. Email: yeletsky@catalysis.ru

6.1 Introduction

Due to the continuing exhaustion of conventional petroleum feedstocks and to CO_2 emissions-induced global warming, looking for sustainable and renewable energy sources for the transport sector is undoubtedly of great importance. The volatility of hydrocarbon fossil fuel prices requires the development of programmes to ensure energy security that consider all existing fuel resources, including those of plant origin. The possibility of converting biomass into

Heterogeneous Catalysis for Energy Applications
Energy and Environment Series No. 27
Edited by Tomas R. Reina and Jose A. Odriozola

Published by the Royal Society of Chemistry, www.rsc.org

alternative fuels/fuel components, fuel additives and other useful products is of great interest due to the renewability and environmentally friendly nature of bio-feedstocks. Today, the use of biomass as an energy source is mainly restricted to either direct combustion or conversion into liquid products. However, liquid products from biomass transformation/extraction demand significant upgrading in order to obtain motor fuel components, fuel additives or chemicals from them – platforms for the synthesis of other biomass-derived products, including fine chemicals. None of these processes are possible without the use of heterogeneous catalysts. In this chapter, we present a brief state-of-the-art literature review on the use of catalysts for biomass conversion in three hot topics: processing of fast pyrolysis bio-oil into upgraded bio-oil; conversion of lipid-based feedstocks into isomerized paraffins – high-quality motor fuel components; and the production of fuel additives from components of both lignocellulosic biomass and vegetable lipids. The main aspects of the use of heterogeneous catalysts in these processes are considered.

6.2 Catalytic Hydrotreatment of Pyrolysis Bio-oil into Valuable Products

Lignocellulosic biomass, including residues from wood processing and harvesting, peat and agricultural residues, is considered a very attractive bioresource for fuel production. Wood components are not only valuable chemical raw materials for producing the majority of petrochemical products, but also can be used for producing unique compounds (*e.g.* biologically active substances).[1,2] Besides this, the production of synthetic fuels from woody biomass is a very promising area. Scientific research and industrial applications of thermochemical and microbial conversion of biomass into alcohols, liquid and gaseous fuels have been implemented in a number of countries.[3] However, many current processes used for the chemical conversion of wood raw materials are inferior to existing oil refinery and petrochemistry technologies in their performance and effectiveness and require large-sized equipment and high energy costs. To overcome these disadvantages, the use of catalysts that increase the rate of chemical reactions and the yield of target products and reduce harmful emissions is essential. The benefit of using catalysts is that all of these effects can be achieved without additional energy costs.

One of the most promising technologies of wood processing is pyrolysis, which allows products such as fuel gas, charcoal and liquid products to be obtained, as well as raw materials for the chemical industry.[4–6] The solid ash residue can be recycled into an environmentally safe pelletized product that can be utilized in road construction and the production of building materials. Modern technologies of biomass pyrolysis can be classified according to heating rate characteristics (fast, slow pyrolysis) and pyrolysis conditions (vacuum, hydropyrolysis, methanopyrolysis, catalytic pyrolysis).[7–11] The distinctive features of fast pyrolysis are the fast heating rate of the feed (up to 1000 °C s^{-1}) and rapid cooling of the generated vapours (residence time 500 ms).[12] Among the fast pyrolysis technologies, the following are to be specified depending on their technical peculiarities: fluidized bed pyrolysis, ablative pyrolysis, pyrolysis in a circulating fluidized bed, stream pyrolysis, *etc.*[13] The majority of pyrolysis technologies have been described in detail in a number of reviews.[7,14–17] Is important to notice that the thermal destruction of the biopolymers of plant biomass during pyrolysis is not complete, and the yield of liquid products can reach up to 70–80%.[4] The liquid fast pyrolysis product, also called bio-oil, is a multi-component mixture of different compounds derived from the depolymerization and fragmentation of cellulose, hemicelluloses and lignin – the main building blocks of woody biomass. Bio-oil has a higher volumetric energy density than wood (21 MJ L^{-1} for bio-oil *vs.* 13.4 MJ L^{-1} for wood pellets) and is more suitable for transportation and processing.[18] In particular, co-processing of pyrolysis bio-oil together with oil fractions in standard oil refinery units is of great interest and is practically relevant. However, this requires pre-treatment of pyrolysis bio-oil to get rid of its undesirable properties: high viscosity and acidity, low thermal stability and tendency to repolymerize. These properties are primarily caused by the high oxygen content in bio-oil. Hence, to improve the quality of pyrolysis liquids, the development of technologies based on deoxygenation reactions is mandatory,[19] and this is best carried out in the presence of hydrogen and a catalyst in a process commonly known as catalytic hydrotreatment or hydrodeoxygenation (HDO).[20] This is a reasonably versatile approach to improve the quality of bio-oil. In general, this process includes a number of chemical transformations that demand hydrogen and/or catalytic functions, such as hydrocracking, decarboxylation, decarbonylation, hydrogenation and hydrodeoxygenation. The overall "reaction" of pyrolysis oil hydrotreatment can be schematically written as follows:

$$\text{Pyrolysis oil} + H_2 \rightarrow \text{Hydrocarbons} + H_2O \tag{6.1}$$

The hydrotreatment of pyrolysis bio-oil is considered to be quite similar to oil refinery industrial processes such as hydrodesulfurization (HDS) and hydrodenitrogenation (HDN). By analogy with petroleum refining, HDO of bio-oil needs high hydrogen pressures to improve its dissolution in the liquid medium. The similarities between HDS/HDN and HDO, as well as high level of oil-refining technologies, determine the initial choice of catalysts for bio-oil hydrotreatment. Thus, the logical step has been to start from commercial sulfided catalysts that are well established in conventional oil refineries. Alumina-supported Ni-Mo and Co-Mo systems became the starting point and are now probably the most widespread in this area.[21–24]

In general, the catalyst has a special place in the catalytic hydrotreatment of pyrolysis oil, and its choice is really decisive. In this chapter, an overview of the historical development of catalysts will be given, with a special focus on the main catalyst generations. As already mentioned, the specificity of bio-oil as a raw material has motivated a number of companies and research institutes to begin the development of new catalysts or to improve traditional ones for the hydrotreatment of biomass pyrolysis liquid. Thus, from a historical viewpoint, we will consider sulfided catalyst applications for bio-oil deoxygenation, consistently turning our attention to other catalytic systems, including catalysts based on noble metals and other non-sulfided systems. We will try to shed some light on catalytic transformations as a function of the catalyst's nature, as well as to consider very important issues related to the activity and stability of catalysts in pyrolysis bio-oil. Since the choice of the catalyst and its performance very much depend on the feedstock, we should first consider the main aspects regarding the chemical composition and properties of typical pyrolysis oil.

6.2.1 Composition and Properties of Pyrolysis Liquids from Lignocellulosic Biomass

The elemental composition of the liquid product (bio-oil) produced by the fast pyrolysis of biomass slightly differs from that of the original biomass and normally is represented by C (55–60 wt%), H (5–7 wt%), O (35–40 wt%) and N (<0.02 wt%). Bio-oil is a multi-component mixture composed of more than 400 different compounds:[25] aliphatic and aromatic alcohols, ethers, ketones, carboxylic acids, sugars, *etc.*, derived from the main biopolymers of plant biomass (cellulose, hemicelluloses and lignin) *via* thermal destruction.

The composition of pyrolysis oil significantly differs from traditional fuels and does not satisfy the requirements for liquid fuels. It has a lower calorific value than diesel fuel and high viscosity and acidity. In addition, it is unstable upon thermal treatment and even when stored, which is called ageing.[12,17,26,27] The low stability of bio-oil results in the formation of higher-molecular-mass compounds and solid carbon residues due to repolymerization reactions; in addition, phase separation might take place due to emulsion destruction. Unlike fossil oil, distillation to fractionate (even under vacuum conditions) appears ineffective in the case of pyrolysis bio-oil, since it also provokes undesirable repolymerization. The high acidity and, as a result, aggressiveness of bio-oil make it unsuitable for use in internal combustion engines. All of these properties are due to the high oxygen content, which is a constituent of almost all of the organic components of bio-oil. In addition, pyrolysis liquid normally contains a lot of water – up to a quarter of its mass – which positively affects fluidity, but negatively affects heating value. It is also significant that the liquid produced by biomass pyrolysis has some advantages over mineral oil, namely a low content of nitrogen and sulfur (<0.01%).[28]

For a more detailed identification of bio-oil components, the extraction scheme is commonly used.[29–32] The following classes of compounds are distinguished on the basis of their solubility: water-soluble compounds – organic acids, alcohols; diethyl ether-soluble compounds – aldehydes, ketones, lignin monomers (O-aromatic species, *etc.*); compounds insoluble in diethyl ether – sugars, hydroxylic acids; *n*-hexane-soluble compounds – fatty acids, extractives; dichloromethane-soluble compounds – low-molecular-weight lignin; and compounds insoluble in dichloromethane – high-molecular-weight lignin, solid residues. It is particularly important that almost all constituents in pyrolysis oil have O-containing functionalities in their structures. Valuable tools that help to characterize the chemical structure of bio-oil are gas chromatography (GC), nuclear magnetic resonance (NMR) and Fourier-transform infrared (FTIR) spectroscopy.[28,30,33–37] However, these methods have limitations and do not allow for the detailed study of the composition of the analysed mixture. GC analysis allows for the identification of only up to 30% of bio-oil components that have sufficient volatility. The rest cannot be identified due to their low volatility, high molecular weight, polarity and high thermal reactivity. NMR and FTIR methods give only a qualitative picture of the presence of certain classes or groups in the composition of the sample and allow for the assessment of their quantity change.

The chemical composition hinders the use of pyrolysis oil as a fuel, but the use of catalysts offers the possibility of upgrading it by reducing the amount of oxygen. Since bio-oil is quite specific in terms of its processing, these catalysts should fulfil some special requirements. The most essential are: (1) sufficiently high activity in the removal of oxygen from a wide range of organic compounds possessing different reactivities; (2) resistance to water at elevated temperatures in an acidic medium; and (3) stability to coking or regeneration *via* burning off carbon deposits.

6.2.2 Sulfided Catalysts: Historical Beginning of Pyrolysis Oil Hydrotreatment

The first generation of catalysts used in bio-oil HDO and its model mixtures was conventional hydrotreating sulfided Ni(Co)Mo systems supported on alumina. This seemed to be quite logical, since commercial hydrotreating catalysts were able to remove heteroatomic species and perform hydrogenation in the case of crude oil fractions, and such chemical transformations were of great importance for bio-oil as well.

The nature of sulfided catalysts determines their specific activity in various hydrogenolysis processes. The activities of sulfided catalysts for oil hydrotreating are given for such processes as HDS, HDN, HDO and hydrogenation. Regarding chemical transformations and their relation to the nature of the sulfided catalyst during HDO of bio-oil, the active sites of conventional sulfided catalysts are sulfur vacancies at the edges of Mo sulfide additionally promoted by Co or Ni atoms.[38] The promoting effect consists in donating electrons to Mo and weakening of the Mo–S bonds, thus giving rise to more active S vacancies. Sulfur vacancies or coordinatively unsaturated active sites promote adsorption of O-containing substrates *via* heteroatomic functionalities. Along with vacancies, S–H groups on the edge or plane sites of the MoS phase can generate activated hydrogen, thus saturating the reacting adsorbed molecule and regenerating sulfur vacancies. In addition, the support material (*e.g.* γ-Al_2O_3) also assists HDO reactions by providing additional "landing" sites for reacting molecules. Obviously, questions arise regarding the stability of these active sites under the influence of different factors (H_2O, coking) during the target reaction.

Indeed, contrary to first impressions, there are issues related to catalyst stability that must be taken into consideration. First, the deactivation of sulfided catalysts for organic HDO can be attributed to sulfur substitution by oxygen in the sulfide structure since the sulfur content in crude bio-oil is low. This results in sulfur losses and conversion of the sulfide phase to an oxide one. The presence of water in the pyrolysis oil favours this process. The addition of sulfiding agents (thiophen, H_2S, CS_2) does not solve the low efficiency problem of this type of catalyst. Moreover, this approach in bio-oil HDO results in the contamination of products with sulfur-containing compounds. In addition, considering the negative effect of water on sulfided catalysts, their use in the HDO of bio-oil with high contents of water becomes highly questionable. Along with active phase deactivation in the presence of water, the irreversible transformation of the γ-Al_2O_3 support into hydroxide has been reported, as well as there being negative effects of acidic species in bio-oil.[39,40] The formation of coke and its (coke) precursors – larger molecules adsorbing onto active centres of the catalyst – is also one of the main reasons for the deactivation of sulfided catalysts. Coke formation is not just the consequence of the low stability of bio-oil constituents, but is also promoted by the acidic surface sites of the γ-Al_2O_3 support. This points to the necessity of non-sulfided catalyst development for bio-oil HDO.

To reduce the negative effect of coking in HDO catalysts, a two-stage scheme of bio-oil processing based on the different reactivity of the pyrolysis oil components was proposed.[41] In the first stage, the HDO process was performed at relatively low temperatures (240–300 °C) with partial separation of target products, and in the second stage – HDO at 320–380 °C – conversion of hydroxy-aromatic compounds was attained. This approach was proposed in order to reduce the effect of the low thermal stability of bio-oil, which can lead to a repolymerization of organic substrates with two or more oxygen-containing functional groups at temperatures of 350–400 °C. Thus, for example, it was shown that the carbonyl group is converted at lower temperatures than methoxy and hydroxyl groups and furane derivatives. A detailed analysis of the reactivities of different oxygenated substrates and of the features of the HDO mechanism for bio-oil model compounds in the presence of sulfided catalysts was provided by Furimsky.[42] Despite the reasonably high activity of sulfided catalysts in the HDO of O-containing constituents, a number of reasons, specified above, prompted further research activities and did not allow the scientific community to consider NiMoS/CoMoS systems as suitable for pyrolysis oil HDO.

6.2.3 Noble Metal-based Catalysts

At around the same time, research activities were directed towards the use of noble metal-based catalysts in the HDO of pyrolysis oils.[43–47] Such catalysts have some advantages over industrial sulfided catalysts because they do not require any additional treatment (such as sulfurizing) either before or during the reaction, with no contamination of products by sulfur. Supported on different materials, such as activated carbon, alumina, zeolite-β, ZSM-5, MCM-41, *etc.*, they exhibited high levels of activity in both HDO and hydrogenation reactions. In general, catalysts based on noble metals demonstrate high hydrogenation ability in aromatic oxygenates: hydrogenation begins at rather low temperatures (100–150 °C), whereas efficient HDO requires temperatures of at least 250–300 °C.

Catalysts based on noble metals proved to be very attractive for mild stabilization of pyrolysis oil.[48,49] It was observed that the H : C atomic ratio increases while the O : C atomic ratio was almost unchanged; this likely was associated with the hydrogenation of highly reactive species, such as carbonyls and sugars, into more stable alcohols.[49] Moreover, it was noted that noble metal additives in bimetallic catalysts increase the selectivity towards hydrogenolysis of the C_{arom}–O bond of aromatic substrates as compared to the rupture of the O–CH_3 bond. It is commonly speculated that such catalysts have a bifunctional nature, with H activation on the metallic phase and organic substrate adsorption on the supporting material. Thus, transfer of active hydrogen atoms from noble metal surfaces to the metal–support interface and then to the substrate by a so-called spill-over phenomenon is assumed.[50] This ability allows for the higher affinity of catalysts based on noble metals for hydrogenation transformations than HDO. Another advantage of catalysts based on noble metals is their relatively high stability due to slower coking as compared to the sulfided catalysts.

Noble metal catalysts have the advantage of avoiding the sulfurizing process, but problems associated with the effect of water on catalyst activity cannot be avoided very easily. For example, Lunzer *et al.*[51] found that, as in the case of sulfided catalysts, water in the reacting system leads to the deactivation of supported Pt catalysts (Pt/TiO_2 and Pt/SiO_2). In general, it is assumed that among the different types of noble metal-based catalysts for the HDO of pyrolysis oil, the most attractive catalyst choice, showing the higher stability relative to water, is carbon-supported catalysts (*e.g.* Pt/C, Pd/C, Ru/C and Rh/C). Nevertheless, the formation of coke deposits, even in the case of the high hydrogenation ability of noble metals, cannot be eliminated entirely, and noble metal-based catalysts also undergo coking. Moreover, the use of carbon as a supporting material places obstacles for their regeneration *via* annealing, raising questions about their feasibility.[45] In addition, a very well-known issue related to the high cost of noble metals causes additional concerns regarding their suitability for the hydrotreatment of pyrolysis oil, especially in large-scale production.

6.2.4 Catalysts Based on Transition Metals

All of the downsides of sulfided catalysts and those based on noble metals stimulated the search for new types of catalysts for the hydrotreatment of bio-oil. One such approach was the use of oxide forms of traditional hydrotreating catalysts in the HDO of biomass pyrolysis products.[52,53] However, in most studies, this approach did not show positive results. In general, the activity of the oxide forms of the catalysts in both hydrogenation and hydrogenolysis was significantly lower than that of their sulfided analogues.

For example, Viljava *et al.*[53] used an industrial hydrotreating catalyst (Ketjenfine 742-1.3Q, Akzo Chemie, 4.4 wt% CoO and 15 wt% MoO_3 on γ-Al_2O_3) in oxide and sulfided forms in the HDO of anisole as a bio-oil model at 1.5 MPa H_2 and 250 °C. Thus, the initial conversion was shown to be higher with the use of the sulfided form (88.2%) than that with the oxide form (63.3%) of the catalyst. The deoxygenation degree (yield of oxygen-free products) for the sulfided catalyst was 10.4%, whereas for the oxide analogue oxygen-free products were not observed. It seemed that HDO in this case did not proceed because at 250 °C NiO is not reduced into Ni^0 to provide catalyst activity in hydrogenation and HDO. In general, independently of the state (sulfided, oxide or reduced) of the bimetallic phase (Ni–Mo or Co–Mo), the catalysts of this type are not promising in the HDO of bio-oil, and so other hydrogenolysis catalysts should be used.

Traditional nickel catalysts of hydrogenation were tested in the hydrotreatment of bio-oil as well. For example, an Ni-1404 catalyst with 68% of nickel shows approximately the same activity in the HDO of bio-oil at 340 °C and 16.8 MPa as sulfided CoMo/γ-Al_2O_3.[54] However, after several hours of reaction, the nickel catalysts became less active than the sulfided ones, very likely due to intense coking. In addition, the use of Ni-1404 led to the formation of up to 15% of undesirable gas-phase products (C_1–C_4), whereas in the case of sulfided catalysts, the yield of C_1–C_4 hydrocarbons was as low as 1–2%. A high yield of gas-phase products was also discussed by Elliott[55] during the HDO of bio-oil at 400 °C in the presence of Raney nickel. Thus, the most common problem of typical nickel-based catalysts of hydrogenation is that they are prone to coking to an even greater extent than sulfided hydrotreating catalysts.[54] The second obstacle relates to the leaching of active metals in the presence of carboxylic acids in pyrolysis oil, especially when keeping in mind that such acids remain quite stable even up to process temperatures of 350 °C.[56] Owing to this, nickel catalysts for the HDO of bio-oil fall into an equivocal situation in which, on the one hand, they should not be used at low temperatures in order to prevent leaching, and on the other hand, their use at high temperatures would lead to fast coking.

In the framework of searching for catalysts for the HDO of biomass pyrolysis products, a series of works devoted to Mo and W carbides and nitrides are of interest.[57] These systems, usually considered as catalysts for hydrotreating mineral oil, were considered to be quite promising for bio-oil hydrotreatment, as they possess the properties of both metals and ceramics. Metal carbides and nitrides are classified as interstitial compounds due to the incorporation of carbon and nitrogen atoms into the metallic phase. In this case, a unique goal of using carbon and nitrogen atoms is to change the electronic metal structure and reduce the density of metal sites at the catalyst surface. Such modification is advantageous in terms of coke formation since metal sites drift apart, and the structurally sensitive reactions of coke formation become unfavourable. Of particular interest in the HDO of organic oxygenates is the use of phosphide catalysts.[58,59] Phosphide catalysts show activity in HDO associated with two types of active sites: Brønsted acidic sites (PO–H) and metallic sites of phosphide species. The former type allows for the activation of reactant molecules, while the latter is proposed to be active in H_2 activation. Despite the fact that phosphides, carbides and nitrides of transition metals are advantageous for the HDO of pyrolysis oils, they are not ready for industrial hydrotreating processes. Nevertheless, deeper understanding of their stability in such specific media as pyrolysis liquids under hydrotreatment conditions is highly desired and remains to be provided.

Particular attention should be given to bimetallic catalysts tested in the HDO of pyrolysis oil. It was shown that bimetallic nickel–copper catalysts possess high activity in both the hydrogenation and HDO of aromatic and aliphatic oxygenates.[60,61] The detailed studies of these catalysts showed that the addition of copper to Ni-containing catalysts results in the formation of a solid solution $Ni_{1-x}Cu_x$, which is one of the reasons for the decrease in the reduction temperature of NiO and thus its easier activation at the temperatures of the target process. It was shown by Ardiyanti *et al.*[62] that the increase in the Ni content in Ni–Cu/Al_2O_3 catalysts leads to an increase in the H:C ratio of the HDO products of bio-oil, which is a positive factor since the products have lower viscosity and molecular weight. However, these catalysts needed further improvement to enhance their stability against coking and sintering of the active component. Additional experiments with different supports under conditions close to those of the HDO of bio-oil (high acidity, presence of water) showed promise for such supports as CeO_2–ZrO_2, ZrO_2 and TiO_2.[39] As for the mechanistic and active species point of view, the catalysts based on transition metals again need to have a bifunctional nature. Thus, activation of molecular hydrogen proceeds on the metallic active phase with O organic species adsorption and activation on the active sites of the supporting material.[20]

The reader should note that, of course, the studies devoted to the hydrotreatment of pyrolysis liquids are quite numerous, and here we provide just a short discussion of some main aspects regarding the catalyst issue. There are a number of recent reviews to which the reader is directed in order to get more detailed information on the topic, since the subject deserves more ample treatment.[55,63–70]

6.3 One-stage Production of Isomerized Alkanes of Fuel Purposes from Vegetable Oils and Other Lipid-based Feedstocks

Along with fast pyrolysis bio-oil obtained from lignocellulosic biomass, another interesting type of biomass-derived renewable energy source is lipid-based feedstocks (LBFs), such as vegetable oils, fats, microalgae lipids, *etc.* The properties of such feedstocks can be found in the appropriate review articles. To date, one of the most utilized biofuels in the world is biodiesel, representing fatty acid methyl esters (FAMEs) obtained *via* transesterification of triglycerides with methanol.[71,72] However, due to its specific properties caused by its oxygen content (*e.g.* water absorption tendency, *etc.*), the use of biodiesel is limited. Therefore, it is utilized mostly in the form of mixtures with fossil diesel fuel at a maximal content of 20%. The HDO of FAMEs and LBFs, on the whole, results in linear diesel fraction alkanes being obtained, known as "green diesel" (GD). There are a number of review articles focused on this subject.[73–78] Despite GD being suitable as a motor fuel component, it has drawbacks, such as high cloud and pour points. Furthermore, linear alkanes are also known to be undesirable components of gasoline due to their low octane index. Thus, an isomerization step is required to limit all of the above drawbacks.

To date, sequential HDO and isomerization of lipid feedstocks have already been realized in the form of proprietary technologies such as NExBTL® (Neste), Ecofining™ (UOP/Eni), *etc.*[79] However, merging the HDO and isomerization stages into a single step would significantly simplify the technology on isomerized fuel alkanes production. Such successful merging, in the first place, depends on selection of proper catalysts active in both HDO and isomerization (*i.e.* having a bifunctional nature).

Since the main features of HDO catalysts were highlighted in Section 6.2, key points of hydrocarbon isomerization as a second key step in the process should be considered here as well.

6.3.1 Basic Aspects of Catalytic Hydroisomerization of Linear Alkanes

Isomerization of linear paraffins is an important process in the petroleum industry. It is usually conducted in a hydrogen atmosphere. Hydroisomerization (HI) of C_4–C_7 paraffins is performed to obtain high-octane-number gasoline, and HI of C_7–C_{15} paraffins is performed to produce diesel fuel with an elevated cetane index and good cold flow characteristics due to reductions in the viscosity and freezing point. Catalysts used for HI are based on metallic active components and acidic supports: the former catalyse hydrogenation/dehydrogenation, meanwhile the latter catalyse skeletal isomerization through carbenium ion formation.[80,81] The HI mechanism of *n*-alkanes over such catalysts occurs in the following consecutive steps:[80]

(1) Dehydrogenation on the metal active site;
(2) Protonation of olefins on the Brønsted acid sites to form a secondary alkylcarbenium;
(3) Rearrangement of the alkylcarbenium ion through the formation of a cyclic alkylcarbenium-type transition state;
(4) Deprotonation;
(5) Hydrogenation.

Schematically, the HI and cracking routes of *n*-alkanes over the HI catalysts are well-established in literature (Figure 6.1).[82]

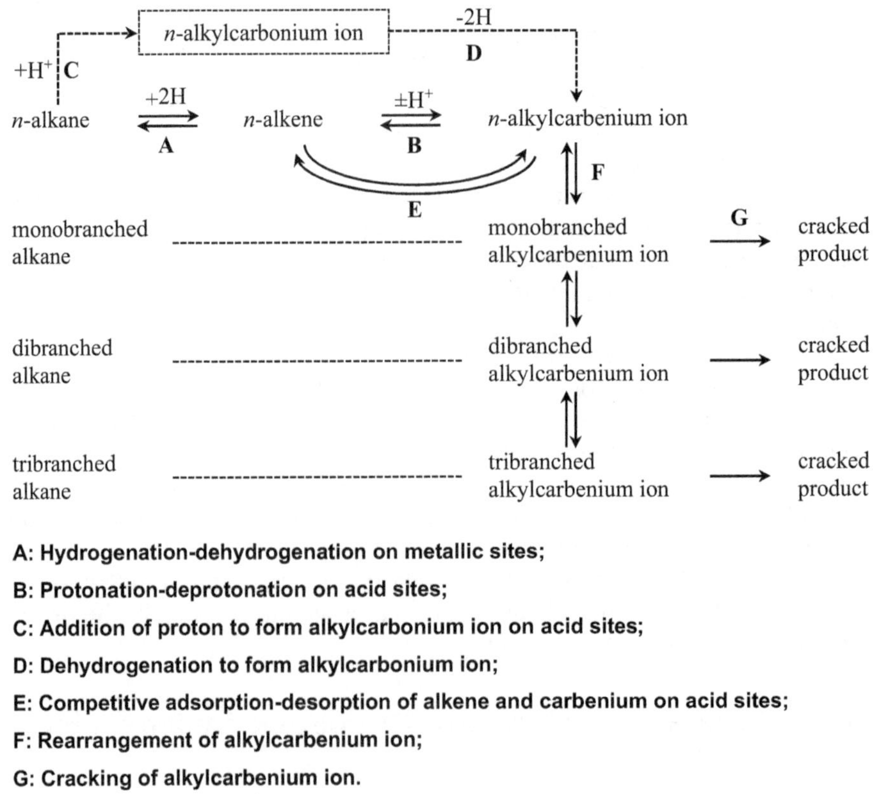

Figure 6.1 Scheme of the pathways of HI and cracking of an *n*-alkane. Reproduced from Ref. 82 with permission from Elsevier, Copyright 2000.

Typical acidic supports used in HI catalysts can be divided into the following groups:[80,83–86]

(1) Amorphous oxides or mixture of oxides (*i.e.* chlorinated Al_2O_3, HF-treated Al_2O_3, SiO_2-Al_2O_3, [Al-]ZrO_2/SO_4^{2-}, WO_3/ZrO_2, WO_3/Al-ZrO_x);
(2) Zeolites (Y, β, mordenite, ZSM-5, ZSM-22, ZSM-23);
(3) Silicoaluminophosphates (SAPO-5, SAPO-11, SAPO-31, SAPO-34, SAPO-41);
(4) Mesoporous materials (MCM-41, AlMCM-41, *etc.*).

The most active in isomerization due to their very high acidity are chlorinated alumina and sulfated zirconia, which allow for the reduction of HI temperatures to 120–180 °C.[81] Such low temperatures are thermodynamically favourable for multibranched alkane formation (Figure 6.2),[87] which are known to have higher octane number ratings compared to monobranched or linear paraffins (*e.g.* *n*-heptane has a research octane number rating of 0, 2-methylhexane has a rating of 53 and 2,3-dimethylpentane has a rating of 93[81]). However, gradual leaching of chlorine and sulfate groups into HI products gives rise to their contamination and the gradual deactivation of the catalysts. Zeolites and silicaaluminophosphates (SAPOs) are less acidic, but more tolerant of admixtures and regenerable. However, zeolite- and SAPO-supported catalysts are used at higher operating temperatures (typically 200–360 °C).[84,88–91] Mesoporous SiO_2-based materials have the lowest acidity, but due to their unique structure, they are usually used in the form of composites with zeolites and SAPOs, having a mostly microporous structure.[88,92]

The active component responsible for hydrogenation/dehydrogenation can be a noble metal (mainly Pt)[80,82,91,93,94] or bicomponent systems such as Ni(Co) Mo(W) in both sulfided[80] and non-sulfided[95] forms. Recently, active components based on phosphides and carbides of transition metals have been considered as promising, as they exhibit platinum-like catalytic activity in hydrogen transfer reactions (including HDS and HDN).[89,90,96–102] On the other hand, they are stable to poisoning by sulfur, unlike Pt.[103]

The best catalysts for the isomerization of *n*-paraffins should have a relatively high hydrogenation activity and low acidity to achieve maximal HI *versus* (hydro)cracking, which always accompanies HI. The hydrocracking reaction causes the degradation of the *n*-paraffins to less valuable and lighter products.[80] Thus, catalysts based on zeolites, SAPOs, other moderate acidic materials and their composites with various mesoporous materials are the most suitable for HI.

Thus, from the above brief review on the main aspects of catalytic HI of linear hydrocarbons, it is seen that HI catalysts are significantly dependent on the support, which should have acidity. The active component of such catalysts is similar to that of HDO catalysts (Pt, Ni[Co]MoS, *etc.*). One can conclude that for the HDO–HI of lipid-based feedstocks, the composition of the catalysts is similar to those of used for HI. However, since the performance of HDO–HI significantly depends on the support, the HDO–HI catalysts are better segregated by support type. The literature review showed that such catalysts can be divided on the ones supported by zeolites/SAPOs and those supported by other porous materials.

6.3.2 Application of Zeolite- and Silicoaluminophosphate-supported Catalysts for HDO–HI of Lipidic Feedstocks

In order to attain satisfactory degrees of HDO and HI simultaneously, a specific temperature balance should be kept because the HDO reaction is favoured at relatively high temperatures (at least 250–300 °C), while the formation of multibranched alkanes is favoured at temperatures below 200 °C. Another aspect of the combined HDO–HI is that in the temperature region where a good HDO degree is achieved, the most suitable catalysts are supported by moderately acidic zeolites and SAPOs. This is why there is a greater quantity

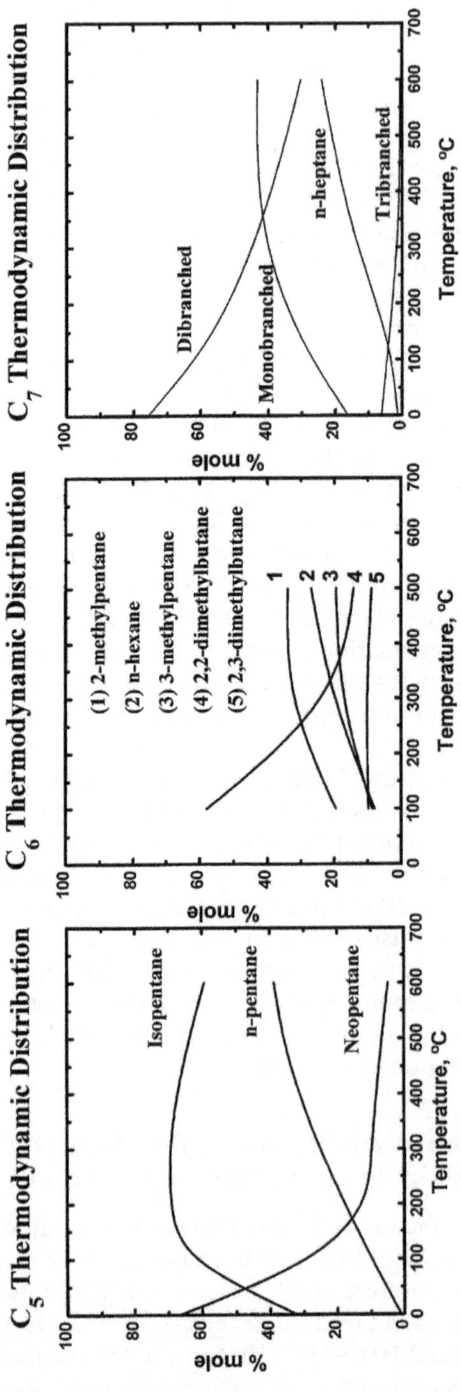

Figure 6.2 Thermodynamic distribution of C_5, C_6 and C_7. Reproduced from Ref. 87 with permission from Elsevier, Copyright 1999.

of successful research studies using such catalysts compared to those where the catalysts are supported by other porous materials. However, if the goal is to obtain bio-jet or even bio-gasoline fractions, higher temperatures are used to carry out HDO–HI in order to intensify the cracking reaction to a certain extent.

On the whole, zeolite- and SAPO-supported catalysts used in the HDO–HI of the lipidic feedstocks and model compounds are mostly based on two types of active components: noble metals (Pt, Pd)[104–112] (which, thus, are typical HI catalysts) and Ni-based components (metallic nickel or containing Ni).[113–119] Several examples of research studies using the above catalysts are presented.

Kikhtyanin *et al.* were among the first to carry out the HDO–HI of a vegetable oil (in this case sunflower oil). They used a Pd/SAPO-31 catalyst. The tests were carried out in a continuous-flow fixed-bed reactor at 310–360 °C at a pressure of 2.0 MPa and weight hourly space velocity (WHSV) of 0.9–1.6 h^{-1}.[104] The hydrocarbon yield in organic liquid products (OLPs) was close to 100%. They were composed mainly of linear and isomerized C_{17} and C_{18} paraffins. The ratio of isomerized paraffins to the linear ones could be higher than 10. However, the used catalyst was gradually deactivated due to inevitable sintering of Pd nanoparticles. Smirnova *et al.* also studied a SAPO-31-supported catalyst – Pt/SAPO-31 in sunflower oil HDO-HI at 330–350 °C.[105,106] They found that the highest isoparaffin yield was in the case of preliminary calcination of the catalyst at 450–550 °C. The catalyst calcinated at 450 °C maintained stability in terms of the isoalkane yield for 40 h of testing.

Wang *et al.* investigated the HDO–HI of soybean oil over Pt-based catalysts (1% Pt) supported by SAPO-11 or ZSM-22, having various levels of acidity.[107] A conversion rate of the substrate of 100% was achieved at 357 °C, a hydrogen pressure of 4 MPa and an LHSV of 1 h^{-1}. The maximum achieved yield of isomerized paraffins was 63%. Herskowitz *et al.* and Rabaev *et al.* studied the hydrotreatment (HT) of soybean oil over Pt/SAPO-11–Al_2O_3 catalyst in a trickle-bed reactor at 375–380 °C, 30 atm and LHSV = 1 h^{-1}.[108,109] The SAPO-11 component was found to gradually deactivate in the process due to the hydrothermal conditions. They showed that addition of a surfactant hexadecylamine (HDA) into the crystallization gel of SAPO-11 at a HDA : Al_2O_3 ratio of ~0.6 leads to a strong increase in the hydrothermal stability of SAPO-11 due to increasing the relative content of Si(nAl[4-n]Si) tetrahedra, with n = 3–4 in its composition. Testing runs for 1000 h of time on stream (TOS) showed that the activity of the catalyst was reduced after 200 h, followed by a stabilization expressed in the achieving of constant values for the cloud point and ratio of isomerized alkanes to linear ones in the OLPs. In the case of the highest HDA : Al_2O_3 ratio of 0.58, the cloud point of the OLPs was minimal (about −35 °C) and the ratio of isomerized alkanes to the linear ones was maximal, at 3.78.[109] The HDO–HI of oleic acid over Pt/SAPO-11 and Pt supported by chlorinated alumina in hydrogen and carbon dioxide media has given rise to the transformation of the substrate to a blend of linear C_6–C_{17} alkanes and dodecylbenzene as a product of dehydrocyclization.[110] The main products were dodecylbenzene and heptadecane.

Chen *et al.* reported on the HDO–HI of methyl palmitate over a PtSn/SAPO-11 system.[111] Promotion of Pt/SAPO-11 with tin improved its activity in terms of both HDO and HI and increased C_{16} alkane yields – products of C–O hydrogenolysis (instead of hydrogenolysis of C–C with formation of C_{15} hydrocarbons). This improvement was due to the fact that addition of tin results in increased dispersion of Pt particles, with their average particle size decreasing down to 4 nm. A part of the tin formed PtSn alloys, and the rest appears to be in the form of SnO_{2-x} oxides ($0 < x < 2$) located at the Pt nanoparticle periphery. These authors have also shown that the Sn content should not be too high (optimum Sn : Pt ratio is 2) to avoid blocking the activation of hydrogen on platinum particles and thus reducing HDO activity. The presence of PtSn alloys is one of the most important reasons for this high activity in HI.

Chen *et al.* investigated Pt/ZSM-22@SiO_2 catalysts in the HT of methyl palmitate. The zeolite component was partially covered with silica to passivate acid sites on the external surface of the support and to direct the pathway of isomerization towards monobranched alkane formation.[112] Monobranched diesel alkanes are known to have a higher cetane number (>85; multibranched <60). The optimal quantity of SiO_2 content was 20 wt%. In this case, the yield of the monobranched alkanes increased from 71 to 95 wt% at 300 °C.

Unlike the HDO of pyrolysis oil, where catalysts based on metallic nickel were shown to tend to coke generation, zeolite- and SAPO-supported catalysts appeared to be quite effective in the HDO–HI of lipid-based feedstocks. Liu *et al.* reported on the HDO–HI of castor oil over Ni-based catalysts supported on most of the available commercial zeolites (ZSM-5, H-β, SAPO-11, USY). Both Ni and Ni_2P were active in the HDO of castor oil, although decarbonylation and dehydration processes are favoured on Ni and Ni_2P, respectively.[113] HT experiments were carried out at 300 °C, 3 MPa and WHSV = 2 h^{-1}. In all cases, the conversion rate of the feedstock was 98–99%. Pathways of substrate transformations and ratios of isomerized to non-isomerized alkanes were found to mainly depend on the acidity of the support: non-modified zeolites with elevated acidity (ZSM-5, USY and H-β) resulted in the formation of C_5–C_7 alkanes and an elevated ratio of isomerized to linear alkanes; supports having a low concentration of acidic sites (SAPO-11, modified USY) predominantly resulted in the formation of long C_{16}–C_{19} alkanes. HT of soybean oil over Ni supported on SAPO-11 and a series of zeolites (ZSM-5, ZSM-22, ZSM-23 and β) was studied by Wang *et al.*[114] The Ni contents were 4, 8 and 12 wt%. The best result was obtained in the case of 8% Ni/SAPO-11 at 370 °C, 4 MPa and WHSW 1 h^{-1}. A conversion rate of soybean oil of 100% was achieved with 85% selectivity to isoalkanes. The excellent conversion, the high liquid yield and the isomerization selectivity are consequences of the mesoporosity, the 1D micropore channels having appropriate size and the moderate acidity. Cao *et al.* studied the HDO–HI of palmitic acid at 200–260 °C over 4% Ni/H-ZSM-22.[115] They used two synthesis routes for preparing the catalysts: impregnation by an aqueous solution of $Ni(NO_3)_2 \cdot 6H_2O$ or by melting the nickel salt at 65 °C (T_{melt} = 56.7 °C). The latter resulted in catalysts with higher Ni dispersions that were more effective in the HDO–HI of palmitic acid due to their elevated activity in both HDO and HI.

Conventional sulfided NiMoS catalysts also appeared to be effective in HDO–HI. Qian *et al.* investigated the effect of the porous structure of the support in sulfided NiMo/SAPO-11 and NiMo/AlSBA-15 catalysts in the HT of methylstearate.[116] They concluded that the porous structure of the support used has a significant effect: SAPO-11-supported catalysts generated mainly monobranched alkanes (pore size = 0.6 nm). In the case of NiMo/AlSBA-15 (pore size 3.3 nm), yields of cracking products and multibranched alkanes were found to be significantly higher due to the higher acidity and wider pores of the support. The HDO–HI of Jatropha oil over NiMoS/SAPO-11 and NiWS/SAPO-11 catalysts at 370–450 °C, 60–80 bar, LHSV 1 h^{-1} allowed for the production of OLPs containing ~40% GD, ~40% biokerosene and ~20% biogasoline. The biokerosene had a freezing temperature of less than −47 °C.[117] The catalysts were stable during testing for at least 6 days, and the sulfur content in the OLPs was relatively low.

Partially reduced Ni-based catalysts promoted by Mo in oxide form as well as NiP_x-based catalysts were also found to be effective in the process. Thus, Chen *et al.* investigated the temperature of reduction of catalysts with an $NiMoO_{3-x}$/SAPO-11 composition in the HDO–HI of methyl laurate.[118] The highest selectivity to isomerized alkanes was found for the catalyst reduced at the lowest temperature (400 °C), which was also the one with the highest acidity. Increasing the reduction temperature gave rise to decreasing Mo^{5+} contents and hence decreased acidity, resulting in a selective decrease of isomerized paraffins. An optimum balance between activity and selectivity in HDO and HI was obtained for the catalyst reduced at 450 °C due to a synergetic effect between Ni and Mo^{4+}. Unsupported Ni_xP and $Ni_{12}P_5$ phosphides and HZSM-22-supported nickel phosphide-based catalysts were studied by Liu *et al.* in the HDO–HI of palmitic acid.[119] It was found that the Ni_xP phase is involved in decarboxylation/decarbonylation reactions, while Brønsted acid sites of the zeolite support are active in hydrocarbon isomerization and Lewis acidic sites are active in cracking.

In addition, Sousa *et al.* have shown the possibility of using the H-form of bare zeolite-β in the HDO–HI of palm kernel oil and palm olein as well as their hydrolisates. The experiments were carried out in a batch reactor at 350 °C and P_{H2} = 10 bar (catalyst content: 15 wt%).[120] The OLP hydrocarbon content varied from 24 to 98 wt% depending on substrate and process time. Significant amounts of isomerization and aromatization products were observed in the hydrocarbon fraction, which resulted in a decrease of the freezing point to between −18 and −30 °C. The catalyst was stable over four consecutive reaction cycles without any notable deactivation.

6.3.3 Application of HDO–HI Catalysts Supported by Other Porous Materials

To obtain isomerized alkanes through one-stage HDO–HI from lipidic feedstocks, many researchers use catalysts consisting of supported noble metal, metallic Ni, NiMoS or Mo_2C active phases as catalysts, as described in Section

6.3.2, but supported on aluminosilicates, SiO_2-based mesoporous materials, SBA-15 doped and non-doped by alumina, γ-Al_2O_3, (*i.e.* modified by WO_3) and porous carbon materials (carbon nanotubes [CNTs], activated carbons) instead of zeolite or SAPO supports.

The HDO–HI of sunflower oil on NiMoS/WO_3-Al_2O_3 catalytic systems at 380 °C, 4.0 MPa and WHSV of 1 h^{-1} was carried out by Nepomnyashchiy *et al.*[121] They found that, on increasing the WO_3 content, the concentration of the support acid sites increased, mainly at the expense of weak and medium acid sites. The OLP yield was 81–83 wt%, with the yield of isomerized alkanes increasing with the WO_3 content. However, this catalyst was not stable enough and gradually deactivated.

The HDO of a mixture of esters, with methyl oleate as a major component (about 70%),[122] on Al_2O_3-modified (Al-SBA-15) and bare SBA-15-supported Ni- and Co-based catalysts was studied by Ochoa-Hernández *et al.*[123] It was found that alumina doping of SBA-15 gave rise to the generation of acid sites in the support and hence to a conversion rate of the ester mixture close to 100% and to a several-fold increase in the content of isomerized paraffins.

Kulikov *et al.* studied the conversion of rapeseed oil on Pt–Pd catalysts supported on mesoporous alumosilicate mixed with 30 wt% boehmite in a Pt–Pd–Al–HMS system (where Al-HMS is aluminum-rich mesoporous aluminosilicate).[124] A 9% gasoline content in OLPs after HDO at 275 °C was found. On increasing temperature, the cracking contribution increases with gasoline and gaseous product yields. At 300 °C and LHSV = 1 h^{-1}, the optimum conditions for maximizing the diesel fraction yield (~94%), a 20% isomerized paraffin content was attained.

Trifunctional catalysts combining hydrogenation, HDO and isomerization/cracking functions were utilized by Liu *et al.* in the HDO–HI of Jatropha oil containing free fatty acids (FFA).[125] Sulfided NiMo/SiO_2, NiMo/γ-Al_2O_3 and NiMo/SiO_2-Al_2O_3 systems were used as catalysts in a continuous-flow reactor at 350 °C and 4 MPa H_2 pressure. Application of catalyst with the mixed support allowed for a mixture of alkanes including 10% C_5–C_{10} and 25% of isoalkanes to be obtained. The freezing point of the OLPs was −10 °C.

A screening of catalysts for the HDO–HI of soybean oil was carried out at 400 °C, 9.2 MPa initial hydrogen pressure by Veriansyah *et al.* in a continuously stirred batch reactor.[126] At these conditions, conversion of feedstock decreases in the following sequence: NiMoS/γ-Al_2O_3 (92.9%) > 4.29 wt% Pd/γ-Al_2O_3 (91.9%) > CoMoS/γ-Al_2O_3 (78.9%) > 57.6 wt% Ni/SiO_2-Al_2O_3 (60.8%) > 4.95 wt% Pt/γ-Al_2O_3 (50.8%) > 3.06 wt% Ru/Al_2O_3 (39.7%). Ni- and Pd-based catalysts were found to generate OLPs mainly composed of linear C_{17} and C_{15} alkanes, while in the case of CoMoS/γ-Al_2O_3 catalysts, elevated yields of isomerized alkanes and lighter cracking products were observed.

Pd/C and sulfided NiMo/γ-Al_2O_3 catalysts were used by Phimsen *et al.* to investigate the HDO–HI of oil extracted from spent coffee grounds in a batch reactor equipped with a shaker.[127] Such oil contains FFA (6.14 wt%). The optimal conditions were 400 °C, 2 h reaction time and initial hydrogen pressure of 40 bar. It was discovered that the catalyst activities were significantly different. The activity of the sulfided catalyst appeared to be higher than that of the Pd/C catalyst. The contents of isoalkanes and olefins in the OLPs were 10.3 and 4.8 wt% and 1.7 and 22.3 wt% for the sulfided and Pd catalysts, respectively.

Han *et al.*[128] used Mo_2C/CNT catalysts for the HDO–HI of several types of vegetable oils and model compounds (decanal, stearic acid and methyl palmitate) at 220–260 °C and 2.0 MPa hydrogen pressure in a batch reactor. The catalytic activity of Mo_2C-based catalysts was compared to those of Pd and Pt catalysts. Unlike the noble metal-based systems, Mo_2C-based catalysts were found to be stable to leaching in these process conditions. A maximal 87% alkane yield (55% of isomerized alkanes) was observed for the 20% Mo_2C/CNT catalyst.

Thus, it has been shown that supported catalysts allow for the combined HDO–HI of lipid-based feedstocks in one stage to obtain isomerized paraffins, despite this being a relatively recent approach. As has been seen, for the LBF conversion, in many cases conventional zeolite (and/or SAPO)-supported catalysts in hydrocarbon isomerization (*e.g.* noble metal-based catalysts) are appropriate. In addition, catalysts based on transition metal (Ni, Cu, Mo, *etc.*) with different phosphide, nitride, carbide additives supported on zeolites, SAPOs or other moderately acidic supports also appeared to be promising catalytic systems. Variation of temperature and/or the texture properties of supports (*e.g.* pore size) allows for flexible tuning of the HDO–HI to obtain more or less branching of the products, as well as heavier or lighter fraction compositions. In any case, this approach has not progressed far, and a rapid growth in its development is expected in the near future.

6.4 Catalytic Production of Fuel Eco-additives from Biomass

This section is focused on using heterogeneous catalysts for obtaining fuel additives from biomass. Global consumption of fuel additives is about 40 million tons per year, and this is constantly increasing. Currently, a number of high-octane additives are known, such as monomethylaniline (MMA), methyl tert-butyl ether (MTBE), Difron R550, Difron R45, Feterol, *etc.* However, the high cost of most of them limits the available additive range. At the same time, the use of inexpensive nitrogen-containing aromatic compounds (MMA and its analogues) results in an increase in the NO_x content in exhaust gases, as well as the formation of various resins, which negatively affect internal combustion engines. It should also be noted that various fuels (gasoline, diesel fuel, jet fuel) are characterized by different susceptibilities to additions; therefore, it is important to have a wide range of fuel additives.

Among the bioresources considered in Section 6.2, bio-oil is not available for use as an additive for gasoline due to its high acidity, viscosity and water content, which does not allow it to be mixed with petroleum products. However, the joint use of bio-oil with diesel fuel can be accomplished *via* the formation of emulsions using surfactants. Ikura *et al.*[129] obtained stable emulsions using 0.8–1.5 wt% surfactant. They showed that the stability of emulsions also depended on the total concentration of bio-oil and power consumption per unit volume for emulsification. Heavy bio-oil fractions were removed by centrifugation and/or emulsion, which increases the cost of this method. Determination of the cetane number of emulsions formed by linear interpolation showed that the pyrolysis liquid has a cetane number of 5.6, but when it is added to diesel fuel, its cetane number (45.7) declines by 4 points for every 10% of bio-oil. Chiaramonti *et al.*[130] also found that the optimal amount of surfactant is 0.5–2.0%, while a higher content leads to high viscosity and the need to use a solvent. Despite the data obtained and the ability to obtain stable emulsions across a very wide range from 5:95 to 75:25, the authors note that the emulsions are stable at a temperature of 70 °C for only 3 days, and the use of these mixtures in injector systems leads to their malfunctioning due to rapid corrosion.[131] Thus, bio-oil can only be used as a diluent of natural fuels, which results in a decrease in their operational properties, and in general, this method is ineffective.

A key step towards successful commercialization of bio-feedstocks is a separate processing of biomass components (Figure 6.3). In the first stage, separation of liquid components (oils) and solid components of biomass, including cellulose, hemicellulose and lignin, is carried out. Triglycerides have long been used as feedstocks for the production of biodiesel, straight paraffins (GD) and isomerized paraffins, which can be used as a mixture with standard fuels. During triglyceride conversion, about 10–11% of crude glycerol is formed. This by-product can be used to obtain glycerol-based fuel additives. Monoacetylglycerol (MAG), diacetylglycerol (DAG) and triacetylglycerol (TAG) are formed *via* transesterification of glycerol with acetic acid using heterogeneous porous materials, and these are widely used in pharmaceuticals, the plastics industry and as eco-additives.[132] Another method for glycerol valorization is the acetalization reaction with ketones and aldehydes.[133] The addition of acetals/ketals in fuels can improve their antiwear properties, reduce the contribution to air pollution from fuel combustion in engines and gum formation and increase the octane number of gasoline.[134]

The processing of biomass components is more difficult compared to the conversion of liquid components. The main disadvantage is the high cost of the separation of bio-components including lignin, hemicellulose and cellulose. The latter has long been used in the production of bioethanol, which is the most popular representative of bio-feedstocks and sees large-scale commercial use. At the same time, several generations of bioethanol are distinguished. These generations are characterized by the nature of the feedstock and include the use of resources such as edible and inedible crops and algal biomass.[135] Bioethanol is already used as a fuel or as a 10% additive to gasoline in the USA and Brazil.[136] However, ethanol has a low C : O ratio and consequently has a low calorific value. Therefore, new approaches are being developed to produce eco-additives based on cellulose, hemicellulose and lignin

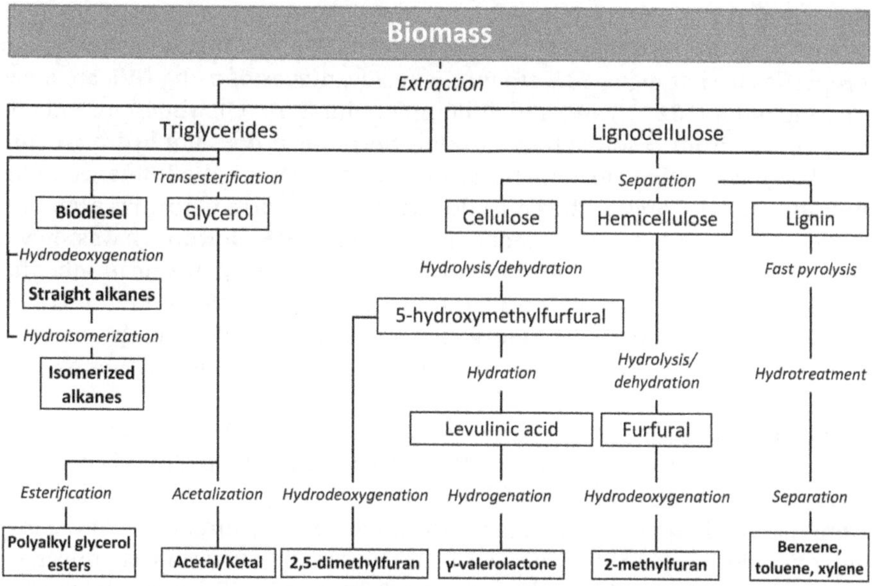

Figure 6.3 Processing of biomass components into fuel additives.

that can not only reduce the load on the oil-refining complex, but also contribute to improving the quality of fossil fuels.

Cellulose is the source of a number of chemical compounds that can be used as fuel additives. For example, conversion of cellulose can lead to the formation of hydroxymethylfurfural (HMF), levulinic acid (LA), 2,5-dimethylfuran (DMF) and γ-valerolactone (GVL).

DMF has excellent physicochemical properties and a high blended research octane number (BRON = 119), as well as low volatility.[137] It also has a higher energy density than ethanol and butanol and is more efficient as a fuel additive. Compounds based on LA (methyl levulinate, ethyl levulinate and butyl levulinate) may be considered as other representatives of additives. The mixture of LA esters with natural fuel improves the fuel economy as well as the physicochemical characteristics of combustion and engine emissions.[138] Esterification of LA over acid catalysts (resins, heteropolyacid, zeolite, *etc.*) using alcohol as a reactant is the most effective way to obtain ester-based additives with almost 100% selectivity.[139] LA is also a platform for a number of other chemical substances. Methyl tetrahydrofuran (MTHF) has an octane number of 87, and it can be mixed with gasoline up to 70% and be produced by dehydration and hydrogenation of LA.[140] Mizugaki *et al.*[141] obtained an 86% yield of MTHF in a one-pot reaction using a Pt-Mo/H-β catalyst in an aqueous solution of LA. The H-β support promotes high dispersion of Pt and MoO_x, leading to 1,4-pentanediol formation, while its acidity promotes the cyclodehydration of 1,4-pentanediol. LA could be transformed into GVL using noble metal-based homogeneous catalysts.[142,143] Horvath *et al.*[144] have proposed the use of GVL as a fuel additive. The authors found that addition of 10% GVL to gasoline with an octane number of 95 leads to a fuel mixture with similar properties to that using 10% ethanol. The advantages of using GVL are a low melting point (−31 °C) and a high boiling point (207 °C), which prevents its evaporation from gasoline fractions. Furthermore, it does not hydrolyse into γ-hydroxypentanoic acid, which makes it stable against the influence of air humidity upon storage. The GVL additive does not form peroxides even during long storage times, which makes it safer than DMF. However, it was shown that the use of low-purity GVL leads to an increase in the peroxide number for 95 octane gasoline from 0.15 to 3.70 and 11.10 mg kg^{-1} with the addition of 3% and 9% GVL, respectively. The authors associated the increase in the peroxide number with the formation of peroxides by impurities present in the composition of the GVL.[145]

Dimethylfuran, levulinic acid and a range of different furan-based additives can be obtained from furfural produced during the acid hydrolysis of hemicellulose. The source of hemicellulose and cellulose may be inedible waste from the wood processing industry and agriculture. Furfural is not only a precursor to biofuels, but also a universal and promising renewable chemical. Its production in the world reaches about 300 000 tons, with an annual market growth of 2%. In addition, furfural can be a platform for various high-octane additives and valuable chemicals: 2-methylfuran (MF; BRON = 131), DMF (BRON = 119), ethyllevulinate (BRON = 110). MF has the highest BRON. To obtain it from furfural, copper–chromium systems are usually used as hydrogenation catalysts. The process is carried out mainly in liquid-phase conditions at 90–120 °C and a hydrogen pressure of 5–6 MPa.[146] As an alternative to chromium-containing catalysts, catalysts based on the transition metals Cu, Ni, Fe, Mo and Mg have been developed.[147–149]

MF is a flammable and water-insoluble liquid. According to studies by Wang *et al.*,[150] it has a higher thermal efficiency compared to gasoline and 2,5-DMF due to its fast burning speed and ability to suppress detonation. In addition, MF can be used as a platform for producing liquid fuels with a high alkane content and low oxygenates.[151] The synthesis of branched alkanes of the diesel fraction from the MF (Figure 6.4) was first carried out by Corma *et al.*[151–153] The process involved hydrolysis of MF to 4-oxo-pentanal (R_1–CH=O) with subsequent hydroxyalkylation and alkylation steps. The yield of the oxidized trimer of 5,5-*bis*(silvan)-2-pentanone was approximately 74% using a 12 wt% aqueous solution of H_2SO_4 at a temperature of 60 °C and a reaction time of 16 h. Then, in an HDO reaction in a fixed-bed reactor using a composite 3 wt% Pt/C-TiO_2 catalyst, a hydrogen pressure of 5 MPa and a space velocity of 1.12 h^{-1}, alkanes of the diesel fraction are formed.[152] As a result of side reactions of oligomerization and cracking, the formation of the C_9–C_{24} diesel fraction was observed with a molar yield exceeding 93%. The group of Zhang has investigated various catalysts, including Ni supported on various acidic supports,[154] Ni-W_xC/C[155] and Pd/C noble metal systems[156] in the HDO reaction of the products of MF condensation with carbonyl compounds. It was shown that conversion of MF

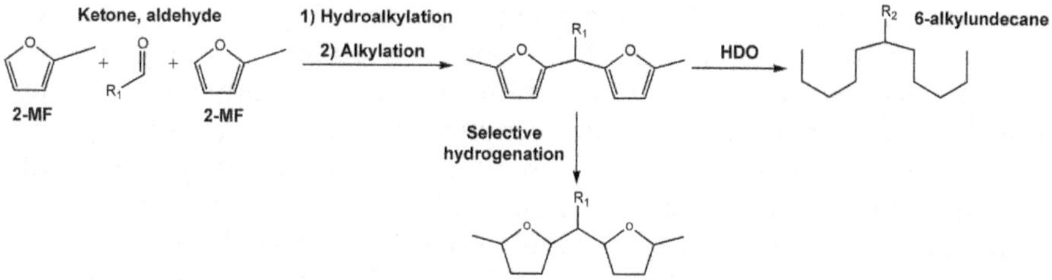

Figure 6.4 Synthesis of branched alkanes by hydroxyalkylation of MF with various substrates and subsequent HDO of intermediate oxygenates and the production of cyclic fuel additives by selective hydrogenation.

to 5,5'-(butane-1,1-diyl)*bis*(2-methylfuran) over Nafion-212 resin followed by HDO over Ni-W_xC/C leads to a 95% yield of diesel fraction.

Balakrishnan *et al.*[157] have studied Pd nanoparticles supported on SiO_2 containing surface-bound ionic liquids in selective hydrogenation of cyclic products formed through MF condensation. At a temperature of 60 °C, a hydrogen pressure of 2.8 MPa, and a reaction time of 30 h, the 92% yield of a full hydrogenated cyclic product, as an additive for diesel, was achieved.

The stoichiometric air demand for MF is close to the value for 95 RON gasoline, which makes it suitable for use in mixtures with fuel. In addition, the high BRON (131) value for MF makes it a promising candidate as an additive to gasoline, and its low boiling point (64 °C) allows for engines to start at low ambient temperatures.

In Lange *et al.*, BRON and boiling point values of the main representatives of eco-additives are compared.[158] This approach allows one to identify components that are compatible with existing fuel additives. The authors note that the presence in a molecule of furfuryl derivatives of aldehyde (furfural) or alcohol (furfuryl alcohol) functional groups leads to a polarity mismatch with fossil fuels. Meanwhile, removal of one or another of the alcohol or aldehyde group results in the formation of components with a lower polarity and increased energy intensity. It was shown that all monomeric derivatives of furfural comply with the boiling range of gasoline.[158] Unsaturated molecules have good octane properties, which often exceed the values for ethanol. Saturation of the furan ring apparently gives rise to a decrease in the octane index by almost 50 points. Thus, it is seen that MF has high rates of use as an additive to motor fuels. Lange *et al.* evaluated the compatibility of 10 vol% MF with the European brand of gasoline, EN228.[158] It was found that over a total travel distance of 90 000 km, the addition of MF did not adversely affect emissions of NO_x, hydrocarbons and CO. The loss of fuel due to the oxygen content is three times less compared to using a mixture of ethanol with gasoline. In addition, the MF additive did not adversely affect the basic characteristics of the engine oil and engine wear.

Recently, the production of aromatic hydrocarbons such as benzene, toluene and xylenes (BTX) from renewable biomass has attracted significant attention from the viewpoint of sustainable BTX production, as these are the main chemicals in the petrochemical industry, and they can be used as fuel additives or solvents and as platforms for the production of polymers. For example, BTX can be obtained by thermochemical transformations of cellulose.[159] Another method of BTX production is rapid pyrolysis in the presence of aromatization catalysts.[160,161] Lignin can be selectively depolymerized into its oxygen-containing aromatic monomers.[162] Further oxygen removal leads to aromatics that can be integrated into the current BTX pool.[163]

Ghorbannezhad *et al.*,[160] when studying the fast pyrolysis of sugarcane bagasse in the presence of HZSM-5 catalysts, showed that a yield of BTX depends on both zeolite acidity and the catalyst : biomass ratio. Under optimal conditions of a temperature of 480 °C, a catalyst : bio-oil ratio of 17 and in the presence of a HZSM-5 catalyst with a Si: Al ratio of 23, the optimal BTX yield was 22%.

In addition to the catalyst : feed ratio in the synthesis of BTX during the pyrolysis of lignin, the acidity of the catalyst, the flow rate of the carrier gas and the process temperature also have significant impacts. Using 2.9% Zr/MCM-48 at 600 °C, Elfadly *et al.* managed to obtain a 49% yield of BTX.[164]

In the HDO of model compounds of lignin pyrolysis liquid (*m*-cresol and guaiacol) under mild reaction conditions, Sirous-Rezaei *et al.*[165] have shown that the highest activity with a 77.2 wt% BTX yield at 350 °C was achieved using a $PdReO_x/ZrO_2$ catalyst. Moreover, the catalytic activities relative to the active component and support decreased in the orders Pd > Ni > Ru > Cu and ZrO_2 > CeO_2 > $ZrCeO_2$, respectively. In this case, the use of Pd, Ru and Cu resulted in the formation of toluene, while the use of Ni led to an additional demethylation reaction with the formation of benzene in addition to toluene.

Thus, along with the fuel additives obtained during the conversion of cellulose and hemicellulose, methods of processing lignin into BTX components are very popular. Conducting more detailed studies in these areas will soon allow for the range of eco-additives to be expanded, as well as creating a base for new chemical raw materials, which will partially reduce the load on the existing oil-refining complex.

6.5 Conclusions

The short literary survey presented in this chapter thus highlights the great importance of the application of heterogeneous catalysts for the transformation of biomass-derived feedstocks into high-quality motor fuel components and for improving the properties of the family of fuel additives. The use of heterogeneous catalysts can significantly increase the yield of target products, as well as improving their quality by increasing selectivity, both in the field of production of fuel components and in the production of high-octane additives. Despite the fact that today there exist problems in the highlighted pathways of biomass conversion (high cost, complexity of the processes and the feedstock), these processes of biomass conversion are developing very rapidly, and significant number of them is expected to be gradually implemented in industry in the near future.

Acknowledgement

The support from Russian Science Foundation (project no. 18-43-08002) is acknowledged.

References

1. V. I. Sharypov, N. G. Beregovtsova, B. N. Kuznetsov, L. Membrado, V. L. Cebolla, N. Marin and J. V. Weber, *J. Anal. Appl. Pyrol.*, 2003, **67**, 325.
2. B. N. Kuznetsov, V. I. Sharypov, S. A. Kuznetsova, V. E. Taraban'ko and N. M. Ivanchenko, *Int. J. Hydrogen Energy*, 2009, **34**, 7051.
3. S. D. Varfolomeev, E. N. Efremenko and L. P. Krylova, *Russ. Chem. Rev.*, 2010, **79**, 491.
4. A. V. Bridgwater, D. Meier and D. Radlein, *Org. Geochem.*, 1999, **30**, 1479.
5. A. Bridgwater and G. V. C. Peacocke, *Renewable Sustainable Energy Rev.*, 2000, **4**, 1.
6. D. K. Shen, S. Gu and A. V. Bridgwater, *Carbohydr. Polym.*, 2010, **82**, 39.
7. V. Dhyani and T. Bhaskar, *Renewable Energy*, 2018, **129**, 695.
8. G. Kabir and B. H. Hameed, *Renewable Sustainable Energy Rev.*, 2017, **70**, 945.
9. S. Yorgun and D. Yildiz, *J. Anal. Appl. Pyrol.*, 2015, **114**, 68.
10. V. K. Guda, P. H. Steele, V. K. Penmetsa and Q. Li, in *Recent Advances in Thermochemical Conversion of Biomass*, ed. A. Pandey, T. Bhaskar, M. Stöcker and R. K. Sukumaran, Elsevier B.V., Amsterdam, 2015, p. 177.
11. A. Aho, T. Salmi and D. Y. Murzin, in *The Role of Catalysis for the Sustainable Production of Bio-fuels and Bio-chemicals*, ed. K. S. Triantafyllidis, A. A. Lappas and M. Stöcker, Elsevier B.V., Amsterdam, 2013, p. 137.
12. Q. Zhang, J. Chang, T. Wang and Y. Xu, *Energy Convers. Manag.*, 2007, **48**, 87.
13 S. Yaman, *Energy Convers. Manag.*, 2004, **45**, 651.
14 E. Butler, G. Devlin, D. Meier and K. McDonnell, *Renewable Sustainable Energy Rev.*, 2011, **15**, 4171.
15. A. V. Bridgwater, *Biomass Bioenergy*, 2012, **38**, 68.
16. D. Mohan, C. U. Pittman and P. H. Steele, *Energy Fuels*, 2006, **20**, 848.
17. R. Venderbosch and W. Prins, *Biofuels, Bioprod. Biorefin.*, 2010, **4**, 178.
18. C. Igathinathane, J. S. Tumuluru, S. Sokhansanj, X. Bi, C.J. Lim, S. Melin and E. Mohammad, *Bioresour. Technol.*, 2010, **101**, 6528.
19. F. de Miguel Mercader, M. J. Groeneveld, S. R. A. Kersten, R. H. Venderbosch and J. A. Hogendoorn, *Fuel*, 2010, **89**, 2829.
20. P. M. Mortensen, J. D. Grunwaldt, P. A. Jensen, K. G. Knudsen and A. D. Jensen, *Appl. Catal., A*, 2011, **407**, 1.
21. P. Grange and X. Vanhaeren, *Catal. Today*, 1997, **36**, 375.
22. R. Durand, P. Geneste, C. Moreau and J. L. Pirat, *J. Catal.*, 1984, **90**, 147.
23. E. Laurent and B. Delmon, *Appl. Catal., A*, 1994, **109**, 77.
24. E. Laurent and B. Delmon, *Stud. Surf. Sci. Catal.*, 1994, **88**, 459.
25. G. W. Huber and A. Corma, *Angew. Chem., Int. Ed.*, 2007, **46**, 7184.

26. Q. Lu, W. Z. Li and X. F. Zhu, *Energy Convers. Manag.*, 2009, **50**, 1376.
27. A. Oasmaa, E. Kuoppala, J. F. Selin, S. Gust and Y. Solantausta, *Energy Fuels*, 2004, **18**, 1578.
28. L. Ingram, D. Mohan, M. Bricka, P. Steele, D. Strobel, D. Crocker, B. Mitchell, J. Mohammad, K. Cantrell and C. U. Pittman, *Energy Fuels*, 2008, **22**, 614.
29. A. Oasmaa, E. Kuoppala, S. Gust and Y. Solantausta, *Energy Fuels*, 2003, **17**, 1.
30. A. Oasmaa, E. Kuoppala and Y. Solantausta, *Energy Fuels*, 2003, **17**, 433.
31. A. Oasmaa and E. Kuoppala, *Energy Fuels*, 2008, **22**, 4245.
32. M. Garcia-Perez, A. Chaala, H. Pakdel, D. Kretschmer and C. Roy, *Biomass Bioenergy*, 2007, **31**, 222.
33. J. H. Marsman, J. Wildschut, F. Mahfud and H. J. Heeres, *J. Chromatogr. A*, 2007, **1150**, 21.
34. C. A. Mullen, G. D. Strahan and A. A. Boateng, *Energy Fuels*, 2009, **23**, 2707.
35. P. K. Kanaujia, Y. K. Sharma, M. O. Garg, D. Tripathi and R. Singh, *J. Anal. Appl. Pyrol.*, 2014, **105**, 55.
36. A. Oasmaa, E. Kuoppala, A. Ardiyanti, R. H. Venderbosch and H. J. Heeres, *Energy Fuels*, 2010, **24**, 5264.
37. M. Staš, D. Kubička, J. Chudoba and M. Pospíšil, *Energy Fuels*, 2014, **28**, 385.
38. Y. Romero, F. Richard and S. Brunet, *Appl. Catal., B*, 2010, **98**, 213.
39. A. R. Ardiyanti, S. A. Khromova, R. H. Venderbosch, V. A. Yakovlev, I. V. Melián-Cabrera and H. J. Heeres, *Appl. Catal., A*, 2012, **449**, 121.
40. L. Wang, D. Li, M. Koike, S. Koso, Y. Nakagawa, Y. Xu and K. Tomishige, *Appl. Catal., A*, 2011, **392**, 248.
41. D. C. Elliott and E. G. Baker, in *Symposium Papers – Energy from Biomass and Wastes*, Institute of Gas Technology, Chicago, IL, 1987, p. 765.
42. E. Furimsky, *Appl. Catal., A*, 2000, **199**, 147.
43. D. C. Elliott, K. L. Peterson, D. S. Muzatko, E. V. Alderson, T. R. Hart and G. G. Neuenschwander, *Appl. Biochem. Biotechnol*, 2004, **115**, 807.
44. M. C. Samolada, W. Baldauf and I. A. Vasalos, *Fuel*, 1998, 77, 1667.
45. J. Wildschut, I. Melián-Cabrera and H. J. Heeres, *Appl. Catal., B*, 2010, **99**, 298.
46. F. H. Mahfud, F. Ghijsen and H. J. Heeres, *J. Mol. Catal. A Chem.*, 2007, **264**, 227.
47. J. Wildschut, F. H. Mahfud, R. H. Venderbosch and H. J. Heeres, *Ind. Eng. Chem. Res.*, 2009, **48**, 10324.
48. R. H. Venderbosch, A. R. Ardiyanti, J. Wildschut, A. Oasmaa and H. J. Heeres, *J. Chem. Technol. Biotechnol.*, 2010, **85**, 674.
49. M. V. Olarte, A. H. Zacher, A. B. Padmaperuma, S. D. Burton, H. M. Job, T. L. Lemmon, M. S. Swita, L. J. Rotness, G. N. Neuenschwander, J. G. Frye and D. C. Elliott, *Top. Catal.*, 2016, **59**, 55.
50. B. Delmon and G. F. Froment, *Catal. Rev. Sci. Eng.*, 1996, **38**, 69.
51. S. Lunzer and R. Kramer, Effect of water vapour on the catalytic activity of supported platinum catalysts, in *Studies in Surface Science and Catalysis*,

ed. A. Corma, F. V. Melo, S. Mendioroz, J. L. G. Fierro, Elsevier, Amsterdam, 2000, vol. 130, p. 2303.
52. O. İ Şenol, T.-R. Viljava and A. O. I. Krause, *Catal. Today*, 2005, **100**, 331.
53. T. Viljava, R. S. Komulainen and A. O. I. Krause, *Catal. Today*, 2000, **60**, 83.
54. D. C. Elliott and E. G. Baker, *Biotechnol. Bioeng. Symp.*, 1984, **16**, 159.
55. D. C. Elliott, *Energy Fuels*, 2007, **21**, 1792.
56. W. Yin, A. Kloekhorst, R. H. Venderbosch, M. V. Bykova, S. A. Khromova, V. A. Yakovlev and H. J. Heeres, *Catal. Sci. Technol.*, 2016, **6**, 5899.
57. E. Furimsky, *Appl. Catal., A*, 2003, **240**, 1.
58. S. T. Oyama, T. Gott, H. Zhao and Y. K. Lee, *Catal. Today*, 2009, **143**, 94.
59. H. Y. Zhao, D. Li, P. Bui and S. T. Oyama, *Appl. Catal., A*, 2011, **391**, 305.
60. V. A. Yakovlev, S. A. Khromova, O. V. Sherstyuk, V. O. Dundich, D. Y. Ermakov, V. M. Novopashina, M. Y. Lebedev, O. Bulavchenko and V. N. Parmon, *Catal. Today*, 2009, **144**, 362.
61. M. V. Bykova, D. Y. Ermakov, V. V. Kaichev, O. A. Bulavchenko, A. A. Saraev, M. Y. Lebedev and V. A. Yakovlev, *Appl. Catal., B*, 2012, **113–114**, 296.
62. A. R. Ardiyanti, S. A. Khromova, R. H. Venderbosch, V. A. Yakovlev and H. J. Heeres, *Appl. Catal., B*, 2012, **117–118**, 105.
63. A. R. K. Gollakota, M. Reddy, M. D. Subramanyam and N. Kishore, *Renewable Sustainable Energy Rev.*, 2016, **58**, 1543.
64. Z. He and X. Wang, *Catal. Sustain. Energy*, 2013, **1**, 28.
65. A. H. Zacher, M. V. Olarte, D. M. Santosa, D. C. Elliott and S. B. Jones, *Green Chem.*, 2014, **16**, 491.
66. E. Furimsky, *Catal. Today*, 2013, **217**, 13.
67. D. C. Elliott, *Curr. Opin. Chem. Eng.*, 2015, **9**, 59.
68. Y. Han, M. Gholizadeh, C.-C. Tran, S. Kaliaguine, C.-Z. Li, M. Olarte and M. Garcia-Perez, *Fuel Process. Technol.*, 2019, **195**, 106140.
69. M. Patel and A. Kumar, *Renewable Sustainable Energy Rev.*, 2016, **58**, 1293.
70. W. Jin, L. Pastor-Pérez, D. Shen, A. Sepúlveda-Escribano, S. Gu and T. Ramirez Reina, *ChemCatChem*, 2019, **11**, 924.
71. F. Ma and M. A. Hanna, *Bioresour. Technol.*, 1999, **70**, 1.
72. H. Bateni, A. Saraeian and C. Able, *Biofuel Res. J.*, 2017, **4**, 668.
73. X. Li, X. Luo, Y. Jin, J. Li, H. Zhang, A. Zhang and J. Xie, *Renewable Sustainable Energy Rev.*, 2018, **82**, 3762.
74. I. Kubičková and D. Kubička, *Waste Biomass Valorization*, 2010, **1**, 293.
75. A. V. Lavrenov, E. N. Bogdanets, Y. A. Chumachenko and V. A. Likholobov, *Catal. Ind.*, 2011, **3**, 250.
76. B. P. Pattanaik and R. D. Misra, *Renewable Sustainable Energy Rev.*, 2017, **73**, 545.
77. D. Singh, D. Sharma, S. L. L. Soni, S. Sharma and D. Kumari, *Fuel*, 2019, **253**, 60.
78. C. Kordulis, K. Bourikas, M. Gousi, E. Kordouli and A. Lycourghiotis, *Appl. Catal., B*, 2016, **181**, 156.
79. S. L. Douvartzides, N. D. Charisiou, K. N. Papageridis and M. A. Goula, *Energies*, 2019, **12**, 809.

80. H. Deldari, *Appl. Catal., A*, 2005, **293**, 1.
81. V. Akhmedov and S. Al-Khowaiter, *Catal. Rev. Sci. Eng.*, 2007, **49**, 33.
82. K. Park and S. Ihm, *Appl. Catal., A*, 2000, **203**, 201.
83. J. Xu and J. Y. Ying, *Angew. Chem., Int. Ed.*, 2006, **45**, 6700.
84. W. Li, K. Chi, H. Liu, H. Ma, W. Qu, C. Wang, G. Lv and Z. Tian, *Appl. Catal., A*, 2017, **537**, 59.
85. H. Yunfeng, W. Xiangsheng, G. Xinwen, L. Silue, H. Sheng, S. Haibo and B. Liang, *Catal. Lett.*, 2005, **100**, 59.
86. A. Melchor, E. Garbowski, M. Mathieu and M. Primet, *J. Chem. Soc. Faraday Trans. Phys. Chem. Condens. Phases*, 1986, **82**, 1893.
87. A. Chica and A. Corma, *J. Catal.*, 1999, **187**, 167.
88. Y. Zhang, D. Liu, Z. Men, K. Huang, Y. Lv, M. Li and B. Lou, *Fuel*, 2019, **236**, 428.
89. G. V. Echevskii, A. V. Toktarev, D. G. Aksenov and E. G. Kodenev, *Catal. Ind.*, 2018, **10**, 135.
90. Y. Yao, Y. J. Tian, H. H. Wu and S. X. Lu, *Fuel Process. Technol.*, 2015, **133**, 146.
91. J. A. Martens, D. Verboekend, K. Thomas, G. Vanbutsele, J. Pérez-Ramírez and J. P. Gilson, *Catal. Today*, 2013, **218–219**, 135.
92. S. V. Konnov, I. I. Ivanova, O. A. Ponomareva and V. I. Zaikovskii, *Microporous Mesoporous Mater.*, 2012, **164**, 222.
93. J. M. Hidalgo, M. Zbuzek, R. Černý and P. Jíša, *Cent. Eur. J. Chem.*, 2014, **12**, 1.
94. M. Busto, C. R. Vera and J. M. Grau, *Fuel Process. Technol.*, 2011, **92**, 1675.
95. G. Xing, S. Liu, Q. Guan and W. Li, *Catal. Today*, 2019, **330**, 109.
96. G. A. Urzhuntsev, A. V. Toktarev, G. V. Echevskii, I. V. Delii, E. N. Vlasova and G. A. Bukhtiyarova, *Catal. Ind.*, 2016, **8**, 32.
97. G. A. Urzhuntsev, E. G. Kodenev and G. V. Echevskii, *Catal. Ind.*, 2016, **8**, 224.
98. G. V. Echevskii, A. V. Toktarev, D. G. Aksenov and E. G. Kodenev, *Catal. Ind.*, 2018, **10**, 57.
99. R. B. Levy and M. Boudart, *Science*, 1973, **181**, 547.
100. R. J. Colton, J.-T. T. J. Huang and J. W. Rabalais, *Chem. Phys. Lett.*, 1975, **34**, 337.
101. M.-L. L. Frauwallner, F. López-Linares, J. Lara-Romero, C. E. Scott, V. Ali, E. Hernández and P. Pereira-Almao, *Appl. Catal., A*, 2011, **394**, 62.
102. C. H. Kim, Y. G. Hur, S. H. Lee and K. Y. Lee, *Fuel*, 2018, **233**, 200.
103. G. V. Echevskii, A. V. Toktarev, D. G. Aksenov and E. G. Kodenev, *Catal. Ind.*, 2017, **9**, 323.
104. O. V. Kikhtyanin, A. E. Rubanov, A. B. Ayupov and G. V. Echevsky, *Fuel*, 2010, **89**, 3085.
105. M. Y. Smirnova, O. V. Kikhtyanin, A. E. Rubanov, L. I. Trusov and G. V. Echevskii, *Catal. Ind.*, 2013, **5**, 253.
106. M. Y. Smirnova, O. V. Kikhtyanin, M. Y. Smirnov, A. V. Kalinkin, A. I. Titkov, A. B. Ayupov and D. Y. Ermakov, *Appl. Catal., A*, 2015, **505**, 524.

107. C. Wang, Z. Tian, L. Wang, R. Xu, Q. Liu, W. Qu, H. Ma and B. Wang, *ChemSusChem*, 2012, **5**, 1974.
108. M. Herskowitz, M. V. Landau, Y. Reizner and D. Berger, *Fuel*, 2013, **111**, 157.
109. M. Rabaev, M. V. Landau, R. Vidruk-Nehemya, A. Goldbourt and M. Herskowitz, *J. Catal.*, 2015, **332**, 164.
110. M. Ahmadi, E. E. Macias, J. B. Jasinski, P. Ratnasamy and M. A. Carreon, *J. Mol. Catal. A Chem.*, 2014, **386**, 14.
111. N. Chen, Y. Ren and E. W. Qian, *J. Catal.*, 2016, **334**, 79.
112. N. Chen, N. Wang, Y. Ren, H. Tominaga and E. W. Qian, *J. Catal.*, 2017, **345**, 124.
113. S. Liu, Q. Zhu, Q. Guan, L. He and W. Li, *Bioresour. Technol.*, 2015, **183**, 93.
114. C. Wang, Q. Liu, J. Song, W. Li, P. Li, R. Xu, H. Ma and Z. Tian, *Catal. Today*, 2014, **234**, 153.
115. Y. Cao, Y. Shi, J. Liang, Y. Wu, S. Huang, J. Wang, M. Yang and H. Hu, *Chem. Eng. Sci.*, 2017, **158**, 188.
116. E. W. Qian, N. Chen and S. Gong, *J. Mol. Catal. A Chem.*, 2014, **387**, 76.
117. D. Verma, B. S. Rana, R. Kumar, M. G. Sibi and A. K. Sinha, *Appl. Catal., A*, 2015, **490**, 108.
118. N. Chen, S. Gong and E. W. Qian, *Appl. Catal., B*, 2015, **174–175**, 253.
119. Y. Liu, L. Yao, H. Xin, G. Wang, D. Li and C. Hu, *Appl. Catal., B*, 2015, **174–175**, 504.
120. F. P. Sousa, L. N. Silva, D. B. de Rezende, L. C. A. de Oliveira and V. M. D. Pasa, *Fuel*, 2018, **223**, 149.
121. A. A. Nepomnyashchiy, E. A. Buluchevskiy, A. V. Lavrenov, V. L. Yurpalov, T. I. Gulyaeva, N. N. Leont'eva and V. P. Talzi, *Russ. J. Appl. Chem.*, 2017, **90**, 1944.
122. Y. Yang, C. Ochoa-Hernández, P. Pizarro, V. A. de la Peña O'Shea, J. M. Coronado and D. P. Serrano, *Top. Catal.*, 2012, **55**, 991.
123. C. Ochoa-Hernández, Y. Yang, P. Pizarro, V. A. de la Peña O'Shea, J. M. Coronado and D. P. Serrano, *Catal. Today*, 2013, **210**, 81.
124. A. B. Kulikov, M. I. Onishchenko, I. A. Sizova, A. L. Maksimov, S. V. Lysenko and E. A. Karakhanov, *Pet. Chem.*, 2016, **56**, 836.
125. Y. Liu, R. Sotelo-Boyás, K. Murata, T. Minowa and K. Sakanishi, *Chem. Lett.*, 2009, **38**, 552.
126. B. Veriansyah, J. Y. Han, S. K. Kim, S.-A. Hong, Y. J. Kim, J. S. Lim, Y.-W. Shu, S.-G. Oh and J. Kim, *Fuel*, 2012, **94**, 578.
127. S. Phimsen, W. Kiatkittipong, H. Yamada, T. Tagawa, K. Kiatkittipong, N. Laosiripojana and S. Assabumrungrat, *Energy Convers. Manag.*, 2016, **126**, 1028.
128. J. Han, J. Duan, P. Chen, H. Lou, X. Zheng and H. Hong, *Green Chem.*, 2011, **13**, 2561.
129. M. Ikura, M. Stanciulescu and E. Hogan, *Biomass Bioenergy*, 2003, **24**, 221.
130. D. Chiaramonti, M. Bonini, E. Fratini, G. Tondi, K. Gartner, A. V. Bridgwater, H. P. Grimm, I. Soldaini, A. Webster and P. Baglioni, *Biomass Bioenergy*, 2003, **25**, 85.

131. D. Chiaramonti, M. Bonini, E. Fratini, G. Tondi, K. Gartner, A. V. Bridgwater, H. P. Grimm, I. Soldaini, A. Webster and P. Baglioni, *Biomass Bioenergy*, 2003, **25**, 101.
132. P. U. Okoye and B. H. Hameed, *Renewable Sustainable Energy Rev.*, 2016, **53**, 558.
133. A. Smirnov, S. Selishcheva and V. Yakovlev, *Catalysts*, 2018, **8**, 595.
134. V. O. Samoilov, D. N. Ramazanov, A. I. Nekhaev, A. L. Maximov and L. N. Bagdasarov, *Fuel*, 2016, **172**, 310.
135. R. K. Prasad, S. Chatterjee, P. B. Mazumder, S. K. Gupta, S. Sharma, M. G. Vairale, S. Datta, S. K. Dwivedi and D. K. Gupta, *Chemosphere*, 2019, **231**, 588.
136. G. W. Huber, S. Iborra and A. Corma, *Chem. Rev.*, 2006, **106**, 4044.
137. Y. Román-Leshkov, C. J. Barrett, Z. Y. Liu and J. A. Dumesic, *Nature*, 2007, **447**, 982.
138. A. Démolis, N. Essayem and F. Rataboul, *ACS Sustain. Chem. Eng.*, 2014, **2**, 1338.
139. K. C. Badgujar, V. C. Badgujar and B. M. Bhanage, *Fuel Process. Technol.*, 2020, **197**, 106213.
140. X. Li, R. Xu, J. Yang, S. Nie, D. Liu, Y. Liu and C. Si, *Ind. Crops Prod.*, 2019, **130**, 184.
141. T. Mizugaki, K. Togo, Z. Maeno, T. Mitsudome, K. Jitsukawa and K. Kaneda, *ACS Sustain. Chem. Eng.*, 2016, **4**, 682.
142. J. Tan, J. Cui, T. Deng, X. Cui, G. Ding, Y. Zhu and Y. Li, *ChemCatChem*, 2015, **7**, 508.
143. C. Xiao, T. W. Goh, Z. Qi, S. Goes, K. Brashler, C. Perez and W. Huang, *ACS Catal.*, 2016, **6**, 593.
144. I. T. Horváth, H. Mehdi, V. Fábos, L. Boda and L. T. Mika, *Green Chem.*, 2008, **10**, 238.
145. V. Fábos, G. Koczó, H. Mehdi, L. Boda and I. T. Horváth, *Energy Environ. Sci.*, 2009, **2**, 767.
146. D. G. Manly and A. P. Dunlop, *J. Org. Chem.*, 1958, **23**, 1093.
147. I. N. Shilov, A. A. Smirnov, O. A. Bulavchenko and V. A. Yakovlev, *Catalysts*, 2018, **8**, 1.
148. A. A. Smirnov, I. N. Shilov, M. V. Alekseeva, S. A. Selishcheva and V. A. Yakovlev, *Catal. Ind.*, 2018, **10**, 228.
149. M. M. Villaverde, T. F. Garetto and A. J. Marchi, *Catal. Commun.*, 2015, **58**, 6.
150. C. Wang, H. Xu, R. Daniel, A. Ghafourian, J. M. Herreros, S. Shuai and X. Ma, *Fuel*, 2013, **103**, 200.
151. A. Corma, O. De La Torre, M. Renz and N. Villandier, *Angew. Chem., Int. Ed.*, 2011, **50**, 2375.
152. A. Corma, O. De La Torre and M. Renz, *ChemSusChem*, 2011, **4**, 1574.
153. A. Corma, O. De La Torre and M. Renz, *Energy Environ. Sci.*, 2012, **5**, 6328.
154. G. Li, N. Li, X. Wang, X. Sheng, S. Li, A. Wang, Y. Cong, X. Wang and T. Zhang, *Energy Fuels*, 2014, **28**, 5112.
155. G. Li, N. Li, J. Yang, A. Wang, X. Wang, Y. Cong and T. Zhang, *Bioresour. Technol.*, 2013, **134**, 66.

156. G. Li, N. Li, S. Li, A. Wang, Y. Cong, X. Wang and T. Zhang, *Chem. Commun.*, 2013, **49**, 5727.
157. M. Balakrishnan, E. R. Sacia and A. T. Bell, *ChemSusChem*, 2014, 7, 2796.
158. J. P. Lange, E. Van Der Heide, J. Van Buijtenen and R. Price, *ChemSusChem*, 2012, **5**, 150.
159. G. Dai, S. Wang, Q. Zou and S. Huang, *Fuel Process. Technol.*, 2018, **179**, 319.
160. P. Ghorbannezhad, M. D. Firouzabadi, A. Ghasemian, P. J. de Wild and H. J. Heeres, *J. Anal. Appl. Pyrol.*, 2018, **131**, 1.
161. A. Zheng, Z. Zhao, S. Chang, Z. Huang, H. Wu, X. Wang, F. He and H. Li, *J. Mol. Catal. A Chem.*, 2014, **383–384**, 23.
162. Z. Sun, B. Fridrich, A. De Santi, S. Elangovan and K. Barta, *Chem. Rev.*, 2018, **118**, 614.
163. Q. Che, M. Yang, X. Wang, Q. Yang, L. Rose Williams, H. Yang, J. Zou, K. Zeng, Y. Zhu, Y. Chen and H. Chen, *Bioresour. Technol.*, 2019, **278**, 248.
164. A. M. Elfadly, I. F. Zeid, F. Z. Yehia, A. M. Rabie, M. M. Aboualala and S.-E. Park, *Int. J. Biol. Macromol.*, 2016, **91**, 278.
165. P. Sirous-Rezaei, J. Jae, K. Cho, C. H. Ko, S. C. Jung and Y. K. Park, *Chem. Eng. J.*, 2019, **377**, 120121.

CHAPTER 7

Catalysis in Modern Bio-refineries: Towards a New Bio-energy Paradigm†

JIE YU

Huazhong University of Science and Technology, State Key Lab of Coal Combustion, Luoyu Road 1037, Wuhan, China
Email: yujie@hust.edu.cn

7.1 Introduction

As a renewable fuel resource for energy production, biomass is likely to play a major role in replacing fossil fuels, helping to meet increasing energy demands and reducing CO_2 emissions. The conversional methods of biomass utilization include pyrolysis for liquid products, gasification for synthesis gas (syngas) and combustion for energy. Regarding the gasification of biomass for syngas, the main technical barrier is the existence of tars in the syngas,[1] which can block downstream facilities and cause serious operational problems. To overcome this problem, gasification of biomass in supercritical water (SCW) was proposed. By means of conversion in SCW, a mixture of H_2, CH_4 and CO can be produced without the drying process of the wet biomass.

†Electronic supplementary information (ESI) available. Colour version of Figure 7.7. See DOI: 10.1039/9781788019576.

Heterogeneous Catalysis for Energy Applications
Energy and Environment Series No. 27
Edited by Tomas R. Reina and Jose A. Odriozola

Published by the Royal Society of Chemistry, www.rsc.org

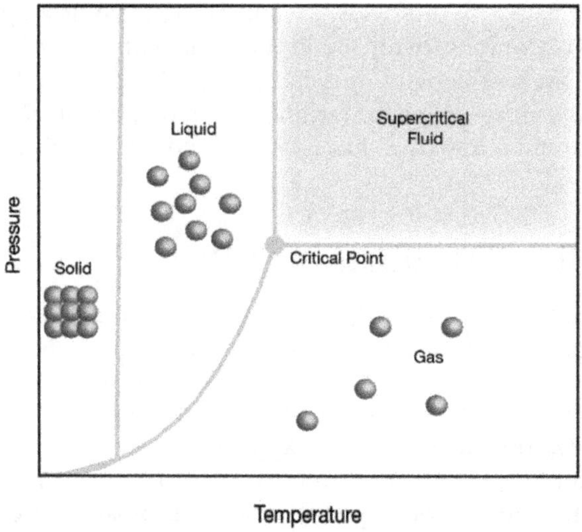

Figure 7.1 Pressure–temperature phase diagram of pure water. Reproduced from Ref. 2 with permission from the Royal Society of Chemistry.

As is shown in Figure 7.1, upon changing temperature and pressure, the phase of water can be changed into solid, liquid or gas. With temperatures and pressures at critical points (T = 374.15 °C, P = 22.1 MPa), various characteristics of water (*e.g.* density, dielectric constant, boiling point, ionic product, thermal conductivity and ionic product) undergoes significant changes. With further increases to the pressure and temperature further, an ideal single-phase environment is created, which can decrease mass transfer limitations to a great extent over other systems. Additionally, the produced gas during reaction can be completely solubilized. Therefore, SCW shows comparable characteristics of ideal miscibility and high solubility analogous to some organic solvents. Under such conditions, a series of chemical reactions such as cracking, hydrogenation, hydrolysis, methanation and water gas shift reactions can readily occur under SCW conditions. Therefore, SCW offered various advantages over conventional process: (1) a high reaction rate; (2) a negligible limit of mass transfer resistance; (3) a higher dispersity and heat transfer; (4) coke precursor clearance; and (5) easy removal of sulfur and nitrogen.

Under SCW conditions, catalysts are commonly used to produce high H_2 yields at comparatively low temperatures by promoting the water gas shift reaction ($CO + H_2O \leftrightarrow H_2 + CO_2$) and minimizing the methanation reaction ($CO_2 + 4H_2 \leftrightarrow CH_4 + 2H_2O$). In contrast, to maximize the production of CH_4, the methanation reaction should be accelerated along with the water gas shift reaction. Therefore, the selectivity of the above reactions should be controlled with the use of different catalysts for the production of target products.

Given the advantages of low cost and better dispersion in SCW, homogeneous catalysts (Na_2CO_3, K_2CO_3, KOH, NaOH, *etc.*) have been widely studied.[3] However, it is relatively difficult to recycle homogeneous catalysts from SCW. Comparatively, heterogeneous catalysts offer a variety of advantages of high selectivity, recyclability and reuse. Zeolites, alumina-based materials, silica-based materials, mixed oxide materials and carbon materials/carbon composites have been used as supports under SCW conditions.[4]

7.2 Stability of Catalysts in SCW

Extensive studies on the heterogeneous catalytic reaction of biomass in SCW have been conducted to achieve a high yield of desired products. The heterogeneous catalysts were generally synthesized by incorporating active metals onto relatively stable porous materials, such as zeolites, Al_2O_3, TiO_2, ZrO_2 and carbon-based supports.[4] The stability of the catalyst support markedly affected the effectiveness of active metals. However, catalysts stable under vapour-phase reactions or subcritical water may not show adequate hydrothermal stabilities in SCW conditions due to the harsh conditions, which can impair their long-term performance as promising catalysts. Therefore, preliminary work on stability tests of catalysts in SCW have been conducted in order to choose ideal catalyst supports resistant to these harsh conditions. The stability of catalysts was correlated with their physical and crystalline structures. The physical structure may be damaged by water molecules under the severe conditions of SCW, and the surface area may be decreased significantly due to physical structure collapse.

Zeolites have been validated as effective catalysts for cracking and hydrodeoxygenation reactions in the vapour phase. The structural changes of zeolites need to be understood under aqueous-phase reactions. In liquid water (>100 °C), zeolites lost stability mainly due to the hydrolysis of Si–O–Si bonds, while in superheated steam, the hydrolysis of Al–O–Al bonds, namely dealumination, led to structural changes in zeolites.[4] For its application in SCW, the hydrothermal stability and physicochemical properties of zeolites were also studied in SCW.[5] In SCW, HY and β-zeolites were completely collapsed, as indicated by significant decreases in the Brunauer–Emmett–Teller (BET) surface area, as shown in Table 7.1. In contrast, ZSM-5 showed higher hydrothermal stability in comparison with other zeolites, even though it also showed a decline in BET surface area, indicating the dependence of stability on zeolite framework type. After SCW treatment, the acidity of zeolites was markedly reduced, along with their micropore areas.

Table 7.1 BET surface area change of catalysts in supercritical water.[5,6]

	ZSM-5	HY	β	Ru/γ-Al_2O_3	Ru/TiO_2	Ru/C
BET surface area (%)	−10.1	−90.6	−83.9	−93.7	−8.3	+1.4

Table 7.2 Stability of catalysts in supercritical water. Reproduced from Ref. 4 with permission from the Royal Society of Chemistry.

Catalyst	Supercritical water			Structural change (Y/N)
	Temperature (°C)	Pressure (atm)	Time (h)	
γ-Al_2O_3	350	197.4	NA	Y
Al_2O_3, SiO_2 and AC	500	276.3	0.25	Y
Mono-, bimetallic Ir, Pt and Pt–Ni on γ-Al_2O_3	450	246.7	0.25	Y
TiO_2, ZrO_2 and Ce–ZrO_2	500	276.3	0.25	N
Pt/CNT	450	246.7	6	N

Refractory metal oxides such as γ-Al_2O_3, TiO_2 and ZrO_2 and carbon materials have been extensively used as catalyst supports in industry. It was reported that metal oxides of γ-Al_2O_3, SiO_2 and activated carbon (AC) all suffered from structural changes in SCW to differing extents, leading to reductions in the BET surface area, as shown in Table 7.2. In contrast, TiO_2, ZrO_2 and carbon nanotubes (CNTs) showed higher stability. Two different forms of monoclinic and tetragonal zirconia were tested for their stability as catalyst supports in SCW.[6] Monoclinic zirconia showed good physical and crystallographic stabilities, while tetragonal ZrO_2 was partially transformed to its monoclinic form. Osada *et al.* evaluated the stability of Ru/TiO_2, Ru/Al_2O_3 and Ru/AC.[6] It was found that Ru/TiO_2 maintained the highest reactivity in SCW and kept their BET surface areas almost unchanged. However, there is still no consensus as to whether the anatase phase of TiO_2 could be transformed into rutile TiO_2.[6,7] In contrast, Ru/Al_2O_3 quickly lost its activity due to the decrease in BET surface area and the structure change from γ-Al_2O_3 to α-Al_2O_3. Even though Ru/AC showed no BET surface area loss in SCW without feed, a significant reduction in BET surface area was observed in lignin gasification, which led to a decrease in gasification reactivity. The characteristic of microporosity dominated within AC, making them susceptible to channel blockage by coke.

Refractory metal oxides of γ-Al_2O_3 are some of the most used catalyst supports in industry due to their low price, superior stability and high porosity. To understand the structure evolution of Al_2O_3 in SCW, Ravenella *et al.* studied structural changes of γ-Al_2O_3 in hot liquid water.[8] It was reported that under mild conditions (200 °C, autogenous pressure), γ-Al_2O_3 was converted into a different phase of hydrated boehmite (AlOOH) and had significantly decreased acidity and surface area, as shown in Figure 7.2. The extent of γ-Al_2O_3 transformation increased with increasing temperature, leading to the phase transformation of alumina ($\gamma \rightarrow \delta \rightarrow \theta \rightarrow \alpha$). The surface area also decreased with the phase transformation and was lowest for α-Al_2O_3. Elliott *et al.* revealed that in subcritical water or SCW, δ- and γ-Al_2O_3 were completely

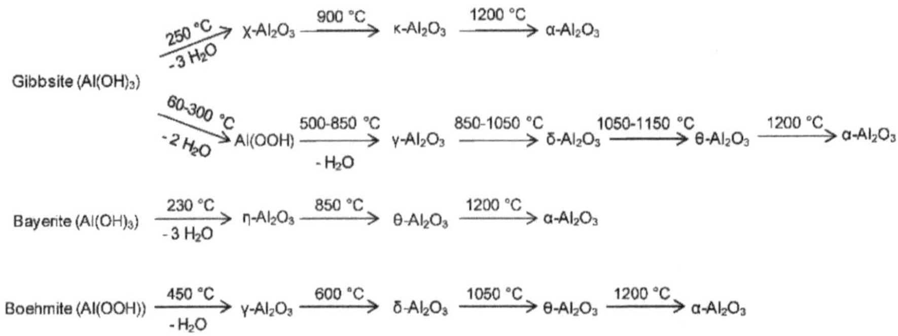

Figure 7.2 Dehydroxylation sequences of alumina hydrates. Reproduced from Ref. 8 with permission from American Chemical Society, Copyright 2011.

transformed into boehmite.[9] The incorporation of metals can effectively delay the transformation rate of γ-Al_2O_3 by preventing the hydration of the alumina surface, thus preserving the stability for a comparatively long period. The surface OH groups were reported to initiate the hydration reaction of γ-Al_2O_3 into boehmite. Therefore, it is possible to enhance the stability of γ-Al_2O_3 by eliminating or capping the OH groups. After treatment in SCW, a decrease in Lewis acid site concentration was also observed. Lewis acid sites in γ-Al_2O_3 are mainly related to tetrahedrally coordinated Al atoms. In aqueous-phase reactions, water can be dissociatively adsorbed onto Lewis acid sites with the formation of octahedrally coordinated Al, which can decrease the Lewis acid sites.

CNTs have also been extensively used for catalyst support in SCW due to their open structure and lack of meso- and micro-pores, which can suppress catalyst deactivation and strengthen resistance to harsh conditions of SCW. Vlieger *et al.* treated CNTs in SCW in order to study the stability of CNTs.[10] The exposure of CNTs to SCW did not change the physical structure of CNTs. In addition, the tube wall was well preserved and the ends of tubes were kept open. Raman analysis also confirmed that the crystallinity of tubes was not affected by SCW. X-ray photoelectron spectroscopy analysis demonstrated that the surface oxygen concentrations did not increase at a result of treatment in SCW. The graphene sheet structure of CNTs explained their high resistance to harsh conditions.

The stability of catalyst supports in SCW is summarized in Figure 7.3. Carbon materials, ZrO_2 and TiO_2 showed the highest stability, being the most suitable for reaction in SCW. Zeolites, Al_2O_3 and other mixed oxides were relatively unstable. The incorporation of active metals can enhance the stability of corresponding catalyst supports.

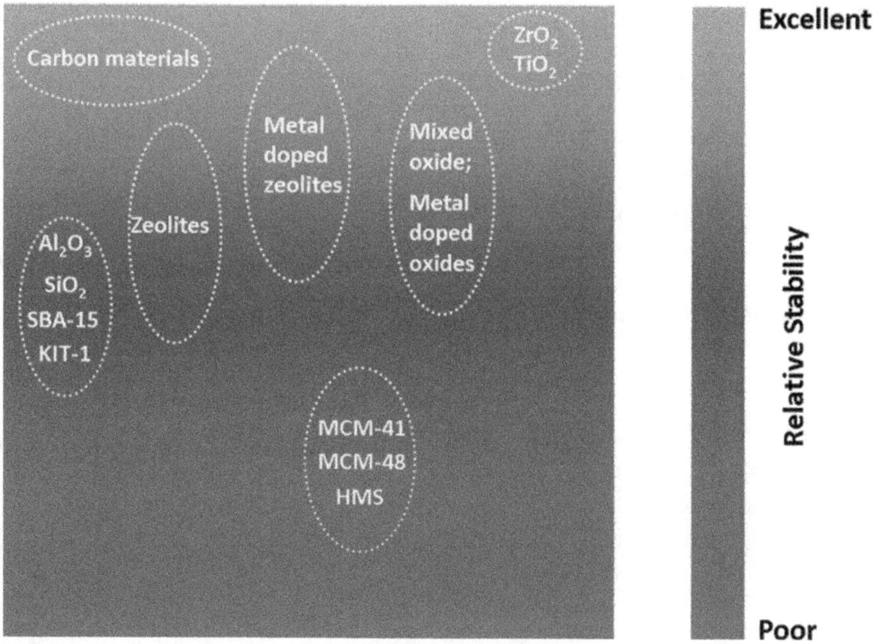

Figure 7.3 Hydrothermal stability of different catalysts in superheated steam (>400 °C). Reproduced from Ref. 4 with permission from the Royal Society of Chemistry.

7.3 Mechanism of Catalytic Reaction in SCW

7.3.1 Reaction Chemistry

Biomass is mainly composed of cellulose, lignin and hemicellulose. The conventional pathway of biomass in SCW is given in Scheme 7.1. Upon treatment in SCW, three components can be decomposed into oxygenated intermediates. The oxygenated intermediates will produce H_2 and CO/CO_2 through the steam reforming reaction and the water gas shift reaction. A maximum theoretical yield of hydrogen can be calculated according to eqn (7.1) and (7.2). The produced CO can be converted into H_2 by the water gas shift reaction following eqn (7.3). The H_2 can be further converted to CH_4 through the methanation reaction. The oxygenated intermediates may also produce CH_4 through the methanation reaction directly.

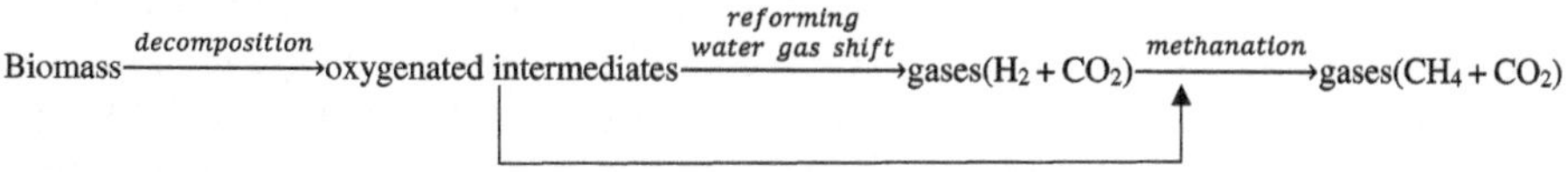

Scheme 7.1

Steam reforming reaction:

$$CH_xO_y + (1-y)H_2O \rightarrow (1-y+x/2)H_2 + CO \quad (7.1)$$

$$CH_xO_y + (2-y)H_2O \rightarrow (2-y+x/2)H_2 + CO_2 \quad (7.2)$$

Water gas shift reaction:

$$CO + H_2O \rightarrow H_2 + CO_2 \quad (7.3)$$

Methanation of CO and CO_2:

$$CO + 3H_2 \rightarrow H_2O + CH_4 \quad (7.4)$$

$$CO_2 + 4H_2 \rightarrow H_2O + CH_4 \quad (7.5)$$

Methanation of char:

$$C + 2H_2 \rightarrow CH_4 \quad (7.6)$$

Boudouard reaction:

$$2CO \leftrightarrow C + CO_2 \quad (7.7)$$

Yakaboylu *et al.* studied the evolution of product gases during the gasification of manure under SCW conditions.[11] The calculations given in Figure 7.4 indicated that CO_2 and CH_4 increased drastically with temperature up to 340 °C. At higher temperatures, the rate of increase slowed down. Above ~380 °C, CH_4 began to decline. Comparatively, H_2 increased with temperature. This trend is consistent with other thermodynamic equilibrium calculations.[12,13] Nikoo *et al.* indicated that lower temperatures and higher concentrations of glucose were beneficial for the methanation reaction and the decomposition of glucose into CH_4, while higher H_2 can be obtained at lower glucose concentrations and higher temperatures.[14]

To achieve high conversion efficiency of the desired reaction without accelerating the severity of reaction conditions and to steer selectivity towards the desired products of H_2 or CH_4, different catalysts are usually used in SCW gasification.[15] Generally, catalysts promote the decomposition of biomass by transforming intermediates (formed from hydrolysis and dehydration) into gases. The gas components can be optimized to CH_4 and H_2O through facilitating the water gas shift reaction and the hydrogenation of CO/CO_2 (methanation reaction). One of the accepted reaction pathways for catalytic gasification of biomass in SCW is shown in Figure 7.5. At the initial stage, the oxygenated intermediates from biomass decomposition undergo the dehydrogenation reaction on the metal surface to form adsorbed intermediates. The adsorbed intermediates are then decomposed into adsorbed C–O* and H_2 (Pathway I) by desired C–C cleavage or to produce alcohols through

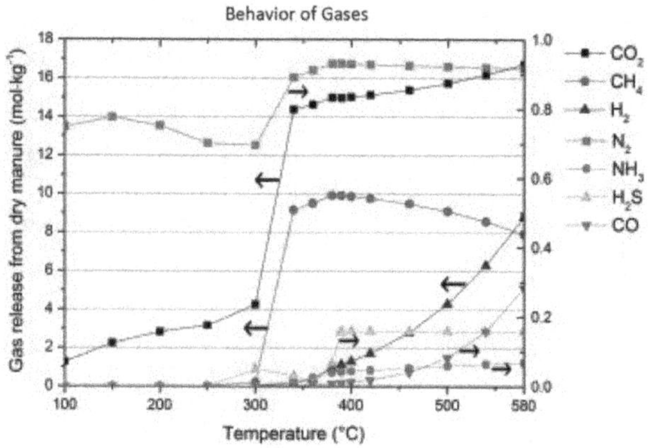

Figure 7.4 Theoretical gas evolution during SCW gasification of manure at 24 MPa and at different temperatures. Reproduced from Ref. 11 with permission from Elsevier, Copyright 2013.

undesired C–O cleavage (Pathway II). The adsorbed C–O* may react with H_2O to produce CO_2 and H_2 by the water gas shift reaction (Pathway III) or form methane by the methanation reaction (Pathway IV). This mechanism illustrates the indirect formation of CH_4 through the methanation reaction. Bjelić *et al.* also proposed details of CH_4 formation on the catalyst surface directly through the following:[16] (1) fragmentation of large molecules to small intermediates by C–C bond breakage; (2) formation of adsorbed C*, O* and H* atoms by breaking C–O and C–H bonds; (3) water molecules split into OH*, O* and H*; (4) accelerating the water gas shift reaction between adsorbed CO* and OH*; and (5) methane formation from adsorbed C* and H* atoms: $C^* + 4H^* \leftrightarrow CH_4^*$. Hence, for a high yield of CH_4, high activities for the cleavage of C–C and C–O bonds and the water gas shift reaction are desired. To maximize the production of H_2, the catalyst should hinder the cleavage of the C–O bond while favouring C–C bond breakage and the water gas shift reaction.

7.3.2 Heterogeneous Catalysts

Generally, the performance of the various catalysts could be attributed to the following aspects: (1) support stability; (2) acidity of support; (3) active metal dispersion; (4) interaction between the active metal and support; and (5) mass transfer issues.[18] Specifically, an ideal catalyst for SCW gasification should have the following characteristics: (1) high hydrothermal stability – the structure of the catalyst should be preserved under SCW conditions; (2) high activity – the catalyst should have high activity to produce the desired gases and avoid the formation of coke; and (3) resistance to deactivation – the catalyst showed have high tolerance to sulfur and low active metal leaching.[19]

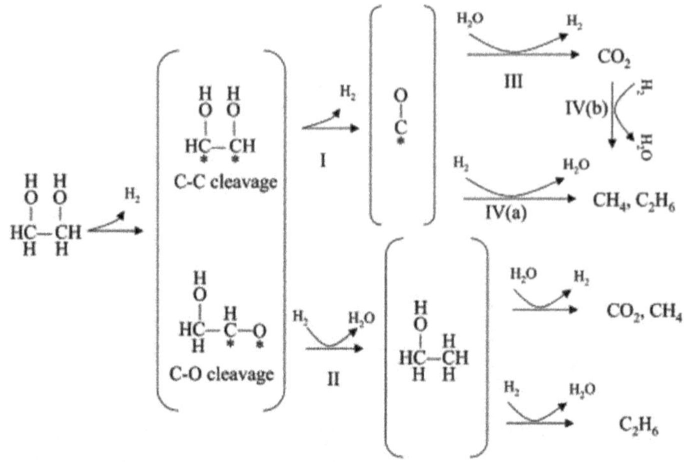

Figure 7.5 Reaction pathways for the conversion of biomass with water. Pathway I is desired C–C cleavage to form C–O*. Pathway II represents C–O cleavage followed by hydrogenation to methane and ethane. Pathway III is the water gas shift reaction. Pathway IV is the methanation and Fischer–Tropsch reactions. Reproduced from Ref. 17 with permission from Elsevier, Copyright 2004.

A high yield of H_2 production, high activity for C–C cleavage, a water gas shift reaction promoting the removal of C–O* along with high hydrothermal stability should be desired, along with low activity for C–O cleavage and hydrogenation of CO and CO_2.[20] To meet the above requirements, heterogeneous catalysts have been extensively studied.

As discussed above, α-Al_2O_3, ZrO_2, carbon nanofibre and TiO_2 were the stable supports used for gasification in SCW, as listed in Table 7.1. Relative to other noble metals, such as Pt, Pd and Rh, Ru has been reported to be more effective for the cracking, water gas shift, methanation and hydrogenation reactions and to be more resistant to sintering. Nickel has also been commonly studied due to its reactivity and recyclability. In contrast, other transition metals are inactive or oxidized. It was proposed that Ni/Al_2O_3 can promote hydrogen formation, while Ru/Al_2O_3 favours methane formation.[21–25] In this chapter, these two active metals will be mainly discussed.

7.4 Effect of Catalysts

Elliott *et al.* reported that the activity of catalysts on the decomposition of 4-cresol in SCW (350 °C) was in the order of Ru/γ-Al_2O_3 > Rh/AC, Ru/AC > Pt/γ-Al_2O_3, Pd/γ-Al_2O_3 and Pd/AC.[26] Azadi *et al.* evaluated the hydrogen production from cellulose, lignin, bark and model carbohydrates in SCW using Ni/α-Al_2O_3, Ni/hydrotalcite, Raney nickel, Ru/C and Ru/γ-Al_2O_3.[27] It was found that the Ni-supported catalysts showed high activity for gasification efficiency and hydrogen production, while Ru-based catalysts and Raney nickel resulted in high CH_4 yields. Sato *et al.* studied the catalytic gasification of alkylphenols in the presence of various catalysts in SCW (400 °C).[28] It was found that catalyst activities decrease in the order Ru/γ-Al_2O_3 > Ru/AC, Rh/AC > Pt/γ-Al_2O_3, Pd/AC and Pd/γ-Al_2O_3. For Ru/γ-Al_2O_3, Ru/AC and Rh/AC, CH_4 and H_2 compositions were ~60% and 10%, respectively. For Pt and Pd catalysts, the H_2 yield was markedly larger than for Ru and Rh. About 50% H_2 yields were obtained for Pd/γ-Al_2O_3 catalysts. Osada *et al.* carried out gasification of lignin and 4-propylphenol in SCW (400 °C).[29] It was found that the turnover number (mol of carbon atoms in gaseous product per mol of surface metal atoms) decreased in the order Ru/TiO_2 > Ru/Al_2O_3 > Ru/C, Pt/C > Pt/Al_2O_3, Rh/C > Pd/C > Pd/Al_2O_3 > Ni/Al_2O_3. The selectivity towards CH_4 follows the order Ru > Rh > Pt > Pd > Ni, while the selectivity towards CO and H_2 follows the order Ni > Pd > Pt > Rh > Ru. Without catalyst, the gas yield was low, and this was mainly composed of CH_4 (69.5%). Ru/TiO_2 promoted the gas yield with lower concentrations of CH_4 and CO, but higher concentrations of CO_2 and H_2, indicating the action of the water gas shift reaction. The group further compared the activity of various transition metals supported on AC for the gasification of lignin in SCW (400 °C).[30] An order of Ru/AC > Rh/AC > Pt/AC > Pd/AC > Ni/AC was observed for gas formation, while H_2 selectivity decreased in the order of Pd/AC > Ru/AC > Pt/AC > Rh/AC > Ni/AC. The authors claimed that the lignin was initially mainly converted to H_2 and CO. In the following stage, CH_4 was formed by the methanation reaction. However, the gas yield and CH_4 composition in the gas product increased simultaneously with reaction time. Therefore, CH_4 could also be directly formed from intermediate decomposition. From the above analysis, it is evident that noble metal-loaded catalysts can promote gas formation. In particular, Ru catalysts are more active than other catalysts for producing CH_4. It was reported that Ru catalysts can also promote H_2 formation. Byrd conducted catalytic gasification of ethanol and glycerol above 600 °C.[31,32] It was found that Ru/Al_2O_3 favoured the carbon conversion to H_2, with small amounts of CH_4 and CO formed. Guan *et al.* gasified lignin using Ni/Al_2O_3-SiO_2, Ru/C, Pt/C and Pd/C in SCW.[33] Ru/C doubled the gaseous and carbon efficiency relative to non-catalytic gasification. H_2, CH_4 and CO_2 were also increased significantly. Without catalyst, the yield of CH_4 was higher than H_2, while H_2 surpassed CH_4 significantly with catalyst. Among the studied catalysts, Ru/C showed the optimal ability of increasing carbon conversion and gas product, especially H_2.

Lu *et al.* incorporated Ni into different catalyst supports (CeO_2/Al_2O_3, La_2O_3/Al_2O_3, MgO/Al_2O_3, ZrO_2/Al_2O_3) for the catalytic gasification of glucose in SCW.[34] Regarding the H_2 yield, the reactivity followed the order of CeO_2/Al_2O_3 > La_2O_3/Al_2O_3 > MgO/Al_2O_3 > Al_2O_3 > ZrO_2/Al_2O_3, while H_2 selectivity followed the order of CeO_2/Al_2O_3 > La_2O_3/Al_2O_3 > ZrO_2/Al_2O_3 > Al_2O_3 > MgO/Al_2O_3. The superior activity of CeO_2 may be due to its resistance to coke formation. Zhang *et al.* synthesized 17 different catalysts with 4 different transition metals (Ni, Ru, Cu or Co) and promoters (Na, K, Mg or Ru) supported on different supports (γ-Al_2O_3, ZrO_2 and AC).[35] It was shown that both metallic and oxidic Ni and Ru result in high activity for producing H_2 under SCW.

The γ-Al_2O_3 support showed better activity, while AC showed inferior performance compared to the other two supports. Promoting Ru with Mg was effective at suppressing coke and tar formation on Ni/γ-Al_2O_3 catalysts, while K or Ca did not exhibit positive effects on reducing coke formation. Azadi *et al.* reported the selectivity and reactivity of 44 different supports with Ni loading for the catalytic gasification of biomass in SCW.[18] Among the catalysts, supports of α-Al_2O_3, CNTs and MgO resulted in the best activities in terms of carbon conversion, while SiO_2, Y_2O_3 and TiO_2 showed modest activities. Zeolites showed negligible activity.

Sato *et al.* examined the activity of Ni/MgO for the gasification of lignin in SCW from 250 to 400 °C.[36,37] MgO can catalyse the reaction, forming fragment intermediates, which could be reformed into gases by Ni or be condensed into tetrahydrofuran-insoluble products. Mastuli *et al.* synthesized NiO/MgO, CuO/MgO and ZnO/MgO catalysts for the catalytic gasification of biomass in SCW.[38,39] Even though ZnO/MgO had larger crystallite sizes and smaller specific surface areas than NiO/MgO and CuO/MgO, ZnO/MgO exhibited the highest H_2 yield. Moreover, to promote the H_2 yields, the strength of basic sites played a more pronounced effect than the density of the basic sites. This means that other factors such as metal dispersion, support acidity, metal–support interaction and support stability may determine catalyst reactivity.

Bimetallic metals catalysts were also proposed to increase selectivity and hinder catalyst deactivation. Huber *et al.* synthesized Ni–Sn catalysts for the catalytic gasification of biomass in SCW. It was found that Sn dramatically decreased CH_4 formation from C–O bond cleavage while improve H_2 selectivity by the cleavage of C–C bonds.[20] Zhang *et al.* revealed that Ru-modified Ni/γ-Al_2O_3 decreased coke in the catalyst without significantly affecting the gas composition.[35] Hossain *et al.* also prepared bimetallic Ni–Ru-Al_2O_3 catalysts for hydrogen production from glucose in SCW.[40] Similarly, Ru was found to hinder coke formation and increase H_2 production.

7.5 Methanation Formation Mechanism

7.5.1 Pathway I: Direct Methane Production

Vogel *et al.* gasified model biomass mixtures and wood in SCW with Ru/C.[41] It was found that CO_2 dominated below 285 °C. With temperature increases up to 305 °C, H_2 and CH_4 gradually dominated. With temperatures increasing further, CH_4 increased with temperature while CO_2 decreased and H_2 remained unchanged. CH_4 was proposed to be formed through the methanation reaction between CO_2 and H_2. Therefore, a sequential gasification–methanation mechanism was proposed here. However, considering that one mole of CO_2 will consume four moles of H_2 for the methanation reaction, it is unreasonable to conclude that CH_4 was formed through the methanation reaction between CO_2 and H_2. Onwudili *et al.* also investigated the catalytic gasification of heavy bio-oil using Ru/Al_2O_3 in SCW between 400 and 500 °C.[42] With higher Ru loading in the catalysts, more gas was formed, while oil decreased accordingly. At the same time, more CH_4 was formed. The authors claimed the CH_4 formation was attributed to the two stages of the water gas shift reaction followed by the methanation reaction. However, it is not clear whether the CH_4 was formed from the decomposition of oil or the methanation reaction.

Ortiz *et al.* gasified glycerol in SCW above 600 °C using Ru/Al_2O_3.[24] It was found that the product gas was mainly composed of CH_4 and CO, while H_2 was quite low. It was reported that Ru catalysts can promote breakage of C–C bonds[42] and thus the methanation reaction of CO/CO_2 and H_2.[22] But at such high temperatures and amounts of water, the methanation reaction of CO/CO_2 with H_2 (reverse reaction of CH_4 reforming) is very unlikely.[24] Therefore, it was proposed that CH_4 may be formed from glycerol decomposition before reaching the catalyst. Similar works were also conducted for the catalytic gasification of glucose and glycerol using Ru/TiO_2 in SCW by Kersten *et al.*[43] Without catalyst, CO dominated the gas products, followed by CH_4, H_2 and CO_2. Given the similar gasification behaviour of glucose and glycerol with respect to carbon conversion and that CH_4, CO and C_2 increased with concentrations of glucose and glycerol while H_2 and CO_2 decreased, it can be inferred that direct thermal cracking of biomass was the main source for CH_4 and CO formation in SCW without catalyst. With Ru/TiO_2, the carbon conversion, CO_2 and H_2 increased markedly, while CO diminished. This means that Ru/TiO_2 is capable of promoting the water gas shift reaction. Interestingly, no CO nor CH_4 was found for 1% glucose at 600 °C and 300 bar using Ru/TiO_2. This indicates that the methanation reaction did not occur or the consumption of CH_4 surpassed the production of CH_4. However, under these conditions without catalyst, CH_4 was detected. Therefore, it can be inferred that CH_4 could be reformed to H_2 and CO_2 by Ru/TiO_2.

Kruse *et al.* carried out methane reforming in SCW (650 °C) with various basic catalysts (KOH, K_2CO_3 and NaOH) and Raney nickel.[44] Without catalyst, around 10% of H_2 was formed through the methane reforming reaction due to the catalytic effect of the reactor wall. In the presence of catalyst, the formation of H_2 was significantly accelerated. At higher concentrations of 5% glucose, CH_4 was markedly increased and H_2 increased slightly, while CO_2 decreased. Based on the fact that the methanation reaction in the gas phase at this condition was negligible, as discussed above, it can be concluded that the CH_4 may be mainly from the glucose directly instead of the methanation reaction between CO_2 and H_2. Cheng *et al.* gasified glycerol using Ni/Al_2O_3 in SCW and also proposed that CH_4 was mainly directly from the hydrogenolysis of glycerol.[45]

7.5.2 Pathway II: Methanation Reaction with Ru Metal

Park and Tomiyasu conducted gasification of biomass model compounds using RuO_2 catalysts under SCW.[46] The production of H_2 was inversely related to that of CH_4. Furthermore, H_2O was replaced with D_2O in order to study the

CH_4 formation mechanism. In the gas product, only CD_4 and D_2 were detected. Therefore, the authors claimed that D_2 was from D_2O and that the CD_4 was produced by the methanation reaction of D_2 and CO/CO_2. The mechanism of catalyst gasification of biomass using Ru was illustrated in Figure 7.6. Biomass was oxidized to CO and H_2O by RuO_2, with Ru^{IV} reduced to Ru^{II}. The reduced Ru_{II} can participate in the reduction of H_2O (or D_2O) to H_2 (or D_2) and oxidized again to Ru^{IV}. The formed H_2 (or D_2) and CO can then lead to the formation of CH_4 though the methanation reaction. Hence, the catalyst gasification of biomass proceeded through the redox cycle between Ru^{IV} and Ru^{II}. The oxide of Ru exhibited outstanding gasification reactivity. Onwudili *et al.* studied the catalytic gasification of solid waste over $RuO_2/\gamma\text{-}Al_2O_3$ catalysts under SCW conditions.[47] Similarly, a redox cycle mechanism of Ru was also proposed for the steam reforming, methanation and hydrolysis reactions of organic compounds. The key step was the redox steam reforming of intermediates compounds to CO and H_2 with RuO_2 reduced as Ru. The reduced Ru can catalyse the methanation reaction forming CH_4 in the gas phase. Therefore, the conversion followed from sequential steam reforming and methanation reactions.

Rabe *et al.* gasified ethanol using Ru/C catalysts in SCW (390 °C, 25 MPa) as assessed this using *in situ* X-ray absorption spectroscopy.[48] During the experiment, it was found that Ru was rapidly reduced to its metallic form at around 150 °C and remained in its metallic form up to 390 °C. This suggests that metallic Ru was stable under SCW conditions, indicating that pure Ru metal is the active phase under SCW, in contrast to the redox cycle of Ru. It was suggested that ethanol was adsorbed onto the catalyst and decomposed into H_2 and acetaldehyde, which can be further decomposed into CO and CH_4. The formed CO remains adsorbed onto the metal surface and can react with H_2O, forming H_2 and CO_2 through the water gas shift reaction. Then, CO_2 could be hydrogenated to CH_4 through the methanation reaction. Therefore, the redox cycle of the Ru mechanism was put into question.

A recent work by Peterson *et al.* presented theoretical and experimental studies aiming to understand the final step of the methanation reaction under SCW.[49] Theoretically, it was predicted when CH_4 and H_2O reacted under SCW conditions, they would be broken down into adsorbents by the

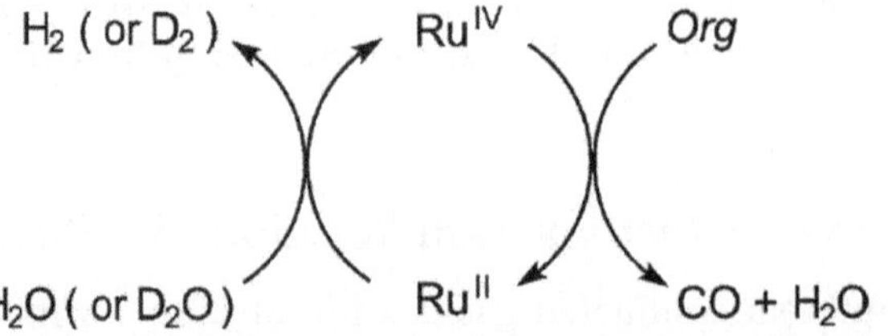

Figure 7.6 Catalytic gasification of organic compounds through redox cycle of Ru. Reproduced from Ref. 46 with permission from the Royal Society of Chemistry.

active metal surface. The adsorbents would undergo scrambling on the Ru surface, interchanging most hydrogen with the surface before being desorbed again as CH_4 and H_2O. An investigation of the CH_4 reaction in D_2O under SCW conditions using Ru/C showed that nearly all reacted CH_4 was converted to CD_4 and CHD_3. Therefore, a new alternative mechanism to the Ru redox cycle mechanism was then proposed: a surface "scrambling" reaction. It was also proven that the metallic form of Ru, other than its oxide state, is active for catalytic gasification under SCW conditions. Dreher *et al.* gasified biomass model compounds of methanol and ethanol to CH_4 in SCW.[50] Using density functional theory calculations and isotope labelling experiments, CH_4 was directed formed by the direct methanation reaction of hydrocarbon compounds on the Ru surface *via* a process that was distinct from the sequential gasification–methanation reaction. At the initial stage, the organic compounds were predominantly decomposed and dehydrogenated to C* on the metal surface. Then the C* was hydrogenated on the Ru surface to form CH_4.

7.5.3 Methane Reforming

Kıpçak *et al.* gasified olive mill waste waters under SCW between 400 and 600 °C at 25 MPa for different reaction times from 30 to 150 s.[51] It was found that CH_4 increased with temperature and reached a maximum of 35% at 550 °C. Moreover, the CH_4 production increased with reaction time below 550 °C, while CH_4 decreased with reaction time and temperature above 550 °C. This indicated that CH_4 could be reformed at elevated temperatures. The authors further employed Ni/Al_2O_3 and Ru/Al_2O_3 catalysts for the catalytic gasification of olive mill waste waters in SCW between 400 and 600 °C at 25 MPa for different reaction times from 30 to 150 s.[52] Reduced CH_4 formation was obtained when using Ni/Al_2O_3 compared with non-catalytic gasification. Comparatively, Ru/Al_2O_3 produced higher amounts of CH_4 than Ni/Al_2O_3. Interestingly, CH_4 also decreased with reaction time at 550 °C, while CO_2 increased accordingly. This indicates that CH_4 could be reformed in the presence of Ru/Al_2O_3. At a higher temperature of 600 °C, CH_4 increased over the first 60 s, but decreased as the reaction proceeded past this point. Therefore, Ru/Al_2O_3 can shorten the time for the decomposition of organic compounds into CH_4, but the formed CH_4 could be reformed, leading to a decrease in CH_4. Kersten *et al.* also indicated that carbon conversion was related to reaction temperatures below 650 °C.[43] Above 700 °C, an excess of H_2 was produced at low concentrations of glucose, indicating the reforming of CH_4. However, a high concentration of glucose produced mainly CH_4. This indicated that the complete gasification was only achieved at diluted concentrations of glucose.

7.6 Deactivation of Catalysts in SCW

Under SCW gasification, catalysts can be deactivated by the following mechanisms: (1) dissolution of active metals into SCW; (2) sintering of active metal and loss of active metal surface area; (3) precipitation of salts or carbonaceous materials onto active metals; and (4) bonding of sulfur to active metals.

7.6.1 Sulfur Deactivation

In biomass, there are always trace amounts of sulfur impurities. During reaction, the sulfur can be released into the gas phase and negatively affect the gasification behaviour of biomass in SCW due to sulfidation of active metals. The mechanism of catalyst poisoning by sulfur is one of the major obstacles in the catalytic gasification of biomass in SCW, which needs to be understood for designing sulfur-resistant catalysts and regeneration procedures. Osada *et al.* gasified organosolv lignin in SCW with or without sulfur.[53] Without the addition of sulfur, lignin can be gasified completely to CH_4, CO_2 and H_2, while the total gas yield decreased in the presence of sulfur. Due to the hindered methanation reaction caused by deactivation of Ru, the formation of CH_4 was markedly decreased, while the production of CO_2 was promoted. Any form of sulfur atom – 4-hydroxythiophenol, 4-methylthiophenol, 2-methyl-1-propanethiol, thiophene and sulfuric acid – can poison Ru catalysts for lignin gasification in SCW. After reaction in SCW, sulfur remained on the catalyst surface in the form of RuS_2, $Ru(SO_3)_2$ and $Ru(SO_4)_2$.

Waldner *et al.* utilized skeletal nickel catalysts and Ru/C for the catalytic gasification of biomass model compounds under SCW conditions (30 MPa, 400 °C).[54] The skeletal nickel catalysts lost reactivity rapidly due to the sintering phenomenon, but Ru/C kept its reactivity for 220 h. It was found that Ru/C was deactivated by the addition of Na_2SO_4, with the proceeding reaction being ascribed to the irreversible sulfidation of Ru^{III} formed in the redox cycle. Osada *et al.* studied the catalytic gasification of lignin, low-molecular-weight model compounds of 4-propylphenol and formaldehyde using Ru/TiO_2 in SCW.[55] It can be concluded that sulfur can inactive the active metal regardless of its form. In the presence of Ru/TiO_2, lignin can be nearly completely gasified to syngas. However, the gas yield in the presence of S-Ru/TiO_2 decreased significantly to 21.0% compared to 97.7% obtained with Ru/TiO_2, and the THF-insoluble yield increased to 3.1% accordingly. In terms of gas compositions, the sulfur-doped catalyst of S-Ru/TiO_2 decreased the formation of methane in comparison with Ru/TiO_2, while it promoted the production of CO_2 and H_2. In addition, the increasing concentration of sulfur compounds further decreased methane, while hydrogen was increased. The formation of methane involved both the breakage of C–O bonds of CO_2 or CO and the breakage of H–H bonds of H_2, which needed ensembles of active metal atoms for the adsorption of aromatic compounds to metal particles. The Ru ensembles, which are indispensable for the methanation reaction, could be decreased by the adsorption of sulfur atoms, while the water gas shift reaction was not related to the ensembles of metals atoms on the metal surface. It was concluded that sulfur could poison the active metal sites for the methanation reaction, while it did not deactivate the sites for the water gas shift reaction.

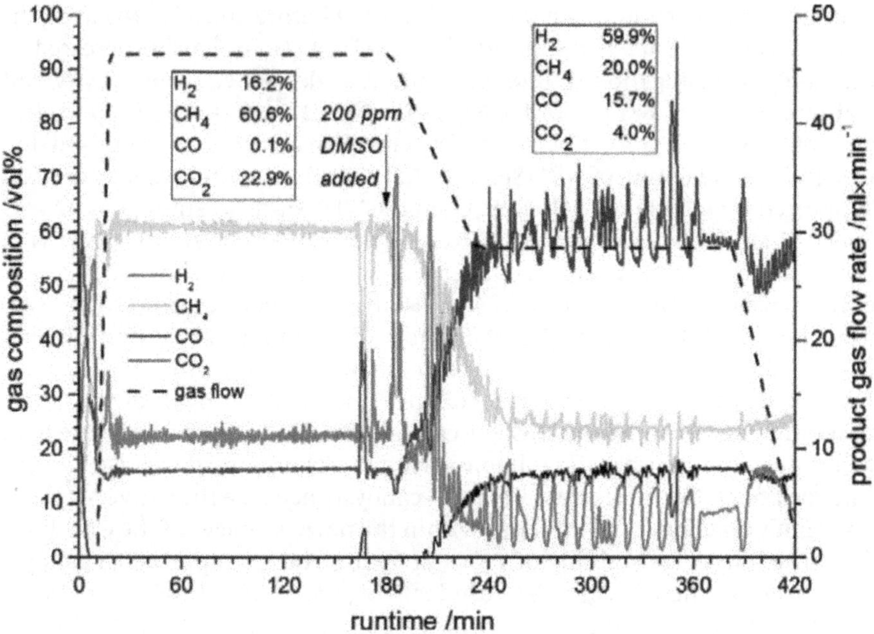

Figure 7.7 Flow rate and composition of product gas during the gasification of 7.5% ethanol in SCW (400 °C and 24.5 MPa). DMSO was added after 180 min and deionized water was fed from 370 min onwards. Reproduced from Ref. 50 with permission from Elsevier, Copyright 2004.

Dreher *et al.* carried out the gasification of ethanol in a continuous-flow reactor using Ru/C.[50] During the experiments, a sulfur model compound of dimethylsulfoxide was added into the system after 180 min. The evolution of product gas is shown in Figure 7.7. Clearly, without the addition of sulfur compounds, ethanol was completely converted into gas, and the product gas consisted of 60% CH_4 and 15% H_2, while CO was not detected. With the addition of sulfur, the total gas yield and concentration of CH_4 decreased significantly, while those of CO and H_2 increased. The conversion also decreased from 99.9% to 29.8%. Even though the sulfur poisoned Ru/C, it retained some activity for the methanation reaction. The researchers also proved that sulfur poisoning of the catalyst is irreversible and that it is impossible to restore activity by treating sulfur-poisoned catalysts in SCW.

7.7 Metal Leaching

During reaction in SCW, active metal may be dissolved into the aqueous phase. In contrast to the active metals supported on catalyst, the leached active metals in the aqueous phase showed low activity, especially at low temperatures.[56] Therefore, the metal leaching under SCW is the issue that needs to be understood by studying the metal content before and after these experiments. Three Ni catalysts supported on CNTs, Al_2O_3 and AC were tested in the catalytic gasification of phenol solution under SCW.[57] The activity and stability trend follows the order Ni/CNT > Ni/AC > Ni/Al_2O_3. The performance of Ni/AC and Ni/Al_2O_3 decreased after 30 h of continuous operation due to significant leaching of Ni^{2+} (Scheme 7.2). The dissolution resistance of Ni supported on α-Al_2O_3, $MgAl_2O_4$-Al_2O_3 and CNTs was also studied at 425 °C and 25.2 MPa.[58] In contrast with Ni/α-Al_2O_3 and Ni/$MgAl_2O_4$-Al_2O_3, Ni/CNTs produced the lowest H_2 yields, but achieved around 80% carbon gasification efficiencies for a steady period of 48 h. Ni dissolution rates from Ni/α-Al_2O_3 were more than 10 times higher than those of the other two catalysts, which may result in lower carbon gasification efficiencies over time. Ni/$MgAl_2O_4$-Al_2O_3 experienced pore structure collapse and crystal growth during reaction under SCW. For Ni/$MgAl_2O_4$-Al_2O_3 and Ni/α-Al_2O_3, parts of Al were lost, which decreased acidity and toluene selectivity. Therefore, the Ni/CNTs were categorized as "hydrothermally stable" catalysts because the active Ni metal was highly dispersed and attached within the narrow spaces of the CNT bundles. The long-term stability of TiO_2-, ZrO_2- and TaO_5-supported Ni catalyst was further studied under SCW.[59] Under SCW, the Ni/TiO_2 lost activity as the reaction proceeded, similarly to Ni/α-Al_2O_3 and Ni/$MgAl_2O_4$-Al_2O_3. For Ni/TiO_2, a new phase of $NiTiO_3$ was formed, which may reduce the BET surface active sites. Similarly, the phase and structure of Ni/ZrO_2 and Ni/TaO_5 were also changed, but these gave higher activities ascribed to the "catalyst *in situ* activation" effect of SCW. Behnia *et al.* carried out gasification of glucose in SCW using $Ni_{20\%}/Al_2O_3$.[60] Similarly, after the experiments, the dissolutions of Ni and Al from the catalyst were up to 76.9% Ni and 41.7% Al after 6 h of reaction. In contrast, both Al and Ni remain unchanged in the used catalyst of $Ni_{20\%}R_{2\%}/Al_2O_3$ due to the stabilizing effect of Ru by increasing the dispersion and reducibility of Ni. Chakinala *et al.* also synthesized Pt–Ni/Al_2O_3 for the catalytic gasification of glycerol in SCW.[61] The leaching of active metals did not occur for this catalyst, but the crystalline structure changed for the support of Al_2O_3. Shabake *et al.* proposed a mechanism of solid reduced Ni oxidation by O_2 and H_2O to form $Ni(OH)_2$.[62] The formed $Ni(OH)_2$ can be dissolved into strongly acidic or basic solutions. A Raney Ni–Sn catalyst was then prepared by adding 9.4% Sn to form a Ni–Sn surface alloy. It was found that Sn can suppress the dissolution of Ni by improving its oxidation resistance.

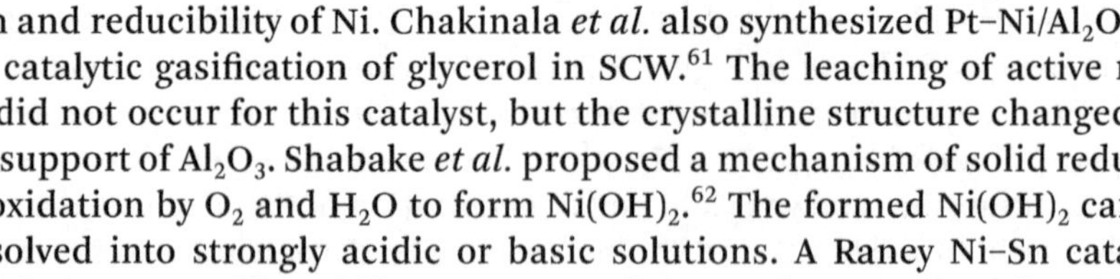

Scheme 7.2

Osada *et al.* evaluated the stability of Ru/TiO_2, $Ru/\gamma\text{-}Al_2O_3$ and Ru/C for lignin gasification under SCW.[6] It was found that $Ru/\gamma\text{-}Al_2O_3$ catalysts showed high activity in the early stages, but lost activity quickly due to the structure change from γ- to $\alpha\text{-}Al_2O_3$ and dissolution of Ru metal into SCW. No leaching of Ru were found for Ru/TiO_2 and Ru/C. Peng *et al.* also studied the catalytic activity and stability of Ru/C catalysts for SCW gasification.[63] The deactivation of catalyst was mainly from progressive coke deposition rather than Ru leaching. In summary, the interaction between active metals and supports and the formation of metal alloys are determining factors for the dissolution of active metals from supports.

References

1. Z. Abu El-Rub, E. A. Bramer and G. Brem, Review of catalysts for tar elimination in biomass gasification processes, *Ind. Eng. Chem. Res.*, 2004, **43**(22), 6911–6919.
2. M. K. M. Lane, and J. B. Zimmerman, Controlling metal oxide nanoparticle size and shape with supercritical fluid synthesis, *Green Chem.*, 2019, **21**(14), 3769–3781.
3. Y. Guo *et al.*, Review of catalytic supercritical water gasification for hydrogen production from biomass, *Renewable Sustainable Energy Rev.*, 2010, **14**(1), 334–343.
4. H. Xiong, H. N. Pham and A. K. Datye, Hydrothermally stable heterogeneous catalysts for conversion of biorenewables, *Green Chem.*, 2014, **16**(11), 4627–4643.
5. Y. Zhang *et al.*, Hydrothermal stability of different zeolites in supercritical water: Implication for synthesis of supported catalysts by supercritical water impregnation, *Korean J. Chem. Eng.*, 2018, **35**(9), 1932–1940.
6. M. Osada *et al.*, Stability of supported ruthenium catalysts for lignin gasification in supercritical water, *Energy Fuels*, 2006, **20**(6), 2337–2343.
7. H. Zöhrer, F. Mayr and F. Vogel, Stability and performance of ruthenium catalysts based on refractory oxide supports in supercritical water conditions, *Energy Fuels*, 2013, **27**(8), 4739–4747.
8. R. M. Ravenelle *et al.*, Structural changes of $\gamma\text{-}Al_2O_3$-supported catalysts in hot liquid water, *ACS Catal.*, 2011, **1**(5), 552–561.
9. D. C. Elliott, L. J. Sealock and E. G. Baker, Chemical processing in high-pressure aqueous environments. 2. Development of catalysts for gasification, *Ind. Eng. Chem. Res.*, 1993, **32**(8), 1542–1548.
10. D. J. M. de Vlieger *et al.*, Carbon nanotubes: A promising catalyst support material for supercritical water gasification of biomass waste, *ChemCatChem*, 2012, **4**(12), 2068–2074.
11. O. Yakaboylu *et al.*, Supercritical water gasification of manure: A thermodynamic equilibrium modeling approach, *Biomass Bioenergy*, 2013, **59**, 253–263.

12. A. Rahbari, M. B. Venkataraman and J. Pye, Energy and exergy analysis of concentrated solar supercritical water gasification of algal biomass, *Appl. Energy*, 2018, **228**, 1669–1682.
13. W. Wan, An innovative system by integrating the gasification unit with the supercritical water unit to produce clean syngas: Effects of operating parameters, *Int. J. Hydrogen Energy*, 2016, **41**(33), 14573–14582.
14. M. K. Nikoo, S. Saeidi and A. Lohi, A comparative thermodynamic analysis and experimental studies on hydrogen synthesis by supercritical water gasification of glucose, *Clean Technol. Environ. Policy*, 2015, **17**(8), 2267–2288.
15. F. L. P. Resende and P. E. Savage, Effect of metals on supercritical water gasification of cellulose and lignin, *Ind. Eng. Chem. Res.*, 2010, **49**(6), 2694–2700.
16. S. Bjelić *et al.*, Deactivation of methanation catalyst (ru/c) under supercritical water by deposition of non-volatile organics: first insights into deposition patterns and chemical properties, *ChemCatChem*, 2019, **11**(6), 1747–1755.
17. J. W. Shabaker, G. W. Huber and J. A. Dumesic, Aqueous-phase reforming of oxygenated hydrocarbons over Sn-modified Ni catalysts, *J. Catal.*, 2004, **222**(1), 180–191.
18. P. Azadi *et al.*, Screening of nickel catalysts for selective hydrogen production using supercritical water gasification of glucose, *Green Chem.*, 2012, **14**(6), 1766–1777.
19. T. Furusawa *et al.*, The evaluation of the stability of Ni/MgO catalysts for the gasification of lignin in supercritical water, *Appl. Catal., A*, 2007, **327**(2), 300–310.
20. G. W. Huber, J. W. Shabaker and J. A. Dumesic, Raney Ni–Sn catalyst for H_2 production from biomass-derived hydrocarbons, *Science*, 2003, **300**(5628), 2075–2077.
21. Q. Q. Guan *et al.*, Catalytic gasification of lignin with Ni/Al_2O_3-SiO_2 in sub/supercritical water, *J. Supercrit. Fluids*, 2014, **95**, 413–421.
22. M. Osada, *et al.*, Low-temperature catalytic gasification of lignin and cellulose with a ruthenium catalyst in supercritical water, *Energy Fuels*, 2004, **18**(2), 327–333.
23. E. Kipcak and M. Akgun, Biofuel production from olive mill wastewater through its Ni/Al_2O_3 and Ru/Al_2O_3 catalyzed supercritical water gasification, *Renewable Energy*, 2018, **124**, 155–164.
24. F. J. G. Ortiz *et al.*, Supercritical water reforming of glycerol: Performance of Ru and Ni catalysts on Al_2O_3 support, *Energy*, 2016, **96**, 561–568.
25. M. Osada *et al.*, Gasification of sugarcane bagasse over supported ruthenium catalysts in supercritical water, *Energy Fuels*, 2012, **26**(6), 3179–3186.
26. D. C. Elliott, L. J. Sealock Jr and E. G. Baker, Chemical processing in high-pressure aqueous environments. 2. Development of catalysts for gasification, *Ind. Eng. Chem. Res.*, 1993, **32**(8), 1542–1548.

27. P. Azadi *et al.*, Hydrogen production from cellulose, lignin, bark and model carbohydrates in supercritical water using nickel and ruthenium catalysts, *Appl. Catal., B*, 2012, **117–118**, 330–338.
28. T. Sato *et al.*, Gasification of alkylphenols with supported noble metal catalysts in supercritical water, *Ind. Eng. Chem. Res.*, 2003, **42**(19), 4277–4282.
29. M. Osada *et al.*, Water Density Effect on Lignin Gasification over Supported Noble Metal Catalysts in Supercritical Water, *Energy Fuels*, 2006, **20**(3), 930–935.
30. A. Yamaguchi *et al.*, Hydrogen production from woody biomass over supported metal catalysts in supercritical water, *Catal. Today*, 2009, **146**(1), 192–195.
31. A. J. Byrd, K. K. Pant and R. B. Gupta, Hydrogen production from ethanol by reforming in supercritical water using Ru/Al_2O_3 catalyst, *Energy Fuels*, 2007, **21**(6), 3541–3547.
32. A. J. Byrd, K. K. Pant and R. B. Gupta, Hydrogen production from glycerol by reforming in supercritical water over Ru/Al_2O_3 catalyst, *Fuel*, 2008, **87**(13–14), 2956–2960.
33. Q. Guan *et al.*, Catalytic gasification of lignin with Ni/Al_2O_3-SiO_2 in sub/supercritical water, *J. Supercrit. Fluids*, 2014, **95**, 413–421.
34. Y. Lu *et al.*, Behavior of nickel catalysts in supercritical water gasification of glucose: Influence of support, *Biomass Bioenergy*, 2014, **67**, 125–136.
35. L. Zhang, P. Champagne and C. Xu, Screening of supported transition metal catalysts for hydrogen production from glucose via catalytic supercritical water gasification, *Int. J. Hydrogen Energy*, 2011, **36**(16), 9591–9601.
36. T. Sato *et al.*, Effect of water density on the gasification of lignin with magnesium oxide supported nickel catalysts in supercritical water, *Ind. Eng. Chem. Res.*, 2006, **45**(2), 615–622.
37. T. Furusawa *et al.*, The evaluation of the stability of Ni/MgO catalysts for the gasification of lignin in supercritical water, *Appl. Catal., A*, 2007, **327**(2), 300–310.
38. M. S. Mastuli *et al.*, Catalytic gasification of oil palm frond biomass in supercritical water using MgO supported Ni, Cu and Zn oxides as catalysts for hydrogen production, *Int. J. Hydrogen Energy*, 2017, **42**(16), 11215–11228.
39. M. S. Mastuli *et al.*, Comparative study between supported and doped MgO catalysts in supercritical water gasification for hydrogen production, *Int. J. Hydrogen Energy*, 2019, **44**(7), 3690–3701.
40. M. Z. Hossain *et al.*, Effect of mesoporosity of bimetallic Ni–Ru-Al_2O_3 catalysts for hydrogen production during supercritical water gasification of glucose, *Fuel Process. Technol.*, 2017, **159**, 55–66.
41. F. Vogel *et al.*, Synthetic natural gas from biomass by catalytic conversion in supercritical water, *Green Chem.*, 2007, **9**(6), 616–619.
42. J. A. Onwudili and P. T. Williams, Catalytic conversion of bio-oil in supercritical water: Influence of RuO_2/gamma-Al_2O_3 catalysts on gasification

efficiencies and bio-methane production, *Appl. Catal., B Environ.*, 2016, **180**, 559–568.
43. S. R. A. Kersten *et al.*, Gasification of model compounds and wood in hot compressed water, *Ind. Eng. Chem. Res.*, 2006, **45**(12), 4169–4177.
44. A. Kruse and E. Dinjus, Hydrogen from methane and supercritical water, *Angew. Chem., Int. Ed.*, 2003, **42**(8), 909–911.
45. C. K. Cheng, S. Y. Foo and A. A. Adesina, Steam reforming of glycerol over Ni/Al_2O_3 catalyst, *Catal. Today*, 2011, **178**(1), 25–33.
46. K. C. Park and H. Tomiyasu, Gasification reaction of organic compounds catalyzed by RuO_2 in supercritical water, *Chem. Commun.*, 2003, (6), 694–695.
47. J. A. Onwudili, Supercritical water gasification of RDF and its components over RuO_2/γ-Al_2O_3 catalyst: New insights into RuO_2 catalytic reaction mechanisms, *Fuel*, 2016, **181**, 157–169.
48. S. Rabe *et al.*, Towards understanding the catalytic reforming of biomass in supercritical water, *Angew. Chem., Int. Ed.*, 2010, **49**(36), 6434–6437.
49. A. A. Peterson *et al.*, Evidence of scrambling over ruthenium-based catalysts in supercritical-water gasification, *ChemCatChem*, 2012, **4**(8), 1185–1189.
50. M. Dreher *et al.*, Catalysis in supercritical water: Pathway of the methanation reaction and sulfur poisoning over a Ru/C catalyst during the reforming of biomolecules, *J. Catal.*, 2013, **301**, 38–45.
51. E. Kıpçak, O. Ö. Söğüt and M. Akgün, Hydrothermal gasification of olive mill wastewater as a biomass source in supercritical water, *J. Supercrit. Fluids*, 2011, **57**(1), 50–57.
52. E. Kıpçak and M. Akgün, Biofuel production from olive mill wastewater through its Ni/Al_2O_3 and Ru/Al_2O_3 catalyzed supercritical water gasification, *Renewable Energy*, 2018, **124**, 155–164.
53. M. Osada *et al.*, Effect of sulfur on catalytic gasification of lignin in supercritical water, *Energy Fuels*, 2007, **21**(3), 1400–1405.
54. M. H. Waldner, F. Krumeich and F. Vogel, Synthetic natural gas by hydrothermal gasification of biomass: Selection procedure towards a stable catalyst and its sodium sulfate tolerance, *J. Supercrit. Fluids*, 2007, **43**(1), 91–105.
55. M. Osada *et al.*, Reaction pathway for catalytic gasification of lignin in presence of sulfur in supercritical water, *Energy Fuels*, 2007, **21**(4), 1854–1858.
56. P. Azadi and R. Farnood, Review of heterogeneous catalysts for sub- and supercritical water gasification of biomass and wastes, *Int. J. Hydrogen Energy*, 2011, **36**(16), 9529–9541.
57. Y. Wang *et al.*, Catalytic performances of Ni-based catalysts on supercritical water gasification of phenol solution and coal-gasification wastewater, *Int. J. Hydrogen Energy*, 2019, **44**(7), 3470–3480.

58. S. Li, P. E. Savage and L. Guo, Stability and activity maintenance of Al_2O_3- and carbon nanotube-supported Ni catalysts during continuous gasification of glycerol in supercritical water, *J. Supercrit. Fluids*, 2018, **135**, 188–197.
59. S. Li, P. E. Savage and L. Guo, Stability and activity maintenance of sol-gel Ni-MxOy (M=Ti, Zr, Ta) catalysts during continuous gasification of glycerol in supercritical water, *J. Supercrit. Fluids*, 2019, **148**, 137–147.
60. I. Behnia *et al.*, Production of methane and hydrogen via supercritical water gasification of renewable glucose at a relatively low temperature: Effects of metal catalysts and supports, *Fuel Process. Technol.*, 2016, **143**, 27–34.
61. A. G. Chakinala *et al.*, Catalytic reforming of glycerol in supercritical water over bimetallic Pt–Ni catalyst, *Ind. Eng. Chem. Res.*, 2013, **52**(15), 5302–5312.
62. J. W. Shabaker *et al.*, Sn-modified Ni catalysts for aqueous-phase reforming: Characterization and deactivation studies, *J. Catal.*, 2005, **231**(1), 67–76.
63. G. Peng *et al.*, Synthesis factors affecting the catalytic performance and stability of Ru/C catalysts for supercritical water gasification, *Catal. Sci. Technol.*, 2014, **4**, 3329–3339.

CHAPTER 8

Catalytic Technologies for the Production of Liquid Transportation Fuels from Biomass

LOLA AZANCOT,[a] LUIS F. BOBADILLA,[a*] FRANCISCA ROMERO-SARRIA[a] AND JOSÉ A. ODRIOZOLA[a,b]

[a]Inorganic Chemistry Department and Institute of Materials Science, University of Sevilla – CSIC, Av. Américo Vespucio 49, Sevilla, 41092, Spain
[b]Department of Chemical & Process Engineering, University of Surrey, Guilford, UK
*Corresponding contributor. Email: bobadilla@icmse.csic.es

8.1 Introduction: A General Overview

Today, crude oil's price volatility together with growing energetic demand have become some of the major concerns of governments around the world. In fact, total daily oil consumption reached 100 million oil barrels worldwide at the end of 2018 and is expected to grow in 2019.[1] The future collapse of fossil fuels and global climate change related to greenhouse gas emissions have motivated the intensification of research efforts to develop more clean and environmentally friendly energetic alternatives.[2] Liquid transportation fuels

Heterogeneous Catalysis for Energy Applications
Energy and Environment Series No. 27
Edited by Tomas R. Reina and Jose A. Odriozola

Published by the Royal Society of Chemistry, www.rsc.org

represent the most high-demand energetic source due to their crucial importance in the transportation sector. To this end, the European Union (EU) has set an ambitious plan for the utilization of biofuels through the EU Biofuels Directive 2009/28/EC.[3] According to this Directive, the European Renewable Energy Council expects that, by the end of 2040, the contribution of renewable sources to energy production will be increased by almost 50%.[4–6]

The technologies for the production of synthetic liquid transportation fuels from carbon-based materials can be collectively referred to feed-to-liquids (XTL) conversion processes and include biomass-to-liquids (BTL), coal-to-liquids (CTL) and gas-to-liquids (GTL) processes, in each instance identifying the raw material that is being converted.[7] As Figure 8.1 shows, these XTL technologies involve a process in which the feed is first converted into synthesis gas (syngas; H_2 + CO) and then liquefied into liquid fuels through Fischer–Tropsch synthesis (FTS).[8] Moreover, since the raw syngas produced can have different types of contaminants, such as particulates, condensable tars and acid gases, which can cause various mechanical problems or deactivate the Fischer–Tropsch (FT) catalyst, the syngas is purified or conditioned by including a gas cleaning unit in the global process.

The CTL and GTL processes are the most important commercial routes to produce higher-molecular-weight derivatives from coal and natural gas, respectively.[9] Today, Sasol is the leading company in the commercialization of CTL processes for producing synthetic fuels. Other companies that have implemented CTL or GTL processes are Shell, Exxon and Statoil.[10] However, the liquid fuels obtained through biomass are more sustainable substitutes than fossil-derived fuels. The BTL process is an attractive option for the production of clean and carbon-neutral diesel from biomass.[11,12] Several analyses on the technical and economic viability of this emerging technology have been reported, concluding that the process has huge potential to become techno-economically competitive.[12,13] In order to advance BTL technology and move it to the next stage, catalytic technologies must be improved to achieve higher product yields under less severe conditions with minimal unit operations to drive down capital costs.[14]

The aim of this chapter is to provide a general overview of the BTL technology by analysing the importance of the catalysis engineering involved in

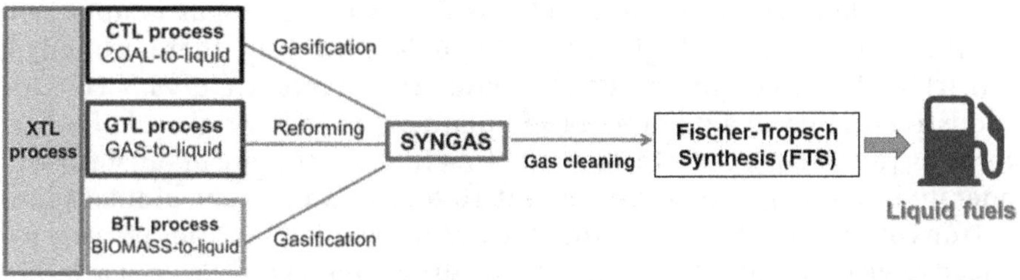

Figure 8.1 General scheme of XTL processes for the synthesis of transportation liquid fuels from different carbon sources.

the global process. To this end, the chapter first discusses recent advances in heterogeneous catalysis for enhancing the direct production of syngas by thermal gasification of biomass or indirect production through the reforming of the biogas produced by anaerobic digestion as well as subsequent transformation into liquid fuels *via* FTS. Furthermore, catalysis advancements for technology commercialization will be examined, with an emphasis on process intensification and the recent development of microreactors.

8.2 Current Scenario of BTL Technology

Depending on the origin and the biomass feedstock type, the liquid fuels produced can be classified into first-, second-, third- or fourth-generation biofuels. First-, second- and third-generation biofuels are obtained from edible feedstocks, non-food competitive feedstocks and aqueous biomass, respectively.[15,16] Meanwhile, fourth-generation biofuels are based on the concept of a circular CO_2 economy regarding carbon capture and storage (CCS) processes.[16] Fourth-generation biofuels symbolize the concept of bioenergy with carbon storage (BECS).[17] This chapter is focused on the production of second-generation biofuels using as raw material biomass not suitable as food, including all forms of cellulosic biomass.

Among the different routes typically used for biomass transformation, both thermal gasification and anaerobic digestion processes are the most commonly utilized to produce a gas mixture that can be further processed and transformed into valuable syngas as a renewable carbon source for the subsequent synthesis of liquid fuels.[18,19] As can be observed in Figure 8.2, the BTL process is typically applied to synthetic fuels produced from biomass through thermochemical routes such as gasification or pyrolysis of biomass.[20] The other route is based on a natural process driven by different microorganisms in which biodegradable feedstocks are fermented in the absence of oxygen, producing a gas mixture composed mainly of methane and CO_2, designated as biogas.[13,21,22] Once the biogas has been obtained, the same technology used in the GTL process is applied, and this pathway is known as a bio-GTL (BGTL) process,[23,24] in which the syngas is produced by means of reforming reactions of biogas and subsequently transformed into liquid fuels through FTS.

The major BTL production technologies are based on gasification and pyrolysis routes in which biomass is pre-treated and converted into syngas by thermochemical processes, and subsequently the syngas is upgraded for the synthesizing of high-quality FT fuels. Currently, there are only a few BTL pilot-scale and large-scale industrial plants. In 1998, CHOREN Industries achieved a demonstration plant on an industrial scale for performing the BTL process. In 2007, the first industrial-scale plant started to operate in Freiberg, converting about 10 tons of anhydrous biomass into 2.5 tons of sun diesel with a planned yearly output of 15 000 tons per year for a feedstock of about 60 000 tons of anhydrous biomass.[20] Several aviation industries such as British Airways, Cathay Pacific Airways and Southwest

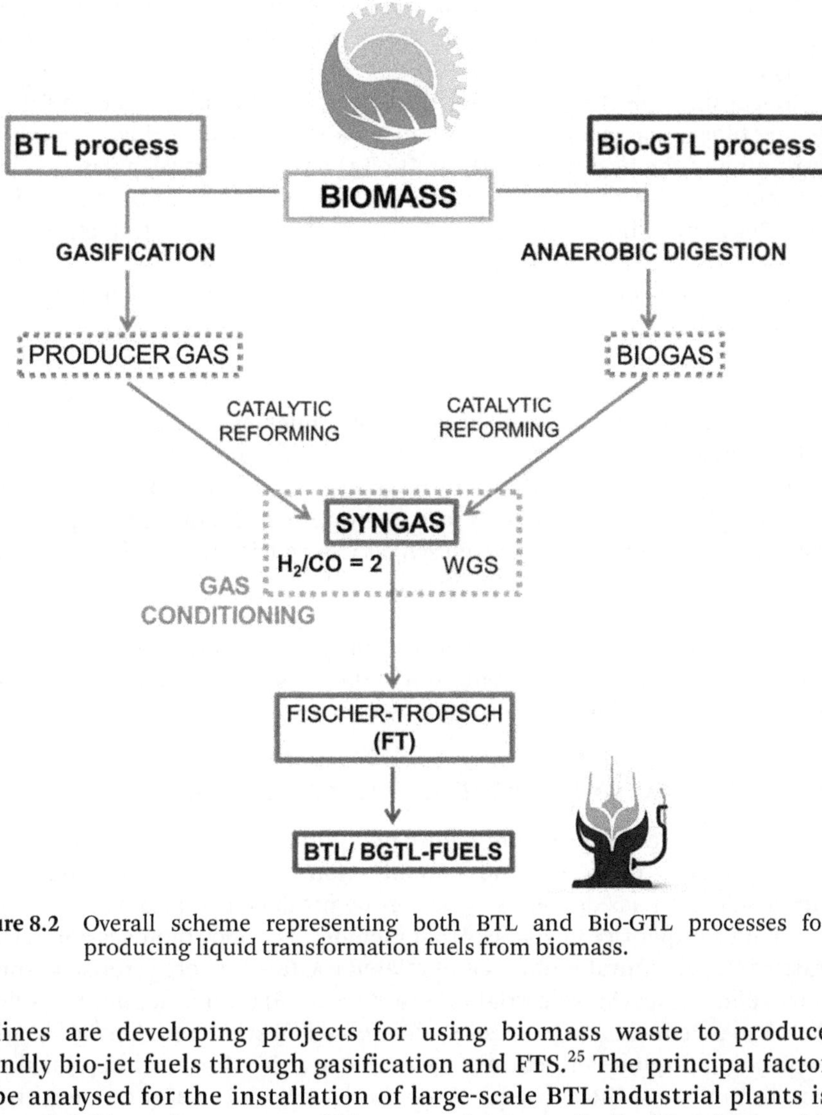

Figure 8.2 Overall scheme representing both BTL and Bio-GTL processes for producing liquid transformation fuels from biomass.

Airlines are developing projects for using biomass waste to produce friendly bio-jet fuels through gasification and FTS.[25] The principal factor to be analysed for the installation of large-scale BTL industrial plants is the availability and transport of biomass, which are the limiting factors for the gasification unit. On the other hand, the BTL system is composed of various processes and thus the stable operation of the integrated system is very important. Kim *et al.*[26] performed a preliminary study in which the technical feasibility was proven for the development of a pilot-scale BTL consisting of a bubbling fluidized bed (BFB), a gasifier (20 kW_{th}), a gas cleaning unit, a syngas compression unit, an acid gas removing unit and an FT reactor. The integrated system was operated for 500 h over several runs, achieving stable operating conditions for each unit. Much effort has been expended on developing economically affordable and environmentally

BTL processes, and it is expected that this technology could be used directly in the existing transportation sector, providing clean, carbon-neutral energy.

Concerning the BGTL process, the anaerobic digestion process for producing biogas has been implemented on a large scale worldwide since the beginning of the twentieth century. The EU has invested deeply in this technology, and most of the anaerobic digestion plants are located in Europe, with Germany being the leading country in global biogas production.[13,27,28]. However, the integration of a process for producing liquid hydrocarbon fuels from biomass through a combined biochemical and thermochemical process has not yet been implemented at the bench scale, although some interesting economic studies have been recently reported. Indeed, the techno-economic study performed by Naqi *et al.*[13] shows that the conversion of biomass to liquid fuels *via* BGTL processes could be profitable under current market conditions only if renewable energy credits are included in the process.

This chapter analyses the current trends in BTL and BGTL processes, focusing particular attention on recent advances in catalyst development for the production of biomass-derived syngas from gasification or biogas reforming and the subsequent production of biofuels *via* FTS. From this perspective, this contribution highlights the relevance of catalysis in the energy and waste sectors in terms of achieving the development of sustainable processes that will contribute to the implementation of the EU's environmental and energy-related policies in the transportation sector.

8.3 Production of Biomass-derived Syngas

The building block for the production of liquid fuels is syngas, which is composed primarily of H_2, CO and a small amount of CO_2.[14] As highlighted in Figure 8.3, biomass-derived syngas can be produced from a wide variety of non-food competitive feedstocks that include: (1) woody biomass and related waste; (2) agricultural biomass and related waste; (3) energy crops; (4) municipal solid waste; (5) industrial bio-waste; and (6) animal manure.[16] Among the biomass conversion processes, anaerobic digestion and thermochemical gasification produce a mixture of gases that are transformed into syngas as a carbon source for the subsequent synthesis of liquid energy carriers. In the context of this chapter, the main focus of this section is on the production of biomass-derived syngas from these mixtures of gas, biogas or producer gas, obtained by anaerobic digestion or biomass gasification, respectively. The following section describes the FTS process in which the syngas produced is transformed into liquid fuels. The processes of the anaerobic digestion and thermal gasification of biomass are outside the scope of this chapter.

8.3.1 Transformation of Biogas into Valuable Syngas

Biogas generated by anaerobic digestion processes from residual biomass is considered one of the most important bio-renewable resources for the next

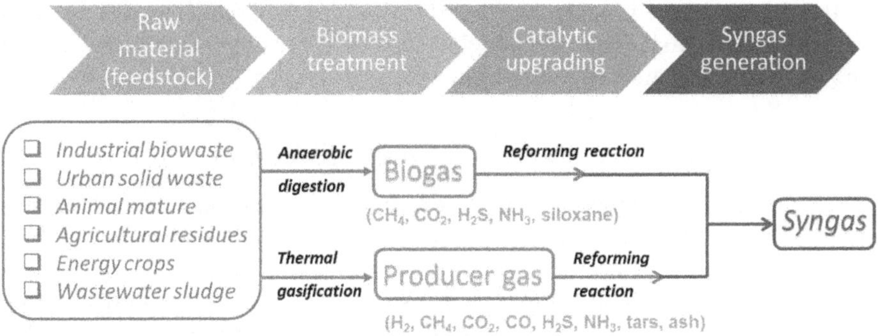

Figure 8.3 Overview of the production of biomass-derived syngas.

generation of biofuels. The composition of biogas varies depending on the fermentation process and the starting raw material, although the major constituents are methane (60%) and carbon dioxide (40%), with some traces of other impurities such as water vapour, ammonia, hydrogen sulfide (H_2S), silicon organic compounds (siloxanes), chlorinated compounds or other volatile organic compounds (VOCs).[29] Several cleaning techniques can be applied to remove these impurities and to upgrade biogas to achieve a suitable quality depending on the application.[30–32] The techniques typically used for raw biogas purification include chemical absorption,[33–35] high-pressure water scrubbing,[36,37] pressure swing adsorption,[38,39] cryogenic separation[40,41] and membrane permeation.[42,43] From a catalytic viewpoint, the separation of H_2S is crucial for preventing catalyst deactivation by metal–sulfur poisoning. Furthermore, H_2S can react with water vapour to produce sulfuric acid (H_2SO_4), which may cause corrosion problems in the reaction system and the pipelines.[44] Other contaminants, such as siloxanes or ammonia, are in very low concentrations or are hardly detectable, and they can be tolerated for the chemical process.

The transformation of biogas into clean syngas is critical for indirectly producing liquid transportation fuels from biogas. Syngas can be obtained from syngas through different biogas reforming processes such as dry, steam or partial oxidative reforming, or from a combination of two or three of these reforming methods.[45] Further, post-conditioning treatments of syngas are required to obtain an optimal $H_2:CO$ molar ratio for producing clean transportation fuels. Of these reforming reactions, dry reforming of methane appears more suitable for the utilization of biogas since no CO_2 separation processes are necessary and both carbon atoms are incorporated into the final product. The $CO_2:CH_4$ molar ratio of close to unity that is required for the stoichiometric dry reforming reaction can be regulated by combusting an adequate quantity of biogas, which at the same time offers the heat required for the highly endothermic reaction.[46] In comparison to steam reforming, dry reforming can be performed in areas where water is not available using simpler installations and reducing operational costs.[46] Furthermore, compared to partial oxidative reforming, it is easier to achieve higher selectivity for syngas in the dry reforming reaction and safety risks are reduced.[47]

8.3.1.1 Importance of CO_2 Reforming of Methane

The production of syngas *via* dry reforming processes has been widely proposed in the literature because of the straightforward utilization of both major constituents from raw biogas (CH_4 and CO_2).[48] The reaction between methane and CO_2 for producing syngas (eqn (8.1)) is highly endothermic, and the equilibrium is influenced by the simultaneous reverse water gas shift (RWGS) reaction (eqn (8.2)), which decreases the $H_2 : CO$ molar ratio to values less than unity.[49–51]

$$CH_4 + CO_2 \leftrightarrow 2CO + 2H_2 \qquad \Delta H^{\circ} = 247\ \text{kJ mol}^{-1} \tag{8.1}$$

$$CO_2 + H_2 \leftrightarrow CO + H_2O \qquad \Delta H^{\circ} = 41\ \text{kJ mol}^{-1} \tag{8.2}$$

Apart from the RWGS reaction, other side reactions such as methane cracking (eqn (8.3)) and the Boudouard reaction (eqn (8.4)) also occur, and these reactions are the main reason for carbon deposition and catalyst deactivation.[52,53] In addition, the steam produced from the RWGS reaction enables steam carbon gasification (eqn (8.5)), and the occurrence of this reaction can decrease the amount of carbonaceous species deposited.[54]

$$CH_4 \leftrightarrow C^* + 2H_2 \qquad \Delta H^{\circ} = 75\ \text{kJ mol}^{-1} \tag{8.3}$$

$$2CO \leftrightarrow CO_2 + C^* \qquad \Delta H^{\circ} = -172\ \text{kJ mol}^{-1} \tag{8.4}$$

$$C^* + H_2O \leftrightarrow H_2 + CO \qquad \Delta H^{\circ} = 131\ \text{kJ mol}^{-1} \tag{8.5}$$

The secondary reactions that take place simultaneously during the dry reforming reaction are strongly dependent on the reaction conditions, including the feed composition, temperature and pressure.[55] Thermodynamically, the optimal conditions for the dry reforming of methane (DRM) reaction are found at $CO_2 : CH_4$ molar ratios higher than unity and temperatures greater than 700 °C in order to minimize carbon formation.[54] Nevertheless, it is difficult to generalize the minimization of coke formation only by calculating thermodynamically the operation conditions, because this is strongly dependent on the type of catalyst.[56] Therefore, it is necessary to analyse the influence of the reaction variables in order to enable the direct use of the resulting syngas in subsequent applications, as well as to design an optimal catalyst that minimizes carbon deposition.

8.3.1.2 Catalysts Design for DRM

The rational design of dry reforming catalysts involves the utilization of the inherent properties of catalyst components and understanding their complex synergistic interactions for optimum performance by combining the state of the art of theoretical methods and characterization techniques.[57] Several extensive reviews have been published in the last decade on recent advances in catalyst development for DRM reactions. Different catalytic formulations with a wide variety of morphologies and topologies have been evaluated in terms of how synergistic component interactions affect the active metal dispersion, surface acidity–basicity properties, mobility of oxygen species, redox properties, confinement effects, reducibility, porosity, surface area and mass transfer limitations of catalysts.[56–61]

The development of robust and efficient catalysts with potential industrial applications requires the control of numerous parameters to ensure optimal activity and long-term durability. To this end, suitable and proper preparation methods must be used for achieving satisfactory dispersion, good metal–support interaction and a high surface area, crucial properties for designing a catalyst that is resistant against sintering.[59] It is desirable for the preparation method to be rapid and efficient, to use a minimal number of steps, to require low amounts of solvent and to employ inexpensive metal precursors.[60] In addition, the activation pre-treatment of the catalysts plays a significant role in the formation of active species, thereby influencing catalytic performance.[56]

Catalysts based on noble metals, especially Ru and Rh, exhibit higher catalytic activity and stronger resistance to coke deposition.[62] Nevertheless, the economic viability of these catalysts for large industrial applications is limited due to the scarcity and high cost of these precious metals. From an industrial perspective, Ni-based catalysts are preferred due to their low cost and high availability.[61] These catalysts have been widely tested in DRM, achieving in many cases catalytic activities comparable with some noble metal-based catalysts. In fact, the use of traditional Ni catalysts has been demonstrated at industrial scale for dry synthesis gas production up to 133 000 $Nm^3\ h^{-1}$ by the company Haldor Topsøe.[63] However, one of the major drawbacks of nickel catalysts is the deactivation caused by coke formation and metal sintering. Diverse approaches have been developed to overcome these deactivation issues and increase the stability of nickel catalysts, such as: (1) optimization of preparation methods to control the nickel size and inhibit carbon growth;[64–66] (2) utilization of basic promoters such as CaO, MgO or K_2O for favouring the gasification of carbon deposits;[67–70] (3) doping with a second metal to create metal alloys to potentially enhance reducibility and avoid coke formation;[71–75] (4) passivation of nickel step-edge sites with sulfur where the growth of carbon whiskers is initiated;[76] (5) utilization of reducible supports or addition of redox promoters to increase oxygen mobility and favour the oxidation of carbon deposits;[77–83] and (6) use of mesoporous supports for stabilizing the Ni nanoparticles through confinement effects and to obtain more accessible sites, thereby limiting diffusional problems.[84–88]

Table 8.1 includes a summary of the combination of properties of each type of support, active metal and promoter, based on an overview of several recently published reviews,[57,60–62] emphasizing the effects of the different combinations and including an example of an optimal Ni-based catalyst for each combination. The choice of an appropriate support is crucial for designing a reforming catalyst with good catalytic activity and stability. As can be observed in Table 8.1, numerous supports including non-reducible metal oxides (Al_2O_3, SiO_2, La_2O_3, MgO), reducible metal oxides (CeO_2, TiO_2, ZrO_2), mixed oxides ($MgAl_2O_4$, mixed oxides derived from hydrotalcites, perovskites) and high-surface mesoporous materials (SBA-15, MCM-41, mesoporous carbon) have been widely explored. Depending of the support, the catalytic properties can be tuned towards the influence of metal–support interactions, acid–base properties, lattice oxygen species and metal dispersion of the supported catalysts.[89]

Table 8.1 Summary of the formulations proposed for Ni-based catalysts for the dry reforming of methane. Adapted from Ref. 60 with permission form Elsevier, Copyright 2018.

Type of support	Active phase		Promoters
	Monometallic	Bimetallic	
Reducible metal oxides	• Good redox properties and oxygen mobility • Major resistance to coke deposition Ni/CeO_2, Ni/ZrO_2, Ni/TiO_2	• Small amounts of noble metal enhance the metallic dispersion of non-noble metals and increase coke resistance Ni–Rh/ZrO_2	• Synergetic effect improves metallic dispersion • Improvement of reducibility of the support $Ni/La_2O_3/ZrO_2$, $Ni/CeO_2/ZrO_2$
Non-reducible metal oxides	• Less prone to metal sintering than reducible metal oxides • Surface basicity improves coke resistance formation Ni/Al_2O_3, Ni/MgO, Ni/SiO_2	• Reduction of coke formation • Enhanced reducibility of Ni • Ni–Co is a desirable combination since it forms alloys and is cheaper than noble metals Ni–Rh/Al_2O_3, Ni–Co/Al_2O_3	• Addition of promoters such as alkaline earth metals prevent Ni sintering and favour coke gasification • Passivation of nickel with sulfur inhibits the sites of carbon filament growth $Ni/MgO/Al_2O_3$, $Ni/CaO/Al_2O_3$, $Ni/K/Al_2O_3$, Ni-S/Al_2O_3

(continued)

Table 8.1 *(continued)*

Type of support	Active phase		Promoters
	Monometallic	Bimetallic	
Mixed oxides	• Important beneficial balance between the metal–support interactions and the basicity of the support • Perovskites are very attractive and increase oxygen mobility $Ni/MgAl_2O_4$, $Ni/LaAlO_3$	• Very good metal–support interactions • High oxygen mobility • Carbon gasification promoted $Co–LaNiO_3$, $Mn–LaNiO_3$	• Improve Ni dispersion and catalyst stability • Better redox properties and oxygen mobility • Major basicity that enhances coke gasification $Ni/La_2O_3/MgAl_2O_4$, $NiCeO_2/MgAl_2O_4$
Mesoporous solids	• Deactivation time can become longer by confining the nickel inside the porous solid, but coke formation is not reduced Ni/Carbon, Ni/SBA-15	• The combination of mesoporous support and bimetallic systems has potential in resistance to carbon poisoning Ni–Rh/SBA-15	• Promotion of Ni dispersion and stronger carbon resistance • More accessible Ni active centres and stabilization of the Ni nanoparticles by the confinement effect Ni/SiO_2/carbon, $NiCeO_2$/SBA-15

The addition of promoters is another feature to be considered when designing an optimal Ni catalyst. The promoters more typically used are based on alkaline and earth metals due to their ability to gasify carbonaceous deposits. For instance, it is well known that potassium is able to enhance steam adsorption and promote the gasification of coke precursors on the catalyst.[63,90] On the other hand, promotion of Ni catalysts with selective doping of atoms such as boron or sulfur on the low coordination sites effectively suppresses the formation of carbon whiskers, although this also decreases catalytic activity. In the sulfur passivated reforming (SPARG) process, sulfur is used to selectively poison the most active sites on the nickel particles, and in this way prevent the growth of carbon filaments while maintaining a good activity for reforming.[63,76]

The other approach proposed for increasing resistance to carbon formation on Ni-based catalysts is the development of bimetallic Ni-based catalysts including noble metals (Pt, Ru, Ir, Rh) and other transition metals (Co, Fe, Cu).[61] In general, the formation of carbon is efficiently supressed by the formation of Ni alloys during the reduction pre-treatment in comparison with single Ni sites. One of the key factors to be controlled in order to achieve excellent catalytic performance with bimetallic catalysts is the preparation method.[91] In general, catalysts synthesized by the sol–gel method give better results than those synthesized by impregnation or those synthesized by the co-precipitation method, since the sol–gel method increases the intimate contact between the metal precursors.[60] Moreover, the order of metal addition in the preparation of bimetallic systems also plays an important role in determining catalyst performance, and the catalysts prepared by simultaneous addition are usually more efficient for the gasification of carbon deposits.[92] On the basis of this general overview, it is evident that optimal catalyst design requires the control of many important variables depending on the synthesis method.

8.3.1.3 Mechanistic Aspects of DRM

Numerous studies have been devoted to understanding the reaction mechanism and elementary steps at the atomic level involved in the dry reforming reaction. Although there is no clear consensus regarding the intrinsic details of the reaction mechanism for DRM on the catalyst surface, it is commonly accepted that the fundamental steps involved in the mechanism can be summarized as follows:[46] (1) dissociative adsorption of methane on the surface metal sites for producing H* and CH_x* species; (2) dissociative adsorption of CO_2 on the metal surface, generating CO* and O* adsorbed species, or associative adsorption on basic sites of the support to create carbonate species; (3) formation of hydroxyl groups *via* hydrogen migration from the metallic sites to the support and O* spill-over to form CH_xO* intermediates by reaction with adsorbed CH* species on the metal sites; and (4) decomposition of CH_xO* intermediates into CO and H_2, which are subsequently desorbed.

Understanding the mechanism of DRM and determining the kinetic parameters is fundamental to designing an effective catalyst and to commercializing the dry reforming technology. The mechanism and the kinetics of DRM depend on the type of the catalyst used as well as the partial pressure range of the reactants involved in the process.[93] Most of the reaction mechanisms proposed in the literature for the dry reforming of methane reaction are based on the Langmuir–Hinshelwood model. This involves a dissociative adsorption of the reactants (CH_4 and CO_2) followed by rate-determining surface reaction of the adsorbed species to the final products. In a very interesting review, Kathiraser *et al.*[54] present a survey of the different rate-determining steps proposed in the literature based on the widely used Langmuir–Hinshelwood model for Ni-catalysed DRM reactions and discuss the importance of the experimental techniques for justifying mechanistic formulation.

Although dry reforming of methane has been extensively studied, the nature of the adsorbed species involved in the process as well as the key intermediates remain under debate. In order to know what exactly is occurring on the catalyst surface during the reaction, it is necessary to combine the direct observation by spectroscopic techniques with computational modelling methods. Spectroscopic methods have also been used to elucidate the reaction mechanism. For instance, a detailed diffuse reflectance Fourier-transform spectroscopy (DRIFTS) study of the mechanism of the DRM reaction over a $Rh/MgAl_2O_4$ catalyst has been performed.[94] Based on the results obtained in this work and the reaction mechanisms postulated in the literature, the elementary reaction steps included in Table 8.2 seem appropriate to describe the mechanism involved in the biogas reforming reaction. Among these steps, methane dissociation could be considered the rate-determining step in the reforming reaction, as it is necessary to activate the highly stable C–H bond.[95] Ceyer *et al.*[96] reported that the dissociation of CH_4 on a Ni (111) surface occurs through a distortion of its tetrahedral geometry to form a trigonal pyramidal structure, and the energetic barrier arises largely from the energy required to deform the molecule sufficiently to allow a strong attractive interaction between the carbon and the Ni surface. In order to achieve C–H cleavage bond, tunnelling of H* atoms through the activation barrier must take place on the Ni surface.

The mechanism of DRM depends largely on the different catalyst formulations. Overall, methane is activated on metal sites and dissociates to form adsorbed H* and CH* species, and this step can be generalized for all metal-based catalysts. By contrast, different types of support can give rise to different reaction mechanisms. For instance, silica-supported catalysts operate through a monofunctional mechanism in which such acidic supports are hardly involved in the mechanism and the activation of both CO_2 and CH_4 only occurs on the metal site. Nevertheless, a bifunctional mechanism has been proposed for other supports such as Al_2O_3, TiO_2, MgO, $MgAl_2O_4$, CeO_2 and La_2O_3, whereby the activation of CO_2 takes place on the support instead, rather than on the metal active site, and methane is activated on the metal.[97]

Table 8.2 Proposed elementary reaction steps for the DRM reaction. Adapted from Ref. 94 with permission from John Wiley and Sons, © 2017 Wiley-VCH Verlag GmbH & Co. KGaA, Weinheim.

Step 1:	$CO_2(g) \rightarrow CO^* + O^*$ (dissociation of CO_2 on metal sites)
Step 2:	$CH_4(g) \rightarrow CH_3^* + H^*$ (methane cracking on metal sites)
Step 3:	$CH_3^* \rightarrow CH_2^* + H^*$
Step 4:	$CH_2^* \rightarrow CH^* + H^*$
Step 5:	$CH^* \rightarrow C^* + H^*$
Step 6:	$CH_3^* \rightarrow CH_2^* + H^*$
Step 7:	$H^* + H^* \rightarrow H_2(g)$ (hydrogen formation and subsequent desorption)
Step 8:	$CO_2(g) + O^* \rightarrow CO_3^*$ (CO_2 adsorption on the support basic sites forming carbonates)
Step 9:	$CO_2(g) + OH^* \rightarrow HCO_3^*$ (formation of bicarbonates on the support)
Step 10:	$HCO_3^* + H^* \rightarrow HCOO^* + OH^*$ (H-assisted formation of formates)
Step 11:	$HCOO^* \rightarrow CO^* + OH^*$ (formate decomposition on the metal–support interface)
Step 12:	$CH_x^* + O^* \rightarrow CH_xO^*$ (oxidation of CH_x fragments on metal sites)
Step 13:	$CH_xO \rightarrow H^* + CO^*$ (CH_xO decomposition on the metal sites)
Step 14:	$CO^* \rightarrow CO(g)$ (CO desorption)
Step 15:	$OH^* + H^* \rightarrow H_2O(g)$ (water production and subsequent desorption)

Regarding the role of the promoter in the mechanism, these additives are usually involved in secondary reactions such as the oxidation of carbonaceous deposits generated on the metal by methane cracking. Future research in this area by combining *in situ/operando* techniques and theoretical calculations for predicting the involved steps is required for achieving the rational design of catalysts.

8.3.2 Catalytic Upgrading of Syngas Obtained by Thermochemical Conversion of Biomass

The thermochemical conversion of biomass by steam gasification or pyrolysis produces a mixture of gaseous effluents that contains H_2, CO, CO_2, methane, water vapour, C_2–C_5 hydrocarbons, tars (aromatic compounds), ammonia and H_2S. This mixture is commonly designated as "producer gas" and it has to be converted into cleaned syngas for producing liquid transportation fuels.[98,99] The upgrading of the producer gas to suitable bio-syngas for producing liquid synthetic fuels requires several complex processes that include: (1) removal of char particulate matter;[100] (2) reduction of tars and carbon dioxide;[99] (3) elimination of traces of alkali metals, chlorine, sulfur and nitrogen compounds;[100] and (4) adjustment of the H_2:CO molar ratio *via* the water gas shift (WGS) reaction (increasing the hydrogen content in the gas mixture through the reaction of water vapour with CO).[101]

The economical clean-up or conditioning of biomass gas effluents is an unsolved problem. Sulfur and tars are the main concerns for many researchers because both components cause severe damage to engines, turbines and pipelines in industrial installations.[102,103] The tar compounds are composed of a mixture of aromatic hydrocarbons that are condensable (benzene and higher) at room temperatures. These tars are well-known precursors of coke and need to be removed from the producer gas before it passes through any downstream processes to avoid rapid deactivation of catalysts and blocking of filters. The elimination of sulfur compound gases is also mandatory in order to decrease corrosion problems and poisoning of catalysts. Therefore, cleaning-up and conditioning treatments of the producer gas are required for further applications. The undesirable tar components can be reduced by downstream gas cleaning methods such as scrubbers and catalytic reforming processes,[99,104] while the most common technologies used for the removal of H_2S are based on solvents or catalytic adsorbents.[105,106] Mostly, Ca-based sorbents with highly basic character such as limestone or dolomite are used for removing acidic sulfur compounds in downstream capture processes.[107] Nevertheless, even after sulfur adsorption processes at high temperatures, some residual sulfur still remains in the effluent.

The catalytic reforming of light hydrocarbons and tars followed by the WGS reaction for adjusting the H_2:CO molar ratio is the most used process for upgrading the biomass-generated gas with the aim of obtaining an enriched syngas that will be further used in the production of fuels *via* FTS.[108] These processes have been established in large-scale facilities, but their application in the upgrading of the gasification-generated gas still requires the design of catalysts that are more active and stable in the presence of these strong contaminants.[109] The complexity of upgrading the syngas is evidenced by the many simultaneous reactions that take place during the producer gas reforming process. Depending on the gas mixture composition, these reactions include CO_2 and steam reforming reactions of methane and tars, CO/CO_2 hydrogenation to methane, tar and methane hydrocracking, the RWGS reaction and the Boudouard reaction.[102] These multiple reactions greatly influence in the product distribution and must be controlled using an adequate catalyst. According to Sutton *et al.*,[110] the criteria that an optimal catalyst must satisfy are the following:

1. High efficiency in removing the tars.
2. The desired product is syngas and the catalyst must be capable of reforming methane.
3. The catalyst should provide a suitable syngas ratio for the intended process.
4. The catalyst must be highly resistant the coke formation and metal sintering.
5. Ability to be easily regenerated.
6. High mechanical resistance and thermal stability.
7. Inexpensive.

Depending on the position of the catalyst with respect to the gasifier in the gasification process, the catalysts can be classified into two distinct groups.[110,111] The first group corresponds to primary catalysts added directly to the biomass prior to gasification either by wet impregnation or by physical dry mixing. Typical primary catalysts includes nature minerals such as dolomite, olivine and limonite, clay minerals such as feldspar or also alkali carbonates, magnesia or alumina.[112–114] These catalysts notably reduce the content of tars in the producer gas, but rarely affect the conversion of methane and other light hydrocarbons. Therefore, this approach is not favourable for obtaining upgraded syngas. The second group of catalysts is located in a secondary reactor downstream of the gasifier and can be operated under different conditions than those of the gasification unit. The secondary catalysts are active in the dry and steam reforming reactions of light hydrocarbons, tars and methane, and this approach is more adequate for conditioning and upgrading the syngas produced by gasification.

Commercially available Ni-based catalysts are highly effective at the removal of hydrocarbons and adjustment of the gas composition to improve syngas quality. Recently, Haldor Topsøe successfully tested a Ni-based catalyst for tar reforming in a commercial large-scale plant.[115] The nickel catalysts act best as secondary catalysts placed in a downstream reactor, which can be operated under different conditions from those of the gasifier.[110] The composition of Ni-based catalysts has been the subject of much research, as the simpler Ni-based catalysts are not optimal due to carbon deposition. The nickel is the catalyst active site, and the addition of promoters increases activity and/or stability, whereas the support provides a high surface area, durability and coking resistance. Promoting the Ni catalysts with other transition metals (Mo, W or Fe)[98,116] or noble metals such as Rh, Ru and Pd can notably enhance the catalytic activity, reducibility, regenerability of spent catalysts and coke resistance.[117] In addition, alkaline and alkaline earth metal promoters such as magnesium, calcium and potassium are usually added to ensure economical operations under severe conditions.[118] Ni doped with alkali and/or alkaline earth promoters shows much promise, especially near or above 900 °C. Moreover, at such temperatures, there is less poisoning of Ni sites by sulfur and irreversible transformations to Ni mixed oxides.[102] Regarding the effect of the support, it should supply a matrix with a high specific surface area, good mechanical properties and resistance to coke and metal sintering.[119] Figure 8.4 includes an overview of the development of Ni-based catalysts for the catalytic reforming of producer gas produced from biomass gasification. In addition to the roles of the active metals, promoters and supports that constitute a Ni-based catalyst, the preparation method and the catalytic pre-treatment are also considered to be some of the most important factors influencing catalytic performance.[119] It is well known that the physicochemical properties such as total surface area, total pore volume, metal distribution and metal size of the catalysts depend directly on the preparation methods.[120] Ion exchange, impregnation and precipitation methods are the common strategies for metal-based catalyst preparation. High metal dispersion can be achieved when the ion exchange method is employed for catalyst preparation.[113] The impregnation method is typically used for dispersing the active metal on the support surface, thereby maintaining the shape and structure of the catalyst. On the other hand, the temperature of calcination and the reduction pre-treatment play key roles in controlling the dispersion of the active phase and the metal–support interaction. To date, many researchers have reported effective catalysts in this area, and Figure 8.4 provides guidance for further research work into the design of novel catalysts. A series of advanced characterizations should be employed for catalyst design and catalytic pathway investigation, which are very helpful for understanding the catalyst structure and active metal state under reaction conditions.

8.4 Production of Liquid Fuels: FTS

8.4.1 Current Status

In the production of liquid transportation fuels, FTS may be considered as the key step for transforming syngas into a mixture of higher hydrocarbons.[121] The obtained mixture is mainly composed of α-olefins and paraffins with different chain lengths, along with small proportions of oxygenated compounds, benzene and other derivatives that may be used as liquid fuels.[122] The synthesis

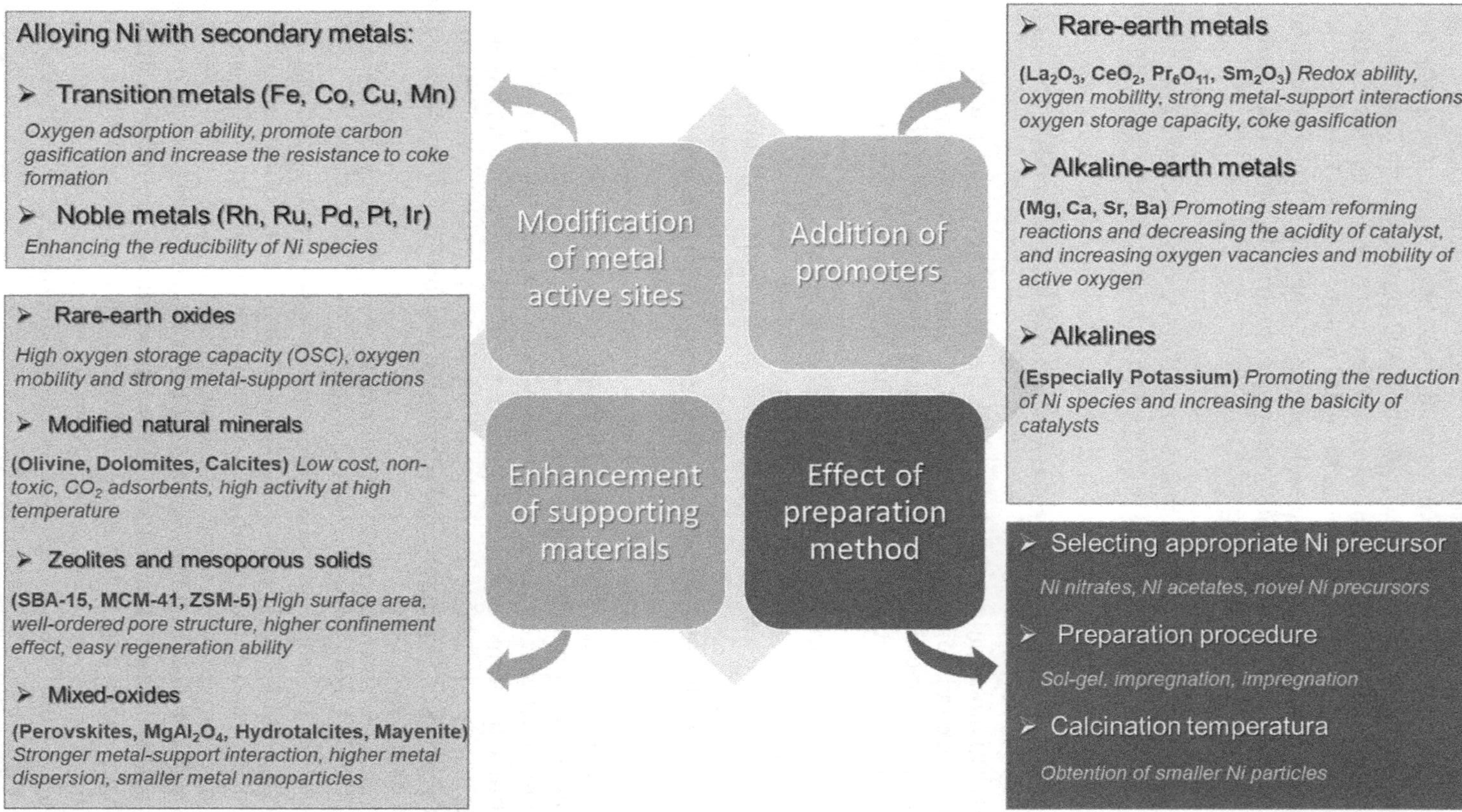

Figure 8.4 Overview of Ni-based catalysts and guidance for further research works in the reforming of producer gas for producing upgraded syngas.[119]

of hydrocarbons from syngas was first developed by Franz Fischer and Hans Tropsch in 1923 and patented in 1925. In 1934, RuhrChemie built the first such industrial plant with a capacity of about 100 tons per year in Oberhausen-Holten (Germany).[123] Since then, the fundamental research on this reaction, although subject to fluctuations in the price of crude oil, has been extensive, and there are several industrial-scale plants in operation. The installations that currently operate in the world use syngas obtained by carbon gasification (CTL) or by reforming natural gas (GTL).

The FTS reaction, which in its simplest form may be represented by eqn (8.6), is an exothermic reaction that is carried out in the presence of a catalyst at temperatures in the 200–350 °C range and pressures of between 20 and 40 bar.[124]

$$CO + 2H_2 \rightarrow -(CH_2)_n- + H_2O \qquad \Delta H^\circ = -165\ \text{kJ mol}^{-1} \qquad (8.6)$$

The H_2 : CO molar ratio in the feed, which depends on the syngas source and production conditions, is between 0.5 and 2.5 and determinates in great part the operating conditions and the employed catalyst in the FTS reactor. In general, to transform a mixture of CO and H_2 into long-chain hydrocarbons, both gases have to be activated on the active sites (metal). As H_2 is easily adsorbed on most metals surfaces giving highly reactive H* species, the activity of a metal in FTS is governed by the strength of the metal–CO interaction. According to Brodén *et al.*,[125] CO is easily dissociated on metals located on the left of the periodic table (strong interaction), while for the transition metals on the right of the periodic table, the CO adsorption is molecular (weak interaction). Considering Sabatier's principle, the metal actives have to exhibit an intermediate interaction strength with the CO, and typically they are Ni, Ru, Co and Fe.[126] Nickel is discarded because it favours methane production (an undesirable product in FTS) over C–C coupling reactions.[127] Ruthenium is the most active metal (working at temperatures above 150 °C) and is very selective to paraffins. However, the high price of this metal and its low availability make it unsuitable for industrial application, and its use is limited to research works.[128] Therefore, industrial catalysts for FTS are fundamentally based on cobalt or iron.

Cobalt, which is cheaper than ruthenium, shows high activity and selectivity to long-chain hydrocarbons waxes and paraffins, and it is relatively resistant to deactivation. The working temperature has to be between 190 and 240 °C (called low-temperature FT [LTFT]) because at higher temperatures the methane production increases and the catalyst is rapidly deactivated. Due to the low activity of this metal in the WGS reaction, the value of the H_2 : CO ratio in the inlet flow has to be very close to 2 (between 2.0 and 2.3), which is accomplished when the syngas is obtained from natural gas (steam reforming or partial oxidation). The main advantage of LTFT is the production of diesel of good quality and free of sulfurs.[129] Co-based catalysts are currently used in some plants operated by Shell and Sasol.[128] Iron, a cheap and widely available metal, may be used either in low- and high-temperature processes (300–350 °C). At high temperatures, the production of α-olefins and oxygenated products is high. Installations using iron-based catalysts are flexible in terms of the H_2 : CO ratio in the feed stream (high activity in the WGS reaction), which can range between 0.5 and 2.5. The main drawback of this metal is its short lifetime.[130]

In addition to the catalyst, the type of reactor also plays a crucial role. The main goal in reactor design is to eliminate the heat produced by this exothermic reaction because the temperature determines the selectivity to desired products. To this end, the most used reactors are multi-tubular fixed beds with internal water cooling systems and the bubble column reactor developed by Sasol.[129,131,132] Currently, many researches and many chemical companies are focusing their efforts on the development of adequate refrigeration systems and novel reactor designs.

8.4.2 Design of FTS Catalysts

The industrially used Co-based catalysts contains between 10 and 30 g of metal per 100 g of support (usually Al_2O_3, SiO_2 or TiO_2) and the average lifetime is between 2 and 3 years depending on the operating conditions. Even though cobalt is cheaper than ruthenium, the optimization of the metallic content, related to the particle size, the intrinsic activity of the metal sites and the analysis of catalyst deactivation are the most important factors to be evaluated from an economical point of view.[133]

Concerning particle size, there is a general consensus on the existence of an optimal value (6–10 nm). Below this optimal value, CO conversion is low, as is selectivity to the C_{5+} fraction, while methane production is significant. For larger particles, FTS turnover rates and selectivity to C_{5+} hydrocarbons and CH_4 are kept approximately constant.[134,135] Therefore, the strict control of the cobalt particle size is an important parameter for optimizing catalytic activity and decreasing the costs associated with catalyst production. To this end, new synthesis methods to obtain a very narrow particle size distribution and high dispersion, such as thermal decomposition, aerosol synthesis, microemulsion methods or modifications of them, are being developed.[129,136–139]

Cobalt catalyst deactivation has been investigated by Rytter and Holmen,[133] who established the sintering, reoxidation of cobalt, formation of stable cobalt–support compounds, surface reconstruction and formation of carbon species on the metal under reaction conditions as the main causes of activity loss. In order to minimize these factors and increase catalytic performance, the use of promoters has been proposed. Thus, the addition of small amounts of noble metals (Ru, Rh, Pt, Pd) to cobalt decreases the reduction temperature of the active phase (spill-over effects), increases its dispersion and decreases the fraction of hardly reducible mixed oxides. That is, noble metal-promoted catalysts show improved activity and longer lifetimes.[129,140,141] Moreover, the addition of rhenium to Co-based catalysts had a positive effect that depended on the pore size distribution of the catalytic system: the effect was small for solids with wide pores.[142] However, the price of noble metal promoters increases the total cost of liquid carbon production by FTS, and thus the addition of metallic oxides such as MnO, CeO_2, ZrO_2 or La_2O_3 as promoters to the cobalt catalysts seems to be a better alternative.[143,144] The presence MnO in a Co/SiO_2 catalyst leads to improved catalytic activity and selectivity to C_{5+} hydrocarbons without modification of the cobalt particle size.[145] The addition of CeO_2 to a Co/ZrO_2 catalyst improves the solid reducibility and cobalt dispersion, thereby increasing the number of surface active sites.[146] The promotion of an Al_2O_3 support with La^{3+} results in higher selectivity to hydrocarbons. When the catalytic support is SiO_2, the addition of La (~3%) improves catalytic activity, probably because the adsorbed CO on the promoted solid is more active and the methane production is suppressed.[147,148]

In the case of Fe-based catalysts, which are typically used in CTL and BTL processes, the addition of promoters improves their catalytic activity. Thus, the commercial catalysts may contain small amounts of potassium, which favours CO dissociation and decreases the selectivity to methane, copper or other transition metals, thereby facilitating the reduction of iron oxides.[149,150] Wang *et al.*[151] studied the addition of Cr, Mn and Zn to an Fe catalyst and concluded that unpromoted and Zn-containing catalysts (form $ZnFe_2O_4$) showed a high initial activity but also a rapid deactivation. In contrast, Mn- and Cr-promoted solids were very resistant to deactivation, although the initial activity was lower than that of other solids. The promoting effect of Mn has also been analysed using steady-state isotopic transient kinetic analysis (SSITKA), and the results reveal that the intrinsic activity of the site is not modified, but the number of active sites is increased.[150]

Considering the positive properties of cobalt in this reaction, bimetallic Fe–Co catalysts have also been tested in FTS. An improved catalytic activity as well as higher selectivities to longer hydrocarbons and lower selectivities to CH_4 are observed in bimetallic solids compared to the monometallic ones.[152] Calderone *et al.*[153] reviewed the use of bimetallic systems in FTS, but the authors found that the comparison of catalytic results reported in the literature is difficult due to the incomplete characterization of the solids and the different reaction conditions or type of reactor employed in the studies. Moreover, the catalysts suffer deep restructuration during the first hours on stream, and most of the published values of conversion and selectivity are measured in this period of time, when the steady state of the reaction has not yet been reached. These authors recommend a reaction time of at least 350 h before measuring activity and selectivity in order to facilitate such comparisons.

In some cases, the design of solids aims not only to improve activity and selectivity, but also to simplify the process of producing liquid fuels. The high temperature FT process (with iron catalysts) produces olefins and aromatic compounds that are mainly used to produce high-value chemicals. At lower temperatures (LTFT), selectivity to long-chain waxes is high, and a further step of hydrotreating is required to obtain liquid hydrocarbons (gasoline: C_5–C_{11}; diesel: C_{12}–C_{20}). In order to avoid post-treatment, the use of bifunctional catalysts with active sites for the FTS (metal, Fe or Co) and cracking/isomerization reactions (acid sites on the support) is being considered. Martínez-Vargas *et al.*[130] analysed the use of different supports (zeolites, clays and metallic oxides) in this "one-stage" process. Zeolites are among the most studied supports, probably due to the possibility of modifying both the number and the strength of the acid sites. For example, some studies demonstrated that high concentrations of strong Brønsted sites decreases the yield of gasoline.[154] In this sense, promising results are obtained with zeolites and other supports, but additional research is essential to set up this process.

8.4.3 Mechanistic Considerations

FTS is considered to be a surface polymerization reaction in which monomers are generated *in situ* by CO hydrogenation. The exact mechanism of this reaction is not completely clear, but there is general agreement on the fact that this is a chain reaction with the following steps: (1) initiation, in which CH_x monomers are formed; (2) propagation, with the formation of CH_x–CH_x fragments; and (3) chain termination by hydrogenation to give paraffins or β-abstraction to form olefins.[155] The relative rates of chain propagation and chain termination (both depending on the chain length) determinate the product distribution.[134,156] The nature of monomeric species and the growth chain step are still controversial, and different reaction mechanisms have been proposed. The main mechanisms are the carbide mechanism, the CO insertion mechanism and modifications of them. The first one, proposed by Fischer and Tropsch, implies the CO dissociation on the metal site to give C* and O* species. These species are hydrogenated to form –CH_2– species, which polymerize, and water, a product of the reaction. In the CO insertion mechanism, the chain growth proceeds by insertion of adsorbed CO (M–CO) into the chain, and this explains the formation of oxygenated compounds *via* FTS. In any case, the CO dissociation on the metal surface is the key step in this reaction given that it determines the formation of monomeric species. This step has been deeply studied by Shetty *et al.*[157] on Ru and Co surfaces with different morphologies. According to this work, the CO may be directly dissociated on the metal ($CO \rightarrow C^* + O^*$) or proceed by so-called H-assisted CO dissociation. In the latter case, the adsorbed CO is hydrogenated, giving HCO* or COH* intermediates before C–O bond scission. It is observed that, on flat surfaces (in both Co and Ru), H-assisted CO dissociation *via* HCO* species shows the lowest activation barrier, being the most probable route for the first step of FTS. However, corrugated and stepped surfaces of both metals favour direct CO dissociation. Furthermore, Ojeda *et al.*[158] highlighted the importance of H-assisted CO dissociation by combining theoretical and experimental results, and they established the pathway in eqn (8.7) for monomer formation as the most favourable:

$$CO^* \rightarrow HCO^* \rightarrow HCOH^* \rightarrow CH^* + OH^* \tag{8.7}$$

In an interesting work, Filot *et al.*[159] described the existence of three different kinetic regimes in FTS as a function of the metal–carbon and metal–oxygen interaction energies. Thus, in metals in which metal–carbon and hydrogen–carbon interactions are weak, the CO dissociation is the rate-limiting step and the α (chain growth probability) values are low. This means that for metals in which the carbon–metal interaction is weak, the selectivity to methane is very high (Regime I). Calculations considering stronger metal–carbon interaction (weak metal–oxygen interaction) indicate that the rate-limiting step is the chain growth and termination steps, and in this case, the methane formation is favoured. This is called Regime III by these authors, and it represents the situation in which oxygen strongly bonds to the metal surface. In this case, water removal is the rate-limiting step and the chain growth probability is increased.

An important aspect to be considered in relation to the reaction mechanism is the effect of the water produced in the reaction or added to the inlet flow, which depends on the catalyst (metal and support).[160] The literature shows that, in the presence of water, CO conversion and selectivity to C_{5+} hydrocarbons are improved,[161] but the presence of water also affects the catalyst's stability due to oxidation of the active phase[162,163] or modification of the support.[164,165]. Other authors affirm that the water filling the pores of the solids affects the diffusion of both reactives and products, in such a way that pore size controls the product distribution.[160,162,166,167] From a mechanistic point of view, Gunasooriya *et al.*[168] analysed the role of water in FTS over Ru-based catalysts using theoretical calculations and concluded that the hydroxyl-assisted CO activation presented the lowest activation barrier, being the most probable method of the formation of monomeric species.

In addition to theoretical calculations, kinetic analysis and spectroscopic techniques have been used to determine the nature of active sites on different catalysts for FTS, as well as the most probable reaction intermediates.[161,169,170] For instance, van der Laan *et al.*[171] carried out a kinetic analysis of Fe-based catalysts and concluded that the active species in FTS are the iron carbides, while the presence of an Fe_3O_4 phase is responsible for activity in the WGS reaction of these solids.

Given that the surface dynamic under reaction conditions is a proven fact, *operando* techniques are the most adequate for this propose. DRIFTS was used to analyse a Co/Al_2O_3-SiO_2 catalyst under reactive flow (H_2 : CO = 2), 220 °C and atmospheric pressure.[172] Under these conditions, metallic carbonyl species with different coordination (linear and bridged) that involve structural modifications with the time on stream were detected. As carbonaceous species were not detected, the authors concluded that the carbon removal rate is higher than the CO dissociation rate, confirming that CO dissociation is the rate-limiting step, as was previously proposed in the literature. Moreover, formate species were observed during the reaction, and due to the different reactivities of both species, the authors proposed that one of these types is associated with the metal. On the other hand, van Ravenhorst *et al.*[173]

investigated a Co/TiO_2 catalyst under reactive conditions using combined microscopy and X-ray absorption, called scanning transmission X-ray microscopy (STXM). The results showed the growth of the hydrocarbon chain on the metal surface, being more significant on the sites where the cobalt oxidation state is higher.

Therefore, it can be concluded in general form that the setting up of the technology to produce liquid transportation fuels from a renewable source requires the development of research works in the fields of catalysts and reactor design.

8.5 Further Trends: Structured Catalysts and Microreactors

The production of fuels from lignocellulosic biomass and wastes very often involves catalytic processes that are characterized by strong heat exchange requirements due to the high thermal effect of the chemical reactions involved, as well as by the difficulty of simultaneously minimizing transport limitations and pressure drops in conventional fixed-bed reactors.[174,175] Sometimes, extremely short contact times are also required. As a result, the conventional catalytic technologies operate under suboptimal conditions. The structured catalytic systems, structured catalysts and microchannel reactors offer excellent opportunities for overcoming those limitations because they efficiently allow for the simultaneous minimization of both transport limitations and pressure drops while improving the radial fluxes of mass and heat and allowing for very short contact times.[176–178] Monoliths with parallel channels, open cell foams and stacked wire meshes can be made of a variety of metallic alloys and cells or pore densities.[179] They can be also coated with any convenient catalyst, thus becoming appropriate for the process of interest. On the other hand, microchannel reactors are capable of providing an incomparable intensification of the process with excellent temperature control and improved product quality and process safety.[180,181] Therefore, the potential of microreactors for the catalytic conversion of biomass derivatives to value-added chemicals and fuels is expected to contribute satisfactorily to accelerating the technological development of biomass conversion processes.

Several reviews have focused on the reactor engineering or process intensification aspects of biomass transformations, and these different aspects have been critically reviewed.[174,182,183] Studies on microstructured reactors have received special attention because of the reduced dimensions of the channels, in which the mass and heat transfer rates are greatly improved and, as a consequence, the yield of liquids is higher and the pressure drop becomes very low.[177,184] A few examples of the application of microreactor technology to the processes examined in this chapter (reforming of biogas, cleaning-up of producer gas and FTS) are described below.

Regarding the advantages of the utilization of monoliths and microreactors in the reforming of biogas, Vita *et al.*[185] investigated syngas production by the steam reforming (SR) and oxy-steam reforming (OSR) of upgraded biogas over cordierite monoliths (400 channels per square inch [cpsi]) coated with Ni, Rh or Pt on a CeO_2 catalyst. They developed a preparation method that allowed uniform and thin-coated layers with high thermal and mechanical strength to be obtained, and they designed a Rh-based structured catalyst that presented optimal performance for the reforming of biogas. Microreactors are reported to have superior intrinsic activity and stability compared to tubular fixed-bed reactors, but there are also problems with microreactors, such as reactor blockage.[186] Izquierdo *et al.*[187] compared biogas steam and oxidative reforming processes for syngas and hydrogen production in conventional and microstructured reactor systems with several Ni-based catalysts and a bimetallic Rh–Ni catalyst supported on different supports. They observed that the catalytic activity measured when operating with a microreactor is an order of magnitude higher than that achieved in fixed-bed reactors. On the other hand, Zhou *et al.*[188] reported the successful design of a corrugated-type structured catalyst based on porous anodic alumina membranes for methane dry reforming, and the catalyst with a strong Ni–$NiAl_2O_4$ interaction showed both excellent activity and stability.

Concerning the application of microstructured systems to the cleaning-up of producer gas, the use of monolithic catalysts and honeycomb structures that can operate with gases containing particulates is very useful in order to avoid the use of hot filters. Corella *et al.*[189] investigated the utilization of Ni-based monoliths for tar elimination in producer gas produced by the gasification of biomass. The intrinsic activity of the monolith for removing tars was compared with that of other competing catalysts, such as dolomites and commercial SR catalysts, and it was concluded that these monoliths are not very high in activity, but they can operate with a producer gas containing particulates, thereby avoiding the use of hot filters, which are problematic when used in biomass gasification. In another work, Wang *et al.*[190] designed a NiO–MgO solid solution cordierite monolith catalyst and observed that producer gas reforming on structured catalysts can improve the quality of syngas obtained as a separate process from adjusting the H_2 : CO ratio. Moreover, the addition of steam notably decreases the carbon deposition on the monolith channels.

A wide variety of works on the utilization of microstructured reactors and monoliths in FTS have been reported during the last two decades. Almeida *et al.*[183] performed an experimental investigation of structured catalysts for FTS using metallic supports of different geometries including aluminium foams, honeycomb monoliths with different cell densities (cpsi) and microchannel block reactors presenting perpendicular channels for reaction and cooling. This study gives evidence of the viability of the use of microstructured catalytic reactors for FTS and provides information about the key parameters in the design and fabrication of monoliths and microchannel blocks. The C_{5+} hydrocarbon selectivity depends on the structured system used, and it was superior in the microchannel reactor. This fact is related to the catalytic layer thickness as well to the better temperature control in the case of the microchannel blocks. Therefore, controlling the coating process and the microreactor configuration are key parameters for good microreactor design for FTS. Visconti *et al.*[191] prepared a prototype of cobalt-based catalysts washcoated onto metallic structured supports with different geometries and evaluated the catalytic properties of such systems in FTS under industrially relevant process conditions. Comparative tests showed that structured catalysts maintained the activity and selectivity of the original powdered catalyst only when the washcoat thickness was sufficiently thin. In spite of these advantages, industrial installations using microreactors are not available today due to technological problems, costs and the need to improve some aspects related to the catalysts, such as to achieve high selectivity towards a selected fraction of hydrocarbons or to increase the stability of the solids.[192,193] Therefore, research in the field of catalyst design for this reaction is currently facing significant challenges.

8.6 Summary and Concluding Remarks

The production of liquid transportation fuels from inedible lignocellulosic biomass is an interesting alternative for substituting fossil fuels with a renewable energy resource and reducing greenhouse gas atmospheric emissions. For the production of synthetic fuels, syngas is first produced from biomass by means of thermochemical routes (gasification and pyrolysis) or through biological processes (anaerobic digestion). In the second stage, the synthetic fuel is produced out of syngas by FTS. Technical barriers to the thermochemical conversion of lignocellulosic biomass into fuels include the upgrading of the producer gas into valuable syngas, with specific emphasis on removing tars and sulfur compounds. In the case of anaerobic digestion, the biogas produced is transformed into valuable syngas by means of dry reforming of the biogas, and the catalytic systems developed for this process require deactivation due to the accumulation of carbonaceous deposits to be minimized. Concerning FTS, despite such a process existing for more than 90 years, several important catalytic challenges deserve special attention. For instance, special attention must be paid to exploring alternative preparation methods that provide a narrow particle size distribution. In addition, more studies regarding stability and the mechanisms of catalyst deactivation are required. This chapter has been devoted to examining recent catalyst developments in each process involved in the production of liquid fuels from biomass. In general form, different requirements regarding porosity, acidity–basicity, metallic dispersion, metal–support interactions, metal particle size, mechanical and thermal stability and high resistance to catalyst deactivation must be considered in order to design an optimal catalyst. The preparation method plays a crucial role in controlling these properties, and understanding of the mechanisms involved in each chemical reaction is fundamental to establishing the structure–property relationships and achieving the rational design of efficient catalysts.

Furthermore, novel reactor technologies such as microstructured reactors and microchannel reactors were presented. In principle, the application of microstructured catalytic reactors can solve many of the problems related to the conventional fixed-bed reactors, particularly concerning the necessity to operate over short diffusion distances, at low pressure drops and with optimal heat transfer control. However, the techno-economical evaluation on the application potential of microreactor technology still presents several challenges to be overcome before such technologies can be effectively applied in large-scale commercial biomass transformation processes.

Acknowledgements

Financial support for this work was obtained from the Spanish Ministerio de Ciencia, Innovación y Universidades co-financed by FEDER funds from the European Union (RTI2018-096294-B-C33). Lola Azancot acknowledges the same institution for their PhD fellowship (BES-2016-0077475), and Luis F. Bobadilla acknowledges the contracts of the Spanish Government, financed in part by the European Regional Development Fund through the "Juan de la Cierva incorporación" programme (IJCI-2015-26348).

References

1. R. G. d. Santos and A. C. Alencar, *Int. J. Hydrogen Energy*, 2019, DOI: 10.1016/j.ijhydene.2019.07.133.
2. R. York and S. E. Bell, *Energy Res. Soc. Sci.*, 2019, **51**, 40–43.
3. P. Bórawski, A. Bełdycka-Bórawska, E. J. Szymańska, K. J. Jankowski, B. Dubis and J. W. Dunn, *J. Clean. Prod.*, 2019, **228**, 467–484.
4. G. Centi and S. Perathoner, *Catal. Today*, 2020, **342**, 4–12.
5. R. Rauch, A. Kiennemann and A. Sauciuc, in *The Role of Catalysis for the Sustainable Production of Bio-fuels and Bio-chemicals*, ed. K. S. Triantafyllidis, A. A. Lappas and M. Stöcker, Elsevier, Amsterdam, 2013, pp. 397–443.
6. R. S. Dhillon and G. von Wuehlisch, *Biomass Bioenergy*, 2013, **48**, 75–89.
7. D. King and A. Klerk, *ACS Symposium Series*, 2011, **1084**, 1–24.
8. K. Y. Koo, H.-S. Roh, Y. T. Seo, D. J. Seo, W. L. Yoon and S. B. Park, *Appl. Catal., A*, 2008, **340**, 183–190.
9. E. F. Sousa-Aguiar, F. B. Noronha and J. A. Faro, *Catal. Sci. Technol.*, 2011, **1**, 698–713.
10. T. Mokrani and M. Scurrell, *Catal. Rev.*, 2009, **51**, 1–145.
11. E. van Steen and M. Claeys, *Chem. Eng. Technol.*, 2008, **31**, 655–666.
12. C. N. Hamelinck, A. P. C. Faaij, H. den Uil and H. Boerrigter, *Energy*, 2004, **29**, 1743–1771.
13. A. Naqi, J. N. Kuhn and B. Joseph, *Biomass Bioenergy*, 2019, **130**, 105395.
14. R. G. Grim, A. T. To, C. A. Farberow, J. E. Hensley, D. A. Ruddy and J. A. Schaidle, *ACS Catal.*, 2019, **9**, 4145–4172.

15. E.-M. Aro, *Ambio*, 2016, **45**, 24–31.
16. X. Zhang, *Green Chem.*, 2016, **18**, 5086–5117.
17. P. Read and J. Lermit, *Energy*, 2005, **30**, 2654–2671.
18. D. Catalán-Martínez, M. E. Domine and J. M. Serra, *Fuel*, 2018, **212**, 353–363.
19. A. Chiodini, L. Bua, L. Carnelli, R. Zwart, B. Vreugdenhil and M. Vocciante, *Biomass Bioenergy*, 2017, **106**, 104–114.
20. P. K. Swain, L. M. Das and S. N. Naik, *Renewable Sustainable Energy Rev.*, 2011, **15**, 4917–4933.
21. J. Kainthola, A. S. Kalamdhad and V. V. Goud, *Proc. Biochem.*, 2019, **84**, 81–90.
22. S. K. Pramanik, F. B. Suja, S. M. Zain and B. K. Pramanik, *Bioresour. Technol. Rep.*, 2019, **8**, 100310.
23. J. P. Gebauer and M. Kaltschmitt, in *Energy from Organic Materials (Biomass): A Volume in the Encyclopedia of Sustainability Science and Technology*, 2nd edn, ed. M. Kaltschmitt, Springer, New York, 2019, pp. 1145–1174.
24. Q. Fei, M. T. Guarnieri, L. Tao, L. M. L. Laurens, N. Dowe and P. T. Pienkos, *Biotechnol. Adv.*, 2014, **32**, 596–614.
25. S. S. Ail and S. Dasappa, *Renewable Sustainable Energy Rev.*, 2016, **58**, 267–286.
26. K. Kim, Y. Kim, C. Yang, J. Moon, B. Kim, J. Lee, U. Lee, S. Lee, J. Kim, W. Eom, S. Lee, M. Kang and Y. Lee, *Bioresour. Technol.*, 2013, **127**, 391–399.
27. S. Achinas, V. Achinas and G. J. W. Euverink, *Engineering*, 2017, **3**, 299–307.
28. F. Di Maria, F. Sisani, O. Norouzi and R. L. Mersky, *Renewable Sustainable Energy Rev.*, 2019, **108**, 347–354.
29. A. M. Yousef, W. M. El-Maghlany, Y. A. Eldrainy and A. Attia, *Fuel*, 2019, **251**, 611–628.
30. F. M. Baena-Moreno, M. Rodríguez-Galán, F. Vega, L. F. Vilches and B. Navarrete, *Int. J. Green Energy*, 2019, **16**, 401–412.
31. S. A. Hosseinipour and M. Mehrpooya, *Renewable Energy*, 2019, **130**, 641–655.
32. L. Yang, X. Ge, C. Wan, F. Yu and Y. Li, *Renewable Sustainable Energy Rev.*, 2014, **40**, 1133–1152.
33. S. E. Hashemi, S. Sarker, K. M. Lien, S. K. Schnell and B. Austbø, *Fuel*, 2019, **245**, 294–304.
34. Q. He, G. Yu, W. Wang, S. Yan, Y. Zhang and S. Zhao, *Fuel Proc. Technol.*, 2017, **166**, 50–58.
35. O. I. Maile, E. Muzenda and H. Tesfagiorgis, *Procedia Manuf.*, 2017, 7, 639–646.
36. S. Gao, C. Bo, J. Li, C. Niu and X. Lu, *Renewable Energy*, 2020, **147**, 2335–2344.
37. R. Noorain, T. Kindaichi, N. Ozaki, Y. Aoi and A. Ohashi, *J. Clean. Prod.*, 2019, **214**, 103–111.
38. R. Augelletti, M. Conti and M. C. Annesini, *J. Clean. Prod.*, 2017, 140, 1390–1398.

39. H. Vogtenhuber, R. Hofmann, F. Helminger and G. Schöny, *Energy*, 2018, **162**, 200–209.
40. Y. Tan, W. Nookuea, H. Li, E. Thorin and J. Yan, *Energy Procedia*, 2017, **142**, 3741–3746.
41. A. M. Yousef, W. M. El-Maghlany, Y. A. Eldrainy and A. Attia, *Energy*, 2018, **156**, 328–351.
42. S. Spitzer, M. Miltner and M. Harasek, *J. Clean. Prod.*, 2019, **231**, 43–53.
43. M. Žák, H. Bendová, K. Friess, J. E. Bara and P. Izák, *Separ. Purif. Technol.*, 2018, **203**, 36–40.
44. L. Meier, D. Stará, J. Bartacek and D. Jeison, *Process Saf. Environ.*, 2018, **119**, 65–68.
45. N. Kumar, M. Shojaee and J. J. Spivey, *Curr. Opin. Chem. Eng.*, 2015, **9**, 8–15.
46. C. Papadopoulou, H. Matralis and X. Verykios, in *Catalysis for Alternative Energy Generation*, ed. L. Guczi and A. Erdôhelyi, Springer, New York, 2012, pp. 57–127.
47. M. Bradford and M. Vannice, *Catal. Rev. Sci. Eng.*, 2007, **41**(**1**), 1–42.
48. A. Moral, I. Reyero, C. Alfaro, F. Bimbela and L. M. Gandía, *Catal. Today*, 2018, **299**, 280–288.
49. P. Cao, S. Adegbite, H. Zhao, E. Lester and T. Wu, *Appl. Energy*, 2018, **227**, 190–197.
50. H. Atashi, J. Gholizadeh, F. Farshchi Tabrizi, J. Tayebi and S. A. H. Seyed Mousavi, *Int. J. Hydrogen Energy*, 2017, **42**, 5464–5471.
51. E. Rezaei and S. Dzuryk, *Chem. Eng. Res. Des.*, 2019, **144**, 354–369.
52. H. Wu, J. Liu, H. Liu and D. He, *Fuel*, 2019, **235**, 868–877.
53. H. Liu, K. Li, R. Zhang, L. Ling and B. Wang, *Appl. Surf. Sci.*, 2017, **423**, 1080–1089.
54. Y. Kathiraser, U. Oemar, E. T. Saw, Z. Li and S. Kawi, *Chem. Eng. J.*, 2015, **278**, 62–78.
55. V. R. Bach, A. C. de Camargo, T. L. de Souza, L. Cardozo-Filho and H. J. Alves, *Int. J. Hydrogen Energy*, 2020, **45**, 5252–5263.
56. W.-J. Jang, J.-O. Shim, H.-M. Kim, S.-Y. Yoo and H.-S. Roh, *Catal. Today*, 2019, **324**, 15–26.
57. A. Abdulrasheed, A. A. Jalil, Y. Gambo, M. Ibrahim, H. U. Hambali and M. Y. Shahul Hamid, *Renewable Sustainable Energy Rev.*, 2019, **108**, 175–193.
58. S. Kawi, Y. Kathiraser, J. Ni, U. Oemar, Z. Li and E. T. Saw, *ChemSusChem*, 2015, **8**, 3556–3575.
59. C.-j. Liu, J. Ye, J. Jiang and Y. Pan, *ChemCatChem*, 2011, **3**, 529–541.
60. N. A. K. Aramouni, J. G. Touma, B. A. Tarboush, J. Zeaiter and M. N. Ahmad, *Renewable Sustainable Energy Rev.*, 2018, **82**, 2570–2585.
61. Z. Bian, S. Das, M. H. Wai, P. Hongmanorom and S. Kawi, *ChemPhysChem*, 2017, **18**, 3117–3134.
62. S. Arora and R. Prasad, *RSC Adv.*, 2016, **6**, 108668–108688.
63. P. M. Mortensen and I. Dybkjær, *Appl. Catal., A*, 2015, **495**, 141–151.
64. R. Debek, M. Motak, T. Grzybek, M. Galvez and P. Da Costa, *Catalysts*, 2017, **7**, 32.

65. J. W. Han, J. S. Park, M. S. Choi and H. Lee, *Appl. Catal., B*, 2017, **203**, 625–632.
66. R. Dębek, M. Motak, M. E. Galvez, T. Grzybek and P. Da Costa, *Appl. Catal., B*, 2018, **223**, 36–46.
67. D. L. Trimm, *Catal. Today*, 1999, **49**, 3–10.
68. Z. Alipour, M. Rezaei and F. Meshkani, *J. Ind. Eng. Chem.*, 2014, **20**, 2858–2863.
69. Q. Zhang, X. Feng, J. Liu, L. Zhao, X. Song, P. Zhang and L. Gao, *Int. J. Hydrogen Energy*, 2018, **43**, 11056–11068.
70. B. M. Nagaraja, D. A. Bulushev, S. Beloshapkin and J. R. H. Ross, *Catal. Today*, 2011, **178**, 132–136.
71. A. Álvarez, M. L. F. Bobadilla, V. Garcilaso, M. A. Centeno and J. A. Odriozola, *J. CO2 Util.*, 2018, **24**, 509–515.
72. U. Guharoy, E. Le Saché, Q. Cai, T. R. Reina and S. Gu, *J. CO2 Util.*, 2018, **27**, 1–10.
73. J. Károlyi, M. Németh, C. Evangelisti, G. Sáfrán, Z. Schay, A. Horváth and F. Somodi, *J. Ind. Eng. Chem.*, 2018, **58**, 189–201.
74. J.-S. Choi, K.-I. Moon, Y. G. Kim, J. S. Lee, C.-H. Kim and D. L. Trimm, *Catal. Lett.*, 1998, **52**, 43–47.
75. D. San-José-Alonso, J. Juan-Juan, M. J. Illán-Gómez and M. C. Román-Martínez, *Appl. Catal., A*, 2009, **371**, 54–59.
76. J. R. Rostrup-Nielsen, *J. Catal.*, 1984, **85**, 31–43.
77. X. Du, D. Zhang, L. Shi, R. Gao and J. Zhang, *J. Phys. Chem. C*, 2012, **116**, 10009–10016.
78. M. Wang, T. Zhao, M. Li and H. Wang, *RSC Adv.*, 2017, **7**, 41847–41854.
79. E. le Saché, L. Pastor-Pérez, D. Watson, A. Sepúlveda-Escribano and T. R. Reina, *Appl. Catal., B*, 2018, **236**, 458–465.
80. G. Valderrama, A. Kiennemann, C. U. de Navarro and M. R. Goldwasser, *Appl. Catal., A*, 2018, **565**, 26–33.
81. R. C. Rabelo-Neto, H. B. E. Sales, C. V. M. Inocêncio, E. Varga, A. Oszko, A. Erdohelyi, F. B. Noronha and L. V. Mattos, *Appl. Catal., B*, 2018, **221**, 349–361.
82. M. Li and A. C. van Veen, *Appl. Catal., B*, 2018, **237**, 641–648.
83. I. Luisetto, S. Tuti, C. Battocchio, S. Lo Mastro and A. Sodo, *Appl. Catal., A*, 2015, **500**, 12–22.
84. A. Mourhly, M. Kacimi, M. Halim and S. Arsalane, *Int. J. Hydrogen Energy*, 2020, **45**, 11449–11459.
85. D. Xia, Y. Chen, C. Li, C. Liu and G. Zhou, *Int. J. Hydrogen Energy*, 2018, **43**, 20488–20499.
86. Z. Li, L. Mo, Y. Kathiraser and S. Kawi, *ACS Catal.*, 2014, **4**, 1526–1536.
87. S. Das, M. Sengupta, A. Bag, M. Shah and A. Bordoloi, *Nanoscale*, 2018, **10**, 6409–6425.
88. G. Zhang, J. Liu, Y. Xu and Y. Sun, *Int. J. Hydrogen Energy*, 2019, **44**, 4809–4820.
89. J.-H. Park, S. Yeo and T.-S. Chang, *J. CO2 Util.*, 2018, **26**, 465–475.

90. J. Rostrup-Nielsen and L. J. Christiansen, Carbon and Sulphur, in *Concepts in Syngas Manufacture*, Imperial College Press, London, 2011, ch. 5, pp. 233–293.
91. B. Abdullah, N. A. Abd Ghani and D.-V. N. Vo, *J. Clean. Prod.*, 2017, **162**, 170–185.
92. A. Álvarez M., M. Á. Centeno and J. A. Odriozola, *Top. Catal.*, 2016, **59**, 303–313.
93. D. Pakhare and J. Spivey, *Chem. Soc. Rev.*, 2014, **43**, 7813–7837.
94. L. F. Bobadilla, V. Garcilaso, M. A. Centeno and J. A. Odriozola, *ChemSusChem*, 2017, **10**, 1193–1201.
95. A. C. Luntz and J. Harris, *Surf. Sci.*, 1991, **258**, 397–426.
96. S. T. Ceyer, Q. Y. Yang, M. B. Lee, J. D. Beckerle and A. D. Johnson, in *Studies in Surface Science and Catalysis*, ed. D. M. Bibby, C. D. Chang, R. F. Howe and S. Yurchak, Elsevier, 1988, **36**, pp. 51–66.
97. M.-S. Fan, A. Z. Abdullah and S. Bhatia, *ChemCatChem*, 2009, **1**, 192–208.
98. S. Anis and Z. A. Zainal, *Renewable Sustainable Energy Rev.*, 2011, **15**, 2355–2377.
99. C. K. Acharya, F. Jiang, C.-h. Liao, P. Fitzgerald, K. S. Vecchio and R. J. Cattolica, *Fuel Process. Technol.*, 2013, **106**, 201–208.
100. W. Zhang, H. Liu, I. Ul Hai, Y. Neubauer, P. Schröder, H. Oldenburg, A. Seilkopf and A. Kölling, *Int. J. Low Carbon Technol.*, 2012, **7**, 69–74.
101. M. P. Aznar, M. A. Caballero, J. Corella, G. Molina and J. M. Toledo, *Energy Fuels*, 2006, **20**, 1305–1309.
102. R. Li, A. Roy, J. Bridges and K. Dooley, *Ind. Eng. Chem. Res.*, 2014, **53**, 7999–8011.
103. T. Baidya and R. J. Cattolica, *Appl. Catal., A*, 2015, **503**, 43–50.
104. M. A. Adnan, S. Adamu, O. Muraza and M. M. Hossain, *Process Saf. Environ.*, 2018, **116**, 754–762.
105. L. Wei, S. Xu, L. Zhang, H. Zhang, C. Liu, H. Zhu and S. Liu, *Fuel Process. Technol.*, 2006, **87**, 863–871.
106. P. R. Westmoreland and D. P. Harrison, *Environ. Sci. Technol.*, 1976, **10**, 659–661.
107. X. Meng, W. de Jong, R. Pal and A. H. M. Verkooijen, *Fuel Process. Technol.*, 2010, **91**, 964–981.
108. S. Albertazzi, F. Basile, D. Barbera, P. Benito, J. Brandin, J. Einvall, G. Fornasari, F. Trifirò and A. Vaccari, *Top. Catal.*, 2011, **54**, 746.
109. J. D. Holladay, J. Hu, D. L. King and Y. Wang, *Catal. Today*, 2009, **139**, 244–260.
110. D. Sutton, B. Kelleher and J. R. H. Ross, *Fuel Process. Technol.*, 2001, 73, 155–173.
111. E. G. Baker, L. K. Mudge and M. D. Brown, *Ind. Eng. Chem. Res.*, 1987, **26**, 1335–1339.
112. J. M. de Andrés, A. Narros and M. E. Rodríguez, *Fuel*, 2011, **90**, 521–527.
113. J. Ren, J.-P. Cao, X.-Y. Zhao, F.-L. Yang and X.-Y. Wei, *Renewable Sustainable Energy Rev.*, 2019, **116**, 109426.

114. N. Berguerand, J. Marinkovic, T. Berdugo Vilches and H. Thunman, *Chem. Eng. J.*, 2016, **295**, 80–91.
115. K. J. Andersson, M. Skov-Skjøth Rasmussen and P. E. Højlund Nielsen, *Fuel*, 2017, **203**, 1026–1030.
116. T. Baidya, R. J. Cattolica and R. Seiser, *Appl. Catal., A*, 2018, **558**, 131–139.
117. D. Laprune, D. Farrusseng, Y. Schuurman, F. C. Meunier, J. A. Z. Pieterse, A. M. Steele and S. Thorpe, *Appl. Catal., B*, 2018, **221**, 206–214.
118. P. Knutsson, V. Cantatore, M. Seemann, P. L. Tam and I. Panas, *Appl. Catal., B*, 2018, **229**, 88–95.
119. Z. Zhang, L. Liu, B. Shen and C. Wu, *Renewable Sustainable Energy Rev.*, 2018, **94**, 1086–1109.
120. I. E. Achouri, N. Abatzoglou, C. Fauteux-Lefebvre and N. Braidy, *Catal. Today*, 2013, **207**, 13–20.
121. D. B. Bukur, B. Todic and N. Elbashir, *Catal. Today*, 2016, **275**, 66–75.
122. O. Eliseev, *Russ. J. Gen. Chem.*, 2009, **79**, 2509–2519.
123. H. H. Storch, in *Advances in Catalysis*, ed. W. G. Frankenburg, V. I. Komarewsky and E. K. Rideal, Academic Press, 1948, **1**, pp. 115–156.
124. J. A. Díaz, M. Calvo-Serrano, A. R. de la Osa, A. M. García-Minguillán, A. Romero, A. Giroir-Fendler and J. L. Valverde, *Appl. Catal., A*, 2014, **475**, pp. 82–89.
125. G. Brodén, T. N. Rhodin, C. Brucker, R. Benbow and Z. Hurych, *Surf. Sci.*, 1976, **59**, 593–611.
126. K. Cheng, J. Kang, D. L. King, V. Subramanian, C. Zhou, Q. Zhang and Y. Wang, in *Advances in Catalysis*, ed. C. Song, Academic Press, 2017, **60**, pp. 125–208.
127. C. G. Okoye-Chine, M. Moyo, X. Liu and D. Hildebrandt, *Fuel Process. Technol.*, 2019, **192**, 105–129.
128. S. S. Ail and S. Dasappa, *Renewable Sustainable Energy Rev.*, 2016, **58**, 267–286.
129. A. Y. Khodakov, W. Chu and P. Fongarland, *Chem. Rev.*, 2007, **107**, 1692–1744.
130. D. X. Martínez-Vargas, L. Sandoval-Rangel, O. Campuzano-Calderon, M. Romero-Flores, F. J. Lozano, K. D. P. Nigam, A. Mendoza and A. Montesinos-Castellanos, *Ind. Eng. Chem. Res.*, 2019, **58**, 15872–15901.
131. N. Majidian and S. Soltanali, *J. Japan Pet. Inst.*, 2016, **59**, 126–139.
132. R. Guettel, U. Kunz and T. Turek, *Chem. Eng. Technol.*, 2008, **31**, 746–754.
133. E. Rytter and A. Holmen, *Catalysts*, 2015, 5, 478–499.
134. E. Iglesia, *Appl. Catal., A*, 1997, **161**, 59–78.
135. J. P. den Breejen, P. B. Radstake, G. L. Bezemer, J. H. Bitter, V. Frøseth, A. Holmen and K. P. de Jong, *J. Am. Chem. Soc.*, 2009, **131**, 7197–7203.
136. T. Kuang, S. Lyu, S. Liu, Y. Zhang, J. Li, G. Wang and L. Wang, *J. Energy Chem.*, 2019, **33**, 67–73.
137. A. Martínez Arias and A. P. Weber, *J. Aerosol Sci.*, 2019, **131**, 1–12.
138. A. N. Pour and M. Housaindokht, *Prog. React. Kinet. Mec.*, 2019, **44**, 316–323.

139. S. Lyu, B. Peng, T. Kuang, K. G. Rappé, Y. Zhang, J. Li and L. Wang, *ACS Appl. Nano Mater.*, 2019, **2**, 2266–2272.
140. F. Diehl and A. Y. Khodakov, *Oil Gas Sci. Technol. Rev. IFP*, 2009, **64**, 11–24.
141. D. Xu, W. Li, H. Duan, Q. Ge and H. Xu, *Catal. Lett.*, 2005, **102**, 229–235.
142. Ø. Borg, N. Hammer, S. Eri, O. A. Lindvåg, R. Myrstad, E. A. Blekkan, M. Rønning, E. Rytter and A. Holmen, *Catal. Today*, 2009, **142**, 70–77.
143. V. Garcilaso, J. Barrientos, L. F. Bobadilla, O. H. Laguna, M. Boutonnet, M. A. Centeno and J. A. Odriozola, *Renewable Energy*, 2019, **132**, 1141–1150.
144. Z. Wang and J. J. Spivey, *Appl. Catal., A*, 2015, **507**, 75–81.
145. G. R. Johnson, S. Werner and A. T. Bell, *ACS Catal.*, 2015, **5**, 5888–5903.
146. X. Zhang, H. Su, Y. Zhang and X. Gu, *Fuel*, 2016, **184**, 162–168.
147. C. Brabant, A. Khodakov and A. Griboval-Constant, *C. R. Chim.*, 2017, **20**, 40–46.
148. M. Adachi, K. Yoshii, Y. Z. Han and K. Fujimoto, *Bull. Chem. Soc. Japan*, 1996, **69**, 1509–1516.
149. D. Martínez del Monte, A. J. Vizcaíno, J. Dufour and C. Martos, *Fuel Process. Technol.*, 2019, **194**, 106102.
150. N. Lohitharn and J. G. Goodwin, *Catal. Commun.*, 2009, **10**, 758–762.
151. H. Wang, Y. Yang, J. Xu, H. Wang, M. Ding and Y. Li, *J. Mol. Catal. A Chem.*, 2010, **326**, 29–40.
152. H. Arai, K. Mitsuishi and T. Seiyama, *Chem. Lett.*, 1984, **13**, 1291–1294.
153. V. R. Calderone, N. R. Shiju, D. C. Ferré and G. Rothenberg, *Green Chem.*, 2011, **13**, 1950–1959.
154. X. Sun, S. Sartipi, F. Kapteijn and J. Gascon, *New J. Chem.*, 2016, **40**, 4167–4177.
155. E. Jiménez-Barrera, P. Bazin, C. Lopez-Cartes, F. Romero-Sarria, M. Daturi and J. A. Odriozola, *Appl. Catal., B*, 2018, **237**, 986–995.
156. O. O. James, B. Chowdhury, M. A. Mesubi and S. Maity, *RSC Adv.*, 2012, **2**, 7347–7366.
157. S. Shetty and R. A. van Santen, *Catal. Today*, 2011, **171**, 168–173.
158. M. Ojeda, R. Nabar, A. U. Nilekar, A. Ishikawa, M. Mavrikakis and E. Iglesia, *J. Catal.*, 2010, **272**, 287–297.
159. I. A. W. Filot, R. A. van Santen and E. J. M. Hensen, *Angew. Chem., Int. Ed.*, 2014, **53**, 12746–12750.
160. D. Shi, J. Faria, T. N. Pham and D. E. Resasco, *ACS Catal.*, 2014, **4**, 1944–1952.
161. D. D. Hibbitts, B. T. Loveless, M. Neurock and E. Iglesia, *Angew. Chem., Int. Ed.*, 2013, **52**, 12273–12278.
162. A. K. Dalai and B. H. Davis, *Appl. Catal., A*, 2008, **348**, 1–15.
163. V. R. R. Pendyala, G. Jacobs, J. C. Mohandas, M. Luo, H. H. Hamdeh, Y. Ji, M. C. Ribeiro and B. H. Davis, *Catal. Lett.*, 2010, **140**, 98–105.
164. Ø. Borg, S. Storsæter, S. Eri, H. Wigum, E. Rytter and A. Holmen, *Catal. Lett.*, 2006, **107**, 95–102.
165. S. Storsæter, Ø. Borg, E. A. Blekkan and A. Holmen, *J. Catal.*, 2005, **231**, 405–419.

166. S. Krishnamoorthy, M. Tu, M. P. Ojeda, D. Pinna and E. Iglesia, *J. Catal.*, 2002, **211**, 422–433.
167. K. Okabe, K. Murata, M. Nurunnabi and Y. Liu, *J. Japan Petrol. Inst.*, 2009, **52**, 139–142.
168. G. T. K. K. Gunasooriya, A. P. van Bavel, H. P. C. E. Kuipers and M. Saeys, *ACS Catal.*, 2016, **6**, 3660–3664.
169. A. I. McNab, A. J. McCue, D. Dionisi and J. A. Anderson, *J. Catal.*, 2017, **353**, 286–294.
170. J. M. González-Carballo, F. J. Pérez-Alonso, M. Ojeda, F. J. García-García, J. L. G. Fierro and S. Rojas, *ChemCatChem*, 2014, **6**, 2084–2094.
171. G. P. van der Laan and A. A. C. M. Beenackers, *Appl. Catal., A*, 2000, **193**, 39–53.
172. A. Paredes-Nunez, D. Lorito, N. Guilhaume, C. Mirodatos, Y. Schuurman and F. C. Meunier, *Catal. Today*, 2015, **242**, 178–183.
173. I. K. van Ravenhorst, C. Vogt, H. Oosterbeek, K. W. Bossers, J. G. Moya-Cancino, A. P. van Bavel, A. M. J. van der Eerden, D. Vine, F. M. F. de Groot, F. Meirer and B. M. Weckhuysen, *Angew. Chem., Int. Ed.*, 2018, **57**, 11957–11962.
174. A. Hommes, H. J. Heeres and J. Yue, *ChemCatChem*, 2019, **11**, 4671–4708.
175. O. Sanz, F. J. Echave, F. Romero-Sarria, J. A. Odriozola and M. Montes, in *Renewable Hydrogen Technologies*, ed. L. M. Gandía, G. Arzamendi and P. M. Diéguez, Elsevier, Amsterdam, 2013, pp. 201–224.
176. X. Yao, Y. Zhang, L. Du, J. Liu and J. Yao, *Renewable Sustainable Energy Rev.*, 2015, **47**, 519–539.
177. F. Kapteijn, R. M. de Deugd and J. A. Moulijn, *Catal. Today*, 2005, **105**, 350–356.
178. O. H. Laguna, M. I. Domínguez, M. A. Centeno and J. A. Odriozola, in *New Materials for Catalytic Applications*, Elsevier, Amsterdam, 2016, pp. 81–120.
179. L. C. Almeida, F. J. Echave, O. Sanz, M. A. Centeno, J. A. Odriozola and M. Montes, in *Studies in Surface Science Catalysis*, ed. E. M. Gaigneaux, M. Devillers, S. Hermans, P. A. Jacobs, J. A. Martens and P. Ruiz, Elsevier, Amsterdam, 2010, vol. 175, pp. 25–33.
180. B. A. Wilhite, *Curr. Opin. Chem. Eng.*, 2017, **17**, 100–107.
181. P. L. Mills, D. J. Quiram and J. F. Ryley, *Chem. Eng. Sci.*, 2007, **62**, 6992–7010.
182. H. J. Venvik and J. Yang, *Catal. Today*, 2017, **285**, 135–146.
183. L. C. Almeida, F. J. Echave, O. Sanz, M. A. Centeno, G. Arzamendi, L. M. Gandía, E. F. Sousa-Aguiar, J. A. Odriozola and M. Montes, *Chem. Eng. J.*, 2011, **167**, 536–544.
184. C. G. Visconti, E. Tronconi, G. Groppi, L. Lietti, M. Iovane, S. Rossini and R. Zennaro, *Chem. Eng. J.*, 2011, **171**, 1294–1307.
185. A. Vita, C. Italiano, M. A. Ashraf, L. Pino and S. Specchia, *Int. J. Hydrogen Energy*, 2018, **43**, 11731–11744.
186. Y. Gao, J. Jiang, Y. Meng, F. Yan and A. Aihemaiti, *Energy Conv. Manag.*, 2018, **171**, 133–155.

187. U. Izquierdo, V. L. Barrio, N. Lago, J. Requies, J. F. Cambra, M. B. Güemez and P. L. Arias, *Int. J. Hydrogen Energy*, 2012, **37**, 13829–13842.
188. L. Zhou, Y. Guo, J. M. Basset and H. Kameyama, *Chem. Commun.*, 2015, **51**, 12044–12047.
189. J. Corella, J. M. Toledo and R. Padilla, *Ind. Eng. Chem. Res.*, 2004, **43**, 2433–2445.
190. C. Wang, T. Wang, L. Ma, Y. Gao and C. Wu, *Energy Conv. Manag.*, 2010, **51**, 446–451.
191. C. G. Visconti, E. Tronconi, L. Lietti, G. Groppi, P. Forzatti, C. Cristiani, R. Zennaro and S. Rossini, *Appl. Catal., A*, 2009, **370**, 93–101.
192. R. Y. Abrokwah, M. M. Rahman, V. G. Deshmane and D. Kuila, *Mol. Catal.*, 2019, **478**, 110566.
193. J. Na, K. S. Kshetrimayum, I. Jung, S. Park, Y. Lee, O. Kwon, Y. Mo, J. Chung, J. Yi, U. Lee and C. Han, *Chem. Eng. Process.*, 2018, **128**, 63–76.

CHAPTER 9

Metal Organic Frameworks: From Material Chemistry to Catalytic Applications

ENRIQUE V. RAMOS-FERNÁNDEZ,[a*] J. C. SERRANO-RUIZ,[b] A. SEPÚLVEDA-ESCRIBANO,[a] J. NARCISO,[a] JESÚS FERRANDO-SORIA[c] AND EMILIO PARDO[c]

[a]Laboratorio de Materiales Avanzados, Departamento de Química Inorgánica – Instituto Universitario de Materiales de Alicante, Universidad de Alicante, Apartado 99, E-03080 Alicante, Spain
[b]Department of Engineering, University Loyola Campus Sevilla, Avda. de las Universidades s/n, 41704 Dos Hermanas, Seville, Spain
[c]Departamento de Química Inorgánica, Instituto de Ciencia Molecular (ICMol), Catedrá tico José Beltrá n Martínez, 2, Universidad de Valencia, 46980 Paterna, Valencia, Spain
*Corresponding contributor. Email: enrique.ramos@ua.es

9.1 Introduction

Metal organic frameworks (MOFs) are a new class of crystalline materials that are made of metals and organic ligands that bind to each other through coordination bonds.[1] In principle, all metals in the periodic table can be used to build MOFs, thus metals ranging from alkalis to rare earths have been used

Heterogeneous Catalysis for Energy Applications
Energy and Environment Series No. 27
Edited by Tomas R. Reina and Jose A. Odriozola

Published by the Royal Society of Chemistry, www.rsc.org

to synthetize MOFs. The metals can form part of the MOF as single metal ions or as clusters of metal ions; these units are normally called metal centres. The metal centres can have different coordinating geometries such as cubic, tetrahedral, octahedral, cubo-octahedral, *etc.*, implying a large number of possible metal centres. The organic part of these materials is constituted by organic ligands, called linkers or connectors, which are usually polycarboxylates,[2–5] imidazolates,[6–15] oxamates,[16–18] *etc*. These linkers can have different geometries such as linear, triangular or square, and the number of linkers is enormous; however, their price can be very high, and this depends on the complexity of the linker. The combination of organic linker and metal centres is almost infinite. The vast amount of possible materials results in a huge amount of opportunities for research.

From a structural point of view, the importance of a large number of metal centre–linker combinations is that a particular MOF structure can be, in principle, rationalized, designed and synthetized. In this way, the pioneering works of O'Keeffe[20–23] and Ferey[21,24,25] showed that by playing with the geometry of both the linker and the metal centre, MOFs structure could be predesigned and then synthesized. As very illustrative example of this approach is MOF-5, which was reported firstly by O'Keefe in 1999.[19] Figure 9.1 shows

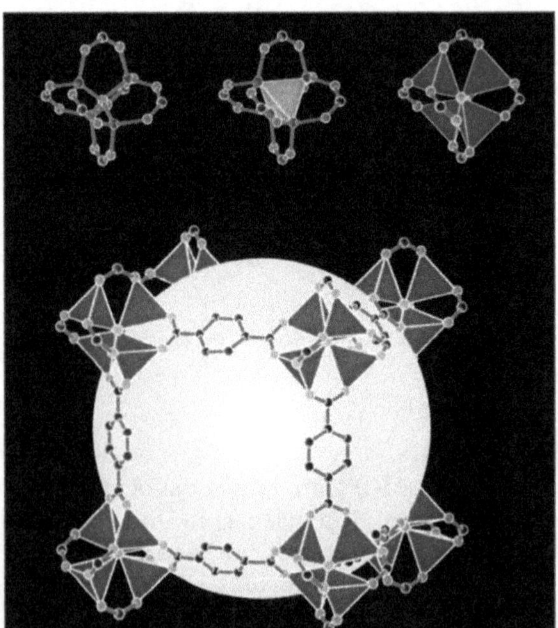

Figure 9.1 (Top) The $Zn_4(O)O_{12}C_6$ cluster. (Left) Ball and stick model. (Middle) Ball and stick model with the $Zn_4(O)$ tetrahedron. (Right) Ball and stick model with the $Zn_4(O)$ tetrahedra. (Bottom) One of the cavities in the $Zn_4(O)(BDC)_3$ MOF-5 framework. Adapted from Ref. 19 with permission from Springer Nature, Copyright 1999.

how MOF-5 is built from an octahedral node based on Zn_4O units where the Zn atoms are tetrahedrally coordinated and the linker is linear. It ends in a cubic structure as expected. If the linker is replaced by other linear linkers, the cubic structure is preserved.[26] This approach is the so-called reticular chemistry.

Another nice example of the design of new frameworks with exceptional properties is that reported by the group of Ferey.[27] The so-called MIL-101 (MIL = Matériaux de l'Institut Lavoisier) is constructed of trimers of octahedral chromium that was computationally designed. These trimers were connected through linear linkers, leading to a predicted MTN topology with exceptionally large pores. Figure 9.2 shows these pre-designed and constructed frameworks.[28] In 2010, more than 6 000 such structures were deposited in the Cambridge Structural Database. Today, there are more than 11 000 such structures, pointing to the importance of this research field and how this field has developed[2,29] in the generation of new structures. These materials may have chemical functionalities. These functionalities can be located in the metal centres (*e.g.* Lewis acidity, redox pairs, *etc.*) or linkers (*e.g.* acid–base). This means that both the inorganic part and the organic part can be modified and designed for specific applications. Later, it will be further shown how the MOF can be functionalized for specific purposes.

MOFs were studied in detail first by Robson, who called them coordination polymers.[30,31] He studied many different coordination polymers, finding that some of them were purely crystalline and could form 1D, 2D and 3D nets. Later, Ferey,[1,24,25] Yaghi[2,22,23,26,32,33] and Kitagawa[32,34–38] promoted the investigation of this type of material, discovering that MOFs had permanent porosity. This implies that potential applications such as the adsorption and separation of gases,[39–43] sensing,[44] and catalysis[29] were plausible.

In heterogeneous catalysis, normally a solid with catalytic active sites is used as a catalyst to accelerate reactions. This is considered a surface phenomenon as the reactants are adsorbed and transformed on the surface of the catalyst. For example, industrial catalysts such as zeolites have Lewis and Brønsted acid sites that are capable of activating hydrocarbon and catalysing transformations such as alkylation or cracking. The active sites in zeolites are intrinsic to the zeolite structure. In the case of MOFs, the situation is different, as many of the MOFs do not have active sites *per se*; a nice example is the family of zeolitic imidazolate frameworks (ZIFs) that are composed of tetrahedrally coordinated Zn cations that are saturated with imidazolate linkers. The coordination bond between the Zn cation and the linker is very strong, making these materials, in principle, catalytically inactive. However, we will see later that ZIFs can be designed to have active sites. There are also other MOFs that have some functional groups that can be potentially catalytically active.[45,46] These functional groups can be located on both the metal centres[47] and linkers.[43] It is not unusual for some MOFs to have one or more of the coordinative positions around the metallic centre not bonded to the linker. These positions are normally occupied by solvent molecules. After removing these guest species, the metal sites is free to be used as an active phase. In general, these sites have Lewis acid site character as they are electron deficient. The linker can also hold pendant

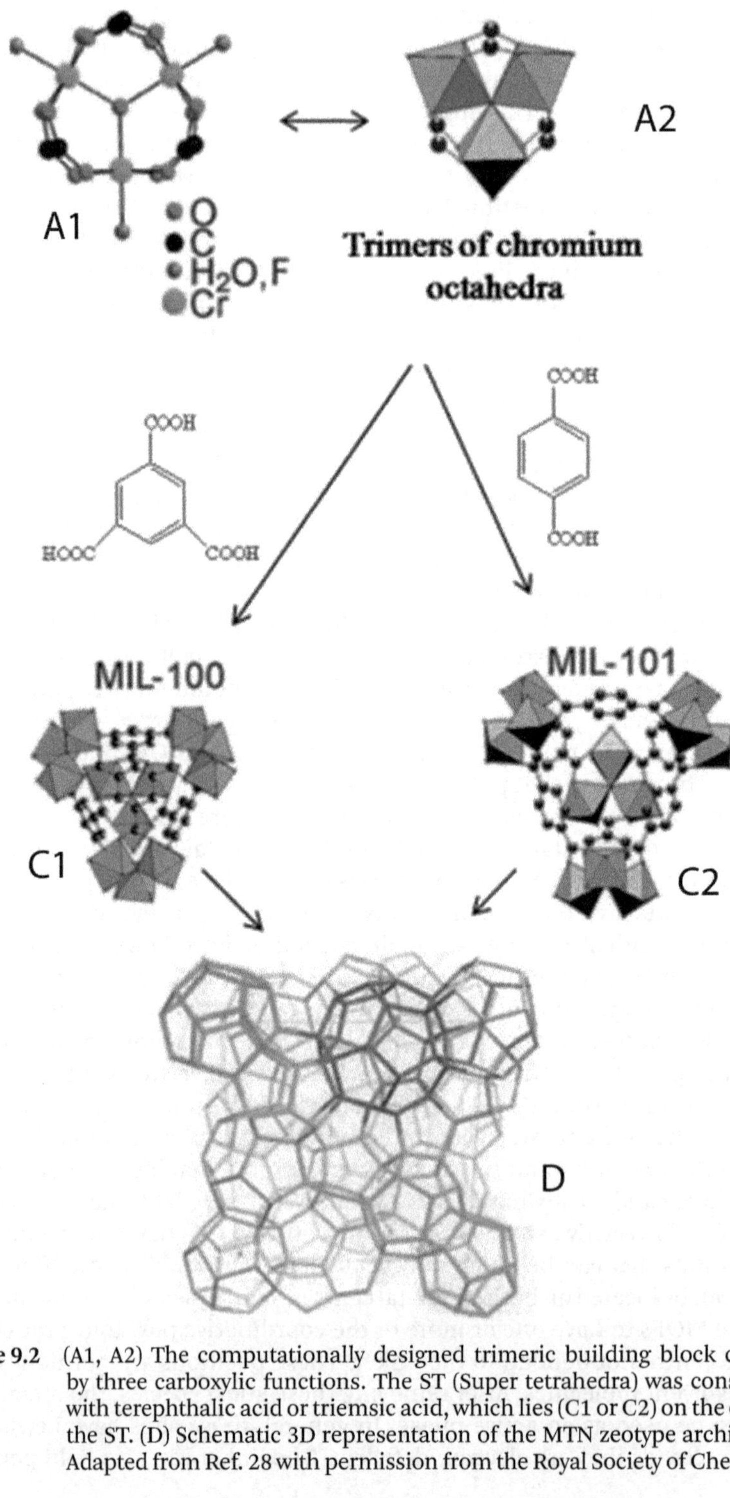

Figure 9.2 (A1, A2) The computationally designed trimeric building block chelated by three carboxylic functions. The ST (Super tetrahedra) was constructed with terephthalic acid or triemsic acid, which lies (C1 or C2) on the edges of the ST. (D) Schematic 3D representation of the MTN zeotype architecture. Adapted from Ref. 28 with permission from the Royal Society of Chemistry.

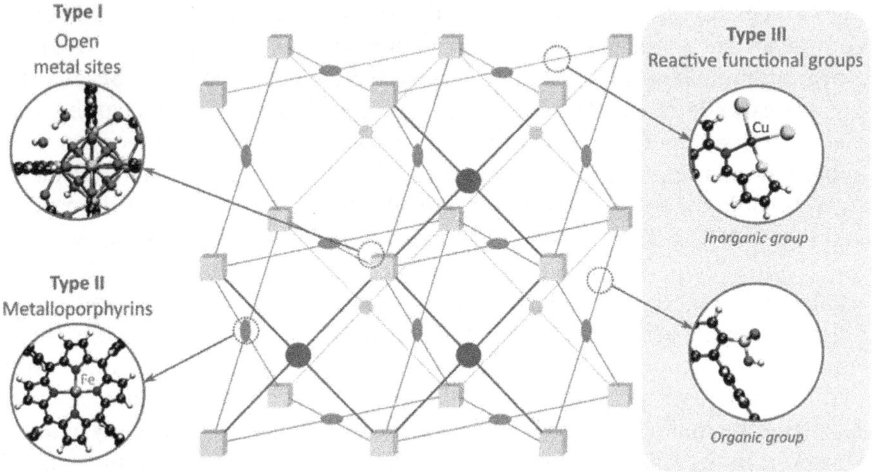

Figure 9.3 Different types of actives that can be generated in different part of the MOFs. Adapted from Ref. 29 with permission from the Royal Society of Chemistry.

functional groups such as amines,[48–50] sulfonic groups,[51,52], nitro groups[53] or pyridinic groups, which can also act as catalytic active sites. The empty spaces of the MOF structures can also be used to immobilize active species. In many cases, the porosity of the MOFs has been used to anchor metal complexes, heteropolyanions,[54] organocatalyts,[55] metal nanoparticles[56–59] or enzymes,[60] using these guest species as active centres for catalysis. The versatility and the tuneability of their chemical and structural properties give them great potential to be applied as catalysts. Lastly, it is also important to bear in mind that MOFs are made of metal centres homogeneously distributed, which, in addition to their use as catalysts, can be used as metal catalyst precursors. They are also made of linkers that can be used as carbon sources to prepare carbon-based materials.[61] Thus, depending on how the MOFs are used as precursors, it is possible to obtain metal or metal oxide nanoparticles[62] or the same particles supported on carbon and finally carbon structures.

MOFs are also ideal platforms for preparing single-sites catalysts, as has been nicely reviewed by Rogge *et al.*[29] Figure 9.3 shows how active sites can be incorporated in the different parts of the MOFs, providing great versatility for designing new catalysts.

The objective of this chapter is not only to review the literature about MOFs and catalysis, but also to describe how the properties of MOFs can be used and tuned to design new catalysts with unprecedented properties. The hybrid nature of MOFs provides them with a large degree of freedom, opening the door for many applications, including catalysis.

9.2 Active Sites in MOFs

As has been mentioned before, the catalytically active sites can be located at linkers, metals and pores. Focusing on the metal centres, one can imagine that

MOFs are just polymers made of metal complexes, and it is known that metal complexes are good catalysts and are widely used as homogeneous catalysts. However, metal complexes can be dissolved and act as a catalysts because the ligands can be detached and attached easily and then reactants can be associated, transformed and dissociated in these free sites in the coordination sphere of the metal. In the case of MOFs, the coordination bond between the ligands (linkers) and the metal is relatively strong, which is the origin of the stability of MOFs. The consequence of this enhanced stability is that MOF ligands cannot be dissociated easily from the metal, which would render MOFs catalytically inactive. Metal complexes based on terephthalate and imidazolate are rarely used in homogeneous catalysis as the complexes that derive from them are poor catalysts in the majority of the chemical transformations of interest. Thus, one could expect that MOFs based on terephthalate and imidazolate (the majority of the known MOF structures) are catalytically inactive. However, many MOFs have shown catalytic activity in many reactions, as will be discussed. The origin of this catalytic activity and how these active sites can be generated will be detailed in the following sections.

9.2.1 Open Metal Sites

When some MOFs are synthetized, the coordination sites of the metal centres may not be fully filled out with the functional groups of the linkers. These unsaturated sites can be occupied by coordinating guest species such as water, hydroxyl groups or the solvent molecules used in synthesis. If the MOF is treated at moderate or high temperatures and/or under vacuum, the coordinating guest molecules can be removed. This generates so-called open metal sites (OMSs; also referred to as unsaturated metal sites, coordinatively unsaturated metal sites or exposed metal centres) that are electron deficient; in other words, they are Lewis acid centres. Figure 9.4 shows how OMSs are generated in the inorganic cluster of MIL-101(Cr).[63]

In terms of the origin of the OMSs, they can be classified into two groups: intrinsic OMSs or induced OMSs. The former are those that are inherent to the MOFs. After the synthesis, these OMSs are already present, and by removing the guest adsorbed molecules, the OMSs are ready to be used. The latter are those that are not inherent to the structure, and they need to be intentionally induced or generated. This can be done by choosing the proper synthesis procedure or using post-synthetic approaches. In both cases, OMSs are electron deficient, and consequently Lewis acid sites, and they may be used as catalysts in Lewis acid-catalysed reactions such as CO_2 cycloaddition,[64] cyanosilylation,[65] acetalization[66] of aldehydes, Knoevenagel condensation,[67] the Mukaiyama-aldol reaction[68] or oxidation reactions.[69–71]

9.2.1.1 MOFs with Intrinsic OMSs

There have been pioneering works describing the catalytic activity of MOFs since late 1990s. One of the most well-known MOFs with OMSs is the so-called

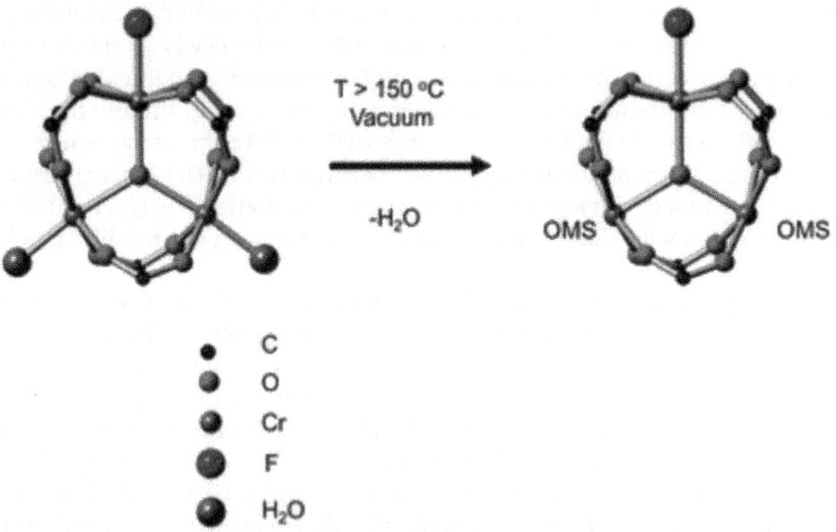

Figure 9.4 Illustrative example of the generation of OMSs. Adapted from Ref. 28 with permission from the Royal Society of Chemistry.

HKUST-1[72] (HKUST = Hong Kong University of Science and Technology), which is a copper trimesate MOF. This MOF is one of the most studied and one of the first that was used as a catalyst.[65] This material has an exceptionally high surface area and thermal stability. In contrast, the Lewis acidity strength is relatively low; furthermore, its water stability is not very high, and this last point is very relevant, as one of the main drawbacks of solid Lewis acid catalysts is their water tolerance. There are only a few solid Lewis acid catalysts that have water stability and tolerance.[73] Another MOF also used for catalysis with OMSs is the one described by Fujita in 1994,[74] which was a Cd bipyridine MOF able to catalyse the clathration of o-dibromobenzene and cyanosililation reactions. After these pioneering studies, the topic has been developed further, and many important results have been obtained. Another study that led to a paradigm shift and demonstrated the importance of OMSs was the one published by Snejko.[66] These authors were able to prepare indium-based MOFs with similar structures, but one of the MOFs had OMSs and the other had indium cations fully saturated with the linker molecules. They found that the MOF with OMSs had catalytic activity that was an order of magnitude higher than the one free of OMSs. This work very elegantly demonstrated the influence of OMSs in catalysis. In the last few years, Dhakshinamoorthy *et al.*[45,75–80] and Ding *et al.*[81] have thoroughly reviewed this topic and have published many reviews showing the applications of OMSs in catalysis. We encourage the reader to go through these relevant works to further complement the content of this chapter.

9.2.1.1.1 Lewis Acidity Strength. Once it has been shown that OMSs can be catalytically active, the next step is to tune the strength and amount of OMSs in the MOFs. Obviously, the Lewis acidity strength is related to the nature of the metal in the cluster. Thus, any modification to the Lewis acidity strength, in principle, occurs *via* the modification of the properties of the MOF's metal centres. There are a few approaches that can be used to modify the Lewis acidity strength: (1) make iso-structural MOFs where the metal centres are made out of different metals; (2) prepare iso-structural MOFs where the metal centres are doped with a second metal; and (3) synthetize iso-structural MOFs having linkers with different functional groups that might modify the electronic properties of the metal centres.

Focusing on the first approach, the preparation of iso-structural MOFs made of different metals is important. The discovery of the MIL-100 and MIL-101 families[70] was a breakthrough in this topic, as these materials are stable under a wide variety of conditions, they are mesoporous and they have large apparent surface areas. Furthermore, these materials can be prepared using a large number of different metals, such as Fe,[83] Sc,[84] V,[85–87] Cr[21,69] and Al.[88] MIL-100 and MIL-101 made of different metals have been evaluated for different catalytic applications. It has been found that there is an appropriate metal for each application; in others words, choosing the proper metal tunes the Lewis acidity strength and consequently the catalytic behaviour. Thus, and considering the works on characterization of Daturi *et al.*,[82,86] it has been found that MIL-100 and MIL-101 made of Cr are very efficient catalysts for oxidation reactions when peroxides are used as the oxidant.

Iron MIL materials are efficient in oxidation reactions when oxygen is used,[71] since iron also supplies redox functionality (Fe^{2+}/Fe^{3+} redox pair). This example is very interesting as is one of the few examples that can be found in the literature of any material where a bi-functional catalyst redox Lewis acid is used. This material has also been characterized spectroscopically, and it has been unambiguously found that Fe^{2+} and Fe^{3+} coexist in the same metallic centres.[82] Figure 9.5 shows the infrared spectra and the amounts of Fe^{2+} and Fe^{3+} present in the metal centres of the MIL-100 structure. These experiments shows that the stretchings of CO adsorbed on Fe^{2+} or Fe^{3+} are located at different wavenumbers, which indicates that the Lewis acidity is also modified by changing the oxidation state of the metal. Finally, it is also important to stress that the most active catalyst for truly Lewis acid-catalysed reactions are those made with the most electropositive metal (Sc) and consequently having the highest acidity strength.[84] Mitchell *et al.* tested 15 different MOF and zeolites catalysts in Friedel–Crafts Michael addition reactions, tandem deprotection carbonylene reactions, reactions between s-methyl styrene and trifluoropyruvate, intermolecular carbonylene reactions, reactions of indoles with vinyl ketone, Michael reactions of pyrroles with methyl vinyl ketone, tert-butylphenylindole reactions with methyl vinyl ketone and imine formation reactions. In all of these cases, the Sc MOF outperformed the others, proving the strength of the Lewis acid sites generated by Sc cations. Particularly efficient was MIL-101(Sc), which, apart from its Lewis acidity, also has a large pore size that favours the diffusion of bulky reactants.

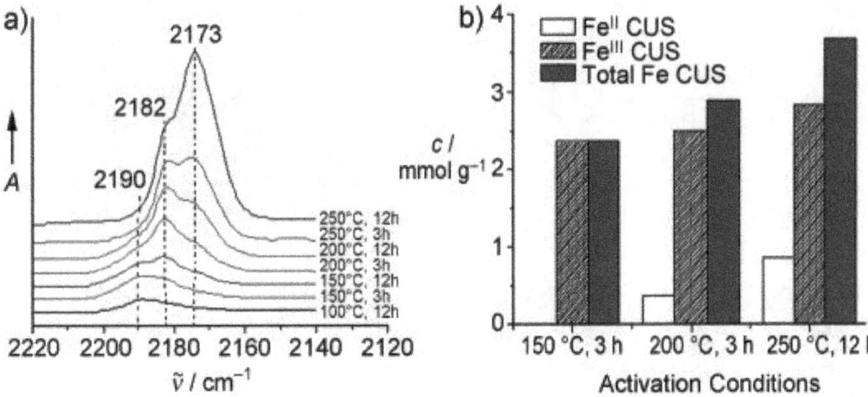

Figure 9.5 (a) Infrared spectra of MIL-100 under a stream of 10% CO at 25 °C after activation under a helium flux at various temperatures over 3 or 12 h. (b) Amount of Fe^{III} CUS and Fe^{II} CUS (coordinative unsaturated sites) detected by infrared analysis upon CO adsorption at −173 °C on MIL-100(Fe) activated under high vacuum at different temperatures. Adapted from Ref. 82 with permission from John Wiley and Sons, Copyright © 2010 Wiley-VCH Verlag GmbH & Co. KGaA, Weinheim.

Another family of iso-structural materials that have been prepared with different metals and that contain intrinsic OMSs is the so-called CPO-27 (Coordination Polymer of Oslo) or MOF-77. These materials have been prepared using Co, Mg, Mn, Ni and Zn. These materials have been used in the oxidation of styrene, cyanosilylation, *etc.* Again, it has been found that the metal nature was crucial to achieving adequate catalysts in each application,[90] as expected. Long and co-workers prepared Fe-MOF-74 with exposed Fe^{2+} sites for the oxidation of cyclohexane. They did not used oxygen as an oxidant, but tert-butylsulfonyl-2-iodosylbenzene. The most relevant achievement of this works is that the authors were able to control the chemical environment of the OMSs and generate significant change in the catalytic activity of the OMSs. This highlights that the Lewis acidity of the OMSs is crucial, but it is not the only factor to be taken into account, as the way the reactants' approach to the active sites also affects the behaviour of the catalysts (see Figure 9.6).[89]

Similar studies have also been carried out with different families of iso-structural materials such as UiO-66 (UiO = Universitetet i Oslo), which can be prepared with Zr,[91] Ce[92] and Hf, or NU-100 (NU = Northwest University), which can be prepared with Zr and Hf.[32] The case of UiO-66 has been deeply studied in the literature, as the material possesses exceptional thermal and chemical stability. It is important to say that in spite of the lack of coordination vacancies in their ideal crystalline structures, the dehydroxylation of the $[Zr_6O_4(OH)_4]^{12+}$ secondary building units (SBUs) to $[Zr_6O_6]^{12+}$ creates μ_3 vacancies. The difference of this material from others is that the removal of the hydroxyl groups requires a higher temperature (>250 °C) than the removal

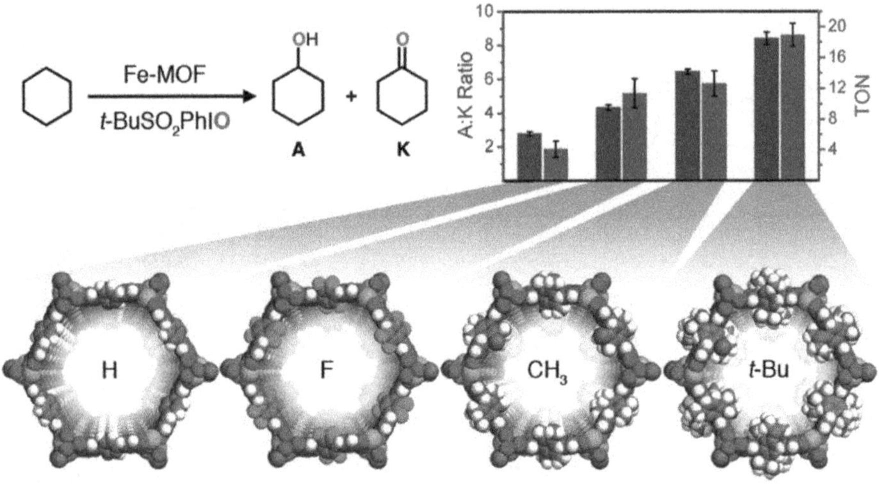

Figure 9.6 Oxidation of cyclohexane to cyclohexanol and cyclohexanone catalysed by Fe-based MOFs. Reproduced from Ref. 89 with permission from American Chemical Society, Copyright 2016.

of guest molecules such as water, ethanol or dimethyl formamide, which are normally removed at below 200 °C. Thermal stability is crucial to creating the OMSs and keeping their structures intact. Furthermore, this structure has a very high density of OMSs, making it especially interesting. The first studies using this material as a catalyst were done by Vermoortelee *et al.* for the synthesis of jasminaldehyde.[93]

The second approach to tuning the Lewis acidity is the preparation of mixed metal MOFs. If a MOF having two different metals in the same cluster is prepared, the Lewis acidity strength can be fine-tuned. There are several methods for preparing mixed metal MOFs, as recently reviewed by Morsali *et al.*,[77] but they can be classified in a very general way into one-pot synthesis approaches and post-synthetic modification. The preparation of MOFs having two metals in the same node is not easy, as there are some prerequisites and not all combinations of metals are possible. Thus, the coulombic charge, the chemical behaviour and the ionic radius must be very similar, or a truly mixed metal MOF becomes difficult to prepare. In this sense, the Zr-based MOFs are very promising, since Zr can be easily replaced by similar atoms, such Ce, Hf or Ti, as occurs in metal oxides where mixed metal oxides based on Zr^{4+} can be easily prepared with Ce,[94] Ti[95,96] or Hf.

The first attempts to prepare bimetallic MOFs were carried out using one-pot synthesis approaches. These consist of adding both metals to the synthesis mixture and controlling the synthesis conditions so that both metals are introduced into the MOF cluster. A pioneering work of Bandosz *et al.*[97] demonstrated that Ce cations can be incorporated into UiO-66(Zr) frameworks by using $CeCl_3$ and $ZrCl_4$ as metal precursors. They claim that Ce^{3+}

cations were incorporated, but they do not explain how they compensated for the charge change due to the change in oxidation state of the cerium cations. Later, Stock *et al.* showed that one-pot synthesis approaches can be used to prepare Ce–Zr mixed metal UiO-66 and MOF-808 structures with different Ce : Zr ratios.[47,92,98,99] In this case, the authors used $(NH_4)_2Ce(NO_3)_6$ and $ZrO(NO_3)_2$ as metal precursors. The similarity in the chemical natures of the metal precursors made it possible to synthetize mixed metal Ce–Zr MOFs with different Ce : Zr ratios. The authors found that the stability of the MOF is greatly influenced by the presence of a second metal in the cluster. Thus, when Ce exceeds 20 wt%, the thermal stability of the MOF decreases substantially. In collaboration with de Vos, they have tested these materials in oxidation reactions, finding that the presence of cerium in the structure is crucial to their catalytic behaviour. Although (2,2,6,6-tetramethylpiperidin-1-yl)oxidanyl (TEMPO) needed to be used to initiate the catalytic cycles, this was ascribed to the difficulty of reducing and oxidizing the Ce^{4+} cations at room temperature. Later, and in collaboration with Bordiga, they were able to demonstrate that, when using TEMPO to initiate the reactions, Ce^{3+} cations were formed in the structure of the MOF.[100,101]

Another approach to prepare mixed metal MOFs is by post-synthetic metal exchange. This was first demonstrated by Cohen[102] and further developed by other authors.[103] This method consists in replacing a metal from the MOF structure in a post-synthetic approach. Thus, the MOF material is suspended in a solution that has a second metal dissolved in it, and by finding the proper conditions, part of the metal of the MOF structure is replaced with the one in solution. This a very versatile method to prepare mixed metal MOFs with tuned Lewis acidity. As with the one-pot method, Zr-based MOFs are the most frequently used for incorporating a second metal. This is how Zr-MOFs doped with different metals, such as Ti or Hf, have been synthesized. An example of the use of these materials in catalysis is the incorporation of Ti cations in UiO-66(Zr) MOFs. Long *et al.*[104] used this Ti-doped UiO-66 for photocatalytic purposes. Another nice example from the same authors[104] involved the development of a two-step method: post-synthetic metathesis and oxidation. They first exchanged the metal of the MOF for a second metal with a low oxidation state, and once this occurred, the second metal was oxidized with air at high temperature.

A third approach is the preparation of iso-structural MOFs having different linkers that, in principle, may change the electronic properties of the metal cluster and consequently the catalytic activity. A pioneering work proving this concept is that of Gascon *et al.*[103] In this paper, the authors prepared iso-reticular MOF-5 materials, where the linkers had different functional groups. They found that there was a nice correlation between the Hammett parameter of the linker and the photocatalytic activity. It is true that photocatalysis is very complex and that MOF-5 does not have OMSs, and the modification of the electronic properties due to the linker properties is difficult to prove. However, this work developed a new concept for the use of MOFs as catalysts and showed the potential of the pendant functional groups in the

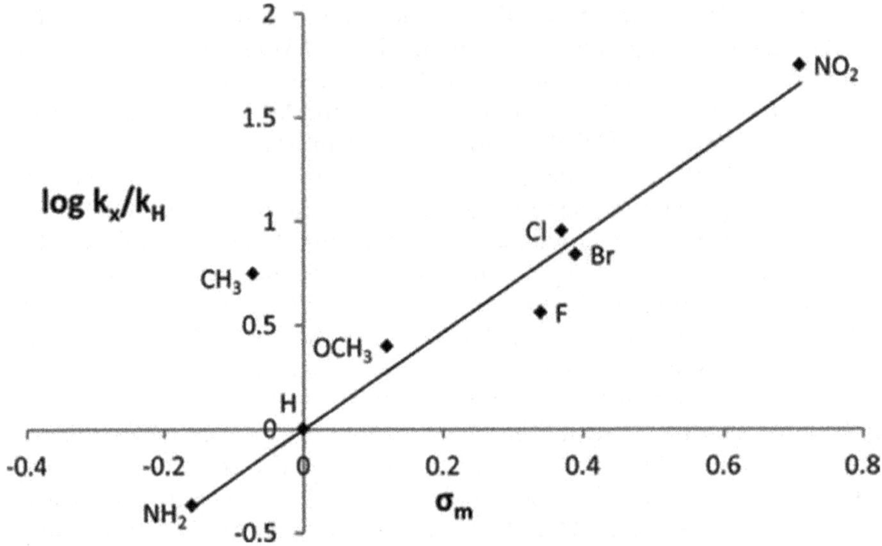

Figure 9.7 Hammett plot for cyclization of citronellal with different UiO-66-X catalysts. Reproduced from Ref. 105 with permission from John Wiley and Sons, Copyright © 2012 Wiley-VCH Verlag GmbH & Co. KGaA, Weinheim.

linker to modify the catalytic properties. This concept was explored in truly Lewis acid-catalysed reactions by de Vos *et al.*,[93,105,106] in which they prepared a series of UiO-66 materials having different functional groups in the linkers. They applied these materials in the cyclization of citronellal to isopulgol, and they found that there was a nice correlation between the electron withdrawing properties of the functional group at the linker and the catalytic activity; in others words, there was a good correlation between the Hammett parameter of the substituent and the catalytic activity. These results were not only spectroscopically demonstrated, but also proven using density functional theory calculations (Figure 9.7) in terms of the linear relation of the Hammett parameter *versus* the kinetic constant of reaction.

9.2.1.2 *Inducing OMSs in MOF Structures*

As has been mentioned before, many MOF structures do not have OMSs after their synthesis, but this is only true if one considers the structure to be free of defects. However, while one can consider a crystal as an "infinite periodic repetition of identical groups of atoms in space",[29,107] is easy to arrive at the conclusion that this is not the case, since real crystals have many defects. There are many types of defects in MOF structures, and many of them can induce OMSs and other types of catalytically active functional groups. There are many classifications of crystal defects. Herein, this will be simplified, and only two types of defects will be considered: surface defects and bulk defects, as we already reviewed.[29] These approaches to generating OMSs have advantages over the OMSs inherent to the structure, as the amount of OMSs is not limited to the structure itself. OMSs can also be generated in MOF structures that, in principle, cannot have OMSs.

9.2.1.2.1 Surface Defects. With regards to a crystal, the most obvious defect is the surface. All crystals end at one point, which can be partially coordinated metals or ligands, depending on the nature of the MOF. Thus, it has been demonstrated that ZIF-8 can catalyse the transesterification reactions of bulky fatty acids, which cannot enter into the pores. The catalytic event happens at the surface of the ZIF-8 crystals.[108] This work was particularly important, since the catalytic activity was an order of magnitude higher than that found for the commercial catalyst. Later, other authors demonstrated that the crystal size is important for the catalytic performance of ZIF-8, and that the surface to volume ratio is crucial when surface defects are used. There are several ways to promote the presence of surface defects.

Preparation of MOF Nanocrystals. When the active sites are only present and the surface or the reactant can access to active sites localized in the inner part of the crystal. The synthesis of MOF nanocrystals is crucial in catalysis. Recently, Barros *et al.*[109] published a nice review of the different strategies for preparing MOF nanocrystals, and we encourage readers to go through this review to gain further insights into this topic. This section will describe the principal methods of generating MOF nanocrystals and the advantages and disadvantages of using MOF nanocrystals as catalysts. The simplest way to create MOF nanocrystals is altering the rate of precipitation. The synthesis of MOFs is normally performed in liquid phase where the metal precursor as well as the linker are dissolved and applying the appropriate conditions for the crystal to start to form and grow. There are many reports showing that by changing the precursors or precursor concentration, temperature, *etc.*, the rate of precipitation can be controlled as well as the crystal size. However, the crystal size distribution is very broad when using this type of method, since it is difficult to stop crystal growth and the synthesis lasts from hours to days. Furthermore, the temperature gradient within the reaction medium is almost impossible to control, and this can have a substantial effect on the particle size.

Another method for preparing MOF nanocrystals is the use of microwaves to provide the heat. The heating is *via* two different mechanisms: dipole rotation excitation and ionic conduction energy transfer.[110] Thus, the heating transfer is not from the wall of the crystallization vessel to the inert part, but rather the reactor vessel is rapidly heated at once. Furthermore, this type of heating also produces an enormous number of hotspots that act as nucleation seeds. Since the first communication was published in 2005[111] on the synthesis of MOFs assisted by microwaves, the number of publications on this topic[112,113] has grown enormously. There are other synthetic methods that efficiently produce MOF nanocrystals, such as surfactant-assisted synthesis in microemulsions,[9] sonochemical synthesis,[114] electrochemical synthesis,[115] non-coordinating modulation approaches [113] and coordination modulation approaches.[116] All of these methods can be considered bottom-up approaches as the synthesis is manipulated to build a nanocrystal from its precursors.

Apart from this direct ways of crystallizing MOFs as nanocrystals (bottom-up approaches), there are post-synthetic approaches to generating nanocrystals of MOFs (top-down approaches). The most relevant method is the one developed by Maspoch's group, in which they performed anisotropic etching of colloidal ZIF-8.[117] They started with large crystals that where selectively etched and the size of the crystal was reduced from micrometres to nanometres. This method is very interesting as it can be used to select the face of the crystal to be etched. The etching method is anisotropic (the same face of the crystal is preferentially etched with respect to the others). In this way, not only can the size of the crystal be tailored, but also the crystal face that is exposed to the reaction. Figures 9.8 and 9.9 show how different crystal sizes and morphologies can be obtained by chemically etching a MOF crystal.

After briefly summarizing the methods for preparing MOF nanocrystals, it is very important to stress that MOF nanocrystals are a nice lab case, but would never be used in real reactors as nanoparticles, as nanocrystal have a large number of inconvenient aspects that hinder their use at large scale; for example, nanocrystals are difficult to handle and recover. In a plug-flow reactor, the presence of nanoparticles would generate a huge pressure drop and so would never be used. One good approach to solve this problem involves the structuring and shaping of the MOF nanocrystal. Ramos-Fernandez *et al.* described a method to prepare MIL-101(Cr) nanoparticles supported on a cordierite monolith, showing that the MIL-101(Cr) monolith worked as well as the suspension of MIL-101(Cr) nanocrystals[118] in the selective oxidation of tetralin using a peroxide as the oxidant. Aguado *et al.* also reported the immobilization of SIM-1, a substituted imidazolate-based MOF, on alumina beads and its application in ketone transfer hydrogenation.[67]

9.2.1.2.2 Bulk Defects. Bulk defects are those that are located randomly anywhere in the crystal of the MOF, not specifically at the surface. There are two ways to generate these defects in MOFs with large numbers of OMSs: one-pot synthesis approaches and by post-synthetic approaches.

Generated during Synthesis. Bulk defects can be created during synthesis. A very simple way to do this is to speed up the crystallization rate. A fast crystallization process produces crystals with very poor quality in terms of crystallinity. This provokes a large number of defects in the lattice that might be used in catalysis. This was first reported by Ravon *et al.*[119] and further developed by Ramos-Fernandez *et al.*[118,120] and Llabrés i Xamena *et al.*[13] A second approach is the use of *dummy linkers* or modulators during synthesis. In the typical synthesis of a MOF, a bidentate (terephthalic acid) linker is used to build up the frameworks. If a portion of monodentate linker (*e.g.* benzoic acid) is added

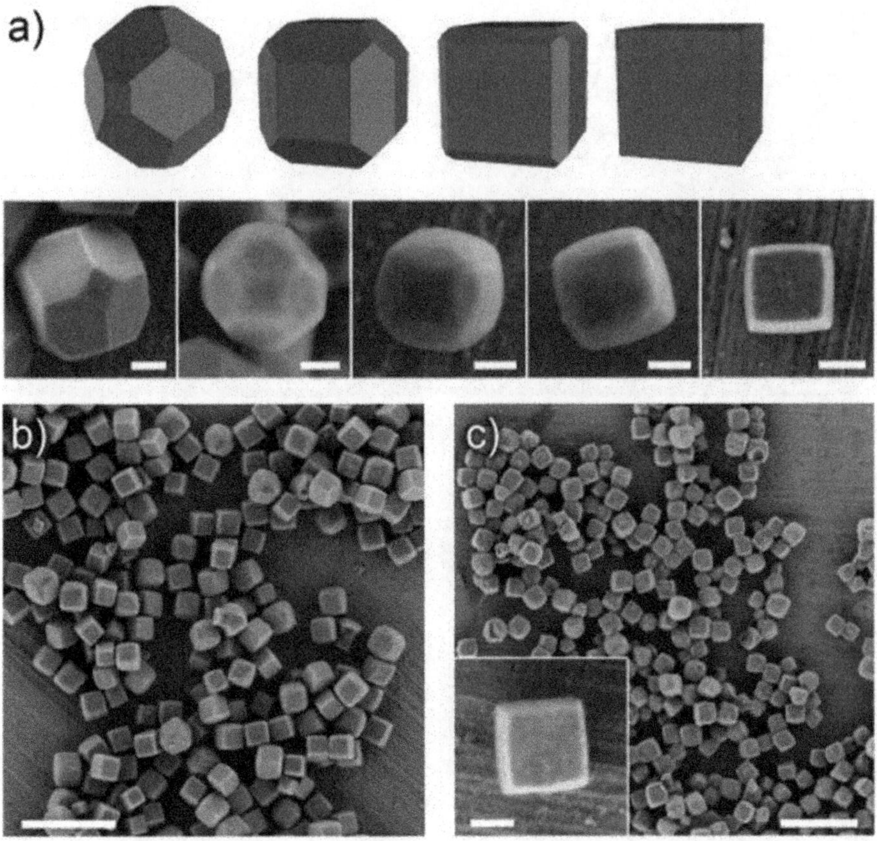

Figure 9.8 (a) Illustration and corresponding FE-SEM images (ZIF-8) of the change in crystal morphology during the etching of truncated rhombic dodecahedral ZIF-8 and ZIF-67 crystals. Note the change in each case from a truncated rhombic dodecahedron to a cube. In each case, etching occurs preferentially in <100> direction. (b, c) Representative FE-SEM images showing general views of the ZIF-8 (b) and ZIF-67 (c) cubes. (Inset) Representative FE-SEM image of a single ZIF-67 cube. Reproduced from Ref. 117 with permission from John Wiley and Sons, Copyright © 2015 Wiley-VCH Verlag GmbH & Co. KGaA, Weinheim.

to the synthesis mixture, both linkers will compete for coordination with the metal. When the MOF crystallizes, the presence of monodentate linkers in the final structure produces defects that are normally called *truncated missing linkers*. This approach is very promising not only for creating defects (OMSs), but also to control the strength of the Lewis acid sites. Fischer's group developed a method to create a defected MOF made of Ru and trimesic acid.[121] The incorporation of dummy linkers generated additional OMSs in the structure, but not only the dummy linker induces the partial reduction of ruthenium. The partially reduced ruthenium OMSs were able to adsorb dissociatively CO_2 or H_2, which are molecules with strong bonds and are difficult to dissociate.

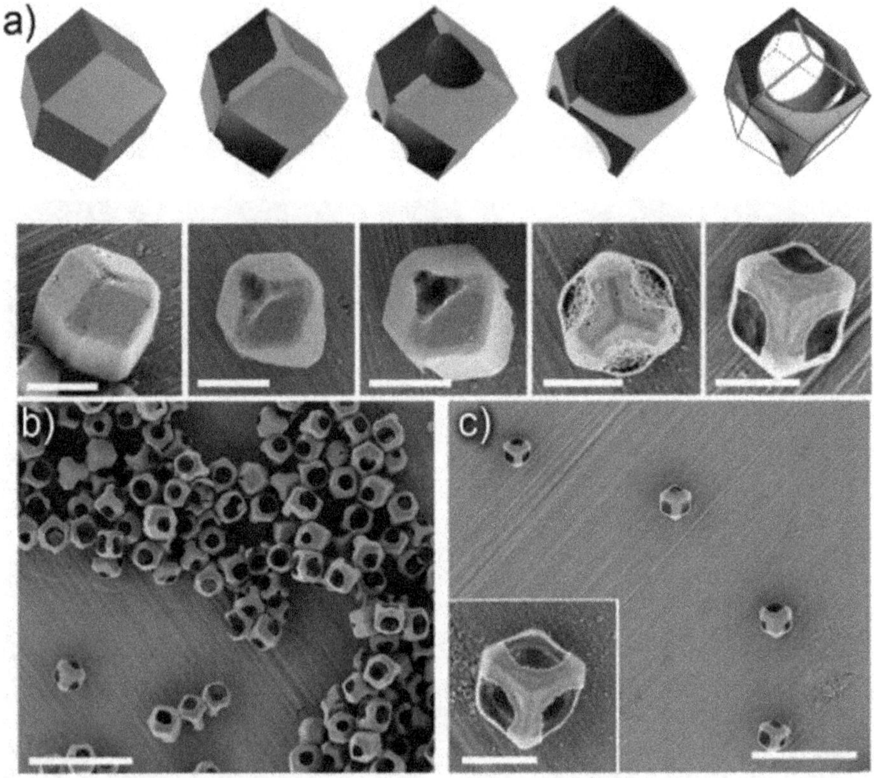

Figure 9.9 (a) Illustration and corresponding FE-SEM images (ZIF-67) of the change in crystal morphology during the etching of rhombic dodecahedral ZIF-8 and ZIF-67 crystals, using an etchant solution at pH 2.5. Note the change from rhombic a dodecahedron to a hollow box. Etching starts preferentially in <211> directions. (b, c) Representative FE-SEM images showing general views of hollow ZIF-67 microboxes. (Inset) Representative FE-SEM image of a single ZIF-67 microbox. Reproduced from Ref. 117 with permission from John Wiley and Sons, Copyright © 2015 Wiley-VCH Verlag GmbH & Co. KGaA, Weinheim.

The authors tested this defected material as a catalyst in the hydrogenation of olefins, and the found that the defected material was catalytically active. Figure 9.10 shows how the defective sites are formed with dummy linkers in the case of pyridine-3,5-dicarboxylate (H_2-pydc).

Post-synthesis Generation of Defects. Post-synthetic approaches to generating OMSs are also very important. Thus, a prepared MOF can be submitted to different treatments to extract or leach out linkers from the framework, thus generating defects. This was first demonstrated by Rosseinsky *et al.*,[122] who prepared an amino acid-based MOF that was later treated with diluted HCl solutions. They found that the treated samples were far more active in the methanolysis of

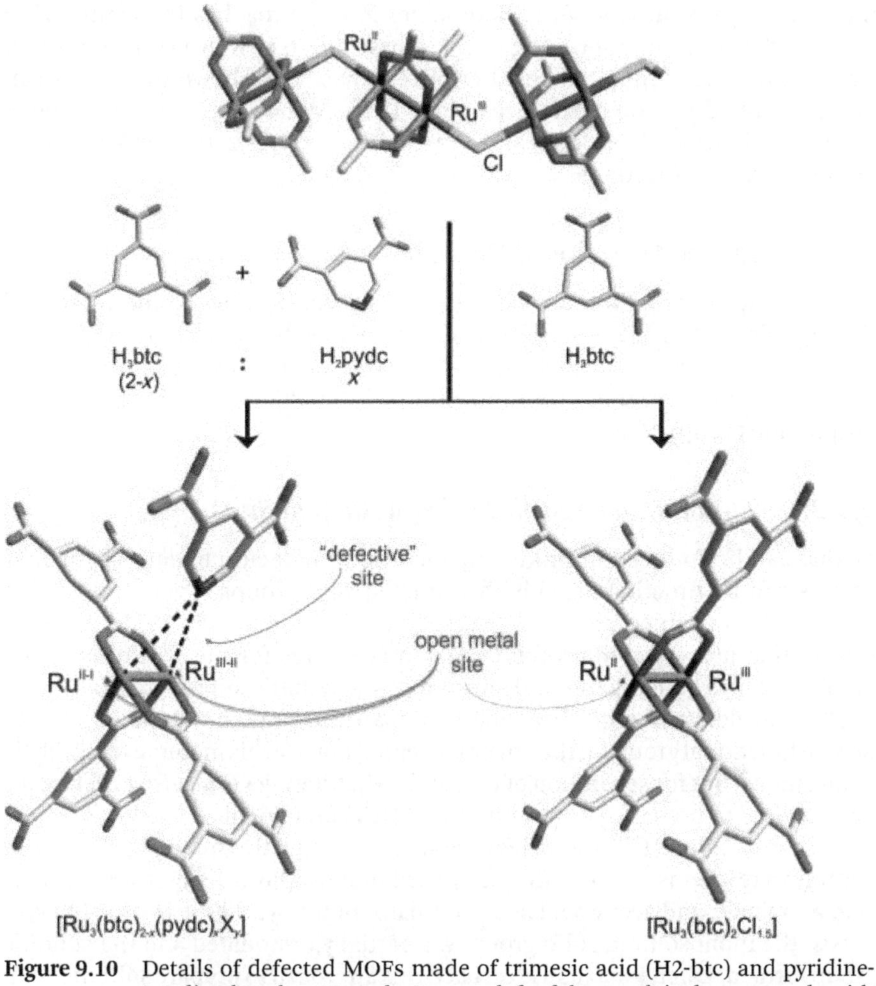

Figure 9.10 Details of defected MOFs made of trimesic acid (H2-btc) and pyridine-3,5-dicarboxylate H_2-pydc. Bottom left of the graph is the one made with a mixture of linker and the bottom right non-defected MOF. Reproduced from Ref. 121 with permission from John Wiley and Sons, Copyright © 2014 Wiley-VCH Verlag GmbH & Co. KGaA, Weinheim.

cis-2,3-epoxybutane than the untreated ones. Later, Vermoortele *et al.*[46] followed the same approach, but in this case with MIL-100(Fe), finding the generation of remarkable numbers of Lewis acid sites.

A different way to induce defects in a MOF is by post-synthetic linker exchange. In this way, Lillerud *et al.* reported that by exchanging a bicarboxylic linker with a monofunctional one, such defects can be generated. They started with UiO-66 and, in a post-synthetic treatment, the linker was exchanged with L-serine or monoethanolamine, inducing functionalities in the MOF.[123] Later, Andreoli *et al.* used a similar approach to functionalize MOFs with picolinic acid or nicotinic acid.[124] In summary, it can be said that there are large numbers of methodologies and approaches for synthetizing MOFs with OMSs and for tuning the acidity strength as well as the amount of sites in this type of material. Furthermore, it has been shown that OMSs can be finely tuned for the desired reaction. Thus, MOFs with OMSs have been used for oxidation,[90] CO_2 cycloaddition,[64,125–134] isomerization or Meerwein–Ponndorf–Verley reactions.

9.2.2 Organic Functional Groups

As we mentioned before, active sites can also be located at the linker (in the organic part of the material). The linkers may be composed of a variety of organic functional groups that can act as active sites, such as sulfonic groups,[51] sulfoxy groups,[135] amino groups,[49,88,128,136] bipyridyl groups,[137,138] porphyrine groups,[132] *etc.*

9.2.2.1 Functional Organic Linker with Acid Sites

Acidity can be induced in linkers by incorporating acid pendant groups, as linkers can be functionalized with Brønsted acid groups. Overall, there are two ways to introduce Brønsted acid sites into the linker of a MOF: (1) using linkers that already contain the Brønsted acid sites for the synthesis of the MOF; or (2) introducing the acid site in a post-synthetic approach.

Thus, van der Voort *et al.*[139] was able to prepare UiO-66 with sulfonate and carboxylate functional groups at the linkers. They did not use this material for catalytic applications, but for separation of CO_2 and CH_4. Kitawaga *et al.* also developed a new method to prepare UiO-66 and MIL-101(Cr) functionalized with very acidic sulfonic groups, and they found high catalytic activity of this material in cellulose hydrolysis reactions.[139] To do this, they used monosodium 2-sulfotherephthalic acid as a linker and acetic acid as a modulator of the synthesis. During the synthesis, the monosodium sulfo group was partially protonated and the sulfonic group formed. Later, Jana Juan *et al.* used this approach to prepare sulfonic acid-functionalized MOFs for esterification reactions.[52] Another application reported for this sulfonic MOF is the alcoholysis of epoxides under ambient conditions.[140] This approach is not straightforward, since the acid functionality of the linker may coordinate the metal, which can lead to undesired structures. This is why they used the monosodium 2-sulfotherephthalic acid, in which the sulfonic group is protected by the sodium cation.

The second approach is generating acid groups in the linker by post-synthetically modifying a MOF structure. Gascon's group also created MOFs with sulfated groups. They developed a post-synthetic modification method to create MIL-101(Cr) with sulfated pendant groups,[135] and they found that this was an efficient catalyst for esterification reactions, as well as being a nice proton-conductive material. Yaghi *et al.* also developed MOFs containing sulfonic groups using a post-functionalization approach.[139] In this work, a pre-functionalized amino MOF was reacted with 1,3 propane-sultone, which led to the formation of sulfonic groups. Later, Stock *et al.* followed the same approach to prepare bismuth-based MOFs with sulfonic groups (CAU-7).[141] The same post-functionalization has been used by Das *et al.* to prepare UiO-66 functionalized with sulfonic groups for proton conductivity.[140]

9.2.2.2 Functional Organic Linkers with Basic Sites

The most used way to induce Lewis basicity in a MOF is the functionalization of the linker with basic sites. If linkers containing nitrogen functional groups are used (*e.g.* $-NH_2$), a MOF with basicity will be obtained. The nitrogen atom in the amine group has five electrons in the valance band, three of them being used for bonding and the other two being unpaired. This causes heavy repulsions with the other bonds of the molecule and, in order to find a more favourable energy level, these groups try to share these electrons with a Lewis acid. Since nitrogen functionalities are very commonly found in the organic linkers of MOFs, nitrogen groups are the most common atoms referred to with regards to Lewis basicity. The basicity of amines and amides is a thoroughly studied topic in organic chemistry[142] and, similarly with Lewis basic groups, they are in a close relationship with the substituents and the surrounding environment of the nitrogen groups. Since MOFs are formed from a large variety of linkers, the functionalization of the basic groups opens up vast possibilities. This trend has already been thoroughly studied[48,143] due to the interesting applications that functionalized nitrogen groups may have.

The interaction of basic sites with acidic sites in the same MOF was demonstrated by Crespo-Serra *et al.*[50] In this work, the authors demonstrated that amine groups and the OMSs of the lattice interact in such a way that the flexibility of MIL-53 is strongly modified due to these interactions.

The basicity in a MOF's linker is normally incorporated by using linker with nitrogen groups, such as amines, pyridines, *etc.* A typical example is the so-called IRMOF-3, which is an iso-structural material of MOF-5. The difference between these materials is that one is made with amino-therephthalic acid and the other is made with therephthalic acid. The IRMOF-3 contains $-NH_2$ functional groups that can be considered as Lewis basic materials. There are other MOF structures that have been synthesized with a mixture of ligands, both functionalized and unfunctionalized. This type of amino-functionalized material has been thoroughly studied and many structures have been reported, such as MIL-101(Al, Fe, Cr)[49,144] and UiO-66-NH_2.[49] However, it is important to remark that the basicity of this amine group is very weak, and only a few reactions are catalysed by these amino-functionalized materials. The most studied reaction is the Knoevenagel condensation between benzaldehyde and malononitrile. Other functional groups that can be incorporated in the framework to promote Lewis basicity are pyridyl groups. Kim *et al.* published a new structure with pyridyl groups, and they used this material for the transesterification reaction, finding very promising results.[145] Zhou *et al.* also prepared MOFs with pyridyl groups by retrosynthesis.[146] These materials were successfully tested in the enantioselective transesterification reaction, showing remarkable results.

Another basic functional group that can be incorporated to induce basicity in MOFs is the amide group. As with the other types of functional groups, amide groups can be introduced during synthesis or in a post-synthetic process. In the first approach, it is interesting to highlight the work of Safarifard *et al.*, who published the iso-reticular structure of interpenetrated MOFs with amide groups.[147] In this pioneering work, the authors found that the basicity is improved, and consequently the capacity for CO_2 adsorption also improved. In the second approach, the work of Cohen is also remarkable.[148] They used this functional groups for the methanolysis of epoxides, showing very high activity. Some time ago, Kitagawa *et al.*[149] published a Cd-based MOF that was made with a complex linker, 1,3,5-benzene tricarboxylic acid tris(N-[4-pyridyl] amide) (4-btapa), a three-connector ligand containing three pyridyl groups that act as coordination sites and amide groups that act as active sites for Knoevenagel condensation. Figure 9.11 shows the linker and how the MOF is constructed. Furthermore, it shows the reaction conversion of benzaldehyde

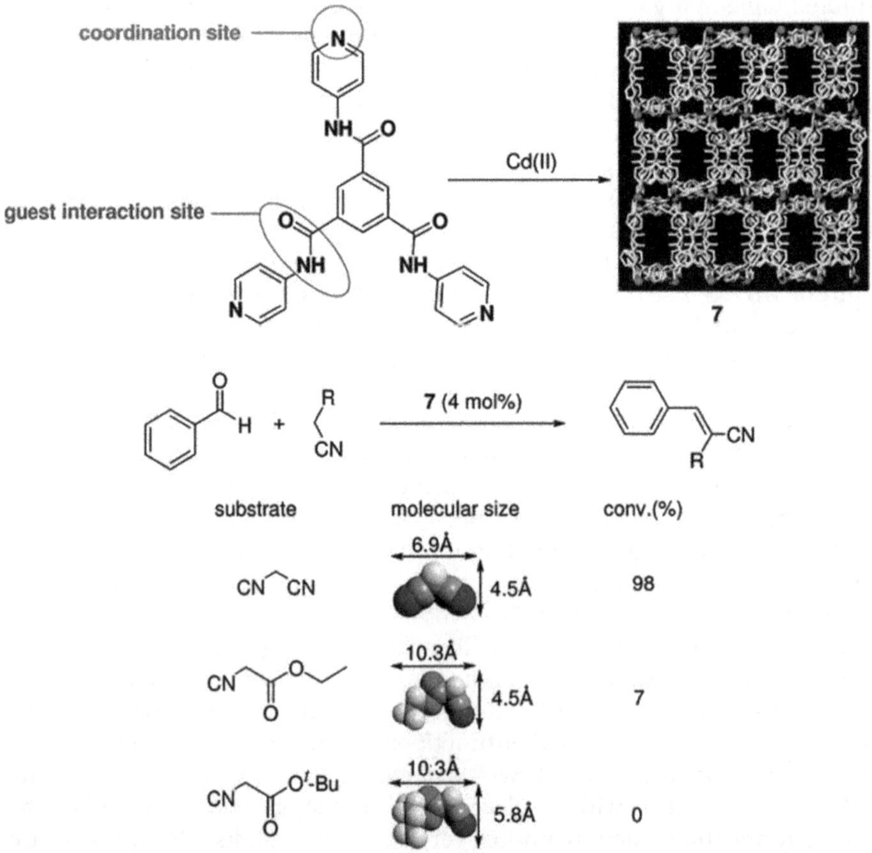

Figure 9.11 Representation of a Cd MOF and its performance in Knoevenagel condensation. Reproduced from Ref. 149 with permission from American Chemical Society, Copyright 2007.[149]

with different nitrile. As can be seen, conversion decreases when the size of the nitrile increases. This is one nice example not only of using amide as a catalyst, but also of inducing size selectivity.

We recommend the detailed review published by Pombeiro *et al.* regarding amide-functionalized MOFs for catalytic applications.[150]

9.2.2.3 Linkers with Metalloporphyrins

Metalloporphyrins are the active sites of many enzymes, such as monooxygenase. These natural catalysts are very efficient in many transformations. Although the metalloporphyrins have been thoroughly studied as homogeneous catalysts, the immobilization of metalloporphyrins has also been studied in traditional porous solids such as activated carbons,[21] mesoporous silicas[138] or carbon nanotubes.[151] The most used metal to immobilize metalloporphyrins in MOFs is the one developed by Hupp and co-workers.[152] They used a mixed ligand approach, involving a tetratopic carboxylic ligand (1,2,4,5-tetrakis[4-carboxyphenyl]benzene) (Figure 9.12) and 5,15-dipyridyl-10,20-*bis*(pentafluorophenyl)porphyrin. They found that a solid material with permanent porosity was generated and half of the linkers were porphyrine structures. After the porphyrin groups were medullated with Zn cations, they used these materials as catalysts in acyl-transfer reactions. Later, they also demonstrated that the synthesis can be extended to others metalloporphyrins.[152] Some of this materials have also been used as catalysts in the oxidation of alkane and alkenes. The same approach has been followed by Zhou[152–154] and Ma[152] to prepare peroxidase-like MOFs showing very high activity and selectivity in oxidation reactions using H_2O_2 as an oxidant. Later, Zhou was also able to prepare porphyrins metallized with cobalt and to demonstrated that this was an efficient catalyst for CO_2 cycloadditions.[155]

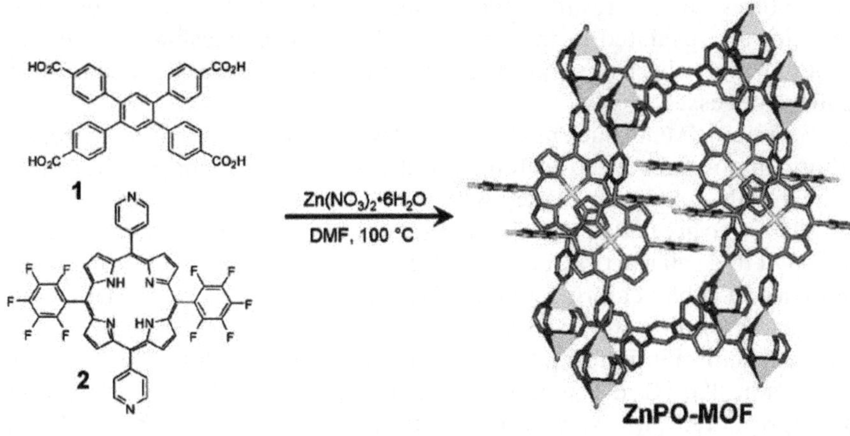

Figure 9.12 ZnPO-MOF prepared by a mixture of 1,2,4,5-tetrakis(4-carboxyphenyl) benzene (1) and 5,15-dipyridyl-10,20-*bis*(pentafluorophenyl))porphyrin (2). Reproduced from Ref. 152 with permission from American Chemical Society, Copyright 2009.

9.2.3 MOFs as Precursors of Catalysts

The particular structures and distributions of chemical species of MOFs can be used to prepare a wide range of heterogeneous catalysts with unique properties. MOFs can be designed with large versatility and extremely high precision, allowing fine control over the distribution of both the active sites (*e.g.* metal ions) and their chemical environments (*i.e.* organic linkers).[156] However, compared with conventional inorganic porous solids, MOFs present significantly lower thermal and chemical stability under the conditions of typical catalytic processes. This drawback of MOFs can, however, be exploited to prepare solid materials by favouring the decomposition of the crystal structures under well-controlled conditions. These strategies, in which MOFs are used as sacrificial templates, will be reviewed in this section, with particular emphasis on those MOF-mediated synthesis (MOFMS) routes leading to catalytically active metal or carbon materials with unique properties.

The MOFMS approach involves the thermal treatment of the MOF at high temperatures and under a controlled atmosphere. Depending on the atmosphere, the organic linkers will be either removed (under oxidizing atmospheres) or carbonized to form a carbon matrix (under inert atmospheres), with the metal ions normally being deposited on the carbon matrix.[157] The main advantage of MOFMS is therefore its simplicity, along with the possibility of generating bulk and supported heterogeneous catalysts with unique properties by finely adjusting the thermal conditions and by selecting the appropriate sacrificial MOF. These most relevant MOF-mediated materials include nanostructured carbons and carbon-supported metal nanoparticles.[158] MOFs can generate carbon materials with well-defined structures by performing carbonization of the framework and subsequent acid leaching of the inorganic moieties. These inorganic species can also be removed by evaporation. Carbonization of a MOF material under an inert atmosphere at high temperatures can also result in the formation of highly dispersed metal nanoparticles encapsulated in the carbon matrix. This encapsulation prevents metal nanoparticles from sintering, which is highly desirable for industrial catalytic processes.

Al-containing MOFs such as CAU-1 and MOF-253 have been used to prepare carbon supports for electrocatalysis and coupling reactions. Several MOF-253 materials, previously loaded with a Pt precursor, were pyrolysed within 700–1050 °C to generate Pt electrocatalysts with similar performance to commercial Pt fuel cell electrocatalysts.[159] MIL-101-NH_2 containing encapsulated thiourea and cobalt chloride was pyrolysed in Ar at 600 °C, acid leached and finally carbonized at 900 °C to prepare oxygen reduction reaction (ORR) catalysts with superior performance to commercial Pt/C catalysts.[160] CAU-1 was pyrolysed at 800 °C to generate an N-doped porous carbon. This material was impregnated with Pd, and the resulting catalyst was used for catalysing Suzuki–Miyaura coupling reactions at room temperature.[161] MIL-101(Cr) was impregnated with Al and K precursors and subsequently calcined at varying temperatures to produce K_2O–Cr_2O_3/Al_2O_3 catalysts with high surface area and pore volume.[162] These materials showed better performance for the dehydrogenation of isobutane to isobutene compared to conventional supported chromia catalysts prepared by the conventional impregnation method. Co-containing MOFs have been used as precursors of dye-sensitized solar cells. ZIF-67 was used as a sacrificial template to prepare dye-sensitized solar cells by pyrolysis under N_2 at high temperature. Temperature was found to have a significant influence on the graphitization degree of the carbon material, with the material carbonized at 850 °C showing optimal electrochemical performance.[163] A Co-containing ZIF-67 was pyrolysed in Ar within 600–900 °C to generate electrocatalysts with excellent ORR activity.[164]

MOFMS has also been exploited to prepare oxidation catalysts for a number of applications. For example, low-temperature CO oxidation was carried out over Co oxides prepared by thermal decomposition of Co- and Ag-containing MOFs.[165] The resulting Ag/Co_3O_4 materials showed outstanding CO oxidation activity and stability compared with other supported Ag conventional catalysts. The oxidation of volatile organic compounds (VOCs) such as benzene was carried out over cobalt–palladium-supported catalysts prepared by calcination of ZIF-67 and subsequent impregnation with a Pd precursor.[166] This material outperformed a similar material prepared using a conventional hydrothermal method. The widely used MOF-199 has been used to prepare a number of catalytic materials by taking advantage of its high surface area and pore volume. Pd–Cu/C catalysts were prepared from MOF-199 by mixing it with a Pd precursor and carbon black.[167] This catalyst showed improved ORR activity *versus* a Pt/C commercial material. MOF-199 was thermally treated above 230 °C to produce CuO nanoparticles with high CO oxidation activity.[168] Important industrial processes such as the Fischer–Tropsch synthesis of hydrocarbons and chemicals have been carried out over MOF-derived materials. Fe-BTC (benzene-1,3,5-tricarboxylic acid) was pyrolysed at 500 °C to generate Fe–C composites with outstanding activity and selectivity towards this important reaction.[169]

Nanostructured carbon materials prepared by pyrolysis of MOF-5 were used as supports of Pd and Ag.[170] The high dispersion of the metallic nanoparticles on the MOF-derived carbon support resulted in excellent activities towards the dehydrogenation of formic acid, well above those of similar catalysts prepared by conventional methods.

9.2.4 Guest Functional Species within the Empty Space of MOFs

9.2.4.1 Introduction

Most of the intriguing properties reported for MOFs so far are intimately connected to the fascinating host–guest chemistry[26,34] of these porous crystalline materials.[171–183] For instance, MOFs were extensively used originally for gas storage and/or separation of mixtures of gases. Indeed, the host–guest interactions established between the guest gas molecules and the empty space of the framework lie at the origin of this efficiency. Such separation properties were studied with other molecules of varying complexity,[184] and even a few examples of simple chemical systems constructed inside MOFs have been reported more recently.[185–188] Overall, the amazing host–guest chemistry of MOFs – which involves the interaction between both MOF channels and a given functional guest molecule – offers endless possibilities for the encapsulation and/or chemical synthesis of functional species within this confined space. In this regard, the empty space can be designed pre- or post-synthetically in terms of size, shape and functionality in order to enhance affinity with guest molecules, which, in addition, can be either inserted "as a whole" or constructed within the channels in different steps. As a consequence, MOFs can be either assembled around the active species intended to be encapsulated ("bottle around a ship") or used as "vessels" to encapsulate directly the pre-formed functional guest species ("ship in a bottle"). In addition, MOFs can be used as "chemical nanoreactors" to synthesize chemically the desired species within their empty spaces by taking advantage of recent advantages in both template-directed synthesis as well as post-synthetic procedures (Figure 9.13).[189,190] Despite the progress that has been made, considerable work remains to be done in terms of molecular recognition,[173,174] encapsulation of more complex molecular systems and, especially, synthetic chemistry inside the tuneable functional channels of the porous material. In this section, we will focus on the main advances related to the use of MOFs as vessels and as chemical reactors and on perspectives and limitations in

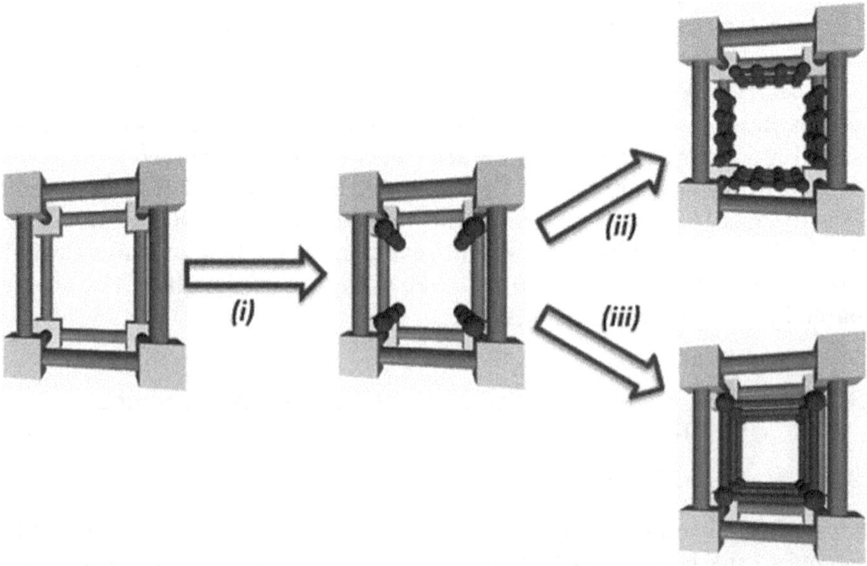

Figure 9.13 Schematic representation of the uses of MOFs as chemical reactors consisting of the use of consecutive PSMs, such as (*i*) insertion of metal cations, (*ii*) *in situ* reduction to form MNPs or MNCs and (*iii*) diffusion of appropriate ligands to self-assemble metal complexes or supramolecular assemblies.

relation to the inclusion or preparation of functional species within MOFs and their application in catalysis. In particular, we will pay special attention to the insertion and/or preparation of polyoxometalates, metal complexes, metal nanoparticles (MNPs) and metal nanoclusters (MNCs) within the empty spaces of MOFs, as they have shown impressive catalytic activity.

9.2.4.2 Encapsulation/Chemical Synthesis of Active Species in the Cavities of MOFs

Supporting catalytically species in MOFs – which may include polyoxometalates, metal complexes, MNPs or MNCs, among others – have experienced rapid development during recent years. The main advantages that these porous materials offer as hosts for such highly active but sometimes unstable species can be summarized as follows: (1) large surface area and high porosity with uniform pore size; (2) easy fine-tuning of the size, shape and functionality of the empty space pre- or post-synthetically; (3) the possibility to use single-crystal X-ray diffraction (SC-XRD) to characterize the resulting host–guest aggregates, which may facilitate studies to elucidate the mechanisms governing catalytic processes; (4) MOF-based materials can be processed and different composites can be also prepared; and (5) the broad and increasing diversity of MOFs that are stable in different solvents and conditions. In turn, in addition to the well-known general limitations of MOFs, such as their lower thermal stability compared to other materials, new challenges arise for this approach, such as preventing the "leaching" of these active species from the channels of the MOF and the large-scale preparation of this novel generation of catalysts, which has been barely studied. In any case, the potential rewards that this chemical avenue offers makes it worth the effort to explore the use of these unique materials that allow, among other advantages, for the support of highly reactive homogeneous catalysts in solid supports with the consequent benefits of heterogeneous catalysis, such as higher stability, reusability or selectivity.

9.2.4.2.1 Polyoxometalates. The rich redox chemistry and chemical diversity of the metal–oxo complexes constituting the polyoxometalates (POMs) – typically transition metals ions of groups 5 and 6 with high oxidation states – together with their strong acidic nature and tuneable size and shape have been intensively exploited in catalysis, especially in reactions where the POM acts as an acid or oxidation catalyst. However, their solubility, small surface area and low stability under working conditions have hampered their catalytic applications. In this context, supporting POMs on MOFs has presented as an efficient approach to improving their stability and catalytic performance.[191–212] The broad chemical versatility and compositional diversity of POMs have enabled different approaches to the construction of open-framework structures based on POMs to develop. For example, a growing amount of work is being devoted to POMs acting as nodes or pillars to form so-called POM-based MOFs (POMOFs).[199,203,206] Despite the interest in this approach, it is beyond the scope of this chapter, and we refer the eager reader to the reviews published on this topic.

In this subsection, we will focus – through relevant examples – on the different approaches to encapsulating POMs within MOF channels and on the new catalytic functions exhibited by these new hybrid materials. The POM-driven formation of HKUST-1 – also known as MOF-199 and Cu-BTC – has been widely used to obtain different examples of POMs encapsulated within MOF channels.[191] The NENU (Northeast Normal University) series, where different preformed dianionic Keggin-type POMs are used as templates, represents one of the most important families of these types of composites (Figure 9.14). These POMOFs have showed increased thermal stability with respect to pristine HKUST-1, high acid–base resistance with a negligible leaching of the incorporated POM and, even more importantly, superior catalytic activity in different hydrolysis reactions compared to the POM and the MOF on their own.

In this respect, for example, NENU-3a[191] exhibited substrate-dependent hydrolytic catalytic activity of ethyl acetate compounds, and nanocrystalline NENU-9 showed nice oxidative activity of dibenzothiophene.[205] Interestingly, in this family of composites, single-crystal X-ray crystallography was used to determine the nature of the interactions between POMs and MOFs in order to increase our understanding of these hybrid assemblies and to provide the possibility of rationalizing their catalytic performance. However, despite the relevant catalytic features of this family of composites, the small size of their

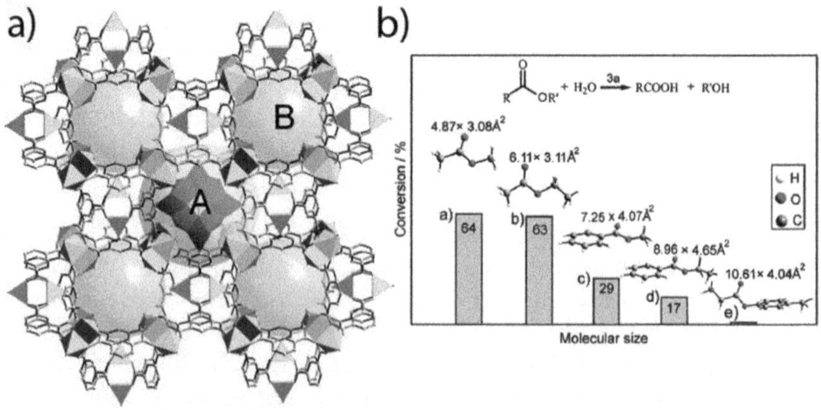

Figure 9.14 (a) Perspective view of the crystal structure of NENU-n (n = 1–6) with two type of pores: pore A, which accommodated the Keggin anions; and pore B, which provides a cuboctahedral space. Cu_2-secondary building unit (blue polyhedral), kegging polyanion (red polyhedral) and C and O atoms (grey and red spheres, respectively). H atoms and solvent molecules are omitted for clarity. (b) Hydrolysis of esters in the presence of NENU-3a. Adapted from Ref. 191 with permission from American Chemical Society, Copyright 2009.

channels has led to diffusion problems, consequently mitigating their potential catalytic activity.

Mesoporous MIL MOFs – MIL-100(M) (M = Cr, Fe) and MIL-101(M) (M = Cr, Fe, Al) – have been widely used to post-synthetically incorporate POMs,[205] mainly by means of impregnation methods. Their structural robustness and large channels have led to different POM@MIL composites with nice catalytic activities in a wide range of reactions, from classical thermal catalysis to light-driven catalytic applications. These studies have nicely reflected the key importance of the total loading amount and the dispersion degree of the POMs within MOF channels on the catalytic performance of the POM@MIL. However, impregnation methods are often associated with POMOFs of reduced stability under working catalytic conditions due to the weak interactions between them. In order to increase their binding, the amino-functionalized version of the MIL series has been used. POM@MILs (Figure 9.15) have also been obtained using POMs as template agents. In this respect, Gascon *et al.* have reported an interesting comparison between both methods for the Knoevenagel condensation reaction between benzaldehyde and ethyl cyanoactetate (Figure 9.15). From this, it was shown that the encapsulated composite showed a higher catalytic activity than the impregnated one. The authors attribute this to both a better dispersion of POMs in the encapsulated sample and the deactivation of POMs during impregnation. Despite the interesting catalytic results obtained with MIL compounds, more work is needed related to the structural elucidation of these hybrid composites using X-ray crystallography. Typically, indirect characterization methods have been used, which have prevented discovering whether the nature of the POM is preserved in the final material, as well as the relevant interactions between POM and MOF.

Zirconium-based MOFs are attracting a growing body of work for hosting POMs. In this context, dealing with microporous materials, Salomon *et al.* reported the use, for the first time, of UiO-67 to encapsulate several tungstate

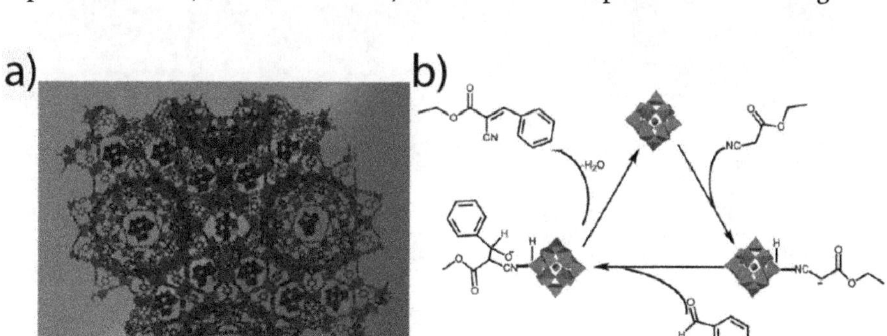

Figure 9.15 (a) Artistic representation of POM@MIL-101(Cr). (b) Proposed reaction mechanism for the Knoevenagel condensation of benzaldehyde with ethyl cyanoacetate. Adapted from Ref. 213 with permission from Elsevier, Copyright 2010.

POMs.[198,207] However, no proper catalytic application has been found for them. Recently, following the same approach, Song *et al.*[198] have reported the encapsulation of polyoxomolybdic cobalt and polyoxomolybdic acid within UiO-67. From the resultant composites, the cobalt one showed excellent catalytic activity for the epoxidation of olefins using H_2O_2 or O_2 as the oxidant. Although the composite was not characterized by single-crystal X-ray crystallography, a plethora of physical characterization techniques have been used to ascertain that the presence of multiple interactions between the encapsulated POMs and the functional groups decorating the UiO-67-type MOF plays a key role in improving the stability and catalytic properties of the hybrid assembly.

Regarding mesoporous Zr-based MOFs, Paille *et al.*[195] have reported the post-synthetic incorporation, using impregnation methods, of a sandwich-type cobalt-POM within the hexagonal channels of the Zr(IV) porphyrinic MOF-545, also known as PCN-222 and MMPF-6. The resulting composite showed nice photocatalytic activity for water oxidation using visible light. Indeed, this is considered the first reported example of a noble metal-free heterogeneous and recyclable photosystem, where both components – the catalyst and the photosensitizer – coexist within the same channel. Following the same approach in NU-1000, a Keggin-type POM has been incorporated (Figure 9.16).[211,212] The resulting hybrid material showed better catalytic

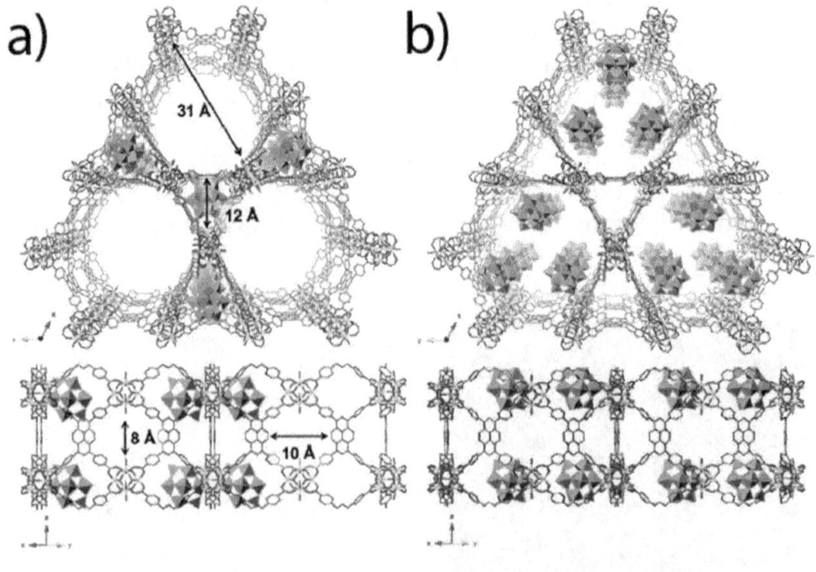

Figure 9.16 Proposed representations by difference envelope density analyses of POM@NU-1000(Zr) with the POMs residing in the microporous (a) or mesoporous (b) channels depending on the chosen activation method. WO_5 (light blue prims), PO_4 (pink prism), Zr (green), O (red) and C (grey); H atoms omitted for clarity. Adapted from Ref. 212 with permission from the Royal Society of Chemistry.

activity for the oxidation of 2-chloroethyl ethyl sulfide using H_2O_2 as the oxidant than either of the individual constituents alone. Interestingly, in a later investigation, a deeper study of this hybrid assembly revealed than the catalytic activity of the POM@NU-1 000 was dependent on the place where it was hosted in the POM, which in turn was related to the method of activation of the resulting material. Heating up to 120 °C led to the POMs being situated in the microporous channels, and supercritical CO_2 drying led to the POMs being hosted in mesoporous channels. Remarkably, this different positioning resulted in different catalytic performance, and the assembly with the POMs within the mesoporous cavities showed a faster reaction rate and higher selectivity.

9.2.4.2.2 Metal Complexes. Metal complexes – including coordination and organometallic compounds – have played a fundamental role as catalysts[214] in a limitless number of reactions, including those mimicking processes naturally occurring in living organisms.[215] However, many of these reactions take place in solution (homogeneous catalysis). Apart from the obvious benefit arising from the higher degree of interaction between reactants and catalysts that homogeneous catalysis offers, it is true that heterogeneous catalysis can lead to great advantages, such as reusability and higher stability and selectivity. Therefore, supporting such metal complexes in solids[214,216–220] constitutes a clear step forwards. In this context, the unique microenvironment of MOF channels and their previously mentioned exceptional features offer clear advantages in comparison to other porous solids: (1) better control of their pore size distribution and functionalities, which allows for a more homogeneous dispersion and stabilization of the inserted metal complexes; (2) higher retention of these active species within their pores, preventing, to a certain extent, leaching processes during the catalytic processes; and (3) the possibility to use SC-XRD for the characterization of the host–guest aggregates,[221–225] which is not possible with other porous materials. Finally, MOFs can also be used as chemical reactors for the *in situ* preparation of metal complexes (or supramolecular aggregates) by the consecutive application of different types of PSMs (post synthetic modifications).

The first attempts to insert such catalytically active guests within MOFs were carried out by Eddaoudi and co-workers by inserting a manganese porphyrin within the cavities of rho-ZMOF (zeolite-like metal–organic framework) (Figure 9.17).[226] This synthetic strategy relied on the use of the "bottle around a ship" method, consisting of the direct incorporation of the complex during the MOF formation process. Thus, the complex is ultimately encapsulated and retained within a large cavity, possessing small openings that prevent the leaching of the complexes. The material showed good results in terms of cyclohexane oxidation with high reusability. However, the crystal structure of the guest functional complex could not be determined, and indirect methods had to be used to characterize the material. In contrast, the "ship in a bottle" strategy was then used by Farruseng[227] and Fedin[228] – in different works – to

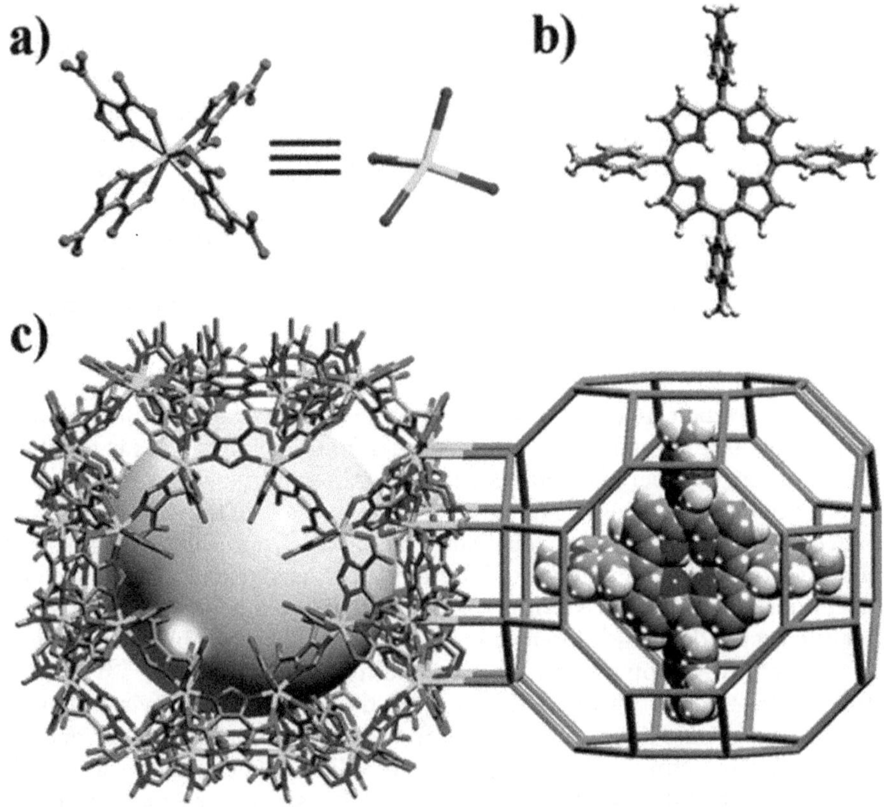

Figure 9.17 Crystal structures of the eight-coordinate SBU (a) and $[H_2TMPyP]^{4+}$ porphyrin (b). (c) Schematic representation showing the encapsulation of the complex within the rho-ZMOF. Adapted from Ref. 226 with permission from American Chemical Society, Copyright 2015.

encapsulate several metal phthalocyanine complexes within MIL-101. The resulting supported complexes showed outstanding behaviour in terms of oxidation catalysis. However, again, no SC-XRD studies could be carried out.

In 2012, Lin *et al.*[229,230] extended the "bottle around a ship" strategy to the highly robust UiO-67 MOF ($Zr_6O_4[OH]_4[bpdc]_6$, where bpdc = para-biphenyldicarboxylic acid) by incorporating a series of Ir, Re and Ru complexes during the MOF formation process under solvothermal conditions. The resulting hybrid materials could be used as catalysts in a plethora of catalytic transformations such as water oxidation, photochemical CO_2 reduction, aerobic amine coupling, the aza-Henry reaction and aerobic thioanisole oxidation, showing excellent catalytic activity and improved reusability. Following a similar strategy, Cohen[231] reported the post-synthetic insertion of a pre-formed organometallic di-iron complex within the channels of the UiO-66 MOF, aiming at mimicking the active site of [FeFe]-H_2ases. Indeed, the

Fe_2 entities supported on the MOF exhibited great efficiency for photochemical H_2 evolution, even surpassing the reference Fe_2 homogeneous system. Several examples consisting of the insertion of metal complexes mimicking hydrogenase active sites and with photocatalytic applications have been subsequently reported.[232–235] Indeed, MOFs are very good candidates for the "heterogenization" of model complexes and for mimicking natural enzymes in general.[215,236–240]

Another original strategy takes advantage of the available coordination sites of the ligands forming the MOF. For example, Rosseinsky and co-workers[241] post-synthetically introduced Zn^{2+} cations within a MOF constructed with the meso-tetra(4-carboxyl-phenyl) porphyrin ligand. The resulting Zn-based porphyrin complex was active for the photocatalytic evolution of H_2 from water with moderately good results.

Overall, following any of the mentioned approaches, a good number of publications reporting the encapsulation of catalytically active metal complexes have been published after these pioneering works.[58,59,242–266] The main advantages of this heterogenizing process consisting of supporting these metal complexes in MOFs are related to the higher stability and reusability they "acquire" from the porous matrix, which includes highly reactive but unstable species[59] with very low applicability in homogeneous catalysis.

The last point of this subsection will be related to the use of MOFs as chemical reactors for the chemical synthesis – by applying consecutive PSMs – of metal complexes (as well as different kinds of supramolecular constructs) within the confined space offered by MOFs. Overall, the few examples reported so far consisted of the post-synthetic insertion of metal cations capable of coordinating the freely available coordination sites of the ligands of the framework.[241] However, there is still much work to do related to the total synthesis of metal complexes within the empty confined spaces of MOFs, which, theoretically, could lead to unprecedented species being forced by these functionally limited spaces. The most relevant example has been reported very recently by some of this chapter's authors. The dual-step strategy – consisting of the two-step consecutive insertion of Pd^{2+} cations and dipyridine ligands – led to the *in situ* heterogeneous self-assembly of a Pd_8 square polygon and a Pd_{16} supramolecular construct (Figure 9.18). The crystal structure of these two unique and original supramolecular coordination complexes (SCCs) could be resolved by SC-XRD. A third Pd–Au heterometallic SCC could also be achieved by inserting Au(III) cations after the assembly of a SCC with free thio-ether groups. The resulting SCC@MOFs constitute a clear advance in the use of MOFs as chemical reactors for the preparation of novel species that are difficult to achieve outside of MOFs. In addition, they efficiently catalysed the coupling of boronic acids and alkynes.

9.2.4.2.3 Metal Nanoparticles. The appearance in the last century of MNPs undoubtedly resulted in a paradigm shift in science and led to a step forward in the manner in which we understand, use and develop metals. This new

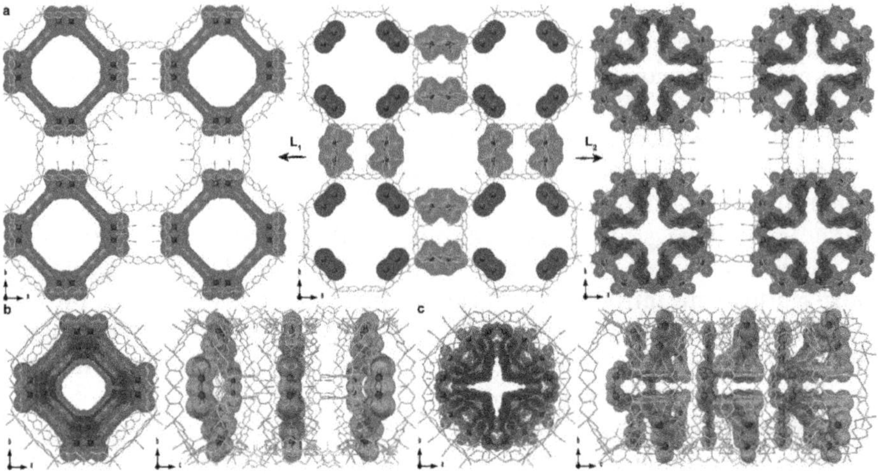

Figure 9.18 Real crystal structures of the Pd_8 (a, left) and Pd_{16} (a, right) SCCs from the starting MOF with Pd^{2+} dinuclear units within its pores, obtained by introducing 1,2-di(pyridn-4-yl)ethyne and methyl 3,5-*bis*(pyridine-4-ylethynyl)benzoate ligands, respectively. (b, c) The resulting SCCs within MOF channels in more detail. Reproduced from Ref. 267 with permission from American Chemical Society, Copyright 2019.

unexploited form of pure metallic matter, with outstanding properties, led to an almost limitless number of publications and patents, and is said to have paved the way to the technological advances of the following next years.[268–270]

However, a main drawback of MNPs is their tendency to aggregate under working catalytic conditions and their thermodynamic instability, which lead to the deactivation of their catalytic properties. Thus, the use of protective coatings with different surface capping agents has been proposed. However, despite the effectiveness of these methods to control the size, shape and dispersion of MNPs, their potential negative influence on their catalytic activities have triggered a quest to find suitable porous materials to encapsulate and stabilize catalytically active MNPs. In this context, a wide range of synthetic strategies have been proposed to obtain hybrid systems consisting of MNPs and MOFs.[45,271–285] Three main approaches have been used: (1) active MNPs are grown and confined, by means of post-synthetic methodologies, within the pores of preformed MOFs; (2) pre-formed well-defined MNPs are used as nucleation seeds to grow a MOF around them; and (3) the concomitant formation of MNPs and MOFs (Figure 9.19). These three different approaches led to a seemingly infinite number of publications reporting on the encapsulation of MNPs within MOFs.[271,280,281,286–326]

Different synthetic approaches – such as impregnation, chemical vapour deposition, solid grinding, co-precipitation, deposition–precipitation or double-solvent approaches – have been used to incorporate MNPs within the channels of pre-formed MOFs. Impregnation, due to its simplicity, has been

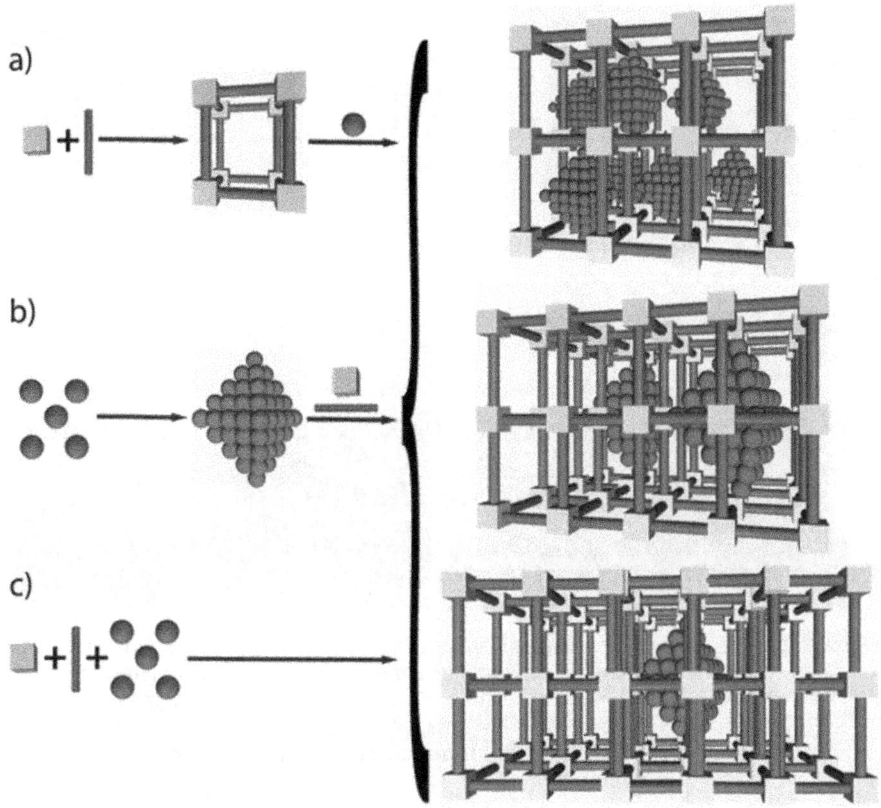

Figure 9.19 Schematic representation of the different approaches (a) "ship in a bottle", (b) "bottle around a ship" and (c) "one-pot" to obtain MNP@MOFs.

the most used approach. It consists of introducing MOF crystals in a solution of a metal ion, followed by isolation of the resulting assembly and reduction of the incorporated metal precursors, leading to MNPs being present within the confined space of the channels. For this approach, mainly the archetypal MIL and UiO families have been intensively studied for the incorporation of different noble metal (*e.g.* silver, gold, platinum, rhodium and palladium) and transition metal (*e.g.* nickel and ruthenium) nanoparticles (Figure 9.20).[313] Among others, it has been possible to obtain ruthenium nanoparticles of 4–5 nm particle size within MIL-101, which have been shown to be efficient catalysts for the conversion – in water media – of furfural to cyclopentanone. Despite this method being widely used, it has the main drawback of it being difficult to avoid deposition on the external surface of the MOF.

Alternatively, the double-solvent approach has been revealed as an appealing approach to preventing this drawback and the aggregation of formed mono- and bi-metallic MNPs. This method consists of the preparation of a MOF dispersion in a hydrophobic solvent, with a volume equal to or less than the pore volume of the MOF, to which a solution of the metal salt prepared in

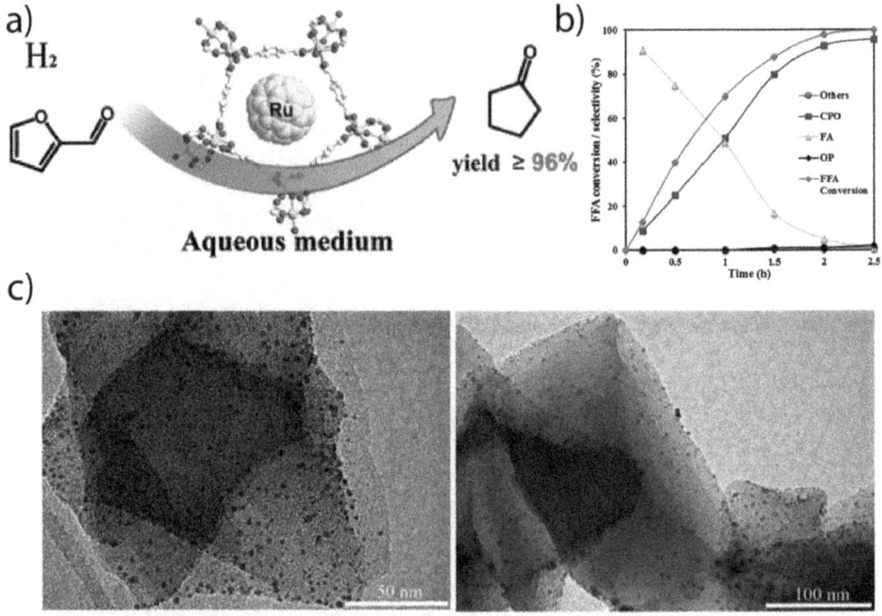

Figure 9.20 (a) Artistic representation of the catalytic activity of Ru NP@MIL-101(Cr) for the selective hydrogenation of furfural to cyclopentanone. (b) Furfural conversion and evolution of product selectivity as a function of reaction time. (c) TEM images of Ru NP@MIL-101(Cr) at different magnifications. Adapted from Ref. 313 with permission from the Royal Society of Chemistry.

a hydrophilic solvent is added. As a consequence of capillary forces and hydrophilic interactions, metallic species enter the MOF pores and, after reduction, lead to MNPs in a more controllable manner. For example, the controlled simultaneous incorporation of palladium and cobalt salts into the pores of MIL-101 and the subsequent reduction with NH_3BH_3 lead to core–shell Pd@Co nanoparticles within MOF channels (Figure 9.21).[297] Interestingly, this hybrid material shows a highly recyclable character and, even more importantly, exhibits synergistically enhanced catalytic activity for hydrolytic dehydrogenation of ammonia borane with respect to the Pd@Co core-shell nanoparticles on the external surface of MIL-101 and their monomeric counterparts and alloys within MOF pores. Despite the formidable advances in the field, the methods reported for the growth of MNPs within MOF channels still have some deficiencies that need to be addressed in order to be able to exploit all of the potential properties of MNP@MOFs. In this sense, more effort should be devoted to accurately controlling the morphology, composition and location of MNPs within MOF channels. To this end, a possible solution may be using the tuneability of MOF pores to generate different chemical environments able to direct and control the nucleation and growth processes.

The incorporation of well-defined MNPs into a solution containing the precursors to build up the MOF has been used as the obvious approach to

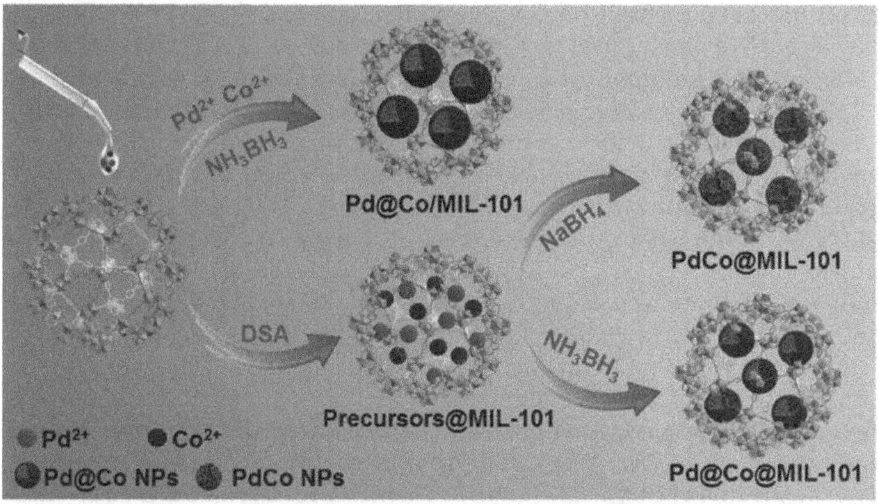

Figure 9.21 Artistic representation of the synthesis of Pd@Co@MIL-101(Cr), Pd@Co/MIL-101(Cr) and PdCo@MIL-101(Cr). Reproduced from Ref. 297 with permission from John Wiley and Sons, Copyright © 2014 Wiley-VCH Verlag GmbH & Co. KGaA, Weinheim.

achieving better control over the physicochemical properties and catalytic performance of MNP@MOFs. A key part of this methodology is the presence in the outer sphere of MNPs of species that are able to promote the heterogeneous growth of the targeted MOF around them, instead of the self-nucleation of the MOF, and to avoid the aggregation of MNPs. However, this is not an easy task, and on top of that, this species could have a non-negligible negative effect on the formation of the targeted MOF and on the accessibility of substrates to the functional MNPs.

The use of ligands or surfactants has frequently been explored for the controlled formation of catalytically active MNP@MOFs. By using polyvinylpyrrolidone to stabilize pre-formed MNPs of different sizes, shapes and compositions, it has been possible to obtain a wide range of functional MNP@ZIF-8.[327] Remarkably, by using this strategy, the encapsulation of different types of nanoparticles has been reported, and even more relevantly, their spatial organization can be controlled by modifying the sequence of addition during the synthesis of the ZIF-8. For example, the synthesis of 3.3 nm Pt NP@ZIF-8 with hydrogenation activity towards *n*-hexene in the liquid phase has been reported. But, conversely, this hybrid material was not able to catalyse the reaction for the bulkier substrate *cis*-cyclooctene. Interestingly, these Pt NP@ZIF-8 show regioselectivity and only possess hydrogenation activity for molecules with terminal alkenes – *n*-hexene – which contrasts with the indiscriminate catalytic activity of bare Pt nanoparticles that were able to catalyse the hydrogenation of *n*-hexene and *trans*-2-hexene. In this investigation, the authors used the MOFs as molecular sieves for performing size selective catalysis. However, more effort has been focused on developing new systems in which the synergy between MNPs and the nodes, metalloligands or organic ligands of the MOFs enables high-performance catalysis to be achieved. In this respect, 18 nm polyvinylpyrrolidone-stabilized Cu nanoparticles wrapped up with UiO-66(Zr) led to a hybrid composite[288] that was able to more efficiently catalyse the CO_2 hydrogenation to methanol than any of the components separately: Cu nanoparticles on UiO-66(Zr) or MIL-101(Cr) and Cu NP@ZIF-8(Zn).[288] On the basis of a complete physical characterization of the composite, the authors assigned this interesting catalytic performance to the influence of zirconium oxides constituting the secondary building units of UiO-66(Zr) on the surrounded Cu nanoparticles.

In addition to organic molecules, other systems such as silica or metal oxides have been used as templating agents to form MNP@MOFs. In this context, in the first stage, a layer of controlled thickness of Al_2O_3 is created around the pre-formed MNPs using different synthetic protocols, such as atomic layer deposition or chemical synthesis. Then, this metal oxide is used as a precursor to synthesize the MOF around the MNPs upon the incorporation of the organic ligand that acts as an etching agent and organic strut of the assembled network. Following this methodology, it has been possible to obtain Au nanorods[328] and Ag nanocrystals[329] encapsulated in two different Al-based MOFs (Figure 9.22). In a similar approach, SiO_2 has been used as a protective shell of different nanoparticles to create composites, whereby after the etching of the SiO_2 layer there is an empty space between the MNPs and the MOFs. For example, it has been reported that the use of SiO_2 as a protective shell for the meta-stable Cu_2O nanoparticles and as sacrificial agent after the formation of the MOF shell around the MNPs led a yolk–shell structure.[288] The resulting Cu_2O NP@ZIF-8(Zn) has been used as an efficient catalyst for the hydrogenation of 4-nitrophenol and has nice reusability, which contrasts with the low stability of meta-stable Cu_2O nanoparticles. The step that needs further attention in this strategy is the etching process. Hard templates with suitable properties are quite limited, and the etching protocols can trigger the decomposition of MOFs. In this context, it is worth noting that MNPs have also been used successfully as sacrificial agents to grow MOFs around them. To this end, some parts of the MNPs are used as sources of the metal ions needed to build up the MOF shell around the remaining MNPs. For example, Pt–Cu NP@HKUST-1(Cu) with different morphologies have been obtained using as precursor copper-rich Pt–Cu alloy nanoparticles capped with polyvinylpyrrolidone.[309] Their dispersion on a dimethylformamide solution containing trimesic acid and iron(III) ions led to the oxidative dissolution of excess copper(II), which reacts with the organic ligand to grow the HKUST-1(Cu) around the Pt–Cu nanoparticles *in situ*. Notably, these materials exhibit remarkable catalytic activity for the hydrogenation of 1-hexene, together with nice reusability and size selectivity for the hydrogenation of other olefins.

Despite the relevant advances using the above-mentioned methodologies to encapsulate single or even multiple MNPs – with the same or different natures – they still lack accurate control of the local position of the MNPs.

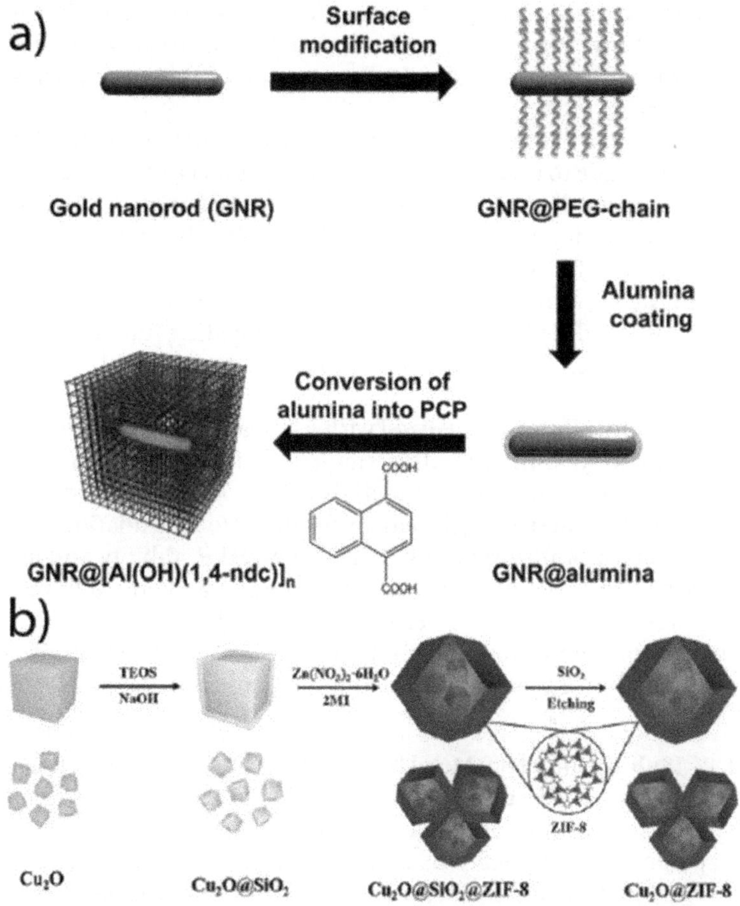

Figure 9.22 (a) Artistic representation of the synthetic approach of using Au nanorods wrapped with Al_2O_3 as seeds to grow around an Al-based MOF. (b) Schematic representation of the use of metastable Cu_2O NPs covered with SiO_2 to obtain yolk–shell Cu_2O@ZIF-8(Zn). Adapted from Ref. 328 with permission from American Chemical Society, Copyright 2013, and from Ref. 329 with permission from John Wiley and Sons, Copyright © 2018 Wiley-VCH Verlag GmbH & Co. KGaA, Weinheim.

In order to achieve better control and circumvent the low stability of MNPs on the external surface of MOFs, the construction of MNPs sandwiched between MOFs has been proposed. Although the synthetic protocol is quite tedious, it offers nice rewards. The core MOF acts as both a support for the MNPs and a promotor for the growth of the MOF shell, and this acts as a protective layer for the MNPs and as a molecular sieve imparting size selectivity in the studied catalytic reactions. Using this methodology, it has been possible to form different MOF/MNP/MOF structures. For example, 2.8 nm Pt nanoparticles have been sandwiched[310] between MIL-101(M) (M = Fe and/or Cr) and studied for the selective hydrogenation of cinnamaldehyde to cinnamyl alcohol (which is thermodynamically unfavoured in the presence of carbon–carbon groups). Overall, this study reveals that such sandwiched structures combine great selectivity, nice reusability and high conversion rates to outperform state-of-the-art catalysts.

The development of efficient one-pot synthetic strategies of MNP@MOFs is highly desirable due to their simplicity, which is associated with reductions in time and cost of production and ease of scale-up. Although this methodology is straightforward, its main challenge is balancing the independent nucleation and growth of MNPs and MOFs with the nucleation and growth of the MOFs surrounding the *in situ* growing MNPs. To this end, the accurate control of the reaction conditions and the presence of chelating functional groups in the linkers have been proposed. To this end, 3.0 nm Pd NP@UiO-67(Zr) have been obtained by combining 2,2'-bipyridine-5,5'-dicarboxylic acid that has been pre-synthetically functionalized with a palladium(II) salt with biphenyl-dicarboxylic acid – in different ratios – and $ZrCl_4$ in dimethylformamide solution.[290] The resulting composite materials exhibit shape selectivity for olefin hydrogenation and high catalytic activity for the aerobic oxidation of alcohols and the reduction of nitrobenzene. Remarkably, the stability and activity of this composite material outperform those obtained when using more traditional impregnation methods. Despite the simplicity of the one-pot methodology, more work is needed to be able to achieve better control of the size, shape, composition and uniform distribution of the resulting MNP@MOFs.

9.2.4.2.4 (Sub-nanometric) Metal Nanoclusters. As described in the previous subsection, the extensive development of new synthetic methods and applications of MNPs at the end of the last century led to a great number of publications, patents and applications and, in addition, many examples of MNPs within MOFs have also been reported.[278,279,281,282] More recently, the metal technological roadmap has presented another form of matter with completely different properties; that is, well-defined small aggregations of metal atoms – close to or within the sub-nanometric regime – so-called MNCs.[330] A series of recent works have demonstrated the exceptional properties of MNCs, such as luminescence,[331] biocompatibility,[332] anti-tumoural activity[333] and superlative catalytic activity.[331,334–336] Focusing on catalysis, the origin of the catalytic activity lies in their small size – less than 1 nm, or at most 2 nm, and with all atoms exposed and interacting between them[330] – and their unique electronic structures.[330] Therefore, it is believed that these novel metal forms will trigger unforeseen technological solutions in catalysis, once their design and large-scale manufacture are properly developed. Unfortunately, despite such importance, the gram-scale synthesis of structurally and electronically well-defined ligand-free MNCs is still far from being a reality.

So far, clearly inefficient ligand stabilization[337,338] and supporting these MNCs in porous solids[339] constitute the main routes to synthesize, stabilize and characterize these MNCs.[340] However, much work remains to be done for: (1) their synthesis and electronic characterization; (2) their stabilization without protective ligands that may reduce or block their catalytic properties, but also prevent their rearrangement into MNPs; and, especially, (3) their synthesis to a multi-gram scale, enabling industrial applications.[330] MOFs offer unique features to overcome most of these limitations, as we describe below.[341]

Again, MOFs emerge as perfect candidates to overcome most of the mentioned limitations as a consequence of their unique features, such as high crystallinity, tuneable functional porosity and fascinating host–guest chemistry, which is helpful for retaining and stabilizing both inserted metal cations and nanoclusters (see below). As a consequence, MOFs are not only excellent materials to act as vessels to host pre-formed MNCs, but also they appear to be the perfect platforms to synthetize chemically, within their pores, such ultra-small metal clusters by taking advantage of PSMs. The reasons for such efficiency are manifold: firstly, the confined state provided by their channels, which, in addition, can be functionalized pre- or post-synthetically, offers an appropriate environment to grow small clusters, taking advantage of both functionality, to anchor metal cations and MNCs to the walls, and also the nanostructured confined space, to limit the size of the formed species. Secondly, they allow exquisite control of the number of metal atoms – which, in addition, can be homogeneously distributed along the channels – that can be inserted within the pores, which undoubtedly influences the nuclearity of the final cluster. Lastly, in contrast to other porous materials, MOFs theoretically provide the possibility to use SC-XRD as a powerful tool to shed light on the crystal structures of MNCs. The resulting hybrid materials are not only high-efficiency nanocatalysts, but they also can show high stability – conferred by the framework to the metal catalyst – and catalytic shape selectivity due to the limited confined space.

Aiming at establishing a clearly defined frontier between MNCs and MNPs, in this section we will focus on those works reporting the preparation or encapsulation of ligand-free, well-defined, ultra-small MNCs – those possessing a small number of metal atoms, thus being close or within the subnanometric regime (less than 2 nm) – and their characterization and catalytic properties. Indeed, the gram-scale preparation and characterization of such small entities is still extremely challenging, and only a few examples have been reported so far. To date, the two main strategies[278] for the preparation of these ultra-small units are as follows: (1) encapsulation of the MNC within the MOF by the self-assembly and growth of the MOF around pre-formed MNCs ("bottle around a ship"); and (2) the use of PSMs that allow for the use of MOFs as chemical nanoreactors for the step-by-step fabrication of MNCs within the confined functional space offered by them. Each of these synthetic approaches shows advantages and disadvantages,[278] which will be described below.

Encapsulation of MNCs ("Bottle around a Ship"). This synthetic strategy basically consists of the *in situ* insertion and encapsulation, in one- or multiple-step processes, of MNCs during the MOF crystallization process. Such a strategy can be a fast and straightforward approach for the preparation of the hybrid MNC–MOFs materials; however, it has substantial problems, such as the agglomeration of MNCs during the crystallization process and the phase separation of the metal species and the MOF matrix, which makes it necessary to use stabilizing agents with the concomitant reduction of the catalytic activity of the MNCs.

As described in Section 9.2.4.2.3, Chen *et al.*[280] reported the one-pot encapsulation of relatively small Pd^0 clusters – with a mean size of 3.0 nm – uniformly distributed along the channels of the MOF UiO-67. The synthetic method is based on anchoring the palladium precursors to the MOF ligands prior to the one-step MOF self-assembly. This strategy requires conscious ligand design. In particular, 2,2-bipyride-5,5′-dicarboxylic acid, which allows for a perfect immobilization of Pd^{2+} cations through the 2,2-bipyride units, was selected. Then, the one-step assembly process that led to Pd^0@UiO-67 takes place when the cluster $Zr_6O_4(OH)_4(CO_2)_{12}$ is added and treated under H_2 at 250 °C (Figure 9.23). The resulting Pd^0 nanoclusters showed great catalytic activity in the aerobic oxidation of alcohols, the reduction of nitrobenzene and olefin hydrogenation, also showing excellent shape selectivity for the former reaction.

A similar strategy was followed by Liu *et al.* for the *in situ* one-step synthesis of ligand-free Pt nanoclusters within the MOF DUT-5.[308] The synthetic procedure consisted of directly mixing $Al(NO_3)_3$, the 4,4'-biphenyldicarboxylic acid ligand and H_2PtCl_6 in DMF (dimethylformamide), which acts as reducing agent, to yield small PT nanoclusters with sizes ranging in the interval 1.5–2.0 nm. Despite the moderately small size of the MNCs formed, they exceeded those of the pores of the MOF (0.7 nm), strongly suggesting inter-network encapsulation and thus leading to a non-homogeneous distribution of Pt nanoclusters. The material showed excellent catalytic performance in the aerobic oxidation of cinnamyl alcohol under base-free conditions.[308] The same authors reported an improved preparation method[342] consisting of the addition, to the previous protocol, of acetic acid as a modulator of the MOF and H_2 as a reducing reagent. As a consequence, greater control of both the reduction rate of the Pt nanoclusters and the growth rate of the MOF UiO-66 was achieved, which allowed for a more homogeneous distribution within the channels. Smaller MNCs of ~1.7 nm were obtained. The Pt@UiO-66

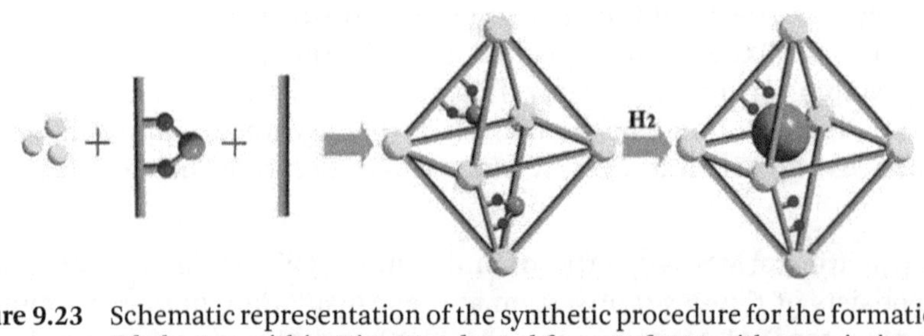

Figure 9.23 Schematic representation of the synthetic procedure for the formation of Pd clusters within UiO-67. Adapted from Ref. 280 with permission from the Royal Society of Chemistry.

composites exhibited exceptional activity and enhanced stability in the aerobic oxidation of cinnamyl alcohol, as well as excellent size selectivity in olefin hydrogenation.

The present strategy offers advantages such as the lack of surfactants or capping agents. However, further examples would be desirable to accept this approach as a general protocol. In addition, the size of the clusters (~3 nm), despite not being too large, is still far from the subnanometric regime, at which superior catalytic performances are expected, and these clusters cannot be considered nanoclusters (2 nm). Other examples, following similar synthetic approaches, gave even larger clusters or MNPs.[314,323]

As an extension of the above-described one-step encapsulation method, the so-called multi-step encapsulation protocol has aroused interest. The synthetic strategy consists of the initial preparation of metal clusters or complexes – by using capping ligands to prevent their agglomeration leading to larger MNPs – and the subsequent growth of the MOF around the pre-formed clusters, to yield the final core–shell cluster@MOF composites. The final stage of this process consists of removing the protecting ligands of the formed clusters in order to enhance catalytic activity. This is perfectly exemplified by Ji and co-workers,[343] who reported the preparation of Ru_3 subnanometric nanoclusters by following a dual-step process consisting first of mixing, in one pot, preformed $Ru_3(CO)_{12}$ complexes with Zn^{2+} cations and 2-methylimidazole ligands – precursors of the MOF ZIF-8 – to yield the hybrid material $Ru_3(CO)_{12}$@ZIF-8. Then, after pyrolysis under H_2 flow, ligands were removed and uniform Ru_3 clusters, anchored by the nitrogen atoms of the framework, were obtained. The Ru_3 nanoclusters exhibited outstanding catalytic activity in the oxidation of 2-aminobenzyl alcohol. This is the only example reporting subnanometric MNCs with such an encapsulation strategy. In fact, apart from the above-described example, this strategy has been rather unsatisfactory so far for the preparation of MNCs below 2 nm. Thus, following this strategy, different hybrid catalysts have been obtained. For example, naked Pd,[318,319] Pt[302,304,310,320,321] and Au[327,344] nanoparticles, with average sizes varying in the range 2–25 nm, were achieved after removing the protecting ligands. The MNPs were dispersed uniformly inside the MOF shells, with no MNPs observed on the external surfaces. The resulting materials are indeed interesting, with great catalytic performance in a myriad of catalytic reactions. However, due to their large size, they cannot be called MNCs, and they will not be described further.

Overall, both one- and multi-step encapsulation approaches lead to some interesting results, but none of them fully exploits the unique features of MOFs, and therefore these methods suffer from most of the traditional metal cluster synthesis disadvantages. Therefore, encapsulation methods have proved ineffective so far in the preparation of ultra-small MNCs.

Post-synthetic Strategies in Pre-formed MOFs. In contrast to the above-described encapsulation methods, PSMs offer the perfect tools for the MOF-driven preparation of electronically and structurally well-defined ultra-small

MNCs. In this sense, a variety of post-synthetic approaches can be used for the MOF-driven preparation of MNCs. They all share a common dual step-by-step strategy consisting of the introduction of the desired metal within the channels of the MOF and the *in situ* reduction to form the zero- or low-valent MNC. This approach allows the size and topology of MNCs to be inherited from the confined functional space where they are synthesized. The first common step consists of the diffusion of the metals – in solid, liquid or gas phase – and the consequent reduction process with an appropriate reducing agent such as ammonia borane or borohydride compounds in gas phase. Both steps are fundamental for the successful formation of ultra-small MNCs. With regards to the first step, the metals must be located within the channels and not on the surfaces of the MOFs in order to profit from the confined space and functionality provided by the MOF channels, thus restricting the size of the MNCs and preventing their agglomeration. Therefore, a proper infiltration method, overcoming the diffusion problems though the pore apertures, must be chosen in order to insert metals within channels and achieve a homogeneous distribution along the pores. On the basis of the different infiltration approaches, four different post-synthetic strategies can be distinguished: (1) gas-phase chemical vapour deposition (CVD) methods; (2) gas-phase solid grinding (SG) methods; (3) wetness impregnation methods; and (4) double-solvent methods (DSMs).

The CVD method consists of introducing, under vacuum, volatile organometallic compounds within the pores of the MOF[345] and concomitant

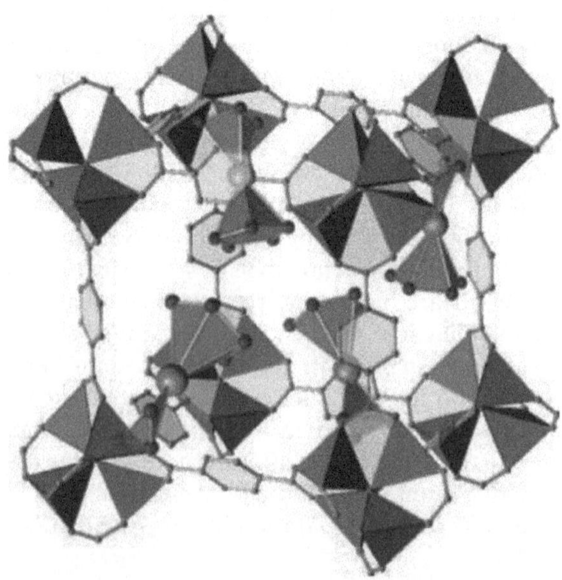

Figure 9.24 Model showing one cage of MOF-5 and four incorporated ($[\eta^5\text{-}C_5H_5]Pd[\eta^3\text{-}C_3H_5]$) complexes prior to the reduction process. Adapted from Ref. 345 with permission from John Wiley and Sons Copyright © 2005 Wiley-VCH Verlag GmbH & Co. KGaA, Weinheim.

reduction with a reducing gas (usually H_2). Different examples have been reported with varying success. The pioneering result was reported by Fischer and collaborators.[345] The desolvated MOF-5 was first exposed to vapours of the organometallic complex $(\eta^5\text{-}C_5H_5)Pd(\eta^3\text{-}C_3H_5)$ in a sealed Schlenk tube (Figure 9.24). Then, after the reduction process, small, well-defined, ligand-free nanoclusters of 1.4 nm were obtained. Despite the success in the preparation of ultra-small MNCs, the resulting Pd-NCs@MOF-5 hybrid material showed only moderate catalytic activity in the hydrogenation of cyclooctene. Other examples have been reported later by Fischer[346] and Kim.[347] They reported the formation of small, naked Au and Ni nanoclusters, with sizes below 2 nm in both cases, within the ZIF-90 and MesMOF-1 MOFs, respectively. Only Ni-NC@MesMOF-1 showed good catalytic activity for both the hydrogenation of styrene and the hydrogenolysis of nitrobenzene. Overall, these results demonstrate that the CVD method allows for the formation of relatively small MNCs. However, despite the fact that this approach shows other remarkable advantages, such as the monetary savings connected to the absence of solvents and a faster synthetic procedure, not many examples have been published to date.[316] The main reasons for this are the limited number of metal precursors that can be sublimated below the temperature of decomposition of MOFs and other problems associated with such organometallic species in terms of stability, synthesis and price.

The gas-phase SG method appears to be a good gas-phase alternative to the CVD method. The main advantage is that it does not require vacuum conditions. It allows for very fast diffusion of the vapour containing the sublimated organometallic precursors, resulting in the more uniform encapsulation of precursors and slightly smaller sizes of MNCs compared to the CVD method. The pioneering work of Ishida *et al.*[348] reported the preparation of a variety of Au clusters within different MOFs, such as CPL-1, CPL-2, Al-MIL53 and MOF-5 (Figure 9.25). The best results were obtained for Al-MIL53 MOFs by introducing by SG $Me_2Au(acac)$ complexes (acac = acetylacetonate), which were then treated with a stream of H_2 to form Au nanoclusters with an average size of 1.5 nm. The Au nanoclusters showed high catalytic activity for different aerobic oxidations of alcohols, for which the activity was highly dependent on

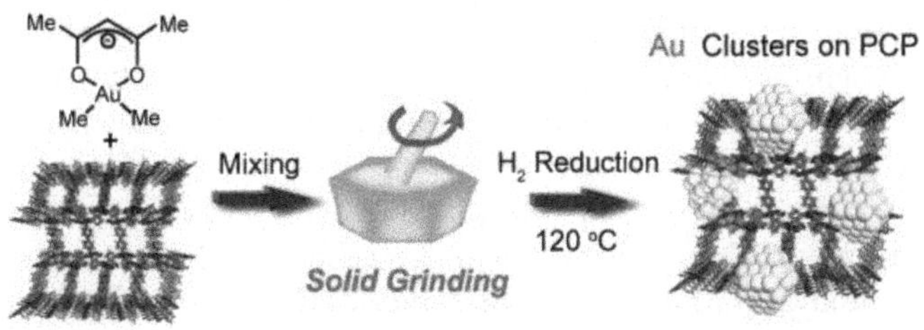

Figure 9.25 Synthetic scheme of the synthetic procedure for the preparation of MNCs within MOFs by the SG method. Adapted from Ref. 348 with permission from John Wiley and Sons, Copyright © 2008 Wiley-VCH Verlag GmbH & Co. KGaA, Weinheim.

the size of the Au particles. Later, Jiang *et al.*[349] followed the same synthetic strategy to form very small Au nanoclusters, below 5 nm. However, catalytic activity was not explored. Despite the very good results with regards to the size of the formed MNCs, very few examples of this have been reported so far, in a similar manner to the CVD approach.

The wetness impregnation method is probably the most explored strategy for the synthesis of MOF-driven ultra-small MNCs, which, in addition, has also achieved the best characterization of these small entities. The synthetic procedure consists of stepwise precursor infiltration in solution, followed by reduction with the selected reducing agent or decomposition to form the ultra-small metal clusters. This approach leads to a uniform distribution of metal atoms, which is one of the key points for the formation of ultra-small MNCs. However, this technique also shows some limitations: for example, the need for solvents makes the process more expensive, and diffusion problems slow down the insertion process and can lead to adsorption onto the outer surface instead of the channels, which can lead to larger MNPs in the absence of a limiting confined space. Different interesting examples have been reported so far using this approach, such as the preparation of Pd nanoclusters of ~1.8 nm by Latroche,[350] Pt nanoclusters with a mean diameter of 1.2 nm by Huang,[351] Pd nanoclusters with a mean size of ~2 nm by Chen *et al.*,[352] mixed Au/Ag nanoclusters of ~2 nm and Pd nanoclusters with an average size of 1.5 nm by Liu *et al.*,[326,353] Pd nanoclusters with sizes varying in the range 0.8–1.1 nm by Coupry [354], Pd^0 and Pt^0 single atoms by Jiang[355] and ultra-small Pd nanoclusters (1.2 nm) by Li and coworkers.[356] Most of these nanoclusters showed exceptional activity different reactions, such as hydrogenations, alcohol oxidations, C–H activations or acetal formations, also with a high degree of recyclability. However, even if different techniques for the characterization of these tiny clusters have been used, such as TEM (transmission electron microscopy) microscopies, EXAFS (extended X-ray absorption fine structure) or X-ray photoelectron spectroscopy, the definitive characterization tool at this atomic level – that of SC-XRD[222,357] – could not be used with all these hybrid materials due to the poor resistance of their crystallinity to PSMs.

This situation changed in 2017 with the publication of the gram-scale preparation of tetranuclear ligand-free Pd_4 clusters within the channels of an anionic oxamato-based MOF.[56] The high levels of robustness and crystallinity of the MOF allowed for the crystal structure to be resolved, even after three consecutive post-synthetic processes that included transmetallation of the framework, insertion of the Pd^{2+} cations within the channels of the MOF and the final reduction with $NaBH_4$ to form the ligand-free $[Pd_4]^{2+}$ cations, supported on the walls of the MOF (Figures 9.26a–d). The MOF plays a key role in the stabilization, formation and characterization of these MNCs. The resulting heterogeneous solid catalyst outperformed the "state-of-the-art" catalysts in different carbene-mediated reactions of diazoacetates, showing high yields (>90%) and turnover numbers close to 100 000. Finally, the MOF-supported Pd_4 clusters showed great recyclability over at least 20 catalytic cycles. The same strategy was followed by the same authors for the preparation of Pt single atoms[358] and

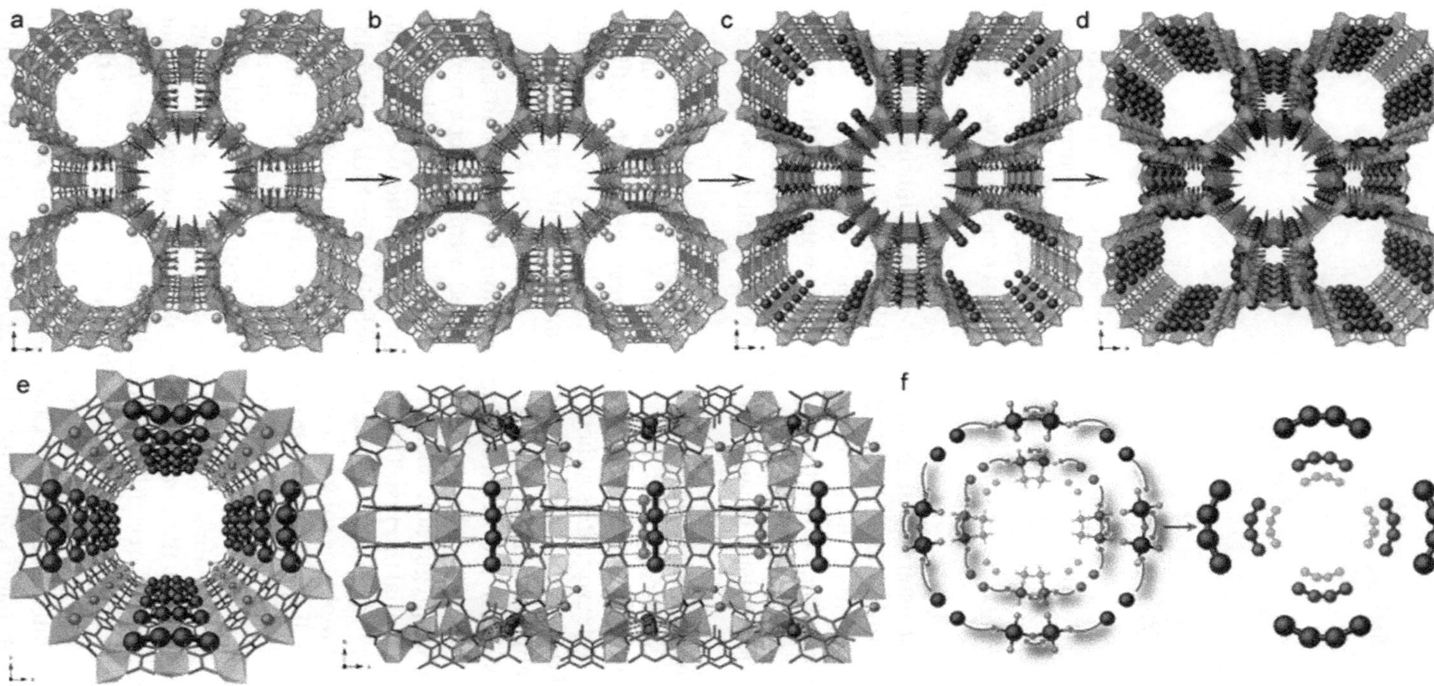

Figure 9.26 Three-step synthesis of $[Pd_4]_{0.5}$@$Na_3(Ni^{II}_4[Cu^{II}_2\{Me_3mpba\}_2]_3)$ consisting of a transmetallation of the starting $(Cu_6Mg_4)Mg_2$ MOF (a) to give $(Cu_6Ni_4)Ni_2$ (b), the exchange of the Ni^{II} cations of the pores by $(Pd^{II}[NH_3]_4)_2$ ones yielding $(Cu_6Mg_4)Pd_2$ (c) and the final reduction process affording $[Pd_4]_{0.5}$@$Na_3(Ni^{II}_4[Cu^{II}_2\{Me_3mpba\}_2]_3)$. (e, f) Perspective views of a channel of the MOF containing the $[Pd_4]^{2+}$ clusters and a proposed formation mechanism for them. Adapted from Ref. 56 with permission from Springer Nature, Copyright 2018.

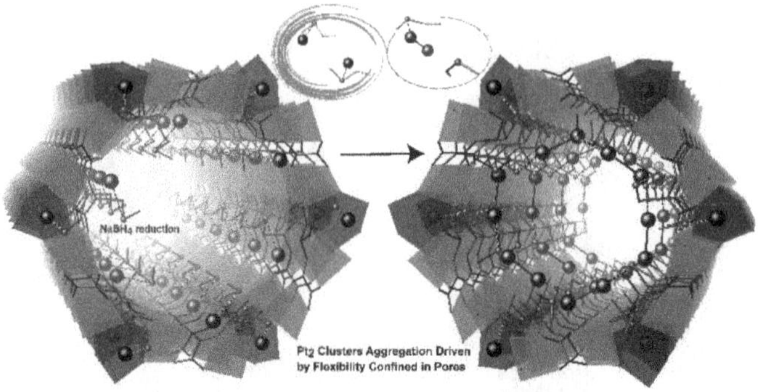

Figure 9.27 Detailed comparison of the crystal structures of a single pore of the MOFs after the Pt^{2+} inclusion (left) and after the reduction process (right), highlighting the proposed mechanism for generating the Pt_2 clusters. Adapted from Ref. 358 with permission from John Wiley and Sons, Copyright © 2018 Wiley-VCH Verlag GmbH & Co. KGaA, Weinheim.

dinuclear Pt nanoclusters[57] within the same MOF and another MOF derived from the natural amino acid L-methionine. These species showed great catalytic activity in important industrial reactions such as cyanide production, ethylene hydrogenation, CO_2 methanation and the water gas shift reaction. In addition, the resolution of the crystal structures revealed, to a certain extent, the formation mechanisms of such small SMNCs (small metal nanoclusters) (Figure 9.27).

Overall, the above-described wetness impregnation method provides the vast majority of MOF-supported MNCs reported so far. The soft conditions used with this method such as the continuous presence of a solvent filling the pores together with exquisite synthetic control, allowing for the retention of the integrity of the crystals, have yielded the only examples reported showing the crystal structure of such tiny MNCs.[56,57,358]

As an extension of the wetness impregnation method, and aiming at improving the ratio of metal atoms incorporated within the channels of the MOF during the impregnation phase, the related improved method of DSM has aroused substantial interest. The strategy relies on the use of a mixture of two solvents, one hydrophilic and one hydrophobic. The MOF is initially suspended in the hydrophobic solvent, and then a solution of the metal precursor dissolved in a small volume of the hydrophilic solvent is added dropwise to the suspension. Then, the metal entities are absorbed totally within the MOF hydrophilic channels by capillarity and the reduction process can take place. This improvement also aims at preventing the formation of larger particles on the surfaces of the MOFs, and a few examples have already been reported. In particular, Qu *et al.*[307] recently reported the immobilization of ultra-small Pt nanoclusters, with an average size of 1.8 nm, within the pores of the MIL-101 MOF by using a mixture of hexane and water as solvents. The formed compounds exhibited outstanding catalytic properties in ammonia borane thermal dehydrogenation, ammonia borane hydrolysis and CO oxidation without aggregation of Pt nanoclusters after catalysis, and they also provided the possibility of reusing the catalysts. Following the same strategy, the same group reported small, ligand-free, heterometallic AuNi[322] and PdAg[359] nanoclusters with mean sizes of ~2 nm in both cases. These heterometallic metal clusters showed great activity for H_2 generation from the catalytic hydrolysis of ammonia borane,[322] and the authors also explored the possibility of carrying out catalytic cascade reactions.[359] In general, these three examples reported by the group of Qu show exceptional results in the preparation of ultra-small MNCs using the DSM. However, more examples are required to validate the strategy. In addition, the use of a hydrophobic solvent to suspend the MOF can damage the crystallinity of the samples, preventing the use of SC-XRD as a powerful characterization tool.

Overall, moderate progress has been achieved over the past 10 years in the synthesis of well-defined naked MNCs. In this context, recent works suggest that MOFs may play a fundamental role in the gram-scale preparation, stabilization and characterization of these unique ultra-small species as a consequence of their unique functional confined state, which, in addition, can be tuned, offering the perfect environment to host, retain and/or synthetize such small moieties. Moreover, important steps have been taken very recently in the characterization at the atomic level of MNCs by means of SC-XRD. As expected, the resulting MNCs exhibit outstanding catalytic activities, paving the way for the real application of these species.

9.2.4.3 Enzymes

Enzymes are the biomolecules involved in almost all chemical reactions in living beings. In a simplified manner, enzymes consist of: (1) an active centre, where the catalytic events take place; and (2) a well-defined external coordination sphere that controls the molecular recognition process of substrates and enables the liberation of products. The efficiency, elegancy and simplicity with which enzymes catalyse the chemical reactions that occur in living beings have been an almost inexhaustible source of inspiration for researchers working in the field of bioenzymatic catalysis, whose eventual goal is to implement artificial systems with the same high degree of sophistication and functionality as enzymes.

The outstanding physicochemical properties of MOFs have enabled their use in obtaining a new class of hybrid biocomposite materials of enzyme@MOFs,[215,236–240,272,360–363] where the enzymes are attached to the external surface of MOFs,[364–372] incorporated into the mesoporous channels of MOFs[178,373–383] or encapsulated.[60,384–394] Notably, there is a considerable amount of work on MOFs able to mimic enzymatic processes. However, this is beyond the scope of this chapter, and we refer the reader to the excellent published reviews.

Adsorption[368–372] and covalent grafting[364–367] on the MOF surface have been exploited in order to obtain biocomposites made up of enzymes and MOFs. The surface adsorption relies on weak non-covalent interactions, such as hydrogen bonding and electrostatic attraction, or partial infiltration of the biomolecules by functionalization of enzymes with different molecules able to enter and interact with MOF channels. Using this approach, for example, *Bacillus subtilis* (BSL2) modified with a surfactant in order to improve interactions with the MOF external surface has been successfully supported on hierarchically porous HKUST-1(Cu) – obtained using a template-free strategy, where nanosized particles of HKUST-1(Cu) were packed to form mesopores (see Figure 9.28).[372] Despite the weakness of these interactions, the amount of BSL2 immobilized after 10 catalytic cycles with respect to the starting material is ~90%, and its catalytic activity – on the esterification reaction of lauric acid and benzyl alcohol – is nearly the same, which speaks of the reliability of this approach. Covalent grafting takes advantage of the carboxylic and amino groups presents in the organic linkers of MOFs to attach, in a more controlled manner, enzymes to MOF surfaces. In this respect, an amino-functionalized UiO-66(Zr) has been used to covalently anchor soybean epoxide hydrolase

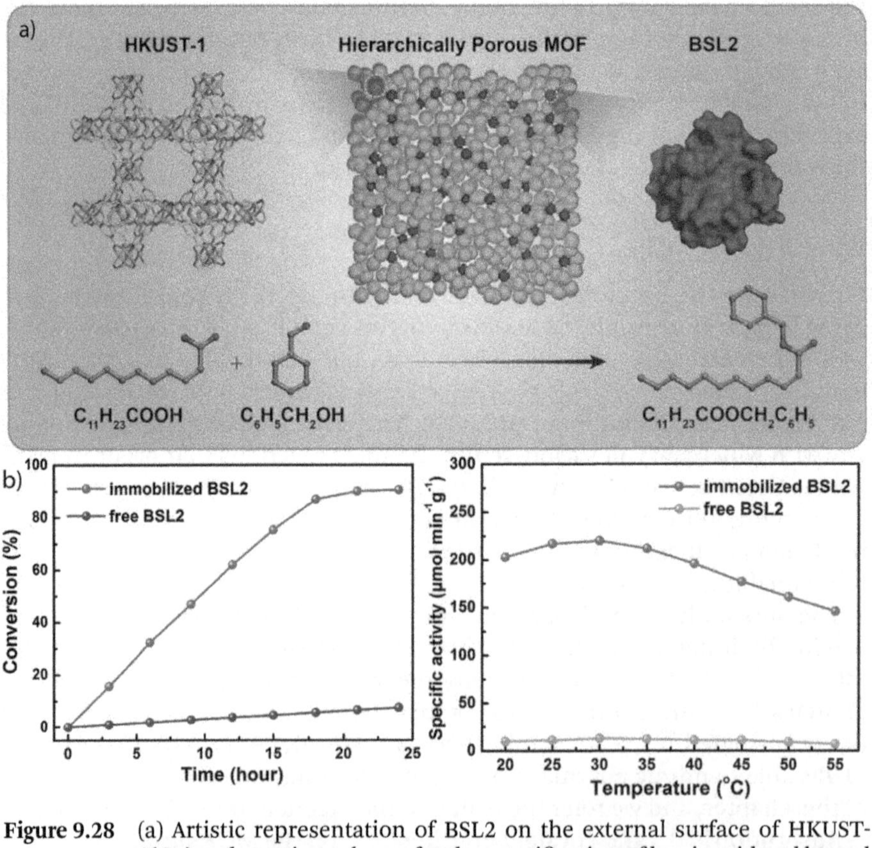

Figure 9.28 (a) Artistic representation of BSL2 on the external surface of HKUST-1(Cu) and reaction scheme for the esterification of lauric acid and benzyl alcohol. (b) Comparative representations of the conversion and specific activity of free BSL2 and immobilized BSL2. Adapted from Ref. 372 with permission from the Royal Society of Chemistry.

(SEH) enzyme.[367] The resulting biomaterial exhibits better pH and temperature stability and tolerance to organic solvents compared to free SEH. Interestingly, SEH@UIO-66(Zr)-NH_2 also showed enhanced enzyme–substrate affinity and higher catalytic efficiency compared to free SEH in the stereospecific epoxide ring opening of 1,2-epoxyoctane to obtain (*R*)-1,2-octanediol in water. The advantaged of this approach is that it does not require mesoporous MOFs. However, this is at the same time one of its weakness, because it does not allow one to take advantage of the porous nature of MOFs.

The great design possibilities of MOFs have allowed for the development of synthetic strategies to obtain stable mesoporous MOFs with appealing physicochemical features to incorporate enzymes post-synthetically.[178,373–383] This approach takes advantage of the high pore volume and void space of MOFs to enable the physical adsorption of enzymes, which considerably improve their stability and reduce the risk of leaching under working catalytic conditions. In addition, pore incorporation allows the size, shape and functionality of MOF channels to be used in order to select the substrates that are able to come into contact with enzymes, which cannot be achieved with surface-immobilized enzymes. In this context, NU-1 000, a mesoporous MOF that has hexagonal and triangular channels with a diameter of 3.1 nm and an edge length of 1.5 nm, respectively, with windows connecting the two channels, has been selected for the post-synthetic incorporation of *Fusarium solani pisi* cutinase.[178] In addition to increased stability of cutinase after incorporation within NU-1 000 channels, in this report the authors explored the influence of having hierarchically mesoporous MOFs with different interconnected channels on the accessibility and activity of incorporated enzymes by comparison with PCN-600[379] – containing only mesoporous hexagonal channels that can communicate *via* channel windows – and CYCU-3[378] – with a hierarchically micro-mesoporous architecture similar to NU-1 000, but with very limited connections between neighbouring channels. From this, the authors have found that NU-1 000 shows the best balance among the compared MOFs between loading capacity, enzymes in channels and accessibility to enzymes. This was attributed to the combination in NU-1 000 of the hierarchical pore structure and the presence of windows between pores allowing for substrate diffusion, and consequently achieving higher catalytic activity (see Figure 9.29).

The encapsulation of enzymes in MOFs has been mainly limited to ZIFs,[60,384,387–393] due to the very mild synthetic conditions needed to obtain them. Different approaches have been used, such as the use of soft and hard templates or the one-pot embedding of enzyme@MOFs. The advantage of this methodology is that if allows enzymes to be stabilized within MOFs, in which the pore windows are smaller than the size of the guest enzyme. This, in turn, offers the advantage of preventing leakage under working catalytic conditions. In addition, it has been reported that the MOF shells play a protective role, conferring the encapsulated enzymes with higher structural and functional stability under high temperatures, in the presence of organic solvents or even in the presence of trypsin, a peptidase enzyme that hydrolyses

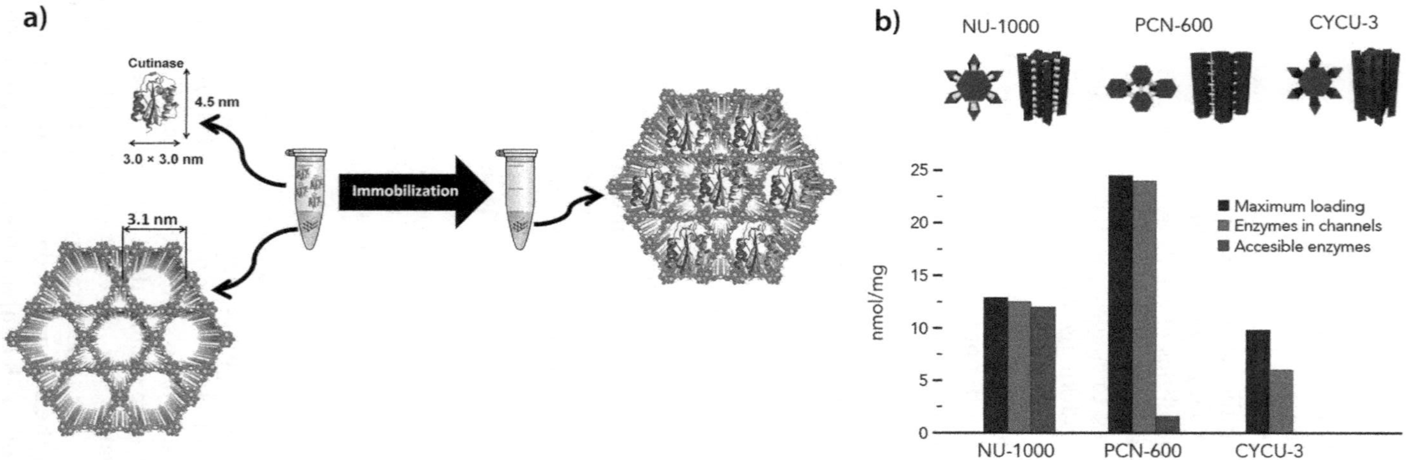

Figure 9.29 (a) Schematic representation of the incorporation of cutinase in the channels of NU-1 000. (b) Comparative graph of the maximum loading, enzymes in channels and accessible enzymes in NU-1 000, PCN-600 and CYCU-3. Reproduced from Ref. 379 with permission from Elsevier, Copyright 2016.

proteins in the digestive systems of many vertebrates. Of these different approaches, biomimetic mineralization has been the one that has attracted the most attention due to the simplicity of it being a one-pot methodology. Although more work still needs to be done, especially dealing with the use of different MOFs and in order to fully understand this biomineralization process, remarkable investigations have been performed in recent years. In this context, the encapsulation of two (glucose oxidase and horseradish peroxidase) or three (β-galactosidase, glucose oxidase and horseradish peroxidase) enzymes or enzyme/cofactors (alcohol deshydrogenase, NAD+ polymer and lactate dehydrogenase) components within ZIF-8(Zn) has recently been reported.[391] Interestingly, the incorporation of different enzymes in ZIF-8(Zn) led to an increase in catalytic cascade reaction studies with regards to the bulk mixture of the catalyst in solution. In addition, it has been observed that the enzyme/cofactor@ZIF-8(Zn) enables biocatalytic cascade reactions, such as the reduction of pyruvic acid to lactic acid by ethanol, to be realized.

References

1. G. Férey, *Chem. Soc. Rev.*, 2008, **37**, 191–214.
2. H. Furukawa, K. E. Cordova, M. O'Keeffe and O. M. Yaghi, *Science*, 2010, **9**, 1230444.
3. A. Czaja, E. Leung, N. Trukhan and U. Müller, in *Metal–Organic Frameworks: Applications from Catalysis to Gas Storage*, Wiley-VCH, Los Angeles, CA, 2011, pp. 337–352.
4. S. Chavan, F. Bonino, J. G. Vitillo, E. Groppo, C. Lamberti, P. D. C. Dietzel, A. Zecchina and S. Bordiga, *Phys. Chem. Chem. Phys.*, 2009, **11**, 9811–9822.
5. M. Tagliabue, C. Rizzo, R. Millini, P. D. C. Dietzel, R. Blom and S. Zanardi, *J. Porous Mater.*, 2011, **18**, 289–296.
6. S. Liu, J. Liu, X. Hou, T. Xu, J. Tong, J. Zhang, B. Ye and B. Liu, *Langmuir*, 2018, **34**, 3654–3660.
7. Y. Pan, Y. Liu, G. Zeng, L. Zhao and Z. Lai, *Chem. Commun.*, 2011, **47**, 2071–2073.
8. M. E. Casco, Y. Q. Cheng, L. L. Daemen, D. Fairen-Jimenez, E. V. Ramos-Fernández, A. J. Ramirez-Cuesta and J. Silvestre-Albero, *Chem. Commun.*, 2016, **52**, 3639–3642.
9. L. Hu, L. Chen, Y. Fang, A. Wang, C. Chen and Z. Yan, *Microporous Mesoporous Mater.*, 2018, **268**, 207–215.
10. M. Goesten, E. Stavitski, E. A. Pidko, C. Gücüyener, B. Boshuizen, S. N. Ehrlich, E. J. M. Hensen, F. Kapteijn and J. Gascon, *Chem. Eur. J.*, 2013, **19**, 7809–7816.
11. I. Stassen, N. Campagnol, J. Fransaer, P. Vereecken, D. De Vos and R. Ameloot, *CrystEngComm*, 2013, **15**, 9308–9311.
12. E. V. Ramos-Fernandez, A. Grau-Atienza, D. Farrusseng and S. Aguado, *J. Mater. Chem. A*, 2018, **6**, 5598–5602.
13. L. Chen, B. Duan, Q. Luo, Z. Gu, J. Liu and C. Duan, *Catal. Sci. Technol.*, 2016, **6**, 1616–1620.

14. A. F. Gross, E. Sherman and J. J. Vajo, *Dalton Trans.*, 2012, **41**, 5458–5460.
15. J. L. Jordá, J. Silvestre-Albero, M. E. Casco, E. V. Ramos-Fernández, F. Rodríguez-Reinoso, F. Fauth, S. Rudić, F. Rey, M. Martínez-Escandell, J. L. Jordá, S. Rudić, F. Fauth, M. Martínez-Escandell, F. Rodríguez-Reinoso, E. V. Ramos-Fernández and J. Silvestre-Albero, *Chem. Sci.*, 2016, 7, 3658–3666.
16. T. Grancha, M. Mon, J. Ferrando-Soria, J. Gascon, B. Seoane, E. V. Ramos-Fernandez, D. Armentano, E. Pardo, T. Grancha, M. Mon, J. Ferrando-Soria, D. Armentano, B. Seoane, E. Pardo and J. Gascon, *J. Mater. Chem. A*, 2017, **5**, 11032–11039.
17. M. Mon, X. Qu, J. J. Ferrando-Soria, I. Pellicer-Carreño, A. Sepúlveda-Escribano, E. V. Ramos-Fernandez, J. C. Jansen, D. Armentano, E. Pardo, I. Pellicer-Carreno, A. Sepulveda-Escribano, E. V. Ramos-Fernandez, J. C. Jansen, D. Armentano, E. Pardo, I. Pellicer-Carreño, A. Sepúlveda-Escribano, E. V. Ramos-Fernandez, J. C. Jansen, D. Armentano and E. Pardo, *J. Mater. Chem. A*, 2017, **5**, 20120–20125.
18. M. A. Rivero-Crespo, M. Mon, J. Ferrando-Soria, C. W. Lopes, M. Boronat, A. Leyva-Pérez, A. Corma, J. C. Hernández-Garrido, M. López-Haro, J. J. Calvino, E. V Ramos-Fernandez, D. Armentano and E. Pardo, *Angew. Chem., Int. Ed.*, 2018.
19. H. Li, M. Eddaoudi, M. O'Keeffe and O. M. Yaghi, *Nature*, 1999, **402**, 276–279.
20. M. O'Keeffe, M. Eddaoudi, H. Li, T. Reineke and O. M. Yaghi, *J. Solid State Chem.*, 2000, **152**, 3–20.
21. J. W. Yoon, H. Chang, S.-J. Lee, Y. K. Hwang, D.-Y. Hong, S.-K. Lee, J. S. Lee, S. Jang, T.-U. Yoon, K. Kwac, Y. Jung, R. S. Pillai, F. Faucher, A. Vimont, M. Daturi, G. Férey, C. Serre, G. Maurin, Y.-S. Bae and J.-S. Chang, *Nat. Mater.*, 2017, **16**, 526–531.
22. H. Deng, S. Grunder, K. E. Cordova, C. Valente, H. Furukawa, M. Hmadeh, F. Gándara, A. C. Whalley, Z. Liu, S. Asahina, H. Kazumori, M. O'Keeffe, O. Terasaki, J. F. Stoddart and O. M. Yaghi, *Science*, 2012, **336**, 1018–1023.
23. H. Furukawa, K. E. Cordova, M. O'Keeffe and O. M. Yaghi, *Science*, 2013, **341**, 1230444–1230444.
24. G. Férey, *J. Solid State Chem.*, 2000, **152**, 37–48.
25. G. Férey, *Science*, 2001, **291**, 994–995.
26. O. M. Yaghi, M. O'Keeffe, N. W. Ockwig, H. K. Chae, M. Eddaoudi and J. Kim, *Nature*, 2003, **423**, 705–714.
27. G. Férey, C. Mellot-Draznieks, C. Serre, F. Millange, J. Dutour, S. Surblé and I. Margiolaki, *Science*, 2005, **309**, 2040 LP – 2042.
28. S. Bhattacharjee, C. Chen and W.-S. Ahn, *RSC Adv.*, 2014, **4**, 52500–52525.
29. S. M. J. J. Rogge, A. Bavykina, J. Hajek, H. Garcia, A. I. Olivos-Suarez, A. Sepúlveda-Escribano, A. Vimont, G. Clet, P. Bazin, F. Kapteijn, M. Daturi, E. V. Ramos-Fernandez, F. X. I. Llabrés Xamena, V. Van Speybroeck, J. Gascon, F. X. Llabrés i Xamena, V. Van Speybroeck and J. Gascon, *Chem. Soc. Rev.*, 2017, **46**, 3134–3184.
30. B. F. Hoskins and R. Robson, *J. Am. Chem. Soc.*, 1989, **111**, 5962–5964.
31. S. R. Batten and R. Robson, *Angew. Chem., Int. Ed.*, 1998, **37**, 1460–1494.

32. F. Carraro, K. Chapman, Z. Chen, M. Dincă, T. Easun, M. Eddaoudi, O. Farha, R. Forgan, L. Gagliardi, F. Haase, D. Harris, S. Kitagawa, J. Knichal, C. Lamberti, J.-S. M. Lee, K. Leus, J. Li, W. Lin, G. Lloyd, J. R. Long, C. Lu, S. Ma, L. McHugh, J. P. H. Perez, M. Ranocchiari, N. Rosi, M. Rosseinsky, M. R. Ryder, V. Ting, M. Van Der Veen, P. Van Der Voort, D. Volkmer, A. Walsh, D. Woods and O. M. Yaghi, *Faraday Discuss.*, 2017, **201**, 369–394.
33. H. Furukawa, N. Ko, Y. B. Go, N. Aratani, S. B. Choi, E. Choi, A. O. Yazaydin, R. Q. Snurr, M. O'Keeffe, J. Kim and O. M. Yaghi, *Science*, 2010, **329**, 424–428.
34. S. Kitagawa, R. Kitaura and S.-I. S. Noro, *Angew. Chem., Int. Ed.*, 2004, **43**, 2334–2375.
35. H.-C. J. Zhou and S. Kitagawa, *Chem. Soc. Rev.*, 2014, **43**, 5415–5418.
36. S. Horike, S. Shimomura and S. Kitagawa, *Nat. Chem.*, 2009, **1**, 695–704.
37. S. Kitagawa and R. Matsuda, *Coord. Chem. Rev.*, 2007, **251**, 2490–2509.
38. S. Kitagawa, S. Noro and T. Nakamura, *Chem. Commun.*, 2006, 701–707.
39. J. E. Bachman, Z. P. Smith, T. Li, T. Xu and J. R. Long, *Nat. Mater.*, 2016, **15**, 845–849.
40. M. E. Casco, J. Fernández-Catalá, Y. Cheng, L. Daemen, A. J. Ramirez-Cuesta, C. Cuadrado-Collados, J. Silvestre-Albero and E. V. Ramos-Fernandez, *ChemistrySelect*, 2017, **2**, 2750–2753.
41. R. F. P. M. Moreira, T. L. P. Dantas, F. M. T. Luna, I. J. Silva Jr., D. C. S. De Azevedo, C. A. Grande and A. E. Rodrigues, in *27th Annual International Pittsburgh Coal Conference 2010, PCC 2010*. Department of Chemical and Food Engineering, Federal University of Santa Catarina, Campus Universitário – Trindade, Florianópolis, 2010, vol. 3, pp. 2332–2343.
42. C. A. Grande, R. P. P. L. Ribeiro and A. E. Rodrigues, *ChemSusChem*, 2010, **3**, 892–898.
43. C. Wang, D. Liu and W. Lin, *J. Am. Chem. Soc.*, 2013, **135**, 13222–13234.
44. M. S. Hosseini, S. Zeinali and M. H. Sheikhi, *Sens. Actuators B Chem.*, 2016, **230**, 9–16.
45. A. Dhakshinamoorthy, Z. Li and H. Garcia, *Chem. Soc. Rev.*, 2018, **47**, 8134–8172.
46. F. Vermoortele, R. Ameloot, L. Alaerts, R. Matthessen, B. Carlier, E. V. R. R. Fernandez, J. Gascon, F. Kapteijn and D. E. De Vos, *J. Mater. Chem.*, 2012, **22**, 10313–10321.
47. M. Lammert, H. Reinsch, C. A. Murray, M. T. Wharmby, H. Terraschke and N. Stock, *Dalton Trans.*, 2016, **45**, 18822–18826.
48. Y. Yang, R. Lin, L. Ge, L. Hou, P. Bernhardt, T. E. Rufford, S. Wang, V. Rudolph, Y. Wang and Z. Zhu, *Dalton Trans.*, 2015, **44**, 8190–8197.
49. P. Serra-Crespo, E. V. Ramos-Fernandez, J. Gascon and F. Kapteijn, *Chem. Mater.*, 2011, **23**, 2565–2572.
50. P. Serra-Crespo, E. Gobechiya, E. V. Ramos-Fernandez, J. Juan-Alcañiz, A. Martinez-Joaristi, E. Stavitski, C. E. A. Kirschhock, J. A. Martens, F. Kapteijn, J. Gascon, A. Martinez-Joaristi, E. V. Ramos-Fernandez, P. Serra-Crespo, J. Juan-Alcañiz, E. Stavitski, F. Kapteijn, C. E. A. Kirschhock and J. Gascon, *Langmuir*, 2012, **28**, 12916–12922.

51. J. Juan-Alcañiz, R. Gielisse, A. B. Lago, E. V. Ramos-Fernandez, P. Serra-Crespo, T. Devic, N. Guillou, C. Serre, F. Kapteijn and J. Gascon, *Catal. Sci. Technol.*, 2013, **3**, 2311–2318.
52. J. Juan-Alcañiz, R. Gielisse, A. B. Lago, E. V. Ramos-Fernandez, P. Serra-Crespo, T. Devic, N. Guillou, C. Serre, F. Kapteijn and J. Gascon, *Catal. Sci. Technol.*, 2013, **3**, 2311.
53. R. Banerjee, A. Phan, B. Wang, C. Knobler, H. Furukawa, M. O'Keeffe and O. M. Yaghi, *Science*, 2008, **319**, 939 LP – 943.
54. M. H. Rosnes, F. S. Nesse, M. Opitz and P. D. C. Dietzel, *Microporous Mesoporous Mater.*, 2019, **275**, 207–213.
55. A. L. W. Demuynck, M. G. Goesten, E. V. Ramos-Fernandez, M. Dusselier, J. Vanderleyden, F. Kapteijn, J. Gascon and B. F. Sels, *ChemCatChem*, 2014, **6**, 2211–2214.
56. F. R. Fortea-Pérez, M. Mon, J. Ferrando-Soria, M. Boronat, A. Leyva-Pérez, A. Corma, J. M. Herrera, D. Osadchii, J. Gascon, D. Armentano and E. Pardo, *Nat. Mater.*, 2017, **16**, 760–766.
57. M. Mon, M. A. Rivero-Crespo, J. Ferrando-Soria, A. Vidal-Moya, M. Boronat, A. Leyva-Pérez, A. Corma, J. C. Hernández-Garrido, M. López-Haro, J. J. Calvino, G. Ragazzon, A. Credi, D. Armentano and E. Pardo, *Angew. Chem., Int. Ed.*, 2018, **57**, 6186–6191.
58. M. Tejeda-Serrano, M. Mon, B. Ross, F. Gonell, J. Ferrando-Soria, A. Corma, A. Leyva-Pérez, D. Armentano and E. Pardo, *J. Am. Chem. Soc.*, 2018, **140**, 8827–8832.
59. M. Mon, R. Adam, J. Ferrando–Soria, A. Corma, D. Armentano, E. Pardo, A. Leyva-Pérez, J. Ferrando-Soria, A. Corma, D. Armentano, E. Pardo and A. Leyva-Pérez, *ACS Catal.*, 2018, **8**, 10401–10406.
60. L. Wang, W. Zhi, J. Wan, J. Han, C. Li and Y. Wang, *ACS Sustain. Chem. Eng.*, 2019, **7**, 3339–3348.
61. M. H. Yap, K. L. Fow and G. Z. Chen, *Green Energy Environ.*, 2017, **2**, 218–245.
62. M. Ronda-Lloret, S. Rico-Francés, A. Sepúlveda-Escribano and E. V. Ramos-Fernandez, *Appl. Catal., A*, 2018, **562**, 28–36.
63. P. Horcajada, T. Chalati, C. Serre, B. Gillet, C. Sebrie, T. Baati, J. F. Eubank, D. Heurtaux, P. Clayette, C. Kreuz, J.-S. Chang, Y. K. Hwang, V. Marsaud, P.-N. Bories, L. Cynober, S. Gil, G. Férey, P. Couvreur and R. Gref, *Nat. Mater.*, 2010, **9**, 172–178.
64. J. Kim, S. N. Kim, H. G. Jang, G. Seo and W. S. Ahn, *Appl. Catal., A*, 2013, **453**, 175–180.
65. K. Schlichte, T. Kratzke and S. Kaskel, *Microporous Mesoporous Mater.*, 2004, **73**, 81–88.
66. B. Gómez-Lor, E. Gutiérrez-Puebla, M. Iglesias, M. A. Monge, C. Ruiz-Valero and N. Snejko, *Chem. Mater.*, 2005, **17**, 2568–2573.
67. S. Aguado, J. Canivet and D. Farrusseng, *J. Mater. Chem.*, 2011, **21**, 7582–7588.
68. S. Horike, M. Dincă, K. Tamaki and J. R. Long, *J. Am. Chem. Soc.*, 2008, **130**, 5854–5855.

69. I. Y. Skobelev, A. B. Sorokin, K. A. Kovalenko, V. P. Fedin and O. A. Kholdeeva, *J. Catal.*, 2013, **298**, 61–69.
70. N. V. Maksimchuk, O. V. Zalomaeva, I. Y. Skobelev, K. A. Kovalenko, V. P. Fedin and O. A. Kholdeeva, *Proc. R. Soc. A Math. Phys. Eng. Sci.*, 2012, **468**, 2017–2034.
71. A. Santiago-Portillo, S. Navalón, F. G. Cirujano, F. X. L. i Xamena, M. Alvaro and H. Garcia, *ACS Catal.*, 2015, **5**, 3216–3224.
72. S. S.-Y. Chui, S. M.-F. Lo, J. P. H. Charmant, A. G. Orpen and I. D. Williams, *Science*, 1999, **283**, 1148–1150.
73. A. Corma, M. E. Domine, L. Nemeth and S. Valencia, *J. Am. Chem. Soc.*, 2002, **124**, 3194–3195.
74. M. Fujita, Y. J. Kwon, S. Washizu and K. Ogura, *J. Am. Chem. Soc.*, 1994, **116**, 1151–1152.
75. I. Stassen, N. Burtch, A. Talin, P. Falcaro, M. Allendorf and R. Ameloot, *Chem. Soc. Rev.*, 2017, **46**, 3185–3241.
76. A. Dhakshinamoorthy, M. Alvaro and H. Garcia, *Chem. Commun. (Camb.)*, 2012, **48**, 11275–11288.
77. M. Y. Masoomi, A. Morsali, A. Dhakshinamoorthy and H. Garcia, *Angew. Chem., Int. Ed.*, 2019, **58**, 15188–15205.
78. A. Dhakshinamoorthy and H. Garcia, *Chem. Soc. Rev.*, 2012, **41**, 5262–5284.
79. A. Dhakshinamoorthy and H. Garcia, *Chem. Soc. Rev.*, 2014, **43**, 5750–5765.
80. A. Dhakshinamoorthy, A. M. Asiri and H. Garcia, *Chem. Soc. Rev.*, 2015, **44**, 1922–1947.
81. Z. Wang, G. Chen and K. Ding, *Chem. Rev.*, 2009, **109**, 322–359.
82. J. W. Yoon, Y.-K. Seo, Y. K. Hwang, J.-S. Chang, H. Leclerc, S. Wuttke, P. Bazin, A. Vimont, M. Daturi, E. Bloch, P. L. Llewellyn, C. Serre, P. Horcajada, J.-M. Grenèche, A. E. Rodrigues and G. Férey, *Angew. Chem., Int. Ed.*, 2010, **49**, 5949–5952.
83. O. A. Kholdeeva, I. Y. Skobelev, I. D. Ivanchikova, K. A. Kovalenko, V. P. Fedin and A. B. Sorokin, *Catal. Today*, 2014, **238**, 54–61.
84. L. Mitchell, B. Gonzalez-Santiago, J. P. S. Mowat, M. E. Gunn, P. Williamson, N. Acerbi, M. L. Clarke and P. A. Wright, *Catal. Sci. Technol.*, 2013, **3**, 606–617.
85. Y. Hou, H. Mao and L. Xu, *Nano Res.*, 2017, **10**, 344–353.
86. B. Van De Voorde, M. Boulhout, F. Vermoortele, P. Horcajada, D. Cunha, J. S. Lee, J.-S. Chang, E. Gibson, M. Daturi, J.-C. Lavalley, A. Vimont, I. Beurroies and D. E. De Vos, *J. Am. Chem. Soc.*, 2013, **135**, 9849–9856.
87. A. Lieb, H. Leclerc, T. Devic, C. Serre, I. Margiolaki, F. Mahjoubi, J. S. Lee, A. Vimont, M. Daturi and J.-S. Chang, *Microporous Mesoporous Mater.*, 2012, **157**, 18–23.
88. E. V. Ramos-Fernandez, C. Pieters, B. Van Der Linden, J. Juan-Alcañiz, P. Serra-Crespo, M. W. G. M. Verhoeven, H. Niemantsverdriet, J. Gascon and F. Kapteijn, *J. Catal.*, 2012, **289**, 42–52.
89. D. J. Xiao, J. Oktawiec, P. J. Milner and J. R. Long, *J. Am. Chem. Soc.*, 2016, **138**, 14371–14379.

90. H.-F. Yao, Y. Yang, H. Liu, F.-G. Xi and E.-Q. Gao, *J. Mol. Catal. A Chem.*, 2014, **394**, 57–65.
91. L. Valenzano, B. Civalleri, S. Chavan, S. Bordiga, M. H. Nilsen, S. Jakobsen, K. P. Lillerud and C. Lamberti, *Chem. Mater.*, 2011, **23**, 1700–1718.
92. S. Smolders, C. Atzori, C. Lamberti, A. L. Bugaev, D. E. De Vos, A. Struyf, N. Stock, K. A. Lomachenko, M. B. J. Roeffaers and B. Bueken, *ChemPhysChem*, 2017, **19**, 373–378.
93. F. Vermoortele, B. Bueken, G. Le Bars, B. Van De Voorde, M. Vandichel, K. Houthoofd, A. Vimont, M. Daturi, M. Waroquier, V. Van Speybroeck, C. Kirschhock and D. E. De Vos, *J. Am. Chem. Soc.*, 2013, **135**, 11465–11468.
94. J. C. Serrano-Ruiz, J. Luettich, A. Sepúlveda-Escribano and F. Rodríguez-Reinoso, *J. Catal.*, 2006, **241**, 45–55.
95. S. Rico-Francés, A. Sepúlveda-Escribano and E. V. Ramos-Fernandez, *Int. J. Hydrogen Energy*, 2017, **42**, 29262–29273.
96. S. Rico-Francés, E. O. Jardim, T. A. Wezendonk, F. Kapteijn, J. Gascon, A. Sepúlveda-Escribano and E. V Ramos-Fernandez, *Appl. Catal., B*, 2016, **180**, 169–178.
97. A. M. Ebrahim and T. J. Bandosz, *ACS Appl. Mater. Interfaces*, 2013, **5**, 10565–10573.
98. M. Lammert, C. Glißmann and N. Stock, *Dalton Trans.*, 2017, **46**, 2425–2429.
99. M. Lammert, M. T. Wharmby, S. Smolders, B. Bueken, A. Lieb, K. A. Lomachenko, D. De Vos and N. Stock, *Chem. Commun.*, 2015, **51**, 12578–12581.
100. K. A. Lomachenko, J. Jacobsen, A. L. Bugaev, C. Atzori, F. Bonino, S. Bordiga, N. Stock and C. Lamberti, *J. Am. Chem. Soc.*, 2018, **140**, 17379–17383.
101. S. Smolders, K. A. Lomachenko, B. Bueken, A. Struyf, A. L. Bugaev, C. Atzori, N. Stock, C. Lamberti, M. B. J. Roeffaers and D. E. De Vos, *ChemPhysChem*, 2018, **19**, 373–378.
102. Z. Wang and S. M. Cohen, *Chem. Soc. Rev.*, 2009, **38**, 1315–1329.
103. S. J. D. Smith, B. P. Ladewig, A. J. Hill, C. H. Lau and M. R. Hill, *Sci. Rep.*, 2015, **5**, 7823.
104. J. Tu, X. Zeng, F. Xu, X. Wu, Y. Tian, X. Hou and Z. Long, *Chem. Commun.*, 2017, **53**, 3361–3364.
105. F. Vermoortele, M. Vandichel, B. Van De Voorde, R. Ameloot, M. Waroquier, V. Van Speybroeck and D. E. De Vos, *Angew. Chem., Int. Ed.*, 2012, **51**, 4887–4890.
106. B. Van de Voorde, I. Stassen, B. Bueken, F. Vermoortele, D. De Vos, R. Ameloot, J.-C. Tan and T. D. Bennett, *J. Mater. Chem. A*, 2015, **3**, 1737–1742.
107. D. S. Sholl and R. P. Lively, *J. Phys. Chem. Lett.*, 2015, **6**, 3437–3444.
108. C. Chizallet, S. Lazare, D. Bazer-Bachi, F. Bonnier, V. Lecocq, E. Soyer, A.-A. Quoineaud and N. Bats, *J. Am. Chem. Soc.*, 2010, **132**, 12365–12377.
109. B. S. Barros, O. J. de Lima Neto, A. C. de Oliveira Frós and J. Kulesza, *ChemistrySelect*, 2018, **3**, 7459–7471.

110. C. O. Kappe, *Angew. Chem., Int. Ed.*, 2004, **43**, 6250–6284.
111. B. Liu, Y. He, L. Han, V. Singh, X. Xu, T. Guo, F. Meng, X. Xu, P. York, Z. Liu and J. Zhang, *Cryst. Growth Des.*, 2017, **17**, 1654–1660.
112. A. Huang, L. Wan and J. Caro, *Mater. Res. Bull.*, 2018, **98**, 308–313.
113. R. Vakili, S. Xu, N. Al-Janabi, P. Gorgojo, S. M. Holmes and X. Fan, *Microporous Mesoporous Mater.*, 2018, **260**, 45–53.
114. D. A. Yang, H. Y. Cho, J. Kim, S. T. Yang and W. S. Ahn, *Energy Environ. Sci.*, 2012, **5**, 6465–6473.
115. S. Motoc, F. Manea, A. Iacob, A. Martinez-Joaristi, J. Gascon, A. Pop and J. Schoonman, *Sensors (Switzerland)*, 2016, **16**, 1719.
116. A. Schaate, P. Roy, A. Godt, J. Lippke, F. Waltz, M. Wiebcke and P. Behrens, *Chem. Eur. J.*, 2011, **17**, 6643–6651.
117. C. Avci, J. Ariñez-Soriano, A. Carné-Sánchez, V. Guillerm, C. Carbonell, I. Imaz and D. Maspoch, *Angew. Chem., Int. Ed.*, 2015, **54**, 14417–14421.
118. E. V. Ramos-Fernandez, M. Garcia-Domingos, J. Juan-Alcañiz, J. Gascon and F. Kapteijn, *Appl. Catal., A*, 2011, **391**, 261–267.
119. U. Ravon, M. Savonnet, S. Aguado, M. E. Domine, E. Janneau and D. Farrusseng, *Microporous Mesoporous Mater.*, 2010, **129**, 319–329.
120. J. Juan-Alcañiz, M. G. Goesten, E. V. Ramos-Fernandez, J. Gascon and F. Kapteijn, *New J. Chem.*, 2012, **36**, 977–987.
121. O. Kozachuk, I. Luz, F. X. Llabrés i Xamena, H. Noei, M. Kauer, H. B. Albada, E. D. Bloch, B. Marler, Y. Wang, M. Muhler and R. A. Fischer, *Angew. Chem., Int. Ed.*, 2014, **53**, 7058–7062.
122. M. J. Ingleson, J. P. Barrio, J. Bacsa, C. Dickinson, H. Park and M. J. Rosseinsky, *Chem. Commun.*, 2008, (11), 1287–1289.
123. G. C. Shearer, J. G. Vitillo, S. Bordiga, S. Svelle, U. Olsbye and K. P. Lillerud, *Chem. Mater.*, 2016, **28**, 7190–7193.
124. A. Koutsianos, E. Kazimierska, A. R. Barron, M. Taddei and E. Andreoli, *Dalton Trans.*, 2019, **48**, 3349–3359.
125. E. E. MacIas, P. Ratnasamy and M. A. Carreon, *Catal. Today*, 2012, **198**, 215–218.
126. O. V. Zalomaeva, A. M. Chibiryaev, K. A. Kovalenko, O. A. Kholdeeva, B. S. Balzhinimaev and V. P. Fedin, *J. Catal.*, 2013, **298**, 179–185.
127. B. Mousavi, S. Chaemchuen, B. Moosavi and K. Zhou, 2017, 674–680.
128. Y.-J. Kim and D.-W. Park, *J. Nanosci. Nanotechnol.*, 2013, **13**, 2307–2312.
129. W. Xiang, Z. Sun, Y. Wu, L. He and C. Liu, *Catal. Today*, 2020, **339**, 337–343.
130. A. Schejn, A. Aboulaich, L. Balan, V. Falk, J. Lalevée, G. Medjahdi, L. Aranda, K. Mozet and R. Schneider, *Catal. Sci. Technol.*, 2015, **5**, 1829–1839.
131. D. Liu, G. Li and H. Liu, *Appl. Surf. Sci.*, 2018, **428**, 218–225.
132. N. Sharma, S. S. Dhankhar and C. M. Nagaraja, *Microporous Mesoporous Mater.*, 2019, **280**, 372–378.
133. M. Tu and R. J. Davis, *J. Catal.*, 2001, **199**, 85–91.
134. Z. Zhou, C. He, J. Xiu, L. Yang and C. Duan, *J. Am. Chem Soc.*, 2013, **137**, 15066–15069.

135. F. Kapteijn, E. V. Ramos-Fernandez, J. Gascon, H. van Bekkum, K. B. Sai Sankar Gupta, E. Stavitski, M. G. Goesten, J. Juan-Alcañiz, E. V. Ramos-Fernandez, K. B. Sai Sankar Gupta, E. Stavitski, H. van Bekkum, J. Gascon, F. Kapteijn, E. V. Ramos-Fernandez, J. Gascon, H. van Bekkum, K. B. Sai Sankar Gupta, E. Stavitski, M. G. Goesten and J. Juan-Alcañiz, *J. Catal.*, 2011, **281**, 177–187.
136. E. Andres-Garcia, A. Dikhtiarenko, F. Fauth, J. Silvestre-Albero, E. V. Ramos-Fernández, J. Gascon, A. Corma and F. Kapteijn, *Chem. Eng. J.*, 2019, **360**, 569–576.
137. H. Fei and S. M. Cohen, *Chem. Commun.*, 2014, **50**, 4810–4812.
138. M. Eddaoudi and J. F. Eubank, in *Metal–Organic Frameworks: Design and Application*, John Wiley and Sons, Tampa, FL, 2010, pp. 37–89.
139. S. Biswas, J. Zhang, Z. Li, Y.-Y. Liu, M. Grzywa, L. Sun, D. Volkmer and P. Van Der Voort, *Dalton Trans.*, 2013, **42**, 4730–4737.
140. Y.-X. Zhou, Y.-Z. Chen, Y. Hu, G. Huang, S.-H. Yu and H.-L. Jiang, *Chem. Eur. J.*, 2014, **20**, 14976–14980.
141. M. Köppen, O. Beyer, S. Wuttke, U. Lüning and N. Stock, *Dalton Trans.*, 2017, **46**, 8658–8663.
142. M. North, *Contemp. Org. Synth.*, 1994, **1**, 475–494.
143. A. Verma, D. De, K. Tomar and P. K. Bharadwaj, *Inorg. Chem.*, 2017, **56**, 9765–9771.
144. Z. Hu and D. Zhao, *CrystEngComm*, 2017, **19**, 4066–4081.
145. J. S. Seo, D. Whang, H. Lee, S. I. Jun, J. Oh, Y. J. Jeon and K. Kim, *Nature*, 2000, **404**, 982–986.
146. S. Yuan, J.-S. Qin, J. Li, L. Huang, L. Feng, Y. Fang, C. Lollar, J. Pang, L. Zhang, D. Sun, A. Alsalme, T. Cagin and H.-C. Zhou, *Nat. Commun.*, 2018, **9**, 808.
147. V. Safarifard, S. Rodr, V. Guillerm, I. Imaz, M. Bigdeli, A. A. Tehrani, J. Juanhuix, A. Morsali, M. E. Casco, J. Silvestre-Albero, E. V Ramos-Fernandez and D. Maspoch, 2016, **16**, 6016–6023.
148. S. J. Garibay, Z. Wang and S. M. Cohen, *Inorg. Chem.*, 2010, **49**, 8086–8091.
149. S. Hasegawa, S. Horike, R. Matsuda, S. Furukawa, K. Mochizuki, Y. Kinoshita and S. Kitagawa, *J. Am. Chem. Soc.*, 2007, **129**, 2607–2614.
150. A. Karmakar and A. J. L. Pombeiro, *Coord. Chem. Rev.*, 2019, **395**, 86–129.
151. A. Benyounes, S. Louisia, R. Axet, Z. Mahfoud, M. Kacimi and P. Serp, *Catal. Today*, 2015, **249**, 137–144.
152. A. M. Shultz, O. K. Farha, J. T. Hupp and S. T. Nguyen, *J. Am. Chem. Soc.*, 2009, **131**, 4204–4205.
153. W. Morris, B. Volosskiy, S. Demir, F. Gándara, P. L. McGrier, H. Furukawa, D. Cascio, J. F. Stoddart and O. M. Yaghi, *Inorg. Chem.*, 2012, **51**, 6443–6445.
154. D. Feng, Z.-Y. Gu, J.-R. Li, H.-L. Jiang, Z. Wei and H.-C. Zhou, *Angew. Chem., Int. Ed.*, 2012, **51**, 10307–10310.
155. D. Feng, W.-C. Chung, Z. Wei, Z.-Y. Gu, H.-L. Jiang, Y.-P. Chen, D. J. Darensbourg and H.-C. Zhou, *J. Am. Chem. Soc.*, 2013, **135**, 17105–17110.

156. M. G. Goesten, F. Kapteijn and J. Gascon, *CrystEngComm*, 2013, **15**, 9249–9257.
157. B. Liu, H. Shioyama, T. Akita and Q. Xu, *J. Am. Chem. Soc.*, 2008, **130**, 5390–5391.
158. L. Oar-Arteta, T. Wezendonk, X. Sun, F. Kapteijn and J. Gascon, *Mater. Chem. Front.*, 2017, **1**, 1709–1745.
159. F. Afsahi, H. Vinh-Thang, S. Mikhailenko and S. Kaliaguine, *J. Power Sources*, 2013, **239**, 415–423.
160. Q.-L. Zhu, W. Xia, T. Akita, R. Zou and Q. Xu, *Adv. Mater.*, 2016, **28**, 6391–6398.
161. L. Zhang, C. Feng, S. Gao, Z. Wang and C. Wang, *Catal. Commun.*, 2015, **61**, 21–25.
162. H. Zhao, H. Song, L. Xu and L. Chou, *Appl. Catal., A*, 2013, **456**, 188–196.
163. H. Jing, X. Song, S. Ren, Y. Shi, Y. An, Y. Yang, M. Feng, S. Ma and C. Hao, *Electrochim. Acta*, 2016, **213**, 252–259.
164. W. Xia, J. Zhu, W. Guo, L. An, D. Xia and R. Zou, *J. Mater. Chem. A*, 2014, **2**, 11606–11613.
165. S. Bao, N. Yan, X. Shi, R. Li and Q. Chen, *Appl. Catal., A*, 2014, **487**, 189–194.
166. X. Zhao, R. Zhang, Y. Liu, J. Deng, P. Xu, S. Lv, S. Li, W. Pei, K. Zhang and H. Dai, *New J. Chem.*, 2019, **43**, 12358–12368.
167. Y. Luo, L. A. Estudillo-Wong, L. Cavillo, G. Granozzi and N. Alonso-Vante, *J. Catal.*, 2016, **338**, 135–142.
168. J. M. Zamaro, N. C. Pérez, E. E. Miró, C. Casado, B. Seoane, C. Téllez and J. Coronas, *Chem. Eng. J.*, 2012, **195–196**, 180–187.
169. V. P. Santos, T. a. Wezendonk, J. J. D. Jaén, a. I. Dugulan, M. a. Nasalevich, H.-U. Islam, A. Chojecki, S. Sartipi, X. Sun, A. a. Hakeem, A. C. J. Koeken, M. Ruitenbeek, T. Davidian, G. R. Meima, G. Sankar, F. Kapteijn, M. Makkee and J. Gascon, *Nat. Commun.*, 2015, **6**, 6451.
170. S. Cheng, N. Shang, X. Zhou, C. Feng, S. Gao, C. Wang and Z. Wang, *New J. Chem.*, 2017, **41**, 9857–9865.
171. V. Duplan, M. Hoshino, W. Li, T. Honda and M. Fujita, *Angew. Chem., Int. Ed.*, 2016, **55**, 4919–4923.
172. S. Urban, R. Brkljača, M. Hoshino, S. Lee and M. Fujita, *Angew. Chem., Int. Ed.*, 2016, **55**, 2678–2682.
173. Y. Inokuma, M. Kawano and M. Fujita, *Nat. Chem.*, 2011, **3**, 349–358.
174. Y. Inokuma, T. Arai and M. Fujita, *Nat. Chem.*, 2010, **2**, 780–783.
175. J. Rabone, Y.-F. Yue, S. Y. Chong, K. C. Stylianou, J. Bacsa, D. Bradshaw, G. R. Darling, N. G. Berry, Y. Z. Khimyak, A. Y. Ganin, P. Wiper, J. B. Claridge and M. J. Rosseinsky, *Science*, 2010, **329**, 1053–1057.
176. Y. Matsuda, T. Mitsuhashi, S. Lee, M. Hoshino, T. Mori, M. Okada, H. Zhang, F. Hayashi, M. Fujita and I. Abe, *Angew. Chem., Int. Ed.*, 2016, **55**, 5785–5788.
177. N. Zigon, M. Hoshino, S. Yoshioka, Y. Inokuma and M. Fujita, *Angew. Chem., Int. Ed.*, 2015, **54**, 9033–9037.
178. P. Li, J. A. Modica, A. J. Howarth, E. Vargas L., P. Z. Moghadam, R. Q. Snurr, M. Mrksich, J. T. Hupp and O. K. Farha, *Chem*, 2016, **1**, 154–169.

179. S. Matsuzaki, T. Arai, K. Ikemoto, Y. Inokuma and M. Fujita, *J. Am. Chem. Soc.*, 2014, **136**, 17899–17901.
180. K. Ikemoto, Y. Inokuma, K. Rissanen and M. Fujita, *J. Am. Chem. Soc.*, 2014, **136**, 6892–6895.
181. R. Matsuda, *Nature*, 2014, **509**, 434–435.
182. Y. Inokuma, S. Yoshioka, J. Ariyoshi, T. Arai, Y. Hitora, K. Takada, S. Matsunaga, K. Rissanen and M. Fujita, *Nature*, 2013, **495**, 461–466.
183. J. Yu, Y. Cui, C. Wu, Y. Yang, Z. Wang, M. O'Keeffe, B. Chen and G. Qian, *Angew. Chem., Int. Ed.*, 2012, **51**, 10542–10545.
184. J. Ferrando-Soria, P. Serra-Crespo, M. de Lange, J. Gascon, F. Kapteijn, M. Julve, J. Cano, F. Lloret, J. Pasán, C. Ruiz-Pérez, Y. Journaux and E. Pardo, *J. Am. Chem. Soc.*, 2012, **134**, 15301–4.
185. A. Corma, M. Iglesias, F. X. Llabrés i Xamena and F. Sánchez, *Chem. Eur. J.*, 2010, **16**, 9789–9795.
186. A. M. Fracaroli, H. Furukawa, M. Suzuki, M. Dodd, S. Okajima, F. Gándara, J. a. Reimer and O. M. Yaghi, *J. Am. Chem. Soc.*, 2014, **136**, 8863–8866.
187. M. H. Zeng, Z. Yin, Y. X. Tan, W. X. Zhang, Y. P. He and M. Kurmoo, *J. Am. Chem. Soc.*, 2014, **136**, 4680–4688.
188. T. M. McDonald, J. A. Mason, X. Kong, E. D. Bloch, D. Gygi, A. Dani, V. Crocellà, F. Giordanino, S. O. Odoh, W. S. Drisdell, B. Vlaisavljevich, A. L. Dzubak, R. Poloni, S. K. Schnell, N. Planas, K. Lee, T. Pascal, L. F. Wan, D. Prendergast, J. B. Neaton, B. Smit, J. B. Kortright, L. Gagliardi, S. Bordiga, J. A. Reimer and J. R. Long, *Nature*, 2015, **519**, 303–308.
189. S. M. Cohen, *Chem. Rev.*, 2012, **112**, 970–1 000.
190. C. K. Brozek and M. Dincă, *Chem. Soc. Rev.*, 2014, **43**, 5456–67.
191. C.-Y. Sun, S.-X. Liu, D.-D. Liang, K.-Z. Shao, Y.-H. Ren and Z.-M. Su, *J. Am. Chem. Soc.*, 2009, **131**, 1883–1888.
192. J. Song, Z. Luo, D. K. Britt, H. Furukawa, O. M. Yaghi, K. I. Hardcastle and C. L. Hill, *J. Am. Chem. Soc.*, 2011, **133**, 16839–16846.
193. S. Ahn, S. L. Nauert, C. T. Buru, M. Rimoldi, H. Choi, N. M. Schweitzer, J. T. Hupp, O. K. Farha and J. M. Notestein, *J. Am. Chem. Soc.*, 2018, **140**, 8535–8543.
194. L. Zhang, S. Li, C. J. Gómez-García, H. Ma, C. Zhang, H. Pang and B. Li, *ACS Appl. Mater. Interfaces*, 2018, **10**, 31498–31504.
195. G. Paille, M. Gomez-Mingot, C. Roch-Marchal, B. Lassalle-Kaiser, P. Mialane, M. Fontecave, C. Mellot-Draznieks and A. Dolbecq, *J. Am. Chem. Soc.*, 2018, **140**, 3613–3618.
196. C. Wang, M. Zhou, Y. Ma, H. Tan, Y. Wang and Y. Li, *Chem. Asian J.*, 2018, **13**, 2054–2059.
197. X. Zhong, Y. Lu, F. Luo, Y. Liu, X. Li and S. Liu, *Chem. Eur. J.*, 2018, **24**, 3045–3051.
198. X. Song, D. Hu, X. Yang, H. Zhang, W. Zhang, J. Li, M. Jia and J. Yu, *ACS Sustain. Chem. Eng.*, 2019, 7, 3624–3631.
199. X.-X. Li, D. Zhao and S.-T. Zheng, *Coord. Chem. Rev.*, 2019, **397**, 220–240.
200. P. Yang, W. Zhao, A. Shkurenko, Y. Belmabkhout, M. Eddaoudi, X. Dong, H. N. Alshareef and N. M. Khashab, *J. Am. Chem. Soc.*, 2019, **141**, 1847–1851.

201. M. Stuckart and K. Y. Monakhov, *Chem. Sci.*, 2019, **10**, 4364–4376.
202. L. Chen and Q. Xu, *Matter*, 2019, **1**, 57–89.
203. D.-Y. Du, J.-S. Qin, S.-L. Li, Z.-M. Su and Y.-Q. Lan, *Chem. Soc. Rev.*, 2014, **43**, 4615–4632.
204. M. C. Wasson, C. T. Buru, Z. Chen, T. Islamoglu and O. K. Farha, *Appl. Catal., A*, 2019, **586**, 117214.
205. M. Samaniyan, M. Mirzaei, R. Khajavian, H. Eshtiagh-Hosseini and C. Streb, *ACS Catal.*, 2019, **9**, 10174–10191.
206. J.-S. Qin, D.-Y. Du, W. Guan, X.-J. Bo, Y.-F. Li, L.-P. Guo, Z.-M. Su, Y.-Y. Wang, Y.-Q. Lan and H.-C. Zhou, *J. Am. Chem. Soc.*, 2015, **137**, 7169–7177.
207. W. Salomon, C. Roch-Marchal, P. Mialane, P. Rouschmeyer, C. Serre, M. Haouas, F. Taulelle, S. Yang, L. Ruhlmann and A. Dolbecq, *Chem. Commun.*, 2015, **51**, 2972–2975.
208. Q. Han, B. Qi, W. Ren, C. He, J. Niu and C. Duan, *Nat. Commun.*, 2015, **6**, 1 0007.
209. J.-W. Sun, P.-F. Yan, G.-H. An, J.-Q. Sha, G.-M. Li and G.-Y. Yang, *Sci. Rep.*, 2016, **6**, 25595.
210. X.-J. Kong, Z. Lin, Z.-M. Zhang, T. Zhang and W. Lin, *Angew. Chem., Int. Ed.*, 2016, **55**, 6411–6416.
211. C. T. Buru, P. Li, B. L. Mehdi, A. Dohnalkova, A. E. Platero-Prats, N. D. Browning, K. W. Chapman, J. T. Hupp and O. K. Farha, *Chem. Mater.*, 2017, **29**, 5174–5181.
212. C. T. Buru, A. E. Platero-Prats, D. G. Chica, M. G. Kanatzidis, K. W. Chapman and O. K. Farha, *J. Mater. Chem. A*, 2018, **6**, 7389–7394.
213. J. Juan-Alcañiz, E. V. Ramos-Fernandez, U. Lafont, J. Gascon, F. Kapteijn, J. Gascon, E. V. Ramos-Fernandez, J. Juan-Alcañiz, U. Lafont, E. V. Ramos-Fernandez, U. Lafont, J. Gascon, F. Kapteijn, J. Gascon, E. V. Ramos-Fernandez, J. Juan-Alcañiz and U. Lafont, *J. Catal.*, 2009, **269**, 229–241.
214. G. B. Shul'pin, D. C. Crans, L. M. Martins and K. T. Mahmudov, *Coord. Chem. Rev.*, 2019, **380**, 600.
215. K. Chen and C.-D. Wu, *Coord. Chem. Rev.*, 2019, **378**, 445–465.
216. D. Farrusseng, S. Aguado and C. Pinel, *Angew. Chem., Int. Ed.*, 2009, **48**, 7502–7513.
217. J. Lee, O. K. Farha, J. Roberts, K. A. Scheidt, S. T. Nguyen and J. T. Hupp, *Chem. Soc. Rev.*, 2009, **38**, 1450.
218. A. Corma, H. García and F. X. Llabrés i Xamena, *Chem. Rev.*, 2010, **110**, 4606–4655.
219. P. Valvekens, F. Vermoortele and D. De Vos, *Catal. Sci. Technol.*, 2013, **3**, 1435.
220. J. Gascon, A. Corma, F. Kapteijn and F. X. Llabrés I Xamena, *ACS Catal.*, 2014, **4**, 361–378.
221. F. Gándara and T. D. Bennett, *IUCrJ*, 2014, **1**, 563–570.
222. W. M. Bloch, N. R. Champness and C. J. Doonan, *Angew. Chem., Int. Ed.*, 2015, **54**, 12860–12867.
223. K. Rissanen, *Chem. Soc. Rev.*, 2017, **46**, 2638–2648.

224. R.-W. Huang, Y.-S. Wei, X.-Y. Dong, X.-H. Wu, C.-X. Du, S.-Q. Zang and T. C. W. Mak, *Nat. Chem.*, 2017, **9**, 689–697.
225. A. Burgun, C. J. Coghlan, D. M. Huang, W. Chen, S. Horike, S. Kitagawa, J. F. Alvino, G. F. Metha, C. J. Sumby and C. J. Doonan, *Angew. Chem., Int. Ed.*, 2017, **56**, 8412–8416.
226. M. H. Alkordi, Y. Liu, R. W. Larsen, J. F. Eubank and M. Eddaoudi, *J. Am. Chem. Soc.*, 2008, **130**, 12639–12641.
227. E. Kockrick, T. Lescouet, E. V. Kudrik, A. B. Sorokin and D. Farrusseng, *Chem. Commun.*, 2011, **47**, 1562–1564.
228. O. V. Zalomaeva, K. A. Kovalenko, Y. A. Chesalov, M. S. Mel'gunov, V. I. Zaikovskii, V. V. Kaichev, A. B. Sorokin, O. A. Kholdeeva and V. P. Fedin, *Dalton Trans.*, 2011, **40**, 1441.
229. C. Wang, J.-L. Wang and W. Lin, *J. Am. Chem. Soc.*, 2012, **134**, 19895–19908.
230. C. Wang, Z. Xie, K. E. DeKrafft and W. Lin, *J. Am. Chem. Soc.*, 2011, **133**, 13445–13454.
231. S. Pullen, H. Fei, A. Orthaber, S. M. Cohen and S. Ott, *J. Am. Chem. Soc.*, 2013, **135**, 16997–17003.
232. K. Sasan, Q. Lin, C. Mao and P. Feng, *Chem. Commun.*, 2014, **50**, 10390.
233. D. Balestri, Y. Roux, M. Mattarozzi, C. Mucchino, L. Heux, D. Brazzolotto, V. Artero, C. Duboc, P. Pelagatti, L. Marchiò and M. Gennari, *Inorg. Chem.*, 2017, **56**, 14801–14808.
234. S. Pullen, S. Roy and S. Ott, *Chem. Commun.*, 2017, **53**, 5227–5230.
235. S. Roy, V. Pascanu, S. Pullen, G. González Miera, B. Martín-Matute and S. Ott, *Chem. Commun.*, 2017, **53**, 3257–3260.
236. M. Zhang, Z.-Y. Gu, M. Bosch, Z. Perry and H.-C. Zhou, *Coord. Chem. Rev.*, 2015, **293–294**, 327–356.
237. I. Nath, J. Chakraborty and F. Verpoort, *Chem. Soc. Rev.*, 2016, **45**, 4127–4170.
238. X. Lian, Y. Fang, E. Joseph, Q. Wang, J. Li, S. Banerjee, C. Lollar, X. Wang and H.-C. Zhou, *Chem. Soc. Rev.*, 2017, **46**, 3386–3401.
239. E. Gkaniatsou, C. Sicard, R. Ricoux, J.-P. Mahy, N. Steunou and C. Serre, *Mater. Horizons*, 2017, **4**, 55–63.
240. C. Doonan, R. Riccò, K. Liang, D. Bradshaw and P. Falcaro, *Acc. Chem. Res.*, 2017, **50**, 1423–1432.
241. A. Fateeva, P. A. Chater, C. P. Ireland, A. A. Tahir, Y. Z. Khimyak, P. V. Wiper, J. R. Darwent and M. J. Rosseinsky, *Angew. Chem., Int. Ed.*, 2012, **51**, 7440–7444.
242. A. Grigoropoulos, G. F. S. Whitehead, N. Perret, A. P. Katsoulidis, F. M. Chadwick, R. P. Davies, A. Haynes, L. Brammer, A. S. Weller, J. Xiao and M. J. Rosseinsky, *Chem. Sci.*, 2016, 7, 2037–2050.
243. X. Dai, M. Liu, Z. Li, A. Jin, Y. Ma, X. Huang, H. Sun, H. Wang and X. Zhang, *J. Phys. Chem. C*, 2016, **120**, 12539–12548.
244. A. W. Peters, Z. Li, O. K. Farha and J. T. Hupp, *ACS Appl. Mater. Interfaces*, 2016, 8, 20675–20681.
245. S. A. Burgess, A. Kassie, S. A. Baranowski, K. J. Fritzsching, K. Schmidt-Rohr, C. M. Brown and C. R. Wade, *J. Am. Chem. Soc.*, 2016, **138**, 1780–1783.

246. M. A. Nasalevich, R. Becker, E. V. Ramos-Fernandez, S. Castellanos, S. L. Veber, M. V. Fedin, F. Kapteijn, J. N. H. Reek, J. I. van der Vlugt and J. Gascon, *Energy Environ. Sci.*, 2015, **8**, 364–375.
247. Y. Lee, S. Kim, H. Fei, J. K. Kang and S. M. Cohen, *Chem. Commun.*, 2015, **51**, 16549–16552.
248. F. Carson, E. Martínez-Castro, R. Marcos, G. G. Miera, K. Jansson, X. Zou and B. Martín-Matute, *Chem. Commun.*, 2015, **51**, 10864–10867.
249. M. I. Gonzalez, E. D. Bloch, J. A. Mason, S. J. Teat and J. R. Long, *Inorg. Chem.*, 2015, **54**, 2995–3005.
250. H. Fei, M. D. Sampson, Y. Lee, C. P. Kubiak and S. M. Cohen, *Inorg. Chem.*, 2015, **54**, 6821–6828.
251. H. Fei and S. M. Cohen, *J. Am. Chem. Soc.*, 2015, **137**, 2191–2194.
252. Y.-C. Luo, K.-L. Chu, J.-Y. Shi, D.-J. Wu, X.-D. Wang, M. Mayor and C.-Y. Su, *J. Am. Chem. Soc.*, 2019, **141**, 13057–13065.
253. M. B. Chambers, X. Wang, N. Elgrishi, C. H. Hendon, A. Walsh, J. Bonnefoy, J. Canivet, E. A. Quadrelli, D. Farrusseng, C. Mellot-Draznieks and M. Fontecave, *ChemSusChem*, 2015, **8**, 603–608.
254. C.-C. Hou, T.-T. Li, S. Cao, Y. Chen and W.-F. Fu, *J. Mater. Chem. A*, 2015, **3**, 10386–10394.
255. C.-Y. Sun, W.-P. To, X.-L. Wang, K.-T. Chan, Z.-M. Su and C.-M. Che, *Chem. Sci.*, 2015, **6**, 7105–7111.
256. H. Fei, J. Shin, Y. S. Meng, M. Adelhardt, J. Sutter, K. Meyer and S. M. Cohen, *J. Am. Chem. Soc.*, 2014, **136**, 4965–4973.
257. S. T. Madrahimov, T. A. Atesin, O. Karagiaridi, A. A. Sarjeant, O. K. Farha, J. T. Hupp and S. T. Nguyen, *Cryst. Growth Des.*, 2014, **14**, 6320–6324.
258. D. T. Genna, A. G. Wong-Foy, A. J. Matzger and M. S. Sanford, *J. Am. Chem. Soc.*, 2013, **135**, 10586–10589.
259. X. Zhang, F. X. Llabrés i Xamena and A. Corma, *J. Catal.*, 2009, **265**, 155–160.
260. S. Roy, A. Bhunia, N. Schuth, M. Haumann and S. Ott, *Sustain. Energy Fuels*, 2018, **2**, 1148–1152.
261. X. Wang, F. M. Wisser, J. Canivet, M. Fontecave and C. Mellot-Draznieks, *ChemSusChem*, 2018, **11**, 3315–3322.
262. S. Yang, B. Pattengale, S. Lee and J. Huang, *ACS Energy Lett.*, 2018, **3**, 532–539.
263. A. Bhunia, B. A. Johnson, J. Czapla-Masztafiak, J. Sá and S. Ott, *Chem. Commun.*, 2018, **54**, 7770–7773.
264. S. Lin, A. K. Ravari, J. Zhu, P. M. Usov, M. Cai, S. R. Ahrenholtz, Y. Pushkar and A. J. Morris, *ChemSusChem*, 2018, **11**, 464–471.
265. W.-M. Liao, J.-H. Zhang, Z. Wang, S.-Y. Yin, M. Pan, H.-P. Wang and C.-Y. Su, *J. Mater. Chem. A*, 2018, **6**, 11337–11345.
266. M. Mon, J. Ferrando-Soria, T. Grancha, F. R. Fortea-Pérez, J. Gascon, A. Leyva-Pérez, D. Armentano and E. Pardo, *J. Am. Chem. Soc.*, 2016, **138**, 7864–7867.
267. R. Adam, M. Mon, R. Greco, L. H. G. Kalinke, A. Vidal-Moya, A. Fernandez, R. E. P. Winpenny, A. Doménech-Carbó, A. Leyva-Pérez, D. Armentano, E. Pardo and J. Ferrando-Soria, *J. Am. Chem. Soc.*, 2019, **141**, 10350–10360.

268. J. Perezjuste, I. Pastorizasantos, L. Lizmarzan and P. Mulvaney, *Coord. Chem. Rev.*, 2005, **249**, 1870–1901.
269. M. Grzelczak, J. Pérez-Juste, P. Mulvaney and L. M. Liz-Marzán, *Chem. Soc. Rev.*, 2008, **37**, 1783.
270. L. Xu, W. Ma, L. Wang, C. Xu, H. Kuang and N. A. Kotov, *Chem. Soc. Rev.*, 2013, **42**, 3114.
271. M. Meilikhov, K. Yusenko, D. Esken, S. Turner, G. Van Tendeloo and R. A. Fischer, *Eur. J. Inorg. Chem.*, 2010, **2010**, 3701–3714.
272. J.-S. Qin, S. Yuan, C. Lollar, J. Pang, A. Alsalme and H.-C. Zhou, *Chem. Commun.*, 2018, **54**, 4231–4249.
273. W. Xu, K. B. Thapa, Q. Ju, Z. Fang and W. Huang, *Coord. Chem. Rev.*, 2018, **373**, 199–232.
274. G. Li, S. Zhao, Y. Zhang and Z. Tang, *Adv. Mater.*, 2018, **30**, 1800702.
275. S. Wang, C. M. McGuirk, A. D'Aquino, J. A. Mason and C. A. Mirkin, *Adv. Mater.*, 2018, **30**, 1800202.
276. F. Chen, K. Shen, J. Chen, X. Yang, J. Cui and Y. Li, *ACS Cent. Sci.*, 2019, **5**, 176–185.
277. D. Yang and B. C. Gates, *ACS Catal.*, 2019, **9**, 1779–1798.
278. A. Dhakshinamoorthy and H. Garcia, *Chem. Soc. Rev.*, 2012, **41**, 5262.
279. H. R. Moon, D.-W. Lim and M. P. Suh, *Chem. Soc. Rev.*, 2013, **42**, 1807–1824.
280. L. Chen, H. Chen, R. Luque and Y. Li, *Chem. Sci.*, 2014, **5**, 3708–3714.
281. P. Hu, J. V Morabito and C. Tsung, *ACS Catal.*, 2014, **4**, 4409–4419.
282. Y. Zhang, Y. Zhou, Y. Zhao and C. Liu, *Catal. Today*, 2016, **263**, 61–68.
283. P. Falcaro, R. Ricco, A. Yazdi, I. Imaz, S. Furukawa, D. Maspoch, R. Ameloot, J. D. Evans and C. J. Doonan, *Coord. Chem. Rev.*, 2016, **307**, 237–254.
284. Q. Yang, Q. Xu and H.-L. Jiang, *Chem. Soc. Rev.*, 2017, **46**, 4774–4808.
285. J. Yu, C. Mu, B. Yan, X. Qin, C. Shen, H. Xue and H. Pang, *Mater. Horizons*, 2017, **4**, 557–569.
286. M. H. Hassan, O. El-Basha, R. R. Haikal, A. H. Ibrahim and M. H. Alkordi, *ACS Appl. Mater. Interfaces*, 2018, **10**, 32942–32945.
287. J. Cure, E. Mattson, K. Cocq, H. Assi, S. Jensen, K. Tan, M. Catalano, S. Yuan, H. Wang, L. Feng, P. Zhang, S. Kwon, J.-F. Veyan, Y. Cabrera, G. Zhang, J. Li, M. Kim, H.-C. Zhou, Y. J. Chabal and T. Thonhauser, *J. Mater. Chem. A*, 2019, 7, 17536–17546.
288. B. Li, J.-G. Ma and P. Cheng, *Angew. Chem., Int. Ed.*, 2018, **57**, 6834–6837.
289. J. Hermannsdörfer, M. Friedrich, N. Miyajima, R. Q. Albuquerque, S. Kümmel and R. Kempe, *Angew. Chem., Int. Ed.*, 2012, **51**, 11473–11477.
290. L. Chen, H. Chen and Y. Li, *Chem. Commun.*, 2014, **50**, 14752–14755.
291. G. Li, H. Kobayashi, J. M. Taylor, R. Ikeda, Y. Kubota, K. Kato, M. Takata, T. Yamamoto, S. Toh, S. Matsumura and H. Kitagawa, *Nat. Mater.*, 2014, **13**, 802–806.
292. Z. Zhang, Y. Chen, X. Xu, J. Zhang, G. Xiang, W. He and X. Wang, *Angew. Chem., Int. Ed.*, 2014, **53**, 429–433.
293. M. Zhao, K. Deng, L. He, Y. Liu, G. Li, H. Zhao and Z. Tang, *J. Am. Chem. Soc.*, 2014, **136**, 1738–1741.

294. L. Lin, T. Zhang, H. Liu, J. Qiu and X. Zhang, *Nanoscale*, 2015, 7, 7615–7623.
295. L. Chen, X. Chen, H. Liu, C. Bai and Y. Li, *J. Mater. Chem. A*, 2015, **3**, 15259–15264.
296. W. Zhou, B. Zou, W. Zhang, D. Tian, W. Huang and F. Huo, *Nanoscale*, 2015, 7, 8720–8724.
297. Y.-Z. Chen, Q. Xu, S.-H. Yu and H.-L. Jiang, *Small*, 2015, 11, 71–76.
298. L. Chen, X. Chen, H. Liu and Y. Li, *Small*, 2015, **11**, 2642–2648.
299. X. Xu, Z. Zhang and X. Wang, *Adv. Mater.*, 2015, **27**, 5365–5371.
300. L. Chen, B. Huang, X. Qiu, X. Wang, R. Luque and Y. Li, *Chem. Sci.*, 2016, 7, 228–233.
301. N. Yuan, V. Pascanu, Z. Huang, A. Valiente, N. Heidenreich, S. Leubner, A. K. Inge, J. Gaar, N. Stock, I. Persson, B. Martin-Matute and X. Zou, *J. Am. Chem. Soc.*, 2018, **140**, 8206–8217.
302. W. Zhang, G. Lu, C. Cui, Y. Liu, S. Li, W. Yan, C. Xing, Y. R. Chi, Y. Yang and F. Huo, *Adv. Mater.*, 2014, **26**, 4056–4060.
303. P. Falcaro, A. J. Hill, K. M. Nairn, J. Jasieniak, J. I. Mardel, T. J. Bastow, S. C. Mayo, M. Gimona, D. Gomez, H. J. Whitfield, R. Riccò, A. Patelli, B. Marmiroli, H. Amenitsch, T. Colson, L. Villanova and D. Buso, *Nat. Commun.*, 2011, **2**, 237.
304. K. M. Choi, K. Na, G. A. Somorjai and O. M. Yaghi, *J. Am. Chem. Soc.*, 2015, **137**, 7810–7816.
305. C. Rösler, A. Aijaz, S. Turner, M. Filippousi, A. Shahabi, W. Xia, G. Van Tendeloo, M. Muhler and R. A. Fischer, *Chem. Eur. J.*, 2016, **22**, 3304–3311.
306. K. Sugikawa, Y. Furukawa and K. Sada, *Chem. Mater.*, 2011, **23**, 3132–3134.
307. A. Aijaz, A. Karkamkar, Y. J. Choi, N. Tsumori, E. Rönnebro, T. Autrey, H. Shioyama and Q. Xu, *J. Am. Chem. Soc.*, 2012, **134**, 13926–13929.
308. H. Liu, L. Chang, L. Chen and Y. Li, 2015, **3**, 8028–8033.
309. Y. Jiang, X. Zhang, X. Dai, Q. Sheng, H. Zhuo, J. Yong, Y. Wang, K. Yu, L. Yu, C. Luan, H. Wang, Y. Zhu, X. Duan and P. Che, *Chem. Mater.*, 2017, **29**, 6336–6345.
310. M. Zhao, K. Yuan, Y. Wang, G. Li, J. Guo, L. Gu, W. Hu, H. Zhao and Z. Tang, *Nature*, 2016, **539**, 76–80.
311. P. Ling, J. Lei, L. Jia and H. Ju, *Chem. Commun.*, 2016, **52**, 1226–1229.
312. C. J. Stephenson, J. T. Hupp and O. K. Farha, *Inorg. Chem.*, 2016, **55**, 1361–1363.
313. R. Fang, H. Liu, R. Luque and Y. Li, *Green Chem.*, 2015, **17**, 4183–4188.
314. F. Ke, L. Wang and J. Zhu, *Nanoscale*, 2015, 7, 8321–8325.
315. L. Sun, Y. Yun, H. Sheng, Y. Du, Y. Ding, P. Wu, P. Li and M. Zhu, *J. Mater. Chem. A*, 2018, **6**, 15371–15376.
316. M. Zahmakiran, *Dalton Trans.*, 2012, **41**, 12690.
317. Y. Liu and Z. Tang, *Adv. Mater.*, 2013, **25**, 5819–5825.
318. X. Wang, M. Li, C. Cao, C. Liu, J. Liu, Y. Zhu, S. Zhang and W. Song, *ChemCatChem*, 2016, **8**, 3224–3228.

319. Q. Yang, Q. Xu, S.-H. Yu and H.-L. Jiang, *Angew. Chem., Int. Ed.*, 2016, **55**, 3685–3689.
320. P. Wang, J. Zhao, X. Li, Y. Yang, Q. Yang and C. Li, *Chem. Commun.*, 2013, **49**, 3330.
321. K. Na, K. M. Choi, O. M. Yaghi and G. A. Somorjai, *Nano Lett.*, 2014, **14**, 5979–5983.
322. Q.-L. Zhu, J. Li and Q. Xu, *J. Am. Chem. Soc.*, 2013, **135**, 10210–10213.
323. L. He, Y. Liu, J. Liu, Y. Xiong, J. Zheng, Y. Liu and Z. Tang, *Angew. Chem., Int. Ed.*, 2013, **52**, 3741–3745.
324. W. Zhang, Y. Liu, G. Lu, Y. Wang, S. Li, C. Cui, J. Wu, Z. Xu, D. Tian, W. Huang, J. S. DuCheneu, W. D. Wei, H. Chen, Y. Yang and F. Huo, *Adv. Mater.*, 2015, **27**, 2923–2929.
325. B. Gole, U. Sanyal and P. S. Mukherjee, *Chem. Commun.*, 2015, **51**, 4872–4875.
326. L. Liu, Y. Song, H. Chong, S. Yang, J. Xiang, S. Jin, X. Kang, J. Zhang, H. Yu and M. Zhu, *Nanoscale*, 2016, **8**, 1407–1412.
327. G. Lu, S. Li, Z. Guo, O. K. Farha, B. G. Hauser, X. Qi, Y. Wang, X. Wang, S. Han, X. Liu, J. S. DuChene, H. Zhang, Q. Zhang, X. Chen, J. Ma, S. C. J. Loo, W. D. Wei, Y. Yang, J. T. Hupp and F. Huo, *Nat. Chem.*, 2012, **4**, 310–316.
328. K. Khaletskaya, J. Reboul, M. Meilikhov, M. Nakahama, S. Diring, M. Tsujimoto, S. Isoda, F. Kim, K. Kamei, R. A. Fischer, S. Kitagawa and S. Furukawa, *J. Am. Chem. Soc.*, 2013, **135**, 10998–11005.
329. Y. Zhao, N. Kornienko, Z. Liu, C. Zhu, S. Asahina, T.-R. Kuo, W. Bao, C. Xie, A. Hexemer, O. Terasaki, P. Yang and O. M. Yaghi, *J. Am. Chem. Soc.*, 2015, **137**, 2199–2202.
330. B. Ni and X. Wang, *Chem. Sci.*, 2016, **7**, 3978–3991.
331. L. Shang, S. Dong and G. U. Nienhaus, *Nano Today*, 2011, **6**, 401–418.
332. Z. Luo, K. Zheng and J. Xie, *Chem. Commun.*, 2014, **50**, 5143–5155.
333. G. Chen, I. Roy, C. Yang and P. N. Prasad, *Chem. Rev.*, 2016, **116**, 2826–2885.
334. J. Oliver-Meseguer, J. R. Cabrero-Antonino, I. Dominguez, A. Leyva-Perez and A. Corma, *Science*, 2012, **338**, 1452–1455.
335. J. Oliver-Meseguer, A. Leyva-Pérez, S. I. Al-Resayes and A. Corma, *Chem. Commun.*, 2013, **49**, 7782.
336. A. Corma, P. Concepción, M. Boronat, M. J. Sabater, J. Navas, M. J. Yacaman, E. Larios, A. Posadas, M. A. López-Quintela, D. Buceta, E. Mendoza, G. Guilera and A. Mayoral, *Nat. Chem.*, 2013, **5**, 775–781.
337. S. L. Benjamin, T. Krämer, W. Levason, M. E. Light, S. A. Macgregor and G. Reid, *J. Am. Chem. Soc.*, 2016, **138**, 6964–6967.
338. T. Murahashi, *Science*, 2006, **313**, 1104–1107.
339. L. Liu, U. Díaz, R. Arenal, G. Agostini, P. Concepción and A. Corma, *Nat. Mater.*, 2016, **16**, 132–138.
340. L. Liu, D. N. Zakharov, R. Arenal, P. Concepcion, E. A. Stach and A. Corma, *Nat. Commun.*, 2018, **9**, 574.
341. N. Wang, Q. Sun and J. Yu, *Adv. Mater.*, 2019, **31**, 1803966.

342. H. Liu, L. Chang, C. Bai, L. Chen, R. Luque and Y. Li, *Angew. Chem., Int. Ed.*, 2016, **55**, 5019–5023.
343. S. Ji, Y. Chen, Q. Fu, Y. Chen, J. Dong, W. Chen, Z. Li, Y. Wang, L. Gu, W. He, C. Chen, Q. Peng, Y. Huang, X. Duan, D. Wang, C. Draxl and Y. Li, *J. Am. Chem. Soc.*, 2017, **139**, 9795–9798.
344. J. Zhou, P. Wang, C. Wang, Y. T. Goh, Z. Fang, P. B. Messersmith and H. Duan, *ACS Nano*, 2015, **9**, 6951–6960.
345. S. Hermes, M.-K. Schröter, R. Schmid, L. Khodeir, M. Muhler, A. Tissler, R. W. Fischer and R. A. Fischer, *Angew. Chem., Int. Ed.*, 2005, **44**, 6237–6241.
346. D. Esken, S. Turner, O. I. Lebedev, G. Van Tendeloo and R. A. Fischer, *Chem. Mater.*, 2010, **22**, 6393–6401.
347. Y. K. Park, S. B. Choi, H. J. Nam, D.-Y. Jung, H. C. Ahn, K. Choi, H. Furukawa and J. Kim, *Chem. Commun.*, 2010, **46**, 3086.
348. T. Ishida, M. Nagaoka, T. Akita and M. Haruta, *Chem. Eur. J.*, 2008, **14**, 8456–8460.
349. H.-L. Jiang, Q.-P. Lin, T. Akita, B. Liu, H. Ohashi, H. Oji, T. Honma, T. Takei, M. Haruta and Q. Xu, *Chem. Eur. J.*, 2011, **17**, 78–81.
350. C. Zlotea, R. Campesi, F. Cuevas, E. Leroy, P. Dibandjo, C. Volkringer, T. Loiseau, G. Férey and M. Latroche, *J. Am. Chem. Soc.*, 2010, **132**, 2991–2997.
351. Z. Guo, C. Xiao, R. V. Maligal-Ganesh, L. Zhou, T. W. Goh, X. Li, D. Tesfagaber, A. Thiel and W. Huang, *ACS Catal.*, 2014, **4**, 1340–1348.
352. G.-J. Chen, J.-S. Wang, F.-Z. Jin, M.-Y. Liu, C.-W. Zhao, Y.-A. Li and Y.-B. Dong, *Inorg. Chem.*, 2016, **55**, 3058–3064.
353. Y.-B. Huang, Q. Wang, J. Liang, X. Wang and R. Cao, *J. Am. Chem. Soc.*, 2016, **138**, 10104–10107.
354. D. E. Coupry, J. Butson, P. S. Petkov, M. Saunders, K. O'Donnell, H. Kim, C. Buckley, M. Addicoat, T. Heine and P. Á. Szilágyi, *Chem. Commun.*, 2016, **52**, 5175–5178.
355. X. Fang, Q. Shang, Y. Wang, L. Jiao, T. Yao, Y. Li, Q. Zhang, Y. Luo and H.-L. Jiang, *Adv. Mater.*, 2018, **30**, 1705112.
356. X. Li, T. W. Goh, L. Li, C. Xiao, Z. Guo, X. C. Zeng and W. Huang, *ACS Catal.*, 2016, **6**, 3461–3468.
357. M. Mon, R. Bruno, J. Ferrando-Soria, L. Bartella, L. Di Donna, M. Talia, R. Lappano, M. Maggiolini, D. Armentano and E. Pardo, *Mater. Horizons*, 2018, **5**, 683–690.
358. M. A. Rivero-Crespo, M. Mon, J. Ferrando-Soria, C. W. Lopes, M. Boronat, A. Leyva-Pérez, A. Corma, J. C. Hernández-Garrido, M. López-Haro, J. J. Calvino, E. V. Ramos-Fernandez, D. Armentano and E. Pardo, *Angew. Chem., Int. Ed.*, 2018, **57**, 17094–17099.
359. Y.-Z. Chen, Y.-X. Zhou, H. Wang, J. Lu, T. Uchida, Q. Xu, S.-H. Yu and H.-L. Jiang, *ACS Catal.*, 2015, **5**, 2062–2069.
360. S. Dutta, J. Kim, P. Hsieh, Y. Hsu, Y. V. Kaneti, F. Shieh, Y. Yamauchi and K. C. -W. Wu, *Small Methods*, 2019, **3**, 1900213.

361. S. M. Cohen, Z. Zhang and J. A. Boissonnault, *Inorg. Chem.*, 2016, **55**, 7281–7290.
362. Y. Chen and S. Ma, *Dalton Trans.*, 2016, **45**, 9744–9753.
363. Q. Qiu, H. Chen, Y. Wang and Y. Ying, *Coord. Chem. Rev.*, 2019, **387**, 60–78.
364. S. Jung, Y. Kim, S.-J. Kim, T.-H. Kwon, S. Huh and S. Park, *Chem. Commun.*, 2011, **47**, 2904.
365. Y.-H. Shih, S.-H. Lo, N.-S. Yang, B. Singco, Y.-J. Cheng, C.-Y. Wu, I.-H. Chang, H.-Y. Huang and C.-H. Lin, *Chempluschem*, 2012, **77**, 982–986.
366. F.-X. Qin, S.-Y. Jia, F.-F. Wang, S.-H. Wu, J. Song and Y. Liu, *Catal. Sci. Technol.*, 2013, **3**, 2761.
367. S.-L. Cao, D.-M. Yue, X.-H. Li, T. J. Smith, N. Li, M.-H. Zong, H. Wu, Y.-Z. Ma and W.-Y. Lou, *ACS Sustain. Chem. Eng.*, 2016, **4**, 3586–3595.
368. W.-L. Liu, S.-H. Lo, B. Singco, C.-C. Yang, H.-Y. Huang and C.-H. Lin, *J. Mater. Chem. B*, 2013, **1**, 928.
369. W.-L. Liu, C.-Y. Wu, C.-Y. Chen, B. Singco, C.-H. Lin and H.-Y. Huang, *Chem. Eur. J.*, 2014, **20**, 8923–8928.
370. W.-L. Liu, N.-S. Yang, Y.-T. Chen, S. Lirio, C.-Y. Wu, C.-H. Lin and H.-Y. Huang, *Chem. Eur. J.*, 2015, **21**, 115–119.
371. S. Patra, T. Hidalgo Crespo, A. Permyakova, C. Sicard, C. Serre, A. Chaussé, N. Steunou and L. Legrand, *J. Mater. Chem. B*, 2015, **3**, 8983–8992.
372. Y. Cao, Z. Wu, T. Wang, Y. Xiao, Q. Huo and Y. Liu, *Dalton Trans.*, 2016, **45**, 6998–7003.
373. J. In, in *Conservation & Exhibitions*, Elsevier, Amsterdam, 1987, vol. 336, pp. 225–240.
374. S. K. Alsaiari, S. Patil, M. Alyami, K. O. Alamoudi, F. A. Aleisa, J. S. Merzaban, M. Li and N. M. Khashab, *J. Am. Chem. Soc.*, 2018, **140**, 143–146.
375. J. Navarro, N. Almora Barrios, B. Lerma Berlanga, J. J. Ruiz-Pernía, V. A. Lorenz Fonfria, I. Tuñón and C. Martí-Gastaldo, *Chem. Sci.*, 2019, **10**, 4082–4088.
376. V. Lykourinou, Y. Chen, X. Wang, L. Meng, T. Hoang, L. Ming, R. L. Musselman and S. Ma, *J. Am. Chem. Soc.*, 2011, **133**, 10382–10385.
377. Y. Chen, V. Lykourinou, C. Vetromile, T. Hoang, L.-J. Ming, R. W. Larsen and S. Ma, *J. Am. Chem. Soc.*, 2012, **134**, 13188–13191.
378. S.-H. Lo, C.-H. Chien, Y.-L. Lai, C.-C. Yang, J. J. Lee, D. S. Raja and C.-H. Lin, *J. Mater. Chem. A*, 2013, **1**, 324–329.
379. P. Li, J. A. Modica, A. J. Howarth, E. Vargas L., P. Z. Moghadam, R. Q. Snurr, M. Mrksich, J. T. Hupp, O. K. Farha, *Chem*, 2016, **1**, 154–169.
380. D. Feng, T.-F. Liu, J. Su, M. Bosch, Z. Wei, W. Wan, D. Yuan, Y.-P. Chen, X. Wang, K. Wang, X. Lian, Z.-Y. Gu, J. Park, X. Zou and H.-C. Zhou, *Nat. Commun.*, 2015, **6**, 5979.
381. P. Li, S.-Y. Moon, M. A. Guelta, S. P. Harvey, J. T. Hupp and O. K. Farha, *J. Am. Chem. Soc.*, 2016, **138**, 8052–8055.
382. X. Lian, Y.-P. Chen, T.-F. Liu and H.-C. Zhou, *Chem. Sci.*, 2016, **7**, 6969–6973.

383. E. Gkaniatsou, C. Sicard, R. Ricoux, L. Benahmed, F. Bourdreux, Q. Zhang, C. Serre, J.-P. Mahy and N. Steunou, *Angew. Chem., Int. Ed.*, 2018, **57**, 16141–16146.
384. F.-K. Shieh, S.-C. Wang, C.-I. Yen, C.-C. Wu, S. Dutta, L.-Y. Chou, J. V. Morabito, P. Hu, M.-H. Hsu, K. C. W. Wu and C.-K. Tsung, *J. Am. Chem. Soc.*, 2015, **137**, 4276–4279.
385. G.-Y. Jeong, R. Ricco, K. Liang, J. Ludwig, J.-O. Kim, P. Falcaro and D.-P. Kim, *Chem. Mater.*, 2015, **27**, 7903–7909.
386. Y. Wang, C. Hou, Y. Zhang, F. He, M. Liu and X. Li, *J. Mater. Chem. B*, 2016, **4**, 3695–3702.
387. K. Liang, R. Ricco, C. M. Doherty, M. J. Styles, S. Bell, N. Kirby, S. Mudie, D. Haylock, A. J. Hill, C. J. Doonan and P. Falcaro, *Nat. Commun.*, 2015, **6**, 7240.
388. F.-S. Liao, W.-S. Lo, Y.-S. Hsu, C.-C. Wu, S.-C. Wang, F.-K. Shieh, J. V. Morabito, L.-Y. Chou, K. C. W. Wu and C.-K. Tsung, *J. Am. Chem. Soc.*, 2017, **139**, 6530–6533.
389. Z. Li, H. Xia, S. Li, J. Pang, W. Zhu and Y. Jiang, *Nanoscale*, 2017, **9**, 15298–15302.
390. Y. Pan, H. Li, J. Farmakes, F. Xiao, B. Chen, S. Ma and Z. Yang, *J. Am. Chem. Soc.*, 2018, **140**, 16032–16036.
391. W.-H. Chen, M. Vázquez-González, A. Zoabi, R. Abu-Reziq and I. Willner, *Nat. Catal.*, 2018, **1**, 689–695.
392. Y. Chu, J. Hou, C. Boyer, J. J. Richardson, K. Liang and J. Xu, *Appl. Mater. Today*, 2018, **10**, 93–105.
393. W. Liang, H. Xu, F. Carraro, N. K. Maddigan, Q. Li, S. G. Bell, D. M. Huang, A. Tarzia, M. B. Solomon, H. Amenitsch, L. Vaccari, C. J. Sumby, P. Falcaro and C. J. Doonan, *J. Am. Chem. Soc.*, 2019, **141**, 2348–2355.
394. J. Huo, J. Aguilera-Sigalat, S. El-Hankari and D. Bradshaw, *Chem. Sci.*, 2015, **6**, 1938–1943.

CHAPTER 10

Application of Ionic Liquids for Sustainable Catalysis

P. GRANGER[a] AND V. I. PARVULESCU[b*]

[a]Univ. Lille, CNRS, Centrale Lille, ENSCL, Univ. Artois, UMR 8181 – UCCS – Unité de Catalyse et Chimie du Solide, F-59000 Lille, France
[b]Director Department of Organic Chemistry, Biochemistry and Catalysis, University of Bucharest 030016, B-dul Regina Elisabeta 4–12, Bucharest, Romania
*Corresponding contributor. Email: vasile.parvulescu@chimie.unibuc.ro

10.1 Concept and Definition of Sustainable Chemistry

According to the definition of the Organisation for Economic Co-operation and Development (OECD), "Sustainable chemistry defines a scientific concept that seeks to improve the efficiency with which natural resources are used to meet human needs for chemical products and services". It encompasses "the design, manufacture and use of efficient, effective, safe and more environmentally benign chemical products and processes".[1] Supplementary support for this has been given by Horváth and co-workers, defining it as follows: "[R]esources including energy should be used at a rate at which they can be replaced naturally and the generation of wastes cannot be faster than the rate of their remediation";[2,3] and by Welton as follows: "[A] sustainable

Heterogeneous Catalysis for Energy Applications
Energy and Environment Series No. 27
Edited by Tomas R. Reina and Jose A. Odriozola

Published by the Royal Society of Chemistry, www.rsc.org

chemical product should be supplied at a price that enables it to be accessed by its users while at the same time being commercially viable for its producers".[4] The concept stimulates innovation across all sectors "to design and discover new chemicals, production processes, and product stewardship practices that will provide increased performance and increased value while meeting the goals of protecting and enhancing human health and the environment".

More than 25 years ago, the problems of wastes in the fine chemicals industry inspired Sheldon to introduce the *E factor* (EF) concept as an instrument for assessing the environmental impact of manufacturing processes.[5] The E(nvironmental) factor is defined as the mass of waste : mass of product ratio (usually expressed as kg kg^{-1}). A second concept has been introduced by Trost and defined as atom economy (AE).[6] It is calculated by dividing the molecular weight of the product by the sum total of the molecular weights of all substances formed in the stoichiometric equation for the reaction involved. AE is a theoretical number that assumes the use of exact stoichiometric quantities of starting materials and a chemical yield of 100%, and it disregards substances such as solvents and chemicals used in the work-up of the reaction mixture that do not appear in the stoichiometric equation.[7] However, the solvent loss represents a major source of waste in the chemical industry, and especially in the pharmaceutical industry. Therefore, in order to provide a better assessment of environmental impact, Sheldon introduced a more complex factor by replacing the simple EF (sEF) with a complete EF (cEF). The new cEF accounts for all process materials, including solvents and water, assuming no recycling, and it is more appropriate for total waste stream analysis.[8]

$$\text{sEF} = \sum m(\text{raw materials}) + \sum m(\text{reagents}) - m(\text{product}) / m(\text{product})$$

$$\text{cEF} = \sum m(\text{raw materials}) + \sum m(\text{reagents}) + \sum m(\text{solvents}) + m(\text{water}) - m(\text{product}) / m(\text{product})$$

Scheme 10.1

To produce at acceptable EF values, it was recommended that: (1) substances from the lithosphere must not systematically increase in the ecosphere; (2) substances produced by society must not systematically increase in the ecosphere; (3) the physical basis for the productivity and diversity of nature must not be systematically deteriorated; and (4) fair and efficient use of resources must match human needs.[9]

Sustainability in chemistry is often associated with the green chemistry and circular economy concepts, of which the objectives are: (1) to reduce the usage and production of hazardous substances for chemical processes concomitantly to a reduction in energy consumption;[10] and (2) to minimize the resource input and waste, emissions and energy leakage by slowing, closing and narrowing energy and material loops.[11]

The general principles operating in this chemistry have already been proposed by Anastas and Werner.[12] Among these, doing reactions with *renewable feedstocks* using *safer solvents and auxiliaries* under high *energy efficiency* are also very important for a sustainable catalytic process.

It thus emerged that sustainable chemistry must take into consideration issues related to accessing different rough and refined resources (carbon, water, metals), problems of access to energy, global warming and the exponential increase in the human population, for which chemistry must allow a serene development.[13] This study also highlights a set of necessary conditions that must be met in order to completely achieve this aim.

10.2 Solvents and Sustainability

In accordance with the fundamentals of sustainable chemistry, which represents a more holistic approach to the development of chemical technologies, sustainability should take into consideration several aspects, such as scalability, economics and regulatory constraints. In this multidisciplinary approach, the role of the solvent is very important. Many organic solvents are harmful, toxic and environmentally damaging, and their use poses risks to both human health and the environment. Therefore, the understanding of solvent properties is a necessary part of sustainable development, and many solvents have therefore also been ranked by their environmental impacts, safety and health characteristics.[14]

Based on the analysis of a series of industrial examples, Welton summarized the ways through which solvents may influence the sustainability of chemical processes.[4] These consider the following possibilities: (1) to act simultaneously as solvent and reagent; (2) to control the quality of the products; (3) to reduce the number of synthetic steps; (4) to reduce the formation of byproducts; and (5) to improve product separation.

At present, the implementation of the sustainability concept in chemistry mainly focuses the valorization of CO_2, the extensive utilization of biomass and greener synthesis, all considering the use of green solvents. However, as an investigation by Jessop showed, there are differences of opinion in the chemical community concerning the type of green solvent that may produce the greatest enhancement in the efficiency of a chemical process and reduction in the environmental damage.[15] Figure 10.1 illustrates the difference between the answers claiming that water and CO_2 will be the most suitable solvents (Figure 10.1a) and the proportion of the scientific papers reporting their utilization (Figure 10.1b). This investigation raised questions again highlighting the differences between academia, which is mostly interested in exploring the opportunities and limits offered by a certain investigation topic, and the more pragmatic view of applied research. In addition, there is a net difference between the use of CO_2 as a solvent (requiring pure, relatively expensive CO_2) and its valorization from different wastes.

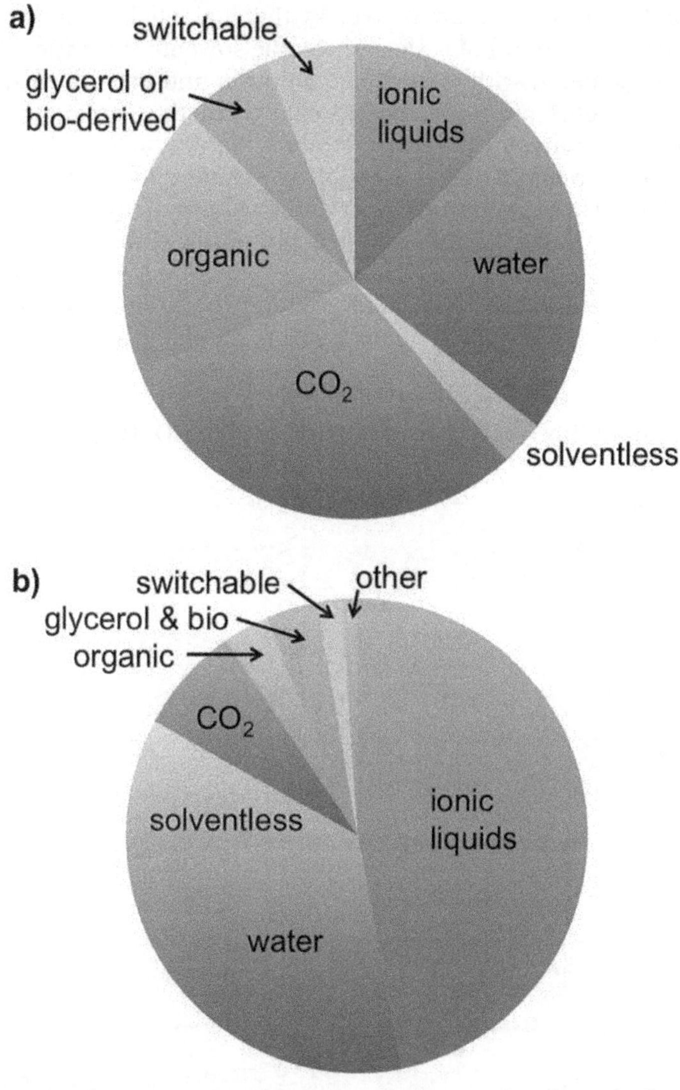

Figure 10.1 (a) The responses to the survey question "... what class of solvents will be responsible for the greatest reduction in environmental damage?". First and second choices were given scores of 1.0 and 0.5, respectively. "Water" includes liquid, supercritical and on-water. "CO_2" includes supercritical, liquid and CO_2-expanded liquid. "Organic" refers to carefully chosen conventional organic solvents. (b) The proportion of papers describing each class of solvent published in the journal *Green Chemistry* in 2010. Reproduced from Ref. 15 with permission from the Royal Society of Chemistry.

Switchable solvents are another class that has elicited interest in the past decade.[16,17] Removal of toxic and volatile solvents is a critical step and can only be achieved with energy consumption and environmental costs. Switchable solvents offer a complete change of the paradigm in which solvent removal from a hydrophobic product could be achieved without distillation if the solvent could be reversibly switched from being hydrophobic to hydrophilic. These are termed switchable-hydrophilicity solvents because they can reversibly change these physical properties abruptly.[18] This unique property is a consequence of a reversible reaction in response to an external stimulus such as a temperature change and/or the addition or removal of a gas. Because of the reversibility of the reaction, the changed solvent can easily be brought back to its original state. Examples in this sense are provided by solvents that have very little miscibility with water in the absence of CO_2, but complete miscibility with water in the presence of CO_2 (Figure 10.2). CO_2 is

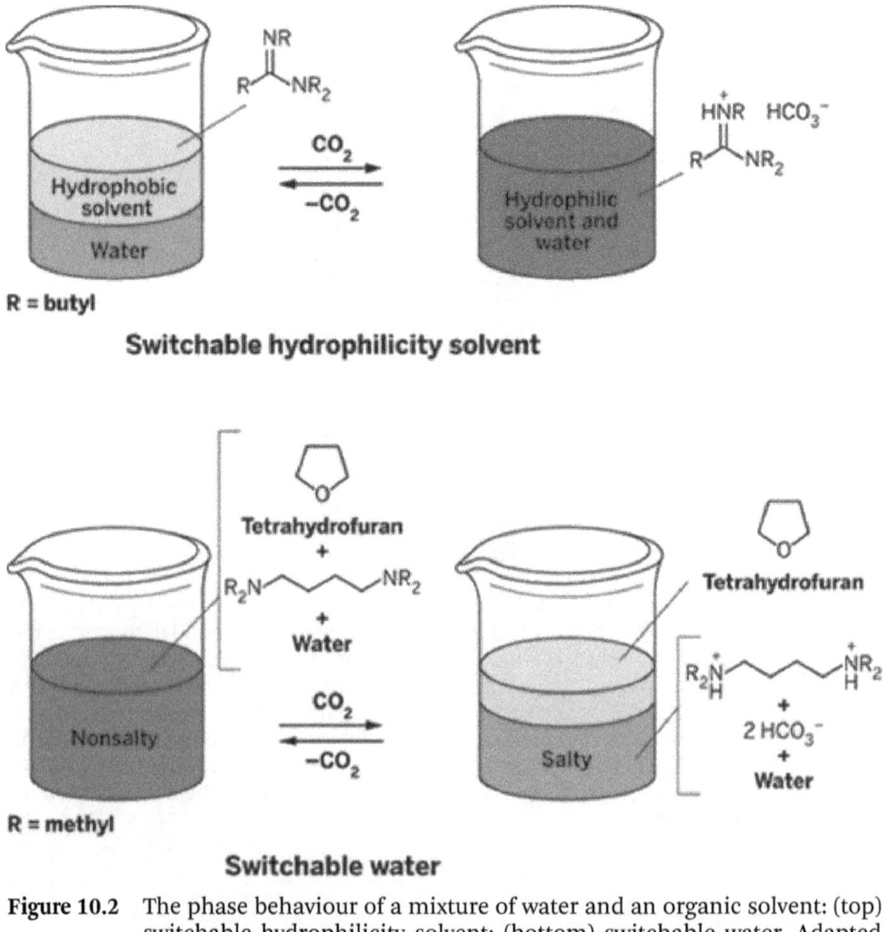

Figure 10.2 The phase behaviour of a mixture of water and an organic solvent: (top) switchable hydrophilicity solvent; (bottom) switchable water. Adapted from Ref. 18 with permission from the Royal Society of Chemistry.

preferred as the trigger for the switching process because it is a removal waste and therefore inexpensive, but also is non-toxic, benign and easily removed. At present, there are two groups of switchable solvents. The first includes amine-derived liquid compounds that, in combination with CO_2 (as amidine/CO_2 or guanidine/CO_2), generate hydrophilic solvents referred to as "switchable-hydrophilicity solvents". The second is named "switchable water" and uses amines that can produce water-soluble carbonates with CO_2 that could easily separate on an immiscible water organic solvent-solubilized reaction product (Figure 10.2).

The influence of the solvents on the position of the chemical equilibriums was reported at the end of the nineteenth century by several chemists.[19–22] Since that time, chemists have only utilized volatile organic solvents with very negative impact on the environment and health. Exoergic solute–solvent interactions and endoergic contributions were evaluated for these solvents, considering parameters for the description of the polarity, the hydrogen bonding and the electron-pair donation and acceptance properties of the solvents.[23] Furthermore, interest in acidic and basic homogeneous catalysis shifted in other directions. This applies to the ideas advanced on the state of the catalyst in solution and to the theory ascribing catalysis to the ionization of the substrate or to the formation of an intermediate product that reacts spontaneously.[24] Water was the solvent that was most utilized in these reactions. Water was also assessed as a green solvent for other catalytic applications, with the huge advantage of its large availability.[25–27] Despite catalysis in water receiving severe criticism, it was stressed that pure water is indeed environmentally friendly, but catalytic organic reactions carried out in water or in the presence of water add complications. Water is only a truly green solvent if it can be directly discharged to a biological effluent treatment plant. Water streams contaminated by organics are very difficult to purify, and therefore may be not released to the environment.[28]

Ionic liquids (ILs) are another class of compounds that emerged as green solvents for environmentally friendly synthesis.[29–35] They are organic salts, usually composed of an organic cation and a polyatomic inorganic anion (Figure 10.3).[36] This combination permits the tuning of the physical and chemical properties (melting point, conductivity, viscosity, density, refractive index, *etc.*) through adequate selection of the cation and anion constituents, thus providing a series of properties that differentiate these compounds from the "classic" organic solvents. At the same time, this allows specific features for a given application to be introduced (hydrophobicity *vs.* hydrophilicity, controlling solute solubility, adding functional groups for catalysis/reactivity purposes, chirality, *etc.*). ILs exhibit an almost negligible vapour pressure, and accordingly have been recognized as environmentally benign alternatives to volatile organic solvents. Indeed, these media have been used as substitutes of volatile organic solvents in a wide variety of chemical processes,[37] such as separation and purification, as well as reaction media in biochemical[38] but also chemical and stereochemical catalysis.[39] The stability of ILs means that they may also act as stable and persistent environmental pollutants.[15] This critical behaviour connected to their poor cost efficiency clearly

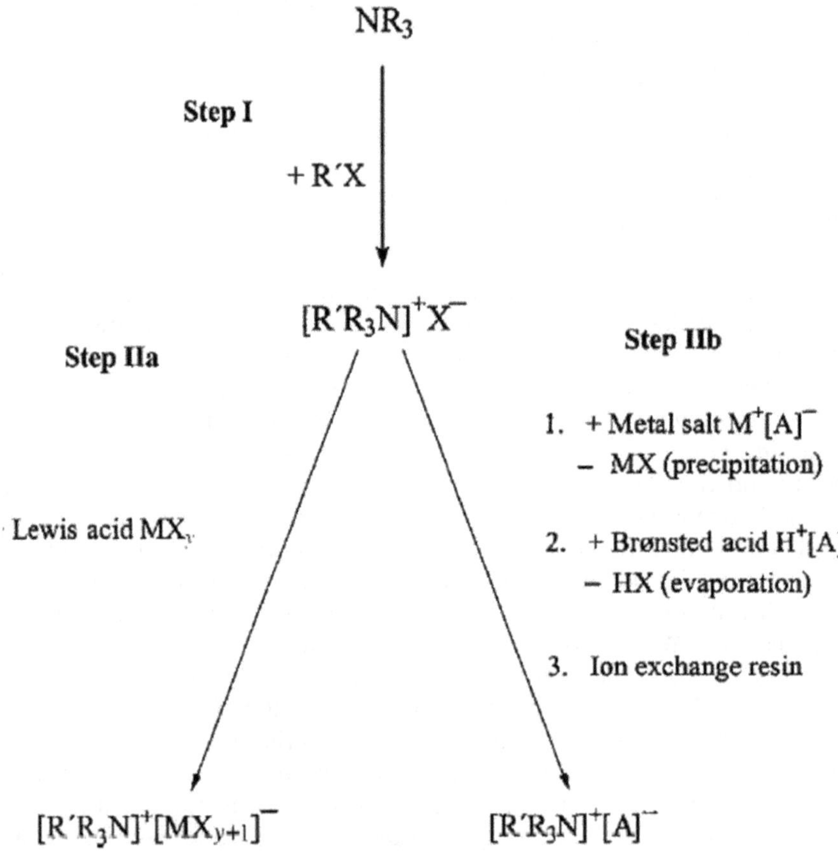

Figure 10.3 Synthesis paths for the preparation of ILs. Reproduced from Ref. 36 with permission from John Wiley and Sons, Copyright © 2000 Wiley-VCH Verlag GmbH, Weinheim, Germany.

recommends their recycling. Moreover, it is worth noting that the synthesis of ILs does not always follow green sustainable routes.

To date, most ILs are typically petroleum derived, which goes strongly against the sustainability concept.[40] However, a consistent number of ILs derived from amino acids, carbohydrates, lignin and other renewable sources have also been reported (Figure 10.4).[41] In this way, the ILs produced from the by-products of different biomass industries and lignocellulose bio-refinery can enable a "closed-loop" bio-refinery that sustains its own needs for process solvents. Proposing this concept, Socha *et al.*[42] reported the production of ILs from lignin and hemicellulose (*i.e.* two lignocellulose pre-treatment by-products) and a series of tertiary amine-based ILs derived from aromatic aldehydes as the major by-products of lignocellulosic biofuel production.

Solvate ILs represent another relatively new class of ILs consisting of a coordinating solvent and salt that give rise to a chelate complex with very

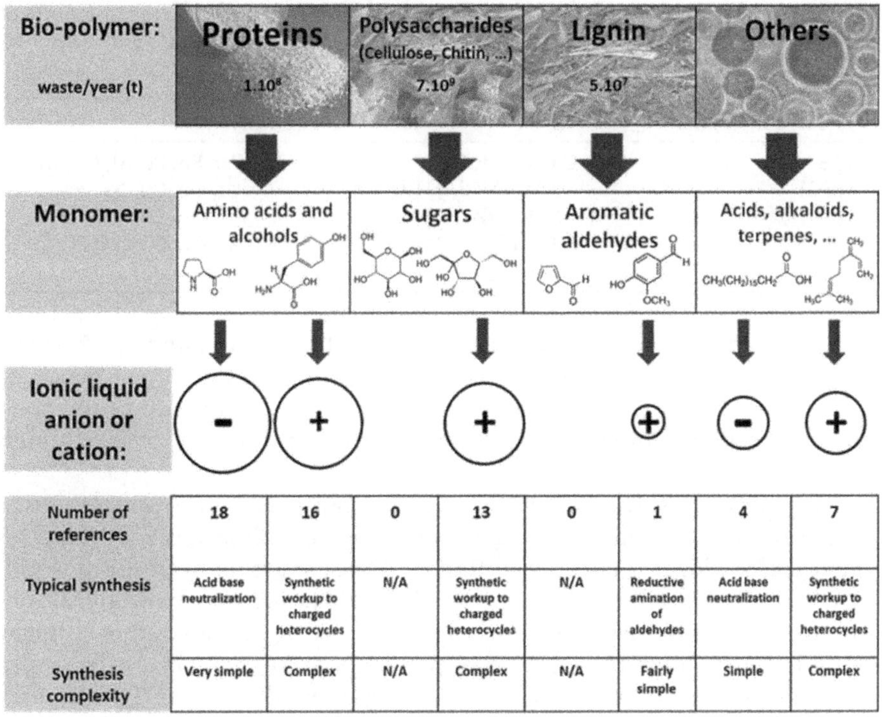

Figure 10.4 Synthesis pathways towards bio-based ILs. For each renewable resource class, typical derived monomers are depicted. The intensity of research (*i.e.* number of published papers) on converting these monomers to IL anions or cations is represented by the radius of the circle around the "−" or "+" signs, respectively. Typical synthetic routes and their complexities are mentioned where available. Reproduced from Ref. 41 with permission from American Chemical Society, Copyright 2016.

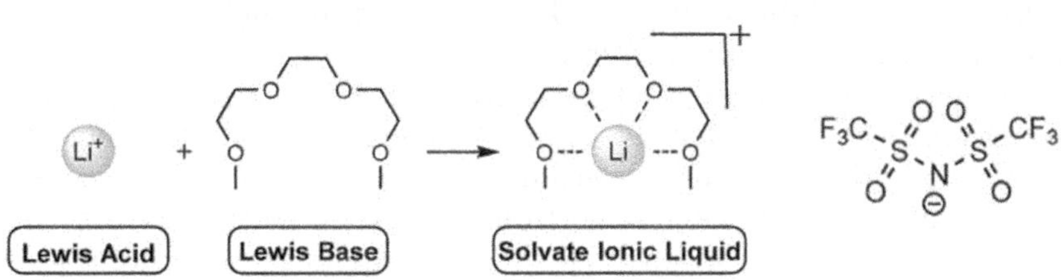

Figure 10.5 Addition of Lewis acid and Lewis base to give a solvate IL. Reproduced from Ref. 43, https://doi.org/10.3389/fchem.2019.00263, under the terms of the CC BY 4.0 licence, https://creativecommons.org/licenses/by/4.0/.

similar properties to ILs (Figure 10.5).[43] They are generated *via* the diffusion of a cationic charge, leading to the sequestering of a hard cation, often in an ethereal solvent. Solvate IL solvents have only been recommended as a reaction medium for organic transformations, including in catalysis for

Table 10.1 General formula for the classification of DESs. Reproduced from Ref. 51, https://doi.org/10.1021/cr300162p, with permission from American Chemical Society, Copyright 2014.

Type	General formula	Terms
Type I	$Cat^+X^-zMCl_x$	M = Zn, Sn, Fe, Al, Ga, In
Type II	$Cat^+X^-zMCl_x{\cdot}yH_2O$	M = Cr, Co, Cu, Ni, Fe
Type III	Cat^+X^-zRZ	$Z = CONH_2$, COOH, OH
Type IV	$MCl_x + RZ = MCl_{x-1}^+{\cdot}RZ + MCl_{x+1}$	M = Al, Zn; $Z = CONH_2$, OH

the Aldol and Kabachnik–Fields reactions or Diels–Alder and [2 + 2] cycloaddition. Replacing the achiral organo-cation with a chiral one may even give rise to asymmetric syntheses that, from the perspective of the complex molecules required by pharmaceutical applications, are extremely important.[44]

Deep eutectic solvents (DESs) have also captured significant scientific attention as a "green" and sustainable class of tailor-made solvents for sustainable biomass transformation.[45–49] They share many properties with ILs, including low vapour pressure, wide liquid range, thermal stability, low flammability and high solvation capability, while being simpler to make from low-cost constituents.[50] DESs are composed of at least one hydrogen bond donor (HBD) and at least one hydrogen bond acceptor (HBA), which can be associated with each other *via* hydrogen bond interactions to form a mixture with a melting temperature far below that of its constituents.[51] DESs can be described by the general formula Cat^+X^-zY where Cat^+ is in principle any ammonium, phosphonium or sulfonium cation (*i.e.* a HBD) and X is a Lewis base, generally a halide anion (*i.e.* a HBA) (Table 10.1).[51] The complex anionic species are formed between X^- and either a Lewis or Brønsted acid Y (*z* refers to the number of Y molecules that interact with the anion). Most of the studies in this area have focused on quaternary ammonium and imidazolium cations.[51]

Liquid polymers, represented by polyethylene glycol (PEG), polypropylene glycol, CO_2 expanded liquids and switchable solvents, were also integrated into the group of the sustainable solvents.[14] There are several properties that differentiate these from the other green solvents. The liquid polymers display an unusual inverse temperature–solubility relationship in water and exhibit the inherent advantage of a negligible volatility. These properties provided advantages over conventional solvents in carbon–carbon bond-forming reactions such as Suzuki, Heck, Stille or Sonogashira coupling reactions.[52] In addition, due to its biphasic nature, PEG is also used as an "alternative reaction medium" for more advanced reaction engineering processes. Reactions carried out in aqueous biphasic reactive extractions can be engineered to control the partitioning of reactants, catalysts and products, affording greater product yields and product selectivities.

10.3 Process Intensification through Catalysis

Sustainability in catalytic processes is also expected to be achieved *via* process intensification. This supposes that the catalytic reaction should be connected with a green technique such as microwave irradiation, ultrasonication or flow microreaction. By using such techniques, the reaction is provided with selective heating (microwave irradiation), increased dispersion and unbundling of the catalyst (ultrasonication) or much better mixing and improved mass transfer of reactants and products (flow microreaction).

Microwaves act as high-frequency electric fields and will generally heat any material containing mobile electric charges, such as polar molecules in a solvent or conducting ions in a solid.[53] Polar constitutive solvents are heated as their component molecules are forced to rotate under the effect of the field and lose energy through collisions. Semiconducting and conducting samples overheat when ions or electrons within them form an electric current and subsequent energy loss is induced by the electrical resistance of the material. The major advantages of using microwaves are: (1) reaction rate acceleration; (2) milder reaction conditions; (3) higher chemical yields; (4) lower energy usage; and (5) different reaction selectivities.

The reactions involving solids (as reactants or substrates) and the use of the heterogeneous catalysts, in addition to the known advantages, may also have several drawbacks in terms of mass transfer resistances generated by the presence of different phases such as a solid, liquid or gas. Another disadvantage is the agglomeration of particles, which may result in lower surface areas and the deposition of impurities on the solid surface, leading to significant decreases in performance.[54,55] The application of ultrasound may overcome these disadvantages and improve reactions that are intrinsically slow and have low product yields and require expensive reagents for progress. Furthermore, ultrasound can be applied to almost all categories of catalytic reactions, such as heterogeneously catalysed reactions, enzyme-catalysed reactions, ion exchange resin-catalysed reactions and homogeneous and phase transfer-catalysed reactions, as well as biphasic reactions. Cavitation generated using ultrasound can also allow for better selectivities based on physical and chemical effects.[55]

The combination of microwaves with ultrasound may lead to synergetic effects on the reaction rate.[56] The reactor design in connection with the type of reaction and the mechanism employed is also important in this area.

10.4 IL Catalysts

10.4.1 Homogeneous IL Catalysts

The physical properties of ILs such as liquid range, vapour pressure, polarity, solubility, viscosity and density are essential factors in controlling their performance in a catalytic process. Therefore, it is necessary to depict the physical properties of ILs and their relationship with their structures in order to obtain useful information on the rational design of ILs for particular processes.[57]

ILs can also act as dual solvent–catalysts for complex synthetic reactions. By way of illustration, commercially available 3-alkyl-1-methyl imidazolium ILs $[C_nC_1Im][X^-]$ was investigated as a dual solvent–organocatalyst for the esterification of α-tocopherol with succinic anhydride, producing satisfactory yields and reaction rates.[58] At the end of the reaction, the IL was recycled by water extraction with very high yields.

ILs were initially used as ionic-type catalysts. In addition, acidic and basic functional ILs have been widely used as acidic and basic catalysts in catalytic transformations. Later, it was realized that, besides solvents, ILs can act as special ligands for metal complexes and can provide a unique stabilization effect for some reaction intermediates, such as free radicals and carboions, and then increase the catalytic activity and selectivity of related catalytic systems.[57] ILs can also provide a smart environment for biocatalysts and increase their stability and activity, and thus optimize their catalytic performance.

Metal-containing ILs or liquid metal catalysts (LMCs) define IL-containing metals in their ionic forms (cationic/anionic, dissolved coordination complex or salt) and not ILs containing metal or metal oxide nanoparticles.[59] They overcome the limitations of using solid metal salts by combining the advantages of ILs with the chemistry of a metal centre (*e.g.* retention of catalytic properties).

LMCs may incorporate p-, d- or f-metal salts and an IL, of which the properties can be tuned by simply changing the nature or ratio of its constituents. After dissolving a Lewis acidic metal salt (*e.g.* metal triflate or halide) in IL, the formation of LMCs occurs *via* a further reaction leading to new unique Lewis and Brønsted Lewis acidic species that may act as active catalysts in organic reactions (Figure 10.6).[60,61] The unique double acidic properties of these catalysts make them more active than individual Brønsted or Lewis acid catalysts, which may be due to the synergetic effects of both acidities. For instance,

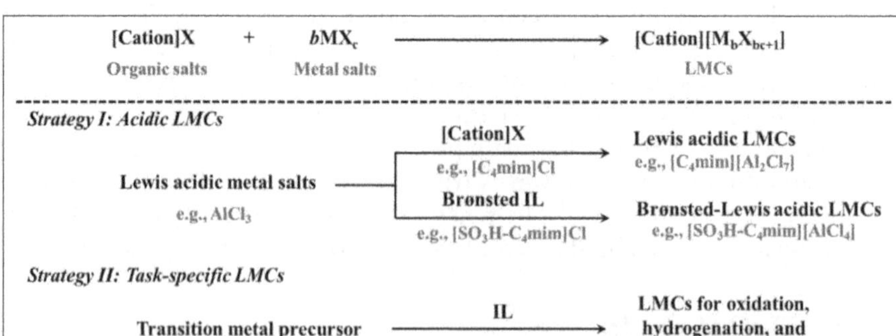

Figure 10.6 Schematic illustration of strategies to design LMCs and their utilization. Reproduced from Ref. 62 with permission from Elsevier, Copyright 2018.

1-butyl-3-methylimidazolium heptachloro-dialuminate ([C4mim]-[Al_2Cl_7]) has superior Lewis acidity than its (C4mim)Cl and $AlCl_3$ constituents.[62]

LMCs have already proved their efficiency in industrially important transformations such as alkene polymerization,[63,64] isomerization and alkylation,[62] but also for other industrial favourites such as C–C bond formation, hydrogenation, hydrochlorination and oxidation reactions (Figure 10.6).

On the other hand, molecular transition metal complexes are extremely powerful catalysts for chemical synthesis, and their use represents a key technology for sustainable manufacturing. However, their use also has a series of drawbacks, such as thermal sensitivity and engineering disadvantages associated with product separation and catalyst recovery that can override the value of the product. Thus, despite the impressive batch performances of these catalysts, industrial exploitation makes this form of homogeneous catalysis uneconomical. In this context, continuous-flow systems that allow steady-state homogeneous turnovers in a stationary liquid phase while at the same time causing integrated product separation at mild process temperatures make these catalysts more attractive.[65]

To overcome these drawbacks, a separation strategy was proposed that was gentle enough to prevent catalyst deactivation yet effective enough to be directly applicable to turnover conditions.[65] This approach assumes an integrated continuous flow (Figure 10.7) that affords a number of engineering advantages, including intensified space–time yields, reduced waste production and enhanced process control through more effective heat management and automation.[66–68]

In recent years, the combination of ILs as both solvents and catalysts with microwave irradiation as an energy source also offered interesting solutions for sustainable processes. Such an approach may allow organic synthesis with less

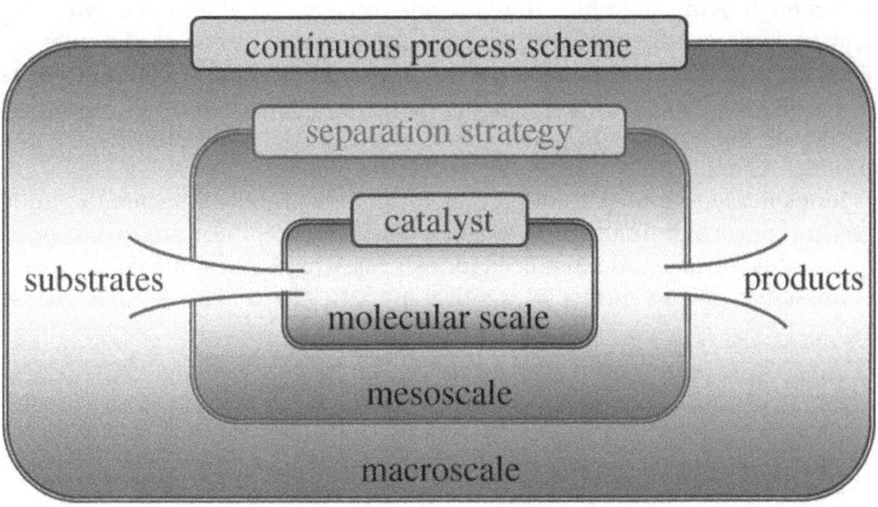

Figure 10.7 Conceptual levels of integrated continuous-flow homogeneous catalysis. Reproduced from Ref. 65 with permission from the Royal Society of Chemistry.

need for protecting/de-protecting steps, with catalysis playing a major role in improving the AE and reducing waste. Examples of already successful reactions are the oxidation of electrophilic alkenes by hydrogen peroxide in the presence of a base,[69] the oxidation of organic sulfides (showing a significant improvement in the rate of reaction and an increase in selectivity),[70] C–C bond formation in Friedel–Craft acylation,[71] the esterification of alcohols[72] and of carboxylic acids,[73] Michael addition of sulfonamides,[74] asymmetric organocatalysis[75] or the thiolene "click" reaction[76] and oxidation and C–C bond formation reactions.[77]

10.4.2 Heterogeneized IL Catalysts

Self-assembly with ILs leads to the formation of ionogels that are soft materials with both high thermal stability and conductivity.[78] Currently, this term comprises a wide variety of materials, including polymeric gels, hydride materials containing inorganic species and, recently, supramolecular gels as well.[79]

Molecular gels may also be formed by other organic salts, such as those comprising diimidazolium organic cations paired with di-, tri-, or tetra-valent anions bearing aliphatic or aromatic spacers of different sizes, symmetries and coordinating abilities, both in "normal" solvents and in ILs. These gelators are able to gel solvents such as conventional solvents and ILs, which are frequently used for their very different properties. While stronger and more crystalline gels are formed in ILs than in organic solvents, both of the entrapped solvents are highly polar and viscous. The properties of ionogels seem to depend mainly on the nature of the IL solvent and of the gelator anion. They also affect the morphology and size of the aggregates.[80] These materials are assumed to have potential for applications in which ILs play a significant role, such as green and sustainable chemistry, catalysis and energy.

Although being solid-like, the ionogels conserve the liquid-like dynamics and an ionic conductivity in the same order of magnitude as the pure IL. Furthermore, these materials can be modified both by chemical modifications of the IL ions (design of task-specific ILs) or by encapsulation of molecular species in the confined IL phase.[81] This strategy offers opportunities for the production of more complex catalysts or drug release systems.

Ionosilicas are ionogels containing siliceous materials in which the ionic groups are covalently attached to silica. Furthermore, there are two types of ionosilicas: surface-functionalized ionosilicas, where the organic ionic group is anchored on the surface of a silica support *via* covalent Si–O–Si bonds

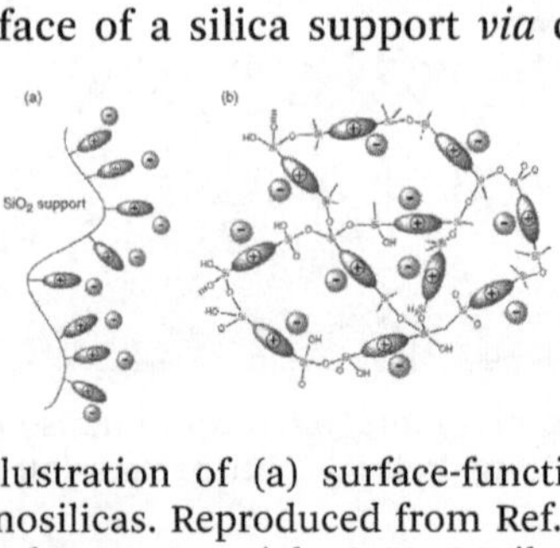

Figure 10.8 Schematic illustration of (a) surface-functionalized ionosilicas and (b) hybrid ionosilicas. Reproduced from Ref. 81 with permission from John Wiley and Sons, Copyright © 2015 Wiley-VCH Verlag GmbH & Co. KGaA.

(Figure 10.8a); and ionosilica hybrid materials that are characterized by the incorporation of the organic entities within the silica scaffold (Figure 10.8b).

The confinement of ILs on solid supports has attracted increased interest from both academia and industry.[81] The confined immobilization in the oxide network (most often silica) endows these ionogels with specific properties, such as ionic conductivity.[82] They have already demonstrated industrial catalytic applications and effectiveness in gas-cleaning technologies. More recently, the interest in IL electrolytes with high ionic conductivity over a wide range of temperatures and electrochemical stability windows has led to progress in the design of advanced electrochemical devices (actuators, lithium batteries, electric double-layer capacitors, dye-sensitized solar cells) and fuel cells, but also as catalysts for the processing and derivatization of cellulose.

The synthesis of surface-functionalized ionosilicas can be achieved by different routes. These include tethering of the cation, the anion or both the cation and the anion. The tethering of zwitterionic species may also produce surface-functionalized ionosilicas (Figure 10.9). "First class" refers to surface-functionalized ionosilicas prepared using cationic precursors from imidazolium, ammonium, pyridinium or phosphonium compounds,[83] "second class" refers to carboxylate-anchored anionic species,[84] while "third class" refers to water-immiscible guanidinium sulfonimide ion pairs.[85] Finally, the chemical grafting of zwitterionic species can be easily produced by reaction of nucleophiles (amines or imidazoles) with sultones.[86]

The catalytic properties of these materials are influenced by a series of properties, such as the hydrophilicity/hydrophobicity of the support, the nature of the anionic/cationic species, the quantity of surface silanol groups and the porosity of the silica support. The synthetic pathway of the support material post-synthesis grafting or co-condensation also influences these properties of the ionosilica materials.[81]

In the context of sustainable processes, base catalysts are highly demanded, especially for fine chemical synthesis.[87,88] To this end, homogeneous catalysts are very efficient but difficult to manipulate in these reactions due to their very high sensitivity to CO_2 and water. Solid bases, although less active and still very sensitive to water, eliminate some of these weaknesses and could be used as heterogeneous catalysts for a large number of industrial processes.[89–91] The mesoporous solid bases derived from hydrophobic polymers offered a step forward in terms of catalysing liquid organic reactions because they benefit from resistance to CO_2 and water.[92,93] In spite of this, they exhibit weak basic sites. Therefore, hydrophobic mesoporous poly(ILs), as a new and improved alternative, better fit the scope of these reactions, providing highly efficient

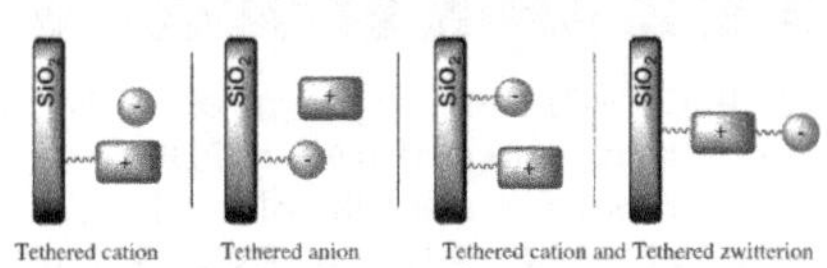

Figure 10.9 Different types of surface-functionalized ionosilicas. Reproduced from Ref. 81 with permission from John Wiley and Sons, Copyright © 2015 Wiley-VCH Verlag GmbH & Co. KGaA.

and contamination-resistant solid-base catalysts with enough strong basicity for these reactions.[94]

10.4.3 Grafting ILs onto Ordered Frameworks

10.4.3.1 Mesoporous Copolymers

Acidity-enhanced heterogenized IL catalysts were prepared by grafting of various strong acidic IL moieties onto the frameworks of mesoporous PDVB–vinylimidazole copolymers.[95] In addition to their strong acidity, the produced solid acids showed abundant nanopores, good hydrophobicity and controllable wettability for various reactants (Figure 10.10). These properties enhanced the degree of exposure of acidic sites and decreased the diffusion resistance of various reactants in catalytic reactions (Figure 10.24), which lead to increased activity compared to their homogeneous counterparts. Such properties were demonstrated for biodiesel production, the depolymerization of crystalline cellulose into glucose, cellobiose and 5-(hydroxymethyl)furfural (5-HMF) or the fixation of CO_2, demonstrating activities even higher than those of mineral acids.[96,97]

10.4.3.2 Mesoporous Oxides

ILs have also been grafted onto different types of large porous silica *via* simple functionalization.[98] They have also been immobilized onto stable structured mesoporous molecular sieves such as SBA-15 or Al-SBA-15 using chemical

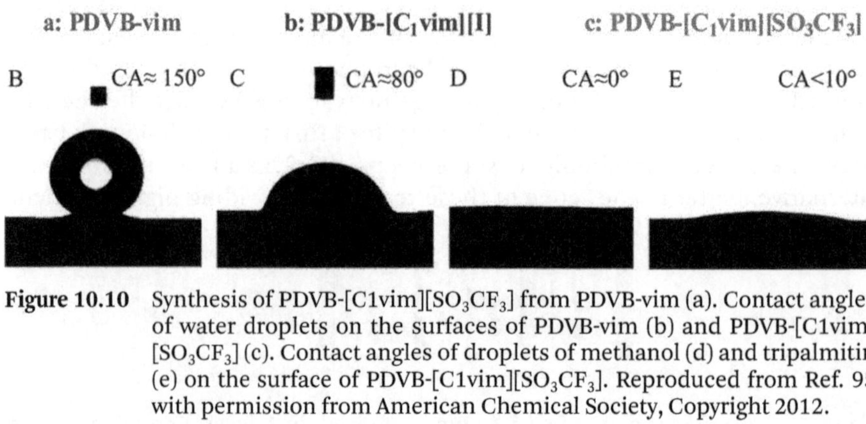

Figure 10.10 Synthesis of PDVB-[C1vim][SO_3CF_3] from PDVB-vim (a). Contact angles of water droplets on the surfaces of PDVB-vim (b) and PDVB-[C1vim][SO_3CF_3] (c). Contact angles of droplets of methanol (d) and tripalmitin (e) on the surface of PDVB-[C1vim][SO_3CF_3]. Reproduced from Ref. 95 with permission from American Chemical Society, Copyright 2012.

grafting methods. These materials were used as catalysts for solvent-free synthesis under relatively mild conditions.[99]

10.4.3.3 Metal–Organic Frameworks

Functionalized iso-reticular metal–organic frameworks (MOFs) were also developed as novel heterogeneous catalysts for both solvent and solvent-less organic synthesis. Solvent-less synthesis merely refers to cyclic carbonates without any co-catalyst.[100] The association of MOFs consisting of coordination bonds between transition metal cations and multidentate organic linkers[101] with ILs combines their advantages through the generation of a synergetic effect between the transition metal centre and the basic sites (anions and cations) of the IL.[102,103] The chemical nature of the cation is important, as has been demonstrated for the catalytic processes using phosphonium cation-based ILs. They also offer the advantage of a superior thermal stability and CO_2 solubility.[104]

10.4.4 Sustainable Catalytic Processes Using ILs

10.4.4.1 Sustainable Biomass Valorization

More and more renewable bio-resources are converted into biofuels, bio-based solvents, platform molecules for fine chemicals and agro-chemicals. Several building blocks derived from renewable resources, such as ethanol, glycerol, lactic acid, succinic and levulinic acids or 5-HMF are already in use and are still considered of potential importance in near-future developments. However, in addition to an active catalyst, these transformations also require an adequate solvent. Therefore, the dual properties of ILs as solvents and catalysts recommended them for such applications.

Catalytic transformation of lignocellulose into chemicals and fuel products in the presence of ILs has been thoroughly investigated.[105] This research has focused on the hydrolysis of cellulose to reducing sugars, the dehydration of glucose and cellulose to 5-HMF, the hydrogenation of cellulose to sugar alcohols, the conversion of glucose and cellulose to levulinic and lactic acids, the conversion of hemicellulose to sugars and of xylose and hemicellulose to furfural and the depolymerization and hydroprocessing of lignin.[106]

The conversion of cellulose into the mentioned platform molecules involves a complex cascade of reactions in which the first step (*i.e.* the solvolysis, which is typically hydrolysis, into glucose or its oligomers) is catalysed by Brønsted acids. Furthermore, under the influence of Lewis acids, glucose can be either isomerized into fructose or undergo a retro-aldol reaction to produce other low-molecular-weight substances.[107] Part of these reactions also requires the presence of the base catalyst.[108] The direction and rate are determined by the nature of the catalysts present and the process conditions, as well as the strength of the synergies between the catalytic sites and solvent systems. The reactor efficiency is also significantly improved in the presence of cooperative

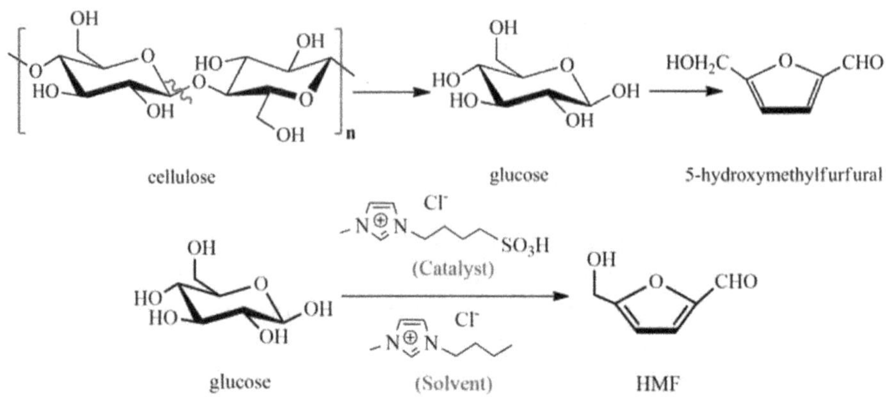

Figure 10.11 General mechanism for the production of 5-HMF from glucose using ILs as both catalyst and solvent. Reproduced from Ref. 111 with permission from American Chemical Society, Copyright 2015.

IL pairs because of dissolution and *in situ* catalytic degradation mechanisms. However, the major limitations of working with ILs lie in the strong interactions of ILs with the reaction products, which must be weakened for efficient recovery of both ILs and products.[109]

10.4.4.2 Dehydration of Glucose to 5-HMF

Dehydration of glucose to produce 5-HMF has been done in the presence of a large number of IL Brønsted acids in association of deep eutectic IL solvents.[110] These solvents can dissolve carbohydrates, even at high concentrations, and can be easily recycled. In addition, they have been shown to act not only as solvents, but also as reaction promoters for carbohydrate dehydration reactions. The reaction mechanism has been proven based on a density functional theory study choosing 1-butyl-3-methylimidazolium chloride as a representative of SO_3H-functionalized ILs (Figure 10.11).[111] Throughout most of the elementary steps, the cation of IL plays a substantial role, functioning as a proton shuttle to promote the reaction. The chloride anion interacts with the substrate and the acidic proton in the imidazolium ring *via* H bonds, as well as providing a polar environment together with the imidazolium cation to stabilize intermediates and transition states.

10.4.4.3 Depolymerization of Crystalline Cellulose into Glucose, Cellobiose and 5-HMF

The depolymerization of cellulose to produce platform molecules is a significant challenge. In this respect, the production of 5-HMF directly from microcrystalline cellulose was also reported using an alkaline IL (1,1,3,3-tetramethyl guanidine [TMG]BF_4) as a catalyst under microwave irradiation. The reaction occurred with good yields at 120 °C and at short reactions times.[112]

Table 10.2 Yield of sugars and dehydration products in the repolymerization of Avicel catalysed by IL moiety-functionalized mesoporous polymers and mineral acids. Reproduced from Ref. 114 with permission from American Chemical Society, Copyright 2017.

Catalysts	Glucose yield (%)[a]	Cellulose yield (%)[a]	TRS (%)[b]
Amberlyst 15	25.1	14.8	56.2
HCl	63.4	13.4	94.1
$[C_3vim][SO_3CF_3]$[c]	66.2	12.2	93.8
H_2SO_4	59.6	9.8	86.8
PDVB-SO_3H	56.8	10.2	82.6
PDVB-SO_3H-$[C_3vim][SO_3CF_3]$	77.0	8.2	99.6
PDVB-SO_3H-$[C_3vim][SO_3CF_3]$[d]	72.7	5.9	94.3
PDVB-$[C_3vim][SO_3CF_3]$	75.9	6.8	98.1
PDVB-SO_3H-$[C_3vim][SO_4H]$	76.8	5.6	98.5
PDVB-SO_3H-$[C_3vim][Cl]$	74.1	6.4	96.3

[a] Measured by high-performance liquid chromatography method; the reaction time was 5 h.
[b] Total reducing sugar (TRS), representing 5-HMF, was measured by the dinitrosalicylic acid (DNS) method.
[c] The same number of acidic sites as in PDVB-SO_3H-$[C_3vim][SO_3CF_3]$.
[d] The catalyst has been recycled five times.

Further combination of an ultrasonic pre-treatment with microwave and stirring treatment led to much higher HMF + furfural production yields compared to regular microwave treatment. These experiments were conducted using a choline chloride/oxalic acid eutectic mixture as the catalyst.[113]

Acidic IL moiety-functionalized mesoporous polymers have also shown activity for the depolymerization of crystalline cellulose (Table 10.2).[114] The specific features of these materials, especially the porosity, acidity and surface polarity, allows for the crystallinity of cellulose to be destroyed, leading to the production of glucose, cellobiose and 5-HMF.

10.4.4.4 Depolymerization of Lignocellulose Biomass to Levulinic Acid

An efficient liquefaction of woody biomass was also reported using various cations (1-butyl-3-methylimidazolium, 1-[3-sulfopropyl]-3-methylimidazolium and 1-[4-sulfobutyl]-3-methyl-midazolium) hydrogen sulfate acidic IL catalysts (Figure 10.12) in polyhydric alcohol, affording liquefaction yields of 99% in the range of catalyst concentration.[115] With similar catalysts (*i.e.* dicationic ILs containing 1,4-*bis*[3-methylimidazolium-1-yl] butane and $[\{2HSO_4\}\{H_2SO_4\}_4]$ counter anions), various types of lignocellulosic biomass such as rubber wood, palm oil fronds, bamboo and rice husk were converted into levulinic acid.[116]

10.4.4.5 Transesterification of Tripalmitin with Methanol

Acidic IL moiety-functionalized mesoporous polymers have shown activity for biodiesel production. The activity of these catalysts was reported to be even

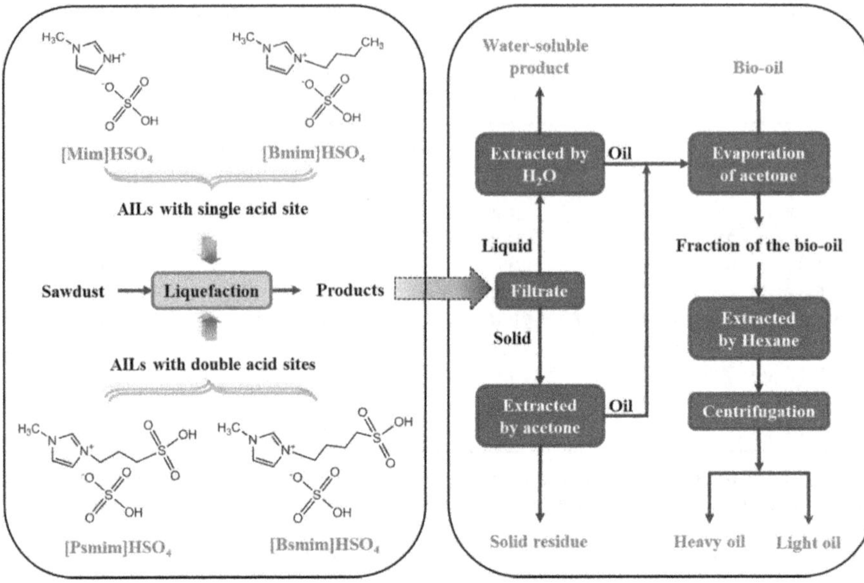

Figure 10.12 Acidic ILs (AILs) and liquefaction procedures. Reproduced from Ref. 115 with permission from Elsevier, Copyright 2015.

higher than that of mineral acids such as HCl and H_2SO_4 (Table 10.3).[95,117] Such activity was attributed to the cumulative contribution of the structural characteristics of these materials, including the high concentration of strong acidic centres, controllable surface wettability, large Brunauer–Emmett–Teller (BET) surface area and large pore widths.

10.4.4.6 Liquefaction of Bagasse Lignins in the Presence of ILs

ILs have also attracted attention as efficient solvents in the dissolution and transformation of lignins. These processes take advantage of ILs as substitutes for proton acids and combine the advantages of both homogeneous and heterogeneous catalysts.[39,118] Previous studies have reported that lignin and its model compounds can be efficiently transformed to liquid alkane and other chemicals in the presence of acidic ILs as solvents and catalysts[119] or in combination with highly dispersed nanoparticle catalysts.[120] As an example, it was demonstrated that the acidic IL [$C_4H_8SO_3$Hmim]HSO_4 acts a powerful catalyst for the direct liquefaction of bagasse lignin (Figure 10.13). This liquefaction competed well with the repolymerization process, leading to a 65.0% degree of liquefaction and a 13.5% yield of phenolic monomer under optimized conditions without char generation.[121]

Table 10.3 Activities of various catalysts in the transesterification of tripalmitin with methanol: 0.05 g of catalyst, 1.04 mmol of tripalmitin, 92.7 mmol of methanol, reaction temperature 65 °C, reaction time 16 h. Reproduced from Ref. 95 with permission from American Chemical Society, Copyright 2012.

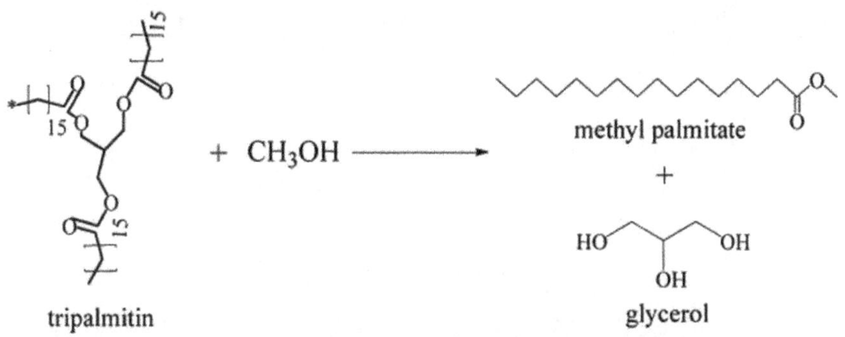

Entry	Catalyst	Yield of methyl palmitate (%)
1	PDVB-[C_1vim][SO_3CF_3]	96.9
2	PDVB-[C_3vim][SO_3CF_3]	>99.9
3	PDVB-[C_4vim][SO_3CF_3]	97.4
4	PDVB-[C_3vpy][SO_3CF_3]	99.5
5	PDVB-[C_4vpy][SO_3CF_3]	92.9
6	PDVB-[C_3vpr][SO_3CF_3]	99.3
7[a]	[C_1vim][SO_3CF_3]	89.1
8	SBA-15-[C_1vim][SO_3CF_3]	24.1
9	Amberlyst 15	28.6
10[b]	PDVB-[C_1vim][SO_3CF_3]	93.5
11[b]	PDVB-[C_3vim][SO_3CF_3]	98.1
12[b]	Amberlyst 15	20.2

[a] The same number of active sites as on PDVB-[C1vim][SO_3CF_3].
[b] Recycled five times.

10.4.4.7 Hydrodeoxygenation of Phenol and Its Derivatives

The use of phenolic bio-oil as feedstock is one of the sustainable solutions envisaged for the production of alkane as fuels. In this respect, the hydrodeoxygenation of phenol to cyclohexane and its derivatives in an aqueous medium is considered as a test reference. Dual-functional catalyst systems consisting of water-soluble, IL-like copolymers offers a very attractive solution in this area. It allows for the stabilization of nanocatalysts and of a mineral acid, H_3PO_4.[122] In the series Ru > Rh > Pt@Pd, Ru provided highest catalytic activity of the metal sites for this application.

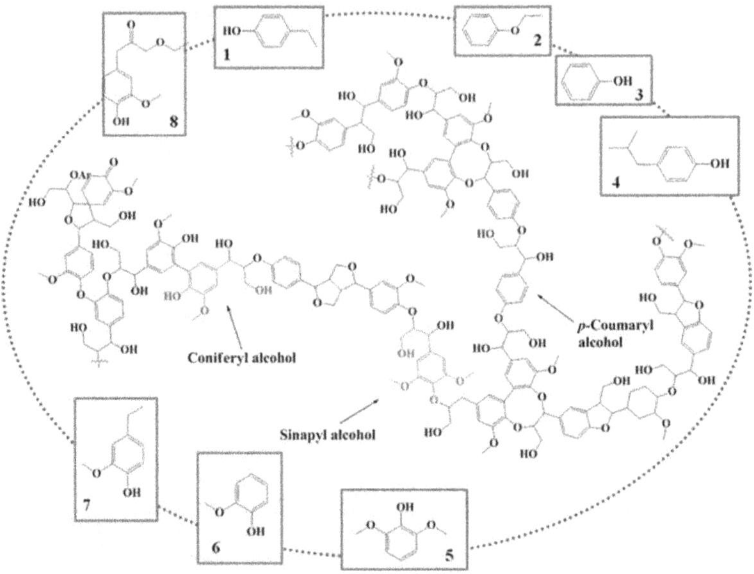

Figure 10.13 Fragmentation of a bagasse lignin using an acidic $[C_4H_8SO_3Hmim]$ HSO_4 catalyst. Reproduced from Ref. 121 with permission from Elsevier, Copyright 2014.

10.4.4.8 Biodiesel and Enzymatic Processes for Biomass Valorisation

The combination of biocatalysts and ILs has also been suggested as promising for industrial use and as an alternative to conventional organic chemistry processes employing organic solvents.[123] The first examples reported on lipase-catalysed transesterification, amidation and epoxidation reactions in ILs for the immobilized *Candida antarctica* lipase B (Novozym® 435) catalysed in anhydrous [bmim][PF_6].

ILs are also able to solubilize biomass and, in the presence of catalysts, convert the biomass into useful platform chemicals. As an example, mixtures of ILs with other solvents and a catalyst provide powerful systems for the selective catalytic transformation of cellulose into 5-HMF (Figure 10.14).[124] Then, the continuous 5-HMF extraction into solvents as methyl-isobutyl-ketone or 1,2-dimethoxyethane, which form a biphase with the IL mixture, makes it easy to further separate 5-HMF in order to achieve high yields.

Mechanistic studies indicate that the nature of the IL anion is critical, contributing to the formation of H bonds.[125] Indeed, good basicity is favourable to the process, whereas weakly basic anions are less effective for cellulose dissolution, and ILs containing non-coordinating anions are unable to dissolve cellulose. An increase in the size of the anion is also important because it lowers

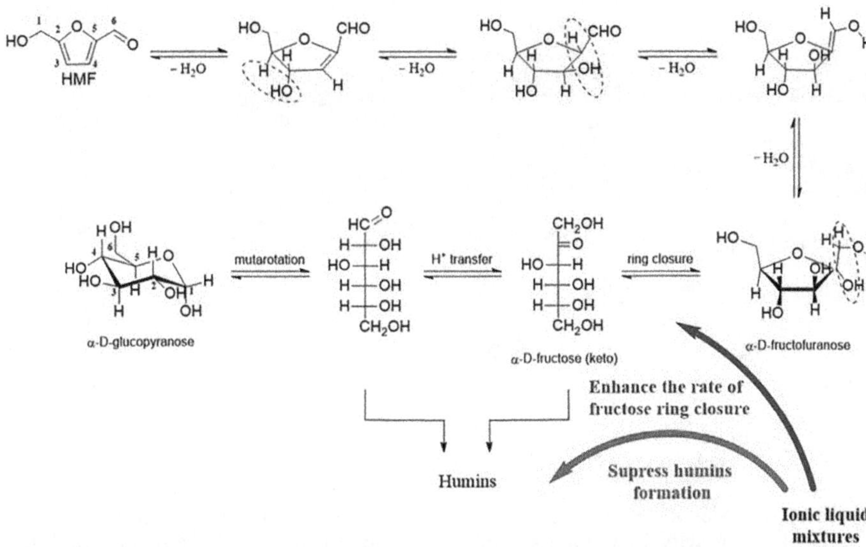

Figure 10.14 Reaction pathways for the dehydration of glucose to fructose and subsequent transformation to 5-HMF through a pathway involving a cyclic fructo-furanosyl intermediate. Reproduced from Ref. 124 with permission from John Wiley and Sons, Copyright © 2016 Wiley-VCH Verlag GmbH & Co. KGaA, Weinheim.

its ability to form H bonds, and by consequence to dissolve cellulose. Although less influential for the anion, the size and functional groups attached to the cation also play significant roles in the solubility of cellulose.[125]

Biocatalysis has also been appealing in the increasing demand for platform chemicals and fuels from renewable sources. However, recalcitrance towards traditional chemical processes and solvents provides a significant barrier to their widespread utility. Therefore, the optimization of the solubility of enzymes in ILs allowed progress in this area. For the particular case of glucosidase, by using [emim][$EtSO_4$], a solvent-induced substrate, promiscuity has been demonstrated for the homogeneous enzyme bioprocessing of cellulose.[126] It produced reducing sugars from cellulose without the need for endo- or exo-cellulases. According to the proposed mechanism, glucosidase indiscriminately attacks glycosidic linkages along the solubilized cellulose strands, and then the smaller oligosaccharides yield glucose (Figure 10.15). The IL facilitates this mechanism by solubilizing both the biocatalyst and the cellulose in a single phase.

Termophilic β-galactosidase is a biocatalyst from *Thermotoga naphthophila* that has also been utilized in combination with a variety of structurally diverse ILs for the production of alkyl galactopyranosides from the milk processing waste lactose as a galactosyl donor. The thermophilic nature of β-galactosidase increased the solubility of the substrate, suppressed

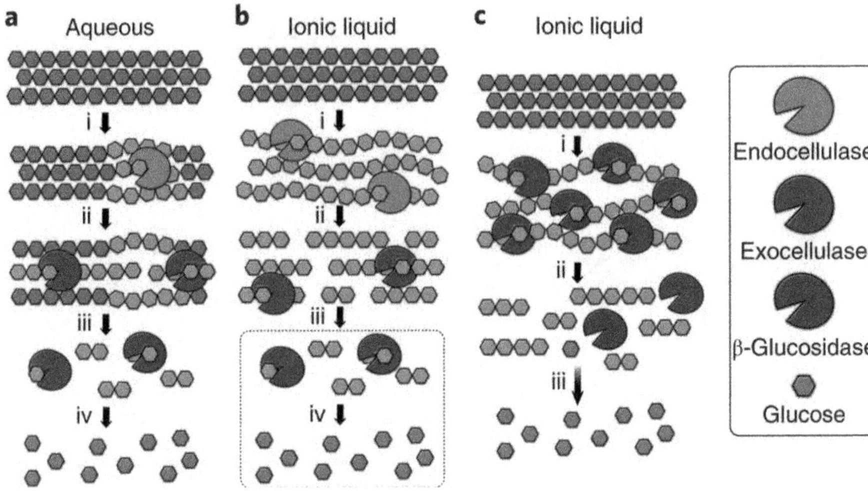

Figure 10.15 Mechanisms of cellulase activity. (a, b) Cartoons depicting cellulose activity in aqueous buffer solution (a) and IL (b) showing swelling (dissolution in appropriate IL) of crystalline cellulose (i), endocellulase-catalysed hydrolysis of amorphous cellulose to reveal individual cellulose chains (ii), exocellulase-catalysed hydrolysis of cellulose chains to oligosachharides (iii) and β-glucosidase-catalysed conversion of oligosachharides (predominantly cellobiose) into glucose (iv). (c) Cartoon detailing the proposed mechanism for cellulose hydrolysis by β-glucosidase solubilized and stabilized in ILs, showing dissolution of cellulose in IL (i) and random hydrolysis of the cellulose chain by activated β-glucosidase, leading to a mixture of oligosaccharides (ii) that will eventually be converted into glucose over time (iii). Reproduced from Ref. 126 with permission from Springer Nature, Copyright 2018.

hydrolysis and increased the biocompatibility of the enzyme.[127] These synthesis processes used *n*-butanol to *n*-tetradecanol as alkyl acceptors and lactose as a galactosyl donor.

10.4.4.9 Fuel Production Using ILs as Solvents and Catalysts

The biocatalytic oxidation of cellulose-derived carbohydrates represents another potential valorization of biomass for energy production in biofuel cells.[128] An example is the combination of biocompatible hydrated choline dihydrogen phosphate ILs containing 35 wt% water as a solvent with cellobiose dehydrogenase as an enzyme catalyst for the oxidation of cellobiose to cellobionolactone (Figure 10.16).

Methanolysis of trioleins to produce biodiesel is another appealing procedure for the sustainable valorization of biomass.[129] However, it is obviously limited by the nature of the catalyst. The homogeneous catalysts usually remain entrapped in the biodiesel, producing corrosion of the engine, while

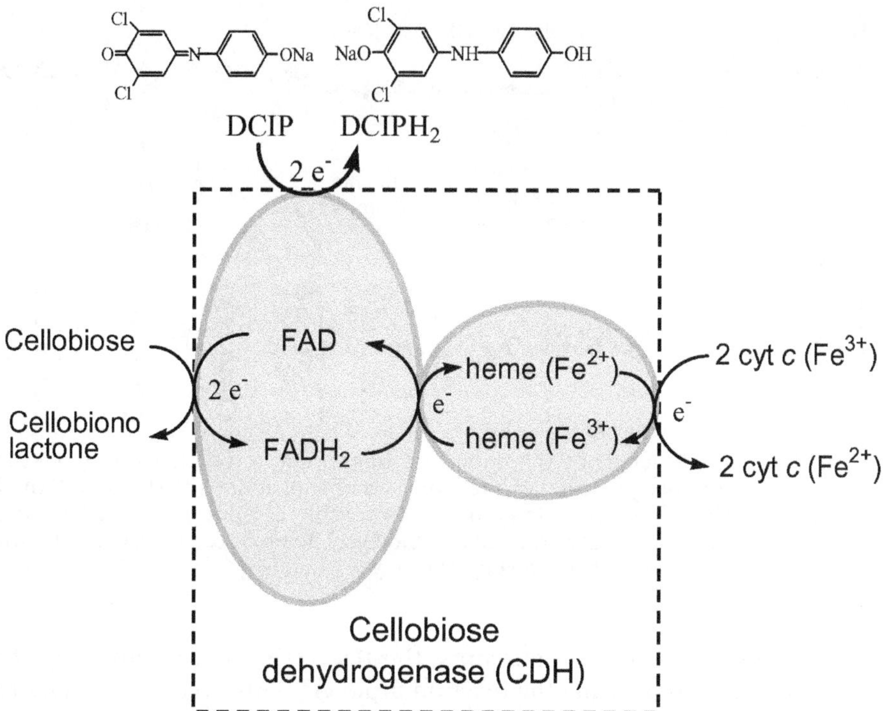

Figure 10.16 CDH-catalysed oxidation of cellobiose by inter- and intra-electron transfer pathways. Reproduced from Ref. 128 with permission from the Royal Society of Chemistry.

the heterogeneous catalysts are less effective and also difficult to be separate. ILs based on cations with long alkyl side-chains such as N-octadecyl-N', N'', N'''-trimethyl-ammonium *bis*(trifuoromethylsulfonyl)imide ([C18tma][NTf2]) behave as sponge-like systems for soaking and releasing (wringing) hydrophobic compounds.[130] These temperature-switchable IL/solid phases acting as reaction/separation media may represent a clean biocatalytic approach for producing biodiesel. As liquid phases, these ILs are excellent monophasic reaction media for the lipase-catalysed methanolysis of triolein, resulting in fast and efficient biodiesel synthesis, while as solid phases, the reaction mixture can easily be fractionated by iterative centrifugations at controlled temperature into three phases: solid IL, glycerol and pure biodiesel (Figures 10.17 and 10.18). This represents a straightforward and sustainable approach for producing biodiesel, allowing full recovery and reuse of the biocatalyst–IL system for successive cycles and suitable for scaling up.[130]

Transesterification of triglyceride-type fish oil containing elevated concentrations of omega-3 fatty acids and ethyl esters by immobilized lipase using imidazolium-based ILs also led to superior performances compared to other reaction systems.[131]

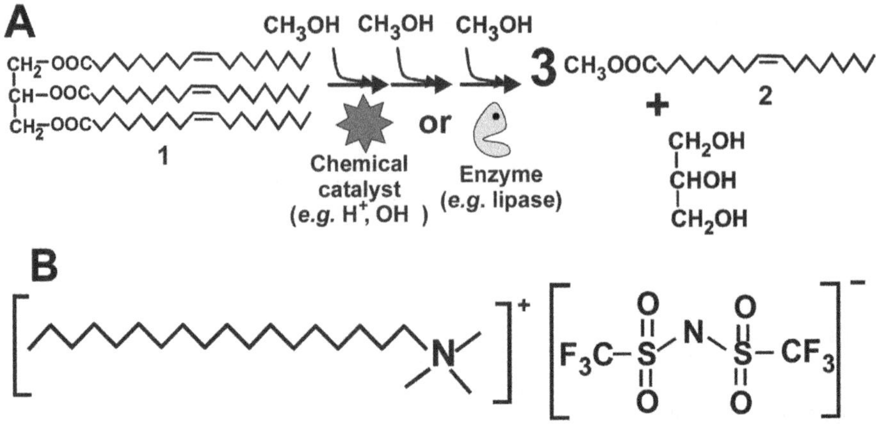

Figure 10.17 (a) Scheme of the immobilized lipase-catalysed synthesis of biodiesel (2, methyl oleate) by methanolysis of triolein (1). (b) The IL [C18tma][NTf2] as an example of the switchable IL/solid phase used for the synthesis and purification of biodiesel. Reproduced from Ref. 130 with permission from the Royal Society of Chemistry.

Despite these successful attempts, the ILs retain their disadvantage of the difficulty of separating the reaction products. This becomes even more difficult when using heterogeneous catalysts. Therefore, a suitable reaction environment that mimics solvents in close proximity to the active sites of the heterogeneous catalysts is necessary. Such an environment should facilitate the separation of the products from high-boiling-point solvents or non-volatile ILs thus simplifying the process. However, the transfer of such complex systems from the liquid phase to a solid state is still challenging because this requires precise control of the spatial continuity and dynamic interactions between the immobilized partners.[132] To achieve this task, several conditions are recommended: (1) amenable synthesis to enable the incorporation of various catalytic components for engineering the chemical microenvironment around the active sites; (2) adjustable chemical composition, allowing for careful positioning of individual functions and forming a well-defined array of non-covalent interactions around a catalytic site; (3) a large surface area with hierarchical porosity to enable fast mass transfer; and (4) robustness under various chemical conditions, allowing for long-term stable performance and potential recycling to simplify the work-up procedures.[132]

Such a microenvironment has been designed by the incorporation of sulfonic acid or imidazole groups in porous polymeric analogues of solvents such as ILs (Figure 10.19). Due to the retained strong solvation ability of the densely populated solvent moieties to fructose, such sulfonic acid groups incorporating porous solid solvents show excellent catalytic performance in the dehydration of fructose to 5-HMF as a sole product, using tetrahydrofuran (THF) as a readily separable solvent.

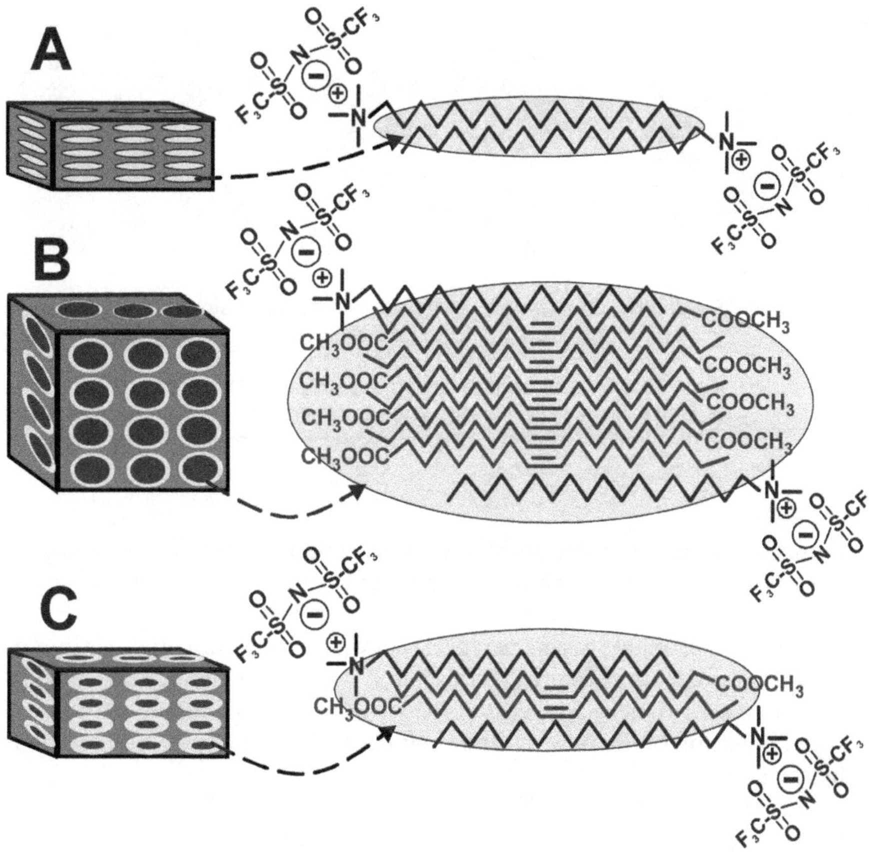

Figure 10.18 Schematic representation of the sponge-like IL hypothesis, showing the [C18tma][NTf2] net as a dry sponge (a), a sponge swollen with methyl oleate (b) and a "wet" sponge after wringing out by centrifugation (c). Reproduced from Ref. 130 with permission from the Royal Society of Chemistry.

10.4.5 CO_2 Valorization

Although non-toxic, carbon dioxide is a major greenhouse gas. Its abundance requires efficient sustainable valorization approaches including recycling to fuels and chemical fixation into useful organic compounds.[133]

10.4.5.1 Coupling of CO_2 with Epoxides

Carbon dioxide is a pollutant greenhouse gas that, owing to its abundance, low cost, non-toxicity and non-flammability, can be regarded as an ideal C_1 chemical feedstock. However, due to its inertness, its subsequent transformation into valuable products requires the use of catalysts and also reaction conditions involving high temperatures and pressures. Within this scope, both homogeneous and heterogeneized ILs have been used.[134]

Figure 10.19 Schematic illustration of reactants as representative of fructose in various systems. (a) A flawlessly engineered enzyme space, (b) a microenvironment with solvent-like behaviour inspired by nature and a solution system and (c) solvent containing HBAs. Reproduced from Ref. 132, https://doi.org/10.1038/s41467-018-05534-5, under the terms of the CC BY 4.0 licence, http://creativecommons.org/licenses/by/4.0/.

From the thermodynamic point of view, the coupling of CO_2 with epoxides using homogeneous IL catalysts is the most accessible reaction, leading to five-membered cyclic carbonates. Ethylene and propylene carbonate are the most common products of this reaction. They have important applications as polar aprotic solvents or intermediates in the production of pharmaceuticals. Due to its performance, coupling of CO_2 with epoxides became a reference for the evaluation of the new ILs and their capability of coupling with various substrates.

Many catalysts have been effectively developed for the fixation of CO_2 and epoxides to produce cyclic carbonates. ILs are among these investigated catalysts,[135–142] and the fixation of CO_2 and ethylene oxide to produce ethylene carbonate was even commercialized by Bayer and Mitsubishi.

Most of the investigated ILs are quaternary ammonium-, phosphonium-, imidazolium- or pyridinium-based cations with inorganic counteranions.[134] In addition to these, Lewis basic ILs derived from 1,4-diazabicycl-[2.2.2] octane (DABCO), 1,8-di-azabicyclo[5.4.0]undec-7-ene (DBU), 1,5,7-triaza-bicyclo[4.4.0]dec-5-ene (TBD) and hexamethylene-tetramine (HMTA) have been investigated as well (Figure 10.20).[141] However, a series of elements differentiate ILs for the conversion of CO_2 to cyclic carbonate with high efficiency, such as their water or air sensitivity,[143,144] their behaviour at high CO_2 pressures[145,146] and their use of expensive transition metal additives.[147–149]

The mechanism for fixation of CO_2 with epoxides catalysed by quaternary onium salts[150] involves ring opening of the epoxide through nucleophilic attack by the anion, which leads to an oxy anion species affording the corresponding cyclic carbonate after reaction with CO_2 (Figure 10.21). The reaction is affected by the chain length and anion of quaternary ammonium.[151]

Mechanistic studies demonstrated that methyl groups on the cation of simple imidazolium salts do not significantly affect the rates and yields of the reaction of CO_2 with epoxide. A similar conclusion has also been drawn for the

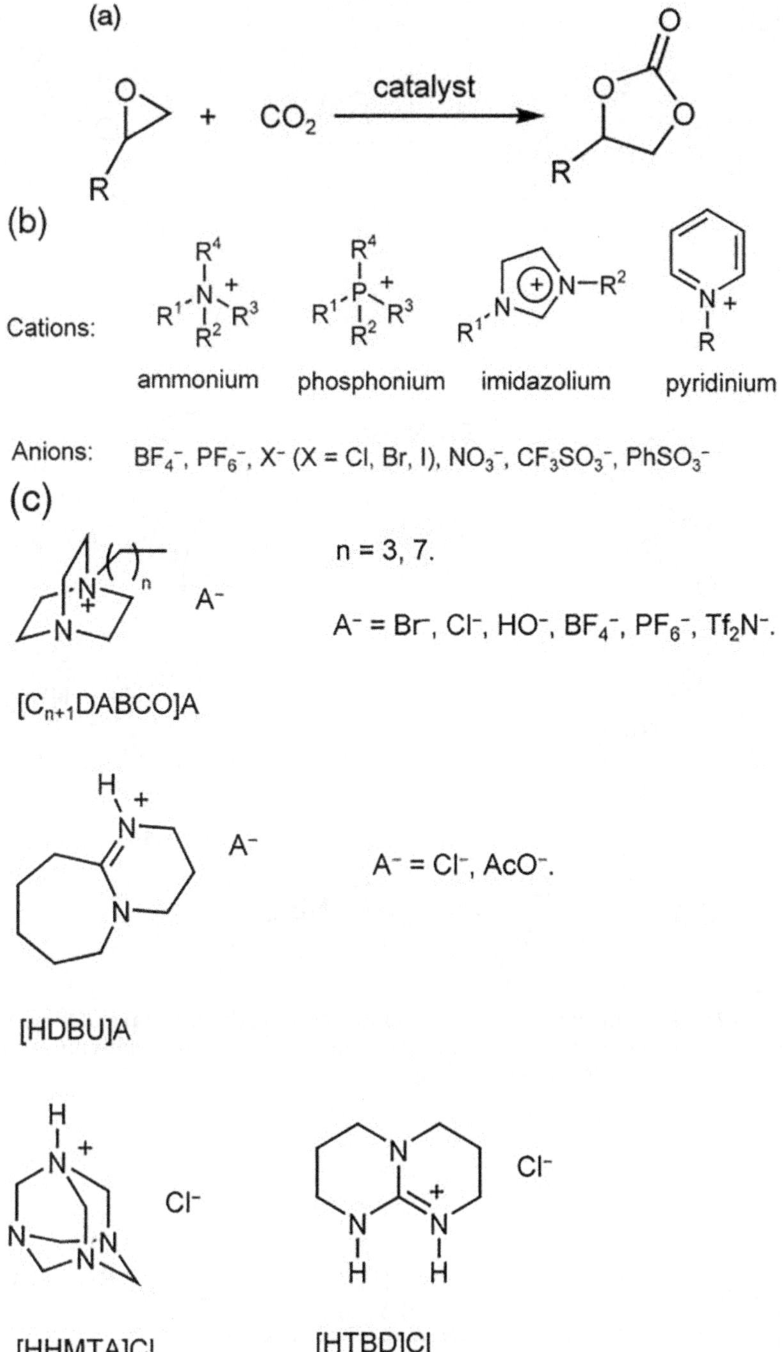

Figure 10.20 (a) Cycloaddition of CO_2 to epoxide. (b) Quaternary onium liquids used in the synthesis of cyclic carbonate. (c) Examples of Lewis basic ILs. Reproduced from Ref. 141 with permission from John Wiley and Sons, Copyright © 2010 Wiley-VCH Verlag GmbH & Co. KGaA, Weinheim.

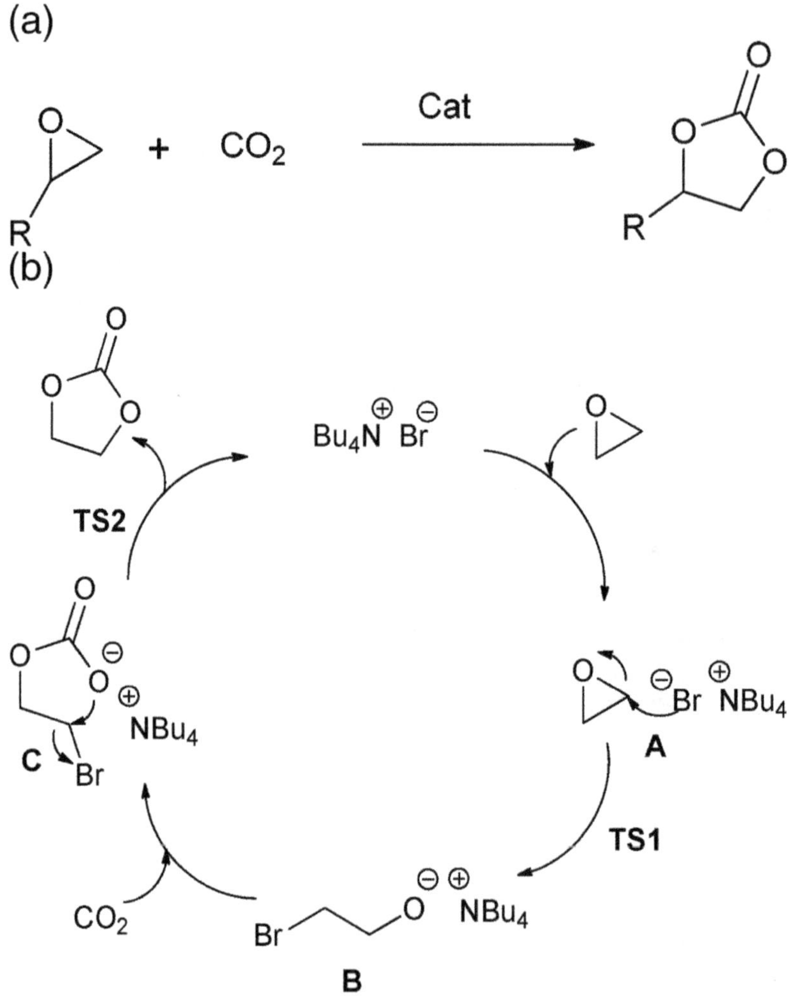

Figure 10.21 Fixation of CO_2 with epoxides into cyclic carbonates (a) and the reaction mechanism (b). Reproduced from Ref. 151 with permission from the Royal Society of Chemistry.

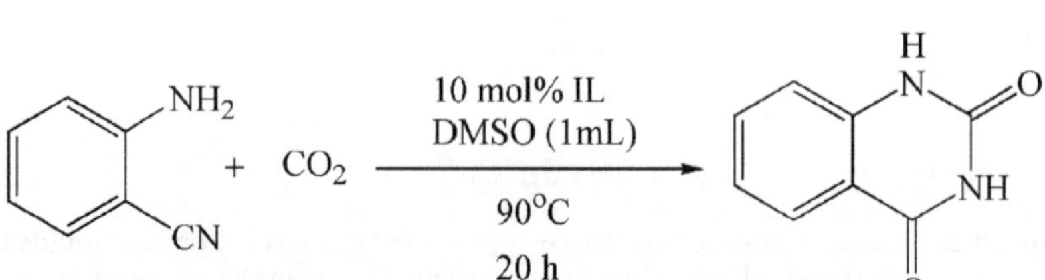

Figure 10.22 Synthesis of quinazoline-2,4(1*H*,3*H*)-dione.

synthesis of quinazoline-2,4(1*H*,3*H*)-dione (Figure 10.22).[152] Instead, it plays a secondary role in controlling the physical properties of the IL. A linear relationship between the pK_a of the IL anion (conjugate acid) and the reaction rate was identified, with maximum catalyst efficiency observed at pK_a >14.7. The base-catalysed reaction is limited by the acidity of the quinazoline-2,4-dione product, which is deprotonated by more basic catalysts, leading to the formation of the quinazolide anion (conjugate acid pK_a = 14.7).[152] The acidic protons in the imidazolium ring of the cation (C_2, C_4, C_5 or a combination thereof) simultaneously activate the epoxide[153] and interact with the anion to reduce its nucleophilicity.[154] Accordingly, in the absence of acidic protons, the epoxide is presumably less well activated, but the nucleophilicity of the anion is higher due to weaker cation–anion pairing, leading to higher activities. Therefore, the preferred anion should be strongly cation dependent.[154] The H-bonding interactions between the cation and substrate are mostly counterbalanced by a decrease in nucleophilicity of the anion, as has been confirmed by the relative activity of three closely related catalysts (*i.e.* *N*,*N*,*N*-triethyl-*N*-hydroxyethylammonium iodide > *N*,*N*,*N*-triethyl-*N*-hydroxyethylammonium bromide > 1-hydroxyethyl-3-methyl-imidazolium chloride).[154]

Obviously, under supercritical CO_2 conditions, activity is enhanced due to the complete miscibility of CO_2 in the ILs under high-pressure conditions. With 1-octyl-3-methylimidazolium tetrafluoroborate as the catalyst, for example, under supercritical conditions (14 MPa CO_2, 110 °C), the cycloaddition of CO_2 and propylene oxide could complete within 5 min with a turnover frequency (TOF) of 516 h^{-1}.[155]

Further functionalization of the ILs by a hydroxyl or carboxylic acid group could even enhance their catalytic performance as an effect of the increase of the polarization of the C–O bond due to the interaction of the acidic hydrogen with the epoxide oxygen (Figure 10.23).[156–160] Alkoxide-functionalized imidazolium betaines exhibit a similar behaviour for the activation and catalytic transformation of CO_2.[161] They have strong tendencies for CO_2 capture, affording a CO_2 adduct, as demonstrated for the reaction of propargylic alcohols with CO_2 under solvent-free reaction conditions (Figure 10.24). Then, microwave-assisted synthesis may also have a positive impact when using these ILs for the production of cyclic carbonates.[162]

The functionalization of the ILs by fluorine exerts a very similar effect to that presented by the hydroxyl group, but it is more stable in various chemical environments. These ILs proved to be efficient and recyclable homogeneous CO_2-soluble catalysts for the solvent-free synthesis of cyclic carbonates from epoxides and CO_2 under supercritical carbon dioxide ($scCO_2$) conditions (Figure 10.25).[163–165]

The presence of water does not deactivate these functionalized ILs, and it can even accelerate the reaction due to the formation of hydrogen bonds.[166]

IL-functionalized PEGs are another family of water-soluble linear polymers with CO_2-philic properties (Figure 10.26). Their catalytic activity has been demonstrated for the chemical fixation of CO_2 with various epoxides, including in solvent-free and supercritical conditions.[167–170] In addition to

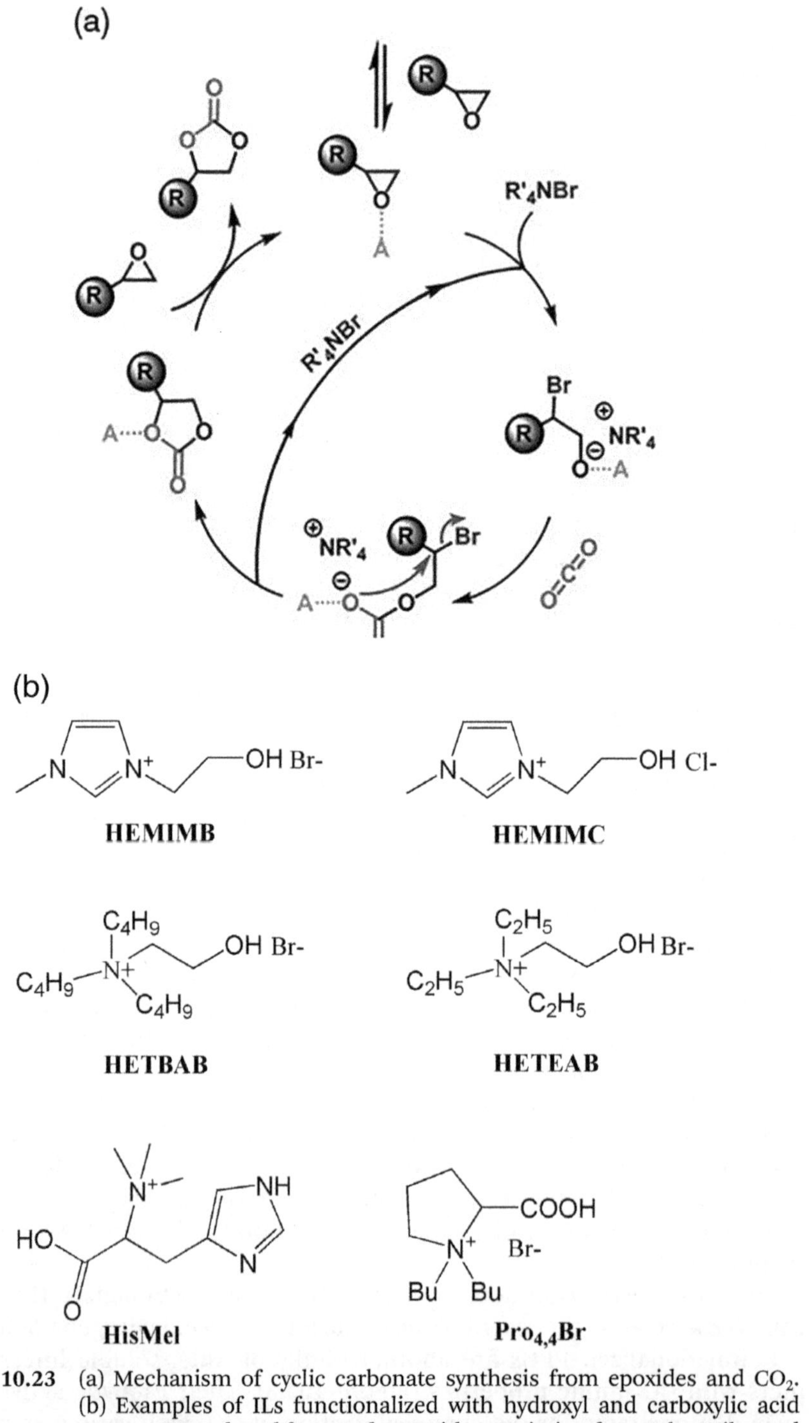

Figure 10.23 (a) Mechanism of cyclic carbonate synthesis from epoxides and CO_2. (b) Examples of ILs functionalized with hydroxyl and carboxylic acid groups. Reproduced from Ref. 156 with permission from John Wiley and Sons, Copyright © 2009 Wiley-VCH Verlag GmbH & Co. KGaA, Weinheim.

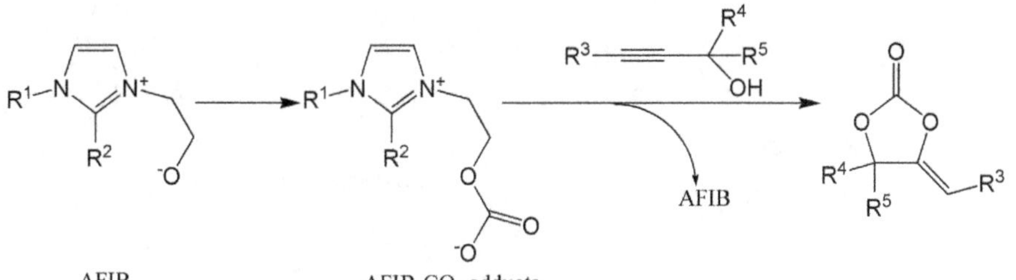

Figure 10.24 Alkoxide-functionalized imidazolium betaine–CO_2 adduct-catalysed carboxylative cyclization of propargylic alcohols with CO.[161]

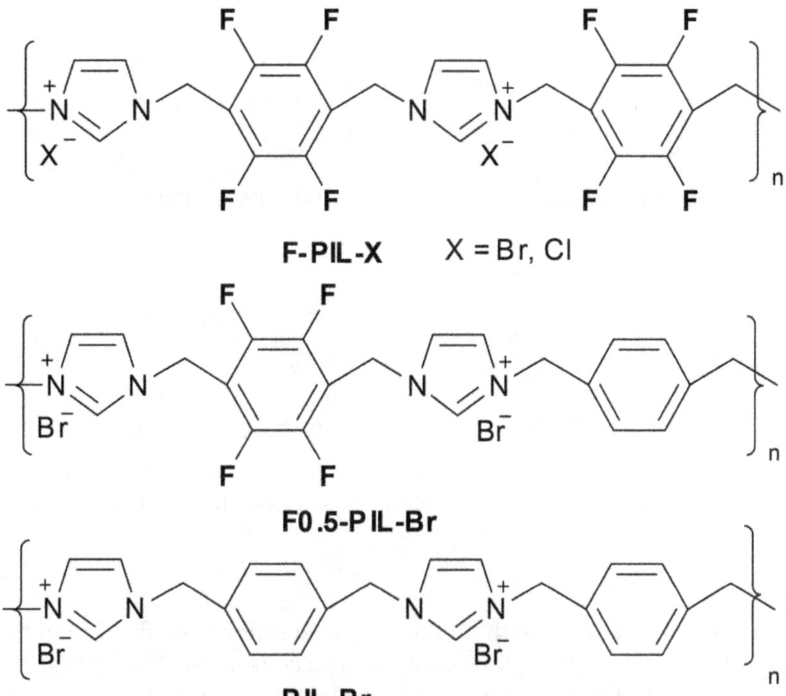

Figure 10.25 Structures of fluorine-functionalized ILs. Reproduced from Ref. 165 with permission from the Royal Society of Chemistry.

this, quaternary ammonium bromide-functionalized PEG has been indicated as a highly efficient and recyclable catalyst for the selective synthesis of 5-aryl-2-oxazolidinones from carbon dioxide and aziridines under solvent-free conditions.[171]

Another reaction of interest catalysed by ILs is the synthesis of quinazoline-2,4(1*H*,3*H*)-dione derivatives through the incorporation of CO_2 into aminobenzonitriles,[172] where the recovery of ILs is performed by simple filtration (Figure 10.27).

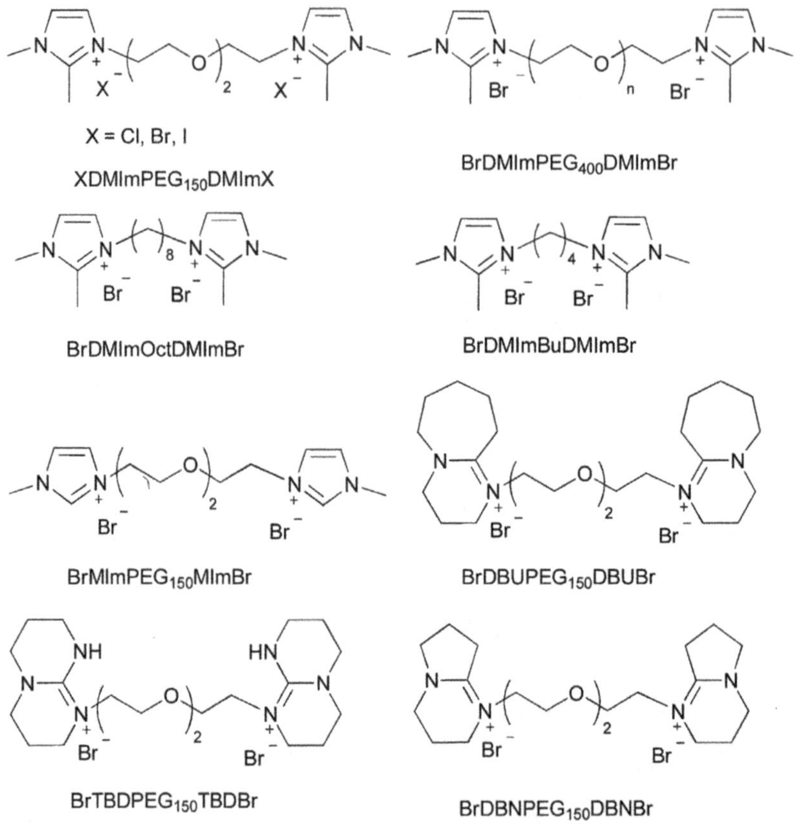

Figure 10.26 Examples of PEG-functionalized ILs for the chemical fixation of CO_2 with various epoxides. Reproduced from Ref. 170 with permission from the Royal Society of Chemistry.

The association of ILs with metals may also expand the fixation of CO_2 for the synthesis of various organic compounds. Thus, a Cu(I) salt/protic IL catalytic system can efficiently catalyse the cycloaddition of CO_2 with propargylic alcohols to produce α-alkylidene cyclic carbonates under mild conditions (Figure 10.28),[173] while a Cu(I) salt and the IL tetrabutylphosphonium imidazole ([P4444][Im]) may act as a catalyst for the three-component reaction of terminal propargyl alcohols, CO_2 and amines.[174]

10.4.5.2 Chemical Fixation of CO_2 into α-alkylidene Cyclic Carbonates

ILs with protic properties such as 1,8-diazabicyclo-[5.4.0]-7-undecenium 2-methylimidazolide may promote the reaction between CO_2 and propargylic alcohol for the synthesis of α-alkylidene cyclic carbonates.[175] Both the cation and anion of the IL play key synergistic roles in promoting the reaction. The

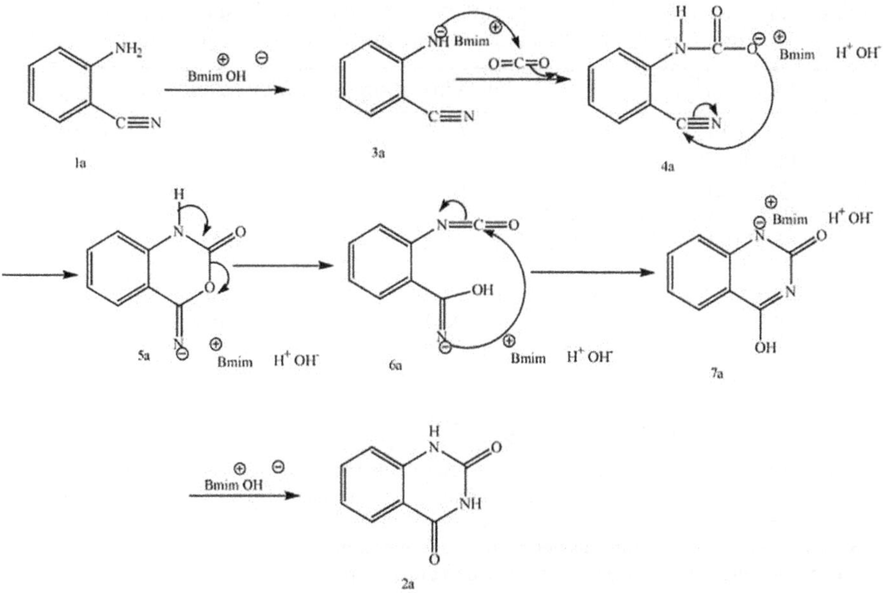

Figure 10.27 The reaction path for the synthesis of quinazoline-2,4(1*H*,3*H*)-dione derivatives through incorporation of CO_2 into aminobenzonitriles. Reproduced from Ref. 172 with permission from Elsevier, Copyright 2009.

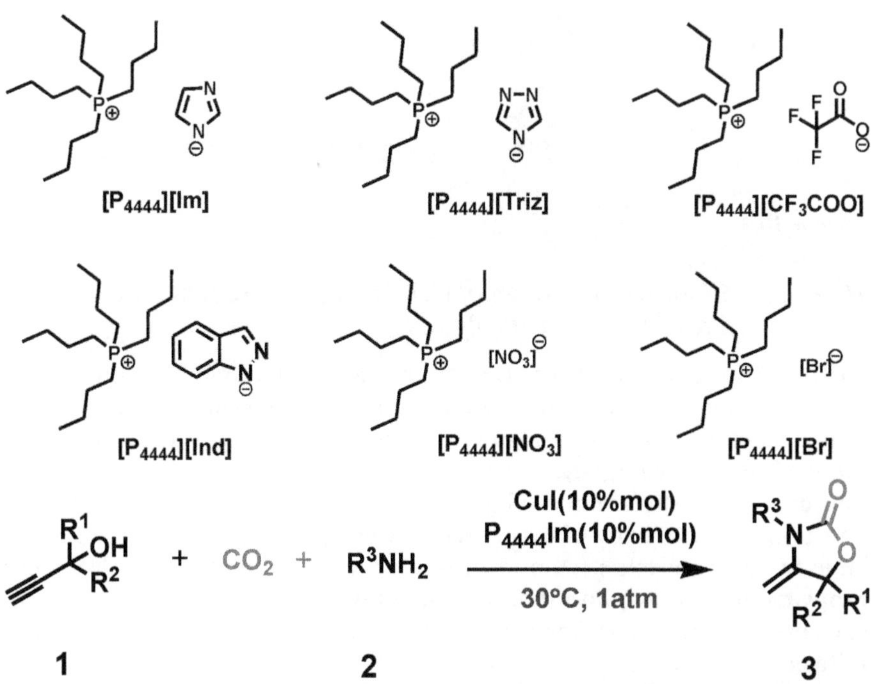

Figure 10.28 Cycloaddition reaction of propargyl alcohol and amines with CO_2. Reproduced from Ref. 174, [https://doi.org/10.3390/molecules24071241], under the terms of the CC BY 4.0 license [https://creativecommons.org/licenses/by/4.0/].

Table 10.4 Catalytic activity for propylene carbonate synthesis.[a] Reproduced from Ref. 176 with permission from the Royal Society of Chemistry.

Entry	Catalysts	Phosphorous[b] (mmol)	Yield (%)	k^c (min^{-1})
1	$P(n\text{-}Bu)_4Cl$	0.57	10	0.18
2	$P(n\text{-}Bu)_4Br$	0.57	4	0.07
3	$P(n\text{-}Bu)_4I$	0.57	5	0.09
4	$PS\text{-}(4\text{-}C_6H_4)CH_2\text{-}P(n\text{-}Bu)_3Cl^d$	0.42	8	0.19
5	$PS\text{-}(4\text{-}C_6H_4)CH_2\text{-}P(n\text{-}Bu)_3Br^d$	0.41	9	0.22
6	$SiO_2\text{-}C_3H_6\text{-}P(n\text{-}Bu)_3Cl$	0.26	29	1.3
7	$SiO_2\text{-}C_3H_6\text{-}P(n\text{-}Bu)_3Br$	0.18	53	4.1
8	$SiO_2\text{-}C_3H_6\text{-}P(n\text{-}Bu)_3I$	0.21	100	26
9	$SiO_2\text{-}C_3H_6\text{-}PEt_3Cl$	0.44	44	1.3
10	$SiO_2\text{-}C_3H_6\text{-}P(n\text{-}Oct)_3Cl$	0.14	12	0.85
11	$SiO_2\text{-}(4\text{-}C_6H_4)CH_2\text{-}P(n\text{-}Bu)_3Cl$	0.20	9	0.44
12	$SiO_2\text{-}(4\text{-}C_6H_4)CH_2\text{-}P(n\text{-}Bu)_3Br$	0.18	37	2.5

[a] 100 °C, 1 h, 10 MPa of CO_2, 4 mL (57.2 mmol) of propylene oxide, 0.57 mmol of homogeneous catalyst or 400–800 mg of immobilized catalyst (0.57 mmol halogen based on the initial halogen content before quaternization).
[b] Absolute amount of phosphorus determined by elemental analyses.
[c] Pseudo-first-order rate constant normalized to phosphorus.
[d] The polystyrene portion consisted of ArgoPore2 resins.

catalytic activity and selectivity of the IL after separation from the reaction mixture can be preserved simply by adding water.

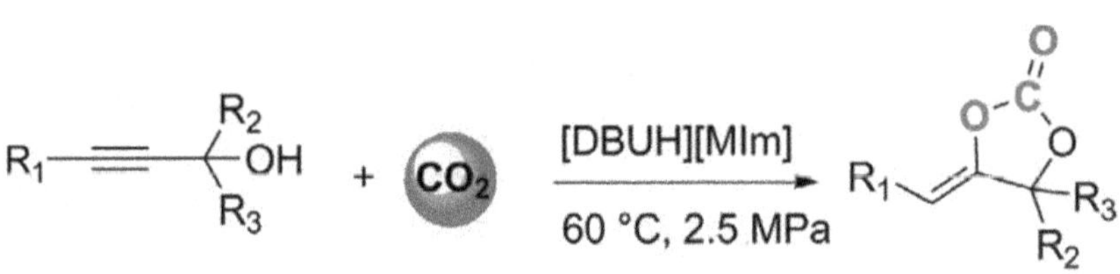

Scheme 10.2

10.4.5.3 *Chemical Fixation of CO_2 Using Heterogeneous Catalysts Containing ILs*

Further immobilization of ILs as surface-functionalized ionosilicas enhances their catalytic performances in this reaction (Table 10.4).[176] This behaviour has been assigned to a possible acceleration mechanism induced by the synergistic hybrid catalyst in association with an activation due to the surface silanol groups (Figure 10.29). A synergistic effect on the catalytic activity of this reaction was also reported for other recyclable hybrid catalysts prepared through the coupling of 3-(triethoxysilyl)propyltriphenylphosphonium bromide and mesoporous silica, where the solvent-free and metal-free reactions proceeded successfully under very mild conditions.[177]

IL-functionalized mesoporous silica resulting from the coupling of 1-(triethoxysilylpropyl)-3-*n*-alkylimidazolium halides with tetraethyl orthosilicate

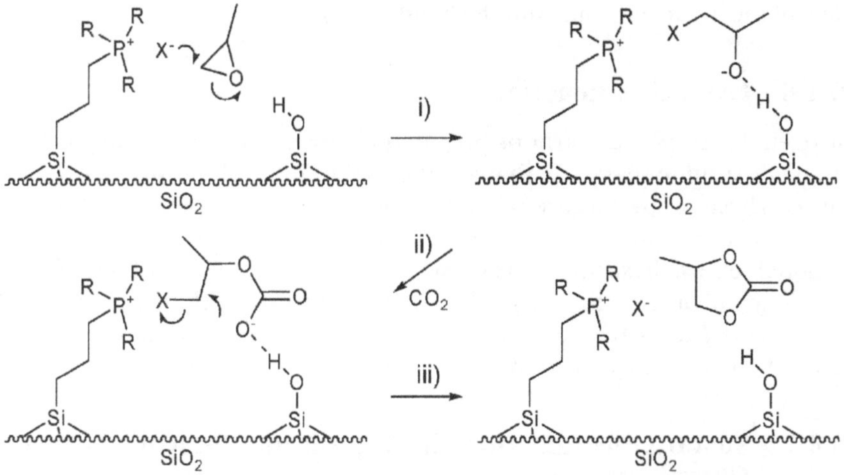

Figure 10.29 Possible acceleration mechanism on the synergistic hybrid catalyst. Reproduced from Ref. 176 with permission from the Royal Society of Chemistry.

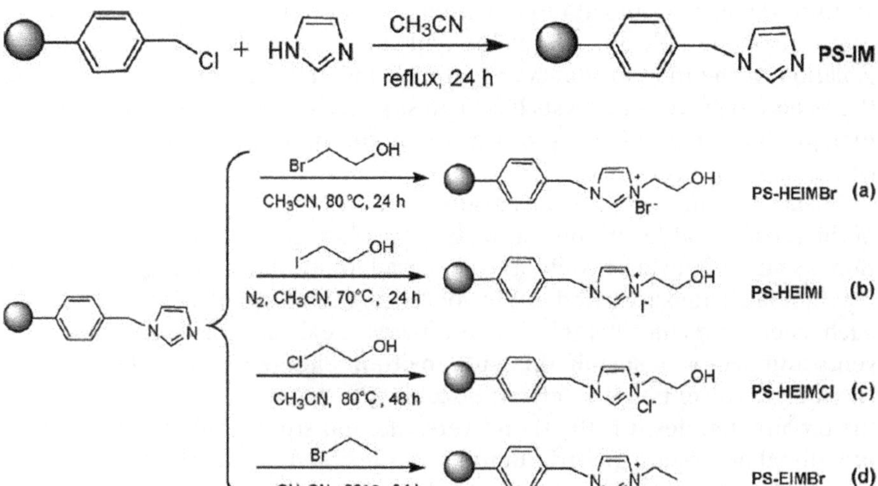

Figure 10.30 Synthesis of polymer-supported ILs. Reproduced from Ref. 179 with permission from Elsevier, Copyright 2009.

are also active catalysts,[178] as proven for the solventless synthesis of cyclic carbonate from allyl-glycidyl ether and carbon dioxide. A longer alkyl chain length and a more nucleophilic nature provided greater conversion of the ether.

1-(2-Hydroxyl-ethyl)-imidazolium-based ILs with both acidic and basic characteristics were also covalently anchored onto cross-linked polystyrene resins (Figure 10.30).[179] Their catalytic activity for the synthesis of cyclic carbonates *via* the cycloaddition reaction of CO_2 has been associated with the synergetic effect of the hydroxyl group and the halide anion.

10.4.6 Oxidation Reactions

Nitrogen-doped porous carbons prepared *via* the conversion of poly(ILs) are rich heteroatomic materials that are thermally stable at temperatures up to 400 °C. These properties are related to their ionic and aromatic nature and the presence of some network-forming groups. By comparison with more conventional routes, these materials show very high CO_2 uptakes with satisfactory selectivity and outstanding reversibility. Equally importantly, they serve excellently as metal-free carbocatalysts for the selective aerobic oxidation of benzyl alcohol to benzaldehyde with high conversion and selectivity.[180]

10.4.7 Sustainable Reactions in Organic and Pharmaceutical Chemistry

10.4.7.1 C–C Coupling

Due to the industrial relevance of the Heck reaction as a very important synthetic tool that enables access to molecular moieties often found in the preparation of pharmaceutical targets, intense effort has been spent on finding protocols using efficient, recyclable and reusable catalytic systems, simple isolation of the final products and minimal waste production.[181,182] Among these, heterogeneous catalysts based on supported IL-like phases ensure very high productivities, allowing the transfer of the application into the flow process (Figure 10.31).[183]

Suzuki–Miyaura coupling has also served over the last two decades as one of the most versatile and powerful strategies for achieving C–C bond formation because of the low toxicity and easy availability of boron reagents.[184] This reaction is obviously carried out in volatile organic solvents. However, a DES such as choline chloride/glycerol eliminates the disadvantages of these solvents and acts as a sustainable and environmentally responsible medium. In eutectic solvents, Pd-catalysed Suzuki–Miyaura cross-couplings between (hetero)aryl halides (Cl, Br, I) and versatile, moisture-stable mono- and bifunctional potassium aryl-tri-fluoroborates proceed efficiently and chemoselectively in air and under generally mild conditions.[185]

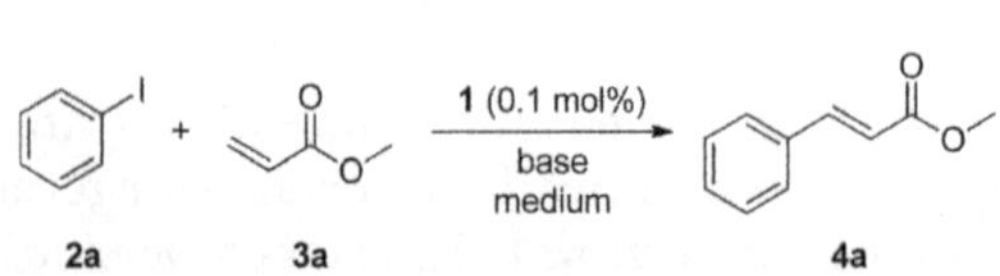

Figure 10.31 Pd-IL-based catalyst and Heck coupling reaction. Reproduced from Ref. 183 with permission from American Chemical Society, Copyright 2014.

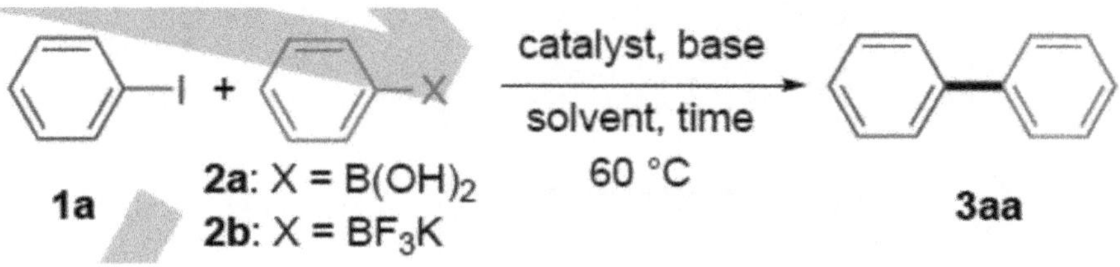

Scheme 10.3

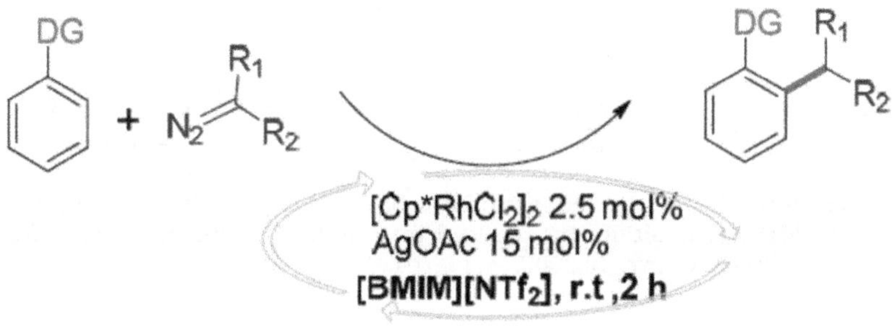

Scheme 10.4

An elegant strategy to balance the high dissociation energy of the C–H bonds (105 kcal mol^{-1} for methane and 110 kcal mol^{-1} for benzene) is to find adequate catalysts and reaction environments. To this end, ILs [BMIM][NTf_2] was found to be an ideal medium in which the catalytic C–H carbenoid coupling occurred efficiently and sustainably. More significantly, the presence of the IL, in addition to recycling the catalyst system, led to improvements in the efficiency of the catalyst. Under these conditions, the carbenoid coupling reactions can occur at room temperature over a short period of time.[186]

The Henry reaction is a base-catalysed C–C bond-forming reaction between nitroalkanes and aldehydes or ketones, and is typically referred to as the nitroaldol reaction. ILs have also been investigated in this reaction. However, *p*-nitro-benzene with nitroethane *bis*-silylated *N,N'*-dialkyl imidazolium showed greater catalytic activity compared to the IL BMIM-PF_6. In addition, functionalized SBA-15 materials showed low catalytic activity. However, passivating the IL-functionalized SBA-15 materials *via* end-capping reactions with 1,1,1,3,3,3-hexamethyldisilazane led to highly active heterogeneous catalysts for Henry reactions. This behaviour was explained by facilitated diffusion of the relatively hydrophobic reactants and substrates through the hydrophobized pores and, as a result, better interaction with the pore surface in the passivated materials.[187]

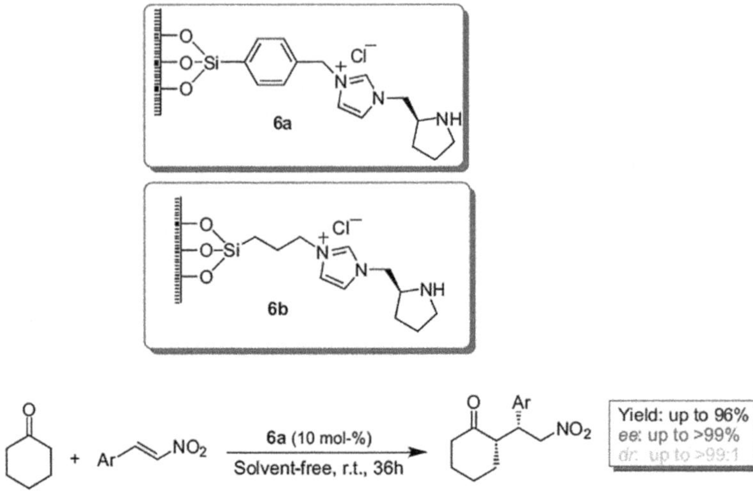

Figure 10.32 Ionosilica-supported pyrrolidine-based chiral IL for stereoselective Michael addition. Reproduced from Ref. 188 with permission from Elsevier, Copyright 2008.

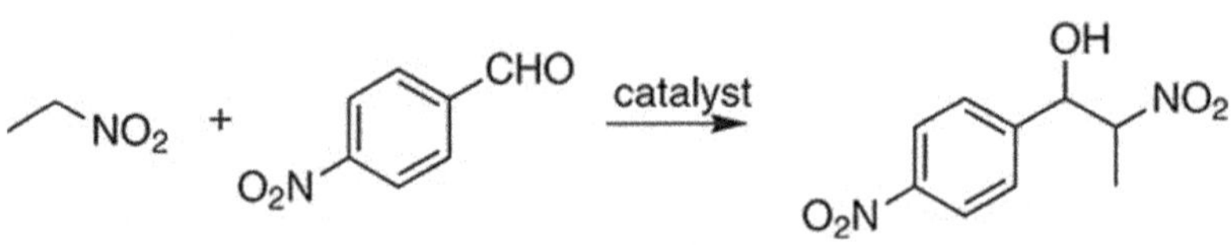

Scheme 10.5

Ionosilica-supported pyrrolidine-based chiral ILs have also been found to be recyclable, highly stereoselective, effective catalysts for the Michael addition of ketones with nitrostyrenes (Figure 10.32). Such reactions generated good yields of the corresponding products (up to 94%), excellent enantioselectivities (up to >99% ee) and high diastereoselectivities (up to 99 : 1 dr).[188]

10.4.8 Other Sustainable Catalytic Reactions

10.4.8.1 Hydration of Alkynes

The catalytic conversion of alkynes into ketones *via* hydration is a route affording important intermediates in the chemical and pharmaceutical industries. Usually, it can be performed *via* rather complicated processes in the presence of liquid acids or highly toxic metal catalysts.[189] Hierarchically porous, strongly acidic poly(IL) solids eliminate these disadvantages, affording the synthesis of a variety of alkynes under much greener and more efficient conditions than using Amberlyst 15, phosphotungstic acid or sulfuric acid (Table 10.5).

Table 10.5 Hydration of various alkynes catalysed by PDVB-[C3VP][SO_3CF_3] (PDVB=polydivinylbenzene; VP=4-vinyl pyridine; C3=1,3-propanesultone; SO_3CF_3 = HSO_3CF_3 anion exchanger). Reaction conditions: alkyne (0.4 mmol), H_2O (0.8 mmol), CF_3CH_2OH (2 mL), catalyst (10 wt% of alkyne), 40 °C, 18 h. Reproduced from Ref. 189 with permission from Elsevier, Copyright 2018.

Entry	Alkyne	Product	Yield (%)
1	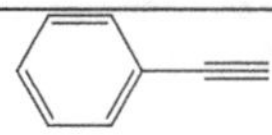	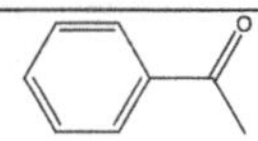	98
2	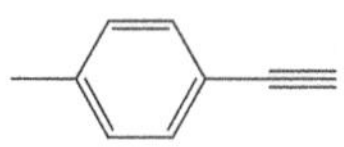	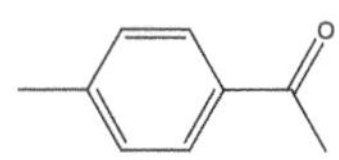	99
3[a]	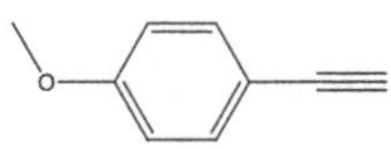	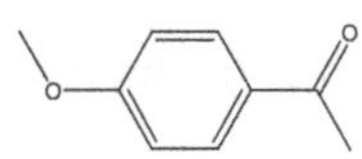	85
4[a]	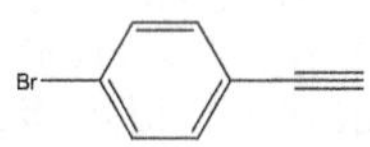	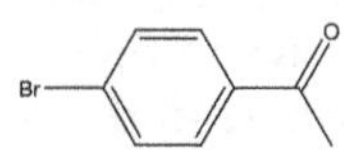	99
5[b]	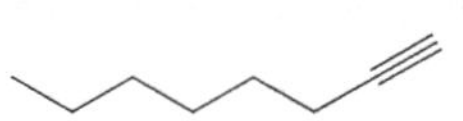	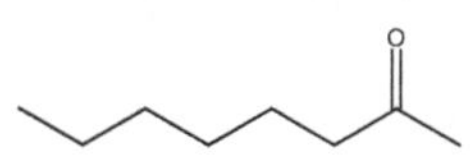	99
6[b]		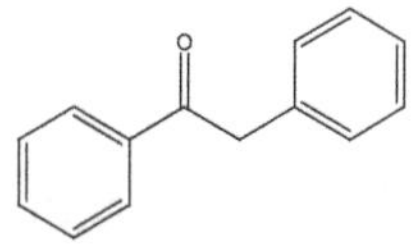	99

[a] 50 °C.
[b] 90 °C, 24 h.

10.4.8.2 Heteropolyanion-based ILs and Applications

Heteropolyanion-based ILs (HPAILs) (Figure 10.33) have emerged as new species of hybrid materials, which were prepared by combining Keggin heteropolyanions with "task-specific" IL cations containing special functional groups.[190–192] Due to their operational simplicity, lack of toxicity, ease of isolation and reusability, they offer an attractive alternative for traditional acid-catalysed[193,194] or oxidative[195–197] organic transformations. More complex molecules such as spirooxindole can also be synthesized with these catalysts (Figure 10.34).[198]

10.4.8.3 Synthesis of Acetamides

Synthesis of acetamides could be carried out with an efficient solid acid IL catalyst such as sulfated choline ILs, which are able to ensure the protection of amine groups by reaction with acetic anhydride under solvent-free grindstone conditions.[199]

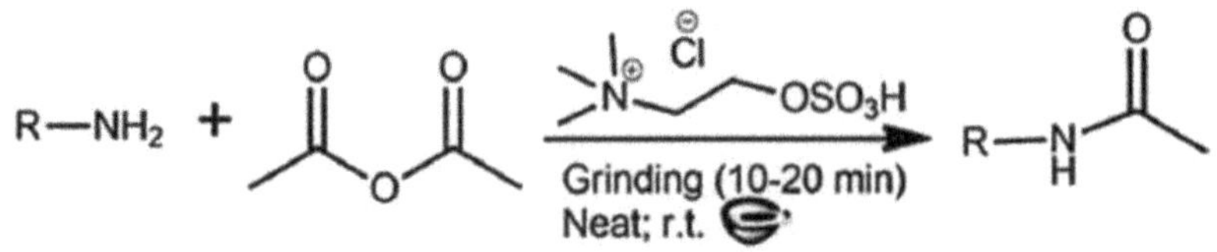

Scheme 10.6

10.4.8.4 *Cyclization between 2-aminobenzamides with Ketones*

Imidazolium Brønsted acidic ILs such as [BSMIm][OTs] or [BSMIm][OTf] have been found to be effective in the formation of 2,2-disubstituted quinazolin-4(1*H*)-one derivatives where both the C_2–H of the imidazolium moiety and an acidic proton are necessary and simultaneously take part in catalysis. This method produced only water as the by-product and represents a green synthetic protocol.[200]

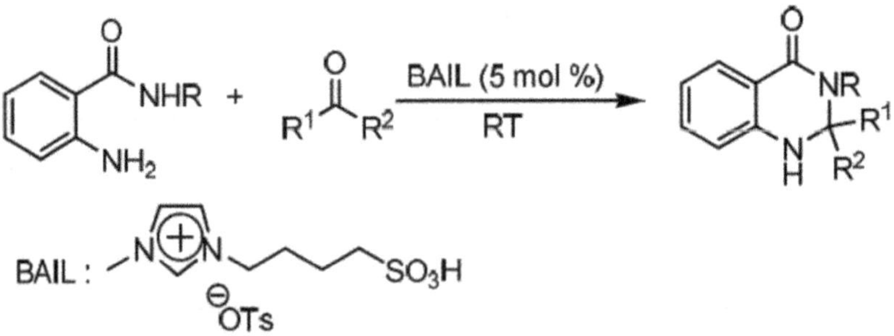

Scheme 10.7

10.4.8.5 *Synthesis of 2-oxazolidinones and Quinazolines*

2-oxazolidinones and quinazoline-2,4(1*H*,3*H*)-diones can be produced without pre-separation and work-up steps in a continuous laminar-flow microreactor in which built-in silicon nanowires (SiNWs) with immobilized IL on the tips of cone-shaped nanowire bundles serve as catalysts (Figure 10.35).[201] A stable gas–liquid interface is maintained between the liquid flow of organo-amines (in the upper part) and the gas flow of CO_2 (in the bottom part of channel) because of the super-amphiphobic SiNWs. This configuration ensures intimate and direct contact of the binary reagents, leading to enhanced mass transfer.

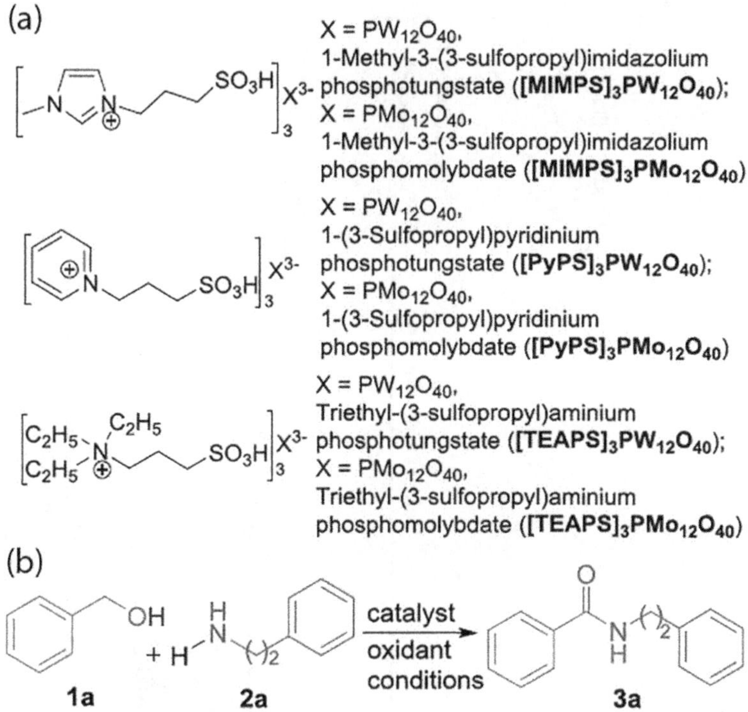

Figure 10.33 (a) N-substituted imidazole-, pyridine- and triethylamine-based HPAILs. Reproduced from Ref. 192 with permission from Elsevier, Copyright 2016. (b) Oxidative amidation of benzyl alcohol with 2-phenylethylamine. Reproduced from Ref. 192 with permission from Elsevier, Copyright 2016.

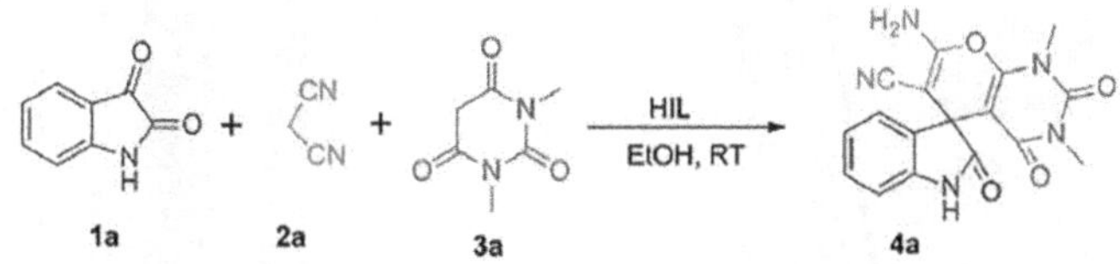

Figure 10.34 Synthesis of spirooxindole using HPAIL catalysts. Reproduced from Ref. 198 with permission from Elsevier, Copyright 2014.

10.4.8.6 Chiral Synthesis

Chiral ILs (CILs) derived from natural amino acids with chirality, biodegradability, reduced toxicity and high biocompatibility have also attracted interest for the synthesis of sustainable ILs.[202] Among these, L-proline- and L-histidine-derived CILs are the most investigated, either as chiral anions or tethered to different standard IL cations.[203] Butyl-imidazolium-proline cations gave high yields and excellent *syn*:*anti* selectivity and enantioselectivity for the

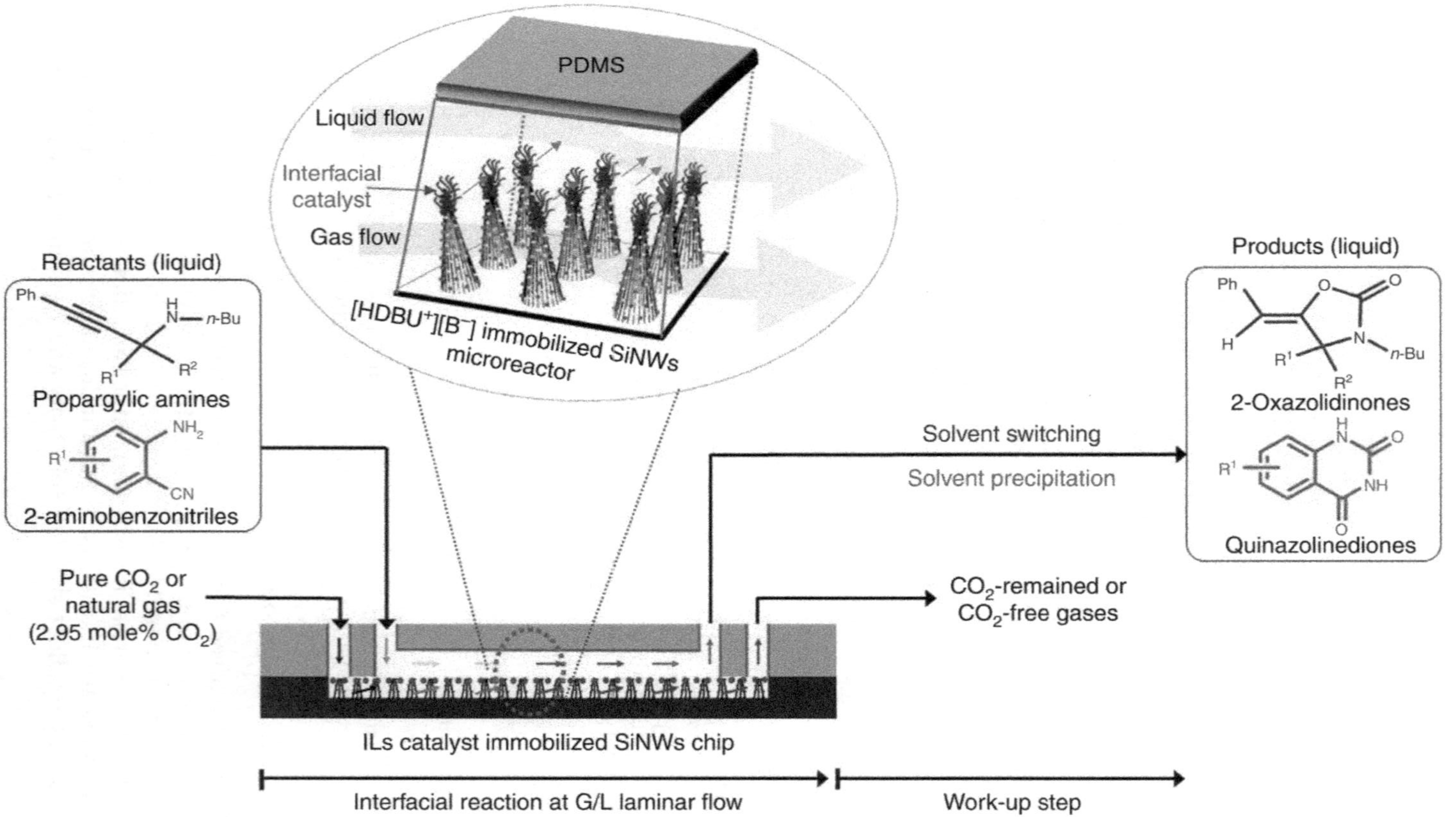

Figure 10.35 An integrated gas–liquid microfluidic system and overall process for the capture and fixation of CO_2 from pure or diluted CO_2 resources to synthesize 2-oxazolidinones and quinazolines-2,4(1*H*,3*H*)-diones, and consecutive work-up for facile product isolation. Reproduced from Ref. 201, https://doi.org/10.1038/ncomms14676, under the terms of the CC BY 4.0 licence, https://creativecommons.org/licenses/by/4.0/.

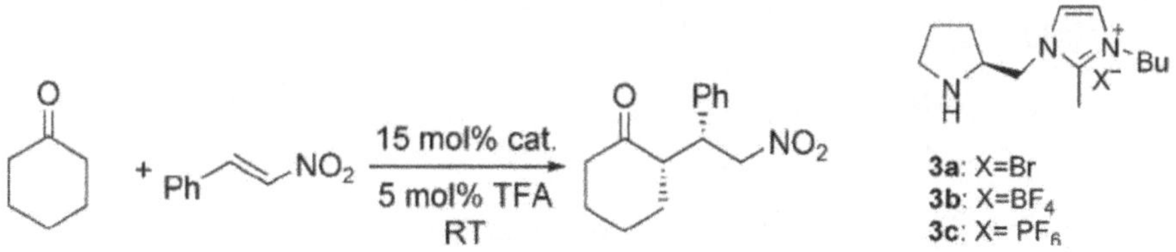

Scheme 10.8

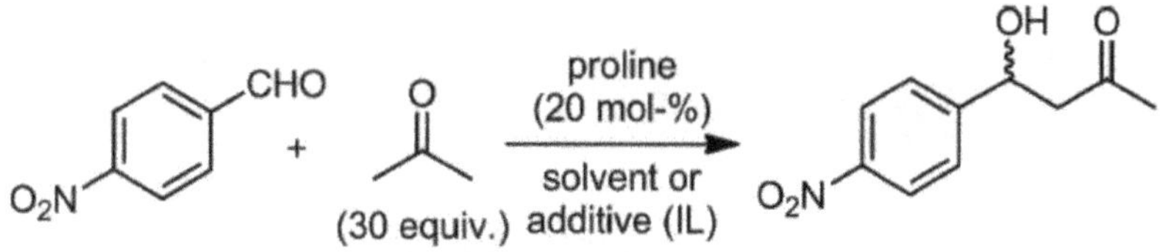

Scheme 10.9

asymmetric Michael addition of cyclohexanone to trans-β-nitrostyrene.[204,205] Then, the combination of the chiral proline with an imidazolium bistriflimide IL that contained a valine-derived amino amide as a second arm led to a high conversion and selectivity for the asymmetric aldol reaction between nitrobenzaldehyde and acetone.[206] However, in this case, the enantioselectivity was less than 80%.

In another example, a methyltrioctylammonium malic acid-based IL[207] was used as a chiral reaction solvent in the aza-Baylis–Hillman reaction of activated alkenes with imines, resulting in an enantiomeric excess of 84%. This is an example of an asymmetric reaction in which the chirality is induced completely by the IL solvent (Figure 10.36). The absence of other solvents in the reaction simplifies product removal and IL regeneration and diminishes thermal degradation of the IL since solvent–solvent separation by distillation is avoided.

10.4.8.7 Synthesis of Organometallic Compounds

The use of ILs also allows for the synthesis of various organometallic compounds widely used in a variety of organic processes or organo-modified solids. Hydrosilylation is a particularly attractive route to produce these compounds since it is clean (*e.g.* no formation of HCl, unlike the condensation pathway) and highly efficient.[208] It requires catalyst loadings at the level of only a few ppm platinum, and therefore attempts to recycle the catalyst are not necessary (Figure 10.37). For these reactions, ILs such as $[C_4py][BF_4]$ and $[C_4\text{-}3\text{-}pic][BF_4]$ offer the optimal compromise between activity and catalyst retention, as is the case with the biphasic hydrosilylation of 1-hexadecene with siloxane. The system also offers advantages with regards to the effects of oxygen, water and chloride impurities.

As another example, organosilicon compounds afford slurry-phase hydrosilylation reactions *via* using a highly active supported ionic liquid phase (SILP) catalyst ([{Rh(cod)(μ-OSiMe3)}2]/[P66614][NTf2] supported on silica). Such a

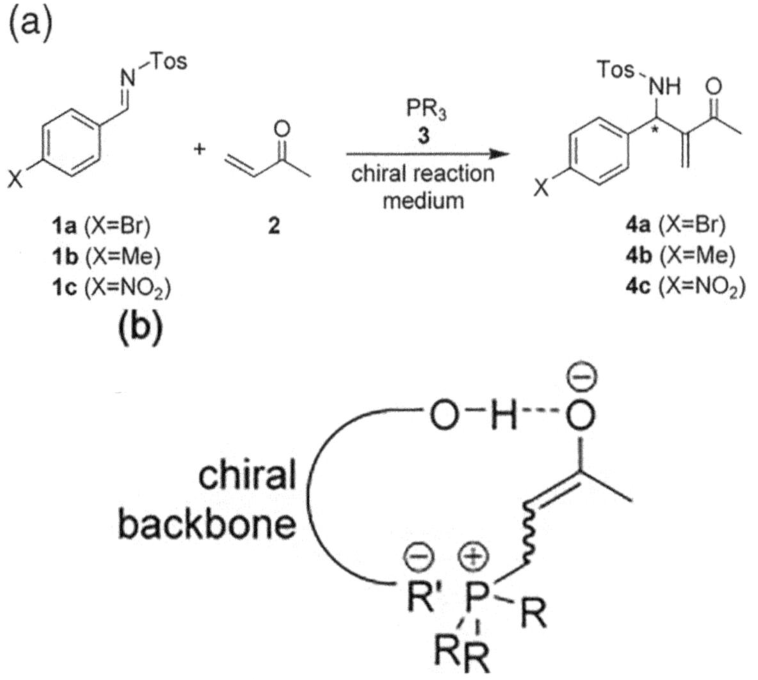

Figure 10.36 (a) The aza-Baylis–Hillman reaction. Tos = 4-toluenesulfonyl. (b) Possible bifunctional interaction of the zwitterionic intermediate of the aza-Baylis–Hillman reaction with the chiral anion of an IL containing a hydrogen bond donor. Reproduced from Ref. 207 with permission from John Wiley and Sons, Copyright © 2006 Wiley-VCH Verlag GmbH & Co. KGaA, Weinheim.

catalyst leads to an increase in turnover of three orders of magnitude in comparison to that obtained with the usual biphasic catalytic system.[209]

Oxyseleno compounds are another class of organometallic compounds of interest. The cyclization of unsaturated alcohols leading to tetrahydrofuran and tetrahydropyran ethers can be achieved by treatment with phenylselenyl chloride in the presence of ILs as solvents/catalysts.[210] These molecules are precursors of many biologically active metabolites. The reaction proceeds very rapidly under mild conditions with absolute regioselectivity using primary, secondary, tertiary and aromatic alcohols, as well as monosubstituted, disubstituted and trisubstituted alkenols.

10.4.8.8 Sustainable Biocatalytic Processes

Whole-cell-based biocatalysis in IL-containing systems has attracted increasing interest due to some major advantages compared to bioreactions catalysed by isolated enzymes.[211] One of the most important is that the whole cell provides a natural intracellular environment for the enzymes to function in, with cofactor regeneration occurring *in situ*. Applications of whole-cell

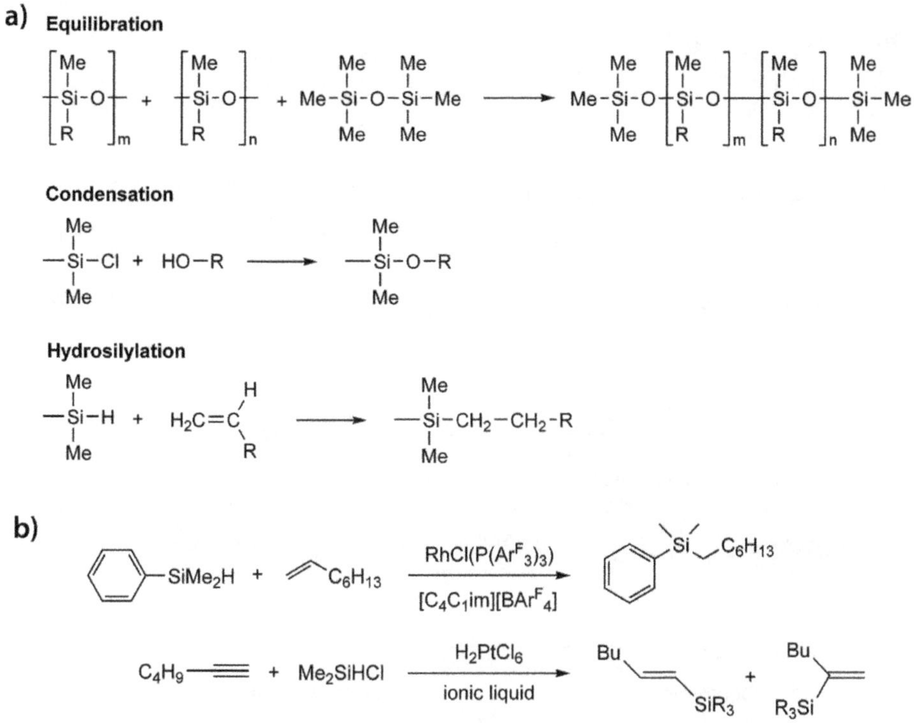

Figure 10.37 Examples of organosilicon compounds obtained *via* hydrosylilation. Reproduced from Ref. 208 with permission from American Chemical Society, Copyright 2006.

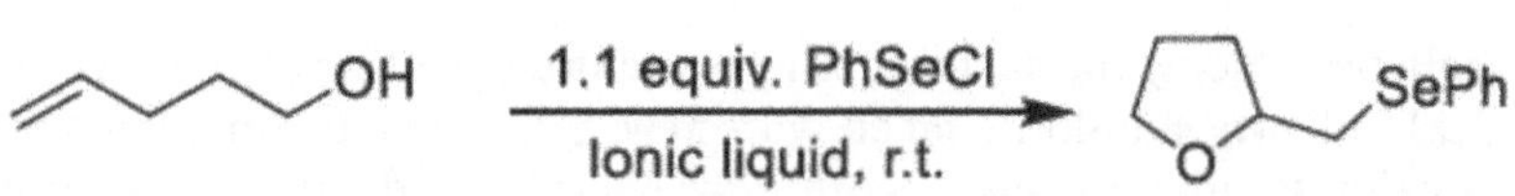

Scheme 10.10

biocatalysis in IL-containing systems have, to date, been focused on the production of valuable compounds, mainly through reduction, oxidation and hydrolytic reactions (Figures 10.38 and 10.39). The mechanism of ILs affecting the whole-cell biocatalysis offers the possibility to effectively integrate ILs with biotransformation. The toxicity of ILs to microbial cells and their biodegradability are, however, two of the crucial factors that could limit this application.[211]

In another approach, the combination of the water-immiscible IL [C_4MIM][PF_6] with choline chloride/urea in a biphasic system enhanced the substrate concentration and led to high product yields and productivity for the biocatalytic synthesis of (*S*)-3-chloro-1-phenylpropanol catalysed by immobilized whole cells of *Acetobacter* sp. CCTCC M209061 cells (Figure 10.40).[211]

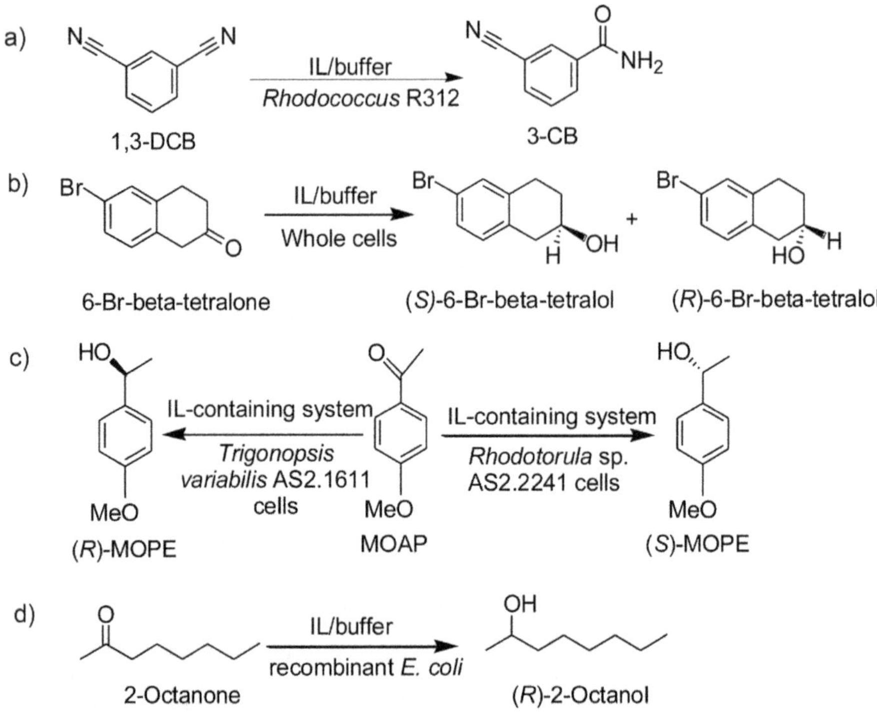

Figure 10.38 Examples of whole-cell biocatalytic reduction reaction in ILs. Reproduced with Ref. 211 with permission from American Chemical Society, Copyright 2015.

10.5 Conclusions

The challenges facing the sustainability of technological advancements in the chemical industry should be faced with new, innovative technologies based on new axioms and premises. Accordingly, chemical industries must support sustainable development by developing green technologies for their molecules and products of interest, supposing not only better performances, but also new reaction pathways, and offering increased adherence to energetic consumption, safety, health and environmental standards. There are already many examples in which ILs alone or functionalized on various supports demonstrated enhanced efficiency. Therefore, the recent interest in biomass valorization and valorization of wastes such as CO_2 as new sources of energy and the sustainable development of the chemical industry, by taking into consideration process intensification concepts, offer new perspectives on the utilization of these compounds.

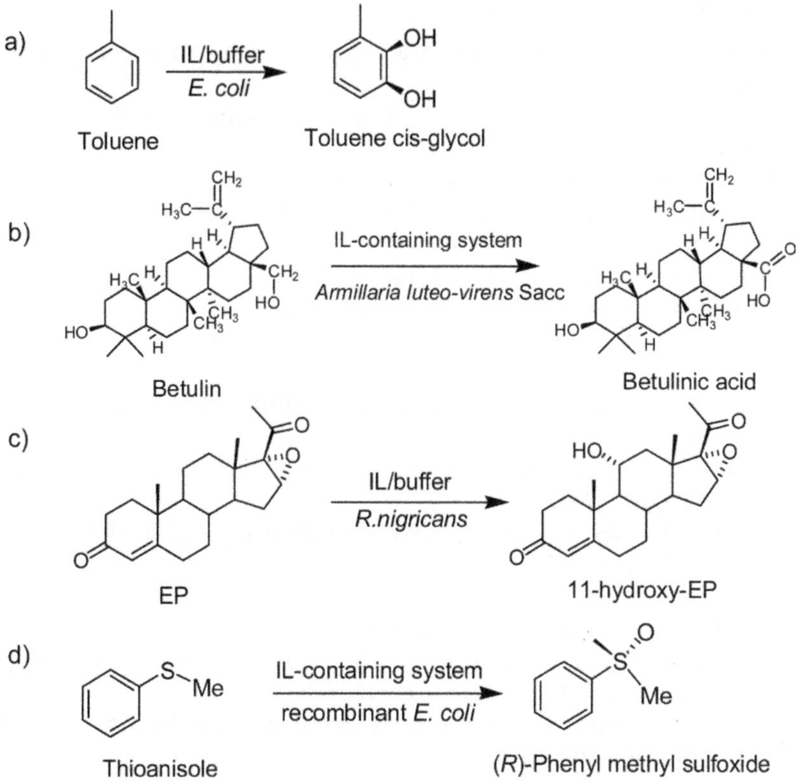

Figure 10.39 Examples of whole-cell biocatalytic oxidation reaction in ILs. Reproduced with Ref. 211 with permission from American Chemical Society, Copyright 2015.

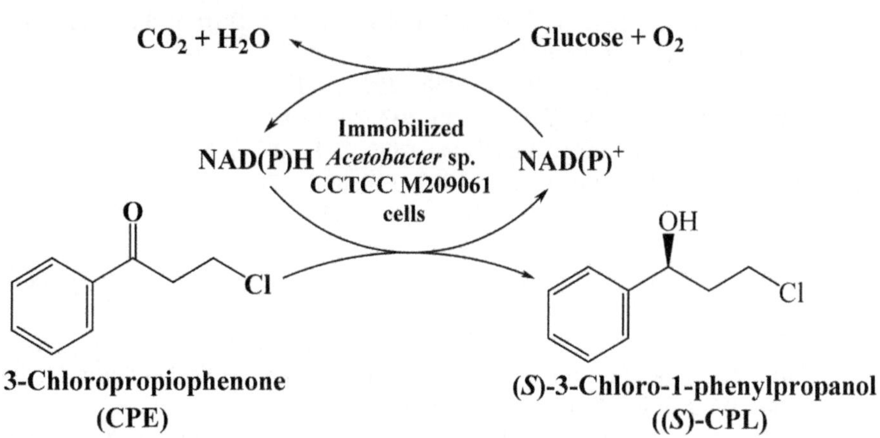

Figure 10.40 Biocatalytic synthesis of (*S*)-3-chloro-1-phenylpropanol catalysed by immobilized whole cells of *Acetobacter* sp. CCTCC M209061. Reproduced with Ref. 211 with permission from American Chemical Society, Copyright 2015.

References

1. http://www.oecd.org/chemicalsafety/risk-management/sustainablechemistry.htm.
2. E. Cséfalvay, G. R. Akien, L. Qi and I. T. Horváth, *Catal. Today*, 2015, **239**, 50–55.
3. I. T. Horváth, *Chem. Rev.*, 2018, **118**, 369–371.
4. T. Welton, *Proc. R. Soc. A*, 2015, **471**, 20150502.
5. R. A. Sheldon, *Chem. Ind.*, 1992, 903–906.
6. B. M . Trost, *Science*, 1991, **254**, 1471–1477.
7. R. A. Sheldon, *Green Chem.*, 2017, **19**, 18–43.
8. F. Roschangar, R. A. Sheldon and C. H. Senanayake, *Green Chem.*, 2015, **17**, 752–768.
9. K.-H. Robert, H. Daly, P. Hawken and J. Holmberg, *Int. J. Sustain. Dev. World. Ecol.*, 1997, **4**, 79–92.
10. C. J. Clarke, W.-C. Tu, O. Levers, A. Brohl and J. P. Hallett, *Chem. Rev.*, 2018, **118**, 747–800.
11. M. Geissdoerfer, P. Savaget, N. M. P. Bocken and E. J. Hultink, *J. Clean. Prod.*, 2017, **143**, 757–768.
12. P. T. Anastas and J. C. Werner, *Green Chemistry: Theory and Practice*, Oxford University Press, Oxford, 1998.
13. P. Marion, B. Bernela, A. Piccirilli, B. Estrine, N. Patouillard and J. Guilbot, F. Jérôme, *Green Chem.*, 2017, **19**, 4973–4989.
14. C. J. Clarke, W.-C. Tu, O. Levers, A. Bröhl and J. P. Hallett, *Chem. Rev.*, 2018, **118**, 747–800.
15. P. G. Jessop, *Green Chem.*, 2011, **13**, 1391–1398.
16. P. G. Jessop, L. Phan, A. Carrier, S. Robinson, C. J. Durr and J. R. Harjani, *Green Chem.*, 2010, **12**, 809–814.
17. P. Pollet, C. A. Eckert and C. L. Liotta, *Chem. Sci.*, 2011, **2**, 609–614.
18. P. G. Jessop, L. Phan, A. Carrier, S. Robinson, C. J. Durr and J. R. Harjani, *Green Chem.*, 2010, **12**, 809–814.
19. L. Claisen, *Liebigs Ann. Chem.*, 1896, **291**, 25–111.
20. L. Knorr, *Liebigs Ann. Chem.*, 1896, **293**, 70–120.
21. W. Wislicenus, *Liebigs Ann. Chem.*, 1896, **291**, 147–216.
22. H. Stobbe, *Liebigs Ann. Chem.*, 1903, **326**, 347–370.
23. Y. Marcus, *Chem. Soc. Rev.*, 1993, **22**, 409–416.
24. J. N. Brønsted, *Chem. Rev.*, 1928, **5**, 231–338.
25. D. Prat, A. Wells, J. Hayler, H. Sneddon, C. R. McElroy, S. Abou-Shehada and P. J. Dunn, *Green Chem.*, 2016, **18**, 288–296.
26. A. Kruse and H. Vogel, in *Handbook of Green Chemistry, Vol. 4: Supercritical Solvents*, ed. W. Leitner and P. G. Jessop, Wiley-VCH Verlag GmbH & Co. KGaA, Weinheim, 2010, pp. 457–475.
27. K. Hartonen and M.-L. Riekkola, in *The Application of Green Solvents in Separation Processes*, ed. F. Pena-Pereira and M. Tobiszewski, Elsevier, Amsterdam, 2017, pp. 19–56.

28. D. G. Blackmond, A. Armstrong, V. Coombe and A. Wells, *Angew. Chem., Int. Ed.*, 2007, **46**, 3798–3800.
29. T. Welton, *Chem. Rev.*, 1999, **99**, 2071.
30. P. Wasserscheid and T. Welton, *Ionic Liquids in Synthesis*, Wiley-VCH Verlag, Weinheim, 2002.
31. P. Wasserscheid and W. Keim, *Angew. Chem., Int. Ed.*, 2000, **39**, 3773.
32. R. D. Rogers and K. R. Seddon, *Science*, 2003, **302**, 792.
33. *Ionic Liquids III A: Fundamentals, Progress, Challenges, and Opportunities Properties and Structure*, ed. R. D. Rogers and K. R. Seddon, American Chemical Society,Washington, DC, 2005.
34. F. V. Rantwijk and R. A. Sheldon, *Chem. Rev.*, 2007, **107**, 2757.
35. *Environmentally Friendly Syntheses Using Ionic Liquids*, ed. J. Dupont, T. Itoh, P. Lozano and S. V. Malhotra, CRC Press, Boca Raton, FL, 2014.
36. P. Wasserscheid and W. Keim, *Angew. Chem., Int. Ed.*, 2000, **39**, 3772–3789.
37. *Catalysis in Ionic Liquids: From Catalyst Synthesis to Application*, ed. C. Hardacre and V. I. Parvulescu, RSC, Cambridge, 2014.
38. T. Itoh, *Chem. Rev.*, 2017, **117**, 10567–10607.
39. V. I. Parvulescu and C. Hardacre, *Chem. Rev.*, 2007, **107**, 2615–2665.
40. C. G. Yoo, Y. Pu and A. J. Ragauskas, *Curr. Opin. Green Sus. Chem.*, 2017, **5**, 5–11.
41. J. Hulsbosch, D. E. De Vos, K. Binnemans and R. Ameloot, *ACS Sustain. Chem. Eng.*, 2016, **4**, 2917–2931.
42. A. M. Socha, R. Parthasarathi, J. Shi, S. Pattathil, D. Whyte, M. Bergeron, A. George, K. Tran, V. Stavil, S. Venkatachalam, M. G. Hahn, B. A. Simmons and S. Singh, *Proc. Natl. Acad. Sci. U. S. A.*, 2014, **111**, E3587–E3595.
43. D. J. Eyckens and L. C. Henderson, *Front. Chem.*, 2019, 7, 263.
44. A. Obregón-Zúñiga, M. Milán and E. Juaristi, *Org. Lett.*, 2017, **19**, 1108–1111.
45. C. Rub and B. Konig, *Green Chem.*, 2012, **14**, 2969–2982.
46. Y. Dai. J. Van Spronsen, G. J. Witkamp, R. Verpoorte and Y. H. Choi, *J. Nat. Prod.*, 2013, **76**, 2162–2173.
47. P. Domínguez de María and Z. Maugeri, *Curr. Opin. Chem. Biol.*, 2011, **15**, 220–225.
48. M. Francisco, A. Van Den Bruinhorst and M. C. Kroon, *Angew. Chem., Int. Ed.*, 2013, 52, 3074–3085.
49. Q. Zhang, K. De Oliveira Vigier, S. Royer and F. Jérôme, *Chem. Soc. Rev.*, 2012, **41**, 7108–7146.
50. S. E. E. Warrag, C. J. Peters and M. C. Kroon, *Curr. Opin. Green Sustain. Chem.*, 2017, **5**, 55–60.
51. E. L. Smith, A. P. Abbott and K. S. Ryder, *Chem. Rev.*, 2014, **114**, 11060–11082.
52. M. Vafaeezadeh and M. M. Hashemi, *J. Mol. Liq.*, 2015, **207**, 73–79.
53. R. Gedye, F. Smith, K. Westaway, H. Ali, L. Baldisera, L. Laberge and J. Rousell, *Tetrahedron Lett.*, 1986, **27**, 279–282.
54. T. J. Mason, *Ultrason. Sonochem.*, 2003, **10**, 175–179.

55. S. V. Sancheti and P. R. Gogate, *Ultrason. Sonochem.*, 2017, **36**, 527–543.
56. Y. Peng and G. Song, *Green Chem.*, 2002, 349–351.
57. J. Wang, A. Zhu and L. Li, in Sustainable Catalytic Processes, ed. B. Saha, M. Fan and J. Wang, Elsevier, Amsterdam, 2015, pp. 61–97.
58. Y. Tao, R. Dong, I. V. Pavlidis, B. Chen and T. Tan, *Green Chem.*, 2016, **18**, 1240–1248.
59. K. Li, H. Choudhary and R. D. Rogers, *Curr. Opin. Green Sustain. Chem.*, 2018, **11**, 15–21.
60. Q. Wang, Y. Geng, X. Lu and S. Zhang, *ACS Sustain. Chem. Eng.*, 2015, **3**, 340–348.
61. R. Kore, P. Berton, S. P. Kelley, P. Aduri, S. S. Katti and R. D. Rogers, *ACS Catal.*, 2017, **7**, 7014–7028.
62. K. Li, H. Choudhary and R. D. Rogers, *Curr. Opin. Green Sustain. Chem.*, 2018, **11**, 15–21.
63. Y. Chauvin, B. Gilbert and I. Guibard, *J. Chem. Soc. Chem. Commun.*, 1990, (23), 1715–1716.
64. R. T. Carlin and J. S. Wilkes, *J. Mol. Catal.*, 1990, **63**, 125–129.
65. G. Franciò, U. Hintermair and W. Leitner, *Phil. Trans. R. Soc. A*, 2016, **373**, 20150005.
66. U. Hintermair, G. Franciò and W. Leitner, *Chem. Commun.*, 2011, **47**, 3691–3701.
67. I. Vural Gürsel, T. Noel, Q. Wang and V. Hessel, *Green Chem.*, 2015, **17**, 2012–2026.
68. S. G. Newman and K. F. Jensen, *Green Chem.*, 2013, **15**, 1456–1472.
69. O. Bortolini, S. Campestrini, V. Conte, G. Fantin, M. Fogagnolo and S. Maietti, *Eur. J. Org. Chem.*, 2003, (24), 4804–4809.
70. V. Conte, F. Fabbianesi, B. Floris, P. Galloni, D. Sordi, I. W. C. E. Arends, M. Bonchio, D. Rehder and D. Bogdal, *Pure Appl. Chem.*, 2009, **81**, 1265–1277.
71. S. Berardi, V. Conte, G. Fiorani, B. Floris and P. Galloni, *J. Organomet. Chem.*, 2008, **693**, 3015–3020.
72. E. C. Gaudino, G. Cravotto, D. Garella, S. Tagliapietra and W. Bonrath, *Org. Prep. Proced. Int.*, 2012, **44**, 175–179.
73. A. Arfan and J. P. Bazureau, *Org. Process Res. Dev.*, 2005, **9**, 745–748.
74. A. Zare, A. Hasaninejad, A. R. Moosavi Zare, A. Parhami, H. Sharghi and A. Khalafi-Nezhad, *Can. J. Chem.*, 2007, **85**, 438–444.
75. J. G. Hernández and E. Juaristi, *Chem. Commun.*, 2012, **48**, 5396–5409.
76. R. Kumar, A. Saima, A. Shard, N. H. Andhare and A. K. Sinha, *Angew. Chem., Int. Ed.*, 2015, **54**, 828–832.
77. B. Floris, F. Sabuzi, P. Galloni and V. Conte, *Catalysts*, 2017, 7, 261.
78. N. Kimizuk and T. Nakashima, *Langmuir*, 2001, **17**, 6759–6761.
79. P. C. Marr and A. C. Marr, *Green Chem.*, 2016, **18**, 105.
80. C. Rizzo, F. D'Anna, R. Noto, M. Zhang and R. G. Weiss, *Chem. Eur. J.*, 2016, **22**, 11269–11282.
81. P. Hesemann, L. Viau and A. Vioux, in *The Sol–Gel Handbook: Synthesis, Characterization, and Applications*, 1st edn. ed. D. Levy and M. Zayat, Wiley-VCH Verlag GmbH & Co. KGaA, Weinheim, 2015, pp. 487–517.

82. J. Le Bideau, L. Viau and A. Vioux, *Chem. Soc. Rev.*, 2011, **40**, 907–925.
83. H. Li, P. S. Bhadury, B. Song and S. Yang, *RSC Adv.*, 2012, **2**, 12525–12551.
84. R. Mouawia, A. Mehdi, C. Reye and R. Corriu, *J. Mater. Chem.*, 2007, **17**, 616–618.
85. A. El Kadib, P. Hesemann, K. Molvinger, J. Brandner, C. Biolley, P. Gaveau, J. J. E. Moreau and D. Brunel, *J. Am. Chem. Soc.*, 2009, **131**, 2882–2892.
86. M. Yang, F. Xu, C. Wang, X. Liu, P. Yan, P. Li and U. Welz-Biermann, *Eur. J. Inorg. Chem.*, 2012, **2012**, 4500–4506.
87. H. Hattori, *Chem. Rev.*, 1995, **95**, 537–550.
88. K. Tanabe and W. F. Hölderich, *Appl. Catal., Aeral*, 1999, **181**, 399.
89. Q. Zhao, P. Zhang, M. Antonietti and J. Yuan, *J. Am. Chem. Soc.*, 2012, **134**, 11852–11855.
90. A. Pourjavadi, S. H. Hosseini, M. Doulabi, S. M. Fakoorpoor and F. Seidi, *ACS Catal.*, 2012, **2**, 1259–1266.
91. E. I. Privalova, E. Karjalainen, M. Nurmi, P. Mäki-Arvela, K. Eranen, H. Tenhu, D. Y. Murzin and J.-P. Mikkola, *ChemSusChem*, 2013, **6**, 1500–1509.
92. F. Liu, W. Li, Q. Sun, L. Zhu, X. Meng, Y.-H. Guo and F.-S. Xiao, *ChemSusChem*, 2011, **4**, 1059–1062.
93. F. Zhang, X. Yang, L. Jiang, C. Liang, R. Zhu and H. Li, *Green Chem.*, 2013, **15**, 1665–1672.
94. X. Wang, J. Li, G. Chen, Z. Guo, Y. Zhou and J. Wang, *ChemCatChem*, 2015, 7, 993–1003.
95. F. J. Liu, L. Wang, Q. Sun, L. F. Zhu, X. J. Meng and F.-S. Xiao, *J. Am. Chem. Soc.*, 2012, **134**, 16948–16950.
96. L. Han, H. J. Choi, D. K. Kim, S. W. Park, B. Liu and D. W. Park, *J. Mol. Catal. A Chem.*, 2011, **338**, 58.
97. B. Zhang, L. Zhang, Q. Wu, Q. Wang, B. Song, W. Wu, B. Lu and T. Ren, *RSC Adv.*, 2014, **4**, 20506–20515.
98. L. Han, S.-W. Park and D.-W. Park, *Energy Environ. Sci.*, 2009, **2**, 1286–1292.
99. W. L. Dai, L. Chen, S. F. Yin, S. L. Luo and C. T. Au, *Catal. Lett.*, 2010, **135**, 295–304.
100. X. Zhou, Y. Zhang, X. Yang and L. Zhao, *J. Mol. Catal. A Chem.*, 2012, **361–362**, 12–16.
101. Y. L. Hu, R. L. Zhang and D. Fang, *Environ. Chem. Lett.*, 2019, **17**, 501–508.
102. L. G. Ding, B. Yao, W. L. Jiang, J. T. Li, Q. J. Fu, Y. A. Li, Z. H. Liu, J. P. Ma and Y. B. Dong, *Inorg. Chem.*, 2017, **56**, 2337–2344.
103. M. Taherimehr, B. Van de Voorde, L. H. Wee, J. A. Martens, D. E. De Vos and P. P. Pescarmona, *ChemSusChem*, 2017, **10**, 1283–1291.
104. W. Dai, Y. Zhang, Y. Tan, X. Luo and X. Tu, *Appl Catal A Gen.*, 2016, **514**, 43–50.
105. Z. Zhang, J. Song and B. Han, *Chem. Rev.*, 2017, **117**, 6834–6880.
106. S. Dutta and S. Pal, *Biomass Bioenergy*, 2014, **62**, 182–197.
107. I. Bodachivskyi, U. Kuzhiumparambil and D. B. G. Williams, *ChemSusChem*, 2017, **11**, 642–660.

108. F. Delbecq and C. Len, *Molecules*, 2018, **23**, 1973.
109. A. M. da Costa Lopes and R. Bogel-Łukasik, *ChemSusChem*, 2015, **8**, 947–965.
110. A. Chinnappan, C. Baskar and H. Kim, *RSC Adv.*, 2016, **6**, 63991–64002.
111. J. Li, J. Li, D. Zhang and C. Liu, *J. Phys. Chem. B*, 2015, **119**, 13398–13406.
112. Y. Qu, Q. Wei, H. Li, P. Oleskowicz-Popiel, C. Huang and J. Xu, *Bioresour. Technol.*, 2014, **162**, 358–364.
113. V. Da Silva Lacerda, J. B. Lopez-Sotelo, A. Correa-Guimeraes, S. Hernandez-Navarro, M. Sanchez-Bascones, L. M. Naras-Gracia, P. Martin-Ramos, E. Perez-Lebena and J. Martin-Gil, *Bioresour. Technol.*, 2015, **180**, 88–95.
114. F. Liu, K. Huang, A. Zheng, F.-S. Xiao and S. Dai, *ACS Catal.*, 2018, **8**, 372–391.
115. Z. Lu, L. Fan, Z. Wu, H. Zhang, Y. Liao, D. Zheng and S. Wang, *Biomass Bioenergy*, 2015, **81**, 154–161.
116. A. S. Khan, Z. Man, M. A. Bustam, A. Nasrullah, Z. Ullah, A. Sarwono, F. U. Shah and N. Muhammad, *Carbohyd. Polym.*, 2018, **181**, 208–214.
117. F. Liu, K. Huang, A. Zheng, F.-S. Xiao and S. Dai, *ACS Catal.*, 2018, **8**, 372–391.
118. Q. Zhang, S. Zhang and Y. Deng, *Green Chem.*, 2011, **13**, 2619–2637.
119. B. J. Cox and J. G. Ekerdt, *Bioresour. Technol.*, 2012, **118**, 584–588.
120. N. Yan, Y. A. Yuan, R. Dykeman, Y. A. Kou and P. J. Dyson, *Angew. Chem., Int. Ed.*, 2010, **49**, 5549–5553.
121. J. Long, W. Lou, L. Wang, B. Yin and X. Li, *Chem. Eng. Sci.*, 2015, **122**, 24–33.
122. J. Chen, J. Huang, L. Chen, L. Ma, T. Wang and U. I. Zakai, *ChemCatChem*, 2013, 5, 1598–1605.
123. R. M. Lau, F. van Rantwijk, K. R. Seddon and R. A. Sheldon, *Org. Lett.*, 2000, **2**, 4189–4191.
124. S. Siankevich, Z. Fei, R. Scopelliti, P. G. Jessop, J. Zhang, N. Yan and P. J. Dyson, *ChemSusChem*, 2016, **9**, 2089–2096.
125. S. Siankevich, Z. Feia, N. Yan and P. J. Dyson, *Chimia*, 2015, **69**, 592–595.
126. A. P. S. Brogan, L. Bui-Le and J. P. Hallett, *Nat. Chem.*, 2018, **10**, 859–865.
127. J. Yang, B. Pérez, S. Anankanbil, J. Li, R. Gao and Z. Guo, *ACS Sust. Chem. Eng.*, 2017, **5**, 2006–2014.
128. K. Fujita, N. Nakamura, K. Igarashi, M. Samejima and H. Ohno, *Green Chem.*, 2009, **11**, 351–354.
129. Z.-L. Huang, T.-X. Yang, J.-Z. Huang and Z. Yang, *Bioenergy Res.*, 2014, 7, 1519–1528.
130. P. Lozano, J. M. Bernal, G. Sanchez-Gomez, G. Lopez-Lopez and M. Vaultier, *Energy Environ. Sci.*, 2013, **6**, 1328–1338.
131. H. Fu, M. Li, R. N and Y. M. Lo, *Food Sci. Nutr.*, 2018, **6**, 2020–2027.
132. Q. Sun, S. Wang, B. Aguila, X. Meng, S. Ma and F.-S. Xiao, *Nat. Commun.*, 2018, **9**, 3236.
133. Q.-W. Song, Z.-H. Zhoua and L.-N. He, *Green Chem.*, 2017, **19**, 3707–3728.

134. S. Zhang, R. Ma and L.-N. He, in *Topics in Organometallic Chemistry, Carbon Dioxide and Organometallics*, ed. X.-B. Lu, 2016, Springer, Amsterdam, vol. 53, pp. 143–170,.
135. J. J. Peng and Y. Q. Deng, *New J. Chem.*, 2001, **25**, 639–641.
136. H. S. Kim, J. J. Kim, H. Kim, H. G. H. Kim and H. G. Jang, *J. Catal.*, 2003, **220**, 44–46.
137. H. Kawanami, A. Sasaki, K. Matsui and Y. Ikushima, *Chem. Commun.*, 2003, (7), 896–897.
138. J. M. Sun, S. I. Fujita, F. Y. Zhao and M. Arai, *Green Chem.*, 2004, **6**, 613–616.
139. L. N. Han, S. W. Park and D. W. Park, *Energy Environ. Sci.*, 2009, **2**, 1286–1292.
140. H. C. Cho, H. S. Lee, J. Chun, S. M. Lee, H. J. Kim and S. U. Son, *Chem. Commun.*, 2011, **47**, 917–919.
141. Z. Z. Yang, L. N. He, C. X. Miao and S. Chanfreau, *Adv. Synth. Catal.*, 2010, **352**, 2233–2240.
142. C. R. Qi, J. W. Ye, W. Zeng and H. F. Jiang, *Adv. Synth. Catal.*, 2010, **352**, 1925.
143. F. W. Li, L. F. Xiao, C. G. Xia and B. Hu, *Tetrahedron Lett.*, 2004, **45**, 8307–8310.
144. J. Palgunadi and H. S. Kim, *Catal. Today*, 2004, **98**, 511–514.
145. H. Kawanami, A. Sasaki, K. Matsui and Y. Ikushima, *Chem. Commun.*, 2003, (7), 896–897.
146. Y. X. Zhou, S. Q. Hu, X. M. Ma, S. G. Liang, T. Jiang and B. X. Han, *J. Mol. Catal. A: Chem.*, 2008, **284**, 52–57.
147. H. S. Kim, J. J. Kim, H. Kim and H. G. Jang, *J. Catal.*, 2003, **220**, 44–46.
148. J. M. Sun, S. I. Fujita, F. Y. Zhao and M. Arai, *Green Chem.*, 2004, **6**, 613–616.
149. Y. J. Kim and R. S. Varma, *J. Org. Chem.*, 2005, **70**, 7882–7891.
150. V. Calo, A. Nacci, A. Monopoli and A. Fanizzi, *Org. Lett.*, 2002, **4**, 2561.
151. J.-Q. Wang, K. Dong, W.-G. Cheng, J. Sun and S.-J. Zhang, *Catal. Sci. Technol.*, 2012, **2**, 1480–1484.
152. M. Hulla, S. M. A. Chamam, G. Laurenczy, S. Das and P. J. Dyson, *Angew. Chem., Int. Ed.*, **56**, 2017,10559–10563.
153. M. H. Anthofer, M. E. Wilhelm, M. Cokoja, I. I. E. Markovits, A. Pöthig, J. Mink, W. A. Herrmann and F. E. Kühn, *Catal. Sci. Technol.*, 2014, **4**, 1749–1758.
154. F. D. Bobbink, D. Vasilyev, M. Hulla, S. Chamam, F. Menoud, G. Laurenczy, S. Katsyuba and P. J. Dyson, *ACS Catal.*, 2018, **8**, 2589–2594.
155. H. Kawanami, A. Sasaki, K. Matsui and Y. Ikushima, *Chem. Commun.*, 2003, (7), 896–897.
156. M. North and R. Pasquale, *Angew. Chem., Int. Ed.*, 2009, **48**, 2946–2948.
157. J. Sun, S. Zhang, W. Cheng and J. Ren, *Tetrahedron Lett.*, 2008, **49**, 3588–3591.
158. Y. Zhou, S. Hu, X. Ma, S. Liang, T. Jiang and B. Han, *J. Mol. Catal. A Chem.*, 2008, **284**, 52–57.

159. Q. Gong, H. Luo, J. Cao, Y. Shang, H. Zhang, W. Wang and X. Zhou, *Aust. J. Chem.*, 2012, **65**, 381–386.
160. X. Wang, L. Wang, Y. Zhao, K. Kodama and T. Hirose, *Tetrahedron*, 2017, **73**, 1190–1195.
161. Y.-B. Wang, D.-S. Sun, H. Zhou, W.-Z. Zhang and X.-B. Lu, *Green Chem.*, 2014, **16**, 2266–2272.
162. K. R. Roshan, T. Jose, D. Kim, K. A. Cherian and D. W. Park, *Catal. Sci. Technol.*, 2014, **4**, 963–970.
163. L.-N. He, H. Yasuda and T. Sakakura, *Green Chem.*, 2003, **5**, 92–94.
164. Q.-W. Song, L.-N. He, J.-Q. Wang, H. Yasuda and T. Sakakura, *Green Chem.*, 2013, **15**, 110–115.
165. Z.-Z. Yang, Y. Zhao, G. Ji, H. Zhang, B. Yu, X. Gao and Z. Liu, *Green Chem.*, 2014, **16**, 3724.
166. J. Sun, J. Ren, S. Zhang and W. Cheng, *Tetrahedron Lett.*, 2009, **50**, 423–426.
167. Y. Du, J.-Q. Wang, J.-Y. Chen, F. Cai, J.-S. Tian, D.-L. Kong and L.-N. He, *Tetrahedron Lett.*, 2006, **47**, 1271–1275.
168. J.-S. Tian, C.-X. Miao, J.-Q. Wang, F. Cai, Y. Du, Y. Zhao and L.-N. He, *Green Chem.*, 2007, **9**, 566–571.
169. L.-N. He, X.-Y. Dou, J.-Q.Wang, Y. Du and E. Wang, *Synlett*, 2007, (18), 3058–3062.
170. Z.-Z.Yang, Y.-N. Zhao, L.-N. He, J. Gao and Z.-S. Yin, *Green Chem.*, 2012, **14**, 519–527.
171. Y. Du, Y. Wu, A.-H. Liu and L.-N. He, *J. Org. Chem.*, 2008, **73**, 4709–4712.
172. Y. P. Patil, P. J. Tambade, K. M. Deshmukh and B. M. Bhanage, *Catal. Today*, 2009, **148**, 355–360.
173. Y. Zhao, L. Tian, J. Qiu, Z. Li, H. Wang, G. Cui, S. Zhang and J. Wang, *J. CO2 Util.*, 2017, **22**, 374–381.
174. J. Qiu, Y. Zhao, Y. Zhao, H. Wang, Z. Li, J. Wang and T. Jiao, *Molecules*, 2019, **24**, 1241.
175. J. Qiu, Y. Zhao, Z. Li, H. Wang, M. Fan and J. Wang, *ChemSusChem*, 2016, **9**, 1–9.
176. T. Takahashi, T. Watahiki, S. Kitazume, H. Yasuda and T. Sakakura, *Chem. Commun.*, 2006, 1664–1666.
177. T. Sakai, Y. Tsutsumi and T. Ema, *Green Chem.*, 2008, **10**, 337–341.
178. H.-L. Shim, S. Udayakumar, J.-I. Yu, I. Kim and D.-W. Park, *Catal. Today*, 2009, **148**, 350–354.
179. J. Sun, W. Cheng, W. Fan, Y. Wang, Z. Meng and S. Zhang, *Catal. Today*, 2009, **148**, 361–367.
180. J. Gong, H. Lin, K. Grygiel and J. Yuan, *Appl. Mater. Today*, 2017, **7**, 159–168.
181. C. Toborg and M. Beller, *Adv. Synth. Catal.*, 2009, **351**, 3027–3043.
182. L.Yin and J. Liebscher, *Chem. Rev.*, 2007, **107**, 133–173.
183. C. Petrucci, G. Strappaveccia, F. Giacalone, M. Gruttadauria, F. Pizzo and L. Vaccaro, *ACS Sust. Chem. Eng.*, 2014, **2**, 2813–2819.
184. F. S. Han, *Chem. Soc. Rev.*, 2013, **42**, 5270–5298.

185. G. Dilauro, S. M. García, D. Tagarelli, P. Vitale, F. M. Perna and V. Capriati, *ChemSusChem*, 2018, **11**, 3495–3501.
186. C. Zhang, X.-M. Chen, Y. Luo, J.-L. Li, M. Chen, L. Hai and Y. Wu, *ACS Sust. Chem. Eng.*, 2018, **6**, 13473–13479.
187. P. Hesemann, B. Motos-Perez, A. Thomas and J. Roeser, *Appl. Organomet. Chem.*, 2013, **27**, 290–299.
188. P. Li, L. Wang, Y. Zhang and G. Wang, *Tetrahedron*, 2008, **64**, 7633–7638.
189. D.-J. Tao, F. Liu, L. Wang and L. Jiang, *Appl. Catal., A*, 2018, **564**, 56–63.
190. A. B. Bourlinos, K. Raman, R. Herrera, Q. Zhang, L. A. Archer and E. P. Giannelis, *J. Am. Chem. Soc.*, 2004, **126**, 15358–15359.
191. P. G. Rickert, M. R. Antonio, M. Firestone, K.-A. Kubatko, T. Szreder, J. F. Wishart and M. L. Dietz, *Dalton Trans.*, 2007, 529–531.
192. R. Fu, Y. Yang, W. Feng, Q. Ge, Y. Feng, X. Zeng, W. Chai, J. Yi and R. Yuan, *Tetrahedron*, 2016, **72**, 8319–8326.
193. Y. Leng, J. Wang, D. Zhu, X. Ren, H. Ge and L. Shen, *Angew. Chem., Int. Ed.*, 2009, **48**, 168–171.
194. W. Zhang, Y. Leng, P. Zhao, J. Wang, D. Zhu and J. Huang, *Green Chem.*, 2011,**13**, 832–834.
195. Y. Leng, J. Wang, D. Zhu, M. Zhang, P. Zhao, Z. Long and J. Huang, *Green Chem.*, 2011, **13**, 1636–1639.
196. Y. Leng, J. Liu, C. Zhang and P. Jiang, *Catal. Sci. Technol.*, 2014, **4**, 997–1004.
197. Z. Y. Long, Y. Zhou, G. J. Chen, W. L. Ge and J. Wang, *Sci. Rep.*, 2014, **4**, 3651–3651.
198. S. P. Satasia, P. N. Kalaria, J. R. Avalani and D. K. Raval, *Tetrahedron*, 2014, **70**, 5763–5767.
199. R. Mohan, N. Kalla, J. Lim, J. Bae and I. Kim, *Tetrahedron Lett.*, 2017, **58**, 1595–1599.
200. S. Das, S. Santra, S. Jana, G. V. Zyryanov, A. Majee and A. Hajra, *Eur. J. Org. Chem.*, 2017, 4955–4962.
201. N. K. Vishwakarma, A. K. Singh, Y.-H. Hwang, D.-H. Ko, J.-O. Kim, A. G. Babu and D.-P. Kim, *Nat. Commun.*, 2017, **8**, 14676.
202. X. Chen, X. Li, A. Hu and F. Wang, *Tetrahedron Asymm.*, 2008, **19**, 1–14.
203. S. Kirchhecker and D. Esposito, *Curr. Opin. Green Sus. Chem.*, 2016, **2**, 28–33.
204. S. Luo, X. Mi, L. Zhang, S. Liu, H. Xu and J.P. Cheng, *Angew. Chem., Int. Ed.*, 2006, **45**, 3093–3097.
205. K. Nobuoka, S. Kitaoka, T. Kojima, Y. Kawano, K. Hirano, M. Tange, S. Obata, Y. Yamamoto, T. Harran and Y. Ishikawa, *Org. Chem. Int.*, 2014, **2014**, 836126.
206. L. Gonzalez, J. Escorihuela, B. Altava, M. I. Burguete and S. V. Luis, *Eur. J. Org. Chem.*, 2014, (24), 5356–5363.
207. R. Gausepohl, P. Buskens, J. Kleinen, A. Bruckmann, C. W. Lehmann, J. Klankermayer and W. Leitner, *Angew. Chem., Int. Ed.*, 2006, **45**, 3689–3692.

208. T. J. Geldbach, D. Zhao, N. C. Castillo, G. Laurenczy, B. Weyershausen and P. J. Dyson, *J. Am. Chem. Soc.*, 2006, **128**, 9773–9780.
209. R. Kukawka, A. Pawlowska-Zygarowicz, J. Dzialkowska, M. Pietrowski, H. Maciejewski, K. Bica and M. Smiglak, *ACS Sust. Chem. Eng.*, 2019, 7, 4699–4706.
210. M. Kostić, P. Verdía, V. Fernández Stefanuto, R. Puchta and E. Tojo, *J. Phys. Org. Chem.*, 2019, **32**, 3928.
211. P. Xu, G.-W. Zheng, P.-X. Du, M.-H. Zong and W.-Y. Lou, *ACS Sust. Chem. Eng.*, 2016, **4**, 371–386.

CHAPTER 11

Structured Catalysts and Non-conventional Reactor Designs for Energy Applications†

MATTEO AMBROSETTI, RICCARDO BALZAROTTI, LAURA FRATALOCCHI, MAURO BRACCONI, GIANPIERO GROPPI AND ENRICO TRONCONI*

Politecnico di Milano, Dipartimento di Energia, Via La Masa 34, 20156 Milano, Italy

*Corresponding contributor. Email: enrico.tronconi@polimi.it

11.1 Introduction

The most important sectors of the worldwide economy such as the petroleum, power generation, chemicals and food industries are mainly dependent on catalytic processes. Therefore, the development of compact, safe, energy-efficient and environmentally friendly processes is a key topic in modern chemical engineering.[1]

Structured catalysts and reactors are topics of growing interest in heterogeneous catalysis, as they represent some of the most promising solutions to

†Electronic supplementary information (ESI) available. Colour version of Figure 11.6. See DOI: 10.1039/9781788019576.

Heterogeneous Catalysis for Energy Applications
Energy and Environment Series No. 27
Edited by Tomas R. Reina and Jose A. Odriozola

Published by the Royal Society of Chemistry, www.rsc.org

overcome the heat and mass transfer limitations of traditional pelletized catalysts.[2,3] The well-known honeycomb substrates consist of solid matrices composed of ceramic (*e.g.* Al_2O_3, cordierite, SiC) or metallic materials (*e.g.* stainless steel, Al, Cu, Ni)[4] that are shaped in regular forms where parallel and segregated channels allow for the flow of the gaseous stream while the catalyst is distributed on the surface of the solid phase. These systems have been introduced for automotive emission control and environmental catalysis thanks to their optimal trade-off between mass transfer and pressure drop,[5] and today they are extensively employed in this field. When manufactured with a thermally conductive bulk material, these structures can be characterized by very high radial and axial effective conductivities, thanks to their continuous solid matrix. In this case, they can be exploited for the intensification of non-adiabatic heat transfer-limited processes.[2,6] However, the main drawback of these structures is the manufacturing constraints associated with extrusion. Metallic monoliths can be manufactured starting from flat and corrugated metal foils that are rolled or piled up, enabling high cell densities (channels per square inch – CPSI – up to 2000) and large voidages (Open Frontal Area – OFA – up to 0.9). Several works at the laboratory micro-scale report the successful use of these materials for the intensification of strongly exo- or endo-thermic reactions.

In the last two decades, open cell foams and, more recently, periodic open cellular structures (POCS) gained interest as potential enhanced catalyst supports. A sketch of their geometrical structure is shown in Figure 11.1. Open cell foams were actually proposed as catalyst supports almost 20 years ago,[8,9] and then received attention at a fundamental level[5,10–13] and for application in selected processes.[14–17] Open cell foams are readily manufactured with different methodologies and with various materials (*e.g.* metals such as stainless steel, copper, aluminium and many others). They can be produced in different forms, typically in long sheets with thicknesses of a few centimetres, and then are shaped to the final geometry (*e.g.* cylinders or pellets). Alternative manufacturing technologies allow for great flexibility in the geometrical and morphological properties of the supports. Open cell foams are cellular materials

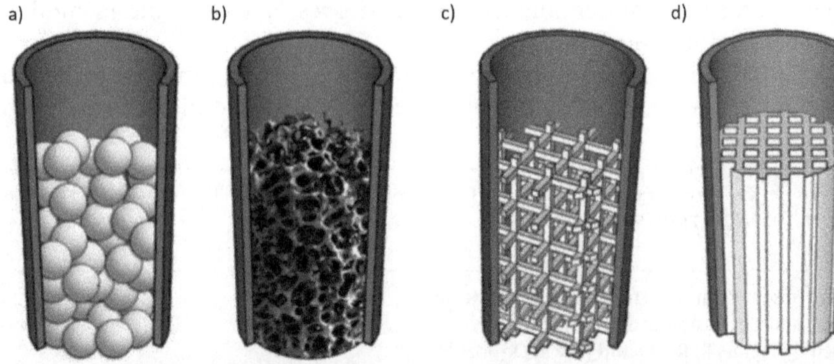

Figure 11.1 Graphical sketch of different reactor layouts: packed bed (a); structured reactors: honeycomb monolith (b), foam (c) and POCS (d). Adapted from Ref. 7 with permission from Elsevier, Copyright 2018.

composed of interconnected solid struts arranged in cells that enclose void regions, with open windows or pores where the flow can stream through.[18] The main morphological properties that characterize open cell foams are the porosity (or the void fraction) the solid fraction and the pore density (*i.e.* the number of pores per linear inch); the smaller the pores, the higher the surface area of the support. An additional interesting feature of open cell foams is the continuous and totally interconnected solid matrix: this provides an alternative mechanism for heat transport, namely heat conduction, which is instead contact limited in random packings of catalytic pellets. By combining this property with foams manufactured in highly conductive metals (Al, Cu), high thermal axial and radial conductive heat transfer rates are expected, making these supports ideal structures for the intensification of catalytic processes where heat management is crucial. The enhanced heat transfer, in fact, can reduce the temperature gradients, minimizing the hot spots and thus surpassing the performance of conventional packed beds.[19]

POCS are characterized by their simple geometry since they are composed of the repetition of regular and ordered unit cells, which can have different shapes.[7,11,20–22] In the literature, cubic, tilted cubic, diamond and tetrakaidecahedron unit cells have been proposed as building blocks for interconnected structures. For these supports, the determination of the geometrical properties is trivial, since they can be derived using analytical approaches.[22,23] These structures are manufactured with several 3D printing techniques. Stereolithography (SLA) can be employed for ceramic samples,[20] whereas selective laser melting (SLM), selective electron beam melting (SEBM) or selective laser sintering (SLS) can be employed for different metallic supports.[24–26] SLA exploits a bath of photo-sensible polymer that is irradiated with a controlled light source and is locally sintered. This technique is quite popular for the manufacturing of resin samples.[27] Recently, it has been proposed as an alternative approach for the manufacturing of ceramic supports for automotive catalysts.[28] Powder bed methods (SLM, SEBM and SLS) are typically employed for the 3D printing of metallic supports. A structure is produced layer by layer performing the sintering – or the melting – of a metal powder. Micro-fusion techniques have also been proposed for the manufacturing of defect-free aluminium supports.[29] This method is based on printing a sacrificial template in wax of the final sample, then the structure is embedded in a heat-resistant slurry and finally the liquid metal is poured into the mould. The deterministic manufacturing process allows for the optimization of the geometry thanks to detailed computational fluid dynamics (CFD) simulations of the principal transport phenomena.

Several techniques are available for the catalytic activation of these structures.[30] The most common ones are washcoating[31] and electrodeposition.[32] A thin layer of catalytic active phase is deposited onto the structure. Open cell foams were employed as structured catalyst supports initially for the intensification of external diffusion-limited adiabatic processes such as catalytic partial oxidation (CPO) of hydrocarbons to synthesis gas (syngas)[33–38] or NO_x removal from exhaust gas from power stations or automotive engines.[39,40] The complex flow field established inside the random and tridimensional geometry increases both the local mixing and the gas-to-solid transport properties,[5,41,42] preventing reactant mal-distribution over the reactor cross-section as well. Thanks to the enhanced thermal conductivity, these structures are suitable solutions both in highly endothermic, dehydrogenation reactions[43–46] and in exothermic systems (*e.g.* selective oxidation,[47,48] methanation,[14,49–51] methanol synthesis[15,52,53] and hydrocarbon hydrogenation reactions[54–58]). Due to their features, POCS can be considered to be advancements over open cell foams and, in principle, they can be applied in the same processes. Papetti *et al.*[20] report the potential of POCS in exhaust after-treatment devices, either as a three-way catalyst (TWC), diesel oxidation catalyst, ammonia slip catalyst and in selective catalytic reduction (SCR) applications. Conductive POCS have been proposed for both exothermic and endothermic applications, including, for example, the heterogeneously catalysed oxidation of methanol to formaldehyde, CO_2 methanation[59,60] and NO oxidation in nitric acid plants.[61]

Washcoated structured supports present some drawbacks with respect to packed bed reactors, as the washcoat management (adhesion, stability to thermal cycles, mechanical stresses) is critical. Furthermore, the catalyst load per unit volume of a washcoated system is much lower than the loads of packed bed systems, limiting the adoption of this reactor configuration.

To overcome these limitations, a new reactor layout was recently proposed, which consists of filling the empty porosity of a structured cellular substrate (open cell foams or monoliths)[62–64] with small catalytic particles. Such a configuration makes it possible to exploit the enhanced heat transfer of the support while increasing the catalyst inventory up to values close to those of a conventional packed bed reactor. The concept of packed foams was first effectively demonstrated for Fischer–Tropsch synthesis (FTS): the introduction of a highly conductive aluminium foam in the FTS reactor prevented thermal runaway at high CO conversions, thus enabling the reactor to run under much more severe conditions than in the packed bed configuration.[63] The concept of packed foams is now being actively investigated, with experimental and computational studies devoted to the analysis of the transport properties[62,65] and to other proofs of concept.[63,66] The extension of the concept to packed POCS is currently also under investigation.[67]

11.2 Heat Transport Properties of Structured Catalyst Substrates

11.2.1 Honeycomb Monoliths

As was pointed out in the introduction to this chapter, one of the main drivers for the adoption of structured catalysts for process intensification is the excellent heat transfer provided by thermally connected conductive structures. Many studies have been devoted to the experimental and numerical assessment of the effective thermal conductivity of honeycomb supports.[2,68–71]

The thermally connected matrix of monolith substrates enables the conductive heat transfer mechanism in both radial and axial directions. However, due to the honeycomb geometry, the axial and the radial effective conductivities are different. The effective conductivity in the axial direction can be estimated according to eqn (11.1), being a function of the solid bulk conductivity and of the solid fraction:

$$k_{\text{eff,ax HC}} = k_{\text{bulk}}\ (1 - \text{OFA}) \tag{11.1}$$

The derivation of an expression for the thermal conductivity in the radial direction is more complex. For square and triangular-channel honeycombs, Tronconi and Groppi derived an expression based on the electrical network analogy of a single channel.[72] This expression was then validated by Hayes and co-workers with numerical simulations of the temperature profiles;[70] the same authors proposed a similar equation based on a different electric network that provided better estimates when compared to the numerical solution. Visconti *et al.*[71] developed a more accurate expression for the effective radial thermal conductivity of honeycombs and compared it to estimates from the temperature profiles obtained by solving the Laplace equation in an idealized honeycomb square cell:

$$k_{\text{eff,rad HC}} = k_{\text{bulk}} \frac{1 - \text{OFA}}{1 + \text{OFA}} \tag{11.2}$$

Figure 11.2 compares the equation proposed by Visconti *et al.*[71] with other models: it clearly represents the best approximation of the numerical solution. The effective radial thermal conductivities of square-channel honeycombs computed by eqn (11.2) are plotted against the void fraction for honeycombs made of different conductive materials. These values should be compared with typical static radial thermal conductivities of packed bed reactors (0.1–1.0 W m^{-1} K^{-1}).

It is obvious that these systems exhibit very high radial and axial thermal conductivities. This enables very small axial and radial temperature gradients even in the case of high heat transfer rates.

When dealing with such highly conductive systems, the coupling between the structure and the tube walls is critical, since it becomes the limiting heat transfer resistance. This resistance can be regarded as a "contact resistance" due to the "gap" between the structure and the monolith. Its reciprocal (*i.e.* the wall heat transfer coefficient) can be expressed accordingly as in eqn (11.3):

$$h_{\text{wall,HC}} = k_{\text{gas}} / d_{\text{gap}} \tag{11.3}$$

The gap is a function of the manufacturing technique and of the loading methodology. Furthermore, it was found that, for exothermic reactions, the gap in reacting conditions would be reduced thanks to the thermal expansion of the materials due to the temperature difference between the bulk solid and the tube walls.[68] Solutions aimed at reducing the gap size have been developed accordingly to improve the wall heat transfer.

The typical configuration of packed honeycomb monoliths is represented in Figure 11.3. In the case of an exothermic reaction, the heat is generated in the catalyst pellets, then it is exchanged with the honeycomb channels and, thanks to the radial conduction, it is transported to the wall and eventually to the external cooling fluid.

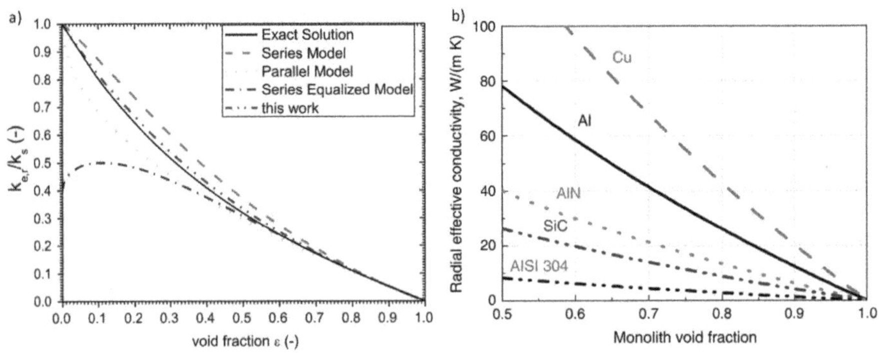

Figure 11.2 Comparison of correlations and exact solutions of the effective radial thermal conductivities of square-channel honeycombs (a) and calculations as a function of bulk metal thermal conductivities and void fractions (b). Part (a) reproduced from Ref. 71 with permission from Elsevier, Copyright 2013. Part (b) reproduced from Ref. 2 with permission from Elsevier, Copyright 2014.

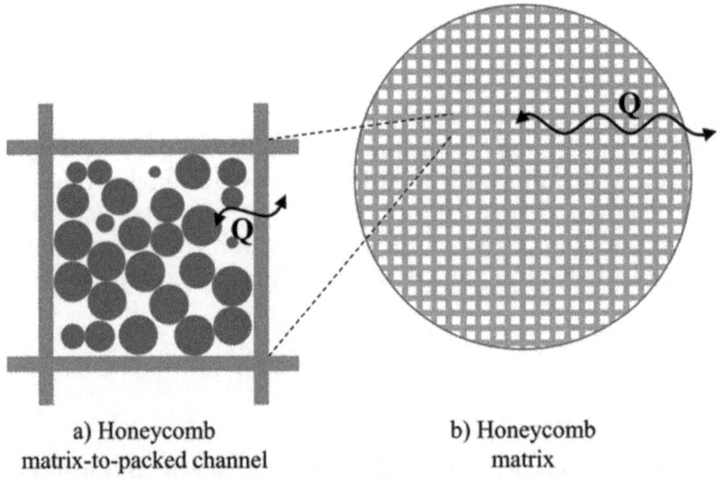

Figure 11.3 Heat transport mechanism of packed honeycombs. Reproduced from Ref. 73 with permission from the Royal Society of Chemistry.

An important piece of information is the evaluation of the heat transport between the packing and the metallic honeycomb. In this regard, Rebughini *et al.*[73] studied by CFD the heat transfer inside the packed microchannels and found that the Wellauer correlation[74] is most suitable for the evaluation of heat transport in this special geometry. A pseudo-continuum heat transport model was developed. Their analysis revealed that the adoption of such a configuration enables the temperature gradients in the reactor to be strongly reduced. In a case study concerning a phtalic anhydride reactor, large operating windows were possible due to the intensified heat transfer allowing for higher o-xylene flow rates, larger tube diameters and higher coolant temperatures without loss of the thermal control of the reaction. Optimal geometries and process parameters of an intensified multitubular reactor for the synthesis of phthalic anhydride have been reported.[73]

11.2.2 Foams

Open cell foams are considered a potential substitute of honeycomb monoliths for non-adiabatic processes thanks to their random interconnected solid structures that enable a high isotropic effective thermal conductivity. The complex geometry of open cell foams, however, does not facilitate the analysis of their transport properties. Experimental work aiming to evaluate the stagnant conductivity of open cell foams has been reported by several researchers. Typically, they loaded foams with different characteristics into tubular reactors, measured the temperature profiles during non-reactive heat transfer flow experiments and fitted the data with numerical heat transfer models aimed at the estimation of effective heat transfer properties.[12,42] In parallel, CFD was used for the prediction of the effective thermal conductivity of commercial foams, relying on micro-computed tomography (μ-CT) scans.[75,76] Our group tackled the same problem by proposing a procedure for the digital reconstruction of open cell foams[77] based on a physically sound geometrical model[78] that considers parabolic strut profiles. With this tool, it is possible to generate synthetic foam structures that essentially retain the geometrical features (topology and surface area) of the real random open cell foams and can therefore be interrogated numerically for the estimation of representative transport properties (pressure drop and mass transfer).[27,79] Simulations of heat conduction in the solid matrix were performed[80] by imposing a temperature difference between two opposite faces of a cubic foam domain and imposing adiabatic boundary conditions on the remaining faces. The heat flux across the solid matrix was calculated and used for the evaluation of the effective thermal conductivity.

The CFD simulation results were compared with a broad set of experimental data, showing very good accordance in the case of high void fractions, whereas the experimental data suffer from large scatter in the case of low void fractions, as shown in Figure 11.4. For high porosities, the results are in line with earlier works by Lemlich, who determined theoretically the electric conductivity of foams in the limit of porosity ε approaching one.[81] On decreasing the void fraction, a more than linear increment of the effective thermal conductivity is

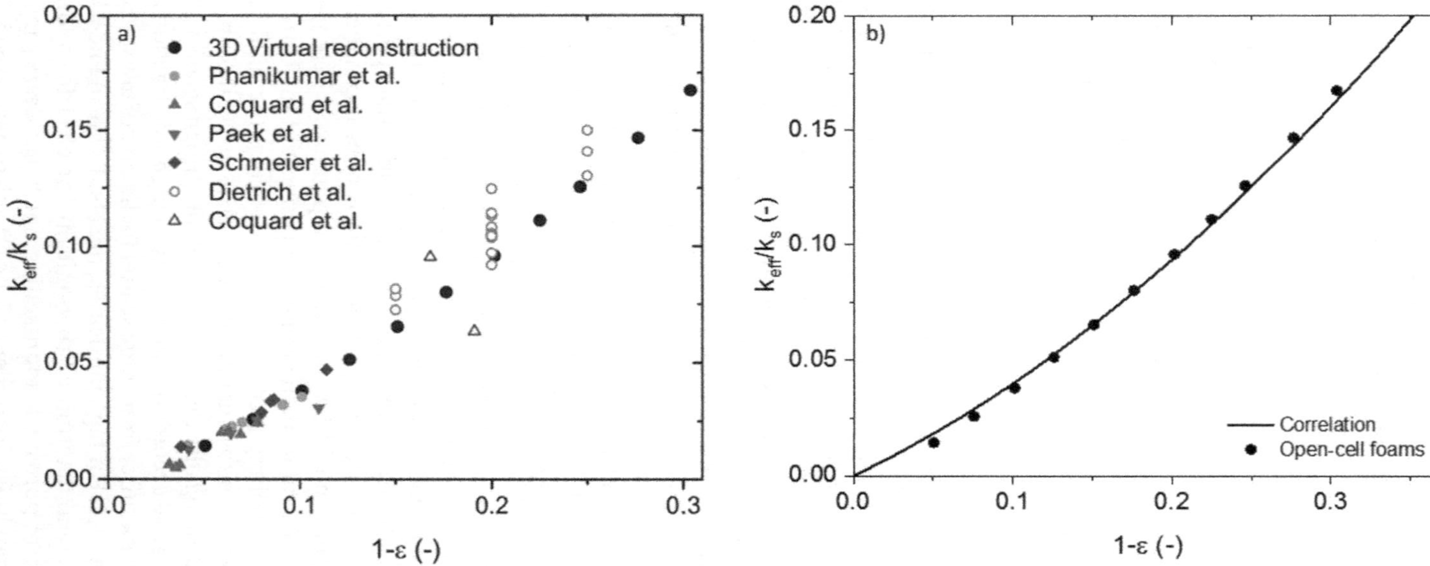

Figure 11.4 Effective thermal conductivity evaluated for the virtually generated foams (black circles) against the solid fraction compared with experimental data for ceramic (open symbols) and metal (filled symbols) foams (a) and effective thermal conductivities of open cell foams (symbols) compared with the correlation by Bracconi *et al.* (continuous line) (b). Reproduced from Ref. 80 with permission from Elsevier, Copyright 2018.

manifest. The effective thermal conductivity of open cell foams is described by eqn (11.4), resulting from the superposition of two asymptotic solutions, namely the Lemlich solution for $\varepsilon = 1$ and the Maxwell solution for $\varepsilon = 0$, which provides a good description of the literature data without any fitting parameter.

$$k_{\text{eff,foam}} = k_{\text{bulk}} \frac{1-\varepsilon}{\left(\frac{2}{3}(1-\varepsilon)+\frac{1}{3}\right)^{-1}} \tag{11.4}$$

As in the case of conductive honeycomb monoliths, the larger heat transfer resistance for conductive open cell foams is represented by the wall coupling. Bianchi *et al.*[42,75] performed experimental and numerical studies to determine the role and extent of the wall heat transfer coefficient. They found that this parameter is inversely proportional to the cell size and proportional to the gas thermal conductivity. The authors proposed a correlation where the Nusselt number, expressed in terms of the cell diameter, is equal to the sum of a constant and of a second term dependent on the Reynolds number. In a more recent study, Aghaei *et al.*[12] evaluated the wall heat transfer coefficient by an equation similar to eqn (11.3), previously derived for honeycombs. In the case of foams, however, an effective gap size is present, which is a function of the cell size:

$$d_{\text{gap,foam}} = 0.00013 + 0.14 \cdot d_{\text{cell}} \tag{11.5}$$

To assess the role of wall coupling, Razza *et al.*[82] evaluated by CFD the effect of "perfect" and "loose" wall contact. The simulations show a very significant role of the wall coupling on the temperature profiles. Experimental heat transfer data were well fitted assuming that only 5% of the struts are in contact with the wall, whereas a 20% contact resulted in almost matching the temperature profiles generated assuming perfect contact by sintering foams to the tube walls.

To develop a multi-scale heat transfer model for open cell foams, it is necessary to consider all of the possible heat transport contributions. In particular, radiative heat transfer in open cell foams may become relevant since they display large surface areas and high void fractions. For this purpose, the radiation model of Gliksmann *et al.*[83] was considered by Aghaei *et al.*[12]

In the case of low pores per inch (PPI) foams, local thermal equilibrium may be not established between gas and solid phases. To account for this heat transfer resistance, thanks to the Chilton–Colburn analogy, the correlation by Bracconi *et al.*[79] may be used to provide a reasonable estimate of the interphase heat transfer.

Packed foams have been proposed as suitable alternatives to washcoated foams for the intensification of several processes. The heat transport properties of these systems have been studied by Visconti *et al.*[62] and more recently by Balzarotti *et al.*[66] by developing a heat transport model based on an electrically equivalent circuit, as shown in Figure 11.5. The two-phase model includes: (1) a solid phase associated with the foam matrix; and (2) a second pseudo-phase lumping the small catalyst particles and the gas, which are supposed to be at thermal equilibrium.

Based on the equivalent circuit in Figure 11.5, a near-wall resistance ($h_{w,eq}$) is present before entering the bed core. This results from the combination of two parallel paths associated with the wall heat transfer resistance of the foam structure and with the wall heat transfer resistance of the packed bed, assumed as a lumped fluid phase-packed particle system. The wall heat transfer coefficient of the foam is evaluated according to Aghaei *et al.*,[12] whereas the packed bed wall heat transfer coefficient is calculated according to Specchia *et al.*[84] Inside the bed core, two parallel heat flow paths are further considered: in the lower branch, two contributions are relevant, the former being associated with the foam interconnected structure (*i.e.* $k_{eff,F}$ – effective radial thermal conductivity of the foam) estimated by eqn (11.4), while the latter is an additional term (*i.e.* $U_{F \rightarrow PB}$) that takes into account the internal resistance associated with heat transfer from the foam struts to the packed bed. The upper branch, instead, represents the thermal resistance of a packed bed, evaluated again by the correlations of Specchia *et al.*[84]

The circuit in Figure 11.5 emphasizes the synergetic effect due to the combination of the two systems (foam + packed bed), as at the wall the packed bed helps to increase the heat transfer, whereas inside the tube the solid matrix contributes to the effective radial heat transport. A detailed description of all the parameters and equations used for the calculation of the components of the heat transfer model is reported in Ref. 66.

11.2.3 Periodic Open Cellular Structures

POCS retain, in principle, all of the advantages of open cell foams, with the additional possibility of an optimized design and enhanced thermal conductivity. Heat transfer in these systems has been addressed by experimental[7,61] and numerical studies.[85]

Bianchi *et al.*[85] studied the effective thermal conductivities of different POCS with different cell shapes, porosities and strut profiles using the same

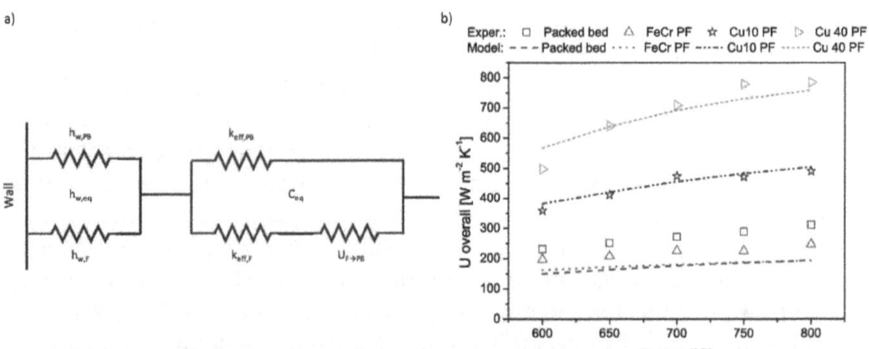

Figure 11.5 Graphical representation of the heat transfer mechanisms within packed foams using an equivalent circuit of resistances. Reproduced from Ref. 66 with permission from Elsevier, Copyright 2019.

approach reported by Bracconi *et al.*[80] The authors exploited the symmetry of the different cells to reduce the computational costs of the numerical simulations. They found that, despite the different morphologies, structures with various cell shapes display very similar conductivities at assigned void fractions. Moreover, structures with struts having a constant cross-sectional area exhibit higher thermal conductivities, resulting in an advantage of up to 30% at high void fractions over foams. A correlation very similar to eqn (11.4) was reported for the effective thermal conductivity of POCS.

Busse *et al.*[7] performed non-reactive heat transfer experiments in POCS under steady-state conditions. The results were analysed according to the same numerical methodology reported by Bianchi *et al.*[75] For poorly conductive structures, the regressions revealed a strong dependence of both the wall heat transfer coefficient and the radial effective conductivity on the flow rates. The authors proposed some engineering correlations to describe the heat transfer performances of cubic cell POCS. They also proposed an interesting comparative study of the overall heat transfer performances of these structures and packed beds as a function of the intrinsic thermal conductivity of the POCS fabrication material λ_S and of the specific mass flow rate G (Figure 11.6b), which clearly highlights the potential of these structures for intensifying the heat transfer in tubular reactors, especially at low flow rates.

11.3 Applications to Methane Steam Reforming

The steam reforming of natural gas and light hydrocarbons (methane steam reforming [MSR]) is the main industrial process for the production of hydrogen and syngas due to the cost effectiveness of the technology.[86] At the industrial scale, steam reforming is mainly carried out in externally heated multi-tubular packed bed reactors, where the endothermic process heat is provided by external burners. The feed stream is typically pre-heated to 425–550 °C and fed to the furnace. Thus, the mixture of steam and methane reacts inside the metallic tubes filled with the catalyst. As the process is strongly endothermic, a large amount of heat has to be supplied: this is achieved by heating the tube walls in a fired radiant furnace, whose typical configurations are shown in Figure 11.7.[87]

In each configuration, the tubes and burners should be ideally arranged in the furnace in such a way that temperature and heat flux along the tube can be easily controlled. These systems are optimized to maximize the convective heat transfer from the tube walls to the catalyst, which implies the adoption of high flow rates at in large-scale production.[88]

In the last few years, the rapidly growing interest in fuel cell and hydrogen technologies has led to increasing demand for low-cost, small-scale fuel processors for distributed hydrogen production systems; this calls for the development of novel reactor configurations, which should pursue the target of clean, sustainable and cost-competitive production processes.[89,90] Additionally, the new configurations will be also asked to manage a new kind of hydrogen demand or production rate, characterized by significant variability in the

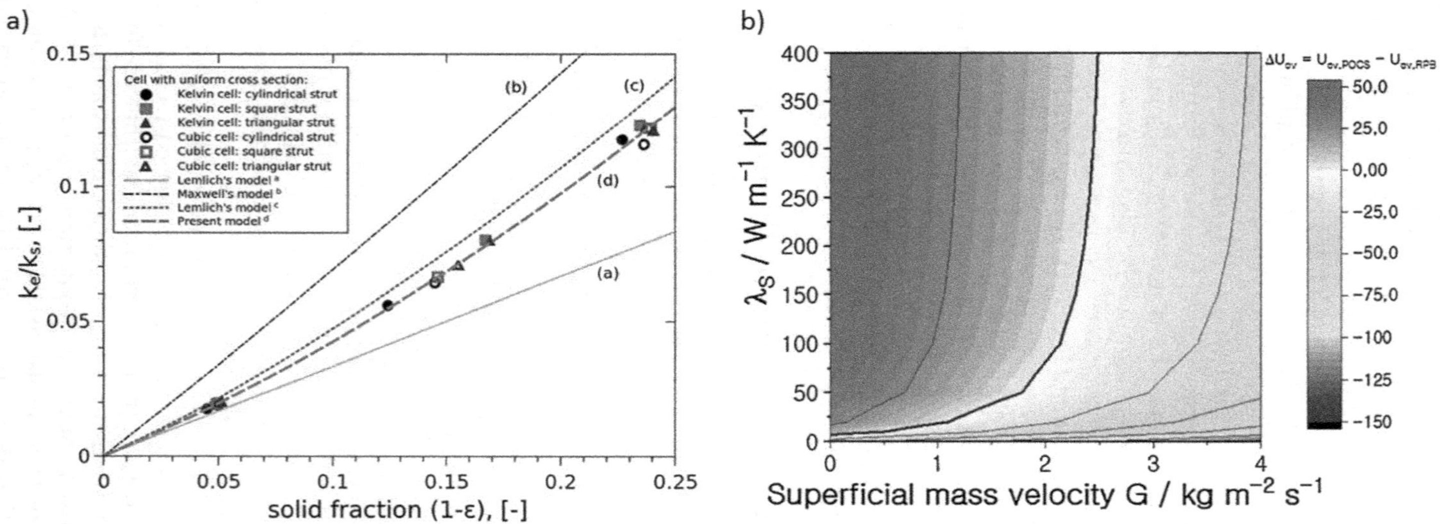

Figure 11.6 Effective thermal conductivity of POCS with different cell shapes (a) and comparison of overall heat transfer coefficients of POCS and packed beds as a function of thermal conductivity and specific mass flow rate (b). Part (a) reproduced from Ref. 85 with permission from John Wiley and Sons, Copyright © 2015 Wiley-VCH Verlag GmbH & Co. KGaA, Weinheim. Part (b) reproduced from Ref. 7 with permission from Elsevier, Copyright 2018.

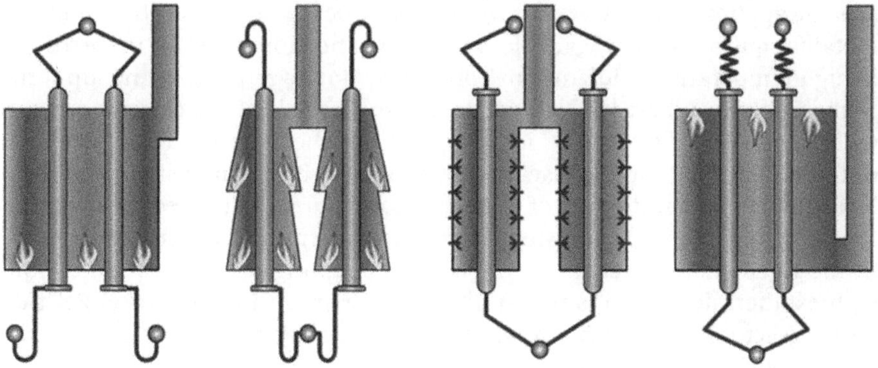

Figure 11.7 Tube and burner arrangements in fire radiant furnaces. Adapted from Ref. 87 with permission from Elsevier, Copyright 1995.

fuel and energy markets.[91] Thus, new research paths have been opened, aiming at meeting the new demand for compact, reliable, flexible and modular reformers.

The new demand for distributed hydrogen production calls for both short-term and long-term solutions to hydrogen production; in the latter case, carbon-free and sustainable production processes need to be designed in order to move beyond the interim solution provided by fossil fuel-based processes. In the case of short-term solutions, however, steam reforming of natural gas (MSR) is still the best compromise between cost, effectiveness and process know-how, which could exploit the already existing methane supply chain, as well as the expertise in the industrial application of the process. Nevertheless, many research and development goals remain to be accomplished due to the difficulties involved in the down-scaling of large industrial reformers. Innovative approaches to intensify the process are required:[92] given the low flow rates of small-scale reformers, heat transfer limitations are likely to be present[3] if the traditional packed bed configuration of industrial reformers is used. Within to this scenario, several technologies are emerging for the distributed production of hydrogen at small scale, such as catalytic heat exchange reactors, membrane reactors, sorption-enhanced reforming, plasma reforming, microchannel reactors and structured reactors.

Among others, washcoated microchannel reactors offer good thermal regulation and control of chemical processes. Indeed, the presence of small-diameter channels (50–5 000 μm) enhances both heat and mass transfer rates, leading to smaller equipment, isothermal behaviour and higher conversion, yield, selectivity and catalyst lifetimes with respect to conventional configurations.[93] This technology is also intrinsically modular: for example, in order to increase plant capacity, it is sufficient to add many small reactors with the same dimensions, instead of increasing the reactor volume of a single unit. Such reactors are typically fabricated using metallic materials (*i.e.* aluminium, copper and stainless steel) according to the requirements of the specific application. However, ceramics may offer advantages with respect to metallic equivalents, as they are less expensive and more resistant to corrosion in high-temperature oxidizing environments. Moreover, the ceramic supports of the catalysts adhere well within the microchannels of the ceramic reactor without additional treatment. At the same time, the low mechanical strength of these materials could cause fractures induced by thermal stresses due to localized hot spots: the risk of thermal shock can be mitigated by adopting longer start-up times.[94] The steam reforming reaction has been investigated in the microchannel reactor configuration coupling the endothermic MSR with exothermic reactions in a multi-flow system; in the literature, FeCrAl-based reactors have been investigated for running the process,[95,96] using noble metal-based catalysts.[97]

An alternative promising approach involves the adoption of structured reactors based on conductive matrices, which have been proven to mitigate the internal heat transfer limitations in endo/exothermic processes:[98] several works have reported the beneficial effect of the introduction of conductive internals in the form of either honeycomb monoliths or open cell foams on overall system performances.[2,98] These structures are typically activated by depositing a thin washcoat layer of catalytic material onto the geometric surface of the support.[30]

MSR is typically operated at high temperatures and, thus, the bulk material of geometric supports needs to be carefully selected in order to meet process specifications. Accordingly, in many works, steel-based materials were reported as valuable candidates for supporting the active phase in the process, thanks to their high thermal resistance and chemical inertness. In this view, Sang *et al.* tested (AISI 316) stainless steel open cell foams as solar absorbers to support Ni- and Ru-based catalysts for the dry reforming of methane,[99] while Yu *et al.* proposed similar materials for the reforming of methanol using copper-based catalysts.[100] Similar steel-based morphological supports were also used for the intensification of steam biogas reforming, paving the way to performance optimization through catalyst formulation and washcoat optimization.[101,102]

One of the most significant limitations in the use of steel-based materials for non-adiabatic processes is the relatively low intrinsic thermal conductivity of steel, especially in comparison with more conductive materials such as copper or aluminium. Accordingly, aluminium has been studied as a bulk material for structured supports, but its use is limited due to its low melting point (~660 °C); thus, Al-based applications in the process intensification of methanol steam reforming[103] and water gas shift[104] have been reported due to the milder operative condition of these processes in comparison with MSR.

As an alternative, other works focused on the use of nickel and nickel alloys as bulk materials for open cell foams, which enables higher thermal conductivity compared to steel-based materials. In particular, promising results were reported for the CO_2 steam reforming of CH_4,[105,106] which highlighted benefits in terms of the heat transfer properties, as the radial heat transfer of the metallic foam catalyst was improved compared to catalytic alumina pellets.

Despite its high thermal conductivity, copper has not been widely investigated as a material of choice for the production of supported reforming catalysts. Efforts have focused on applying copper foams to the methanol reforming process:[107,108] in all cases, benefits were reported in terms of improvements in heat transfer properties, which directly translated into a reduction in the intensity of the cold spot. Moreover, copper-supported materials were found to provide improvements over commercial catalysts in the form of pellets.[109] Additionally, structured copper catalysts have been applied in the solar reforming of methane for supporting Ni-based Mg/Al oxide catalysts.[110,111]

In addition to overcoming the internal heat transfer limitations by adopting conductive internals, improvements in catalytic activity can be achieved by using noble metals, as is widely established in the literature and was recently reviewed by Farrauto *et al.*[112] This approach allows for an increase in the kinetic performance of the system thus meeting the need for a smaller catalyst inventory. In this view, Yang *et al.* investigated the use of both conductive substrates (namely nickel and copper open cell foams) and noble metal-based catalysts (*i.e.* Pd/Al_2O_3), providing insights into both the washcoat deposition process and the role of Ni and Cu metal foams on overall process performance.[113]

11.3.1 Packed Bed *versus* Washcoated Foams

Based on this background, a recent work has been reported in the literature focusing on the intesification of MSR thanks to the adoption of a highly active Rh/Al_2O_3 catalyst and the use of highly conductive copper foams.[64] Forty PPI copper foams were catalytically activated by finely tuning the Rh/Al_2O_3 washcoat layer onto the support surface, in terms of both metal load and layer thickness, by using the slurry coating method. Two different washcoated samples (*i.e.* 80 μm thick with 1 wt% Rh and 35 μm thick with 2.2 wt% Rh, labelled as WF 1 and WF 2.2 in the following; Figure 11.8) were prepared, aiming at replicating the properties of a home-made eggshell pellet catalyst.

A packed bed configuration was also tested for comparison. In particular, pellets were prepared by dry impregnation of a rhodium nitrate solution on alumina particles with 600 μm nominal diameter (Puralox by Sasol), based on

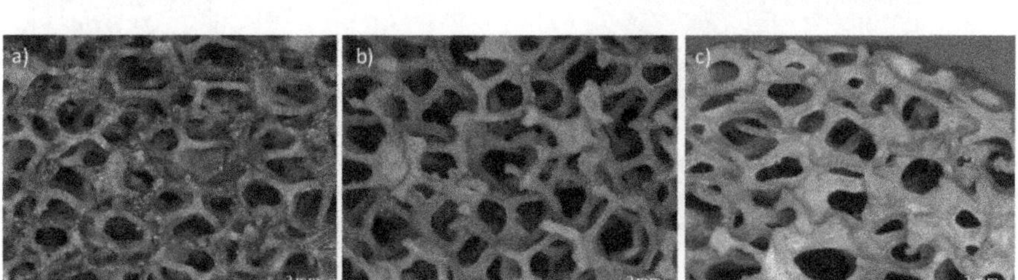

Figure 11.8 Optical microscopy analysis of the bare foam (a) and washcoated samples of 2.2 wt% Rh WF (b) and 1 wt% Rh WF (c) after flash drying. Adapted from Ref. 64 with permission from the Royal Society of Chemistry.

procedures reported in the literature,[114] aiming at a final Rh metal content of 0.3 wt% with respect to alumina carrier mass and to an eggshell thickness of 35 μm, which corresponds to 1 wt% metallic Rh with respect to the alumina eggshell mass.

Catalytic tests were performed in a tubular reactor with an internal diameter of 29.5 mm, as shown in Figure 11.9c, where a schematic diagram of the system is included.

Focusing on the layout, an inert FeCrAl foam was placed at the bottom of the reactor as a space holder. Then, the catalytic bed was loaded into the system and, finally, the upper part of the reactor was filled with inert silicon carbide particles in order to allow water evaporation and reactant mixing. The different reactor layouts were tested at two gas hourly space velocities (GHSV), namely 5 000 and 10 000 h^{-1}. In order to properly compare the different systems, the same mass of metallic rhodium per catalytic bed volume unit was loaded. Catalytic tests were performed at atmospheric pressure over a broad range of temperatures (600–800 °C range, with 50 °C steps), using a tubular furnace. The reactants feed consisted of a steam/CH_4 mixture with an S:C ratio of 3.5; reactants were fed separately into the reactor, and water evaporation and reactant mixing took place in the SiC pre-mixing bed. Temperature profiles were recorded across the

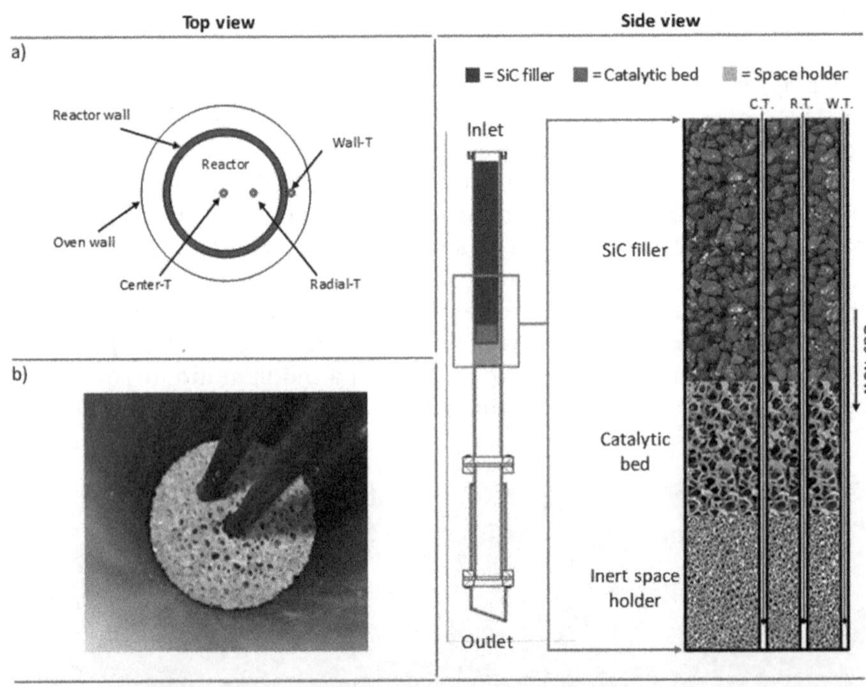

Figure 11.9 Graphical sketch of the reactor layout: top view (a), top view of the washcoated sample loaded in the reactor (b) and cross-section view of the reactor (c). Adapted from Ref. 66 with permission from Elsevier, Copyright 2019.

catalytic bed with a resolution of 5 mm on a total length of 7.5 cm at three radial positions, as sketched in Figures 11.9a and 11.9c.

The results in terms of the maximum radial temperature difference across the catalytic bed and methane conversion are summarized in Figure 11.10.

The washcoated configuration induced significant improvements in terms of reductions of the maximum temperature difference across the catalytic bed (Figure 11.10a). These results confirmed that the solid matrix has a positive effect on the heat transport mechanism from the external duty (*i.e.* furnace) to the reactor centre in the investigated temperature range.

As far as methane conversion is concerned (Figure 11.10b), the WF 1 configuration showed good performances for oven temperatures higher than 700 °C, outperforming the packed bed system. Accordingly, washcoated copper foams are promising supported catalysts for enhancing the performance of MSR, as the adoption of a copper substrate improves and optimizes both heat transfer and temperature distribution along the catalytic bed. However, despite the advantageous presence of a solid matrix that flattens the temperature profiles in both axial and radial directions, some limitations are still present: in the case of the WF 1 sample, the washcoat layer is too thick, and this likely induces internal mass transfer limitations. Indeed, the WF 2.2 sample is characterized by insufficient metal dispersion, whose direct effect is the lowering of the number of active sites available for gaseous reagents.

11.3.2 Packed Foam Configuration

Despite the encouraging results reported so far, washcoated catalysts may also be affected by other intrinsic drawbacks, such as issues related to washcoat adhesion and catalyst loading/unloading, which have discouraged the application of this technology at the industrial scale.[62] Moreover, washcoated catalysts are characterized by limitations in terms of catalyst inventory, which is usually smaller than in packed bed systems.

As far as the MSR process is concerned, eggshell pellets are typically employed in packed bed reactors to maximize the activity at fixed active species contents, enabling tuning of the thickness of the active layer according to the internal mass transfer limitations: in this case, the catalyst load is in trade-off with the effectiveness factor. By contrast, in the case of washcoated systems, for a given catalyst load, the thickness of the layer is a function of the geometric surface area of the support: thus, foams with high pore density (in terms of PPI) should be used to decrease the thickness.

According to this scenario, the packed foam configuration was investigated for the intensification of the MSR process. As discussed in previous sections, this reactor configuration consists of filling the empty porosity of open cell foams with small catalyst particles.[62] The introduction of a highly conductive and interconnected metallic internal enhances the overall heat transfer properties of the system, thanks to the efficient conductive transport of the reaction heat. This is also valid at low flow velocities, which allows the design of short tube reactors.[65] The concept of packed foams was first effectively applied to

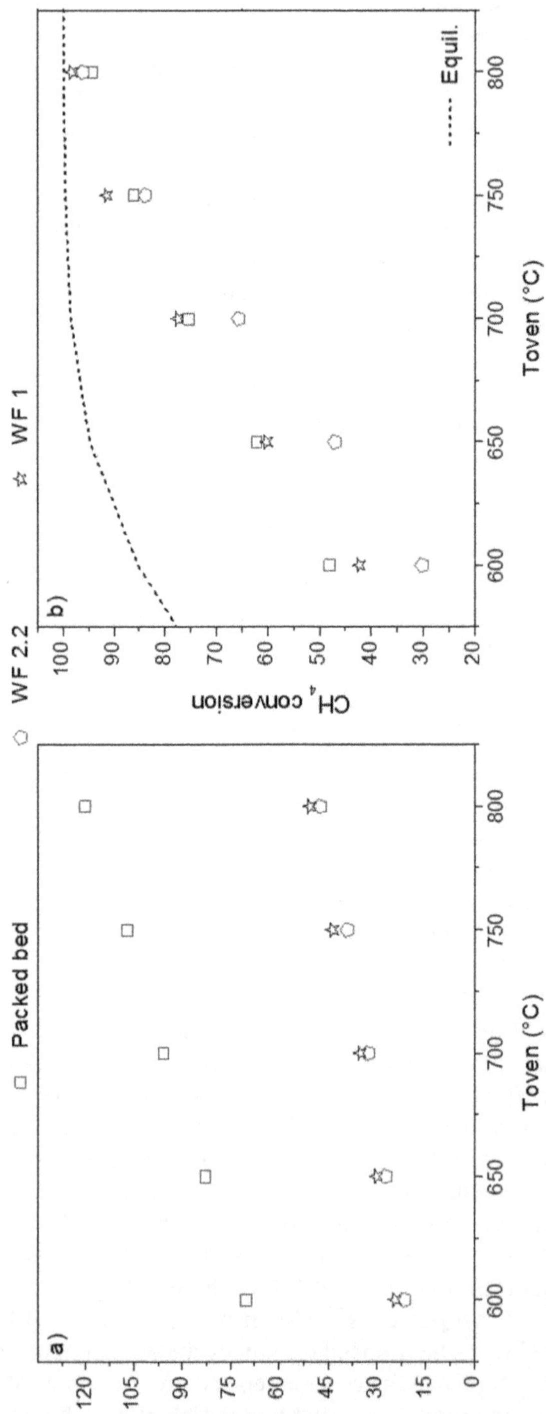

Figure 11.10 Maximum difference of temperature across the catalytic bed (a) and methane conversion (b) as a function of oven temperature.

the strongly exothermic FTS, where the introduction of highly conductive aluminium internals prevented the occurrence of thermal runaway, thus enabling the test reactor to run under much more severe conditions than in the packed bed configuration.[63] A detailed description of the use of aluminium foams for the FTS process will be reported in the following paragraphs.

In the case of the MSR process, different metallic supports were tested to intensify radial and axial heat transfer using the packed foam configuration, as summarized in Table 11.1.

A steel-based support (FeCrAlY open cell foams of 12 PPI with 5.2 mm cell size and 0.92 void fraction [ε_F]) was compared to copper open cell foams (two different cell densities, namely 10 and 40 PPI). In order to guarantee fully comparable conditions, all of the packed foam configurations were tested with the same procedure used for the washcoated systems. As reported for the washcoated systems, the packed bed configuration was assumed as a benchmark of the traditional steam reforming process, using the same reactor layout reported in Figure 11.9, as well as the same operative conditions. Based on a similar approach to the one used for washcoated foams, the maximum radial temperature difference across the catalytic bed (Figure 11.11a) and the conversion (Figure 11.11b) were evaluated as a function of the oven temperature to assess the performance of each reactor configuration.

The presence of the FeCrAl matrix apparently did not bring about any improvement in process performances with respect to those of the packed bed system, probably due to both the beneficial effect of the conductive diluent (SiC particles) in the present packed bed configuration and the low thermal conductivity of FeCrAl. By contrast, the highly conductive internals (Cu matrices) led to remarkable enhancements in the heat transfer system performances, which directly translate into a reduction of the maximum radial temperature difference across the catalytic bed.

Table 11.1 Geometrical properties of open cell foams. In all cases, foam height and diameter are equal to 25 and 29 mm, respectively. Adapted from Ref. 66 with permission from Elsevier, Copyright 2019.

	FeCr12	Cu10	Cu40
	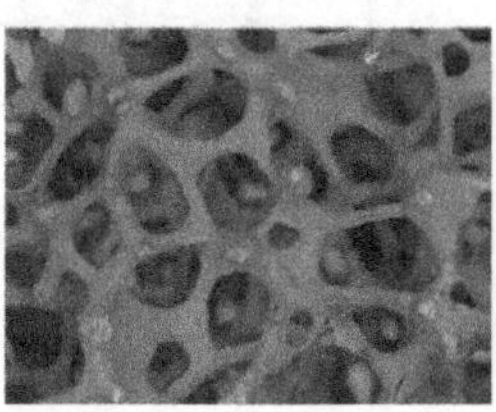	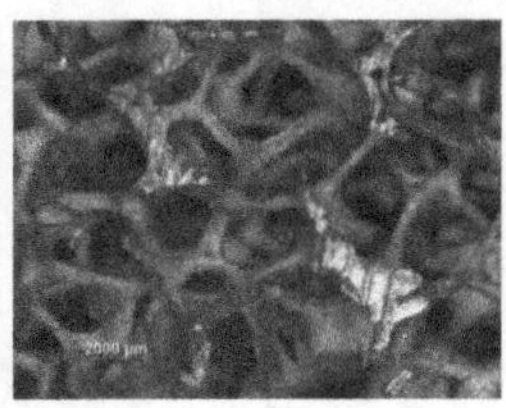	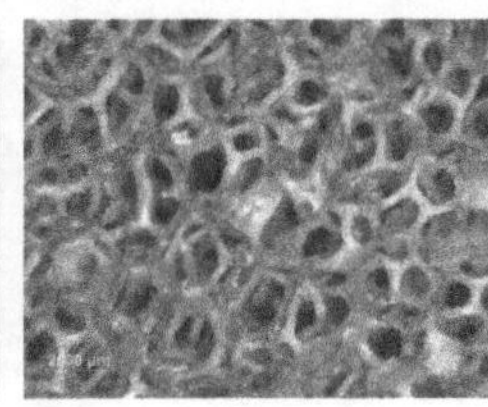
PPI	12	10	40
ε_F (–)	0.92	0.91	0.88
$S_{V,F}$ (m^{-1})	500	600	1220
d_{cell} (mm)	5.2	4.6	2
d_{strut} (μm)	745	705	285

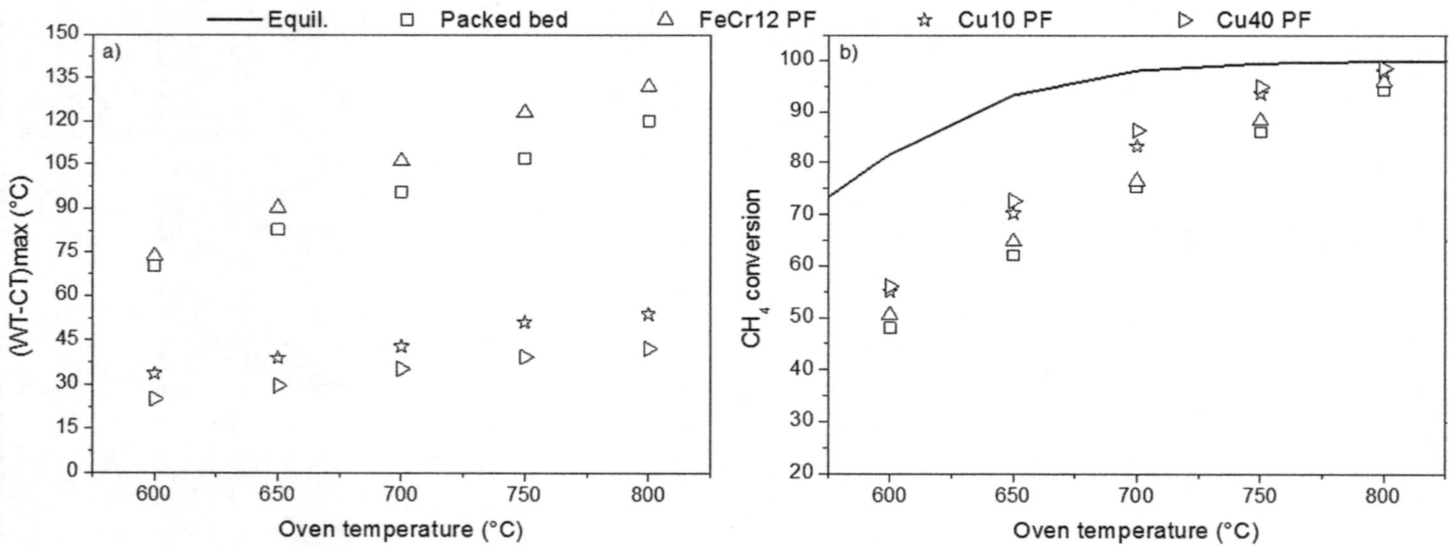

Figure 11.11 Maximum radial temperature difference between the wall thermocouple (WT) and the central thermocouple (CT) plotted against the furnace temperature at GHSV = 10 000 h^{-1} (a). Methane conversion as a function of the oven temperature (b): the solid line corresponds to the chemical equilibrium computed at the oven temperature. Adapted from Ref. 66 with permission from Elsevier, Copyright 2019.

The catalytic performance was quantified by comparing methane conversions at fixed furnace temperatures (Figure 11.11b), where the thermodynamic equilibrium calculated at the oven temperature (continuous line) is also plotted in order to highlight the maximum limit of the system. The introduction of highly conductive internals allowed for a remarkable increase in methane conversion at any oven temperature. The authors ascribed this phenomenon to the beneficial effect on system performance induced by the flattening of the temperature profiles inside the catalytic bed due to the adoption of the conductive Cu internals. As a result, the reactor operated at higher average catalyst temperatures, thus having improved reaction kinetics. The thermal and H_2 efficiencies of the new reactor design were thus both improved, being strictly interconnected performance indicators.

Figure 11.12 illustrates the good accordance between the experimental global heat transfer coefficient and model predictions based on the equivalent thermal circuit depicted in Figure 11.5. Additionally, the model allows for the estimation of each component of resistance in the electrical equivalent circuit. As a result, in these conditions, the wall coupling is the dominant heat transfer resistance for conductive (Cu) packed foams, whereas in the cases of the packed bed and of the FeCrAl packed foam configurations, the internal resistance becomes the limiting term. Therefore, in the case of copper foams, the wall coupling is identified as a crucial parameter to ensure good heat transfer performances. This allows for pinpointing of a strategy to further improve the heat transfer properties, which consists in using foams with external skins: the latter would guarantee bulk material continuity and a large contact area between the reactor tube and the structured internals, thus improving wall heat transfer.

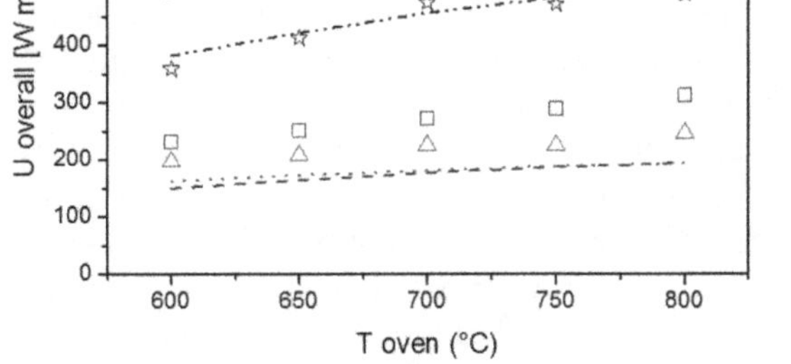

Figure 11.12 Overall heat transfer coefficient as a function of oven temperature: model predictions and experimental results. Reproduced from Ref. 66 with permission from Elsevier, Copyright 2019.

11.4 FTS over Conductive Structured Catalyst Supports

FTS is a viable alternative to petroleum sources for the production of liquid fuels. FTS converts syngas (CO and H_2) into a mixture of hydrocarbons, which is further processed to maximize the yield of diesel fuels. These products are highly appreciated by the market due to their high performances (high cetane number) and low environmental impact (being free from of sulfur and aromatics).[115]

The Fischer–Tropsch (FT) reaction competes with the methanation reaction (reverse of steam reforming) and with those reactions that lead to the production of propane and butane (liquefied petroleum gas [LPG]). In order to promote FTS with respect to CH_4 and LPG reactions, the synthesis is carried out at low temperatures (~210–240 °C) and high pressures (~20–30 bar) over selected catalysts (*i.e.* supported cobalt-based catalysts).

To date, low-temperature FT (LTFT) synthesis is run in two types of commercial reactors: the slurry bubble column reactor (SBCR) and the externally cooled multi-tubular fixed bed reactor (MTFBR). In large-scale gas to liquids (GtL) commercial plants (6 000–15 000 barrels per day of GtL products), the size of these reactors is huge, with reactor diameters in the range of 7–10 m and lengths up to 60 m.[116,117]

In SBCRs, the well-mixed liquid phase results in nearly isothermal operation that allows the process to be run at higher CO conversions per pass. However, the SBCR technology has low specific productivity, which makes it convenient only at a very large scale.

MTFBRs are most often preferred[118] due to their several advantages, such as the quasi-plug flow behaviour, high catalyst hold-up, no need for catalyst separation and less difficult scale-up. However, due to the highly exothermic nature of FTS (standard reaction enthalpy of −165 kJ mol_{CO}^{-1}), relatively poor heat removal is one of the main drawbacks of using MTFBR technology.[118] The dominant pathway of the heat transfer in a MTFBR is associated with thermal convection, typical of packed bed reactors operated at high flow velocities. Heat transfer by solid thermal conduction is insignificant as few contact points are present between the catalyst particles and between the particles and the reactor walls.[119] Inefficient temperature control in the reactor may result in the formation of hot spots and strong axial and radial temperature gradients within the catalyst bed, thus degrading the process selectivity, with the FT product distribution shifted towards low-molecular-weight hydrocarbons (*e.g.* methane, which is strongly undesired). In addition, poor temperature management may cause faster catalyst deactivation and, in the worst case, thermal runaway of the reactor.[119] At the industrial scale, the heat removal issue of the MTFBRs is overcome by recycling a fraction of the tail gas into the reactor to limit per-pass CO conversions. This helps to increase the gas velocity, thus improving the heat transfer coefficients, as well as decreasing the heat generation per unit of volume. Recycling of produced liquid hydrocarbons at high flow rates is also adopted as a strategy to boost heat removal. However, this increases pressure drops and makes the reactor less flexible for scaling.[120]

Today, interest in small-scale GtL technology has been considerably renewed in view of exploiting both associated and remote natural gas fields to produce synthetic liquid fuels and valuable chemicals.[121] The compact GtL approach (up to 3000 barrels per day of syncrude) does not involve the direct on-site production of high-value distillates, since the synthetic crude oil produced by the compact FT units should then be exported to conventional refineries for onward processing. One of the main advantages of this technology is obtaining feed gas at essentially no cost, as flaring or re-injection, with the associated costs, are avoided with this technology.

The compact GtL technology requires the down-scaling of conventional MTFBRs into a compact and modular FT reactor technology. The need for short reactors (~1–3 m long) rules out the conventional multi-tubular packed-bed designs, since heat transfer by thermal convection becomes limited by the low flow velocity associated with operation at a constant space velocity.[122] Indeed, a decreased bed contact time would result in decreased CO conversion together with reduced selectivity to long-chain hydrocarbons.[123] Accordingly, the heat removal from the reactor becomes a key issue for the development of an intensified compact FT reactor technology.[124]

Today, several research groups are focusing on the development of structured reactors with improved thermal management suitable for small-scale GTL applications.[119,124–134] In this regard, a microchannel reactor design is now commercially offered by Velocys.[134] In these reactor configurations, the catalyst is packed within microchannels formed by wave-like metallic fin structures and the pressurized water coolant flowing inside cross-flow microchannels effectively removes the reaction heat.[134] However, the microchannel design introduces a totally new reactor technology, which is intrinsically more complex and expensive in comparison to the conventional MTFBR.

As an alternative approach, the adoption of structured catalysts with different geometries was proposed in order to efficiently overcome the heat transfer issue in compact FT packed bed reactors.[63,119,124–133] In this regard, promising results were obtained when the catalytic material was washcoated onto a spatially structured substrate made of conductive materials.[119,124,125,127,135] This enabled more isothermal operation of the reactor, thus reducing the presence of hot spots.

The potential of coated conductive monoliths in FTS was actively investigated by our group at Politecnico di Milano.[124,135] Visconti *et al.*[135] reported an experimental investigation of the deposition of catalytic washcoats active in FTS onto aluminium supports with different geometries, such as spirals and honeycomb monoliths. The spirals were prepared by rolling 2.5 cm × 25 cm aluminium slabs and prototype extruded aluminium honeycombs (27 cpsi) were supplied by Corning, Inc.[135] The authors analysed the effects of the washcoating process parameters with the goal of obtaining adherent washcoat layers of controlled thickness.[135] Limiting the thickness to 40–50 μm was found to be crucial in order to avoid the onset of intra-catalyst mass transport limitations, which would lead to decreases in both reaction rate and C_{5+} selectivity.[136–139] Furthermore, the authors assessed the catalytic properties of the prepared structured catalysts in FTS under relevant process conditions (T = 220–235 °C, P = 20 bar, H_2 : CO inlet molar ratio = 2.1, GHSV = 5 000 cm^3[STP]‡ h^{-1} g_{cat}^{-1}), both in terms of activity (measured in terms of CO conversion) and selectivity (measured in terms of olefin and paraffin selectivity from C_1 to C_{49}). The activity and selectivity data obtained over the structured catalysts compared well with those obtained over the same catalyst in powder form.[135] The latter was sieved in the range 75–100 μm so as to ensure a kinetically controlled regime.

Before testing the structured systems, the authors verified the chemical inertia of the aluminium supports under the adopted process conditions. To this end, uncoated aluminium spirals were loaded into the reactor, pretreated with the classical reduction step adopted for Co-based FT catalysts (400 °C for 12 h, under a flow of pure hydrogen) and then tested in FTS. The absence of products in the gas phase confirmed the chemical inertia of the aluminium supports.[135]

The slight differences observed in both FTS activity and selectivity over the washcoated spirals and monoliths with respect to the powdered catalysts were attributed to the presence of modest internal mass transfer limitations in the case of the structured samples.[135] The diffusion lengths of the latter (23 μm in the case of the spirals and 43 μm in the case of the monoliths) were in fact significantly longer than the characteristic length of the powders (between 12.5 and 15.0 μm). In particular, in the case of the monoliths, the washcoat thickness was close to the threshold value of 40–50 μm for the onset of intraporous mass transfer limitations.[135]

Afterwards, Visconti *et al.*[124] demonstrated by numerical simulation the ability of the washcoated honeycomb monoliths to tolerate up to a 10-fold increment in catalytic activity (Figure 11.13). With a 10-fold increment of the catalyst activity, the temperature gradients were still limited (*i.e.* less than 6 °C), whereas the productivity was of course drastically increased. It was thus shown computationally that heat conduction in the aluminium honeycomb substrates was able to remove effectively the heat generated by the strongly exothermic FT reaction and to guarantee excellent temperature control.[124] Flat axial and radial temperature profiles were found along the catalytic bed, showing the strong potential of the adopted structured catalysts to manage the heat removal issue and to enable excellent temperature control without recycling a fraction of the liquid reaction products as a result of the static heat conduction within the thermally connected solid matrix of the honeycomb monolith.[124]

The potential advantages associated with the adoption of conductive structured reactors, however, are limited by the fact that the thin washcoated layer (up to 40–50 μm) results in a very low catalyst inventory that is insufficient for industrial applications. Indeed, this limits the productivity per reactor volume of the FT reaction.

In addition, the catalyst loading and unloading in the reactor, as well as the replacement of the spent catalyst, are challenging aspects in washcoated reactor technologies that require the development of dedicated technologies.[62]

‡STP: standard temperature and pressure.

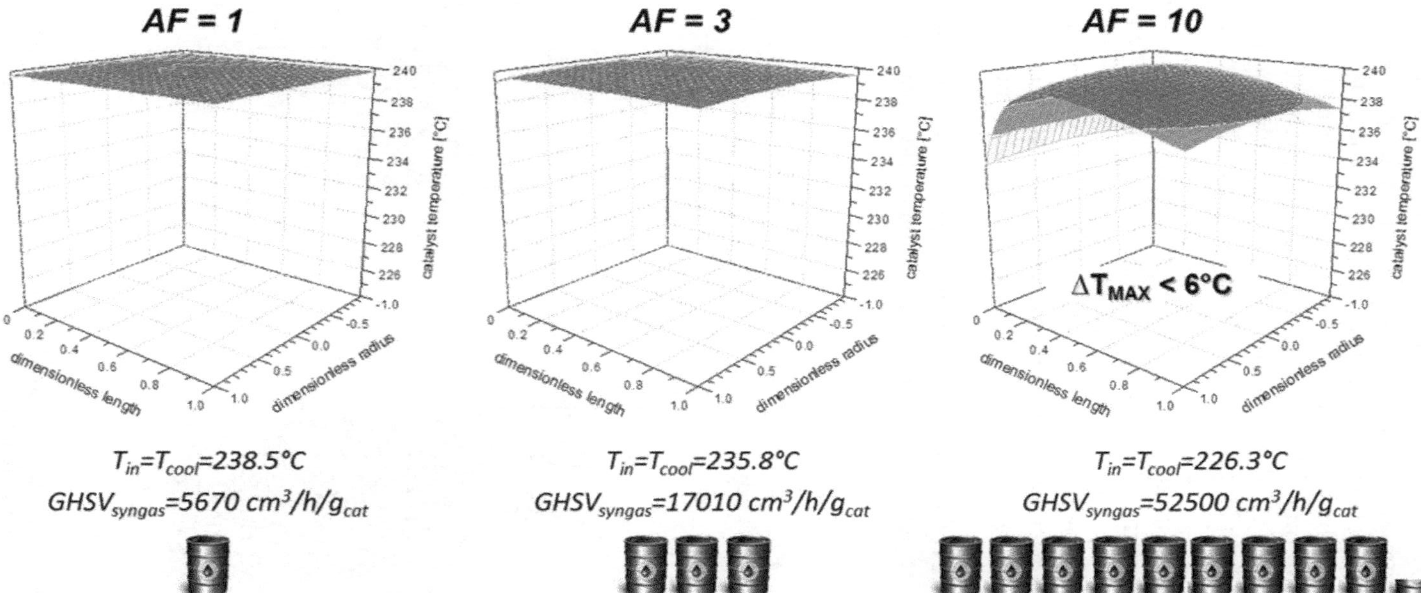

Figure 11.13 Simulated solid-phase temperature profiles of a 1 in. O.D. monolith (27 cpsi) in the case of: (a) AF = 1, GHSV = 5670 $cm^3\ h^{-1}\ g_{cat}^{-1}$, $T_{in} = T_{cool}$ = 238.5 °C; (b) AF = 3, GHSV = 17 010 $cm^3\ h^{-1}\ g_{cat}^{-1}$, $T_{in} = T_{cool}$ = 235.8 °C; and (c) AF = 10, GHSV = 52 500 $cm^3\ h^{-1}\ g_{cat}^{-1}$, $T_{in} = T_{cool}$ = 226.3 °C. Adapted from Ref. 124 with permission from Elsevier, Copyright 2011.

To overcome all of these constraints, the adoption of "packed structured reactors" was proposed by our group in collaboration with Eni.[140,141] In this reactor configuration, the catalyst is loaded in the form of pellets randomly packed in the voids of the structured internals, thus boosting the productivity per reactor volume. The potential of this reactor configuration was recently investigated by our group[63] for FTS. For the first time in the scientific literature, Fratalocchi *et al.*[63] have demonstrated experimentally that the adoption of an FT tubular reactor loaded with a highly conductive (aluminium) open cell foam packed with catalyst pellets enabled strongly enhanced heat exchange.

The open cell foam used in Ref. 63 was purchased from ERG Aerospace. It was made of aluminium (Al6101) with a nominal pore density of 40 PPI ($\varepsilon_{foam} \approx 0.906$; $d_{cell} \approx 2$ mm). The shape of the foam was cylindrical with a length of 4 cm and a diameter of 2.78 cm, which matched the internal diameter of the stainless steel reactor tube used in the activity tests so as to ensure good contact and to avoid dead space ("gap") between the tube and the structure (Figure 11.14).

The performances of a home-made, highly active Co-based catalyst packed into the metallic structure were assessed at industrially relevant operating

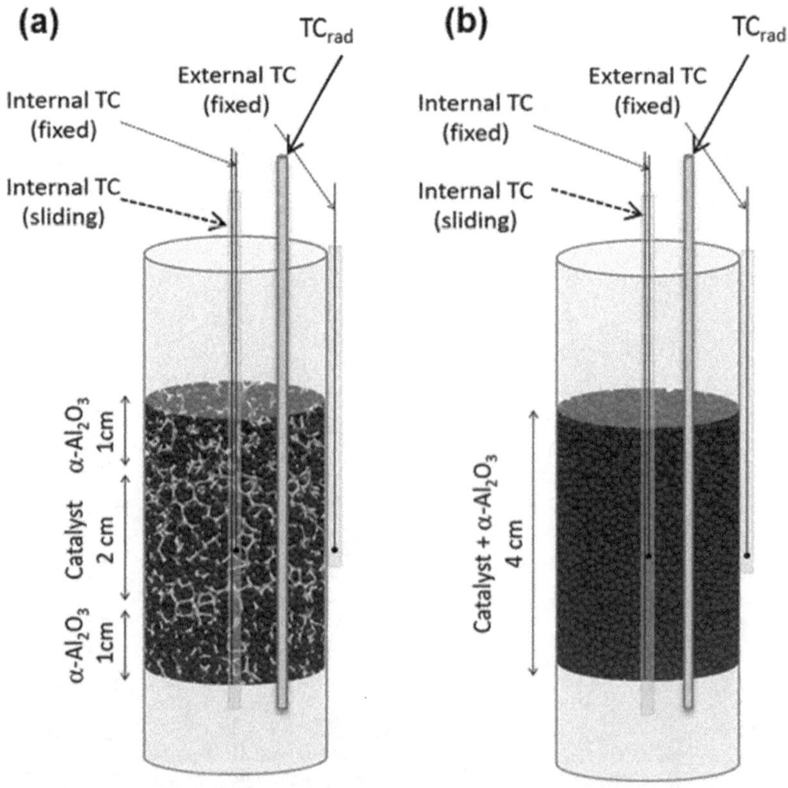

Figure 11.14 Scheme of the packed foam (a) and packed bed reactors (b) (TC = thermocouple).

conditions (200–240 °C, 25 bar, H_2:CO inlet molar ratio = 2, GHSV = 6410 cm^3 [STP] h^{-1} g_{cat}^{-1}, inerts [N_2 + Ar] in the feed = 24 vol%) and compared with those obtained in a conventional randomly packed fixed-bed reactor.[63]

A sketch of the packed foam and packed bed reactors configuration is shown in Figure 11.14.

The structured catalyst reached outstanding performances (CO conversion >65%) with a remarkable temperature control.[63] As shown in Figure 11.15, the axial temperature profiles measured along the catalytic bed were almost flat, even under the most severe process conditions (*i.e.* high CO conversions ≈ high reaction duties), showing the excellent ability of this reactor concept to manage the strong exothermicity of the reaction. In this regard, the ΔT_{cat}, defined as the difference between the maximum and minimum temperatures recorded along the catalyst bed, was negligible at 180 and 190 °C, in line with the very low catalyst activity. ΔT_{cat} was also small (≤4.5 °C) in the 195–225 °C temperature range, although the CO conversion increased by up to 50%. The ΔT_{cat} was still acceptable at 230 and 240 °C (5 and 6 °C, respectively) with significant CO conversions of 55% and 67.5%, respectively.[63]

In contrast, when the same experiment was carried out over the same catalyst randomly packed in the reactor, an abrupt increase in the catalyst temperature occurred already at low temperatures (195 °C, CO conversion of 12%), eventually leading to thermal runaway.[63] The catalyst bed in the packed

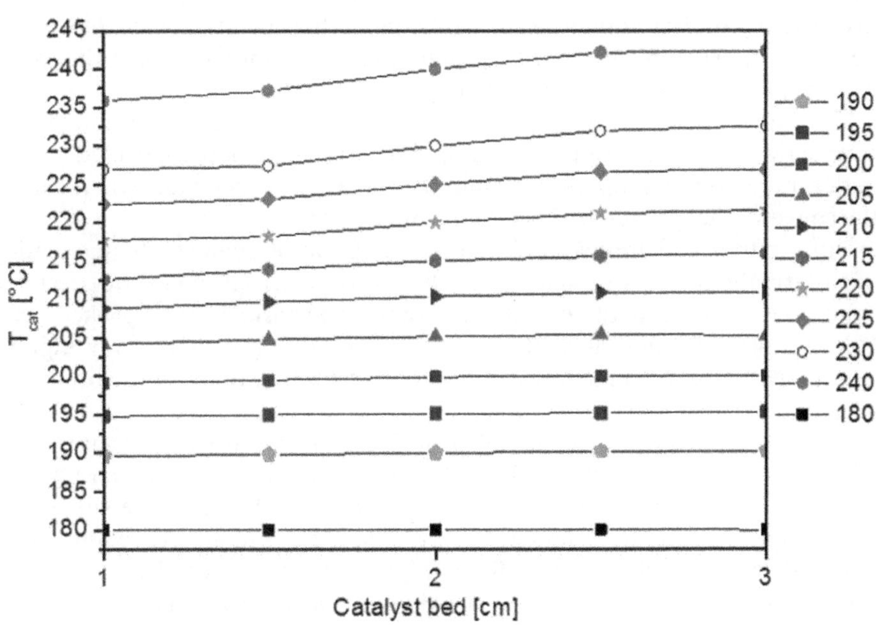

Figure 11.15 Axial temperature profiles measured at different temperatures in the packed foam reactor by sliding the thermocouple located at the centreline of the foam. P = 25 bar, H_2/CO^{in} = 2 mol/mol, GHSV = 6410 cm^3(STP) h^{-1} g_{cat}^{-1}, inerts = 24 vol%.

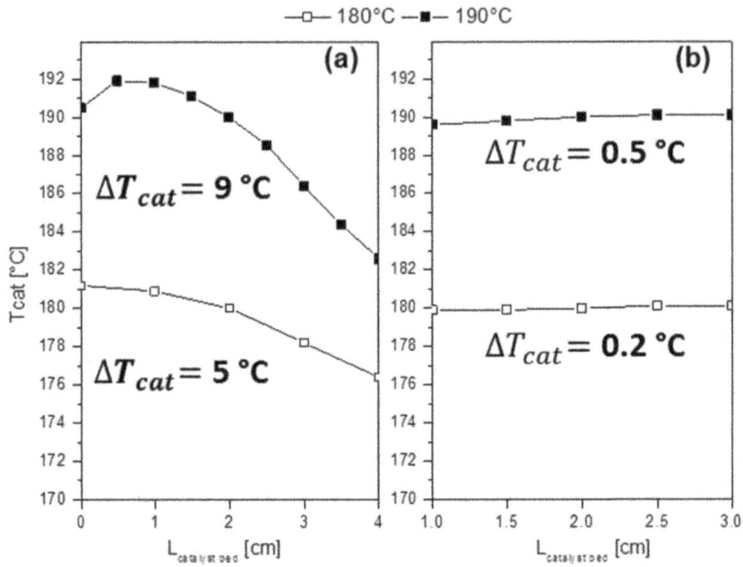

Figure 11.16 Axial temperature profiles of (a) the packed bed reactor measured at 180 and 190 °C. The results obtained with the (b) packed foam reactor are also shown for comparison purposes. P = 25 bar, H_2/CO^{in} = 2 mol/mol, GHSV = 6410 cm^3(STP) h^{-1} g_{cat}^{-1}, inerts = 24 vol%.

bed reactor configuration was strongly non-isothermal even at the low temperatures of 180 and 190 °C, with axial temperature gradients along the catalyst bed (ΔT_{cat}) of 5 and 9 °C, respectively (Figure 11.16).[63]

The data reported in Ref. 63 clearly show that, thanks to the adoption of the conductive foam, the mean temperature inside the reactor is controlled much better, providing new operating windows that are not accessible when using the conventional packed bed reactor technology.[63]

Our group first demonstrated that the adoption of open cell foams made of a thermally conductive material as the reactor internal can be an effective solution to enhancing the overall heat transfer performances of packed bed FTS reactors.[63] Furthermore, these results prove that the adoption of highly conductive packed foams is also an innovative strategy to boost the productivity of the FT reaction.

In our recent work,[142] we have also clarified an aspect that is often raised when discussing the structured reactor concept: what is the actual advantage of packing the catalyst within an interconnected conductive substrate over simply diluting the catalyst bed with "particles" made of a conductive material? Accordingly, Fratalocchi *et al.*[142] compared the catalytic and heat transfer performances of the aforementioned packed foam and packed bed reactor technologies with those obtained with a packed bed reactor diluted with crushed fragments of the foam (average size of 0.3–0.4 cm). Interestingly, a slight improvement in heat exchange was noted when diluting the catalyst bed with conductive fragments with respect to the conventional packed bed

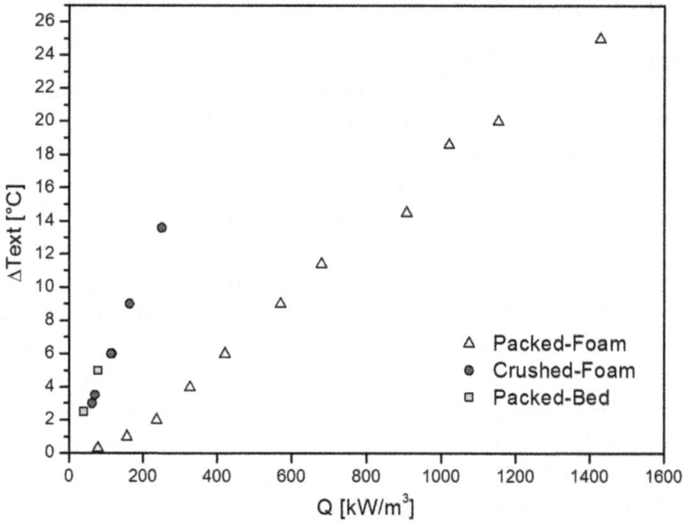

Figure 11.17 External (ΔT_{ext}) temperature difference as a function of the volumetric heat duty (Q) calculated during FTS experiments at different temperatures. Adapted from Ref. 142 with permission from Elsevier, Copyright 2019.

reactor technology. The catalytic performances of the crushed foam reactor could be investigated over a wider range of reaction temperatures (up to 205 °C) with respect to the conventional packed bed reactor. In this regard, the thermal runaway of the crushed foam reactor was shifted to higher temperatures (210 °C) than in the packed bed reactor (195 °C).[142] This was explained by looking at the temperature profiles of the crushed foam reactor, which revealed the presence of weaker hot spots along the catalyst bed than in the packed bed reactor. In particular, the ΔT_{cat} was found to be 5 °C at 190 °C, which was about half that obtained in the absence of the conductive fragments.[142] By increasing the reaction temperature, the ΔT_{cat} increased up to 22 °C at 205 °C.

However, this was not sufficient to match the heat transfer performance of the packed foam reactor configuration, which was operated under more severe conditions (up to 240 °C).[142] This is clearly shown in Figure 11.17, where the external temperature difference (ΔT_{ext}), calculated as the difference between the temperature reading at the centre of the catalyst bed and that at the outer surface of the reactor tube in correspondence to the mid-point of the catalyst bed, is plotted as a function of the volumetric heat duty (Q) calculated during the FTS experiments over the packed foam, packed bed and crushed foam reactors.[142] It is clear from Figure 11.17 that the packed foam configuration outperforms both the packed bed and the crushed foam configurations, highlighting the dominant role of the highly conductive thermally connected solid matrix in managing the heat removal issue in FTS.[142]

11.5 Conclusions

In this chapter, a comprehensive review of the heat transfer properties of conductive washcoated and packed cellular substrates/internals has been reported. With these structures, it is possible to increase by almost an order of magnitude the overall heat transfer coefficient in tubular reactors with respect to the conventional packed bed configuration.

According to the results reported so far in the literature[64,66] and herein reviewed, conductive metallic (Al, Cu) foams and POCS have great potential as improved reactor internals due to their ability to enhance process yields and to approach the isothermal behaviour in the case of both endothermic and exothermic heat transfer-limited reactions. Indeed, such structures are advantageous since they combine excellent effective radial thermal conductivity with good mechanical resistance, high void fractions and large surface-to-volume ratios. In fact, their properties make them good candidates both as substrates of washcoated structured catalysts and as conductive reactor internals packed with catalyst microparticles.

Concerning specific process applications, results for the endothermic MSR pointed out that washcoated Cu foams provide enhancements in terms of both methane conversion and heat transfer at high temperatures, thanks to the presence of a highly conductive and interconnected solid matrix. However, washcoated catalysts may be affected by internal mass transfer limitations and by issues with metal dispersion and with coating adhesion. These drawbacks were effectively overcome by adopting the packed foam configuration: in this view, copper foams were proven to be promising reactor internals for the intensification of the steam reforming process, as they allow for higher conversions and reduced thermal gradients inside the reactor with respect to the packed bed configuration, thus offering the possibility of operating the reformer with more flexibility and better thermal efficiency.

Similar results were reported for the exothermic FTS, where aluminium-based structured internals have been successfully used to boost the heat transfer performance of a FTS tubular reactor, resulting in unprecedented efficiency in temperature control. Indeed, the conductive packed foam reactor enabled the FT process to be run under very severe conditions (*i.e.* high CO conversions and large heat duties). In comparison, the conventional packed bed reactor experienced thermal runaway at very low temperatures and CO conversions (*i.e.* with a limited release of reaction heat), even if operated under milder conditions (*i.e.* the catalyst density was halved with respect to the foam-based reactor).

Conductive POCS, fabricated by additive manufacturing techniques, for example, today represent the emerging third generation of enhanced substrates/internals, whose many more degrees of freedom can be exploited to further boost the intensification of heat transfer-limited key catalytic processes for energy applications.

Acknowledgements

The research leading to these results has received funding from the European Research Council under the European Union's Horizon 2020 research and innovation programme (Grant Agreement no. 694910/ INTENT).

References

1. A. I. Stankiewicz and J. A. Moulijn, *Chem. Eng. Prog.*, 2000, **96**, 22–34.
2. E. Tronconi, G. Groppi and C. G. Visconti, *Curr. Opin. Chem. Eng.*, 2014, **5**, 55–67.
3. J. Gascon, J. R. van Ommen, J. A. Moulijn and F. Kapteijn, *Catal. Sci. Technol.*, 2015, **5**, 807–817.
4. G. Groppi and E. Tronconi, *Catal. Today*, 2005, **105**, 297–304.
5. L. Giani, G. Groppi and E. Tronconi, *Ind. Eng. Chem. Res.*, 2005, **44**, 4993–5002.
6. G. Groppi, E. Tronconi, C. Cortelli and R. Leanza, *Ind. Eng. Chem. Res.*, 2012, **51**, 7590–7596.
7. C. Busse, H. Freund and W. Schwieger, *Chem. Eng. Process. Process Intensif.*, 2018, **124**, 199–214.
8. J. Richardson, D. Remue and J.-K. Hung, *Appl. Catal., A*, 2003, **250**, 319–329.
9. J. T. Richardson, Y. Peng and D. Remue, *Appl. Catal., A*, 2000, **204**, 19–32.
10. G. Incera Garrido, F. C. Patcas, S. Lang and B. Kraushaar-Czarnetzki, *Chem. Eng. Sci.*, 2008, **63**, 5202–5217.
11. A. Inayat, M. Klumpp, M. Lämmermann, H. Freund and W. Schwieger, *Chem. Eng. J.*, 2016, **287**, 704–719.
12. P. Aghaei, C. G. Visconti, G. Groppi and E. Tronconi, *Chem. Eng. J.*, 2017, **321**, 432–446.
13. B. Dietrich, *Chem. Eng. Sci.*, 2012, **74**, 192–199.
14. M. Frey, T. Romero, A. Roger and D. Edouard, *Catal. Today*, 2016, **273**, 83–90.
15. A. Montebelli, C. G. Visconti, G. Groppi, E. Tronconi, S. Kohler, H. Johnsen and R. Myrstad, *Applied Catal. A, Gen.*, 2014, **481**, 96–103.
16. D. Park, D. J. Moon and T. Kim, *Fuel Process. Technol.*, 2014, **124**, 97–103.
17. V. Palma, D. Pisano and M. Martino, *Int. J. Hydrogen Energy*, 2017, **42**, 23517–23525.
18. L. J. Gibson and M. F. Ashby, *Cellular Solids: Structure and Properties*, Cambridge University Press, Cambridge, 1999.
19. B. R. Coquard, M. Loretz and D. Baillis, 2008, **10**, 323–337.
20. V. Papetti, P. Dimopoulos Eggenschwiler, A. Della Torre, F. Lucci, A. Ortona and G. Montenegro, *Int. J. Heat Mass Transf.*, 2018, **126**, 1035–1047.
21. M. Lämmermann, C. Busse, W. Schwieger and H. Freund, *Chem. Ing. Tech.*, 2016, **88**, 1209.
22. M. Lämmermann, G. Horak, W. Schwieger and H. Freund, *Chem. Eng. Process. Process Intensif.*, 2018, **126**, 178–189.

23. M. Klumpp, A. Inayat, J. Schwerdtfeger, C. Körner, R. F. Singer, H. Freund and W. Schwieger, *Chem. Eng. J.*, 2014, **242**, 364–378.
24. T. Selvam, F. Warmuth, M. Klumpp, K. Warnick, M. A. Lodes, C. Körner and W. Schwieger, *Chem. Eng. J.*, 2016, **288**, 223–227.
25. E. Hansjosten, A. Wenka, A. Hensel, W. Benzinger, M. Klumpp and R. Dittmeyer, *Chem. Eng. Process. Process Intensif.*, 2018, **130**, 119–126.
26. C. Hutter, D. Büchi, V. Zuber and P. Rudolf von Rohr, *Chem. Eng. Sci.*, 2011, **66**, 3806–3814.
27. M. Bracconi, M. Ambrosetti, O. Okafor, V. Sans, X. Zhang, X. Ou, C. P. Da Fonte, X. Fan, M. Maestri, G. Groppi and E. Tronconi, *Chem. Eng. J.*, 2018, **377**, 120123.
28. O. Santoliquido, G. Bianchi, P. Dimopoulos Eggenschwiler and A. Ortona, *Int. J. Appl. Ceram. Technol.*, 2017, **14**, 1164–1173.
29. A. Anglani and M. Pacella, *Procedia CIRP*, 2018, **67**, 504–509.
30. A. Montebelli, C. G. Visconti, G. Groppi, E. Tronconi, C. Cristiani, C. Ferreira and S. Kohler, *Catal. Sci. Technol.*, 2014, **4**, 2846–2870.
31. M. Ambrosetti, R. Balzarotti, C. Cristiani, G. Groppi, E. Tronconi, *Catalysts*, 2018, **8**, 510.
32. G. Do, T. Stiegler, M. Fiegl, L. Adler, C. Körner, A. Bösmann, H. Freund, W. Schwieger and P. Wasserscheid, *Ind. Eng. Chem. Res.*, 2017, **56**, 13402–13410.
33. O. Korup, C. F. Goldsmith, G. Weinberg, M. Geske, T. Kandemir, R. Schlögl, R. Horn, *J. Catal.*, 2013, **297**, 1–16.
34. P. Benito, G. Nuyts, M. Monti, W. De Nolf, G. Fornasari, K. Janssens, E. Scavetta and A. Vaccari, *Applied Catal. B Environ.*, 2015, **179**, 321–332.
35. E. Verlato, S. Barison, S. Cimino, F. Dergal, L. Lisi, G. Mancino, M. Musiani and L. Vàzquez-Gòmez, *Int. J. Hydrogen Energy*, 2014, **39**, 11473–11485.
36. J. E. P. Navalho, J. M. C. Pereira and J. C. F. Pereira, *Int. J. Hydrogen Energy*, 2014, **39**, 3666–3680.
37. A. Donazzi, M. Maestri, B. C. Michael, A. Beretta, P. Forzatti, G. Groppi, E. Tronconi, L. D. Schmidt and D. G. Vlachos, *J. Catal.*, 2010, **275**, 270–279.
38. R. Horn, N. J. Degenstein, K. A. Williams and L. D. Schmidt, *Catal. Lett.*, 2006, **110**, 169–178.
39. J. Ochoń ska-Kryca, M. Iwaniszyn, M. Piątek, P. J. Jodłowski, J. Thomas, A. Kołodziej and J. Łojewska, *Catal. Today*, 2013, **216**, 135–141.
40. J. Kryca, P. J. Jodłowski, M. Iwaniszyn, B. Gil, M. Sitarz, A. Kołodziej, T. Łojewska and J. Łojewska, *Catal. Today*, 2016, **268**, 142–149.
41. L. Giani, G. Groppi and E. Tronconi, *Ind. Eng. Chem. Res.*, 2005, **44**, 9078–9085.
42. E. Bianchi, T. Heidig, C. G. Visconti and H. Freund, *Catal. Today*, 2013, **216**, 121–134.
43. H. Ba, Y. Liu, X. Mu, W. Doh, J. Nhut, P. Granger and C. Pham-huu, *Applied Catal. A Gen.*, 2015, **499**, 217–226.
44. N. Xiao, Y. Zhou, Z. Ling, Z. Zhao and J. Qiu, *Carbon NY*, 2013, **60**, 514–522.
45. J. P. Bortolozzi, L. B. Gutierrez and M. A. Ulla, *Applied Catal. A Gen.*, 2013, **452**, 179–188.

46. Z. Zhang, L. Han, R. Chai, Q. Zhang, Y. Li, G. Zhao, Y. Liu and Y. Lu, *CATCOM*, 2017, **88**, 90–93.
47. A. Sirijaruphan, J. G. Goodwin, R. W. Rice, D. Wei, K. R. Butcher, G. W. Roberts and J. J. Spivey, *Appl. Catal., A*, 2005, **281**, 11–18.
48. A. Sirijaruphan, J. G. Goodwin, R. W. Rice, D. Wei, K. R. Butcher, G. W. Roberts and J. J. Spivey, *Appl. Catal., A*, 2005, **281**, 1–9.
49. M. Frey, D. Edouard and M. C. Roger, *C. R. Chim.*, 2015, **18**, 283–292.
50. Y. Li, Q. Zhang, R. Chai, G. Zhao, F. Cao, Y. Liu and Y. Lu, *Applied Catal. A Gen.*, 2016, **510**, 216–226.
51. C. Wang, D. Ping, X. Dong, Y. Dong and Y. Zang, *Fuel Process. Technol.*, 2016, **148**, 367–371.
52. A. Montebelli, C. Giorgio, G. Groppi, E. Tronconi and S. Kohler, *Chem. Eng. J.*, 2014, **255**, 257–265.
53. A. Montebelli, C. G. Visconti, G. Groppi, E. Tronconi, C. Ferreira and S. Kohler, *Catal. Today*, 2013, **215**, 176–185.
54. J. N. Tourvieille, R. Philippe and C. de Bellefon, *Chem. Eng. J.*, 2015, **267**, 332–346.
55. F. J. Méndez, O. Sanz, M. Montes, J. Guerra, C. Olivera-Fuentes, S. Curbelo and J. L. Brito, *Catal. Today*, 2017, **289**, 151–161.
56. F. Lali, G. Böttcher, P. Schöneich, S. Haase, S. Hempel and R. Lange, *Chem. Eng. Res. Des.*, 2014, **94**, 365–374.
57. L. Truong-Phuoc, T. Truong-Huu, L. Nguyen-Dinh, W. Baaziz, T. Romero, D. Edouard, D. Begin, I. Janowska and C. Pham-Huu, *Applied Catal. A Gen.*, 2014, **469**, 81–88.
58. Y. Liu, Y. Luo, G. Chu and J. Chen, *Catal. Today*, 2016, **273**, 34–40.
59. S. Danaci, L. Protasova, J. Lefevere, L. Bedel, R. Guilet and P. Marty, *Catal. Today*, 2016, **273**, 234–243.
60. S. Danaci, L. Protasova, F. Snijkers, W. Bouwen, A. Bengaouer and P. Marty, *Chem. Eng. Process. Process Intensif.*, 2018, **127**, 168–177.
61. N. F. Bastos Rebelo, K. A. Andreassen, L. I. Suarez Ríos, J. C. Piquero Camblor, H.-J. Zander and C. A. Grande, *Chem. Eng. Process. Process Intensif.*, 2018, **127**, 36–42.
62. C. G. Visconti, G. Groppi and E. Tronconi, *Catal. Today*, 2015, **273**, 178–186.
63. L. Fratalocchi, C. G. Visconti, G. Groppi, L. Lietti and E. Tronconi, *Chem. Eng. J.*, 2018, **349**, 829–837.
64. R. Balzarotti, A. Beretta, G. Groppi and E. Tronconi, *React. Chem. Eng.*, 2019, **4**, 1387–1392.
65. M. Ambrosetti, M. Bracconi, M. Maestri, G. Groppi and E. Tronconi, *Chem. Eng. J.*, 2020, **382**, 122801.
66. R. Balzarotti, M. Ambrosetti, A. Beretta, G. Groppi and E. Tronconi, *Chem. Eng. J.*, 2019, 123494.
67. L. Fratalocchi, G. Groppi, C. G. Visconti, L. Lietti and E. Tronconi, *Chem. Eng. J.*, 2020, **386**, 123988.
68. T. Boger and A. K. Heibel, *Chem. Eng. Sci.*, 2005, **60**, 1823–1835.

69. M. Flytzani-Stephanopoulos, G. E. Voecks and T. Charng, *Chem. Eng. Sci.*, 1986, **41**, 1203–1212.
70. R. E. Hayes, A. Rojas and J. Mmbaga, *Catal. Today*, 2009, **147**, 113–119.
71. C. G. Visconti, G. Groppi and E. Tronconi, *Chem. Eng. J.*, 2013, **223**, 224–230.
72. G. Groppi and E. Tronconi, *AIChE J.*, 1996, **42**, 2382–2387.
73. S. Rebughini, M. Bracconi, A. G. Dixon and M. Maestri, *React. Chem. Eng.*, 2018, **3**, 25–33.
74. T. Wellauer, D. L. Cresswell and E. J. Newson, *ACS Symp. Ser.*, 1982, 527–542.
75. E. Bianchi, T. Heidig, C. G. Visconti, G. Groppi, H. Freund and E. Tronconi, *Chem. Eng. J.*, 2012, **198–199**, 512–528.
76. X. Fan, X. Ou, F. Xing, G. A. Turley, P. Denissenko, M. A. Williams, N. Batail, C. Pham and A. A. Lapkin, *Catal. Today*, 2016, **278**, 350–360.
77. M. Bracconi, M. Ambrosetti, M. Maestri, G. Groppi and E. Tronconi, *Chem. Eng. J.*, 2017, **315**, 608–620.
78. M. Ambrosetti, M. Bracconi, G. Groppi and E. Tronconi, *Chem. Ing. Tech.*, 2017, **89**, 915–925.
79. M. Bracconi, M. Ambrosetti, M. Maestri, G. Groppi and E. Tronconi, *Chem. Eng. J.*, 2018, **352**, 558–571.
80. M. Bracconi, M. Ambrosetti, M. Maestri, G. Groppi and E. Tronconi, *Chem. Eng. Process. Process Intensif.*, 2018, **129**, 181–189.
81. R. Lemlich, *J. Colloid Interface Sci.*, 1978, **64**, 107–110.
82. S. Razza, T. Heidig, E. Bianchi, G. Groppi, W. Schwieger, E. Tronconi and H. Freund, *Catal. Today*, 2016, **273**, 187–195.
83. L. R. Glicksman, in *Low Density Cellular Plastics*, ed. N. C. Hilyard and A. Cunningham, Springer, Amsterdam, 1994, p. 369.
84. V. Specchia, G. Baldi and S. Sicardi, *Chem. Eng. Commun.*, 1980, **4**, 361–380.
85. E. Bianchi, W. Schwieger and H. Freund, *Adv. Eng. Mater.*, 2016, **18**, 608–614.
86. L. Baharudin and M. J. Watson, *Rev. Chem. Eng.*, 2017, **34**, 43–72.
87. I. Dybkjaer, *Fuel Process. Technol.*, 1995, **42**, 85–107.
88. O. Sanz, I. Velasco, I. Reyero, I. Legorburu and G. Arzamendi, *Catal. Today*, 2016, **273**, 131–139.
89. S. Specchia, *Int. J. Hydrogen Energy*, 2014, **39**, 17953–17968.
90. G. Kolb, *Chem. Eng. Process. Process Intensif.*, 2013, **65**, 1–44.
91. S. Ahmed and M. Krumpelt, *Int. J. Hydrogen Energy*, 2001, **26**, 291–301.
92. J.-H. Ryu, K.-Y. Lee, H. La, H.-J. Kim, J.-I. Yang and H. Jung, *J. Power Sources*, 2007, **171**, 499–505.
93. Z. Anxionnaz, M. Cabassud, C. Gourdon and P. Tochon, *Chem. Eng. Process. Process Intensif.*, 2008, **42**, 2029–2050.
94. D. M. Murphy, A. Manerbino, M. Parker, J. Blasi, R. J. Kee and N. P. Sullivan, *Int. J. Hydrogen Energy*, 2013, **38**, 8741–8750.
95. A. K. Avcı, D. L. Trimm and M. Karakaya, *Catal. Today*, 2010, **155**, 66–74.

96. G. Kolb, V. Hessel, V. Cominos, H. Pennemann, J. Schürer, R. Zapf and H. Löwe, *J. Mater. Eng. Perform.*, 2006, **15**, 389–393.
97. R. J. Farrauto, Y. Liu, W. Ruettinger, O. Ilinich, L. Shore and T. Giroux, *Catal. Rev.*, 2007, **49**, 141–196.
98. L. Baharudin and M. J. Watson, *Rev. Chem. Eng.*, 2018, **34**, 481–501.
99. L. Sang, B. Sun, H. Tan, C. Du, Y. Wu and C. Ma, *Int. J. Hydrogen Energy*, 2012, **37**, 13037–13043.
100. H. Yu, H. Chen, M. Pan, Y. Tang, K. Zeng, F. Peng and H. Wang, *Appl. Catal., A*, 2007, **327**, 106–113.
101. P. S. Roy, N.-K. Park and K. Kim, *Int. J. Hydrogen Energy*, 2014, **39**, 4299–4310.
102. P. S. Roy, A. S. K. Raju and K. Kim, *Fuel*, 2015, **139**, 314–320.
103. F. J. Echave, O. Sanz, I. Velasco, J. A. Odriozola and M. Montes, *Catal. Today*, 2013, **213**, 145–154.
104. V. Palma, D. Pisano, M. Martino and P. Ciambelli, *Chem. Eng. J.*, 2016, **304**, 544–551.
105. D. Park, D. J. Moon and T. Kim, *Fuel Process. Technol.*, 2014, **124**, 97–103.
106. D. Park, D. J. Moon and T. Kim, *Fuel Process. Technol.*, 2013, **112**, 28–34.
107. W. Zhou, Y. Ke, Q. Wang, S. Wan, J. Lin, J. Zhang and K. S. Hui, *Fuel*, 2017, **191**, 46–53.
108. C. C. Shen, T. Y. Jian and Y. T. Wang, *Fuel Cells*, 2013, **13**, 965–970.
109. S. Catillon, C. Louis and R. Rouget, *Top. Catal.*, 2004, **30**, 463–467.
110. J. Qi, Y. Sun, Z. Xie, M. Collins, H. Du and T. Xiong, *J. Energy Chem.*, 2015, **24**, 786–793.
111. J. T. Jang, K. J. Yoon and G. Y. Han, *Sol. Energy*, 2014, **101**, 29–39.
112. R. J. Farrauto, *Chem. Eng. J.*, 2014, **238**, 172–177.
113. H. Yang, J. Li, H. Yu, F. Peng and H. Wang, *Int. J. Chem. React. Eng.*, 2015, **13**, 83–93.
114. A. Porta, L. Falbo, C. G. Visconti, L. Lietti, C. Bassano and P. Deiana, *Catal. Today*, 2020, **343**, 38–47.
115. L. P. Dancuart, R. de Haan and A. de Klerk, *Stud. Surf. Sci. Catal.*, 2004, **152**, 482–532.
116. R. L. Espinoza, A. P. Steynberg, B. Jager and A. C. Vosloo, *Appl. Catal., A*, 1999, **186**, 13–26.
117. M. J. V. D. Burgt, C. J. Van Leeuwen, J. J. Del'Amico and S. T. Sie, *Stud. Surf. Sci. Catal.*, 1988, **36**, 473–482.
118. M. E. Dry and A. P. Steynberg, *Stud. Surf. Sci. Catal.*, 2004, **152**, 406–481.
119. D. Merino, O. Sanz and M. Montes, *Chem. Eng. J.*, 2017, **327**, 1033–1042.
120. S. T. Sie, M. M. G. Senden and H. M. H. Van Wechem, *Catal. Today*, 1991, **8**, 371–394.
121. L. Fratalocchi, C. G. Visconti, L. Lietti, G. Groppi, E. Tronconi, E. Roccaro and R. Zennaro, *Catal. Sci. Technol.*, 2016, **6**, 6431–6440.
122. S. Saeidi, M. K. Nikoo, A. Mirvakili, S. Bahrani, N. A. Saidina Amin and M. R. Rahimpour, *Rev. Chem. Eng.*, 2015, **31**, 209–238.
123. E. Iglesia, S. C. Reyes, R. J. Madon and S. L. Soled, *Adv. Catal.*, 1993, **39**, 221–302.

124. C. G. Visconti, E. Tronconi, G. Groppi, L. Lietti, M. Iovane, S. Rossini and R. Zennaro, *Chem. Eng. J.*, 2011, **171**, 1294–1307.
125. L. C. Almeida, F. J. Echave, O. Sanz, M. A. Centeno, G. Arzamendi, L. M. Gandía, E. F. Sousa-Aguiar, J. A. Odriozola and M. Montes, *Chem. Eng. J.*, 2011, **167**, 536–544.
126. L. C. Almeida, O. Sanz, D. Merino, G. Arzamendi, L. M. Gandía and M. Montes, *Catal. Today*, 2013, **215**, 103–111.
127. A. Egana, O. Sanz, D. Merino, X. Moriones and M. Montes, *Ind. Eng. Chem. Res.*, 2018, **57**, 10187–10197.
128. N. Hooshyar, D. Vervloet, F. Kapteijn, P. J. Hamersma, R. F. Mudde and J. R. van Ommen, *Chem. Eng. J.*, 2012, **207–208**, 865–870.
129. D. Vervloet, M. R. Kamali, J. J. J. Gillissen, J. Nijenhuis, H. E. A. van den Akker, F. Kapteijn and J. R. van Ommen, *Catal. Today*, 2009, **147**, 138–143.
130. K. Pangarkar, T. J. Schildhauer, J. R. van Ommen, J. Nijenhuis, J. A. Moulijn and F. Kapteijn, *Catal. Today*, 2009, **147**, 2–9.
131. B. Kaskes, D. Vervloet, F. Kapteijn and J. R. van Ommen, *Chem. Eng. J.*, 2016, **283**, 1465–1483.
132. M. Sheng, H. Yang, D. R. Cahela and B. J. Tatarchuk, *J. Catal.*, 2011, **281**, 254–262.
133. M. Sheng, H. Yang, D. R. Cahela, W. R. Yantz, C. F. Gonzalez and B. J. Tatarchuk, *Appl. Catal., A*, 2012, **445–446**, 143–152.
134. S. Leviness, S. R. Deshmukh, L. A. Richard and H. J. Robota, *Top. Catal.*, 2014, **57**, 518–525.
135. C. G. Visconti, E. Tronconi, L. Lietti, G. Groppi, P. Forzatti, C. Cristiani, R. Zennaro and S. Rossini, *Appl. Catal., A*, 2009, **370**, 93–101.
136. A.-M. Hilmen, E. Bergene, O. A. Lindvåg, D. Schanke, S. Eri and A. Holmen, *Catal. Today*, 2001, **69**, 227–232.
137. R. M. de Deugd, *Catal. Today*, 2005, **105**, 350–356.
138. L. Fratalocchi, C. G. Visconti, L. Lietti, E. Tronconi and S. Rossini, *Appl. Catal., A*, 2016, **512**, 36–42.
139. L. Fratalocchi, C. G. Visconti, L. Lietti, E. Tronconi, U. Cornaro and S. Rossini, *Catal. Today*, 2015, **246**, 125–132.
140. M. Iovane, R. Zennaro, P. Forzatti, G. Groppi, L. Lietti, E. Tronconi, C. G. Visconti, S. Rossini and E. Mignone, *Reactor for Exothermic or Endothermic Catalytic Reactions*, WO 2010/ 130399 A1, 2010.
141. G. Groppi, E. Tronconi, C. G. Visconti, A. Tasso and R. Zennaro, *Packed-bed Tubular Reactor for Heterogeneous Exothermic or Endothermic Catalytic Reactions*, WO 2015/ 033266 A1, 2015.
142. L. Fratalocchi, G. Groppi, C. G. Visconti, L. Lietti and E. Tronconi, *React. Chem. Eng.*, 2019, **4**, 1917–1921.

CHAPTER 12

Catalytic Conversion of CO_2 to Fuels and Value-added Chemicals

A. ÁLVAREZ,[a]* O. H. LAGUNA[b] AND S. MURCIA-LÓPEZ[c]

[a]Universidad de Boyacá, Departamento de Química y Bioquímica, Carrera 2ª Este No. 64-169, 150003, Tunja, Colombia
[b]Universidad de Jaén, Escuela Politécnica Superior de Linares, Campus Científico Tecnológico de Linares, Departamento de Ingeniería Química, Ambiental y de los Materiales, Av de la Universidad s/n, 23700, Linares (Jaén), Spain
[c]Institut de Recerca en Energía de Catalunya (IREC), Energy Storage Group-Advanced Materials for Energy Area, C/ Jardins de les Dones de Negre 1, 08930, Barcelona, Spain
*Corresponding contributor. Email: andalvarez@uniboyaca.edu.co

12.1 Introduction

Humanity has always relied on fuels for energy generation. At the beginning of civilization, wood allowed our ancestors to eat cooked food and avoid freezing. Many centuries later, in about 1780, the "star fuel" became coal, as it was responsible for powering the steam machines that characterized the Industrial Revolution. A century later, a new fuel appeared in the energy scenario when the first oil well was drilled, changing forever the energy paradigm when paired with the new-born automotive industry. More recently, after

Heterogeneous Catalysis for Energy Applications
Energy and Environment Series No. 27
Edited by Tomas R. Reina and Jose A. Odriozola

Published by the Royal Society of Chemistry, www.rsc.org

effective pipelines began to be built in the twentieth century, the use of natural gas started to be implemented regularly in homes for heating and cooking. Today, one thing is certain: the energy scenario is ruled by fossil fuels – coal, oil and natural gas.

The last report of the International Energy Agency (IEA)[1] indicates that the global primary energy demand for fuel still relies on approximately 27% coal, 32% oil and 22% natural gas (Figure 12.1). The data show that the overall share of fossil fuels in global primary energy demand has not changed over the last 25 years and is not likely to change in the future. Oil, coal and gas account for more than 81% of the world's primary energy, and the forecasts show that, even including the actual policies and targets announced by governments (years 2025 and 2040 in Figure 12.1), 75% of the total energy will still come from fossil fuels 20 years from now.

Fossil fuels have some advantages; for example, they are able to store and deliver large quantities of energy more effectively and consistently than current alternative energies,[2] and their versatility is also a huge plus. Coal, which is mainly used for electricity production, can be gasified to produce natural gas for lower-emission consumption. Natural gas can provide space heating, drive combined-cycle turbines for efficient electricity generation and, with additional infrastructure (gas-to-liquid [GTL] technologies), provide fuel for road vehicles. Oil can be refined to yield large fractions of gasoline, diesel and aviation fuels. Similarly, oil is essential to the petrochemical industry, which provides plastics, fibres, synthetic materials, paints, pharmaceuticals and fertilizers to humanity (Figure 12.2).[2]

However, even though fossil resources are incredibly versatile, there are two main drawbacks associated with their use. Firstly, fossil fuels are not renewable resources, implying that, sooner or later, their supply will end. Secondly, they are mainly responsible for the climate change that the world is currently experiencing.[3] Climate change is mostly induced by the greenhouse effect, and carbon dioxide (CO_2) accounts for a dominant proportion of this issue.[4,5] The fossil fuel-based economy contributes nearly 80% of the CO_2 emitted in the atmosphere,[6,7] and it does not seem that its generation will be mitigated anytime soon (Figure 12.1).

Fortunately, fossil fuels are not the only versatile feedstock for achieving heating or generating electricity, liquid fuels or even value-added chemicals derived from the petrochemical industry, as CO_2 can be just as versatile. Carbon dioxide can be transformed into a wide variety of products, from polymers, carboxylates, lactones, carbamates, isocyanates, formaldehyde and formic acid to renewable fuels, such as synthesis gas (syngas), methane, methanol and dimethyl ether (DME) (Figure 12.2).[8–10] Today, a new scenario is being explored, where CO_2 is not assumed to be a waste, but an alternative carbon feedstock.

However, the challenges associated with the conversion of CO_2 are primarily related to both its kinetic and its thermodynamic stability.[11] CO_2 is inert, and its chemical transformation is thermodynamically highly unfavourable. Its inertness is due to it being the most oxidized state of carbon, implying that

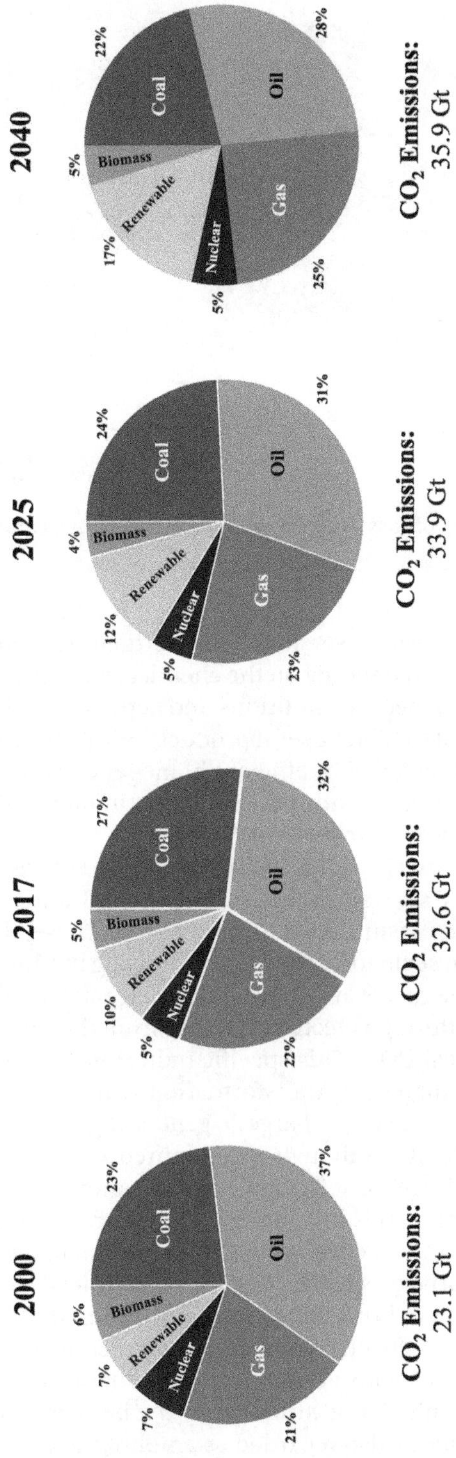

Figure 12.1 World primary energy demand by fuel. Data extracted from Ref. 1.

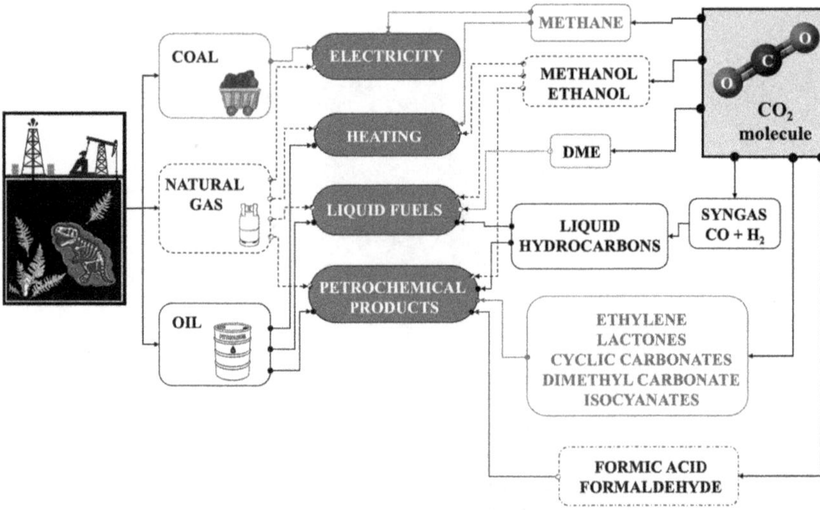

Figure 12.2 Versatility of fossil fuels *versus* versatility of fuels and chemicals derived from CO_2.

it is a molecule in its lowest energy level, which is a major obstacle in establishing industrial process conversion.[10,12] Nevertheless, it can be transformed with notable reactivity depending on the chemical environment. A large input of energy, optimized reaction conditions and active catalysts are required for any transformation of CO_2 into useful products. While transforming CO_2 into polymers is generally enabled by efficient homogeneous catalysts, large-scale products of bulk chemicals and fuels need continuous heterogeneous catalysts to activate CO_2 over their surfaces.[13]

The activation of the CO_2 molecule over heterogeneous catalyst surfaces can be performed by several methods: thermal activation, photoactivation, electroactivation and plasma activation.[13] Between these, thermal activation and electroactivation seem to be the most promising methods in order to generate value-added chemicals and fuels.[14] Thermal activation of CO_2 can occur over metal surfaces through electron transfer from the metal to CO_2 in order to generate the radical CO_2^-. This specific radical has been identified as the key intermediate in many CO_2 transformation reactions.[13] In this state, the carbon atom bears a negative charge,[15] generating nucleophilic behaviour and increasing reactivity. Although the required energy for this process is very high due to the large work function of some metal surfaces, higher reaction temperatures can result in a larger number of electrons with sufficient energy to take part in CO_2 activation. Surface intermediates such as CO, formate (HCOO) and carboxyl species (COOH)[13] are typically found in thermally activated, metal-catalysed reactions. Similarly, thermal activation can occur over acid–base sites. Following the definition of acidity from the Lewis concept (acid = electron acceptor; base = electron donor), CO_2 is recognized as a mildly acidic molecule that is attracted by the basic sites of solid surfaces. However, the molecule is also regarded as amphoteric owing to its polarized character due to the more electron-deficient C atom and the more electron-rich O atoms. In this manner, CO_2 can be activated over an acid surface when the oxygen atoms interact with the surface through the metal cation (M^{n+}) or through oxygen vacancies, and it can also be activated over a basic surface when the carbon atom interacts through the O^{2-} sites.[16] Both situations induce charge repartition states that enhance the reactivity of the CO_2 molecule.[13] Surface intermediates such as carbonates (CO_3^{2-}) and bicarbonates (HCO_3^-) are classically found in the acid–base activation of CO_2.

Very similar to thermally induced metal activation, electroactivation is also based on a charge transfer from the metal's Fermi level to CO_2 in order to generate the radical $CO_2^{\cdot-}$. This charge transfer occurs across the interface between the surface of the electrode (often metallic) and adsorbed CO_2.[13] As occurs in thermal activation, the process is energy intensive owing to the thermodynamic stability of the CO_2 molecule. Most setups for CO_2 electrocatalytic reduction are electrolytic cells where a cathode and an anode are immersed in aqueous electrolyte and gaseous CO_2 is continuously bubbled thought the solution. Reaction engineering plays a key role in electrocatalytic CO_2 conversion to improve the contact between CO_2 and the cathode with optimized mass transport properties. Reactor engineering can provide outstanding CO_2 reactivity, even in continuous operations.[13]

Between these two activation methods, thermal catalysis is a more mature technology than electrocatalysis, mostly because it has been used for longer by the current fossil fuel industry; therefore, catalytic thermal conversion of CO_2 towards fuels is a more promising approach than electrocatalytic conversion.

Technologies that could be considered for CO_2 based fuels using thermocatalytic conversion include the Sabatier reaction for the production of methane (CH_4) and the Fischer–Tropsch synthesis (FTS) of higher hydrocarbons from syngas. Both technologies provide access to existing infrastructure for utilization in transportation, but also for energy storage and later reconversion to grid electricity (Figure 12.3). Similarly, novel alternative fuels and fuel components are emerging with potentially improved physical and chemical properties. Oxygenates, such as methanol and DME, are of particular interest in this context.[17] Due to their oxygen content and resulting changes in the radical pathways of combustion mechanisms, such structures have the potential to reduce soot formation during engine operation significantly. Furthermore, their miscibility with diesel as well as their low toxicity and corrosiveness can facilitate integration into today's transportation infrastructure.[17]

On the other hand, electrocatalytic conversion of CO_2 seems quite promising in generating value-added chemicals such as CO, formic acid and ethylene. These last two products depend directly on the petrochemical industry; therefore, establishing a viable strategy for their production from CO_2 could boost the "circular economy" concept associated with the recycling of CO_2 towards value-added chemicals.

The grand challenge facing converting CO_2 waste streams into useful products is the development of processes that require minimal amounts of non-renewable energy, are economically competitive and provide substantial

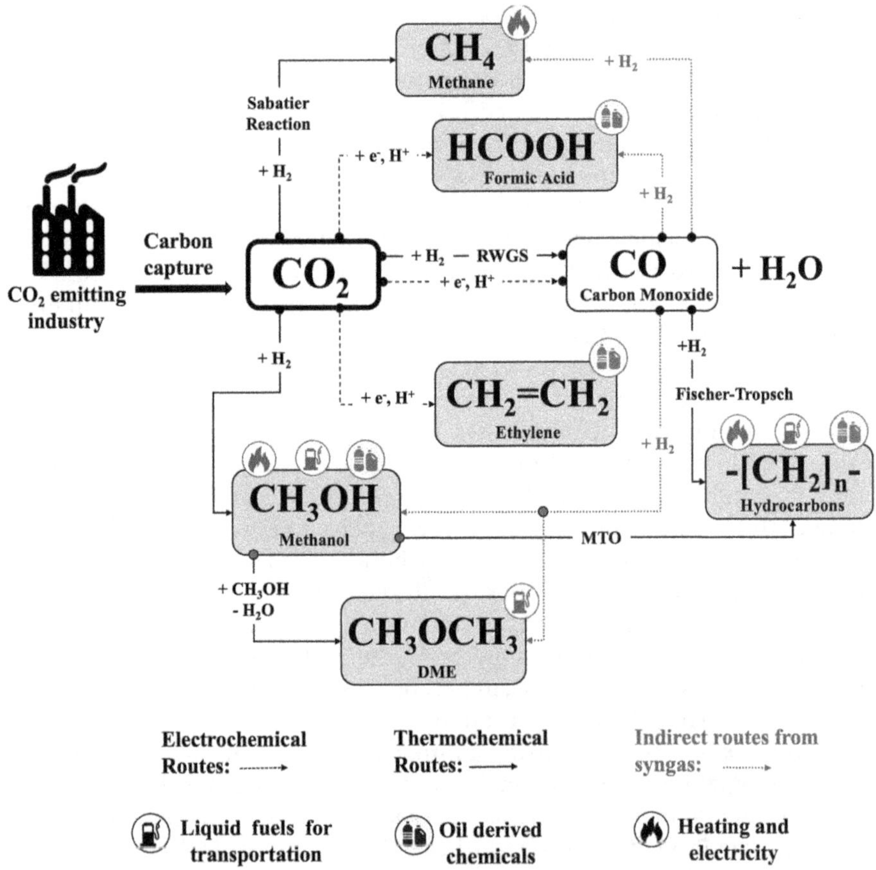

Figure 12.3 Fuels and value-added chemicals derived from CO_2 by thermochemical routes and electrochemical routes addressed in this chapter.

reductions in greenhouse gas emissions compared to existing technologies.[11] The most sustainable proposed strategy is the "carbon capture and utilization" (CCU) technology.[18–20] With CCU technology, CO_2 emissions from point sources are captured or redirected to another reaction system where they meet an energy-intensive reactant such as H_2 in order to generate fuels and chemicals. The approach is quite reasonable, as the vast majority of fuels or bulk chemicals are organic compounds that contain essentially carbon and hydrogen (Figure 12.3). In addition, forecasts show that CO_2 is a feedstock that is in continuous generation (Figure 12.1), implying that its supply will not be an issue. However, the economic perspectives of this strategy are largely dependent on the cost of producing renewable H_2[21–24] and the enhancement of CO_2 capture/separation technologies.

This chapter presents the latest advances in fuel generation from CO_2 hydrogenation, focusing on the thermocatalytic processes which generate methanol, DME, methane and hydrocarbons as the major products.

Additionally, new alternatives for generating value-added chemicals are explored, describing the electrochemical generation of CO, formic acid and ethylene from CO_2.

12.2 Thermochemical Conversion of CO_2 to Fuels

12.2.1 Methanol

Methanol (CH_3OH) is one of the most important bulk chemicals produced worldwide, with a current production rate of 95 million metric tons per year.[25] It is a primary raw material for the chemical industry since it is an intermediate for the production of a variety of chemicals, including formaldehyde, methyl tert-butyl ether (MTBE), methylamines, acetic acid and DME (Figure 12.4). Most of these chemicals are the basic building blocks of many commodity products in our daily lives, including paints, plastics, resins, adhesives and antifreezes.[26] Furthermore, in addition to being an excellent fuel on its own, methanol can be blended with gasoline as an oxygenated additive (fuel M85),[27] and it can also be the feedstock to generate hydrocarbons through methanol-to-hydrocarbon (MTH) processes: methanol to olefin (MTO), methanol to propene (MTP) and methanol to gasoline (MTG) (Figure 12.4).

In the early days (*ca.* 1920), methanol was commercially produced from syngas ($CO + H_2$)[26] by the direct hydrogenation of CO (eqn (12.1)).

$$CO + 2H_2 \leftrightharpoons CH_3OH \qquad \Delta H_r^\circ(298\ K) = -90.6\ kJ\ mol^{-1} \qquad (12.1)$$

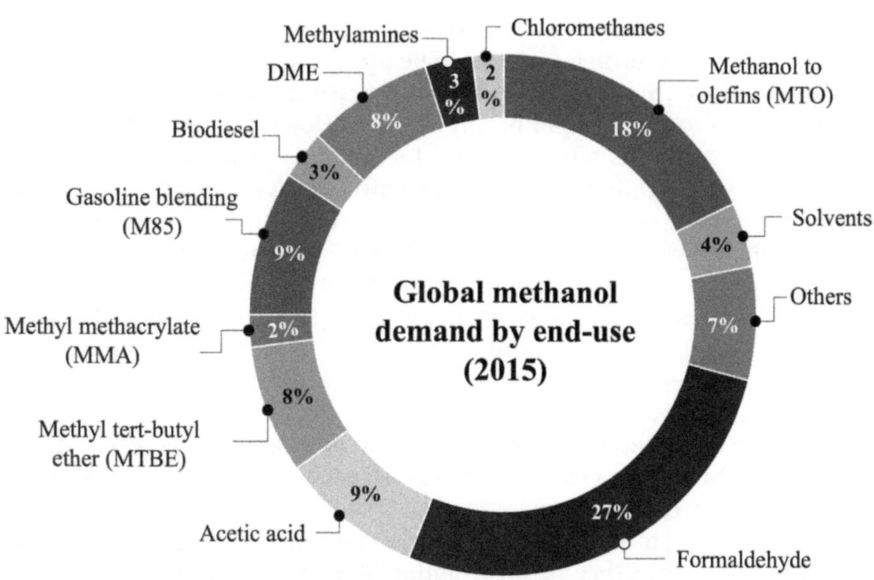

Figure 12.4 Global methanol demand in 2015. Data extracted from Ref. 28.

The situation has not changed much, and currently, the industrial methanol production is still based on syngas that is generated from the steam reforming of natural gas or coal gasification, with the only variation being that, today, a mixture of CO, H_2 and CO_2 is used.[28–30] The latter implies that the actual methanol production depends directly on fossil fuel resources. However, since the "methanol economy" concept was introduced by Nobel Laureate George Olah[27] in 2006, the idea of capturing CO_2 from fossil fuel-powered plants in order to use it as a feedstock for methanol generation has intensified research into the direct hydrogenation of CO_2 towards methanol (eqn (12.2)).

$$CO_2 + 3H_2 \leftrightharpoons CH_3OH + H_2O \qquad \Delta H_r°(298\ K) = -49.3\ kJ\ mol^{-1} \quad (12.2)$$

The direct hydrogenation of CO_2 has a couple of drawbacks. The first is that there is competition with the reverse water gas shift (RWGS) reaction that generates CO and water (eqn (12.3)), and the second is that H_2O generated in the previous reaction has been reported to inhibit CO_2 adsorption in Cu-based catalysts, leading to catalyst deactivation.

$$CO_2 + H_2 \leftrightharpoons CO + H_2O \qquad \Delta H_r°(298\ K) = 41.2\ kJ\ mol^{-1} \quad (12.3)$$

In order to reduce the inhibiting effect of water, the CAMERE process[31] (CArbon dioxide hydrogenation to form MEthanol *via* a REverse-water–gas-shift reaction) is sometimes implemented. The method is based on a two-step process. First, the feed gas ($CO_2 + H_2$) is passed into a RWGS reactor and water is then removed from the first reactor. The composition of the feed gas at the second reactor is then closer to that of the syngas ($CO + H_2$) used in conventional methanol synthesis. Results show three times higher productivity than the single-stage process due to the removal of water and the shifting of the equilibrium.[24,31] Nevertheless, it would be preferable to develop direct routes for converting CO_2 to methanol in order to overcome the thermodynamic constrains related to RWGS equilibrium reaction. However, in order to achieve this goal, new types of catalysts, specific for methanol synthesis from CO_2, not sensitive to water inhibition and possibly involving a different type of mechanism, not passing through the first stage of the RWGS reaction, are necessary.

Researchers have developed several catalysts and reactors for direct hydrogenation of CO_2 to methanol, but high rates and high methanol selectivities have only been possible using high pressures (>300 bar) owing to the mole-reducing character of the reaction (eqn (12.2)).[32–34] However, some research shows that the cost of this technology is not competitive with the cost of methanol synthesis from syngas.[11]

Regarding the heterogeneous catalysts that are used for this reaction, one can identify three main groups: (1) Cu-based catalysts; (2) noble metal-based catalysts; and (3) oxygen-deficient materials such as indium oxide.[26,35]

Cu-based catalysts are by far the most widely studied catalytic systems for CO_2 hydrogenation as they are an evolution of the $Cu/ZnO/Al_2O_3$ catalysts that were originally developed by Imperial Chemical Industries (ICI) for methanol synthesis from syngas. Although it is recognized that Cu acts as the active phase of these catalysts, ZnO is described as a promoter enhancing the dispersion of Cu, and Al_2O_3 provides the thermal and chemical stability of the catalyst.[26] The interactions of the components in this catalyst have been a matter of debate for years. The "active site" component has been widely discussed. While some research groups suggest the direct relationship between the surface area of metallic copper (Cu^0) and the catalytic activity of the catalyst,[7,36,37] others suggest that there is a special solid-state interaction between ZnO and Cu that is responsible for this outstanding activity. Regarding this interaction, one possibility proposes that there is an intimate synergy between Cu and ZnO at the interface, where ZnO could act as a structural modifier, hydrogen reservoir or direct promoter for bond activation.[38,39] The other possibility argues that a highly active ZnCu alloy forms by partial reduction of ZnO, and even a Cu surface decorated with metallic Zn has been proposed.[40,41] Generally, this Cu–ZnO synergy, where the surface area of Cu^0 is maximized and the Cu–ZnO interaction is enhanced, is often determined by the catalyst synthesis method.[26,37,42,43]

Despite the fame of the Cu/ZnO/Al_2O_3 catalysts, unsatisfactory stability performance has been reported in direct CO_2 hydrogenation owing to the presence of water and the strong hydrophilic character of alumina.[44] In order to address this issue, the use of less hydrophilic supports, such as ZrO_2, has been proposed.[44–46] To this end, Cu/ZrO_2 catalysts have been studied extensively. Due to the abundant availability of oxygen at the surface and its good ion exchange capacity, zirconia is considered to be a good candidate for hydrogenation reactions.[47] ZrO_2 added on Cu-based catalysts increases the Cu^0 surface area and improves Cu particle dispersion.[48] However, special care should be taken regarding the synthesis method, as the final activity of the Cu/ZrO_2 catalysts has been reported to be highly dependent on the crystallographic structure of ZrO_2.[45]

Concerning noble metal-based catalysts in direct CO_2 hydrogenation, Pd systems are the most commonly used. Pd is widely applied as a major catalytic component in CO_2 hydrogenation to methanol because it can efficiently activate the reactants due to its spill-over effect.[7,49] Other systems based on Pd have also been described. Pd–ZnO catalysts have been reported as very selective catalysts towards methanol, avoiding the RWGS reaction owing to the formation of the PdZn alloy.[50] Pd–Cu catalysts have also been reported. However, even if lower metal loadings are required to achieve high methanol production, an increase of CO is evidenced owing to the RGWS reaction.[51,52]

Apart from the traditional Cu-based catalysts and noble metal catalysts, many CO_2 hydrogenation catalysts have been developed in recent years. Between these, the In_2O_3 oxide-based catalysts have shown remarkable behaviour. Ye *et al.*[53] reported a density functional theory study that predicted the high selectivity of In_2O_3 for CO_2 hydrogenation without passing though the RWGS reaction. Based on these predictions, Oliver *et al.*[54] prepared a series of In_2O_3 catalysts supported on ZrO_2 that exhibited extraordinary activity and 100% selectivity towards methanol, surpassing the activity of the benchmark Cu/ZnO/Al_2O_3 catalyst.

As for the CCU approach for methanol production, the Methanol Institute[55] has published its latest report addressing this subject. Herein, they established the category "renewable methanol" for methanol generated from carbon dioxide and hydrogen produced from renewable electricity.[56] Of the mentioned institutions, Carbon Recycling International (CRI) and the Innogy initiative stand out. CRI is considered to run the "world's largest CO_2 methanol plant".[57] This is the George Olah Renewable Methanol Plant in Svartsengi, near Grindavik. In 2015, CRI achieved a CCU milestone when the plant's capacity was scaled from 1 300 to 4 000 metric tons per year, which resulted in a CO_2 recycling capacity of 5 500 tons.[57] In their technology, the use of hydrogen obtained by electrolysis using renewable energy from a geothermal power plant is a key point.[56,57] The Innogy initiative captures CO_2 in order to generate methanol as well; however, the process is based on enzymatic transformation rather than thermochemical transformation.

12.2.2 Dimethyl Ether

DME (CH_3OCH_3) is a clean, colourless gas that is easy to liquefy and transport. Physically, is very similar to liquefied petroleum gas (LPG), and owing to its clean combustion (no soot formation), ease in transportation and energetic density (Figure 12.5), it has incredible potential to replace the conservative fossil fuels used for transportation (diesel and gasoline).[58,59]

Today, primary DME applications include: supplements in cooking gas, aerosol spray-can propellants, solvents and chemical feedstocks. However, developing markets are exploring DME as a transportation fuel. Auto

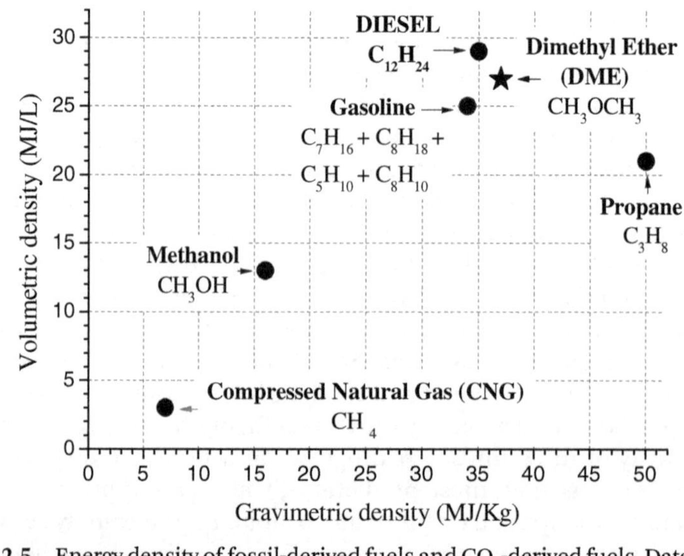

Figure 12.5 Energy density of fossil-derived fuels and CO_2-derived fuels. Data extracted from Refs 30, 60 and 61.

manufacturers such as Shanghai Diesel Co., AB Volvo, Isuzu Truck and Nissan Diesel are actively developing DME fuelled heavy-duty vehicles.[62]

At an industrial level, DME is synthesized from syngas following two main strategies (eqn (12.4) and (12.5)).

$$2CO + 4H_2 \leftrightharpoons CH_3OCH_3 + H_2O \qquad \Delta H_r^{\circ}(298\ K) = -49.0\ kJ\ mol^{-1} \quad (12.4)$$

$$3CO + 3H_2 \leftrightharpoons CH_3OCH_3 + CO_2 \qquad \Delta H_r^{\circ}(298\ K) = -58.0\ kJ\ mol^{-1} \quad (12.5)$$

Eqn (12.4) is a combination of methanol synthesis from syngas (eqn (12.1)) and methanol dehydration (eqn (12.6)), while eqn (12.5) combines both previous reactions with the water gas shift reaction (eqn (12.3)).

$$2CH_3OH \leftrightharpoons CH_3OCH_3 + H_2O \qquad \Delta H_r^{\circ}(298\ K) = -23.4\ kJ\ mol^{-1} \quad (12.6)$$

The strategy of eqn (12.4) is used by Haldor-Topsoe,[27] while the strategy of eqn (12.5) is used by JFE Corporation.[63] In the case of JFE Corporation, their syngas comes from coal gasification or from methane dry reforming (eqn (12.7)). Although they claim to have a process with zero CO_2 emissions (because the CO_2 generated is then used for the dry reforming of methane), this reaction is highly energy intensive and requires natural gas as feedstock.

$$CH_4 + CO_2 \leftrightharpoons 2CO + 2H_2 \qquad \Delta H_r^{\circ}(298\ K) = 247.3\ kJ\ mol^{-1} \quad (12.7)$$

Therefore, in order to fit DME synthesis into the methanol economy scenario, syngas as feedstock should be avoided, and the implementation of a process that involves the direct hydrogenation of CO_2 towards methanol, followed by the dehydration of this molecule to yield DME (eqn (12.6)) should be taken into consideration.

In order to perform the direct transformation of CO_2 into DME, a bifunctional catalyst that is able to conduct the two reactions (methanol synthesis and methanol dehydration) is required. This type of catalyst is usually referred to as "hybrid" and includes an active phase in the methanol synthesis, such as Cu, Cu–ZnO, Cu–ZnO–ZrO_2 or Pd and a solid acid catalyst capable of the dehydration reaction, such as γ-Al_2O_3, or acid zeolites such as H-ZSM5, HV, SAPO-18, MCM or ferrierite.[24,26,64–69]

The two functions – namely, methanol synthesis and methanol dehydration of a hybrid catalyst – can be combined mainly in two ways. The first is a physical mixture in which a methanol synthesis catalyst and a solid acid catalyst are simply mixed together. In this case, special attention should be given to how the two catalysts are placed in a reactor, because this has a substantial influence on catalytic performance.[26] The second option of combination is an integrated mixture, in which the catalytically active components for the reactions are intentionally placed in close proximity from the synthesis stage using co-precipitation or impregnation methods. Recent examples of this type of catalyst have been presented by Atakan *et al.*,[70] who reported the synthesis of an active catalyst for DME generation from CO_2 where Cu nanoparticles where integrated into SBA-15 mesoporous supports previously doped with Zr in order to modulate the acidity of the surface. In addition, reports of Cu nanoparticles supported over Al_2O_3 were described as active in CO_2 hydrogenation to DME.[64] Recent reviews can be found in the literature regarding the most innovative catalytic systems developed in this field.[68,71]

In terms of commercialization plans for CO_2 to DME, two initiatives stand out. The Korea Gas Corporation (KOGAS) has developed an indirect strategy for DME synthesis from CO_2 that is performed on a scale of 100 tons per day. It involves the tri-reforming of methane, CO_2 and H_2O into syngas. The syngas is then converted into DME. Although this route is scalable, it is not economically viable and does not involve the direct conversion of CO_2 as the only feedstock for DME production.[11] Similarly, BASF recently outlined a project in collaboration with Linde[72] in which a novel one-step route from CO_2 was announced. Even though they proposed a very clear plan of scaling up the project with significant CO_2 emissions reductions, the concept is based on the dry reforming of methane with CO_2 and the direct conversion of syngas to DME. Unfortunately, to the best of our knowledge, no reports regarding a CCU approach in which CO_2 from a highly CO_2-emitting industry is used as a feedstock to generate DME has yet been reported.

12.2.3 Hydrocarbons

The production of hydrocarbons directly from CO_2 can be considered as a step ahead in the aim of reusing this renewable carbon source, especially for the option of obtaining value-added products with more than one carbon in their molecules. However, few studies are currently focused on this due to the extreme inertness of the CO_2 molecule and the high C–C coupling barrier.[73–75] Generally, the direct transformation of CO_2 into hydrocarbons is carried out through a modified FTS route (MFTSR) by means of the inclusion of a previous step of CO production through the RWGS reaction, followed by the Fischer–Tropsch reaction for obtaining short-chain olefins.[24,75] Furthermore, there is another indirect route that implies the methanol synthesis followed by the methanol-to-olefin (MTO) pathway (methanol-mediated route [MMR]).[76] Probably the most attractive route, but at the same time the most challenging, is direct hydrogenation over promoted catalysts.[76,77]

CO_2 presents low adsorption rates over the catalyst surfaces due to its thermodynamic and chemical stability, resulting in high H:C ratio products over the catalyst surface during the hydrogenation process. This is why methane production is favoured instead of chain growth. Within the first systems studied for the production of olefins from CO_2, we found iron-based catalysts with a formulation that is active in FTS.[78,79] However, the main challenge is still the promotion of the high selectivity of light olefins while methane production is avoided, and for that, catalysts that are able to simultaneously conduct the following two tasks must be developed: CO_2 activation and selective

C–C coupling. In this sense, the approach of developing bifunctional catalysts arises as a suitable option, such as that proposed by Gao *et al.*,[75] where an In–Zr mixed oxide is combined with SAPO-34, aiming to perform the CO_2 activation and C–C coupling in every component, respectively. In this case, the authors reported that C_2–C_4 olefins reached selectivity levels near to 80% and CO_2 conversion was above 35%.

A novel strategy proposed by Guo *et al.*[77] considered the use of an enzyme-like integrated catalyst comprising iron carbides and alkali promoters from calcined corncob ash, which generate valuable products such as linear α-olefins that contain a terminal C=C double bond. In this case, the results were interesting not only in terms of obtaining molecules with C=C bonds, as observed in Figure 12.6, but also in terms of the approach of using mineral elements from residual biomass named biopromoters for the synthesis of the catalysts.[77]

Regarding the catalysts used for obtaining olefins from CO_2, recently, Ronda-Lloret *et al.*[80] presented a comprehensive review that included an interesting scheme (presented in Figure 12.7) that allows for identifying the most common components of the catalysts tested from 2016 to 2019.

From Figure 12.7, it can be inferred that, in the case of the MFTSR, the use of a simple catalyst configuration (support + active phase) has been assumed, where the inclusion of promoters enhances the abilities of the supports and their interaction with the active phase. These systems are inspired by the traditional Fe-based FTS catalysts.

As stated above, the MFTSR combines the RWGS and the FTS. In the case of the RWGS step, two main mechanisms have been proposed, depending on the nature of the catalysts.[81] The first is the redox mechanism by which CO_2 adsorbs and dissociates on reduced active sites. This results in the oxidation

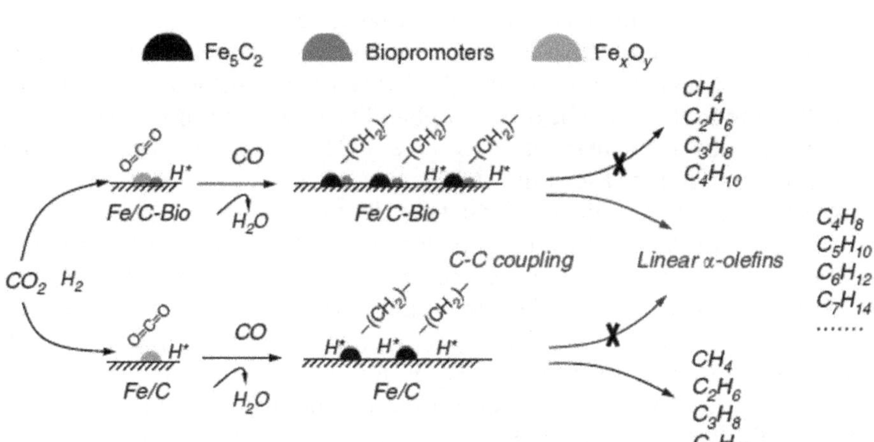

Figure 12.6 Reaction scheme for CO_2 hydrogenation to linear α-olefines over bio-promoted catalysts that weaken secondary hydrogenation of olefins and strengthen the reaction activity of C–C coupling. Reproduced from Ref. 77, https://doi.org/10.1038/s42004-018-0012-4, under the terms of the CC BY 4.0 licence, https://creativecommons.org/licenses/by/4.0/.

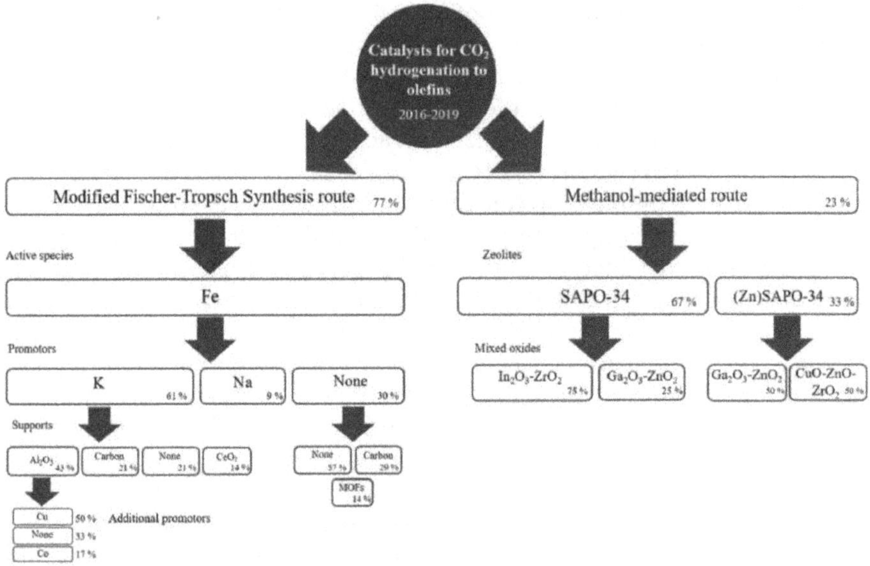

Figure 12.7 The main components of the catalysts used for CO_2 hydrogenation to olefins, considering two routes: MFTSR and MMR. The percentages included in the scheme correspond to the frequency of use of every component and were calculated by the authors based on the reviewing of 26 papers from the previous four years. Reprinted from Ref. 80.

of the active sites that are subsequently reduced by hydrogen, producing H_2O. The second mechanism is the associative one by which adsorbed CO_2 reacts with adsorbed H_2, resulting in the creation of carbon-containing intermediates that can be C–C linked or directly hydrogenated.[81]

In the second step, which is FTS, an important parameter is the product distribution, which can be described by Anderson–Schulz–Flory (ASF) theory. ASF theory models the chain growth probability (α) depending on the features of the catalysts and the reaction conditions.[82–85] However, ASF theory is just an approximation in the case of the MFTSR, as replacing the feed-stream CO/H_2 by CO_2/H_2 induces significant product distribution changes.[86]

Another important aspect to consider in the MFTSR is that it combines RWGS and FTS, an endothermic and an exothermic process, respectively. Therefore, variables such as the temperature, pressure, $H_2:CO_2$ ratio and the presence of H_2O in the feed stream may strongly influence not only the CO_2 conversion, but also the product distribution. In this sense, Yao *et al.*[87] presented a thermodynamic study of this process, and they concluded that long-chain olefins were favoured at lower temperatures, while short-chain olefins were favoured at higher temperatures. Furthermore, they established that at around 673 K, the highest production of light olefins is achieved.[87]

In the case of MMR, it seems that a bifunctional catalyst approach is preferred, according to Figure 12.7.[80] MMR combines CO_2 hydrogenation to methanol and MTO reactions. The CO_2 hydrogenation to methanol was presented in Section 12.2.1, and thus it is more illustrative in this case to highlight some aspects of the MTO reaction (eqn (12.8)). Firstly, methanol can be converted into olefins using acid catalysts and zeolitic materials (ZSM-5 and SAPO-34), which are used industrially as catalysts for the MTO reaction.[88,89]

$$2CH_3OH \rightarrow CH_3OCH_3 + H_2O \rightarrow CnH_{2n} + nH_2O \qquad (12.8)$$

The MTO reaction is thermodynamically favoured at high temperatures. Nevertheless, the previous methanol production presents the opposite behaviour. The high temperatures enhance CO generation as a by-product due to the promotion of the competing RWGS reaction. Therefore, a compromise between these two main processes must be established in every catalytic application. Hence, the catalysts require a combination of a metal oxide and a zeolite/molecular sieve. In this way, the CO_2 conversion to methanol is performed by the metal oxide, while the zeolite component is responsible for the subsequent methanol conversion through selective C–C coupling. This means that it is clear where the processes occur, but determining how they take place remains a challenge.[80]

As for the product distribution, the role of the acidic component of the catalysts is critical during the MTO reaction. This is not only due to the strength of the acid sites of the catalysts, but also due to their porous structure, which may influence the type of olefins produced. For instance, the ZSM-5 that is used in the MTO reaction presents straight channels (5.1 × 5.4 Å) and intersecting sinusoidal channels (5.4 × 5.6 Å), forming a large space with a diameter of ~8.9 Å, as discussed elsewhere.[90,91] These large channels do not constrain the growth of hydrocarbons from forming longer-chain hydrocarbons and aromatics, and this is why such materials are mainly used for the production of gasoline.[80,92]

As can be observed from what has been described above, CO_2 hydrogenation to olefins is a complex process as, regardless of the applied route (MFTSR or MMR), different steps must be combined, and this implies the occurrence of competing reactions that divert the process from the desired path and generate by-products that lower overall performance. Consequently, this process is presently being studied only at the laboratory scale, with no commercial applications developed yet.

12.2.4 Methane

The production of methane by means of carbon dioxide methanation has received great attention in recent years as it is a highly versatile process.[24,93–98] Firstly, it generates a valuable fuel called substitute natural gas (SNG) that can be stored and distributed through the well-developed and already existing natural gas infrastructure. Hence, no further investments in the design and development of a new distribution technology are required.[99] Moreover, the H_2 requirement for CO_2 methanation makes this process an appropriate strategy for energy storage in renewable processes for the generation of electric power, where intermittency in power generation is one of the biggest disadvantages. This is why CO_2 methanation is one of the power-to-gas (PtG) technologies that combines the use of H_2 produced by water electrolysis feed with renewable energy (as wind or solar) and subsequent CO_2 hydrogenation through the Sabatier reaction (eqn (12.9)).[24,97]

$$CO_2 + 4H_2 \leftrightharpoons CH_4 + 2H_2O \qquad \Delta H_r^{\circ}(298\ K) = -165\ kJ\ mol^{-1} \qquad (12.9)$$

The Sabatier reaction presents the same drawback as the hydrogenation process for methanol production, which is competition with the endothermic RWGS reaction that generates CO (eqn (12.3)). Furthermore, considering that methane production is an exothermic process, CO_2 is mainly promoted at low temperatures (~473–573 K), which implies some kinetic limitations.[93,97] The exothermicity of the Sabatier reaction makes this process hardly adiabatic, which lead to in considerable challenges in reactor design because ensuring the stabilization of the reaction temperature is difficult without hot spot production or quenching of the reaction.

Due to the kinetic limitations, the use of heterogeneous catalysts to promote the reaction is mandatory, and so the nature of the catalyst strongly influences the efficiency of the process. Therefore, variables such as the active phase, the nature of the support, the presence of promoters and other components must be deeply analysed in order to ensure the contribution of the catalyst to the enhancement of the reaction. Different reviews[97,100,101] have summarized the most common active phases used in the formulation of catalysts for CO_2 methanation, including noble metals such as Rh, Ru, Pd and other transition metals such as Co, Fe and Mo. Regarding the support, the most common systems already tested are Al_2O_3 or SiO_2, although catalysts using CeO_2 and TiO_2 have also been reported.[100] The huge amount of analysed systems that include the combination of several active phases and supports, with or without the addition of promoters, makes it hard to compile a concise analysis of all of the possible systems in this brief section. However, there are some active phases and supports that are more commonly used. Therefore, in Table 12.1, some relevant highlights of the typical components of methanation catalysts are presented.

Regarding the reaction path for the CO_2 methanation, Jalama[97] recently presented a comprehensive review of the most relevant studies, particularly focused on Ni- and Ru-based catalysts, which are the most active systems in such reactions. He highlighted that the early literature suggested two main pathways: (1) the formation of CO as an intermediate; and (2) the direct hydrogenation of CO_2 by activated hydrogen to formic acid, structural rearrangement and stepwise hydrogenation of intermediates.[97]

Despite different studies supporting both approaches, subsequent studies generated unfavourable evidence regarding the first option of using CO as an intermediate, such as the fact that CO_2 hydrogenation in Ru- and Fe-based catalysts mainly produces CH_4, unlike CO hydrogenation, or that CO_2 is not hydrogenated in the presence of CO. Therefore, most of the current studies

Table 12.1 Common active phases and supports used to design CO_2 methanation catalysts.

Commonly used active phase	Relevant highlights
Rh	The particle size of Rh nanoparticles influences the catalytic performance (larger Rh particles exhibited superior catalytic activity than smaller ones).[102,103] Small amounts of O_2 may promote methane production, avoiding CO formation.[104] Density functional theory studies suggest a CO_2 and H_2 co-adsorption over Rh atoms, where a subsequent interaction results in the production of CO as an intermediate prior to the CH_4 generation.[105,106]
Ru	Ru is the most efficient active phase in the Sabatier reaction, although the price of this noble metal hinders its use in massive applications. The particle size of Ru nanoparticles strongly influences the performance of the catalysts. Large Ru nanoparticles present superior performance,[94,95,97] as this allows for the spill-over of H_2 that produces the hydrogenation of the intermediate CO species. Controlling the particle size allows for a certain level of control over kinetic parameters such as the reaction orders of some species.[94] The contribution of the hydroxyl groups of the support may influence the adsorption of the CO_2 molecules over the Ru nanoparticles. In this sense, the presence of additives may alter the adsorption capacity of the catalytic surface and consequently modify its performance.[95]
Pd	Pd presents a considerable catalytic activity in the Sabatier reaction. Shape-controlled Pd nanoparticles embedded in mesoporous silica demonstrated superior performance than other systems obtained by traditional methods. This is because the preferential exposition of some crystallographic planes – (100) and (111) – renders superior activity, thus suggesting that the catalytic activity also depends on the shape of the active phase.
Ni	Ni-supported catalysts are the most studied systems in the Sabatier reaction due to their high efficiency and relatively low cost.[107] Therefore, Ni has been combined with many supports.[97,100] In fact, the most important companies focused on the commercial production of methanation catalysts include Ni as the main component in their formulations. Despite the high efficiency of Ni-supported catalysts, their main drawback is their low resistance to sintering, which results in the reduction of their catalytic performance. This makes the role of the support in this sort of system more relevant in order to prevent the sintering of the active phases.[97,100] Smaller-sized nanoparticles of Ni present higher catalytic activities than larger ones.[97] The interaction between Ni and basic supports enhances catalytic performance and prevents the deactivation by deposition of carbonaceous species.[108]

(*continued*)

Table 12.1 (*continued*)

Commonly used supports	Relevant highlights
Al_2O_3	Alumina is the most widely used support for the synthesis of methanation catalysts,[107] and depending on the phase of this oxide, important differences in the interaction between this material and the different active phases may be observed.[109] The role of the hydroxyl groups of this material is determinant not only in the interaction with the active phases, but also during the activation of the reactant molecules on the periphery of active sites.[95]
SiO_2	Different synthesis procedures may result in SiO_2 oxides with controlled morphologies that may influence the catalytic activity of catalysts supported on this oxide.[110,111] Controlling the porosity of SiO_2 may result in an enhanced interaction between the active phase and the support, which generates more stable catalysts.[110–112]
CeO_2	The oxygen exchange and storage abilities of this oxide are the most relevant contribution to the enhancement of CO_2 conversion.[109,113] Both exchange and storage capacities may be enhanced in this oxide through the addition of promoters that mainly increase the amount of oxygen vacancies within the structure of the oxide.[113–115] The combination of ceria with zirconia as a mixed oxide is often used due to the enhancement of the oxygen exchange abilities and thermal stability promoted by the synergy between these two systems.[100,114,116] The interphase between the active phase and ceria influences the activation of the reactants during the reaction.[116]
TiO_2	The presence of oxygen vacancies in TiO_2 strongly influences the interaction of TiO_2 with the active phase and adsorption.[117] The phase of the TiO_2 also determines the performance of catalysts based on this oxide.[100] In this sense, controlling the morphology of TiO_2 may result in a preferential orientation of the crystallographic planes, which promotes a more suitable interaction with the active phase, aiming to increase the catalytic activity.[118]
ZrO_2	The main interest in the use of this oxide in CO_2 methanation is due to its acidic/basic properties, as well as its CO/CO_2 adsorption capabilities.[100,119,120] The phase of the ZrO_2 may influence the performance of the catalysts prepared with this oxide as a support.[100,119] The performance of ZrO_2 as a support of CO_2 methanation catalysts may be improved through the inclusion of promoters such as ceria.[100]

in the analysis of Sabatier reaction mechanisms are focused on identifying the active intermediate, such as formate species on the catalyst surfaces, or even including in some cases CO combined with other intermediates.[97]

The findings are still controversial, not only due to the different possible intermediates, but also due to the influence of the nature of the catalyst components. This includes, of course, the active phase and the support, which play a critical role in the adsorption process. In this sense, Navarro *et al.*[95] studied Ru/Al_2O_3 catalysts obtained by different synthesis procedures, and they demonstrated that the remains of Na and B in one of the cases blocked the hydroxyl groups in the alumina. This forced the adsorption of CO_2 on the active phase and the subsequent production of formyl (HCO*) intermediates, which were hydrogenated until the generation of methane. However, in Ru/Al_2O_3 without the blocking of hydroxyl groups in the alumina by Na or B, the reaction proceeded *via* formate intermediates.

Nevertheless, beyond understanding the reaction pathway based on a detailed study of different catalytic systems at the laboratory scale, the new information must be rapidly transferred to scalable procedures aimed at generating economic profit. The relevant recently launched initiatives demonstrate the interest in promoting PtG technology, such as the European Gas Platform, which explores the viability of PtG projects in European countries, where Germany is clearly a front runner.[121] There are some remarkable PtG demonstration plants worldwide, such as the pioneer pilot plant built in the Tohoku Institute of Technology (Japan, 2003),[122] which is able to produce 4 $Nm^3\ h^{-1}$ of H_2 and 1 $Nm^3\ h^{-1}$ of CH_4. Another example is the Audi e-gas plant (Germany), which is the biggest PtG plant worldwide.[122]

12.3 Alternative Technologies: CO_2 to Value-added Chemicals

In addition to the well-known thermocatalytic approaches for CO_2 conversion into value-added chemicals, alternative processes that aim to work under milder conditions have been considered, attaining different levels of technological development. Although, for instance, all "photo" approaches are normally grouped into the artificial photosynthesis concept, which aims to mimic natural photosynthesis, a deep relationship between electrochemical, photoelectrochemical and photocatalytic processes can be found, and the difference practically lies in the integration degree, as shown in Figure 12.8. In the photoelectrochemical and photocatalytic approaches, a photoactive material, normally a metal oxide, absorbs sunlight, generating hole–electron pairs that directly oxidize and reduce some species on the surface. The sites at which these redox reactions take place depend on the system used: on the same particle for the photocatalytic approach, and on different electrodes at two half-cells for the (photo)electrochemical approach. However, while direct absorption and conversion of solar energy is an interesting concept, the intrinsic low efficiency of the light absorbers used under these architectures limits their practical application. As alternatives, the development of electrochemical systems with, for instance, coupled or even integrated photovoltaics seems to be more reliable after considering the individual efficiencies of solar cells and electrolysers. For this reason, and because today only the

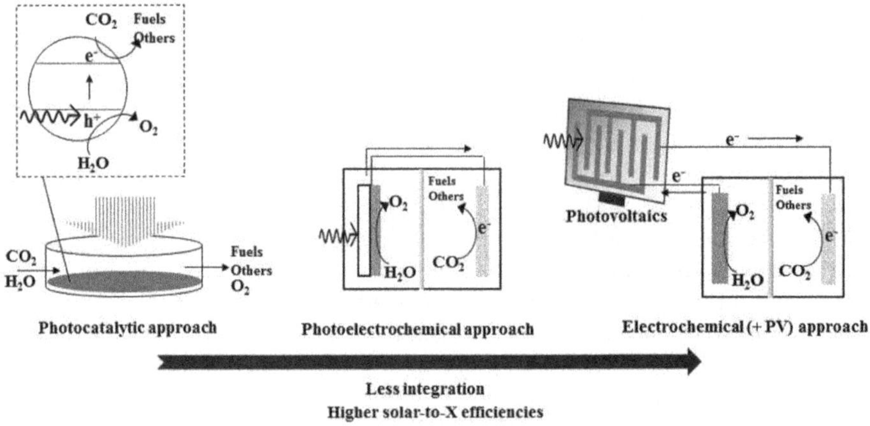

Figure 12.8 Scheme showing the working principle of photocatalytic, photoelectrochemical and "photovoltaics + electrolysis" CO_2 valorization processes, from the more integrated on the left side to the less integrated on the right.

electrochemical approach can be considered as an alternative to thermocatalysis, this is the only technology that will be analysed in this chapter.

12.3.1 Electrocatalytic Conversion of CO_2

With the deployment of renewable energy sources (*i.e.* solar and wind energy) as alternatives to the traditional fossil fuel technologies, and because of their intrinsic intermittency, the development of large-scale energy storage approaches has become mandatory. A promising alternative is the use of the energy surplus by generating energy carriers that allow power release when needed. In this sense, CO_2 and H_2O are ideal sources for this purpose in the so-called power-to-X concept. This concept, which aims at the generation of fuels or value-added products from renewable power sources, can be addressed through several pathways, including electrocatalysis and thermal catalysis approaches.[123] The generation of hydrogen through water electrolysis is probably the most paradigmatic example of a PtG process, in which the excess of renewable electricity can be used for producing H_2 from water splitting in an electrolyser. Further coupling to a thermal hydrogenation unit with a thermal catalytic reactor and a CO_2 stream is a promising alternative for obtaining value-added products and fuels. Certainly, this approach offers the advantage of using the commercial alkaline or Polymer Electrolyte Membrane (PEM) electrolysers and thermal catalysis processes described above (*e.g.* Sabatier reaction, Fischer–Tropsch), reaching a high technological level. However, the electrochemical CO_2 reduction reaction (CO_2RR) has emerged as alternative for directly converting CO_2 and H_2O into fuels or value-added chemicals and, although still at the laboratory scale, it offers some advantages *versus* the thermocatalytic approach: (1) the reaction can be carried out

Table 12.2 Standard redox potentials for different electrolysis reactions at pH 7.

Reaction	E (V *versus* SHE)
$2H^+ + 2e^- \rightarrow H_2$	−0.41
$2H_2O \rightarrow O_2 + 4H^+ + 4e^-$	+0.82
$CO_2 + e^- \rightarrow {}^*COO^-$	−1.90
$CO_2 + 2H^+ + 2e^- \rightarrow HCOOH$	−0.61
$CO_2 + 2H^+ + 2e^- \rightarrow CO + H_2O$	−0.53
$CO_2 + 4H^+ + 4e^- \rightarrow HCHO + H_2O$	−0.48
$CO_2 + 6H^+ + 6e^- \rightarrow CH_3OH + H_2O$	−0.38
$CO_2 + 8H^+ + 8e^- \rightarrow CH_4 + 2H_2O$	−0.24
$12CO_2 + 12H^+ + 12e^- \rightarrow C_2H_4 + 4H_2O$	+0.06

under mild conditions, even at ambient temperature; (2) the reaction rate can be controlled by tuning the applied bias; and (3) the oxidation and reduction products are obtained separately on different electrodes and compartments, minimizing the cost of further separation processes.[124]

Despite these inherent advantages, the CO_2RR is a thermodynamically uphill reaction requiring high external bias in order to drive the overall reaction, especially when considering the one-electron reaction from CO_2 to the CO_2^- radical (−1.9 V *versus* the Standard Hydrogen Electrode (SHE)),[124] as shown in Table 12.2. As an alternative, the proton (H^+)-assisted reactions are thermodynamically more favourable, varying in a range of around −0.2 to −0.6 V *versus* SHE. However, the small potential differences among the possible redox reactions not only imply lower selectivity for specific products, but also the competitive hydrogen evolution reaction (HER) occurring at a lower thermodynamic potential, making this process challenging, especially when working in aqueous media. These aspects demonstrate the importance of controlling the reaction conditions and designing proper electrocatalysts for specific CO_2RR products.

12.3.2 Electrolysers for CO_2 Reduction

In general terms, the electrochemical CO_2RR is based on reaction systems consisting of two compartments, with the anode and cathode separated by an ionic membrane. In the standard configuration, reactors for electrochemical CO_2 reduction are based on the oxygen evolution reaction (OER) on the anode side and CO_2 reduction on the cathode side, as shown in Figure 12.9a. Although most of the research in this area has focused on the development of the electrocatalytic material, the reactor configuration has also been demonstrated to have a significant influence on the efficiencies attained, and some intrinsic aspects, such as CO_2 diffusion, can determine the performance. In fact, although many studies have been done in traditional H-cell reactors with dissolved CO_2 in the electrolyte, the intrinsic low current density limitations of such systems have drawn focus to systems inspired by water electrolysers. Three types of configurations can be described: liquid-phase, gas-phase and solid oxide electrolysers.[125] Among these, and because of their more extensive use, only the first two will be described in this chapter.

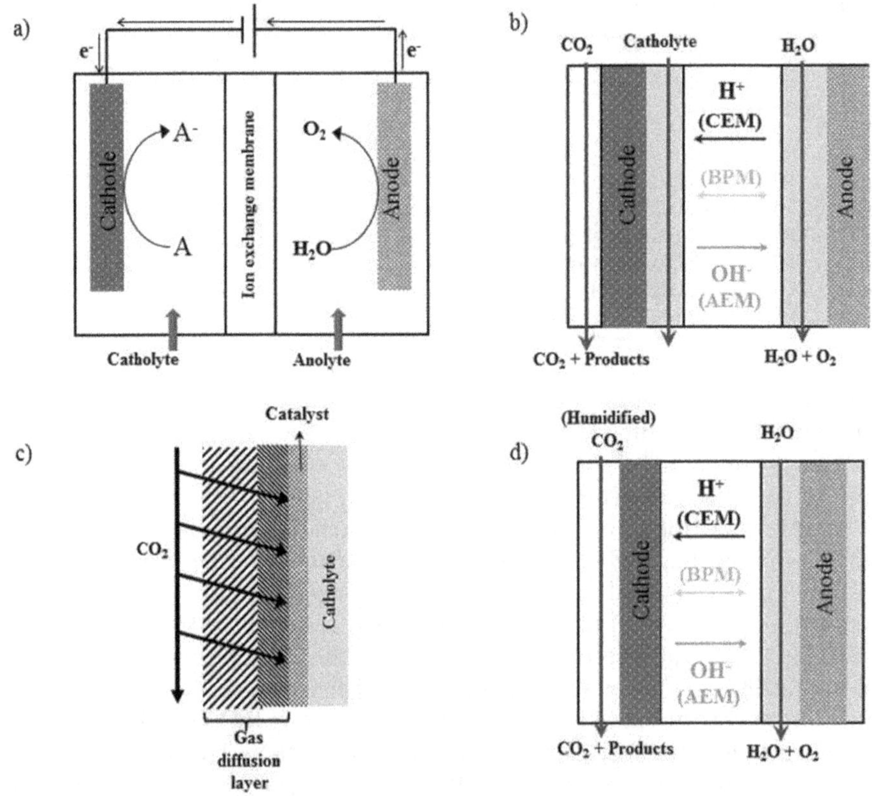

Figure 12.9 General description of an electrolysis system (a), liquid-phase electrolyser (b), gas-diffusion electrode (c) and gas-phase electrolyser for CO_2RR (d). The schemes show the three possible types of membranes: cation exchange membranes (CEMs), anion exchange membranes (AEMs) and bipolar membranes (BPMs).

Liquid-phase electrolysers represent the most commonly used flow-cell configuration, directly resembling alkaline electrolysers. These systems consist of three flow channels: two for the catholyte and anolyte, separated by an ion exchange membrane, and one for the CO_2 gas, as shown in Figure 12.9b. On the cathode side, gas diffusion electrodes (GDEs) are frequently used in order to improve the contact at the triple-phase boundary between the CO_2, electrolyte and catalyst. Therefore, a catalyst layer is deposited on the side facing the catholyte, as represented in Figure 12.9c. Ion exchange membranes are used for avoiding cross-over of products between the anode and cathode side, which could in turn reoxidize CO_2RR products on the anode or favour competitive reactions, such as the oxygen reduction reaction. The selection of the membrane type depends on specific parameters such as working pH and targeted product, and this is particularly necessary when aiming to obtain liquid products. Although high current densities and faradaic efficiencies (FEs) have been achieved for CO, formate and multicarbon products, liquid electrolysers might have two important drawbacks: (1) flooding issues caused by penetration of the liquid electrolyte through the GDE; and (2) possible formation of bicarbonates/carbonates and salt precipitation when passing CO_2 into alkaline electrolytes.

Gas-phase electrolysers are similar to PEM electrolysers. In this configuration, the catalyst is directly deposited onto the membrane on the cathode side and the CO_2 is humidified before entering the reactor, as seen in Figure 12.9d. In this way, greater compression can be achieved, decreasing ohmic losses, while the intrinsic disadvantages of having a liquid electrolyte are overcome. Thanks to this, some systems showing higher FEs and working at higher current densities have been developed. However, additional issues such as acidification of the cathode side and higher HER generation during long-term operation have also been found, making proton transport an important parameter to control. A simple way to avoid these limitations is working at high pH conditions on the cathode side, which is not possible in liquid-phase electrolysers, as CO_2 could in turn form carbonates, which is perfectly acceptable in gas-phase configurations.[126]

Certainly, this reaction system has attracted attention, and several strategies have been proposed for improving not only selectivity and efficiency, but also particularly stability. There is no doubt that further development of the individual components, from type of membrane (*i.e.* cationic, anionic, bipolar) to GDEs, is also necessary, while a parallel effort is devoted to the electrocatalytic system and to understanding the intrinsic reaction mechanisms.

12.3.3 Electrocatalysts for CO_2RR

From the beginning, mostly bulk transition metals have been used as electrocatalysts for CO_2 reduction, and the intrinsic selectivity of specific CO_2RR products has been strongly related to the CO binding energy on the metal.[127] Based on this, several subgroups can be identified for different products, as shown in Table 12.3.

Table 12.3 Classification of the most used electrocatalytic materials for CO_2RR according to the specific product.

Product	Metallic catalysts	Non-metallic catalysts	Hybrid catalysts
CO	Noble metals: Au, Ag, Pd. Alloys: Zn	Heteroatom (*e.g.* S, N, B)-doped metal-free carbon structures	—
Formate ion	p-block metals: Sn, In, Pb, Ti, Hg, Cd, Bi. Others: Pd	Boron-doped diamond, SnO_2	—
Multiple-electron transfer products (*e.g.* C_2H_4, CH_3OH, CH_4)	Cu	Heteroatom (*e.g.* S, N, B)-doped metal-free carbon structures	MOF-derived Cu clusters

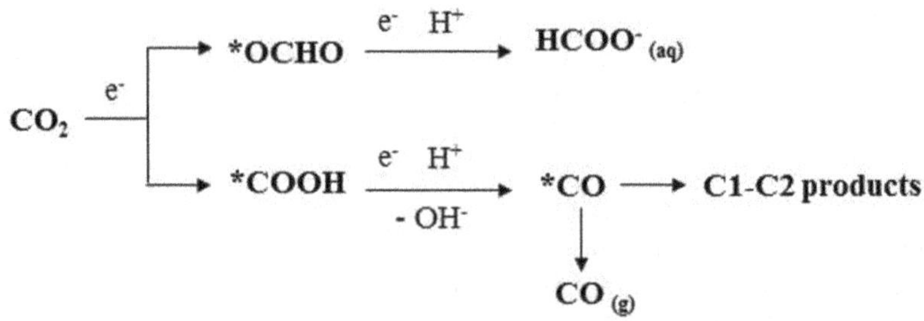

Figure 12.10 General reaction pathways for electrochemical CO_2RR in aqueous electrolyte.

In terms of reaction mechanism, electrochemical CO_2RR is a multistep reaction with multiple electron/proton transfer steps. The one-electron transfer to CO_2 to form $*CO_2^-$ is the initial step (often considered the rate-determining step). The interaction between this adsorbed intermediate and the metal electrocatalyst surface determines further product evolution,[128] as shown in Figure 12.10. If the intermediate is bound *via* an oxygen atom (*OCHO), it might later be protonated/reduced at the C-position and lead to the formation of formate ion. By contrast, if the intermediate binds to the metal *via* the carbon atom (*COOH), protonation/reduction occurs at the O-position, forming the *CO intermediate.[126] Later on, depending on the *CO adsorption strength, CO can be desorbed as product, or further proton/reduction and/or C–C coupling reactions can lead to the formation of alternative C_1–C_2 and even longer-chain products. While noble metals such as Ag and Au show poor *CO adsorption, Cu is a prominent metal for other C_1 and longer-chain products, as the *CO intermediate is mildly enough adsorbed and might undergo further reactions.

Although a certain number of products have been derived from electrochemical CO_2RR, from two-electron transfer products such as CO and formate ($HCOO^-$) to multiple-electron products such as CH_4, CH_3OH, C_2H_4 and even C_3H_8O, both syngas and formate have been the most developed products, not only because of the intrinsic lower energy requirements favouring their formation, but also because of economic aspects. In fact, some techno-economic analyses carried out by considering several electrochemical CO_2RR products have shown that CO and formate are the most competitive targets,[125] and that some longer-chain products, such as ethanol, ethylene and propanol, might also become attractive because of their higher market prices compared to methane and methanol. Based on this, as previously mentioned, this chapter presents some of the latest advances in CO, formate and C_2H_4 generation from electrochemical CO_2RR.

12.3.3.1 CO Generation

Without doubt, CO is the most technologically developed product from electrochemical CO_2RR in terms of target indicators such as selectivity, current density and overpotentials. In fact, most of the medium-scale demonstrators based on this technology reaching a pre-commercial state aim to obtain syngas from CO_2 and H_2O for further coupling to alternative upgrading processes such as FTS or fermentation. For example, Siemens and Evonik have recently developed a (photo)electrolyser for syngas generation coupled to a fermentation unit for obtaining butanol and hexanol as final products.[129,130] This system has operated with pure CO_2 on the cathode side (in a gas-phase configuration) and used an Ag-based GDE. In the end, the electrolysis unit was able to work at 300 mA cm^{-2}, reaching a FE to CO of around 70% (30% to H_2) with stable operation for up to 1200 h.

Other examples of ongoing European projects have been summarized by Gutiérrez *et al.*,[130] among which at least three are aiming to carry out syngas electrochemical generation coupled to non-electrochemical (most of them thermocatalytic) generation of value-added products.

Concerning more fundamental and lower-scale systems, and as previously stated, noble metals such as Au, Ag and Pd are the most studied metal electrocatalysts for CO generation.[128] In general terms, in recent years, there has been a shift from bulk and polycrystalline metals to nanostructured materials, such as nanoporous surfaces, nanoparticles and nanowires. The so-called oxide-derived catalysts have also shown tremendous improvement in terms of overpotential *versus* polycrystalline metals, associated with increased stabilization of the $*CO_2^-$ intermediate.[131] Some examples can also be found by using Zn-based materials as non-expensive alternatives to noble metals, although these show lower activity and thus lower FEs to CO. Cu alloys with transition metals such as Zn or Sn have shown promising performances: in the former case, for instance, electrodeposited Zn on Cu foam has led to an FE to CO of around 85% with a −0.8 V *versus* RHE;[132] in the latter case, CO or formate generation has shown thickness dependency on Cu/SnO_2 core–shells, reaching an FE of around 93% to CO at −0.7 V *versus* the Reversible Hydrogen Electrode (RHE).[133]

12.3.3.2 Formate Generation

Similar to CO, formate generation from electrochemical CO_2RR has achieved some targeted performance values in order to be economically viable,[125] and several examples of large-scale demonstrators have been fabricated in the last decade.[130] The ECFORM demonstrator was manufactured by Det Norske Veritas in 2011, aiming to convert 1 kg of CO_2 per day into formate/formic acid with 600 cm^2 electrodes based on Sn-covered carbon paper.[134] Another example is the system developed by Dioxide Materials, Inc., who have developed a three-compartment electrochemical cell for formic acid generation with GDE, combining Sn nanoparticles and imidazole-based ionomers on the cathode side. This demonstrator has operated at around 140–200 mA cm^{-2} and 3.5–4.0 V, reaching an FE to formic acid of over 90%.[135] Other examples of current projects based on formate/formic acid generation have been collected by Gutiérrez *et al.*[130]

As seen in Figure 12.10, the electrochemical CO_2RR to formate requires the *CO_2^- intermediate to be bonded on the O-site before protonation/reduction.[128] For this reason, the most effective metals for formate generation are those weakly binding most CO_2RR intermediates and *H, such as p-block ones: Sn, Hg, Pb and Bi.[126] Among these, Sn and Bi are the most extensively used metals as electrocatalysts for this reaction, with the former being the most studied one, either in metal or in derived metal oxide form. In addition to these metals, Pd nanoparticles are also selective for formate production with bicarbonate electrolyte. The mechanism in this case is proposed to be related to the initial formation of PdH_x on the electrocatalyst surface, which later reduces CO_2 to *OCHO.[128] Pd–Pt alloying with a low Pt : Pd ratio (30 : 70) has been proposed to further improve the performance of such systems.[136]

12.3.3.3 Ethylene Generation

Ethylene is an important chemical material obtained from the petrochemical industry, and it is considered to be a building block of the chemical industry,[137] especially for the fabrication of plastics. In addition to this, the industrial production of ethylene is a highly energy-demanding process. For this reason, it is important to find alternative production routes, such as electrochemical CO_2RR.

As previously indicated, Cu is the only metal that is able to catalyse electrochemical CO_2RR towards C_2 and longer-chain products. However, there is a lack of understanding about the underlying mechanisms, and a deep correlation between the exposed surface and the product distribution has been observed. In early experiments, the Cu (100) facet showed, for instance, higher selectivity towards C_2H_4, in comparison to the Cu (111) facet, which leads to preferential formation of CH_4.[126] This effect was initially associated with the lower dimerization energy on the (100) terraces. However, further studies trying to emulate the cubic motif through different preparation and pre-treatment methods, such as O_2 plasma, have shown the formation of Cu^+ species, which are later reduced under operation conditions. Some authors have attributed the higher C_2H_4 selectivity to remaining Cu^+ species rather than to structural aspects on the surface.[138] At the same time, the local pH increase related to the higher electrochemical surface area and to the major consumption of protons has been demonstrated to increase the FE to C_2H_4. Other authors suggest that the initial conditions of the Cu electrocatalysts are not the determining aspect, rather than the applied voltage window during operation.[139] Thus, some of the highest FE values reported are around 65–70%, which are lower than those commonly reached for CO and formate.

New approaches using metal–organic frameworks (MOFs) are also becoming attractive for the generation of multicarbon products. MOF-regulated Cu dimers, for instance, have served to stabilize the Cu active sites, and FE values of around 45% for C_2H_4 have been reported.[128]

12.3.4 Electrochemical CO_2RR: Future Perspectives

There has been significant progress in recent years in terms of FE, current density and stability as key performance indicators for electrocatalytic reduction of CO_2, particularly to C_1 products such as CO and formate, making this process more economically compelling. Moreover, with the significant decrease in the cost of renewable energy (wind and solar energy), the development of processes integrating renewable sources has become even more attractive. Despite this, significant effort should be devoted not only to the individual components (electrocatalysts, membranes, reactor configurations), but also to integrated devices, especially for large-scale demonstrators.

In parallel with the research on the different components and electrolyser configurations, a few examples of coupled and/or integrated systems with photovoltaics for CO_2 electrochemical reduction without additional voltage bias have been reported, demonstrating solar-to-fuel efficiencies in the range of 2–5%.[132,140,141] However, a record value of 13.4% has also been obtained by using higher-efficiency solar cells and a selective bifunctional electrocatalyst based on SnO_2–CuO nanowires.[142] These examples, in turn, demonstrate the alternative possibility of developing integrated or directly coupled devices without losing efficiency from the electrolyser point of view. This relatively new field aims to design photoelectrodes with specific configurations in terms of photovoltaics and electrocatalysts. In a similar manner, the development of new configurations with alternative anodic reactions to OER, such as chloride and organic compound oxidation, constitutes an attractive alternative for decreasing the overall voltage requirements and/or obtaining value-added compounds in both compartments.[130] These are some examples of the additional possibilities relating to electrochemical CO_2RR.

12.4 Conclusion

The continuous growth of energy consumption worldwide has generated a scenario of resource depletion and global warming whose effects must be minimized. Hence, it is very important to reduce the emission of greenhouse gases and especially the emission of CO_2. The international agreements that have laid the foundations for addressing this global problem propose comprehensive strategies such as the transition to sustainable models of energy consumption and resource use. These are based on principles such as those of the circular economy, which seeks to maximize the use of resources by minimizing waste production and promoting recycling. From this point of view, CO_2 can be seen as a source of renewable carbon instead of a residue that is only emitted into the atmosphere, and this has generated great interest in all kinds of processes that allow for the use of this compound. Therefore, many processes are being proposed based on the principle that CO_2 is no longer a waste, but a C_1 building block. Thermocatalytic and electrocatalytic routes have been demonstrated to be able to generate fuels such as methanol, methane, olefins and DME. Similarly, important bulk chemicals such as formic acid and ethylene have been proven to be efficiently generated form CO_2. The catalytic conversion of CO_2 towards CO through the RWGS reaction and the following transformation to methanol/DME or hydrocarbons *via* FTS and the subsequent production of gasoline or of diesel are well stablished; however, there is still need for further development due to the change of feed composition that comes from starting from CO_2 instead of syngas. These challenges are not simple, but they ensure that the processes are really friendlier to the environment and generate economic benefits. We must bear in mind that the transition to a more sustainable model does not have to impede economic growth.

The future of the valorization of CO_2 relies on some important aspects that require further advances: (1) the development of the renewable and efficient production of hydrogen using clean energy sources such as solar or wind power; (2) the development of catalytic applications able to perform complex processes through a selective pathway aiming to maximize the production of the desired products while side reactions are avoided; and (3) the scaling of catalytic applications to profitable production levels that allow high thermal efficiency in endo- or exo-thermic processes. These applications must also have versatile designs that can be coupled to traditional production systems in order to make their transition to the use of CO_2 as a raw material viable.

References

1. IEA, *World Energy Outlook 2018*, 2018, IEA, Paris, https://www.iea.org/reports/world-energy-outlook-2018.
2. M. J. Economides and D. A. Wood, *J. Nat. Gas Sci. Eng.*, 2009, **1**, 1–13.
3. M. Grasso, *Energy Res. Soc. Sci.*, 2019, **50**, 106–115.
4. G. A. Olah, G. K. S. Prakash and A. Goeppert, *J. Am. Chem. Soc.*, 2011, **133**, 12881–12898.
5. M. Liu, Y. Yi, L. Wang, H. Guo and A. Bogaerts, *Catalysts*, 2019, **9**.
6. T. A. Boden, G. Marland and R. J. Andres, *Global, Regional, and National Fossil-Fuel CO_2 Emissions*, Oak Ridge National Laboratory, U.S. Department of Energy, Oak Ridge, 2010.
7. C. Le Quéré, R. M. Andrew, P. Friedlingstein, S. Sitch, J. Pongratz, A. C. Manning, J. I. Korsbakken, G. P. Peters, J. G. Canadell, R. B. Jackson, T. A. Boden, P. P. Tans, O. D. Andrews, V. K. Arora, D. C. E. Bakker, L. Barbero, M. Becker, R. A. Betts, L. Bopp, F. Chevallier, L. P. Chini, P. Ciais, C. E. Cosca, J. Cross, K. Currie, T. Gasser, I. Harris, J. Hauck, V. Haverd, R. A. Houghton, C. W. Hunt, G. Hurtt, T. Ilyina, A. K. Jain, E. Kato, M. Kautz, R. F. Keeling, K. Klein Goldewijk, A. Körtzinger, P. Landschützer, N. Lefèvre, A. Lenton, S. Lienert, I. Lima, D. Lombardozzi, N. Metzl, F. Millero, P. M. S. Monteiro, D. R. Munro, J. E. M. S. Nabel, S. I. Nakaoka, Y. Nojiri, X. A. Padin, A. Peregon, B. Pfeil, D. Pierrot, B. Poulter, G. Rehder, J. Reimer, C. Rödenbeck, J. Schwinger, R. Séférian, I. Skjelvan, B. D. Stocker, H. Tian, B. Tilbrook, F. N. Tubiello, I. T. van der Laan-Luijkx, G. R. van der Werf, S. van Heuven, N. Viovy, N. Vuichard, A. P. Walker, A. J. Watson, A. J. Wiltshire, S. Zaehle and D. Zhu, *Earth Syst. Sci. Data*, 2018, **10**, 405–448.

8. J. Wu and X.-D. Zhou, *Chinese J. Catal.*, 2016, **37**, 999–1015.
9. A. Agarwal, E. Rode, N. Sridhar and D. Hill, 2015, in *Handbook of Climate Change Mitigation and Adaptation*, ed. W. Y. Chen, T. Suzuki and M. Lackner, Springer, New York, pp. 1–40.
10. A. A. Olajire, *J. CO2 Util.*, 2013, **3–4**, 74–92.
11. National Academies of Sciences and Medicine, *Gaseous Carbon Waste Streams Utilization: Status and Research Needs*, National Academies Press, Washington, DC, 2019.
12. M. Aresta, A. Dibenedetto and A. Angelini, *Chem. Rev.*, 2014, **114**, 1709–1742.
13. A. Álvarez, M. Borges, J. J. Corral-Pérez, J. G. Olcina, L. Hu, D. Cornu, R. Huang, D. Stoian and A. Urakawa, *ChemPhysChem*, 2017, **18**, 3135–3141.
14. E. A. Quadrelli, G. Centi, J.-L. Duplan and S. Perathoner, *ChemSusChem*, 2011, **4**, 1194–1215.
15. S. Nakamura, M. Hatakeyama, Y. Wang, K. Ogata and K. Fujii, in *Advances in CO_2 Capture, Sequestration, and Conversion*, American Chemical Society, Washington, DC, 2015, vol. 1194, ch. 5, pp. 123–134.
16. H. Metiu, S. Chrétien, Z. Hu, B. Li and X. Sun, *J. Phys. Chem. C*, 2012, **116**, 10439–10450.
17. J. Artz, T. E. Müller, K. Thenert, J. Kleinekorte, R. Meys, A. Sternberg, A. Bardow and W. Leitner, *Chem. Rev.*, 2018, **118**, 434–504.
18. R. M. Cuéllar-Franca and A. Azapagic, *J. CO2 Util.*, 2015, **9**, 82–102.
19. J. Klankermayer and W. Leitner, *Phil. Trans. R. Soc. A*, 2016, **374**, 20150315.
20. T. Bruhn, H. Naims and B. Olfe-Kräutlein, *Environ. Sci. Policy*, 2016, **60**, 38–43.
21. G. Centi and S. Perathoner, *Catal. Today*, 2009, **148**, 191–205.
22. A. Sternberg and A. Bardow, *ACS Sust. Chem. Eng.*, 2016, **4**, 4156–4165.
23. A. Sternberg, C. M. Jens and A. Bardow, *Green Chem.*, 2017, **19**, 2244–2259.
24. G. Centi, E. A. Quadrelli and S. Perathoner, *Energy Environ. Sci.*, 2013, **6**, 1711–1731.
25. R. Kajaste, M. Hurme and P. Oinas, *AIMS Energy*, 2018, **6**, 1074–1102.
26. A. Álvarez, A. Bansode, A. Urakawa, A. V. Bavykina, T. A. Wezendonk, M. Makkee, J. Gascon and F. Kapteijn, *Chem. Rev.*, 2017, **117**, 9804–9838.
27. G. A. Olah, A. Goeppert and G. K. S. Prakash, *Beyond Oil and Gas: The Methanol Economy*, Wiley-VCH, Weinheim, 2006.
28. I. Markit, *Methanol – Chemical Economics Handbook*, Methanol Report, IHS Markit, London, 2017.
29. A. Behr, *Angew. Chem., Int. Ed.*, 2014, **53**, 12674–12674.
30. Z. Jiang, T. Xiao, V. Kuznetsov and P. Edwards, *Phil. Trans. R. Soc. A*, 2010, 368, 3343–3364.
31. O.-S. Joo, K.-D. Jung, I. Moon, A. Y. Rozovskii, G. I. Lin, S.-H. Han and S.-J. Uhm, *Ind. Eng. Chem. Res.*, 1999, **38**, 1808–1812.
32. B. Tidona, C. Koppold, A. Bansode, A. Urakawa and P. Rudolf von Rohr, *J. Supercrit. Fluids*, 2013, **78**, 70–77.
33. B. Tidona, A. Urakawa and P. Rudolf von Rohr, *Chem. Eng. Process. Process Intensif.*, 2013, **65**, 53–57.

34. R. Gaikwad, A. Bansode and A. Urakawa, *J. Catal.*, 2016, **343**, 127–132.
35. H. Yang, C. Zhang, P. Gao, H. Wang, X. Li, L. Zhong, W. Wei and Y. Sun, *Catal. Sci. Technol.*, 2017, 7, 4580–4598.
36. S. Li, Y. Wang, B. Yang and L. Guo, *Appl. Catal., A*, 2019, **571**, 51–60.
37. M. Behrens and R. Schlögl, *Z. Anorg. Allg. Chem.*, 2013, **639**, 2683–2695.
38. S. Kattel, P. J. Ramírez, J. G. Chen, J. A. Rodriguez and P. Liu, *Science*, 2017, **355**, 1296.
39. T. Lunkenbein, J. Schumann, M. Behrens, R. Schlögl and M. G. Willinger, *Angew. Chem., Int. Ed.*, 2015, **54**, 4544–4548.
40. S. Kuld, M. Thorhauge, H. Falsig, C. F. Elkjær, S. Helveg, I. Chorkendorff and J. Sehested, *Science*, 2016, **352**, 969.
41. M. Behrens, F. Studt, I. Kasatkin, S. Kühl, M. Hävecker, F. Abild-Pedersen, S. Zander, F. Girgsdies, P. Kurr, B.-L. Kniep, M. Tovar, R. W. Fischer, J. K. Nørskov and R. Schlögl, *Science*, 2012, **336**, 893.
42. S. A. Kondrat, P. J. Smith, P. P. Wells, P. A. Chater, J. H. Carter, D. J. Morgan, E. M. Fiordaliso, J. B. Wagner, T. E. Davies, L. Lu, J. K. Bartley, S. H. Taylor, M. S. Spencer, C. J. Kiely, G. J. Kelly, C. W. Park, M. J. Rosseinsky and G. J. Hutchings, *Nature*, 2016, **531**, 83.
43. C. Baltes, S. Vukojević and F. Schüth, *J. Catal.*, 2008, **258**, 334–344.
44. F. Arena, K. Barbera, G. Italiano, G. Bonura, L. Spadaro and F. Frusteri, *J. Catal.*, 2007, **249**, 185–194.
45. K. Samson, M. Śliwa, R. P. Socha, K. Góra-Marek, D. Mucha, D. Rutkowska-Zbik, J. F. Paul, M. Ruggiero-Mikołajczyk, R. Grabowski and J. Słoczyński, *ACS Catal.*, 2014, **4**, 3730–3741.
46. Y.-W. Suh, S.-H. Moon and H.-K. Rhee, *Catal. Today*, 2000, **63**, 447–452.
47. I. U. Din, M. S. Shaharun, M. A. Alotaibi, A. I. Alharthi and A. Naeem, *J. CO2 Util.*, 2019, **34**, 20–33.
48. R. Raudaskoski, M. V. Niemelä and R. L. Keiski, *Top. Catal.*, 2007, **45**, 57–60.
49. X.-M. Liu, G. Q. Lu, Z.-F. Yan and J. Beltramini, *Ind. Eng. Chem. Res.*, 2003, **42**, 6518–6530.
50. H. Bahruji, M. Bowker, G. Hutchings, N. Dimitratos, P. Wells, E. Gibson, W. Jones, C. Brookes, D. Morgan and G. Lalev, *J. Catal.*, 2016, **343**, 133–146.
51. A. G. Saputro, R. I. D. Putra, A. L. Maulana, M. U. Karami, M. R. Pradana, M. K. Agusta, H. K. Dipojono and H. Kasai, *J. Energy Chem.*, 2019, **35**, 79–87.
52. X. Jiang, Y. Jiao, C. Moran, X. Nie, Y. Gong, X. Guo, K. S. Walton and C. Song, *Catal. Commun.*, 2019, **118**, 10–14.
53. J. Ye, C. Liu, D. Mei and Q. Ge, *ACS Catal.*, 2013, **3**, 1296–1306.
54. O. Martin, A. J. Martín, C. Mondelli, S. Mitchell, T. F. Segawa, R. Hauert, C. Drouilly, D. Curulla-Ferré and J. Pérez-Ramírez, *Angew. Chem., Int. Ed.*, 2016, **55**, 6261–6265.
55. The Methanol Institute, https://www.methanol.org/ (accessed 9 September 2019).
56. C. Hobson, in *Renewable Methanol Report*, ed. C. Márquez, ATA Markets Intelligence S.L, Madrid, 2018, pp. 13–17.

57. Carbon Recycling International, https://www.carbonrecycling.is/.
58. J. A. Martens, A. Bogaerts, N. De Kimpe, P. A. Jacobs, G. B. Marin, K. Rabaey, M. Saeys and S. Verhelst, *ChemSusChem*, 2017, **10**, 1039–1055.
59. C. Arcoumanis, C. Bae, R. Crookes and E. Kinoshita, *Fuel*, 2008, **87**, 1014–1030.
60. R. J. Pearson and J. W. G. Turner, in *Alternative Fuels and Advanced Vehicle Technologies for Improved Environmental Performance*, ed. R. Folkson, Woodhead Publishing, Sawston, 2014, pp. 19–51.
61. L. J. France, P. P. Edwards, V. L. Kuznetsov and H. Almegren, in *Carbon Dioxide Utilisation*, ed. P. Styring, E. A. Quadrelli and K. Armstrong, Elsevier, Amsterdam, 2015, pp. 161–182.
62. Methanol Institute, http://www.methanol.org/wp-content/uploads/2016/06/DME-An-Emerging-Global-Guel-FS.pdf).
63. O. Yotaro, Y. Masahiro, S. Tsutomu, I. Osamu, O. Takashi and I. Norio, *New Direct Synthesis Technology for DME and its Application Technology,* JFE TECHNICAL REPORT No. 8, 2006, pp. 34–40.
64. E. Lam, J. J. Corral-Pérez, K. Larmier, G. Noh, P. Wolf, A. Comas-Vives, A. Urakawa and C. Copéret, *Angew. Chem., Int. Ed.*, 2019, **58**, 13989–13996.
65. S. Ren, W. R. Shoemaker, X. Wang, Z. Shang, N. Klinghoffer, S. Li, M. Yu, X. He, T. A. White and X. Liang, *Fuel*, 2019, **239**, 1125–1133.
66. G. Bonura, M. Migliori, L. Frusteri, C. Cannilla, E. Catizzone, G. Giordano and F. Frusteri, *J. CO2 Util.*, 2018, **24**, 398–406.
67. G. Bonura, C. Cannilla, L. Frusteri and F. Frusteri, *Appl. Catal., A*, 2017, **544**, 21–29.
68. Z. Azizi, M. Rezaeimanesh, T. Tohidian and M. R. Rahimpour, *Chem. Eng. Process. Process Intensif.*, 2014, **82**, 150–172.
69. F. Frusteri, M. Migliori, C. Cannilla, L. Frusteri, E. Catizzone, A. Aloise, G. Giordano and G. Bonura, *J. CO2 Util.*, 2017, **18**, 353–361.
70. A. Atakan, J. Keraudy, P. Mäkie, C. Hulteberg, E. M. Björk and M. Odén, *J. Colloid Interface Sci.*, 2019, **546**, 163–173.
71. E. Catizzone, G. Bonura, M. Migliori, F. Frusteri and G. Giordano, *Molecules*, 2018, **23**, 1–28.
72. I. D. Association, https://www.aboutdme.org/index.asp?sid=97, 2019.
73. T. Sakakura, J.-C. Choi and H. Yasuda, *Chem. Rev.*, 2007, **107**, 2365–2387.
74. X. Zhang, X. Zhu, L. Lin, S. Yao, M. Zhang, X. Liu, X. Wang, Y.-W. Li, C. Shi and D. Ma, *ACS Catal.*, 2017, **7**, 912–918.
75. P. Gao, S. Dang, S. Li, X. Bu, Z. Liu, M. Qiu, C. Yang, H. Wang, L. Zhong, Y. Han, Q. Liu, W. Wei and Y. Sun, *ACS Catal.*, 2018, **8**, 571–578.
76. Z. Ma and M. D. Porosoff, *ACS Catal.*, 2019, **9**, 2639–2656.
77. L. Guo, J. Sun, X. Ji, J. Wei, Z. Wen, R. Yao, H. Xu and Q. Ge, *Commun. Chem.*, 2018, **1**, 11.
78. J. Wei, J. Sun, Z. Y. Wen, C. Y. Fang, Q. J. Ge and H. Y. Xu, *Catal. Sci. Technol.*, 2016, **6**, 4786–4793.
79. C. G. Visconti, M. Martinelli, L. Falbo, A. Infantes-Molina, L. Lietti, P. Forzatti, G. Iaquaniello, E. Palo, B. Picutti and F. Brignoli, *Appl. Catal., B*, 2017, **200**, 530–542.

80. M. Ronda-Lloret, G. Rothenberg and N. R. Shiju, *ChemSusChem*, 2019, **12**, 3896–3914.
81. S. Navarro-Jaen, M. A. Centeno, O. H. Laguna and J. A. Odriozola, *J. Mater. Chem. A*, 2018, **6**, 17001–17010.
82. A. Egana, O. Sanz, D. Merino, X. Moriones and M. Montes, *Ind. Eng. Chem. Res.*, 2018, **57**, 10187–10197.
83. D. Merino, I. Perez-Miqueo, O. Sanz and M. Montes, *Top. Catal.*, 2016, **59**, 207–218.
84. D. Merino, O. Sanz and M. Montes, *Fuel*, 2017, **210**, 49–57.
85. D. Merino, O. Sanz and M. Montes, *Chem. Eng. J.*, 2017, **327**, 1033–1042.
86. G. P. Van der Laan and A. Beenackers, *Catal. Rev. Sci. Eng.*, 1999, **41**, 255–318.
87. B. Z. Yao, W. J. Ma, S. Gonzalez-Cortes, T. C. Xiao and P. P. Edwards, *Greenh. Gases*, 2017, **7**, 942–957.
88. C. D. Chang, W. H. Lang and R. L. Smith, *J. Catal.*, 1979, **56**, 169–173.
89. U. Olsbye, S. Svelle, M. Bjorgen, P. Beato, T. V. W. Janssens, F. Joensen, S. Bordiga and K. P. Lillerud, *Angew. Chem., Int. Ed.*, 2012, **51**, 5810–5831.
90. S. Xu, Y. Zhi, J. Han, W. Zhang, X. Wu, T. Sun, Y. Wei and Z. Liu, *Adv. Catal.*, 2017, **61**, 37–122.
91. D. Mores, E. Stavitski, M. H. F. Kox, J. Kornatowski, U. Olsbye and B. M. Weckhuysen, *Chem. Eur. J.*, 2008, **14**, 11320–11327.
92. Y.-K. Park, K.-C. Park and S.-K. Ihm, *Catal. Today*, 1998, **44**, 165–173.
93. A. Ricca, L. Truda and V. Palma, *Chem. Eng. J.*, 2019, **377**, 120461.
94. S. Navarro-Jaén, J. C. Navarro, L. F. Bobadilla, M. A. Centeno, O. H. Laguna and J. A. Odriozola, *Appl. Surf. Sci.*, 2019, **483**, 750–761.
95. S. Navarro-Jaen, A. Szego, L. F. Bobadilla, O. H. Laguna, F. Romero-Sarria, M. A. Centeno and J. A. Odriozola, *ChemCatChem*, 2019, **11**, 2063–2068.
96. J. C. Navarro, M. A. Centeno, O. H. Laguna and J. A. Odriozola, *Catalysts*, 2018, **8**, 25.
97. K. Jalama, *Catal. Rev.*, 2017, **59**, 95–164.
98. C. Janke, M. S. Duyar, M. Hoskins and R. Farrauto, *Appl. Catal., B*, 2014, **152–153**, 184–191.
99. M. A. A. Aziz, A. A. Jalil, S. Triwahyono and A. Ahmad, *Green Chem.*, 2015, **17**, 2647–2663.
100. P. Frontera, A. Macario, M. Ferraro and P. Antonucci, *Catalysts*, 2017, **7**, 59.
101. G. Du, S. Lim, Y. Yang, C. Wang, L. Pfefferle and G. L. Haller, *J. Catal.*, 2007, **249**, 370–379.
102. M. Jacquemin, A. Beuls and P. Ruiz, *Catal. Today*, 2010, **157**, 462–466.
103. A. Karelovic and P. Ruiz, *Appl. Catal., B*, 2012, **113–114**, 237–249.
104. A. Beuls, C. Swalus, M. Jacquemin, G. Heyen, A. Karelovic and P. Ruiz, *Appl. Catal., B*, 2012, **113–114**, 2–10.
105. K. Fukui, T. Yonezawa and H. Shingu, *J. Chem. Phys.*, 1952, **20**, 722–725.
106. S. Ma, W. Song, B. Liu, H. Zheng, J. Deng, W. Zhong, J. Liu, X.-Q. Gong and Z. Zhao, *Catal. Sci. Technol.*, 2016, **6**, 6128–6136.
107. D. Türks, H. Mena, U. Armbruster and A. Martin, *Catalysts*, 2017, **7**, 152.

108. H. Takano, Y. Kirihata, K. Izumiya, N. Kumagai, H. Habazaki and K. Hashimoto, *Appl. Surf. Sci.*, 2016, **388**, 653–663.
109. H. Liu, X. Zou, X. Wang, X. Lu and W. Ding, *J. Nat. Gas Chem.*, 2012, **21**, 703–707.
110. R. P. Ye, W. Gong, Z. Sun, Q. Sheng, X. Shi, T. Wang, Y. Yao, J. J. Razink, L. Lin, Z. Zhou, H. Adidharma, J. Tang, M. Fan and Y. G. Yao, *Energy*, 2019, **188**, 116059.
111. G. Zhou, H. Liu, Y. Xing, S. Xu, H. Xie and K. Xiong, *J. CO2 Util.*, 2018, **26**, 221–229.
112. T. A. Le, J. K. Kang and E. D. Park, *Top. Catal.*, 2018, **61**, 1537–1544.
113. H. C. Yao and Y. F. Y. Yao, *J. Catal.*, 1984, **86**, 254–265.
114. O. H. Laguna, A. Pérez, M. A. Centeno and J. A. Odriozola, *Appl. Catal., B*, 2015, **176–177**, 385–395.
115. O. H. Laguna, F. Romero Sarria, M. A. Centeno and J. A. Odriozola, *J. Catal.*, 2010, **276**, 360–370.
116. F. Ocampo, B. Louis, L. Kiwi-Minsker and A.-C. Roger, *Appl. Catal., A*, 2011, **392**, 36–44.
117. Z. Qin, X. Wang, L. Dong, T. Su, B. Li, Y. Zhou, Y. Jiang, X. Luo and H. Ji, *Chem. Eng. Sci.*, 2019, **210**, 115245.
118. S. Chai, Y. Men, J. Wang, S. Liu, Q. Song, W. An and G. Kolb, *J. CO2 Util.*, 2019, **33**, 242–252.
119. E.-M. Köck, M. Kogler, T. Bielz, B. Klötzer and S. Penner, *J. Phys. Chem. C*, 2013, **117**, 17666–17673.
120. M. Yamasaki, H. Habazaki, T. Yoshida, E. Akiyama, A. Kawashima, K. Asami, K. Hashimoto, M. Komori and K. Shimamura, *Appl. Catal., A*, 1997, **163**, 187–197.
121. C. Wulf, J. Linßen and P. Zapp, *Energy Procedia*, 2018, **155**, 367–378.
122. M. Bailera, P. Lisbona, L. M. Romeo and S. Espatolero, *Renewable Sustainable Energy Rev.*, 2017, **69**, 292–312.
123. B. Rego de Vasconcelos and J. M. Lavoie, *Front. Chem.*, 2019, 7, 392.
124. Q. Lu and F. Jiao, *Nano Energy*, 2016, **29**, 439–456.
125. M. G. Kibria, J. P. Edwards, C. M. Gabardo, C. T. Dinh, A. Seifitokaldani, D. Sinton and E. H. Sargent, *Adv. Mater.*, 2019, **31**, e1807166.
126. M. B. Ross, P. De Luna, Y. Li, C.-T. Dinh, D. Kim, P. Yang and E. H. Sargent, *Nat. Catal.*, 2019, **2**, 648–658.
127. K. P. Kuhl, T. Hatsukade, E. R. Cave, D. N. Abram, J. Kibsgaard and T. F. Jaramillo, *J Am Chem Soc*, 2014, **136**, 14107–14113.
128. Z. Yin, G. T. R. Palmore and S. Sun, *Trends Chem.*, 2019, **1**, 739–750.
129. T. Haas, R. Krause, R. Weber, M. Demler and G. Schmid, *Nat. Catal.*, 2018, **1**, 32–39.
130. O. G. Sánchez, Y. Y. Birdja, M. Bulut, J. Vaes, T. Breugelmans and D. Pant, *Curr. Opin. Green Sust. Chem.*, 2019, **16**, 47–56.
131. Y. Chen, C. W. Li and M. W. Kanan, *J. Am. Chem. Soc.*, 2012, **134**, 19969–19972.
132. F. Urbain, P. Tang, N. M. Carretero, T. Andreu, L. G. Gerling, C. Voz, J. Arbiol and J. R. Morante, *Energy Environ. Sci.*, 2017, **10**, 2256–2266.

133. Q. Li, J. Fu, W. Zhu, Z. Chen, B. Shen, L. Wu, Z. Xi, T. Wang, G. Lu, J. J. Zhu and S. Sun, *J. Am. Chem. Soc.*, 2017, **139**, 4290–4293.
134. A. Agarwal, Y. Zhai, D. Hill, S. Guan, E. Rode and N. Sridhar, *Clean Technol.*, 2011, 303–306.
135. J. J. Kaczur, H. Yang, Z. Liu, S. D. Sajjad and R. I. Masel, *Front. Chem.*, 2018, **6**, 263.
136. R. Kortlever, I. Peters, S. Koper and M. T. M. Koper, *ACS Catal.*, 2015, **5**, 3916–3923.
137. J.-M. Chen, B. Yu and Y.-M. Wei, *Appl. Energy*, 2018, **224**, 160–174.
138. J. Kim, W. Choi, J. W. Park, C. Kim, M. Kim and H. Song, *J. Am. Chem. Soc.*, 2019, **141**, 6986–6994.
139. D. Ren, J. Fong and B. S. Yeo, *Nat. Commun.*, 2018, **9**, 925.
140. T. N. Huan, D. A. Dalla Corte, S. Lamaison, D. Karapinar, L. Lutz, N. Menguy, M. Foldyna, S. H. Turren-Cruz, A. Hagfeldt, F. Bella, M. Fontecave and V. Mougel, *Proc. Natl. Acad. Sci. U. S. A.*, 2019, **116**, 9735–9740.
141. Gurudayal, J. Bullock, D. F. Srankó, C. M. Towle, Y. Lum, M. Hettick, M. C. Scott, A. Javey and J. Ager, *Energy Environ. Sci.*, 2017, **10**, 2222–2230.
142. M. Schreier, F. Héroguel, L. Steier, S. Ahmad, J. S. Luterbacher, M. T. Mayer, J. Luo and M. Grätzel, *Nat. Energy*, 2017, **2**, 17087.

CHAPTER 13

In Situ *Characterization of Metal/Oxide Catalysts for CO_2 Conversion: From Fundamental Aspects to Real Catalyst Design*

MAUSUMI MAHAPATRA,[a] LUIS E. BETANCOURT,[a] ZONGYUAN LIU,[a] DIMITRIY VOVCHOK,[a] JUAN P. SIMONOVIS,[a,b] JOSÉ A. RODRIGUEZ[a,c]* AND SANJAYA D. SENANAYAKE[a]*

[a]Chemistry Department, Brookhaven National Laboratory, Upton, NY 11973, USA
[b]National Synchrotron Light Source II, Brookhaven National Laboratory, Upton, NY 11973, USA
[c]Chemistry Department, Stony Brook University, Stony Brook, NY 11794, USA
*Corresponding contributor. Email: rodrigez@bnl.gov; ssnanay@bnl.gov

13.1 Introduction

CO_2 is a major greenhouse pollutant with increasing presence in the atmosphere, directly associated with global warming and the impacts of climate change.[1–3] Today, great effort has been devoted to the development of new strategies such as CO_2 capture,[4] sequestration[5] and chemical conversion[6–8]

Heterogeneous Catalysis for Energy Applications
Energy and Environment Series No. 27
Edited by Tomas R. Reina and Jose A. Odriozola

Published by the Royal Society of Chemistry, www.rsc.org

for reducing its concentration in the atmosphere. A large number of researchers have focused their scientific attention on developing methods for efficiently capturing CO_2 directly from the emission source or from ambient air.[4,9] Although the captured CO_2 can be sequestered underground in geological formations, a more sustainable approach is to directly utilize the CO_2 to produce value-added products.[6–8] In this aspect, the chemical conversion of CO_2 to CH_3OH has gained major attention over the last few decades since methanol can be used as a liquid fuel or fuel additive and is a precursor in many industrial organic synthesis processes. The utilization of CO_2 to produce clean fuels such as CH_3OH results in an overall carbon-neutral cycle and is economically attractive.[10,11]

The development of a catalytic process for the large-scale production of methanol with high activity and selectivity from CO_2 hydrogenation to methanol is very challenging due the molecule's high stability and strong bonds (bond energy of C=O in CO_2 = 127 kcal mol^{-1}). Traditional methanol synthesis catalysts are heterogeneous and requires high temperatures and pressures.[12–14] In addition, the product selectivity is highly dependent on the nature of the catalyst and its active sites.[15–19] Typically, the Cu-based catalysts supported on oxides show high selectivity for yielding methanol from CO_2 and H_2.[15–17] On the other hand, Pt-, Au- and Ag-based catalysts on an oxide support show high selectivity for CO formation through the reverse water gas shift reaction.[18,19] The discovery of effective and active catalysts for this reaction is ongoing, and the strategy of trying different combinations of metals and oxides has been suggested in order to optimize the catalytic conditions for methanol yield. Traditionally, the metal nanoparticles dispersed on oxide supports (metal/oxide) were heavily used for this reaction.[20–22] However, it has been shown in recent years that the so-called "inverse catalysts", where oxide nanoparticles are dispersed on a metal support (oxide/metal), show superior catalytic performance compared to the traditional metal/oxide configuration.[17,23,24] The inverse oxide/metal catalysts have special structural and electronic properties due to the strong interaction between the oxide and the metal, which is difficult to attain in traditional metal/oxide catalysts. The synergy between the oxide nanoparticles and the metal in the interfacial oxide/metal sites is highly active for CO_2 activation and is considered to be the active site for this reaction.[23,25] We explore several such systems in more detail in this chapter.

The majority of the research in this area of catalyst optimization focuses on the fundamental, atomic-scale understanding of the physical and chemical properties of catalytic materials. Since real catalysts and catalytic processes are extremely complex, the development of techniques for the characterization of catalytic systems as they evolve in time with a changing chemical environment is a high priority. Over the years, the general trend in this field has been the multimodal characterization of the catalytic systems (both powders and models) by using various spectroscopic, diffraction and imaging techniques both *in situ* and under *operando* conditions. The information regarding the electronic properties, chemical compositions and structures obtained by these techniques is crucial to identifying the catalyst active site, reaction mechanism and overall conversion processes.

This chapter will focus on the use of various spectroscopic, diffraction and imaging techniques for characterizing the interaction of CO_2 and its hydrogenation reaction on both model and powder catalytic systems. The reaction that happens on powder catalytic surfaces under ambient conditions is extremely complex. To simplify the complexity and to gain an atomic-scale understanding regarding the catalyst structure and reaction mechanism, it has been common to employ model studies performed on well-defined single-crystal catalyst surfaces, typically under ultrahigh vacuum (UHV) conditions. However, the pressure gap between the powder and UHV studies can have a huge impact on the actual catalytic process, which can be bridged by performing experiments under near-ambient pressure (NAP) conditions, which will also be addressed in this chapter. We will start by showing how scanning tunnelling microscopy (STM) has been useful for characterizing the morphology of well-defined oxide–metal catalysts active for the hydrogenation of CO_2 and for studying the type of overlayer formed upon exposing an oxide to CO_2 at regular pressures. This will be followed by a discussion of studies of ambient-pressure X-ray photoelectron spectroscopy (AP-XPS) and infrared spectroscopy (IR), where surface reactions have been examined while exposing different types of metal–oxide interfaces to CO_2 or CO_2/H_2 mixtures. Finally, we will present studies focused on the *in situ* or *operando* characterization of catalyst structure and composition with X-ray diffraction (XRD) and X-ray absorption fine structure (XAFS). From these works, it is clear that we need to use a multi-technique approach when aiming at a full characterization of a catalyst or its associated catalytic processes.

13.2 Scanning Tunnelling Microscopy

Since the invention of STM by Binning and Rohrer in 1983, it has proved itself as a very powerful imaging technique in modern surface science.[26,27] STM is commonly used for studying the surface structure and local electronic properties of conductive sample surfaces at the atomic scale. By correlating the STM topographic images with other surface science techniques with chemical specificity (*e.g.* X-ray photoelectron spectroscopy [XPS]), important information regarding the clean and adsorbate-covered surface structure can be determined.

This chapter briefly explains the basic concept and applications of STM for the *in situ* characterization of catalytic surfaces. For in-depth knowledge about the theory and instrumentation, readers are referred to the literature.[28–30] According to classical physics, an electron cannot pass through a barrier if its kinetic energy is smaller than the potential within the barrier, but its quantum nature allows it to be simultaneously on both sides of the barrier as a result of the Heisenberg uncertainty principle. STM is based on the concept of quantum tunnelling in which an electron can tunnel through a barrier, which is classically forbidden. In this technique, a sharp metallic tip is brought very close (in the order of few ångströms) to a conductive surface and a small bias voltage is applied to the tip sample junction. The overlap of the electron wave functions permits quantum mechanical tunnelling and a current (*I*) will flow across the gap that is inversely proportional to the distance (*d*) between the tip and the sample, as shown below in eqn (13.1).

$$I \propto e^{-2Kd} \tag{13.1}$$

K is the decay constant whose value is given by $=(2m\Phi/\hbar^2)^{1/2}$, where *m* is the mass of the electron and Φ is the work function of the tip/sample material. For a work function (Φ) of ~5 eV (a typical metal), $K = 1.0$ Å^{-1}, and the current decreases by an order of magnitude when the gap *d* increases by 1 Å. This effect of the exponential dependence between the tunnelling current and the tip–sample distance provides the basis for STM and the cause of its remarkable vertical resolution.

There are two basic modes used in the scanning process, namely constant-current or constant-height modes. In the constant-current mode of operation, first the tip is brought close to the surface such that a measuring tunnelling current (pico-nano ampere) can be detected for an applied bias voltage (commonly used 2 mV to 2 V). To maintain a constant tunnelling current, a feedback loop moves the tip up or down in the *Z* direction while the tip is rastered across the surface and essentially follows the surface topography. The applied bias voltage determines the number of electronic states from which electron can tunnel and the polarity of the bias determines the direction of the flow of electrons. When the tip is positively biased with respect to the sample, electrons tunnel from the occupied states of the sample surface to the unoccupied states of the tip, and with negative bias the electrons tunnel from the occupied states of the tip to the unoccupied states of the sample surface. Alternatively, in the constant-height mode of operation, the tip is scanned across the surface at constant height and voltage while the current is measured. Generally, the constant-current mode is more frequently used for STM data acquisition because, with the constant-height mode of operation, it is possible that, when scanning on an uneven surface, the tip will be damaged as a result of contact between the tip and sample.

The state-of-the-art applications of STM include the *in situ* characterization of model catalytic surfaces and their interactions with adsorbate molecules under UHV as well as NAP conditions. In this section, we will detail some such *in situ* STM studies that focus on the structural characterization of various oxide/metal inverse model catalytic surfaces relevant for CO_2 hydrogenation to methanol synthesis reactions, particularly emphasizing the catalyst "structure property" relationship.[31–35]

Mahapatra *et al.*[32,33] investigated the growth and structure of ZnO/CuO_x/Cu(111) inverse model catalysts by using UHV-STM, and then explored their catalytic activity for CO_2 hydrogenation to methanol reaction. The experimental results suggest that the surface structure of the ZnO/CuO_x/Cu(111)

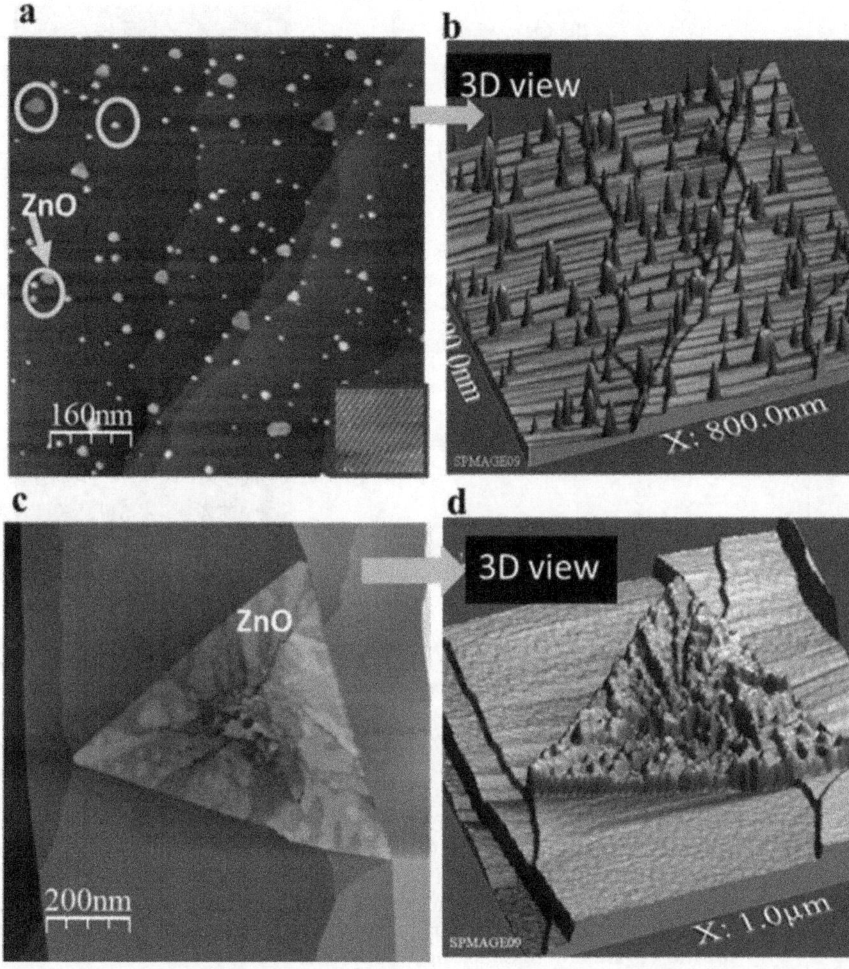

Figure 13.1 STM images $ZnO/CuO_x/Cu(111)$: (a) 0.2 ML Zn was vapour deposited onto Cu(111) at 300 K followed by oxidation at 550 K; the inset shows the structure of $CuO_x/Cu(111)$ substrate. (b) 3D view of (a). (c) 0.2 ML Zn was vapour deposited onto $CuO_x/Cu(111)$ at 550 K in a 5×10^{-7} torr O_2 background. (d) 3D view of (c). (a,b) cover an area of 800 nm × 800 nm; (c,d) 1 000 nm × 1 000 nm $V_t = 1.2$ V, $I_t = 0.1$ nA. Adapted from Ref. 32 with permission from American Chemical Society, Copyright 2018.

system is highly sensitive to the preparation conditions that subsequently affect catalytic performance. Oxidation of a Zn–Cu(111) alloys with O_2 at 550 K produces $ZnO/CuO_x/Cu(111)$ systems where the size of the ZnO islands is in the range of 5–20 nm (Figure 13.1a,b). On the other hand, the $ZnO/CuO_x/Cu(111)$ surfaces prepared by vapour deposition of Zn at 550 K in O_2 background exhibit triangular islands of ZnO that are extremely large (400–500 nm) and contain a minimum of three to four layers of the oxide (Figure 13.1c,d). The catalytic activity of the $ZnO/CuO_x/Cu(111)$ surfaces prepared by the above-mentioned

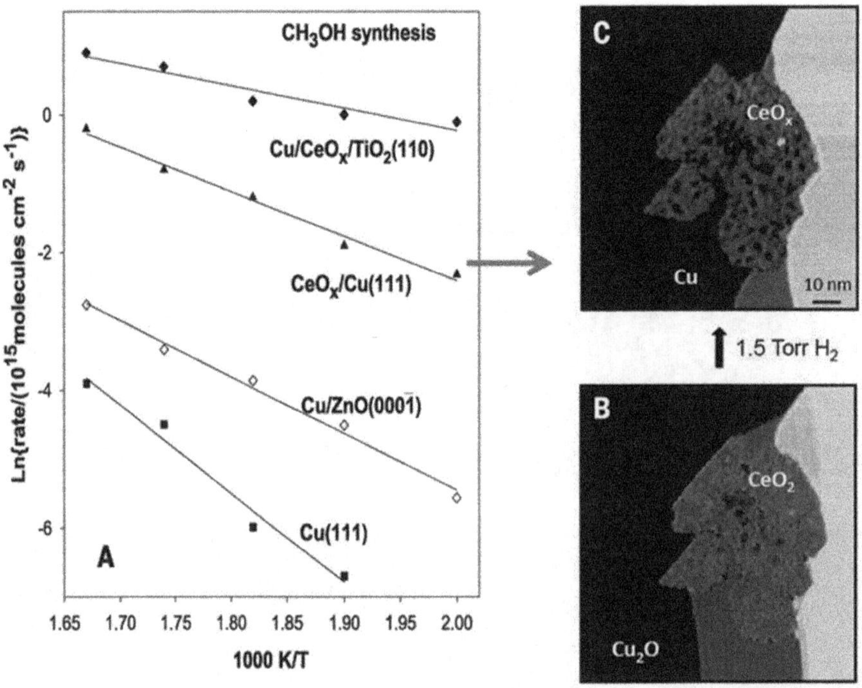

Figure 13.2 Kinetics and STM studies. (a) Arrhenius plot for methanol synthesis on Cu(111), a 0.2 ML of Cu on ZnO(000$\bar{1}$), a Cu(111) surface covered 20% by ceria and a 0.1 ML of Cu on a TiO_2(110) surface 15% pre-covered with ceria. In a batch reactor, the catalysts were exposed to 0.5 atm of CO_2 and 4.5 atm of H_2. The reported values are steady-state rates measured at 600, 575, 550, 525 and 500 K. (b) STM image of a CeO_x/Cu(111) surface as prepared. (c) *In situ* STM image taken during exposure to 1.5 torr of H_2 at 300 K after 26 h of reaction. Scanning parameters: 0.3 nA, 1.0 V. Reproduced from Ref. 16 with permission from American Association for the Advancement of Science, Copyright 2014.

methods was measured in a batch reactor for the CO_2 hydrogenation reaction. The small ZnO islands (Figure 13.1a,b) are not very stable at the high temperatures (550–600 K) used for the CO_2 hydrogenation reaction; above 500 K, the ZnO particles get reduced to metallic Zn and migrate to the Cu bulk. On the other hand, the large ZnO islands (Figure 13.1c,d) have high thermal stability and provide ZnO/Cu interfaces that are efficient as catalysts for the CO_2 hydrogenation reaction at 550–625 K. In a similar context, the authors have also explored the catalytic performance of those ZnO/CuO_x/Cu(111) surfaces for CO oxidation and the water gas shift reaction.[33] Those studies reported that the small ZnO particles are highly active for the CO oxidation reaction at moderate temperatures (400–500 K), whereas the large ZnO islands are suitable for water gas shift reactions at 550–625 K owing to their high stability at higher temperatures.[32,33] These studies provide good examples of the correlation between the catalyst structure, stability and activity and how STM is useful to fine tune the catalyst structure in the rational design of catalytic materials.

With modern advances in experimental techniques, STM can now also be performed under NAP conditions, which is essential for studying the structural evolution of the catalyst surface under reaction conditions.[16,36,37] This has been done by using a unique instrumentation offered by SPECS Aarhus HT-NAP STM, which contains a sealed ambient pressure cell housed in a UHV chamber.[16] This experimental setup allows the pressurization of the ambient pressure cell at up to ~2 torr of reactant gases and the simultaneous monitoring of the surface evolution of the catalyst surface under reaction conditions by *in situ* STM measurement.

Figure 13.2 shows an example of such an *in situ* STM study, where the surface evolution of the $CeO_2/Cu_2O/Cu(111)$ surface, relevant for CO_2 hydrogenation to methanol synthesis, was observed under 1.5 torr H_2.[16] The activity for methanol production was compared for a series of Cu-based catalysts (Figure 13.2a), which demonstrated that $CeO_2/Cu(111)$ shows high turnover for methanol production. Figure 13.2b shows the STM image of a $CeO_2/Cu_2O/Cu(111)$ surface prepared by vapour depositing Ce onto a pre-oxidized Cu(111) surface ($Cu_2O/Cu(111)$) at 600 K in the presence of a 5×10^{-7} torr oxygen background. The STM image shows the formation of large CeO_2 islands on the $Cu_2O/Cu(111)$ step edges. The CeO_2 islands have a rough morphology and are one layer thick. After exposing the surface to 1.5 torr H_2, the complete reduction of Cu_2O to Cu and the formation of clusters of O vacancies in the ceria island were observed. Reduction in H_2 increased the surface roughness of the ceria particles and expanded the ceria–Cu interface. The XPS study suggests that the active phase of the catalyst is the interface between the reduced ceria and Cu, which agrees well with the structural transformation observed in Figure 13.2b,c.

A recent publication by Hamlyn *et al.*[36] reported the structure of CO_2 adsorption on a $TiO_2(110)$ surface by using ambient pressure STM. It is well known that the high stability and non-polar nature of CO_2 leads to weak bonding with well-defined surfaces of metals and oxides. Previous studies show that CO_2 does not adsorb onto the $TiO_2(110)$ surface at 300 K under UHV conditions owing to its low adsorption energy, requiring cooling down of the surface to liquid nitrogen temperatures.[38] In this report, for the first time, Hamlyn *et al.* imaged CO_2 on $TiO_2(110)$ at 300 K under an ambient pressure of CO_2. Images of STM for a $TiO_2(110)$ surface exposed to 0.53 torr of CO_2 are shown in Figure 13.3. CO_2 is imaged as single symmetrical protrusions along the direction of the Ti_{5c} rows (Figure 13.3a) at an apparent height of about 1.0 Å with respect to the Ti atoms, as shown in the line profile measurements in Figure 13.3c,e. The coverage 0.2 monolayer (ML) is variable over different regions, with some (1×3) and (1×4) arrangements corresponding to about 0.2 monolayer (ML) (Figure 13.3b), and other regions that exhibit a coherent (1×2) superstructure (Figure 13.3d). The maximum coverage seen at 0.5 ML is attributed to these latter areas, giving an intermolecular distance of about 6.0 Å. Under these conditions, there is no aggregation of CO_2 into islands, but rather a distributed wetting of the surface to form a 2D condensate or film.

Upon evacuation of the background gas, CO_2 on the Ti_{5c} rows desorbs from the surface. This study provides an excellent use of ambient pressure STM to show that, in spite of this weak binding, a substantial amount of CO_2 can reside on a $TiO_2(110)$ surface at room temperature, forming 2D ordered films.

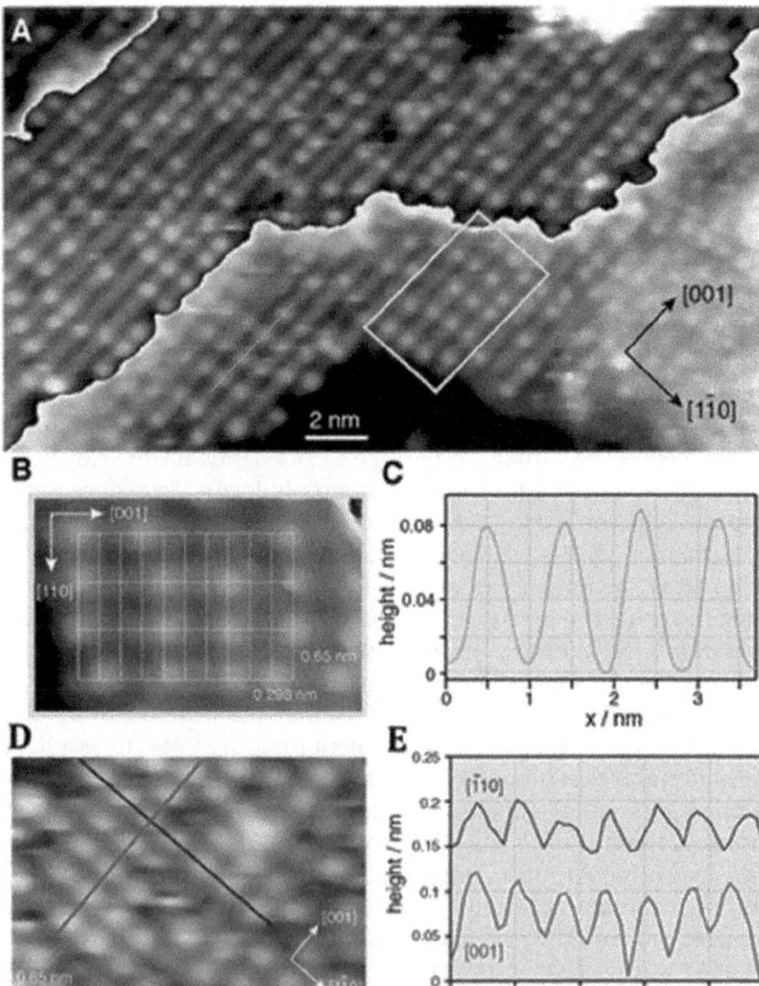

Figure 13.3 (a) STM image of the rutile $TiO_2(110)$ surface under 0.53 torr CO_2 at 300 K. (24×14 nm^2, V = +1.73 V, I = 0.11 nA). Individual terrace contrast was adjusted independently for ease of viewing. (b,d) Small areas demonstrating the adsorbate spacing relative to the titania substrate. (b) is taken from the rectangle in (a) and includes an overlay of the rutile $TiO_2(110)$ unit cell. (d) depicts the presence of (1×2) domains of adsorbates on the titania surface. (6.5×4.3 nm^2, V = +1.73 V, I = 0.07 nA). (c) Line profile from (a). (E) Line profiles from (d). Reproduced from Ref. 36 with permission from the PCCP Owner Societies.

This study opens up the possibilities for further investigations of the interaction of CO_2 on other oxide surfaces.

13.3 Ambient Pressure XPS

XPS is a well-known surface-sensitive technique that is both elementally specific and quantitative. It is capable of detecting surface compositions and chemical/electronic states and differentiating between varied environments of surface adsorbates. The surface sensitivity of XPS is derived from a limited mean free path of photoelectrons ejected from the surface as a result of the photoelectric effect. The conventional XPS system requires operation in a UHV environment to minimize the scattering of the photoelectrons by gases/liquids. Fortunately, the development of ambient pressure (AP) XPS allowed for the application of this powerful technique to *in situ* measurements in an ambient regime (up to a few torrs), and henceforth helped to bridge the pressure gap between UHV and working pressure conditions.[39,40] Both lab-source (Al and Mg K-α) and synchrotron-based X-rays have been utilized in the implementation of the AP-XPS technique. With lab-source X-rays, AP-XPS experiments can be performed conveniently on a daily basis. These are able to provide a sufficient signal-to-noise ratio for the determination of the elemental compositions and electronic states of most transition metals that are commonly used in catalysis studies (2p, 3d and 4f regions). Synchrotron-based AP-XPS systems require the support of large synchrotron facilities, but they have particular advantages in the studies of surface chemistry (C 1s, N 1s, O 1s, *etc.*), as well as depth profiling owing to the intense, focused and tuneable X-rays. In the field of heterogeneous catalysis, where reactions mainly occur at the gas–solid interface, AP-XPS has demonstrated its ability to unravel the catalytic active phases, surface species and associated reaction mechanisms under reaction conditions, providing *in situ* and surface-sensitive information that cannot be achieved by other "bulk" characterization techniques. One good example of such an application is the *in situ* studies of CO_2 hydrogenation on Cu-based model and powder catalytic systems using AP-XPS.

Copper is the essential element in methanol synthesis catalysts. However, CO_2 interacts poorly with copper alone.[15,41] Both experimental and theoretical calculations have shown that combining copper with ZnO or CeO_2 can significantly enhance the activation of CO_2 and its subsequent conversion to methanol with H_2.[16,17] Investigation of the inverse ZnO/Cu(111) and CeO_2/Cu(111) model catalysts is an important methodology for establishing the effect of metal–oxide interfaces on the activation of CO_2 and their catalytic performance. The morphology of these systems has been studied using STM (see Figures 13.1 and 13.2c), and the corresponding surface chemistry under CO_2 hydrogenation reaction conditions was examined using AP-XPS. Figure 13.4 displays O 1s and C 1s AP-XPS spectra of an as-prepared CeO_2/CuO_x/Cu(111) surface under pure CO_2 and with the addition of H_2 at different temperatures.[25] When exposed to 30 mtorr

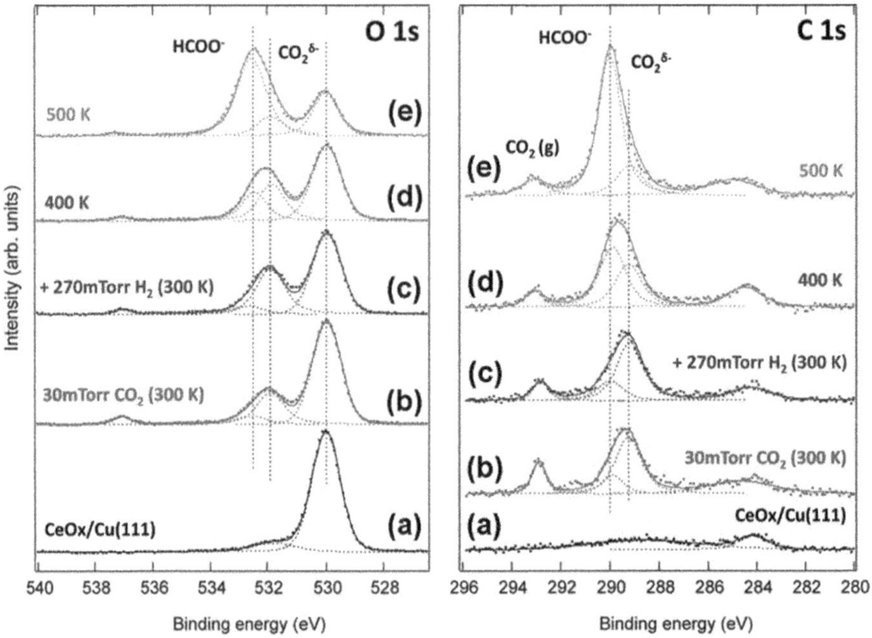

Figure 13.4 O 1s and C 1s XPS spectra of an as-prepared $CeO_2/CuO_x/Cu(111)$ surface with ~0.2 ML of CeO_x (a), under 30 mtorr of CO_2 at 300 K (b), with an addition of 270 mtorr of H_2 (c) and subsequent heating to 400 K (d) and 500 K (e). Reproduced from Ref. 25 with permission from American Chemical Society, Copyright 2016.

of CO_2 at 300 K, peaks that can be attributed to $CO_2^{\delta-}$ were observed with centres at 531.9 eV (O 1s) and 289.3 eV (C 1s), respectively. Weak features of formate ($HCOO^-$) species were also present at 300 K. Subsequently adding 270 mtorr of H_2 and heating up to 500 K led to a gradual increase of formate species, while the $CO_2^{\delta-}$ feature diminished. AP-XPS experiment carried out as a function of ceria coverages showed that a maximum in the amount of $CO_2^{\delta-}$ species was reached at 0.2–0.3 ML of ceria coverage, and no $CO_2^{\delta-}$ was detected on plain Cu surfaces, indicating that the generation of a ceria–copper interface is crucial for the activation of CO_2.[25] The AP-XPS study on the $ZnO/CuO_x/Cu(111)$ system showed a similar surface chemistry under the reaction conditions, as shown in Figure 13.5.[42] Starting with $CO_2^{\delta-}$ species on the surface at 300 K, a strong signal for formate species appeared at 400 K. Heating from 400 to 575 K induces a slow decrease in the formate/$CO_2^{\delta-}$ signals, and both features eventually disappeared at 575 K. The Zn 2p and Zn LMM regions of the AP-XPS data also indicate that the formed ZnO overlayers were stable and did not reduce to metallic Zn during the reaction.[42] A comparison of the surface chemistry of plain Cu(111) with oxide-deposited Cu(111) demonstrated the important role played by

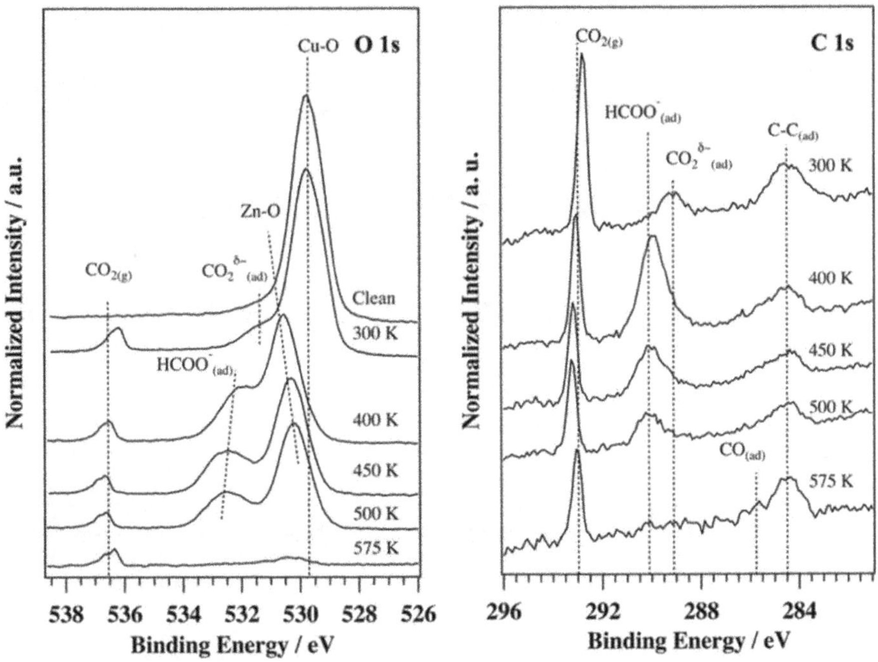

Figure 13.5 O 1s and C 1s XPS spectra of a $ZnO/CuO_x/Cu(111)$ surface under 100 mtorr of CO_2 and 500 mtorr of H_2 at 300–575 K. The ZnO covered ~25% of the copper substrate. Reproduced from Ref. 42 with permission from American Chemical Society, Copyright 2018.

the metal–oxide interface: it accelerates the binding and transformation of CO_2 to the possible active species, including $CO_2^{\delta-}$ and formate.

AP-XPS could also be a useful tool in the study of high-surface-area catalysts, as it provides more surface-sensitive information to complete the understanding from bulk characterizations such as XRD and XAFS. Figure 13.6 shows the Ce 3d AP-XPS spectra of two types of Cu/CeO_2 catalysts.[24] A quantitative analysis of the deconvoluted Ce 3d spectra depicts the reduction degree of ceria in association with oxygen vacancy formation during the CO_2 hydrogenation reaction. The reduction degree of the supported copper ceria nanorods and ceria nanospheres was very similar after the H_2 pre-treatment (Figure 13.6c).[24] When exposed to a mixture of H_2/CO_2 at room temperature, a reoxidation of the ceria surface was observed, suggesting the recovery of oxygen vacancies on ceria due to the activation and dissociation of CO_2. Increasing the reaction temperature to 350 and 400 °C reduced the surface again as a consequence of reacting with hydrogen. However, a clear difference in the reduction extent on the two morphologies of ceria was seen. The Cu supported on ceria nanorods exhibited a stronger affinity for CO_2 with a tendency to generate surface intermediate species or O adatoms during the reaction, which is probably the reason for the higher catalytic performance of Cu/CeO_2 nanorods.[24]

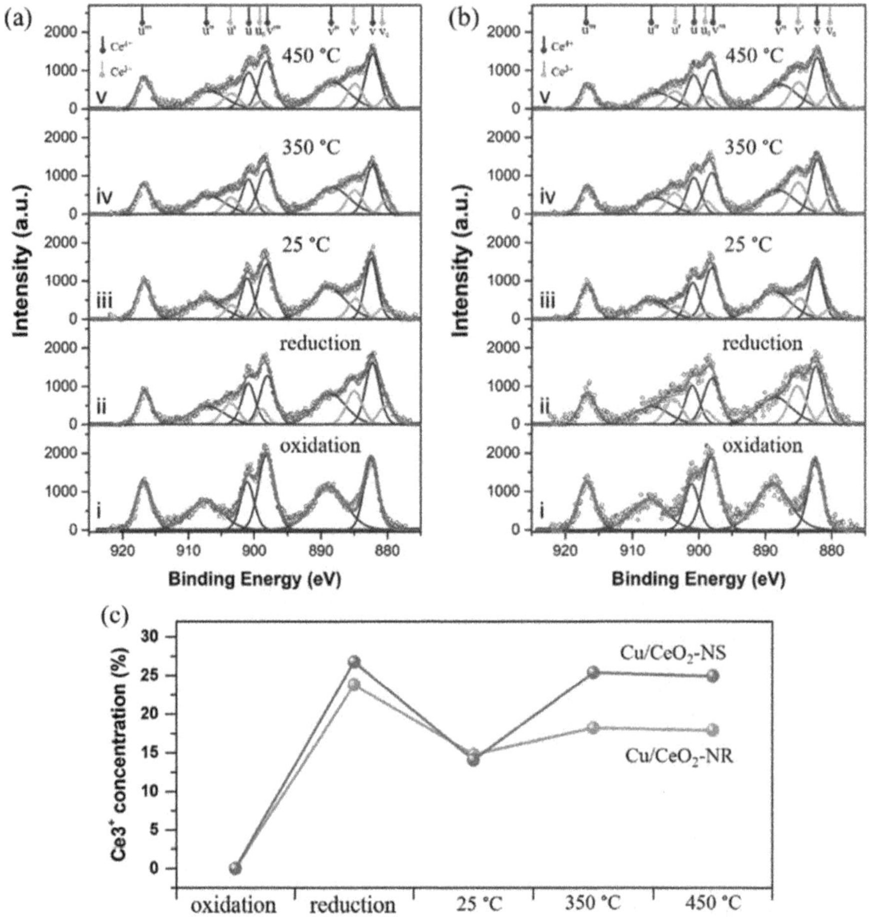

Figure 13.6 Ce 3d profiles of (a) Cu/CeO_2 nanorod and (b) Cu/CeO_2 nanosphere catalysts under different reaction conditions. (c) Ce^{3+} concentration of the two catalysts under different reaction conditions. Reproduced from Ref. 24 with permission from American Chemical Society, Copyright 2018.

13.4 Vibrational Spectroscopy

In catalysis, Fourier-transform infrared spectroscopy (FTIR) has been widely used because it is a robust technique that has allowed for the identification of adsorbed surface species and, with the use of probe molecules such as CO and NO, has aided in identifying and quantifying surface active sites. The region of most relevance for surface science and catalysis is the mid-infrared (200–4000 cm^{-1}), as most transitions in the discrete vibrational modes of chemisorbed molecules are excited in this range. The study of the absorption frequencies of adsorbates yields information about molecular geometry, the strength of absorption forces and the nature of their bonding to the surface.[43]

13.4.1 Reflection Absorption Infrared Spectroscopy

As a surface-sensitive technique, reflection absorption infrared spectroscopy (RAIRS) is a specialized use of infrared spectroscopy, as it requires flat, reflective surfaces and it only probes adsorbates in a small area of the sample.[44] The adsorption of infrared radiation by chemisorbed molecules on the surface is markedly enhanced at higher angles of incidence and is effectively limited by p-polarized radiation, since it is parallel to the incident plane and can interact with molecules that have dynamic dipoles that are perpendicular to the surface. This is commonly known as the surface selection rule.[30]

The uniqueness of RAIRS has been applied to a variety of pure metal systems and has also been employed in studies of adsorbates on metal oxides for the activation and conversion of hydrocarbons. In particular, it has been used to track the activation and conversion of CO_2 into methanol over CeO_x/Cu(111) and to explore the role of the metal–oxide interface in this reaction.[25] Figure 13.7 shows the ambient pressure RAIRS (AP-RAIRS) spectra gathered during exposure of a low CeO_x coverage (~0.2 ML) and medium CeO_x coverage (~0.5 ML) supported on Cu(111) to a reaction mixture composed of 1 torr of

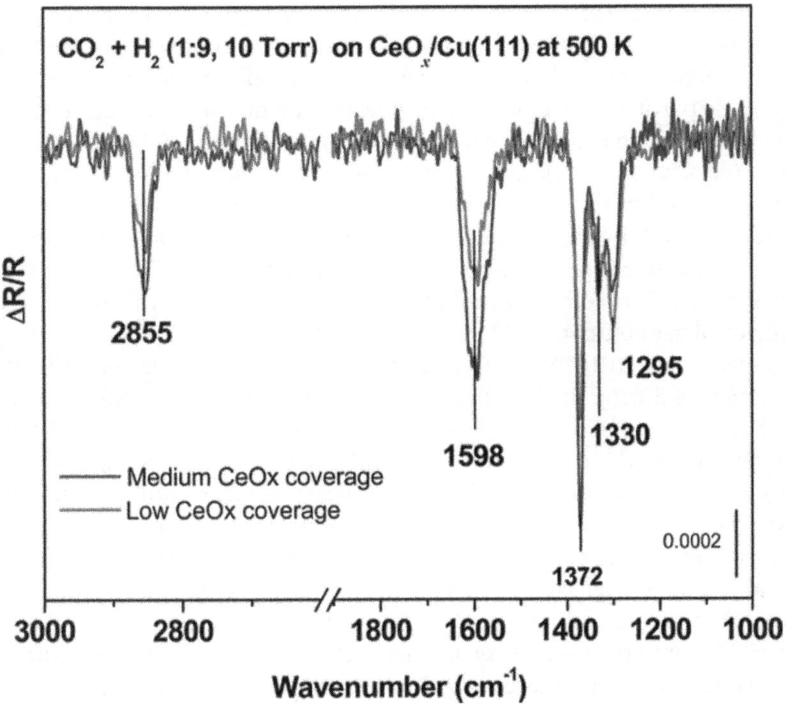

Figure 13.7 AP-RAIRS spectra obtained during the exposure of low (~0.2 ML) and medium (~0.5 ML) coverages of CeO_x on Cu(111) to CO_2 (1.0 torr) and H_2 (9.0 torr) at 500 K. Reproduced from Ref. 25 with permission from American Chemical Society, Copyright 2016.

CO_2 plus 9 torr of H_2 at 500 K. The band observed at 1295 cm^{-1} corresponds to carboxylate ($CO_2^{\delta-}$),[45] while the rest have been previously attributed to formate species ($HCOO^-$):[16] 1330 and 1372 cm^{-1} to the symmetric OCO stretch, v_s(OCO); 1598 cm^{-1} to the asymmetric OCO stretch, v_a(OCO); and 2855 cm^{-1} to the CH stretch, v(CH). The use of RAIRS in these experiments helped confirm ambient pressure XPS (AP-XPS) results carried out in the same work, where it was found that there is a substantial decrease in coverage of the carboxylate species with higher coverages of CeO_x. This was commensurate with a decrease in catalytic activity, supporting the role of ($CO_2^{\delta-}$) as a reaction intermediate and confirming that $HCOO^-$, despite its increasing surface concentration with greater amounts of CeO_x, can simply be regarded as a spectator species. This is a typical example of how the study of model systems with FTIR can help determine and elucidate key mechanistic steps of industrially and environmentally relevant processes.

13.4.2 Diffuse Reflectance Infrared Fourier-transform Spectroscopy

Amongst the available FTIR technologies for the characterization of technical catalysts (*i.e.* supported catalysts), diffuse reflectance infrared Fourier-transform spectroscopy (DRIFTS) stands out as one of the most robust and versatile because it requires very little sample preparation and can be applied to powders, crystals and rough surfaces. This makes DRIFTS especially suited for solids that are opaque and particularly fragile for handling in transmission infrared (TIR).[46,47] In this setup, the solid is placed in a receptacle, and when the IR beam illuminates the sample particles, it is scattered rather than transmitted. The scattered beam is then collected in a hemispherical mirror and focused onto the detector. This feature increases the signal intensity (compared to traditional TIR setups), presumably because of the numerous internal reflections of the light in the vicinity of the surface before collection, enhancing the sensitivity towards surface species *versus* gas-phase compounds.[44]

Lin *et al.* used DRIFTS to carry out the *in situ* characterization of Cu/CeO_2 nanocatalysts during the hydrogenation of CO_2. For this particular study, Cu/CeO_2 nanorods and Cu/CeO_2 nanospheres were used, and the DRIFTS spectra collected during the reverse water gas shift reaction (RWGS) can be seen in Figure 13.8.[24] In the presence of the gas-phase reactants, multiple absorption peaks can be observed in the range between 700 and 1800 cm^{-1} at room temperature, confirming the presence of surface carbonate and other oxo-species (see Table 13.1 for IR assignments). These oxygenates occupy previous oxygen vacancies in the ceria, causing the disappearance of the Ce^{3+} sites that were observed by time resolved X-ray diffraction (TR-XRD) and AP-XPS in the same study. The temperature increase to 250 °C shows the formation of surface formate, evidenced by the absorption bands at 2848 and 2945 cm^{-1}, which are typical of the C–H stretch of this species. In addition, features for bidentate formate can also be observed at 1565 and 1374 cm^{-1}.[48–52] The DRIFTS at 350 and 450 °C show a gradual decrease in the intensity of the aforementioned signals, which suggested that the bidentate formate could be an active surface

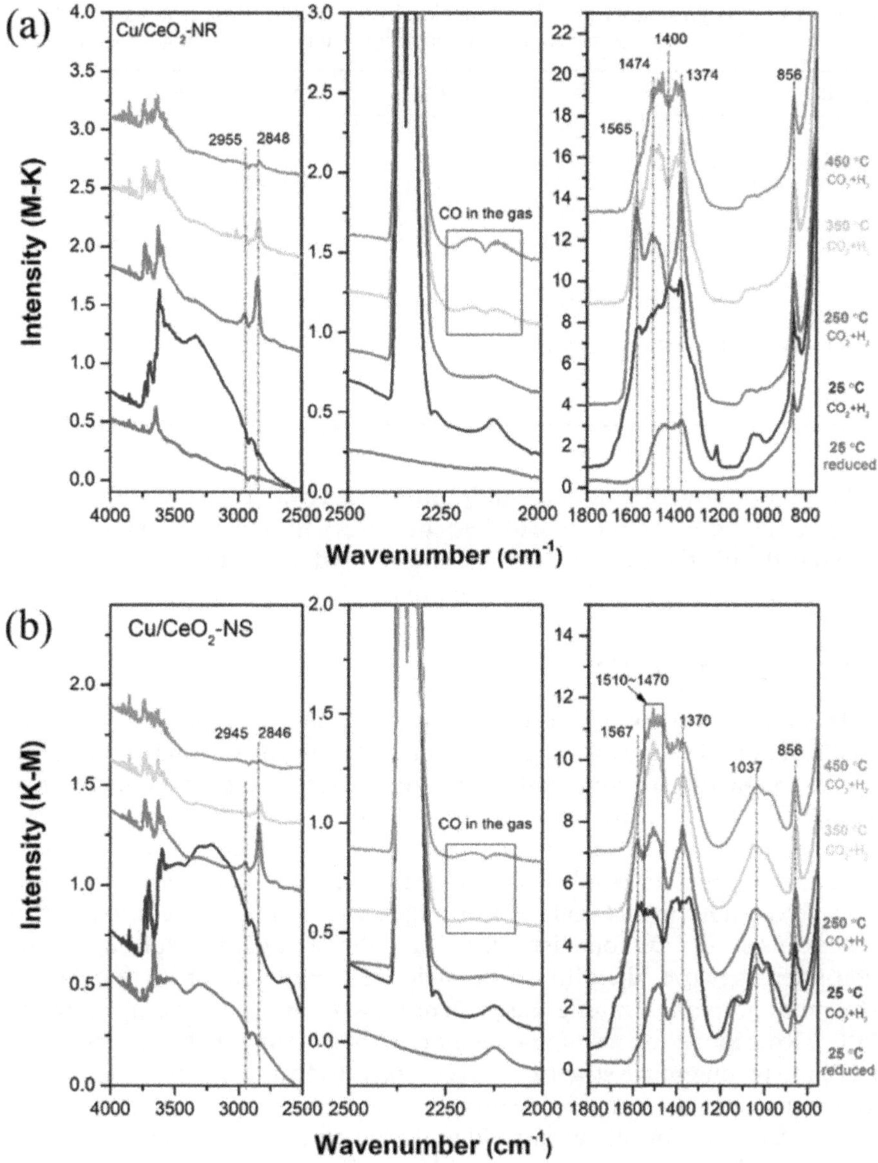

Figure 13.8 *In situ* DRIFTS spectra collected for (a) Cu/CeO_2 nanorod and (b) Cu/CeO_2 nanosphere catalysts during the RWGS reaction (10% CO_2/50% H_2/He) at different temperatures. Reproduced from Ref. 24 with permission from American Chemical Society, Copyright 2018.

intermediate. This was a compelling result, as formate species are mainly seen to be adsorbed after the hydrogenation of carbon dioxide.[15,35,53,54] Finally, the DRIFTS analysis also revealed higher concentrations of more stable species at 450 °C. These bands were assigned to polydentate surface carbonates, which were strongly bound to the surface and were harder to hydrogenate.

Table 13.1 Assignment of absorbance peaks observed on FTIR spectra. Adapted from Ref. 24 with permission from American Chemical Society, Copyright 2018.

Species	Frequency (cm^{-1})	Vibration mode
	2845	$\nu(CH)$
Bidentate formate	1565	$\nu(CO_2)$
	1372	$\nu(CO_2)$
	771	$\delta(CO_2)$
	1580	$\nu(CO_3)$
Bidentate carbonate	1292	$\nu(CO_3)$
	851	$\delta(CO_3)$
	1474	$\nu(CO_3)$
Polydentate carbonate	1367	$\nu(CO_3)$
	1037	$\nu(CO_3)$
	856	$\delta(CO_3)$

The use of DRIFTS in this study provided a way to correlate changes observed in the electronic state of the active sites with chemical changes associated with adsorbates on the surface and allowed for detailed description of the reaction pathway for the hydrogenation of CO_2.

13.5 *In Situ* XRD

XRD is a powerful and well-established structural probe, particularly in its application to the study of polycrystalline materials. Diffraction is produced by the interaction of X-ray light with the ordered crystallographic planes within a crystalline material. Incident radiation of wavelength λ onto a crystallographic plane with interplanar spacing d may produce diffraction at incident angle θ when the condition of Bragg's law ($n\lambda = 2d\sin\theta$) is satisfied.

In polycrystalline materials, the arrangement of crystallographic faces in space consists of a random distribution, and the sums of the resulting diffraction intensities over the diffraction angle are recorded in order to measure structural properties among the present crystallites. Consequently, powder XRD is considered an averaging technique that is suitable for determining medium- to long-range structures, in contrast to local structural probes such as transmission electron microscopy.

XRD data are typically used to identify the crystalline phases present in the material, to determine composition ratios between multiple crystalline phases, to calculate lattice parameters for each phase and to determine average particle size and strain broadening for each phase based on isotropic or multiaxial models. Diffraction data may also be used to obtain information about atomic vacancies, substitutions, lattice defects and thermal parameters. This type of information may be obtained by the application of Rietveld refinement,[55] where a diffraction pattern for a refined theoretical model is fitted to existing data.

State-of-the-art application of XRD to the study of catalysts is primarily composed of *in situ* or *operando* methodologies that are attainable at synchrotron

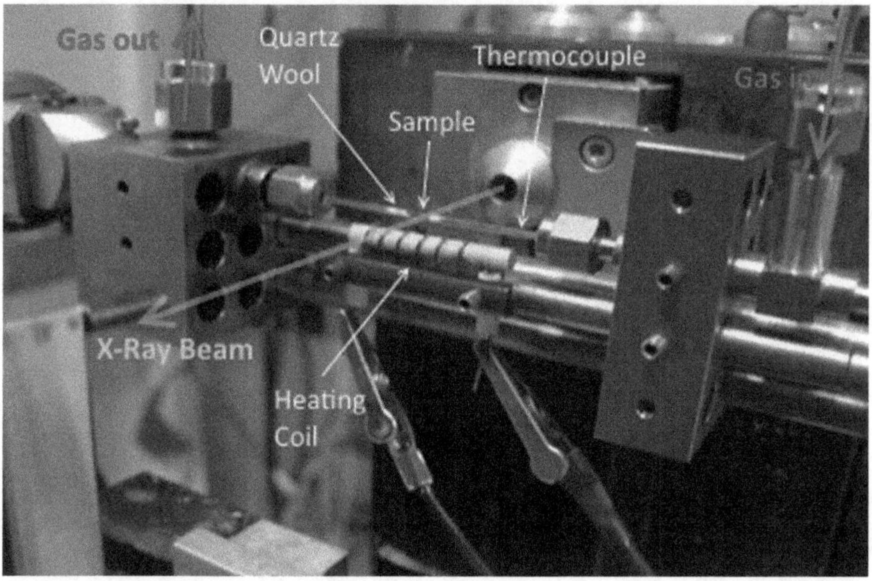

Figure 13.9 Flow cell for *in situ* XRD.

light source facilities such as the National Synchrotron Light Source II (NSLS-II) at Brookhaven National Laboratory and the Advanced Photon Source (APS) at Argonne National Laboratory. Synchrotron radiation enables very fast, high-throughput measurement of diffraction data using flat-plane charge-coupled device detectors.

A wide variety of sample environments have been developed for XRD experiments that enable continuous measurement under a wide range of conditions to facilitate research in areas including, but not limited to, materials synthesis, solid–gas-phase catalysis, solid–liquid/triphase catalysis, photochemical and electrochemical systems and thin films, as well as materials related to engineering applications.

Here, we detail the use of XRD to study the structural properties of catalysts under reactant gas flow conditions. For this application, the reaction environment used[56] was a gas flow cell that utilizes a quartz capillary with an outer diameter of 1.1 mm and an inner diameter of 0.9 mm to hold approximately 4 mg of sample between two pieces of quartz wool. This sample environment allows for the flow of gas through the catalyst material, as well as local heating of the catalyst material up to 500 °C (or 700 °C with shielding). The flow cell used is depicted in Figure 13.9.

Lin *et al.*[24] investigated the use of Cu/CeO_2 catalysts for the RWGS reaction ($CO_2 + H_2 \rightarrow CO + H_2O$). Two catalysts were examined for this application, consisting of a 5 wt% loading of copper deposited on two morphologically distinct CeO_2 nanostructures: nanorods (Cu/CeO_2-NR) and nanospheres (Cu/CeO_2-NS). The purpose of the study was to evaluate the impact of variable ceria morphologies on the catalytic properties of the materials. Catalytic activity studies

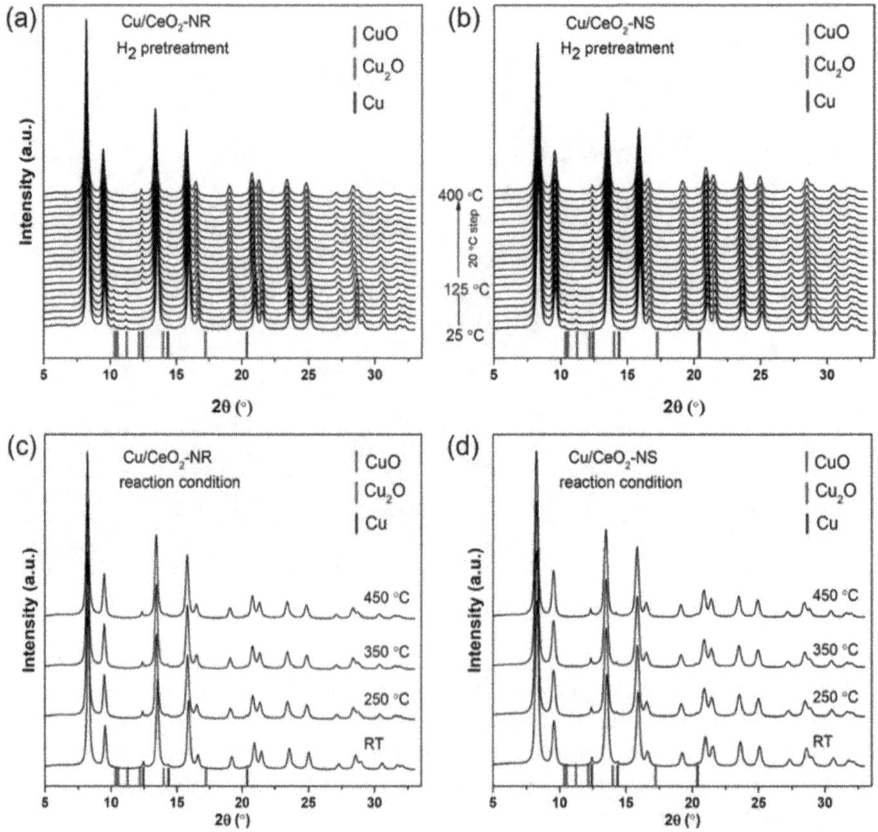

Figure 13.10 Time-resolved XRD patterns for (a) CuO/CeO_2-NR and (b) CuO/CeO_2-NS catalysts with H_2 pre-treatment (50% H_2/He), and pre-reduced (c) Cu/CeO_2-NR and (d) Cu/CeO_2-NS under RWGS reaction conditions at different temperatures (10% CO_2/50% H_2/He). Reproduced from Ref. 24 with permission from American Chemical Society, Copyright 2018.

were carried out for these materials in the temperature range between 150 and 450 °C, with an H_2:CO_2 ratio of 5. The Cu/CeO_2-NR catalyst was more active, reaching a CO_2 conversion of approximately 50% at 450 °C, which is near the RWGS equilibrium value of approximately 53%. The Cu/CeO_2-NR catalyst displayed lower activity, reaching a 42% CO_2 conversion at 450 °C.

In situ XRD was carried out at beamline 17-BM at the APS at Argonne National Laboratory, with an X-ray wavelength of 0.445226 Å. After pre-treatment under 50% H_2/He, the XRD measurements were collected under RWGS reactant gas flow (10% CO_2/50% H_2/He) at between 250 and 450 °C in 100 °C stepwise increments, with a 30 min hold at each step. The XRD patterns were collected using a PerkinElmer amorphous silicon detector and integrated using GSAS-II software.[57] GSAS-II was also used to carry out the Rietveld refinement necessary to determine phase compositions, particle sizes and lattice parameters.

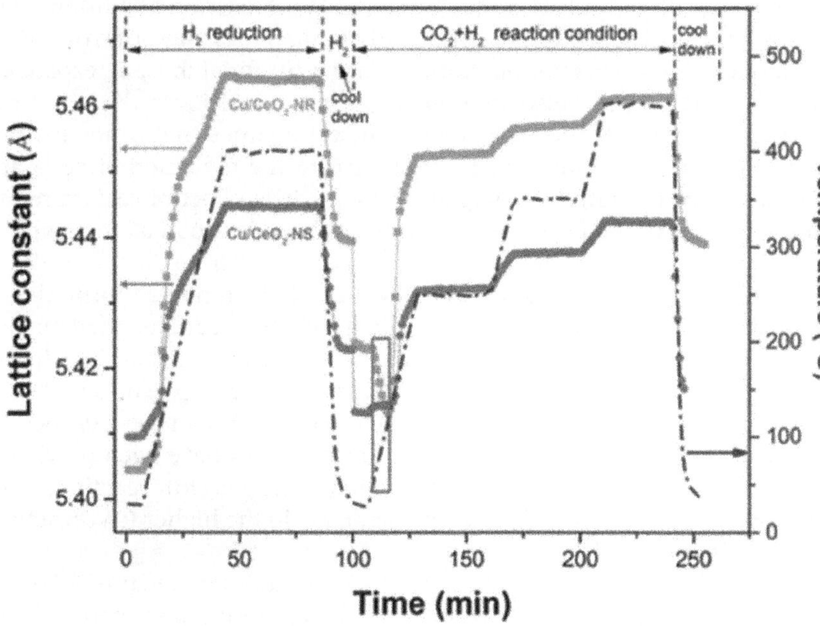

Figure 13.11 CeO_2 lattice constant for Cu/CeO_2-NR and Cu/CeO_2-NS catalysts reduced in molecular hydrogen or under the reaction conditions for the hydrogenation of CO_2. Calculated through Rietveld refinement of *in situ* XRD data. Reproduced from Ref. 24 with permission from American Chemical Society, Copyright 2018.

The resulting diffraction patterns are presented in Figure 13.10, which was reproduced from Lin *et al.*[24]

The fluorite CeO_x phase was clearly evident in the measured diffraction patterns for the Cu/CeO_2-NS and -NR catalysts. Additionally, two crystalline copper phases could be identified during the pre-treatment of the catalysts under H_2 gas flow conditions (Figures 13.10a,b). As-prepared materials initially presented a crystalline CuO phase. As both materials were heated to 125 °C under hydrogen, reduction of the copper oxide to crystalline metallic copper was observed as the CuO phase attenuated and a Cu phase evolved. The metallic copper phase persisted throughout the remainder of the pre-reduction process, as well as under RWGS conditions, indicating that the catalyst remains in its robust active state in the presence of the oxidant CO_2.

While the behaviour of copper under the reaction conditions remained stable, the CeO_2 phase underwent subtle changes with regards to diffraction peak positions. In order to more closely probe the structural characteristics of the evolving CeO_2 phase, Rietveld refinement was carried out. The CeO_2 lattice parameter was determined from each measured diffraction pattern, and the results are plotted in Figure 13.11, which is reproduced from Lin *et al.*[24] The lattice constant of the cubic fluorite CeO_2 cell was observed to expand upon heating under both H_2 reduction and RWGS reaction conditions. The portion of this expansion that is linear with temperature may be attributed to thermal expansion. This porion, however, accounts for at most an expansion of approximately 0.02 Å between room temperature and 450 °C.[53]

The >50% portion of the expansion of the ceria lattice that is not attributable to thermal expansion may be attributed to the reduction of ceria and the subsequent formation of oxygen vacancies. The effect of cerium reduction on the size of the lattice may be intuitively understood as the result of the increase in size from the oxidized Ce^{4+} ion (r = 0.97 Å) to the reduced Ce^{3+} ion (r = 1.14 Å). This effect is amplified by the well-documented formation of oxygen vacancies in partially reduced oxides, of which ceria is often considered among the most reducible and amenable to vacancy formation.[53] In this study, the Cu/CeO_2-NR catalyst demonstrated a greater degree of ceria lattice reduction under hydrogen conditions (Δa = 0.0352 Å) than the Cu/CeO_2-NS catalyst (Δa = 0.0135 Å). Because oxygen vacancy sites have been previously observed to serve as active sites in the related water gas shift reaction,[53] and this greater degree of ceria reducibility may explain the higher RWGS activity displayed by Cu/CeO_2-NR.

Upon introduction of RWGS reactant gases at room temperature following reduction (around t = 100 min), the ceria lattice was observed to sharply shrink for both catalysts. This shrinkage may be attributed to the recovery of oxygen vacancies by the oxidative agent CO_2 in the RWGS reactant gas feed. When the temperature was increased from ~45 to 110 °C under RWGS reaction conditions, the ceria lattice in the Cu/CeO_2-NR catalyst was observed to sharply shrink even further (rectangle in Figure 13.11). This effect was not observed for Cu/CeO_2-NS, where the ceria lattice instead increased slightly during the same portion of the experiment. This phenomenon suggests that CO_2 activation over Cu/CeO_2-NR is more facile than over Cu/CeO_2-NS at these lower temperatures. At higher temperatures, the ceria lattices in both catalysts achieved and maintained a degree of lattice expansion comparable to their respective maxima under the reductive H_2 gas flow conditions. This observation confirms that the RWGS reaction conditions in this experiment are sufficient to keep these catalyst materials reduced (activated) despite the presence of the oxidant CO_2.

This study by Lin *et al.*[24] serves as a typical example of the application of *in situ* XRD to the study of catalysts under working conditions. Phase analysis yielded information regarding the chemical state of the copper component of the catalysts, while Rietveld refinement yielded phase information that was used for informing the chemical and structural state of two similar but distinct materials at all points during the reduction and reaction components of the experiment. *In situ* XRD serves as a powerful structural characterization tool that is capable of probing multiple phases present within dynamic crystalline systems under transient conditions. Combined with other *in situ* characterization techniques such as X-ray absorption near-edge structure (XANES)/extended absorption fine structure (EXAFS), DRIFTS or AP-XPS, the *in situ* XRD technique is an indispensable fixture in any comprehensive characterization study of catalysts and other crystalline materials.

13.6 *In Situ* XAFS Spectroscopy

Chemical reactions, particularly in heterogeneous catalysis, are characterized by having short timescales (milliseconds to picoseconds); therefore, determination of reaction mechanisms is an ongoing challenge during catalytic transformations.[58] X-ray absorption spectroscopy (XAS) is a powerful tool that is able to extract structural and electronic information from an absorbing element. As an experimental probe, information such as the number and type of neighbouring atoms, interatomic distances and structural disorders can be obtained by careful analysis of the absorbing element's chemical environment. *In situ* techniques provide tools for examining catalytic reactions in real time under reaction conditions. XAS remains the method of choice to resolve the structural environment and electronic properties of active sites in a catalytic material due to its excellent spatial, temporal and energy resolution.[59]

The EXAFS region refers to the oscillations above 1000–1500 eV of the absorption energy edge concerning the excitation energy of the core-level electron. The signal obtained or the fine structure refers to the photoelectron scattering of neighbouring atoms. EXAFS measurements can probe the atomic structure within the distance range of approximately 6–8 Å from the resonant absorbing atom.[60] The X-ray absorption near-edge structure (XANES) portion of the XAS spectra is approximately 30 eV below to 40 eV past the absorption edge and its signal corresponds to the energies between the first symmetry-allowed unoccupied state and the continuum states of an absorbing atom. The XANES spectra contains information regarding the electronic structure, the density of occupied states and the bonding geometry of the absorbing atom. While crystallographic techniques are often considered complementary, offering superior quantitative and sample structures, EXAFS is not limited to ordered phases, allowing useful information on dilute, strongly disordered or dispersed structures to be obtained.

Bersani *et al.*[61] examined nano copper-based catalysts for CO_2 hydrogenation (H_2:CO_2:N_2 = 4:1:1 vol% weight hourly space velocity (WHSV) = 120 000 L kg^{-1} h^{-1}) in an effort to highlight relationships between the structural evolution monitored by *in situ* EXAFS on the Cu K-edge with subsequent DRIFTS analysis for the determination of surface products. To fully understand the Cu behaviour during the reaction and to remove any external contribution, no catalyst support was used for any of the measurements. Structural evolution of Cu was monitored using EXAFS with linear combination analysis where the structural state of Cu was found to be strongly related to the formation of reaction products: formate, methane and carbon monoxide. Specifically, with the reduction of CuO to Cu_2O, characterized by the distinctive emergence of a 1s–4s peak at the edge, formate was observed; while the further reduction of Cu_2O to Cu, similarly to the pure metal, determined the concurrent onset of CH_4 and CO production. The structural evolution and the surface product dependence of the products obtained after CO_2 hydrogenation were used to propose a reaction mechanism consistent with the reaction studied over a wide range of temperatures by combining multiple characterization tools.

The incorporation of a catalyst support greatly varies the reaction mechanism, granting additional active sites where the reaction proceeds in synergy. Lin *et al.*[24] examined the effects of different ceria morphologies deposited onto copper metal for CO_2 hydrogenation. EXAFS was implemented in order to examine nanoparticle size based on the hemispherical model where the atom packing and the geometric arrangement of atoms influence the coordination number of an absorbing element and ultimately impact the resultant spectra.[62] Figure 13.12 shows the EXAFS data of the three catalysts used consisting of copper metal deposited into three different ceria morphologies: nanorods (Cu/CeO_2-NR), nanocubes (Cu/CeO_2-NC) and nanoparticles (Cu/CeO_2-NP). *In situ* X-ray absorption proceeded by loading the sample into a custom-made microchannel reactor with graphite paper windows where the XAFS spectra were collected in fluorescence-yield mode. For the tested catalyst, the powder material was reduced in a flow of 50% H_2 in He for a total flow rate of 20 mL min^{-1} at 400 °C for 2 h and cooled down to room temperature under H_2. The white line intensity for the Cu/CeO_2-NR, Cu/CeO_2-NC and Cu/CeO_2-NP closely resembled that of Cu Foil. The EXAFS R-space further confirmed the Cu–Cu bond (2.53 Å) path as the predominant signal obtained for the three samples. In the EXAFS fitting results (Table 13.2), it can be seen that the Cu coordination number (C.N.) of Cu/CeO_2-NR and Cu/CeO_2-NS (Cu–$Cu_{C.N}$ = 6.7) after pre-treatment were similar, whereas the Cu/CeO_2-NC (Cu–$Cu_{C.N}$ = 10.3) was coordinated mostly to Cu, resulting in different coordination environments, so it was excluded from further analysis. Higher activity was obtained for the Cu/CeO_2-NR, demonstrating that the RWGS reaction (H_2:CO_2:N_2 = 5:1:1 vol% WHSV = 150 000 L kg^{-1} h^{-1}) is a structure-sensitive reaction, and complementary DRIFTS was used to conclude that CeO_2(110) terminations are responsible for the improved performance.

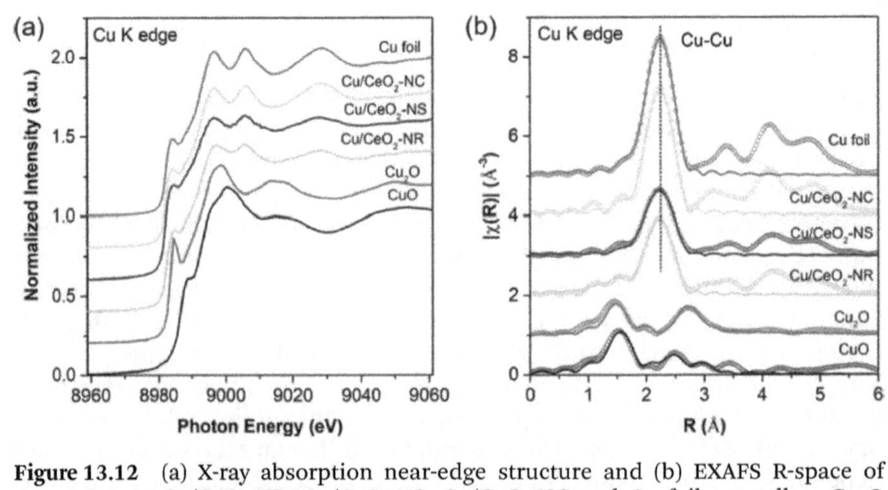

Figure 13.12 (a) X-ray absorption near-edge structure and (b) EXAFS R-space of Cu/CeO_2-NR, Cu/CeO_2-NS, Cu/CeO_2-NC and Cu foil, as well as Cu_2O and CuO. Reproduced from Ref. 24 with permission from American Chemical Society, Copyright 2018.

Table 13.2 Cu K-edge EXAFS fitting results for Cu-based catalysts. Adapted from Ref. 24 with permission from American Chemical Society, Copyright 2018.

Sample	Shell	Bond length (Å)	C.N.	σ^2 (Å^2)	E_0 shift	R factor
Cu/CeO_2-NS	Cu–Cu	2.53 ± 0.01	6.7 ± 1.8	0.010	2.9	0.009
Cu/CeO_2-NR	Cu–Cu	2.53 ± 0.01	6.7 ± 1.3	0.008	2.7	0.008
Cu/CeO_2-NC	Cu–Cu	2.54 ± 0.01	10.3 ± 1.0	0.008	3.6	0.001
Cu foil	Cu–Cu	2.54 ± 0.001	12	0.009	3.7	0.002
Cu_2O	Cu–O	1.85 ± 0.01	4	0.003	9.5	0.030
	Cu–Cu	3.02 ± 0.01	12	0.015		
	Cu–O	3.54 ± 0.01	12	0.015		
CuO	Cu–O	1.95 ± 0.03	4	0.003	7.8	0.040
	Cu–O	2.80 ± 0.03	2	0.003		
	Cu–Cu	2.91 ± 0.03	4	0.010		
	Cu–Cu	3.10 ± 0.03	4	0.010		
	Cu–Cu	3.19 ± 0.03	2	0.010		

CO_2 can also be converted into methanol, providing a carbon-neutral source of fuel given adequate temperatures and high pressures. As such, an In_2O_3 catalyst has been studied for CO_2 conversion into methanol. Tsoukalou *et al.*[63] was able to probe the structural evolution and dynamics of In_2O_3 *in situ*. A spatially resolved EXAFS at the In-K edge (27 940 eV) was used to follow the state of the catalyst during CO_2 reduction on an In_2O_3 nanoparticle catalyst. At the reaction temperature conditions (300 °C, 20 bar, 10 mL min^{-1}, $H_2:CO_2:N_2 = 3:1:1$ vol% WHSV = 1 000 000 L kg^{-1} h^{-1}), it can be seen how, as the reaction progresses, the intensity of the In white line of each XAS spectrum decreases, indicating that, at the beginning of the reaction, In is partially oxidized, and as the reaction progresses, it becomes fully reduced. Therefore, when CO_2 conversion decreases, oxidized In is converted to reduced In, suggesting that partially oxidized In is the more active phase. Three distinctive steps were carefully scrutinized using EXAFS analysis: activation, performance and deactivation. The activation step is characterized by a small decrease in the In–O coordination number, consistent with the formation of oxygen vacancies, leading to the formation of an In_2O_{3-x} phase. At this point, the highest methanol production rate is observed, and this is associated with the change in structure and oxidation state from In_2O_3 to In_2O_{3-x} calculated through linear curve fitting and multivariate curve resolution alternating least square analysis, whereas the deactivation of the catalyst is characterized by the formation of In^0 species, suggesting that the partially oxidized In is the more active phase. In summary, these space resolved EXAFS studies allowed the demonstration of the dynamic structure of In_2O_3 in a plug flow reactor for the CO_2 hydrogenation reaction.

13.7 Outlook and Future Directions

The studies described above illustrate how *in situ* techniques can advance our understanding of CO_2 activation and hydrogenation over oxide and metal–oxide systems. On the one hand, it is clear that the structure and composition of a catalyst usually evolve under CO_2 hydrogenation conditions. The oxidizing ability of CO_2 and the reducing ability of H_2 induce different effects on the physical properties of the sample, and how they balance depends on the pressure of the reactants and the temperatures used for the conversion of CO_2. *A priori*, without *in situ* measurements, it is very difficult to predict the evolution or behaviour of the catalytic system.

Much more needs to be done in this area. The techniques described above require specific sample setups or reaction cells, and they cannot all be applied simultaneously. In the future, the goal is to perform simultaneous characterization of the catalyst compositions/properties and the surface chemistry for CO_2 hydrogenation under the high-pressure conditions used in industrial operations. This will require an integration of techniques such as IR or Raman spectroscopy with XRD or XAFS. Some progress has been made in this area, but much more systematic work applied to CO_2 hydrogenation is necessary. The STM images shown in Figure 13.3 indicate that CO_2 can form ordered overlayers on top of an oxide surface. It is not known how the internal arrangement of the CO_2 overlayer changes with the symmetry of the oxide substrate or when transition or alkali metals are present. This information is necessary in order to understand the first steps in CO_2 activation on metal–oxide catalysts.

Acknowledgements

This manuscript has been authored by employees of Brookhaven National Laboratory supported by the US Department of Energy (DOE), Office of Science, Office of Basic Energy Sciences, Division of Chemical Sciences, Biosciences and Geosciences, under contract No. DE-SC0012704.

References

1. M. Aresta, *Carbon Dioxide as Chemical Feedstock*, vol. 416, Wiley-VCH, Weinheim, 2010.
2. T. R. Karl and K. E. Trenberth, Modern global climate change, *Science*, 2003, **302**(5651), 1719–1723.
3. B. Obama, The irreversible momentum of clean energy, *Science*, 2017, **355**(6321), 126–129.
4. N. MacDowell *et al.*, An overview of CO_2 capture technologies, *Energy Environ. Sci.*, 2010, **3**(11), 1645–1669.
5. K. S. Lackner, A guide to CO_2 sequestration, *Science*, 2003, **300**(5626), 1677–1678.
6. M. Aresta and A. Dibenedetto, Utilisation of CO_2 as a chemical feedstock: Opportunities and challenges, *Dalton Trans.*, 2007, (28), 2975–2992.
7. E. A. Quadrelli *et al.*, Carbon dioxide recycling: Emerging large-scale technologies with industrial potential, *ChemSusChem*, 2011, **4**(9), 1194–1215.
8. E. V. Kondratenko *et al.*, Status and perspectives of CO_2 conversion into fuels and chemicals by catalytic, photocatalytic and electrocatalytic processes, *Energy Environ. Sci.*, 2013, **6**(11), 3112–3135.

9. K. Z. House *et al.*, Economic and energetic analysis of capturing CO_2 from ambient air, *Proc. Natl. Acad. Sci. U. S. A.*, 2011, **108**(51), 20428–20433.
10. G. A. Olah, A. Goeppert and G. S. Prakash, *Beyond Oil and Gas: The Methanol Economy*, John Wiley & Sons, Hoboken, NJ, 2018.
11. A. Goeppert, G. A. Olah and G. S. Prakash, *Toward a sustainable carbon cycle: The methanol economy*, in *Green Chemistry*, 2018, Elsevier, Amsterdam, pp. 919–962.
12. M. Behrens *et al.*, The active site of methanol synthesis over Cu/ZnO/Al_2O_3 industrial catalysts, *Science*, 2012, **336**(6083), 893–897.
13. G. Chinchen *et al.*, Mechanism of methanol synthesis from $CO_2/CO/H_2$ mixtures over copper/zincoxide/alumina catalysts: Use of ^{14}C-labelled reactants, *Appl. Catal.*, 1987, **30**(2), 333–338.
14. A. B. Vidal *et al.*, CO_2 activation and methanol synthesis on novel Au/TiC and Cu/TiC catalysts, *J. Phys. Chem. Lett.*, 2012, **3**(16), 2275–2280.
15. Y. Yang *et al.*, Fundamental studies of methanol synthesis from CO_2 hydrogenation on Cu (111), Cu clusters, and Cu/ZnO (0001), *Phys. Chem. Chem. Phys.*, 2010, **12**(33), 9909–9917.
16. J. Graciani *et al.*, Highly active copper–ceria and copper–ceria–titania catalysts for methanol synthesis from CO_2, *Science*, 2014, **345**(6196), 546–550.
17. S. Kattel *et al.*, Active sites for CO_2 hydrogenation to methanol on Cu/ZnO catalysts, *Science*, 2017, **355**(6331), 1296–1299.
18. X. Yang *et al.*, Low pressure CO_2 hydrogenation to methanol over gold nanoparticles activated on a CeOx/TiO_2 interface, *J. Am. Chem. Soc.*, 2015, **137**(32), 10104.
19. S. Kattel *et al.*, CO_2 Hydrogenation over oxide-supported PtCo catalysts: The role of the oxide support in determining the product selectivity, *Angew. Chem., Int. Ed.*, 2016, **28**(128), 8100–8105.
20. J. M. Thomas and W. J. Thomas, *Principles and Practice of Heterogeneous Catalysis*, John Wiley & Sons, Hoboken, NJ, 2014.
21. S. D. Jackson and J. S. Hargreaves, *Metal Oxide Catalysis*, Wiley-VCH, Weinheim, 2008.
22. J. R. Ross, *Heterogeneous Catalysis: Fundamentals and Applications*, Elsevier, Amsterdam, 2011.
23. J. A. Rodriguez *et al.*, Inverse oxide/metal catalysts in fundamental studies and practical applications: A perspective of recent developments, *J. Phys. Chem. Lett.*, 2016, 7(13), 2627–2639.
24. L. Lin *et al.*, *In situ* characterization of Cu/CeO_2 nanocatalysts for CO_2 hydrogenation: Morphological effects of nanostructured ceria on the catalytic activity, *J. Phys. Chem. C*, 2018, **122**(24), 12934–12943.
25. S. D. Senanayake *et al.*, Hydrogenation of CO_2 to Methanol on CeOx/Cu (111) and ZnO/Cu (111) Catalysts: Role of the metal–oxide interface and importance of Ce^{3+} Sites, *J. Phys. Chem. C*, 2016, **120**(3), 1778–1784.
26. G. Binnig *et al.*, Surface studies by scanning tunneling microscopy, *Phys. Rev. Lett.*, 1982, **49**(1), 57.

27. G. Binnig and H. Rohrer, Scanning tunneling microscopy, *Surf. Sci.*, 1983, **126**(1–3), 236–244.
28. C. J. Chen, *Introduction to Scanning Tunneling Microscopy*, vol. 4, Oxford University Press on Demand, Oxford, 1993.
29. H.-J. Güntherodt and R. Wiesendanger, *Scanning Tunneling Microscopy I: General Principles and Applications to Clean and Adsorbate-covered Surfaces*, Springer, Berlin, 1992.
30. D. Woodruff and T. Delchar, *Modern Techniques of Surface Science*, Cambridge University Press, Cambridge, 1994.
31. J. A. Rodriguez, Gold-based catalysts for the water–gas shift reaction: Active sites and reaction mechanism, *Catal. Today*, 2011, **160**(1), 3–10.
32. M. Mahapatra *et al.*, Growth, Structure, and Catalytic Properties of ZnO_x Grown on CuO_x/Cu (111) Surfaces, *J. Phys. Chem. C*, 2018, **122**(46), 26554–26562.
33. M. Mahapatra *et al.*, The behavior of inverse oxide/metal catalysts: CO oxidation and water–gas shift reactions over ZnO/Cu (111) surfaces, *Surf. Sci.*, 2019, **681**, 116–121.
34. F. Yang *et al.*, Determining the behavior of RuO_x nanoparticles in mixed-metal oxides: Structural and catalytic properties of $RuO_2/TiO_2(110)$ Surfaces, *Angew. Chem., Int. Ed.*, 2011, **50**(43), 10198–10202.
35. J. A. Rodriguez *et al.*, Ceria-based model catalysts: Fundamental studies on the importance of the metal–ceria interface in CO oxidation, the water–gas shift, CO_2 hydrogenation, and methane and alcohol reforming, *Chem. Soc. Rev.*, 2017, **46**(7), 1824–1841.
36. R. C. Hamlyn *et al.*, Imaging the ordering of a weakly adsorbed two-dimensional condensate: Ambient-pressure microscopy and spectroscopy of CO_2 molecules on rutile TiO_2 (110), *Phys. Chem. Chem. Phys.*, 2018, **20**(19), 13122–13126.
37. I. Waluyo *et al.*, Potassium-Promoted Reduction of Cu_2O/Cu (111) by CO, *J. Phys. Chem. C*, 2018, **123**(13), 8057–8066.
38. D. C. Sorescu *et al.*, CO_2 adsorption on $TiO_2(110)$ rutile: Insight from dispersion-corrected density functional theory calculations and scanning tunneling microscopy experiments, *J. Chem. Phys.*, 2011, **134**(10), 104707.
39. H. Siegbahn and K. Siegbahn, ESCA applied to liquids, *J. Electron Spectros. Relat. Phenomena*, 1973, 2(3), 319–325.
40. K. Siegbahn, Electron spectroscopy – an outlook, *J. Electron Spectros. Relat. Phenomena*, 1974, 5(1), 3–97.
41. J. A. Rodriguez *et al.*, Hydrogenation of CO_2 to methanol: Importance of metal–oxide and metal–carbide interfaces in the activation of CO_2, *ACS Catal.*, 2015, 5(11), 6696–6706.
42. R. M. Palomino *et al.*, Hydrogenation of CO_2 on ZnO/Cu (100) and ZnO/Cu (111) catalysts: Role of copper structure and metal–oxide interface in methanol synthesis, *J. Phys. Chem. B*, 2018, **122**(2), 794–800.
43. L. H. Little, *Infrared Spectra of Adsorbed Species*, Academic Press, Washington, DC, 1966.

44. F. Zaera, New advances in the use of infrared absorption spectroscopy for the characterization of heterogeneous catalytic reactions, *Chem. Soc. Rev.*, 2014, **43**(22), 7624–7663.
45. K. Mudiyanselage *et al.*, Importance of the metal–oxide interface in catalysis: In situ studies of the water–gas shift reaction by ambient-pressure X-ray photoelectron spectroscopy, *Angew. Chem., Int. Ed.*, 2013, **52**(19), 5101–5105.
46. G. Busca, The use of vibrational spectroscopies in studies of heterogeneous catalysis by metal oxides: An introduction, *Catal. Today*, 1996, **27**(3–4), 323–352.
47. J. P. Blitz and S. M. Augustine, Characterization of heterogeneous catalysts by FT-IR diffuse reflectance spectroscopy, *ChemInform*, 1995, **26**(30), 347–359.
48. D. Mei, First-principles characterization of formate and carboxyl adsorption on stoichiometric CeO_2 (111) and CeO_2 (110) surfaces, *J. Energy Chem.*, 2013, **22**(3), 524–532.
49. C. Li *et al.*, CO disproportionation at mild temperatures over partially reduced cerium oxide, *J. Chem. Soc. Chem. Commun.*, 1991, (6), 410–411.
50. T. Shido and Y. Iwasawa, Regulation of reaction intermediate by reactant in the water–gas shift reaction on CeO_2, in relation to reactant-promoted mechanism, *J. Catal.*, 1992, **136**(2), 493–503.
51. K. Yoshikawa *et al.*, Synthesis and analysis of CO_2 adsorbents based on cerium oxide, *J. CO2 Util.*, 2014, **8**, 34–38.
52. G. N. Vayssilov *et al.*, Reassignment of the vibrational spectra of carbonates, formates, and related surface species on ceria: A combined density functional and infrared spectroscopy investigation, *J. Phys. Chem. C*, 2011, **115**(47), 23435–23454.
53. X. Wang *et al.*, *In situ* studies of the active sites for the water gas shift reaction over Cu–CeO_2 catalysts: Complex interaction between metallic copper and oxygen vacancies of ceria, *J. Phys. Chem. B*, 2006, **110**(1), 428–434.
54. J. Stubenrauch, E. Brosha and J. Vohs, Reaction of carboxylic acids on CeO_2 (111) and CeO_2 (100), *Catal. Today*, 1996, **28**(4), 431–441.
55. A. Albinati and B. Willis, The Rietveld method in neutron and X-ray powder diffraction, *J. Appl Crystallogr.*, 1982, **15**(4), 361–374.
56. P. J. Chupas *et al.*, A versatile sample-environment cell for non-ambient X-ray scattering experiments, *J. Appl Crystallogr.*, 2008, **41**(4), 822–824.
57. B. H. Toby and R. B. Von Dreele, GSAS-II: The genesis of a modern open-source all purpose crystallography software package, *J. Appl Crystallogr.*, 2013, **46**(2), 544–549.
58. B. S. Clausen *et al.*, A combined QEXAFS/XRD method for on-line, in situ studies of catalysts: Examples of dynamic measurements of Cu-based methanol catalysts, *Catal. Lett.*, 1993, **20**(1–2), 23–36.
59. J. Sinfelt, G. Via and F. Lytle, Application of EXAFS in catalysis. Structure of bimetallic cluster catalysts, *Catal. Rev. Sci. Eng.*, 1984, **26**(1), 81–140.
60. G. Vlaic, D. Andreatta and P. Colavita, Characterisation of heterogeneous catalysts by EXAFS, *Catal. Today*, 1998, **41**(1–3), 261–275.

61. M. Bersani *et al.*, Combined EXAFS, XRD, DRIFTS, and DFT study of nano copper-based catalysts for CO_2 hydrogenation, *ACS Catal.*, 2016, **6**(9), 5823–5833.
62. A. M. Beale and B. M. Weckhuysen, EXAFS as a tool to interrogate the size and shape of mono and bimetallic catalyst nanoparticles, *Phys. Chem. Chem. Phys.*, 2010, **12**(21), 5562–5574.
63. A. Tsoukalou *et al.*, Structural evolution and dynamics of an In_2O_3 catalyst for CO_2 hydrogenation to methanol: An *operando* XAS-XRD and *in situ* TEM study, *J. Am. Chem. Soc.*, 2019, **141**(34), 13497–13505.

CHAPTER 14

Catalytic Aspects of Fuel Cells: Overview and Insights into Solid Oxide Fuel Cells

MOHAMMADMEHDI CHOOLAEI AND BAHMAN AMINI HORRI*

Chemical and Process Engineering Department, University of Surrey, Guildford, GU2 7XH, UK
*Corresponding contributor. Email: b.aminihorri@surrey.ac.uk

14.1 Introduction

Fuel cells are electrochemical devices that convert chemical energy into electricity at a low pollution level. Since fuel cells do not have the limitations of internal combustion engines (Carnot cycle), they can generate electricity with much higher efficiency. In addition, the progressive increase in utilizing hydrogen as a clean energy carrier has led to great interest in developing more efficient fuel cells as a prospective power source from both research and manufacturing communities. In this regard, fuel cells have already shown great potential in providing electricity for rural areas with no or limited access to the public grid, removing the huge cost of wiring and electricity transfer to these areas. As is presented in Table 14.1, when compared to common power generation systems, fuel cells provide the highest efficiency, reaching efficiencies as high as 80% when used in combined heat and power (CHP) applications.

Heterogeneous Catalysis for Energy Applications
Energy and Environment Series No. 27
Edited by Tomas R. Reina and Jose A. Odriozola

Published by the Royal Society of Chemistry, www.rsc.org

In general, according to the choice of fuel and electrolyte, fuel cells are categorized into six major groups: (1) alkaline fuel cells (AFCs); (2) phosphoric acid fuel cells (PAFCs); (3) solid oxide fuel cells (SOFCs); (4) molten carbonate fuel cells (MCFCs); (5) proton exchange membrane fuel cells (PEMFCs); and (6) direct methanol fuel cells (DMFCs). A comparison between different fuel cell technologies along with their schematic illustrations can be found in Table 14.2 and Figure 14.1. PEMFCs and SOFCs are the two most promising

Table 14.1 Comparison of fuel cells with other power-generating systems.[1]

	Reciprocating engine: diesel	Turbine generator	Photovoltaic	Wind turbine	Fuel cells
Capacity range	500 kW–50 MW	500 kW–5 MW	1 kW–1 MW	10 kW–1 MW	200 kW–2 MW
Efficiency	35%	29–42%	6–19%	25%	40–85%
Capital cost (US$ kW^{-1})	200–350	450–870	6600	1000	1500–3000
Operation and maintenance cost (US$ kW^{-1})	0.005–0.015	0.005–0.0065	0.001–0.004	0.01	0.0019–0.0153

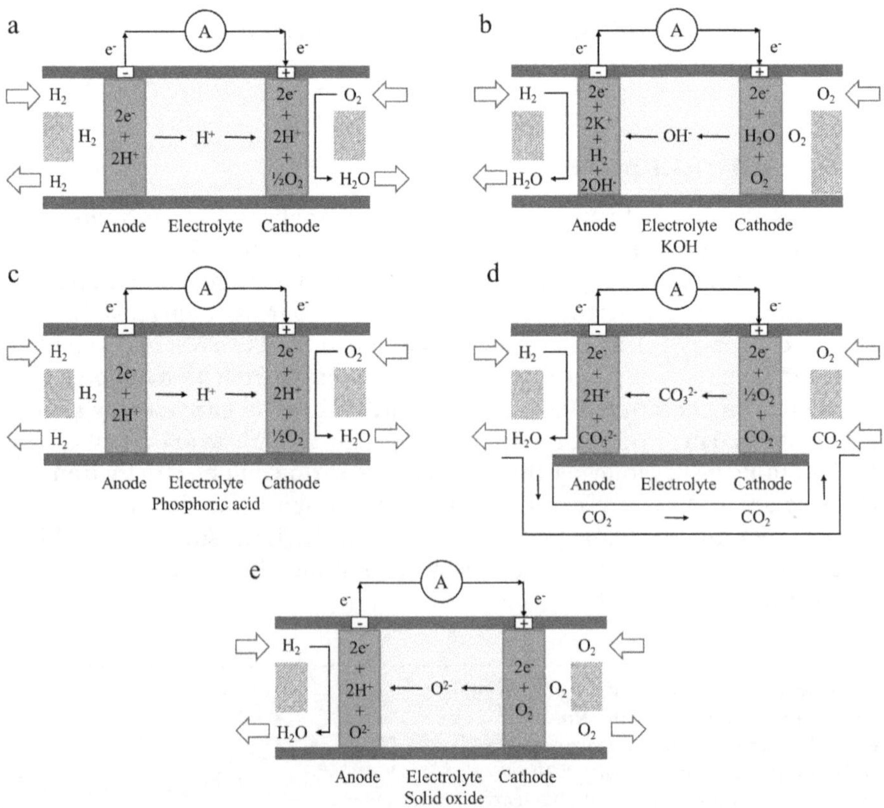

Figure 14.1 (a) A PEMFC, (b) an AFC, (c) a PAFCs, (d) a MCFC and (e) a SOFC.[9]

types of fuel cells that have already passed early commercialization stages. Nonetheless, further development is required to overcome some existing limitations surrounding common fuel cell technologies. Reducing cost, improving durability and further optimizing performance has been the focus of most fuel cell research at the individual cell level, stack level and general system level. The interdisciplinary nature of this technology, in which thermodynamics, chemistry, material science and engineering, electrochemistry, engineering economics and electrical engineering all combine, makes this development a difficult task. However, great breakthroughs are yet to come, and positive progress has been witnessed in the recent years.

Due to the mentioned complexity surrounding each fuel cell, it is not possible to cover all six categories in this chapter. Thus, more emphasis is placed on SOFCs, as some of the most promising fuel cells in providing efficient electric power generation and substantial environmental benefits with high levels of fuel flexibility. It is this fuel flexibility that has made it possible to operate SOFCs with today's conventional fuels.[2] Another important aspect of SOFCs is there capability to operate in reverse mode, turning energy and water back into H_2. By using the energy from renewables when they are not feeding into the grid, SOFCs can run in reverse, producing H_2 gas through electrolysis (SOECs). H_2 allows a huge amount of energy to be stored for long periods, so the energy from the sun could be used in summer to create H_2, which becomes a fuel source in winter, unlocking greater renewable energy adoption through long-term energy storage and grid balancing. Figure 14.2 presents a schematic image of a SOFC in operation. The electrolyte transfers the reduced oxygen atoms formed in the cathode to a fuel-rich anode zone, where the oxygen ions react with the fuel, say hydrogen. This reaction yields the production of heat and water and a release of electrons. Since the electrolyte is not electron conductive, the generated electrons are pushed towards an external circuit, connecting the anode to the cathode. The flow of electrons in this circuit results in the generation of electrical power. In general,

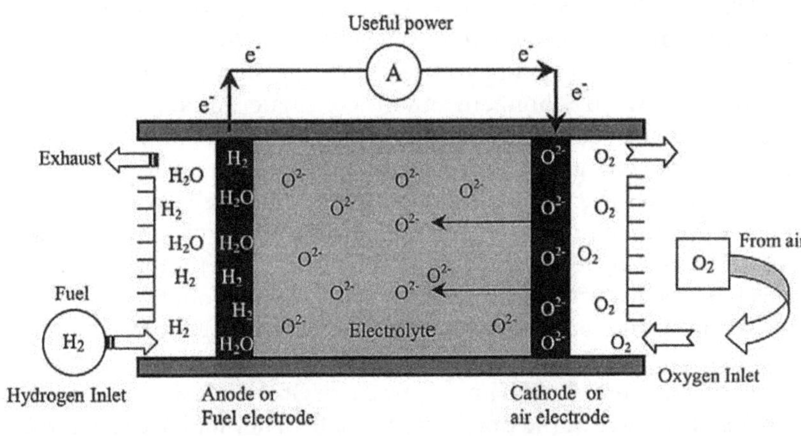

Figure 14.2 Schematic illustration of a SOFC. Reproduced from Ref. 3 with permission from Elsevier, Copyright 2002.

each SOFC is composed of three major components: electrolyte, cathode and anode.

The electrolyte determines the operating temperature of the SOFC and is used to prevent the two electrodes from coming into electronic contact by blocking the electrons. However, standard electrolyte materials, based on stabilized zirconia, require that SOFCs are operated at 800–1000 °C to ensure sufficient ionic conductivity and, therefore, output current density. This high temperature causes several issues, such as cell degradation due to thermochemical instability of the electrode constituents, thermal expansion mismatch between the individual cell layers and a higher cost of auxiliary materials (*e.g.* sealants and interconnects) suitable at such high temperatures.[4,5] Therefore, a great deal of interest has been placed on reducing the operating temperatures to around 500 °C.

This chapter focuses on the current development and challenges regarding anode materials for low- and intermediate-temperature SOFCs operating in hydrocarbon fuels. Possible approaches for achieving high coking resistance, improved electrocatalytic activity and superior long-term durability of the anode layer will be evaluated. Starting with the fundamental aspects of a working SOFC and presenting the current state-of-the-art materials, this discussion will be based on the principles of improving ionic conductivity and electrochemical activity in the low- and intermediate-temperature ranges. At the end, recent progresses and strategies for overcoming carbon deposition over Ni-based cermet anodes operating in hydrocarbon fuels will be discussed.

14.1.1 History

There has been quite the controversy regarding the discovery of fuel cell principles. The first published paper addressing the scientific work conducted on the phenomenon of fuel cells was published by a German scientist, Friedrich Schönbein, in *Philosophical Magazine* issued in January 1839.[6,7] In addition, the Department of Energy of the United States has officially reported Schönbein as the first researcher working on fuel cells.[8] However, in Schönbein's paper, he himself referenced Sir William Robert Grove as the one who introduced the concept of hydrogen fuel cells for the first time.[9] A set of letters written between Schönbein and his colleague, Faraday, between 1836 and 1862 reflect the problems they were facing in understanding how Grove had been able to generate power through oxidation of a positive electrode.[10] This itself indicates the fact that the origin of the fuel cell came from Grove. Although Grove developed the first fuel cell (Figure 14.3) – or, as he called it, the *gaseous voltaic battery* – in 1842, it took almost a century for the development of fuel cell to reach a practical device.[11] During this period, researchers such as Ludwig Mond and Car Langer focused on *new forms of gas batteries*, solving the problem of liquid electrolyte flooding in Grove's design. The result of their work in 1889 was more than an improvement of Grove's design, and was more like the first prototype of the practical fuel cell.[12] They had managed to fix in place the liquid electrolyte (sulfuric acid) in Grove's design using a matrix,

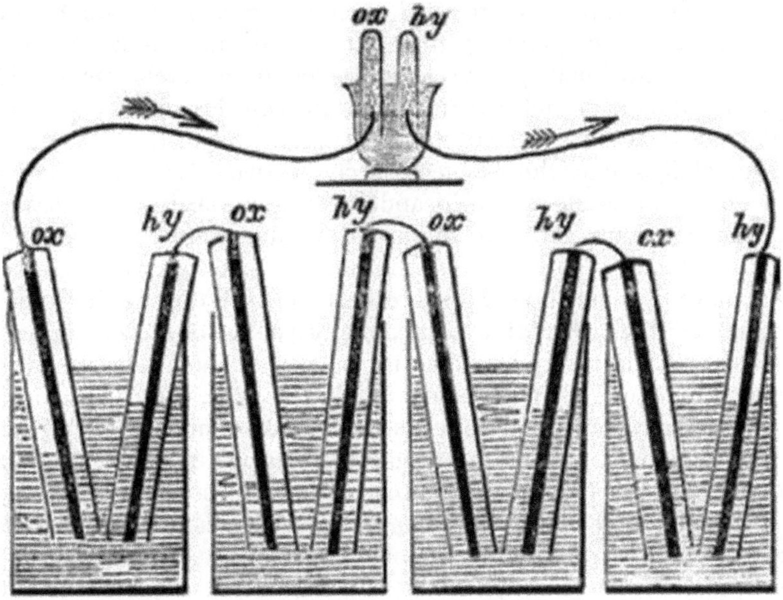

Figure 14.3 Grove's gaseous voltaic battery. Reproduced from Ref. 6 with permission from Elsevier, Copyright 1843.

also called a diaphragm. The theoretical performance of fuel cells was mostly described by Friedrich Wilhelm Ostwald, studying the interactions between different parts such as electrolytes, electrodes, cations and anions.[3] It was he that related the chemical and physical reactions taking place in Grove's gas battery, through which he explained the chemistry of fuel cells. It was in 1896 that William W. Jacques introduced the first fuel cell that generated electricity and could be used for practical applications.[13] Jacques was also able to practically prove that, unlike Grove's belief that only pure hydrogen could be used as a fuel, it was possible to use coal as a source of hydrogen for fuel cells. By the end of the nineteenth century, Mond, Langer and Ostwald firmly believed that the twentieth century would be the era of *electrical combustion*, where fuel cells would replace thermodynamic engines operating on Rankine's cycle. However, even today, in the beginning of the twenty-first century, this prediction has not yet been realized.

In the beginning of the twentieth century, both William W. Jacques and E. Baur were the leading researchers in the development of fuel cells. By then, Jacques had been able to build the first high-power fuel cell system containing of a stack of 100 fuel cell units each capable of generating 1.5 kW.[14] Like Jacques, Baur and Ehrenberg were also focused on studying the possibility of directly using coal as a fuel, leading them to study the effect of molten carbonates in fuel cells.[13] The first molten carbonate fuel cell (Appendix A) was built by Baur in 1921. He based his design on the use of a mixture of molten salts acting as the electrolyte. The cell operated at 1000 °C and the results were highly satisfactory; however, the high level of corrosion and problems associated with the high operating temperature were showstoppers.[15] The need to deal with a more manageable electrolyte led to the development of SOFCs (Appendix A) by Baur and Preis in 1937.[16] In their design, they used a solid compound already developed by Nernst as the electrolyte. In 1899, Nernst had experimentally studied the change in the conductivity of solid materials at different temperatures. Back then, incandescent lamps had just been developed, and filaments such as carbon, osmium and tantalum were mainly used in their structure. Although filaments were chosen from materials with high melting points, most of them showed a short lifespan, since they would vaporize at high temperatures. In addition, since the resistance of metals dramatically increases with increasing temperature, metal filaments could not reach the temperature required to release radiation in the visible spectrum.[17] However, Nernst had realized that unlike metals, melted salts demonstrated higher conductivity at higher temperatures. In addition, during his work on the development of a new type of electrical lamp, using solid electrolytes, he discovered the surprisingly high conductivity of mixed metal oxides at high temperatures. In particular, he found that mixtures of Mg, Si, Zr and rare earth elements gave conductivities beyond 4 siemens, which was significantly higher than the most conductive electrolyte tested at the time (sulfuric acid with 0.74 siemens).[18] Although Nernst's lamp showed no effect of decomposition even after 100 h of continuous work, the complicated design of the lamp and the introduction of tungsten filaments lowered the level of interest in his invention.[19]

Thanks to the work done by Bacon in the early 1930s, fuel cells reached their adolescence. Bacon began his work on studying alkaline fuel cells (Appendix A) and soon developed cells using oxygen and hydrogen as fuels with practical use.[20] During the Second World War, he developed the first fuel cell that could power submarines in the Royal Navy. In 1958, he proposed an alkaline fuel cell with 25.4 mm (in diameter) electrodes to Britain's National Research Development Corporation.[21] The designed fuel cell was capable of producing 6 kW and, as part of the demonstration, it was used to power welding equipment, a circular saw and a lift truck.[22] Although the developed fuel cell ended up being much more expensive than expected, it drew the attention of Pratt and Whitney Company, which later on applied Bacon's patent in the Apollo spacecraft.[23]

By this time, acids, carbonates, oxides and alkalis had been used as electrolytes in fuel cells. The availability of polytetrafluoroethylene (PTFE; Teflon) in 1950 greatly influenced the development of fuel cells using aqueous electrolytes and soon introduced a new type of electrolyte. In 1960, a solid ion exchange polystyrene sulfated electrolyte was developed by Grubb and Niedrach, introducing the first polymer membrane fuel cell (Appendix A).[24] The membrane proved to have a conductivity similar to a 0.5 M sulfuric acid solution; however, in order to operate at its highest level of conductivity, 100% humidity at 25 °C was required. A solution to this was to bubble the inlet gas through water and soak the membrane with water. An important aspect of this approach was the fact that it would not dilute the acidity of the membrane since the excess water was rejected by the membrane whenever it reached saturation.[25] Later on, General Electric Company (GE) developed the technology of the ion exchange membrane fuel cells (later called PEMFCs) (Appendix A) for NASA and McDonnell Aircraft for the Gemini space mission. In the final design of the fuel cell used in that mission, fuels (hydrogen and oxygen) were cryogenically stored, and the produced water during the reaction was used by the crew on long missions. Gemini 5 was the first mission to use that fuel cell, launching on 21 August 1965.[26]

In 1960, Broers and Ketelaar reported testing a fuel cell operating at high temperatures (650 °C) over a 6 month period using a mixture of alkali metal carbonates (lithium, sodium and/or potassium) as an electrolyte impregnated within a magnesium oxide porous disc. It was understood that charge transfer with in the molten electrolyte occurs by means of the carbonate (CO_3^{2-}).[27] Elmore and Tanner reported the development of a new type of fuel cell, which they called *intermediate-temperature fuel cells* (operating at temperatures around 150–200 °C). The electrolyte was a mixture of 65% silicon powder and 35% phosphoric acid attached to a Teflon gasket, reporting successful continuous operation of 6 months at 90 mA cm^{-2} and 0.25 V.[28] An important aspect in their fuel cell was the possibility of using air instead of oxygen in the cathode. Later on, this type of fuel cell was mostly known as a phosphoric acid fuel cell (Appendix A).

In 1957, a paper published by Kiukkola and Wagner inspired global research in the field of solid-state electrochemistry.[29] Shortly afterwards, a growing number of papers and patent applications appeared in the field of SOFCs. In this development, a great deal of interest was shown towards zirconia-based electrolytes, which immediately found different applications (*e.g.* the use of $Zr_{0.85}Ca_{0.15}O_{1.85}$ by Weissbart and Ruka for measuring oxygen concentration in a gas phase).[30] After 1961, many companies began applying for patents on SOFCs and started to invest in developing the technology. Westinghouse Electric Corporation gathered a team of 16 researchers under the direct supervision of Archer to develop SOFC stacks. In this work, tubular and parallel flat-plate cells of platinum electrodes were connected in series and their performance was experimentally tested.[31] The results indicated poor performance for flat disc plates, showing bulky stakes and indicating sealing difficulties. On the other hand, the tubular design using bell-and-spigot joints showed satisfactory results, except for their low durability. The problem seemed to be the tendency of the electrolyte layer to peel off from the anode layer due to the formation of water vapour between the two layers. By early 1970, the high level of research carried out on the applications and properties of solid oxide electrolytes makes it very difficult to survey. However, there still seemed to be great obstacles to the fabrication of practical SOFCs. The main problems that had to be solved by then were finding reliable cathode and interconnection materials, along with achieving a suitable technique to fabricate gastight super-thin films of interconnection and electrolyte.[32] These issues resulted in intensive studies into the topic of SOFCs, and many laboratories were focused on resolving them. A leading laboratory in this field was

Brown Boveri research laboratories, which between 1969 and 1973 synthesized and experimentally studied more than 100 different oxide compositions.[33] The results indicated a good performance for Bi_2O_3-doped $LaNO_3$ and SrO-doped $LaMnO_3$ as cathode materials. Today's most commonly used cathodes, and specifically the exclusive use of $La_{0.84}Sr_{0.16}MnO_3$ after 1973, are directly related to the results obtained from that experimental research. Along with the investigation into finding suitable materials for cathodes, profound research was also devoted to finding compatible materials for interconnections.[34] By around 1974, single cells were successfully tested for a period of 3 years at temperatures of around 1000 °C, and modules consisting of stacks containing 25 unit cells were being reviewed. However, since the costs of the available fabrication processes did not seem to be economically adequate, the development of SOFCs was not continued after 1975.[35]

The urge to decrease the fabrication cost of the cells resulted in the introduction of modern fabrication techniques. In 1978, Westinghouse applied the electrochemical vapour deposition (EVD) method developed by Isenberg in 1970 to fabricate a more efficient and cost-effective multi-tubular SOFC.[36] Although tubular cells have proven to have higher stability against thermal and mechanical stresses, the development of modern fabrication techniques, such as screen printing, tape casting, wet spraying, vapour deposition, plasma spraying and others, has brought about lower costs for the fabrication of planar cells. This matter led to much greater development of SOFCs with planar designs during the 1980s and 1990s.[37] In 2000, Ceramic Fuel Cells Ltd in Australia developed a 25 kW SOFC system containing 3840 planar electrolyte-supported cells, which was capable of operating on natural gas due to the use of internal reforming anodes.[37] A summary of the historical developments of fuel cells can be seen in Figure 14.4.

Although until the beginning of the twenty-first century great effort was devoted to reducing the fabrication cost of SOFCs, their costs still remained high due to their high operational temperature (850–1000 °C). This high-temperature operating condition is still causing several issues, such as cell degradation, thermal expansion mismatches between different layers, high operating costs and a higher cost of auxiliary materials.[4,5,38,39] Therefore, developing cells capable of operating at lower temperature ranges (500–700 °C) could improve cell stability and greatly decrease fabrication costs through the use of less expensive materials.[40]

14.2 Reaction Kinetics at Electrodes

In a SOFC, oxygen reduction reaction takes place at the cathode, resulting in the formation of O^{2-} ions as shown in eqn (14.1).[41,42]

$$\frac{1}{2}O_2(\text{gas}) + 2e^- \leftrightarrow O^{2-} \tag{14.1}$$

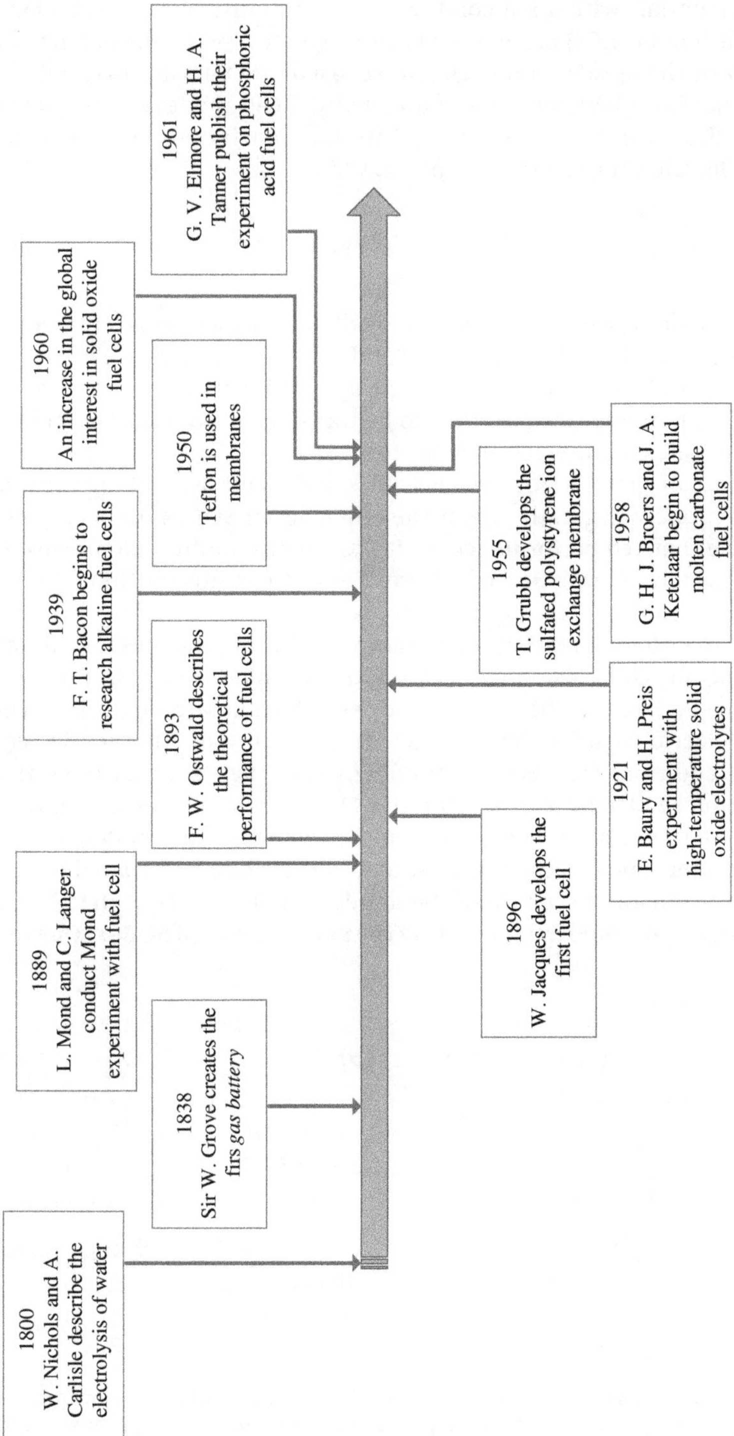

Figure 14.4 Historical review of fuel cells.

Ceramic materials with good conductivities and fairly high thermal expansion coefficients (TECs) are mainly used as electrolyte. Yttria-stabilized zirconia (YSZ) is the most common electrolyte used in SOFCs due to its relatively high ionic and very low electronic conductivity.[43] The anode layer is where the fuel is oxidized (*e.g.* H_2) by combining with the transferred oxygen ions from the cathode, releasing electrons (eqn (14.2)).[3,41]

$$H_2 + O^{2-} \leftrightarrow H_2O + 2e^- \tag{14.2}$$

A porous anode provides sites for the oxidation of fuel, allows the delivery and removal of fuel and by-products to and from the triple-phase boundaries (TPBs; interface between electrode, electrolyte and fuel shown in Figure 14.5) and provides pathways for electrons to be transferred from the electrochemical reaction sites to the electron collectors.

The choice of material for the electrodes and electrolytes along with their structural properties greatly affects the efficiency of SOFCs. Hence, in order to achieve enhanced performance for SOFCs, many studies have focused on the synthesis and fabrication of different anode composites, in addition to exploring alternative electrolyte materials capable of providing high ionic conductivity at low operating temperatures.[45–47] The thickness of the electrolyte and the electrochemical activity of the anode towards fuel oxidation mainly dictate the efficiency of the cell. Thus, anode-supported SOFCs with thin electrolyte layers have attracted a great deal of attention in recent years.[48] In addition, the electrical conductivity of the anode for the transfer of the produced electrons away from the TPBs and to the current collectors also plays an important role in improving the generated power density of the cell. Parameters such as the composition, particle size and spatial distribution of each constituent phase of the anode can greatly affect the electrical conductivity of the anode, thus altering the performance of SOFCs. Due to the

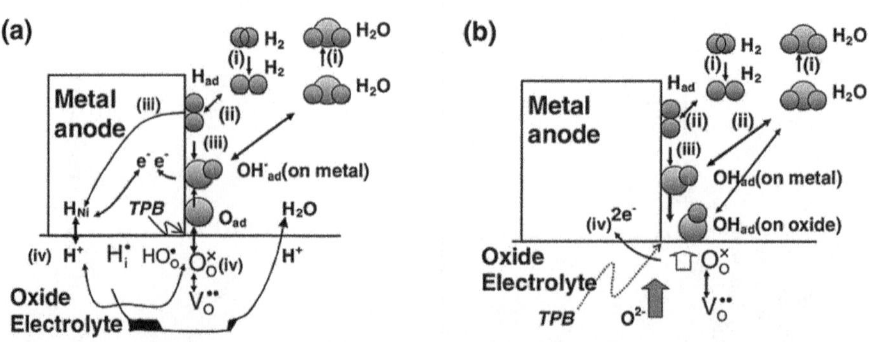

Figure 14.5 Schematic visualization of the TPBs and possible reaction processes for H_2 (fuel oxidation). Reproduced from Ref. 44 with permission from Elsevier, Copyright 2006.

mentioned importance of the anode layer, the anode will be the main focus of this chapter.

To lower the polarization losses, anode materials should possess two main characteristics: (1) high electronic conductivity; and (2) sufficient electrocatalytic activity towards complete fuel oxidation. Although the choice of material itself can greatly influence these requirements, other aspects, such as the microstructure and morphology of the anode, exposed area, level of porosity, distribution and size of the pores, have also been shown to play important roles in optimizing the polarization losses. In addition to the electrochemical necessities, the anode layer should provide sufficient thermal and chemical stability, high mechanical strength and minimal TEC mismatch with other cell components. Furthermore, material selection and cermet composition for the anode layer should support the high level of fuel flexibility required from SOFCs. Thus, a high level of tolerance towards carbon deposition, reoxidation and sulfur poisoning is required when operating under common fuel gases (*e.g.* hydrogen, CO, natural gas and other hydrocarbons).[49,50] In addition to the mentioned criteria, reaching a fundamental understanding regarding the possible charge and mass transfer mechanisms taking place in the bulk, across interfaces and along the surfaces of SOFCs should also be considered as a critical challenge regarding the development of SOFC anode materials.

14.3 Temperature-based SOFC Classification

Based on the operating temperature of SOFCs, they can be categorized as high-, intermediate- or low-temperature SOFCs. A significant decrease in the material cost and higher durability are guaranteed by moving towards lower operating temperatures for SOFCs; hence, many studies have focused on the fabrication of low-temperature operating anodes.[51]

14.3.1 High-temperature SOFC Anodes

High-temperature anodes mainly consist of materials with high melting points, which possess very high electrocatalytic activity at temperatures above 800 °C.[52] High-temperature SOFCs exhibit great fuel flexibility and have high efficiencies. They can operate on hydrocarbon fuels, showing high carbon deposition and sulfur poisoning resistance.[53,54] However, the high operating temperature results in high degradation levels and high operating costs. Further details on high-temperature anodes and their required temperature for operation along with their advantages and drawbacks can be found in the literature.[55–57]

14.3.2 Intermediate-temperature SOFC Anodes

Due to the mentioned limitations in the choice of material and high level of mechanical degradation when operating at high temperatures, it is desirable to decrease the temperature of SOFCs below 800 °C. This approach would also greatly decrease the cost, reduce energy losses and benefit the market penetration of SOFCs.[58] However, moving towards lower operating temperatures has been shown to decrease reaction rates and greatly increase the ohmic losses in the anode. Hence, high-performance anodes with continuous gas diffusion networks need to be development.[59] So far, there have been numerous attempts to overcome the high electrode overpotential and electrolyte resistance observed in intermediate-temperature SOFCs.[60–65] Approaches such as decreasing the electrolyte thickness and searching for alternative active materials with higher ionic conductivities at lower operating temperatures can be seen as noteworthy attempts to encounter the mentioned shortcomings of intermediate-temperature SOFCs.[66,67]

14.3.3 Low-temperature SOFC Anodes

SOFCs operating at temperatures below 650 °C are known as low-temperature SOFCs. Operating at such low temperatures greatly expands the choice of material, reduces corrosion and degradation rates, removes sealing problems and leads to higher energy conversions under partial load circumstances.[68] Low-temperature SOFCs are mostly targeted to powering mobile devices by providing high power densities, short start-up and shut-downs and reduced costs for material and cell fabrication.[69] However, the increased polarization loss in the anode and low ionic conductivity of the electrolyte are still the main drawbacks of operating at such low temperatures. Further details on the recent developments in low-temperature anodes along with their advantages and drawbacks can be found in the literature.[49,51,70–72]

14.4 Structure-based Anodes

Different structures, such as fluorite, perovskite, pyrochlore, rutile, tungsten bronze, spinel, *etc.*, have been used to improve the performance of anode materials in SOFCs.[73] For instance, fluorite's structure, having coordination numbers of four and eight for anions and cations, respectively, has proven to be highly effective in improving the ionic conductivity of the anode composites (Figure 14.6a). Ceria and some types of doped zirconia are well known to exhibit such structures. It is the possibility of the improved oxygen ion mobility of such structures that has made them extremely interesting for SOFC researchers. This is due to the formation of added oxygen vacancies to the fluorite structure when cations with lower valance are introduced into the lattice structure. Furthermore, the addition of multivalence elements (*e.g.* transition metals) to the fluorite structure can result in the formation of mixed conductors that are known as high-performance anodes for SOFCs.[74] Rutile's structure can be explained as close-packed hexagonal anion arrays with cations occupying its octahedral sites, sharing edges between each two octahedra (Figure 14.6b). It is believed that the high electronic conductivity observed in some materials with rutile structure is due to these edge-sharing octahedra, facilitating the electron transfer between the octahedra. However, it is rarely possible to see materials with such structures exhibit high ionic conductivity. To improve its ionic conductivity, attempts have been made to introduce oxygen vacancies into the rutile structure; however, for the oxygen vacancies to jump from one octahedron to another, the octahedra need to rotate. It has been shown that this rotation of edge-sharing octahedra is much more difficult than that of corner-sharing octahedra, observed in perovskite structures. Thus, the anode performance of rutile structures (*e.g.* Nb_2TiO_6) has been reported to be quite poor, despite illustrating high electronic conductivity.[75] Perovskite oxides are a group of mixed oxides with a general formula of ABO_3, where A with a coordination number of 12 and B with a coordination number of 6 present the large and small cations in the material structure, respectively (Figure 14.6c). Their unique structure enables them to accommodate a large number of oxygen vacancies, showing very high levels of oxygen ion conductivity. In addition, the B-site in the perovskite structure allows for the insertion of multi-valence elements (*e.g.* first row transition elements) into its lattice, providing a source of high electronic conductivity.[74] Therefore, high ionic and mixed conductivities can be observed in many perovskite oxides, making them potential candidates for SOFC anodes. Chromites and titanates have proven to be the most promising SOFC anode materials.[76,77]

Tungsten bronze-type materials have also been studied as possible anode materials. They have almost the same structure as perovskites, except for a rotation of some Ti/NbO_6 octahedra (Figure 14.6d). In this regard, a size increase is observed in 40% of the A-sites by their transformation from tetracapped square prisms to pentacapped pentagonal prisms; 40% remain unchanged and the rest of the remaining sites reduce in size.[78] Thus, the general formula of tungsten bronze-type materials can be written as $A_{0.6}BO_3$. It is important to note that the distortion of the octahedra results in the presence of B–O bonds with different bond lengths, some being shorter and some longer than the average B–O length. It has been suggested that it is the short B–O bond connections that provide a percolation path for the transfer of charge, which can lead to high electronic conductivity.[79] However, the conductivity of most tungsten bronze-type materials is significantly related to the level of pO_2. For instance, $Sr_{0.2}Ba_{0.4}Ti_{0.2}Nb_{0.8}O_3$ has been proven to have acceptable conductivity (10 S cm^{-1}), but only at $pO_2 = 10^{-20}$ atm and 930 °C.[80] Since the number of oxygen vacancies significantly depends on the operating conditions, it is the partial reduction of Ti^{4+}/Nb^{5+} at pO_2 lower than 10^{-17} atm at 930 °C that releases the required electrons for charge transfer. Nevertheless, $Sr_{0.2}Ba_{0.4}Ti_{0.2}Nb_{0.8}O_3$ still fails to provide acceptable performance when used as an anode material for SOFCs.

Pyrochlore materials with a general formula of $A_2B_2O_7$ are another type of anode material, and they have been extensively studied for their potential use in SOFCs.[81] Due to the presence of small empty A-sites in their structure, they show great similarity to the tungsten bronze structure, especially when observed from the (001) direction (Figure 14.6e). On the other hand, they can also resemble a perovskite structure with extra oxygen atoms in their lattices. Some materials with a pyrochlore structure (*e.g.* $[Gd_{0.98}Ca_{0.02}]_2Ti_2O_7$)

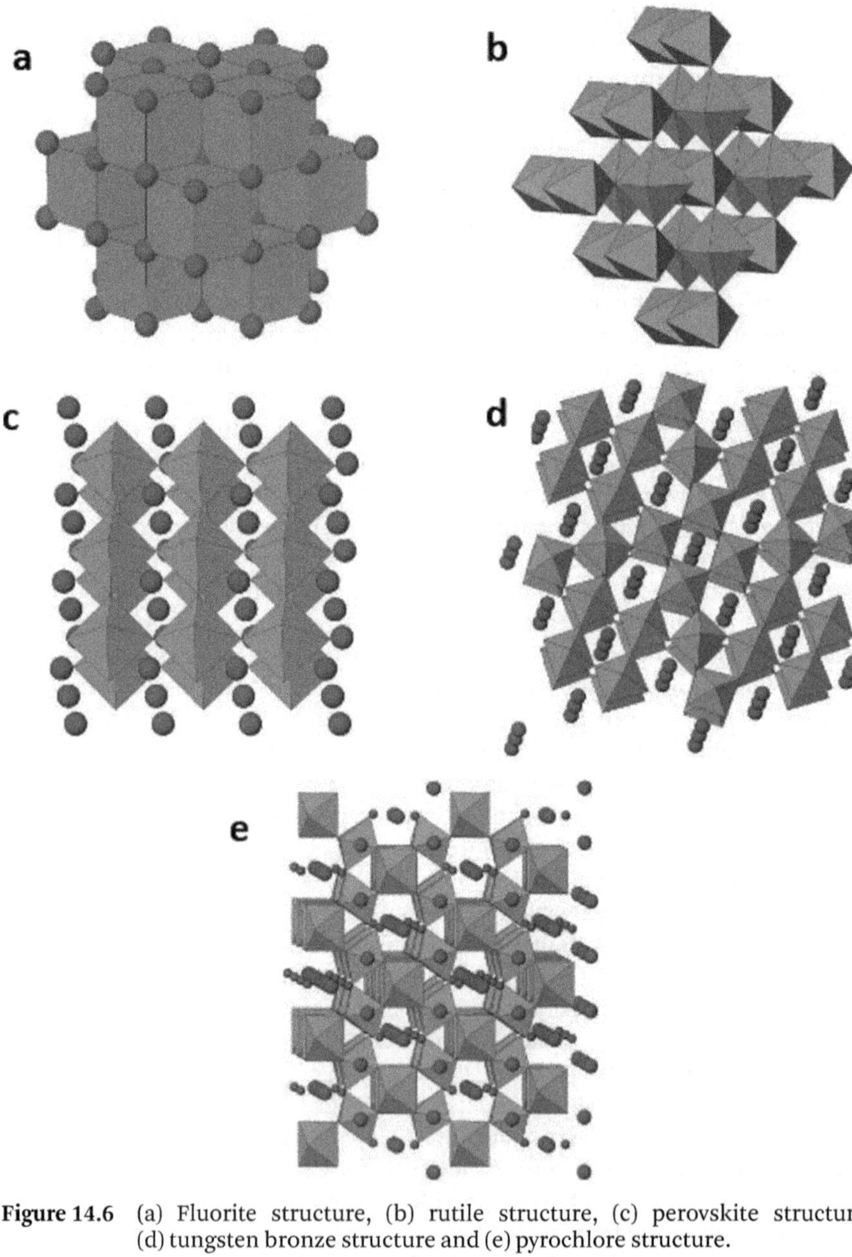

Figure 14.6 (a) Fluorite structure, (b) rutile structure, (c) perovskite structure, (d) tungsten bronze structure and (e) pyrochlore structure.

have been reported to exhibit high levels of mixed ionic/electronic conductivity, making them promising candidates for anode materials in SOFCs.[82] Although $(Gd_{0.98}Ca_{0.02})_2Ti_2O_7$ has been proven to exhibit ionic conductivities as high as YSZ, its electronic conductivity still needs to be improved if it is to be used as a SOFC anode. In this regard, it was assumed that the addition of multi-valance transition elements (*e.g.* Mn) into the $Gd_2Ti_2O_7$ lattice structure could improve its electronic conductivity.[83] However, it was further revealed that Mn with p-type conductivity only shows improvement at high pO_2, and it has almost no effect on conductivity at low pO_2. Thus, the addition of Mo as an n-type-favouring dopant was further studied. It was shown that $Gd_2Mo_2O_7$ exhibits metallic-like electrical conductivities as high as 10^2 S cm^{-1} at room temperature; nevertheless, this phase was very much unstable at non-reducing atmospheres.[81] Although successful attempts have been made to improve the stability of pyrochlore phases, they have been at the expense of reduced conductivity. From the pyrochlore phases investigated thus far, it seems difficult to reach both redox stability and high conductivity in these systems.

14.5 Hydrocarbon-fuelled SOFCs

The potential of SOFCs in directly using hydrocarbon fuels has given them a great advantage over conventional hydrogen-based fuel cells.[84] In this regard, removing the need for external fuel reformers and water gas shift reactors, especially when using relatively cheap and clean fuels such as natural gas, can greatly decrease the size, complexity and cost of an operating fuel cell system. By considering methane as the main component of natural gas, the reforming reactions in eqn (14.3) and (14.4) can directly take place inside a SOFC cermet anode:[85]

$$CH_4 + H_2O \leftrightarrow 3H_2 + CO \tag{14.3}$$

$$O + H_2O \leftrightarrow H_2 + CO_2 \tag{14.4}$$

However, the process of carbon deposition has long been recognized as the main consequence of internal reforming of hydrocarbon fuels. Thus, it is not surprising that many SOFC studies using hydrocarbon fuels have reported the gradual loss of cell performance due to the vast coverage of anode active sites by carbon compounds, especially when using Ni as part of the anode cermet.[86,87] A common method used in conventional steam reformers to overcome this issue is the use of high steam : carbon (S : C) ratios (*e.g.* up to 3). However, these high S : C ratios can greatly lower the electrical efficiency of SOFCs by steam dilution of the fuel.[88] In addition, since the steam reforming reaction (eqn (14.3)) has an endothermic nature, the unavoidable local cooling points can greatly affect the durability of SOFCs by causing possible mechanical damage to the anode layer.[89] Nevertheless, a combination of internal and external reforming activity or applying a process known as gradual internal reforming (GIR) of methane have been proven to greatly reduce the damaging effect of this endothermicity.[90–92] In the following, the advantages and disadvantages of commonly used anode cermets along with recent developments in anode materials will be described and compared to provide an overview of the ongoing search for new generations of anode materials for hydrocarbon-fuelled SOFCs.

14.5.1 Ni–$ZrO_2(Y_2O_3)$ Cermet

Currently, the most common material used as an anode in SOFCs is Ni/YSZ cermet. Its low cost and high chemical stability in reducing atmospheres along with its close TEC to that of YSZ electrolytes has made it a suitable anode material for SOFC applications. To facilitate the transport of the fuel and the product gases to and from the TPBs, the presence of more than 30% continuous porosity is an important requirement for the anode layer in which, in a compromise between a high level of porosity and high mechanical stability, Ni/YSZ cermets have been proven to provide a very good level of balance. Nickle itself has long been considered as an excellent electrocatalyst and reforming catalyst for electrochemical oxidation of hydrogen, in addition to its eminent electronic conductivity for an anode. In the anode cermet, YSZ constitutes a framework for the dispersion of Ni particles and greatly broadens the TPBs by offering a significant part of the ionic contribution to the overall conductivity in the anode. In addition, the arbitrary mixing ratio of YSZ with Ni also makes it possible to control the TEC of the anode layer to match with those of other SOFC components.[89] Since YSZ and Ni are almost immiscible and non-reactivate over a wide temperature range, it is possible to fabricate cells by conventional sintering methods, where the developed thin microstructure can be maintained for a long period of time when in service. Previously, a two-step fabrication process was used to apply the Ni/YSZ anode cermet on the electrolyte surface. In the first step, a Ni slurry was deposited and dried over the electrolyte surface, then YSZ was grown around the Ni particles using the EVD method. However, the development of more cost-effective fabrication methods, involving simultaneous deposition of Ni/YSZ followed by adequate final sintering, has greatly reduced fabrication costs and has enhanced the commercialization of SOFCs.

Despite all of the mentioned advantages, it is well known that carbon deposition is the main issue of Ni-based cermets when operating under hydrocarbon fuels. In addition, the high tendency of Ni in these anodes to dissolve carbon can lead to considerable volume expansion, leading to severe structural failures.[93] The problem with YSZ regarding this issue arises from its inertness and inability to lessen the consequences of solid coke formation on nickel. The very low electronic conductivity and low electrochemical oxidation activity of YSZ practically leads to a complete failure of the whole cell once the nickel deactivates. Thus, much research has focused on finding alternative materials to YSZ, in which ceria-based electrolytes have shown great potentials, especially when using liquid oxygenated hydrocarbon fuels.[94,95] Liu *et al.* studied the overall performance and degradation rate of Ni–Sm-doped CeO_2 (Ni-SDC) anodes operating under methanol, methane and ethanol.[96] While relatively stable performance was achieved for direct methanol-fuelled cells, a rapid degradation rate with severe carbon deposition was reported for cells operating under methane and ethanol. Although in this study an acceptable stability was reported when using methanol, the overall electrocatalytic activity of Ni-SDC for methanol oxidation was much lower than its activity when operating under H_2 fuel.[97] In addition, the low durability of Ni-SDC in ethanol fuel shows that the coking resistance of the Ni-SDC anode still needs to be improved.

To overcome the issue of carbon deposition on the Ni-YSZ anode, we first need to look at the possible chemical reactions and conditions in which carbon is formed on the anode surface: (1) methane cracking happening at high temperatures and at low ratios of carbon to steam in thermodynamic equilibrium, shown in eqn (14.5); (2) carbon monoxide reaction, shown in eqn (14.6); and finally (3) CO disproportionation taking place at low temperatures and high concentrations of CO, shown in eqn (14.7).[98]

$$\mathrm{H_4 \rightarrow C + 2H_2},\ \Delta H_{298\,K} = -172\,\mathrm{kJ\,mol^{-1}} \tag{14.5}$$

$$\mathrm{O + H_2 \rightarrow C + H_2O},\ \Delta H_{298\,K} = -131\,\mathrm{kJ\,mol^{-1}} \tag{14.6}$$

$$\mathrm{2CO \rightarrow C + CO_2},\ \Delta H_{298\,K} = -172\,\mathrm{kJ\,mol^{-1}} \tag{14.7}$$

In general, the anode material, fuel composition and operation temperatures determine the morphology of the deposited carbon.[98–100] The mechanism in which carbon induces corrosion on the Ni-based anode of SOFCs has been extensively studied by Chun *et al.*[101] It was suggested that at high temperatures (~750 °C), it is the metal dusting corrosion of nickel that acts as a key factor in the cell performance deterioration of hydrocarbon-fuelled SOFCs. It was further discussed that this phenomenon is caused by a combination of carbon precipitation and diffusion in the bulk metal followed by Ni atom migration through surface carbon deposition. At high operating temperatures, carbon is initially transferred from the fuel environment to the nickel surface, and then graphite is produced and precipitated on the Ni surface *via* the catalytic graphitization mechanism. Through this process, regional basal graphite planes were found to be formed in perpendicular orientations to the Ni surface. It is in the channels formed between these planes that small Ni particles are shaped due to the intercalation and coalescing of Ni cations in these regions. The presence of these small Ni particles on the outer surface of the anode layer further catalyses the surface carbon formation, transporting the metal particles away from the surface and causing severe metal dusting corrosion (Figure 14.7). Therefore, it is the unrestricted diffusion of Ni atoms *via* the channels between the basal planes of graphite that greatly decreases the conductivity and overall cell performance of SOFCs at high operational temperatures (~750 °C), whereas it is suggested that at lower operating temperatures (~650 °C), it is mainly the vast carbon deposition on the Ni-based anode material that restricts the fuel diffusion to the TPBs, greatly reducing the electrochemical reaction sites on the anode.[101] Therefore, many researchers have focused on improving anode performance and carbon tolerance by decreasing the working temperature (below 700 °C) and increasing the current density to facilitate carbon oxidation on the anode surface.[102]

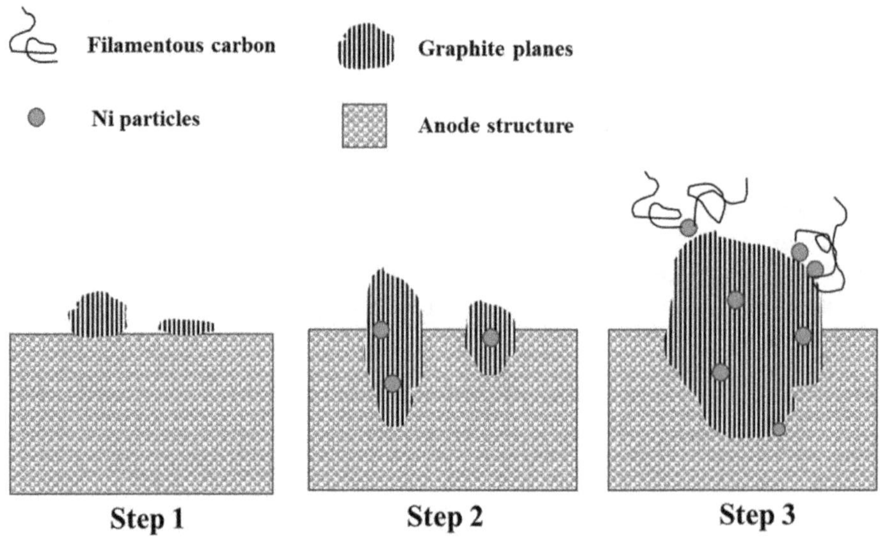

Figure 14.7 Schematic representation of Ni-dusting corrosion at high operating temperatures.

When using oxygenated hydrocarbons as fuel in SOFCs, the O : C ratio in the chemical structure of the fuel has been proven to play an important role in the durability of the cell. In this regard, SOFCs using oxygenated hydrocarbon fuels with high O : C ratios (*e.g.* formic acid or ethylene glycol [EG]) have shown superior durability, even when used with traditional Ni-based cermets.[103,104] Qu *et al.* have used oxygen-temperature programmed oxidation to study the carbon deposition rates of different carbon-containing fuels on Ni-YSZ anodes and have reported EG as a highly coking-resistant fuel for SOFCs.[103] The tested Ni-YSZ anode operating under EG steam (1 : 1) fuel not only showed very low performance loss over a 200 h test, but also presented attractive peak power densities (PPDs) as high as 1.2 W cm^{-2} at 750 °C, being about 86% of the PPD of H_2 fuel.

14.5.2 Alloying Ni with Metals

Partial replacement of Ni with more materials that are more inert towards the hydrocarbon cracking reaction has been proven to supress coke formation over Ni. The improved coke tolerance of Ni alloys can occur through decreasing the formation rate of carbon–carbon bonds and/or facilitating competing reactions, such as carbon oxidation.[105–108] However, reducing the amount of Ni can greatly decrease the electrocatalytic activity of the anode towards oxidation reactions. Jeong *et al.* have reported the synthesis of heterostructured nickel and ruthenium bimetallic catalysts *via* atomic layer deposition (ALD) as an active anode material for direct methanol low-temperature SOFCs.[105] The compared PPDs between the synthesized Ni/ALD Ru and

Pt/ALD Ru (known as the best catalyst for direct methanol fuel cells) showed less than 7% difference at 300 °C. In this study, the higher electrocatalytic activity of Ni/ALD Ru than the Ni anodes was attributed to the absence of Ni agglomeration and very low level of coke formation. Although the development of such high-performance anodes can find applications in different types of electrochemical energy conversion devices, the scaling up of such heterostructured anodes prepared by the ALD technique requires much more development in order to reach commercialization.

The idea of reducing the cost of SOFCs has directed many studies towards finding cheaper alternatives when it comes to material selection. So far, transition metals, such as cobalt and iron, have shown great potential in decreasing the level of graphitic carbon formation on the anode active sites. Huang *et al.* have investigated the effect of different Fe to Ni ratios on the overall performance and carbon deposition levels of Fe–Ni/Scandia-stabilized zirconia (ScSZ) composite anodes running on ethanol fuel.[106] Although their work reveals the positive effect of Fe on improving the coke resistance of Ni-based cermet anodes, its excessive addition has also been shown to greatly reduce the electrocatalytic activity of the anode towards the full oxidation of ethanol fuel. In this study, a 1 : 1 ratio between Fe and Ni ($Fe_{0.5}Ni_{0.5}$/ScSZ) was proven to provide a good balance between the coke resistivity and electrocatalytic activity of the anode when operating under ethanol. Nevertheless, the very poor long-term stability of Ni–Fe cermet anodes is considered the main drawback for their future market penetration. Ni–Co alloy composites have also proven to be promising anode materials for hydrocarbon-fuelled SOFCs. A recent study done by Ding *et al.* has investigated the performance of a $Ni_{1-x}Co_xO$-$Ce_{0.8}Sm_{0.2}O_{1.9}$(SDC) composite as a promising anode material for methanol-fuelled SOFCs.[107] In this study, $Ni_{0.9}Co_{0.1}$-SDC was found to deliver the lowest polarization resistance, highest PPD (675 mW cm^{-2}, 700 °C) and highest coking resistance under a certain current density. Additionally, it was confirmed that the addition of Co to the anode composition not only expanded the TPBs through enhancing the interaction between NiO and SDC, but also greatly reduced the anode particle size and improved the $Ni_{1-x}Co_xO$ dispersion by hindering the grain growth of NiO particles. In addition to Fe and Co, relatively high-performance Ni-based anode alloys with low precious metal loadings, such as Pt, Pd and Ru, have also been proven to greatly increase coke tolerance in SOFCs.[109,110] In this regard, Nabae *et al.* studied the electrocatalysis of Pd–Ni anodes for the direct oxidation of dry CH_4 at 800–900 °C.[110] The results revealed stable power generation in a dry methane atmosphere with negligible levels of cell degradation and carbon deposition. Tin (Sn) has also been proven to be a promising addition to Ni-based anodes for increasing both carbon and sulfur tolerance in SOFCs.[108,111,112] However, most studies have only observed significant suppression of coke formation for Ni alloys containing low levels of Sn. These studies stress the importance of controlling the amount of Sn as the secondary metal in Ni–Sn alloys, since excess addition of Sn can greatly decrease the electrocatalytic activity and stability of Ni-based anodes.[111]

14.5.3 Surface Decoration of Ni-based Anodes with Metals and Metal Oxides

It has been reported that decorating Ni-based anodes with other metal/metal oxide materials, forming a second phase in the anode structure, could suppress coke formation in the anode. The second phase is usually added on the surface of the Ni particles in very small quantities.[113,114] For instance, Wang *et al.* have reported the deposition of Cu on a Ni-based anode for methane-fuelled SOFCs.[115] In this work, Cu was impregnated into an already prepared Ni/SDC porous matrix and the cell was operated at intermediate temperatures (600 °C). The final molar ratio of Cu/Ni was 1 : 19 and it was simply controlled by the number of impregnation steps carried out. Over a 12 h operation, only 2% loss was observed in the power density, presenting an effective suppression of coke formation on the anode. It was suggested that the improved carbon tolerance is due to the decrease in the exposed Ni surface area and the level of carbon monoxide disproportionation. Pd has also been proven to be a useful modifier for Ni-based SOFCs, especially when operated in ethanol. Babaei *et al.* have studied the electrochemical performance and carbon tolerance of Pd-promoted Ni-based anodes in both methane and alcohol fuels.[116] The presence of Pd nanoparticles greatly promoted the electrocatalytic activity of the anode operating under methane and significantly enhanced the coking resistance of Ni-based anodes in ethanol-fuelled SOFCs. The absence of filament carbon in the methanol- and ethanol-fuelled SOFCs suggested a possible correlation between the presence of hydroxide groups in the molecular structure of the fuel, here ethanol and methanol, and the level of carbon removal from the surface of the anode. Many studies have reported successful surface modification of Ni-based cermet anodes using various oxides, such as $BaZrO_3$, $PrBaMn_2O_{5+\delta}$ and $BaCe_{0.9}Yb_{0.1}O_{3-\delta}$, as a promising method for improving both coking and sulfur tolerance in SOFCs.[117–119] Cimenti *et al.* have decorated the surface of Ni-YSZ anodes using $Zr_{0.35}Ce_{0.65}O_{2-\delta}$ (ZDC) to increase their coking resistance when operated in methanol fuel.[120] It was reported that the addition of ZDC significantly increased the coking resistance and fuel electro-oxidation of the modified anode, possibly due to the high oxygen storage capacity of ZDC and the improved conductivity of the anode. It was further revealed that the addition of ZDC resulted in the formation of a much more reactive type of carbon, making it possible to remove the deposited carbon without destroying the anode microstructure. Recently, the addition of a metal/metal oxide mixture of Mo/MoO_3 at very small quantities has been proven to improve both the coking resistance and electrocatalytic activity of Ni-based anodes operating in methanol.[121] The durability of the cell was reported to be improved by tailoring the ratio of Ni to Mo (1 : 0.03 molar); however, the effect of this ratio on the valance state of Mo needs to be further studied due to the significant effect it can have on the $Mo:MoO_3$ ratio in the anode.

14.5.4 Modification of the Ceramic Phase in Ni-based Anodes

In both hydrogen- and hydrocarbon-fuelled anodes, water is produced as a by-product of fuel oxidation. It was suggested that by improving the water storage capacity of the anode, especially when using hydrocarbon fuels, the *in situ* formed water can be used to facilitate carbon removal from the surface, significantly decreasing the required ratio of steam to fuel for the steam reforming process. Following this hypothesis, a considerable research effort has been devoted to increasing the water storage capability of Ni-based cermet anodes by modifying their ceramic phase.[122–124] For instance, Lei *et al.* have reported the successful modification of Ni-based anode surfaces using nanostructured barium oxide operating in C_3H_8, CO and gasified carbon fuels at 750 °C.[125] It was revealed that the high level of absorbed water at the interface of BaO–Ni particles facilitates water-mediated carbon removal reactions, dramatically suppressing carbon deposition over the Ni-YSZ anode. BaO nanoparticles with high levels of water storage capacity have been proven to promote the dissociation of water to OH and H species, which further facilitates the gasification of carbon at the Ni–BaO interface. However, one downside to the use of nanosized BaO is its high tendency to agglomerate, evaporate and react with fuel impurities at high temperatures. In addition, the inherent insulating property of BaO makes it necessary to only apply very low BaO coverage on the anode, since over-coverage of the Ni surface would block charge transfer within the anode. Consequently, the water adsorption capability of the anode would be very limited, and so high-precision coating methods are required.

To overcome this shortcoming, proton-conducting perovskite oxides with a strong bulk interaction with Ni and high water storage capacity, such as $BaZr_{0.4}Ce_{0.4}Y_{0.2}O_{3-\delta}$ (BZCY), have been introduced as the ceramic phase for Ni-based anodes to inhibit coke formation.[126] BZCYs have long been studied due to their wide range of applications in high-temperature electrochemical devices. Their great advantage is that although they do not have the host constituents to liberate conducting protons, they can take the protons from the dissociation of the captured water molecules in the environment.[127,128] Figure 14.8 illustrates a possible mechanism of water-mediated carbon removal on the Ni + BZCY anode surface, as proposed by Wang *et al.*[122] As has been presented, the presence of BZCY particles triggers water incorporation or storage, promoting the dissociation of water to OH species. Then, the formed OH species further react with the absorbed C atoms near the Ni–BZCY interface, producing CO and H. Finally, the produced CO and H are later electrochemically oxidized in the TPBs, thus increasing the coking resistance and power density of the Ni-based anode. To further improve the coking tolerance, water storage capacity and electrochemical activity of Ni + BZCY anodes, Wang *et al.* studied the effect of decreasing the amount of Zr and adding Yb to the structure.[129] The results showed that doping barium zirconate-cerate with rare earths Y and Yb can greatly enhance the resistance of Ni-based anodes to coking and even to the presence of traces of hydrogen sulfide.[87] It was further suggested that this enhanced catalytic activity for sulfur oxidation and hydrocarbon cracking and

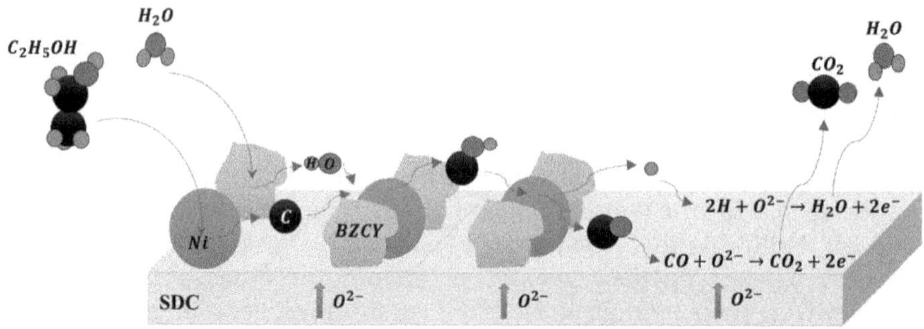

Figure 14.8 Proposed mechanism for the water-mediated carbon removal process on the Ni + BZCY anodes.

reforming is greatly linked to the mixed proton and oxygen ion conductivity of $BaZr_{0.1}Ce_{0.7}Y_{0.2-x}Yb_xO_{3-\delta}$ (BZCYYb). The improved durability and catalytic activity towards the steam reforming reaction and high coking resistance of Ni + BZCYYb cermets has made them promising candidates for tackling the carbon deposition problems associated with Ni-based anode materials in SOFCs, especially when operated in ethanol.

14.5.5 Copper and Other Noble Metal Cermets

Cu has been found to be almost inactive to coke formation when used as an anode in hydrocarbon-fuelled SOFCs.[130] However, its low melting point and the high sintering temperatures required in the conventional solid-state fabrication processes make it impractical to use Cu in the anode. In SOFCs, infiltration of a porous structure with metal/metal oxide nanoparticles was first used for the preparation of Cu-based anodes. Applying this method not only made it possible to use materials with low melting points in the anode, but also simultaneously maximized the surface area and suppressed carbon deposition by limiting the area of graphitic carbon growth.[131,132] However, in addition to the added complexity of anode preparation, there are many problems associated with the use of infiltration for Cu-based anodes, especially the issues related to their durability and scaling-up processes to meet the requirements of industrial-sized anodes. A study done by Lu *et al.* concluded that Cu does not present any catalytic activity towards the breaking of hydrocarbons in SOFCs, and it solely acts as an electron conductor in the anode.[133] It was further discussed that it is this poor catalytic activity of Cu in breaking C–H and C–C bonds that has resulted in its high stability in the environment of hydrocarbon fuels. However, this behaviour also leads to a great increase in the activation polarization of the cell. Thus, the combination of Cu with other materials with high catalytic activity towards the oxidation of hydrocarbon fuels would make it possible to enhance both stability and oxidation activity in the anode.

Kim *et al.* have studied the performance of different compositions of Cu–Ni alloys as anode materials at 800 °C in dry methane.[52] It was reported that the presence of Cu led to greatly decreased carbon deposition and a significant increase in the recorded power density. The authors attributed this enhancement to the increased electronic conductivity of the anode due to the presence of Cu. The excellent catalytic activity of ceria-based materials has made them ideal candidates for anode materials, especially when combined with Cu. Tsipis *et al.* have studied the performance of both Cu-gadolinium-doped ceria (Cu-GDC) and Ni-GDC cermets as potential anode materials for SOFCs.[134] They used a three-electrode technique to study the current density and overpotential characteristics of both anodes in a wet 10% H_2 and 90% N_2 gas mixture (600–800 °C). The results indicated a lower overpotential for the Ni-GDC anode, which was linked to the high sinterability and very low catalytic activity of Cu when compared to Ni. However, when compared with Ni-YSZ anodes (operating in H_2 and CO), Cu–CeO_2-YSZ composite anodes demonstrated significant improvements in overall cell performance.[130] While different maximum power densities were observed for Ni-YSZ anodes operating under H_2 (136 mW cm^{-2}) and CO (73 mW cm^{-2}) fuel atmospheres, Cu–CeO_2-YSZ-based anodes presented similar PPDs in both H_2 and CO fuels at 700 °C (305 mW cm^{-2}). It was also demonstrated that the addition of cobalt to Cu–CeO_2-YSZ anode cermets can further enhance catalytic activity, increasing the PPD of the cell up to 310 mW cm^{-2} with H_2 and 370 mW cm^{-2} with CO fuel. Further extensive research on Cu-based anodes can be found in the literature.[131,132,135]

14.5.6 Ceria and Other Compounds

Acting as a mixed ionic–electronic conductor, CeO_2 has been extensively studied as an important additive in anode composites for SOFCs. Its significant catalytic activity towards fuel oxidation combined with its high ionic conductivity has helped to expand the reaction zones beyond TPBs. Furthermore, the addition of dopant oxides such as CaO, Y_2O_3, Gd_2O_3 and Sm_2O_3 has been proven to improve its ionic conductivity, and it shows good level of compatibility with adjoining cell component materials.[136,137] Studies have confirmed the attractive catalytic activity of cerium-based materials towards the direct oxidation of hydrocarbon fuels (*e.g.* methane, ethane, 1-butene, *n*-butane and toluene).[138] Park *et al.* have studied the direct electrochemical oxidation of hydrogen and different hydrocarbon fuels on Cu–CeO_2 anode composites tested at 700–800 °C in which Cu only provides electronic conductivity in the anode, while ceria acts as a mixed ion conductor and catalyst towards fuel oxidation. The results indicated a reasonable PPD for both *n*-butane- and H_2-fuelled anodes (0.18 and 0.31 W cm^{-2} at 800 °C, respectively), and the final oxidation products were reported to be CO_2 and H_2O. Infiltration of YSZ scaffolds with various metal/metal oxide composites has been proven to be a useful technique for studying the effects of different materials on anode stability and electrochemical performance.[139] Kurokawa *et al.* reported a high level of sulfur tolerance when conventional Ni-YSZ anodes were infiltrated with ceria nanoparticles.[140] An appreciable sulfur tolerance was observed when the cell was operated on humidified H_2 fuel containing 40 ppm H_2S at 700 °C for 500 h.

In similar works, infiltration of $La_{0.75}Sr_{0.25}Mn_{0.5}Cr_{0.5}O_3$ and Nb-doped $SrTiO_3$ with GDC and nanosized GDC, respectively, presented enhanced performance when compared to conventional Ni-YSZ anodes.[141–143] Although considerable enhancement in the overall cell performance and carbon deposition/sulfur poisoning resistance has been observed for most infiltrated Ni-YSZ anodes, the significant initial drop of cell voltage still needs to be addressed in these anodes. It has been suggested that the difficulty in achieving a uniform and homogeneous layer during the infiltration of the anode surface can contribute to this sudden initial decay in the cell voltage. In addition, attempts at forming a continuous dense layer through impregnation of the anode surface can possibly lead to the blockage of electron pathways, decreasing the TPBs in the modified Ni-YSZ anodes.

The idea of using ceramics with mixed oxide ion and electronic conductivity was applied in order to expand the charge transfer reaction along the entire gas–electrode interface. Mixed conducting ceramics such as doped $LaCrO_3$, $La_{0.6}Sr_{0.4}Co_{0.2}Fe_{0.8}O_3$ (LSCF), $ZrO_2/TiO_2/Y_2O_3$ and La-, Fe- and Y-doped $SrTiO_3$ are some of the noteworthy endeavours in the literature.[92,144–146] Although they have been proven to improve the stability of anodes towards carbon deposition and sulfur poisoning, they hold certain performance limitations (*e.g.* lower catalytic activity is observed for $La_{1-x}Sr_xCr_{1-y}Ni_yO_{3-\delta}$ when compared to Ni composites operating in H_2/CH_4 fuels).[147] In addition, some (*e.g.* LSCF) possess a high level of instability in reducing atmospheres and weight loss at high operating temperatures, which greatly limit their use, despite presenting promising performances.[148] Thus, modified mixed conducting cermets such as $La_{0.8}Sr_{0.2}Cr_{0.97}V_{0.03}O_3$ (LSCV) have been developed, showing significant stability at high operating temperatures (800 °C), in addition to their great level of coke resistance and catalytic activity towards CH_4 steam reforming.[92]

As an attempt to further improve the long-term stability of the anode, Skarmoutsos *et al.* studied the effect of Ti on the ionic conductivity of YSZ and further evaluated its effects in terms of an anode substrate or anode material in H_2-fuelled SOFCs.[149] Cermets of Ni–$Y_{0.25}Zr_{0.60}Ti_{0.15}O_{2-x}$ (Ni-YZT) with different Ni contents were synthesized and exposed to 1000 °C for up to 1000 h and their stability and electrochemical performance was compared to that of Ni/yttria (8 mol%)-stabilized zirconia (Ni/8YSZ). The results indicated an improvement in the long-term stability of the anode in terms of microstructure and electronic conductivity, suggesting it to be the result of a change in the bonding strength between Ni and the cermet interface due to the presence of Ti. However, a higher electrochemical performance was reported for Ni/8YSZ when compared to Ni-YZT, which was further attributed to the lower ionic conductivity of YZT than that of 8YSZ.[149] In a similar work, Tsai *et al.* attempted to decrease the interfacial resistance between the 8YSZ electrolyte and the Ni-YSZ anode by introducing a thin layer of yttria-doped CeO_2 (YDC) and YZT between the two layers.[150] Impedance spectroscopy measurements at temperatures ranging from 600 to 750 °C confirmed a reduction in the interfacial resistance between the anode and electrolyte layer, where the lowest resistances were obtained using 0.5 μm-thick YDC interfacial layers and were reported to be 0.13 and 0.29 Ω cm^2 at 750 and 600 °C, respectively. It was suggested that the enhanced charge transfer between the two layers was a result of the improved redox reaction and/or mixed conductivity rate of the interfacial layer.[150]

The oxidation of hydrogen fuel on a Ni-based anode can be explained by the following steps: (1) transfer of the fuel to the anode region; (2) diffusion of the fuel gas through the porous structure of the anode; (3) physical adsorption on the nickel surface; (4) diffusion of the hydrogen species on the surface to the TPBs; and (5) oxidation reaction involving charge transfer.[151,152] It has been proven that restriction in the surface diffusion rates and limitations in the material conductivity play major roles in controlling anode polarization resistance. Hence, controlling the microstructure of the anode can play an important role in minimizing the surface diffusion resistance and the resulting polarization.[152,153] In this regard, Wu *et al.* have reported the synthesis and application of flake-shaped NiO-YSZ particles as anode materials and observed a significant enhancement in cell performance when compared with conventional anode powders in Ni-YSZ/YSZ/LSM single cells operating in hydrogen as the fuel.[154] In addition, the catalytic activity of flake-shaped NiO-YSZ particles towards CH_4 reforming with both CO_2 and steam was also improved when compared with that of commercial powder anodes. It was further discussed that the flake shape and uniform structure of the synthesized anode material significantly improved the surface diffusion rate and electrical conductivity of the anode layer, resulting in lower polarization resistance.[155] Further details of nanostructured and other noteworthy materials investigated for SOFC anodes, along with their advantages and drawbacks, can be found in the literature.[156,157]

14.6 Conclusion

In this chapter, a brief overview of the working principles and developments regarding anode materials for hydrocarbon-fuelled SOFCs is presented. It is shown that Ni-based anode cermets still exhibit higher PPDs when compared with nickel-free anodes without precious metals. However, significant modification needs to be done on the microstructure and composition of Ni-based anodes to overcome the severe coke formation observed when operating in hydrocarbon fuels. External reforming of hydrocarbon fuels to CO and H_2 is still considered the main solution for this shortcoming. However, due to the added complexity and the loss in efficiency of hydrocarbon-fuelled SOFCs, a great deal of effort is being placed on removing the need for an external reforming step.

In this chapter, four main strategies are reviewed as possible solutions to the problems associated with direct conversion of hydrocarbon fuels in SOFCs. First is modifying the operating conditions of conventional Ni-based anode SOFCs. Based on the type of hydrocarbon fuel, it was shown that it is possible to lower the carbon deposition and cell degradation rates by decreasing the operating temperatures to the intermediate range. It was illustrated that when using fuels with low rates of carbon deposition (*e.g.* methane), the rate of steam formation and oxygen ion transfer between both electrodes would be enough to remove the deposited carbon at temperatures below 650 °C. Although operating at such low temperatures eliminates the degradation of conventional Ni-based anodes through Ni dusting processes, the increased polarization resistance and reduced ionic conductivity significantly decrease the generated power density. The second mentioned strategy consists of modifying Ni-based anodes through surface decoration, Ni alloy formation or partial substitution of Ni with more inert materials towards carbon formation reactions (*e.g.* Cu and Ce). Although all of the mentioned methods have been proven to significantly increase the coke formation resistance of Ni-based anodes, high agglomeration rates, instability in reduced atmospheres and low electrocatalytic activity towards fuel oxidation are still the main drawbacks of such modified anodes. The third strategy for avoiding external reforming involves the use of mixed and electronic conductors as anode materials. The use of mixed La-doped strontium titanate with ceria-doped $SrTiO_3$ is an example of such anodes, which have presented resealable performance when operated in methane.[158,159] The fourth and final approach herein discussed is the modification of the anode ceramic phase and the use of alternative hydrocarbon fuels, such as liquid oxygenated hydrocarbons with high O : C ratios in their chemical structures. This approach is mainly based on acknowledging the formation of steam as the main oxidation product in the anode layer, which can be further used as a valuable source for direct fuel reforming. In summary, by observing the current rate of development in SOFC anode materials, we can be hopeful of seeing highly durable hydrocarbon-fuelled SOFCs in the near future.

Appendix A. Types of Fuel Cells

Due to the wide power generation range in fuel cells, they can be employed in almost any application requiring power. However, with different fuel cell technologies available, it is essential to indicate the type of technology best suited to a specific application. Regarding small range power devices, fuel cells can be employed in personal electronic equipment such as personal computers and mobile phones. Medium scale power generating fuel cells have found widespread commercial applications, including fuel cell vehicles and domestic appliances. Finally, fuel cells capable of generating large power range (1–10 MW) are used in distributed power systems and grid-quality alternating current electricity. Table 14.2 presents such applications, key features, and main advantages/disadvantages of different fuel cell technologies.

Table 14.2 Comparison of fuel cell technologies.[160]

Fuel cell type	Common electrolyte	Operating temperature	Efficiency	Applications	Advantages	Disadvantages
PEMFC	Perfluoro sulfonic acid	<120 °C	60% direct H_2; 40% reformed fuel	• Backup power • Portable power • Distributed generation • Transportation • Speciality vehicles	• Solid electrolyte reduces corrosion and electrolyte management problems • Low temperature • Quick start-up	• Expensive catalysts • Sensitive to fuel impurities • Low temperature waste heat
AFC	Aqueous solution of potassium hydroxide soaked in a matrix	<100 °C	60%	• Military • Space	• Cathode reaction faster in alkaline electrolyte, leads to high performance • Low-cost components	• Sensitive to CO_2 in fuel and air • Electrolyte management
PAFC	Phosphoric acid soaked in a matrix	150–200 °C	40%	• Distributed generation	• Higher temperature enables CHP • Increased tolerance to fuel impurities	• Pt catalyst • Long start-up time • Low current and power
MCFC	Solution of lithium, sodium and/or potassium carbonates soaked in a matrix	600–700 °C	50%	• Electric utility • Distributed generation	• High efficiency • Fuel flexibility • Can use a variety of catalysts • Suitable for CHP	• High-temperature corrosion and breakdown of cell components • Long start-up time • Low power density
SOFC	Yttria-stabilized zirconia	500–1000 °C	60%	• Auxiliary power • Electric utility • Distributed generation	• High efficiency • Fuel flexibility • Can use a variety of catalysts • Solid electrolyte • Suitable for CHP combined hydrogen, heat and power (CHHP) • Hybrid/gas turbine (GT) cycle	• High-temperature corrosion and breakdown of cell components • High-temperature operation requires long start-up time and limits

References

1. A. Kirubakaran, S. Jain and R. Nema, A review on fuel cell technologies and power electronic interface, *Renewable Sustainable Energy Rev.*, 2009, **13**(9), 2430–2440.
2. M. Chourashiya and L. Jadhav, Synthesis and characterization of 10% Gd doped ceria (GDC) deposited on NiO-GDC anode-grade-ceramic substrate as half cell for IT-SOFC, *Int. J. Hydrogen Energy*, 2011, **36**(22), 14984–14995.
3. A. B. Stambouli and E. Traversa, Solid oxide fuel cells (SOFCs): A review of an environmentally clean and efficient source of energy, *Renewable Sustainable Energy Rev.*, 2002, **6**(5), 433–455.
4. V. Kharton, F. Marques and A. Atkinson, Transport properties of solid oxide electrolyte ceramics: A brief review, *Solid State Ionics*, 2004, **174**(1), 135–149.
5. A. Lashtabeg and S. J. Skinner, Solid oxide fuel cells – a challenge for materials chemists?, *J. Mater. Chem.*, 2006, **16**(31), 3161–3170.
6. W. Grove, On a gaseous voltaic battery, *J. Franklin Inst.*, 1843, **35**(4), 277–280.
7. G. Hoogers, *Fuel Cell Technology Handbook*. CRC Press, Boca Raton, FL, 2002.
8. D. Bose, Fuel cells: The fuel for tomorrow, *J. Energy Res. Environ. Technol.*, 2015, **2**(2), 71–75.
9. J. M. Andújar and F. Segura, Fuel cells: History and updating. A walk along two centuries, *Renewable Sustainable Energy Rev.*, 2009, **13**(9), 2309–2322.
10. F. Schönbein, *The Letters of Faraday and Schönbein*. Universidad de California 1899; Available from: http://www.archive.org/stream/lettersoffaraday00fararich#page/32/mode/2up.
11. W. R. Grove, *LXXII*. On a gaseous voltaic battery, *Philosophical Magazine Series 3*, 1842, **21**(140), 417–420.
12. P. Chaurasia, Y. Ando and T. Tanaka, Regenerative fuel cell with chemical reactions, *Energy Convers. Manag.*, 2003, **44**(4), 611–628.
13. A. Appleby, From Sir William Grove to today: Fuel cells and the future, *J. Power Sources*, 1990, **29**(1–2), 3–11.
14. W. W. Jacques, *Electricity direct from coal*. Harpers New Monthly Magazine, 1896, **96**, 144–150.
15. C. Stone and A. E. Morrison, From curiosity to "power to change the world®", *Solid State Ionics*, 2002, **152**, 1–13.
16. E. Baur and H. Preis, Über Brennstoff-Ketten mit Festleitern, *Ber. Bunsen-Ges.*, 1937, **43**(9), 727–732.
17. N. Walther, *Electrical Incandescent Lamp*, US Patent App. 0/906,550, 1897.
18. W. Nernst, *Material for Electric-lamp Glowers*, US Patent 685730, 1901.
19. S. C. Singhal and K. Kendall, *High-temperature Solid Oxide Fuel Cells: Fundamentals, Design and Applications*, Elsevier, Amsterdam, 2003.
20. F. T. Bacon, Research into the Properties of the Hydrogen-Oxygen Fuel Cell, *British Electrical and Allied Manufacturers Association Journal*, 1954, **61**(6), 6–8.

21. M. N. Eisler, "A Modern Philosopher's Stone": Techno-analogy and the Bacon Cell, *Technology and Culture*, 2009, **50**(2), 345–365.
22. F. Bacon, Fuel cells, past, present and future, *Electrochim. Acta*, 1969, **14**(7), 569–585.
23. F. T. Bacon, The fuel cell: Some thoughts and recollections, *J. Electrochem. Soc.*, 1979, **126**(1), 7C–17C.
24. L. Niedrach and H. Alford, A new high-performance fuel cell employing conducting-porous-teflon electrodes and liquid electrolytes, *J. Electrochem. Soc.*, 1965, **112**(2), 117–124.
25. W. Grubb and L. Niedrach, Batteries with solid ion-exchange membrane electrolytes II. Low-temperature hydrogen–oxygen fuel cells, *J. Electrochem. Soc.*, 1960, **107**(2), 131–135.
26. J. M. Grimwood, B. C. Hacker and P. Vorzimer, *Project Gemini Technology and Operations – A Chronology*, 1969, National Aeronautics and Space Administration (NASA), Washington, DC.
27. A. L. Dicks, Molten carbonate fuel cells, *Curr. Opin. Solid State Mater. Sci.*, 2004, **8**(5), 379–383.
28. G. Elmore and H. Tanner, Intermediate temperature fuel cells, *J. Electrochem. Soc.*, 1961, **108**(7), 669–671.
29. K. Kiukkola and C. Wagner, Measurements on galvanic cells involving solid electrolytes, *J. Electrochem. Soc.*, 1957, **104**(6), 379–387.
30. J. Weissbart and R. Ruka, Oxygen gauge, *Rev. Sci. Instrum.*, 1961, **32**(5), 593–595.
31. D. Archer *et al.*, *An Investigation of Solid-Electrolyte Fuel Cells*, Westinghouse Electric Corp., Pittsburgh, PA, 1963.
32. W. Feduska and A. Isenberg, High-temperature solid oxide fuel cell – technical status, *J. Power Sources*, 1983, **10**(1), 89–102.
33. W. Fischer *et al.*, Hochtemperatur-Brennstoffzellen mit keramischen Elektrolyten zur Umsetzung billiger Brennstoffe. Teil II: Aufbau und Wirtschaftlichkeit von HTBZ-Aggregaten und deren Anwendungsmöglichkeiten, *Chem. Ing. Tech.*, 1972, **44**(11), 726–732.
34. W. Baukal *et al.*, Das elektrodenverbindungsmaterial, ein zentrales bauelement in hochtemperatur-brennstoffzellen – Teil II: Untersuchung ausgewählter hochleitfähig, *J. Power Sources*, 1976, **1**(2), 203–213.
35. F. Rohr, P. Hagenmuller and W. Van Gool, *Solid Electrolytes*, Material Science Series, Academic Press, New York, 1978, p. 431.
36. A. Isenberg, Recent advancements in solid electrolyte fuel cell technology, in *National Fuel Cell Seminar Abstracts*, Courtesy Associates, Washington, DC, 1982, pp. 154–156.
37. K. Foger and B. Godfrey, System demonstration program at Ceramic Fuel Cells Ltd in Australia, in *4th European SOFC Forum, Switzerland*, ed. A. J. McEvoy, 2000, pp. 167–173.
38. B. Steele, Appraisal of $Ce_{1-y}Gd_yO_{2-y/2}$ electrolytes for IT-SOFC operation at 500°C, *Solid State Ionics*, 2000, **129**(1), 95–110.
39. J. B. Goodenough, Oxide-ion electrolytes, *Ann. Rev. Mater. Res.*, 2003, **33**(1), 91–128.

40. B. Steele, Materials for IT-SOFC stacks: 35 years R&D: The inevitability of gradualness?, *Solid State Ionics*, 2000, **134**(1), 3–20.
41. L. Carrette, K. A. Friedrich and U. Stimming, Fuel cells: Principles, types, fuels, and applications, *ChemPhysChem*, 2000, **1**(4), 162–193.
42. C. Sun, R. Hui and J. Roller, Cathode materials for solid oxide fuel cells: A review, *J. Solid State Electrochem.*, 2010, **14**(7), 1125–1144.
43. R. M. Ormerod, *Solid oxide fuel cells, Chem. Soc. Rev.*, 2003, **32**(1), 17–28.
44. T. Horita *et al.*, Materials and reaction mechanisms at anode/electrolyte interfaces for SOFCs, *Solid State Ionics*, 2006, **177**(19–25), 1941–1948.
45. A. Rafique *et al.*, Significance enhancement in the conductivity of core shell nanocomposite electrolytes, *RSC Adv.*, 2015, **5**(105), 86322–86329.
46. Z. Yang *et al.*, Recent progress in doped carbon nanomaterials as effective cathode catalysts for fuel cell oxygen reduction reaction, *J. Power Sources*, 2013, **236**, 238–249.
47. A. Majid *et al.*, Ti_{Ga}–V_N complexes in GaN: A new prospect of carrier mediated ferromagnetism, *RSC Adv.*, 2015, 5(106), 87437–87444.
48. D. Sarantaridis and A. Atkinson, Redox cycling of Ni-based solid oxide fuel cell anodes: A review, *Fuel Cells*, 2007, **7**(3), 246–258.
49. S. P. Jiang and S. H. Chan, A review of anode materials development in solid oxide fuel cells, *J. Mater. Sci.*, 2004, **39**(14), 4405–4439.
50. N. Q. Minh, *Ceramic fuel cells, J. Am. Ceramic Soc.*, 1993, **76**(3), 563–588.
51. C. Ding *et al.*, Synthesis of NiO–Ce 0.9Gd 0.1O 1.95 nanocomposite powders for low-temperature solid oxide fuel cell anodes by co-precipitation, *Scripta Materialia*, 2009, **60**(4), 254–256.
52. H. Kim *et al.*, Cu–Ni cermet anodes for direct oxidation of methane in solid-oxide fuel cells, *J. Electrochem. Soc.*, 2002, **149**(3), A247–A250.
53. S. Tao and J. T. Irvine, Synthesis and Characterization of $(La_{0.75}Sr_{0.25})Cr_{0.5}Mn_{0.5}O_{3-\delta}$, a Redox-Stable, Efficient Perovskite Anode for SOFCs, *J. Electrochem. Soc.*, 2004, **151**(2), A252–A259.
54. G. S. Kim *et al.*, Highly active and stable $Sr_{0.92}Y_{0.08}Ti_{1-x}Ru_xO_{3-d}$ in dry reforming for hydrogen production, *Int. J. Hydrogen Energy*, 2019, **44**(1), 202–212.
55. W. Wang *et al.*, Core–shell structured $Li_{0.33}La_{0.56}TiO_3$ perovskite as a highly efficient and sulfur-tolerant anode for solid-oxide fuel cells, *J. Mater. Chem. A*, 2015, **3**(16), 8545–8551.
56. V. Vorontsov *et al.*, Synthesis and characterization of new ternary transition metal sulfide anodes for H_2S-powered solid oxide fuel cell, *J. Power Sources*, 2008, **183**(1), 76–83.
57. N. Ni *et al.*, Degradation of $(La_{0.6}Sr_{0.4})_{0.95}(Co_{0.2}Fe_{0.8})O_{3-\delta}$ Solid Oxide Fuel Cell Cathodes at the Nanometer Scale and below, *ACS Appl. Mater. Interfaces*, 2016, **8**(27), 17360–17370.
58. J. Cao *et al.*, An investigation of metal ion diffusion in ceria-based solid oxide fuel cell with barium-containing anode, *J. Am. Ceramic Soc.*, 2018, **101**(12), 5791–5800.
59. H. Uchida, H. Suzuki and M. Watanabe, High-performance electrode for medium-temperature solid oxide fuel cells effects of composition and microstructures on performance of ceria-based anodes, *J. Electrochem. Soc.*, 1998, **145**(2), 615–620.

60. J. Qu *et al.*, Stable direct-methane solid oxide fuel cells with calcium-oxide-modified nickel-based anodes operating at reduced temperatures, *Appl. Energy*, 2016, **164**, 563–571.
61. Y. Okawa and Y. Hirata, Sinterability, microstructures and electrical properties of Ni/Sm-doped ceria cermet processed with nanometer-sized particles, *J. Eur. Ceramic Soc.*, 2005, **25**(4), 473–480.
62. S. Ohara *et al.*, High performance electrodes for reduced temperature solid oxide fuel cells with doped lanthanum gallate electrolyte: I. Ni–SDC cermet anode, *J. Power Sources*, 2000, **86**(1–2), 455–458.
63. X. Yao *et al.*, Hydrothermally synthesized NiO-samarium doped ceria nano-composite as an anode material for intermediate-temperature solid oxide fuel cells, *Int. J. Hydrogen Energy*, 2017, **42**(34), 22192–22200.
64. X. Zhang *et al.*, Ni-SDC cermet anode for medium-temperature solid oxide fuel cell with lanthanum gallate electrolyte, *J. Power Sources*, 1999, **83**(1–2), 170–177.
65. H. Kan and H. Lee, Sn-doped Ni/YSZ anode catalysts with enhanced carbon deposition resistance for an intermediate temperature SOFC, *Appl. Catal., B*, 2010, **97**(1–2), 108–114.
66. Y. Leng *et al.*, Low-temperature SOFC with thin film GDC electrolyte prepared in situ by solid-state reaction, *Solid State Ionics*, 2004, **170**(1–2), 9–15.
67. Z. Gao, E. C. Miller and S. A. Barnett, A high power density intermediate-temperature solid oxide fuel cell with thin $(La_{0.9}Sr_{0.1})_{0.98}(Ga_{0.8}Mg_{0.2})O_{3-\delta}$ electrolyte and nano-scale anode, *Adv. Funct. Mater.*, 2014, **24**(36), 5703–5709.
68. J. Sun *et al.*, The performance of cathode material of perovskite-like La–Ni–M–O for low temperature solid oxide fuel cell, *Int. J. Energy Res.*, 2011, **35**(12), 1032–1036.
69. C. Xia and M. Liu, Low-temperature SOFCs based on $Gd_{0.1}Ce_{0.9}O_{1.95}$ fabricated by dry pressing, *Solid State Ionics*, 2001, **144**(3–4), 249–255.
70. T. Fukui *et al.*, Morphology control of Ni–YSZ cermet anode for lower temperature operation of SOFCs, *J. Power Sources*, 2004, **125**(1), 17–21.
71. S. Zha, W. Rauch and M. Liu, Ni-$Ce_{0.9}Gd_{0.1}O_{1.95}$ anode for GDC electrolyte-based low-temperature SOFCs, *Solid State Ionics*, 2004, **166**(3–4), 241–250.
72. T. Suzuki *et al.*, Performance of Ni-based Anode-Supported SOFC s with Doped Ceria Electrolyte at Low Temperatures Between 294 and 542 °C, *Int. J. Appl. Ceramic Technol.*, 2015, **12**(2), 358–362.
73. S. Tao and J. T. Irvine, A redox-stable efficient anode for solid-oxide fuel cells, *Nat. Mater.*, 2003, **2**(5), 320.
74. S. Tao and J. T. Irvine, Discovery and characterization of novel oxide anodes for solid oxide fuel cells, *Chem. Rec.*, 2004, **4**(2), 83–95.
75. C. Reich, A. Kaiser and J. Irvine, Niobia based rutile materials as SOFC anodes, *Fuel Cells*, 2001, **1**(3–4), 249–255.
76. S. Primdahl *et al.*, Sr-doped $LaCrO_3$ anode for solid oxide fuel cells, *J. Electrochem. Soc.*, 2001, **148**(1), A74–A81.
77. G. Pudmich *et al.*, Chromite/titanate based perovskites for application as anodes in solid oxide fuel cells. *Solid State Ionics*, 2000, **135**(1–4), 433–438.

78. P. Jamieson, S. Abrahams and J. Bernstein, Ferroelectric tungsten bronze-type crystal structures. II. Barium sodium niobate $Ba_{(4+x)}Na_{(2-2x)}Nb_{10}O_{30}$, *J. Chem. Phys.*, 1969, **50**(10), 4352–4363.
79. P. Slater and J. Irvine, Niobium based tetragonal tungsten bronzes as potential anodes for solid oxide fuel cells: Synthesis and electrical characterisation, *Solid State Ionics*, 1999, **120**(1–4), 125–134.
80. A. Kaiser *et al.*, Tetragonal tungsten bronze type phases $(Sr_{1-x}Ba_x)_{0.6}Ti_{0.2}Nb_{0.8}O_{3-\delta}$: Material characterisation and performance as SOFC anodes, *Solid State Ionics*, 2000, **135**(1–4), 519–524.
81. M. Subramanian, G. Aravamudan and G. S. Rao, Oxide pyrochlores – a review, *Prog. Solid State Chem.*, 1983, **15**(2), 55–143.
82. S. Kramer, M. Spears and H. Tuller, Conduction in titanate pyrochlores: Role of dopants, *Solid State Ionics*, 1994, **72**, 59–66.
83. O. Porat *et al.*, Modelling and characterization of mixed ionic-electronic conduction in $Gd_2(Ti_{1-x}Mn_x)_2O_{7+y}$, *Solid State Ionics*, 1996, **86**, 285–288.
84. K. Sato, K. Sakamaki and Y. Inoue, Electrocatalytic activity of $Co_{0.5}Ni_{0.5}O$ particles for the anodic oxidation of methane, *Mater. Trans. JIM*, 2000, **41**(12), 1621–1625.
85. A. Weber *et al.*, Oxidation of H_2, CO and methane in SOFCs with Ni/YSZ-cermet anodes, *Solid State Ionics*, 2002, **152**, 543–550.
86. A. L. Dicks, Hydrogen generation from natural gas for the fuel cell systems of tomorrow, *J. Power Sources*, 1996, **61**(1–2), 113–124.
87. J.-H. Koh *et al.*, Carbon deposition and cell performance of Ni-YSZ anode support SOFC with methane fuel. *Solid State Ionics*, 2002, **149**(3–4), 157–166.
88. M. L. Faro *et al.*, Investigation of Ni-based alloy/CGO electro-catalysts as protective layer for a solid oxide fuel cell anode fed with ethanol, *J. Appl. Electrochem.*, 2015, **45**(7), 647–656.
89. U. Stimming, S. Singhal and H. Tagawa. *Proceedings of the Fifth International Symposium on Solid Oxide Fuel Cells (SOFC-V).* The Electrochemical Society, Pennington, NJ, 1997.
90. W. Dollard, Solid oxide fuel cell developments at Westinghouse, *J. Power Sources*, 1992, **37**(1–2), 133–139.
91. P. Vernoux, J. Guindet and M. Kleitz, Gradual internal methane reforming in intermediate-temperature solid-oxide fuel cells, *J. Electrochem. Soc.*, 1998, **145**(10), 3487–3492.
92. P. Vernoux *et al.*, Alternative anode material for gradual methane reforming in solid oxide fuel cells, *Solid State Ionics*, 2000, **135**(1–4), 425–431.
93. J. Y. Won *et al.*, Glycerol as a bioderived sustainable fuel for solid-oxide fuel cells with internal reforming, *ChemSusChem*, 2009, **2**(11), 1028–1031.
94. A. Da Silva *et al.*, Effect of the type of ceria dopant on the performance of Ni/CeO_2 SOFC anode for ethanol internal reforming, *Appl. Catal., B*, 2017, **206**, 626–641.
95. B. L. Augusto *et al.*, Nickel/gadolinium-doped ceria anode for direct ethanol solid oxide fuel cell, *Int. J. Hydrogen Energy*, 2014, **39**(21), 11196–11209.

96. M. Liu *et al.*, Direct liquid methanol-fueled solid oxide fuel cell, *J. Power Sources*, 2008, **185**(1), 188–192.
97. X. Meng *et al.*, Low-temperature ceria-electrolyte solid oxide fuel cells for efficient methanol oxidation, *J. Power Sources*, 2011, **196**(23), 9961–9964.
98. H. Sumi *et al.*, Effect of carbon deposition by carbon monoxide disproportionation on electrochemical characteristics at low temperature operation for solid oxide fuel cells, *J. Power Sources*, 2011, **196**(10), 4451–4457.
99. T. Chen *et al.*, Evaluation of carbon deposition behavior on the nickel/yttrium-stabilized zirconia anode-supported fuel cell fueled with simulated syngas, *J. Power Sources*, 2011, **196**(5), 2461–2468.
100. T. Takeguchi *et al.*, Study on steam reforming of CH_4 and C_2hydrocarbons and carbon deposition on Ni-YSZ cermets, *J. Power Sources*, 2002, **112**(2), 588–595.
101. C. Chun, J. Mumford and T. A. Ramanarayanan, Carbon-induced corrosion of nickel anode, *J. Electrochem. Soc.*, 2000, **147**(10), 3680–3686.
102. X.-Y. Zhao *et al.*, Studies on the carbon reactions in the anode of deposited carbon fuel cells, *J. Power Sources*, 2008, **185**(1), 104–111.
103. J. Qu *et al.*, Ethylene glycol as a new sustainable fuel for solid oxide fuel cells with conventional nickel-based anodes. *Appl. Energy*, 2015, **148**, 1–9.
104. Y. Chen *et al.*, Coke-free direct formic acid solid oxide fuel cells operating at intermediate temperatures, *J. Power Sources*, 2012, **220**, 147–152.
105. H. Jeong *et al.*, Bimetallic nickel/ruthenium catalysts synthesized by atomic layer deposition for low-temperature direct methanol solid oxide fuel cells, *ACS Appl. Mater. Interfaces*, 2016, **8**(44), 30090–30098.
106. B. Huang *et al.*, Preparation and performance characterization of the Fe–Ni/ScSZ cermet anode for oxidation of ethanol fuel in SOFCs, *J. Power Sources*, 2007, **167**(2), 288–294.
107. G. Ding *et al.*, Carbon-resistant $Ni_{1-x}Co_x$-$Ce_{0.8}Sm_{0.2}O_{1.9}$ anode for solid oxide fuel cells fed with methanol, *Catal. Today*, 2017, **298**, 250–257.
108. B. Farrell and S. Linic, Direct electrochemical oxidation of ethanol on SOFCs: Improved carbon tolerance of Ni anode by alloying, *Appl. Catal., B*, 2016, **183**, 386–393.
109. C. Chun and T. A. Ramanarayanan, Mechanism and control of carbon deposition on high temperature alloys, *J. Electrochem. Soc.*, 2007, **154**(9), C465–C471.
110. Y. Nabae *et al.*, Catalytic behavior of Pd–Ni/composite anode for direct oxidation of methane in SOFCs, *J. Electrochem. Soc.*, 2006, **153**(1), A140–A145.
111. F. Wang *et al.*, Aluminum oxide as a dual-functional modifier of Ni-based anodes of solid oxide fuel cells for operation on simulated biogas. *J. Power Sources*, 2014, **268**, 787–793.
112. H. Chen *et al.*, H2S poisoning effect and ways to improve sulfur tolerance of nickel cermet anodes operating on carbonaceous fuels, *Appl. Energy*, 2016, **179**, 765–777.
113. I. Gavrielatos, V. Drakopoulos and S. G. Neophytides, Carbon tolerant Ni–Au SOFC electrodes operating under internal steam reforming conditions, *J. Catal.*, 2008, **259**(1), 75–84.

114. J. Qu *et al.*, One-pot synthesis of silver-modified sulfur-tolerant anode for SOFCs with an expanded operation temperature window, *AIChE J.*, 2017, **63**(10), 4287–4295.
115. Z. Wang *et al.*, Catalytic modification of Ni–Sm-doped ceria anodes with copper for direct utilization of dry methane in low-temperature solid oxide fuel cells, *J. Power Sources*, 2008, **179**(2), 541–546.
116. A. Babaei *et al.*, Performance and carbon deposition over Pd nanoparticle catalyst promoted Ni/GDC anode of SOFCs in methane, methanol and ethanol fuels, *Int. J. Hydrogen Energy*, 2012, **37**(20), 15301–15310.
117. B. Hua *et al.*, Grafting doped manganite into nickel anode enables efficient and durable energy conversions in biogas solid oxide fuel cells, *Appl. Catal., B*, 2017, **200**, 174–181.
118. M. Li *et al.*, Enhancing sulfur tolerance of Ni-based cermet anodes of solid oxide fuel cells by ytterbium-doped barium cerate infiltration, *ACS Appl. Mater. Interfaces*, 2016, **8**(16), 10293–10301.
119. O. Kwon *et al.*, In situ surface modification of Ni-YSZ with $BaZrO_3$ for enhancing the sulfur tolerance of Ni-YSZ anode, *J. Electrochem. Soc.*, 2016, **163**(9), F1055–F1058.
120. M. Cimenti, V. Alzate-Restrepo and J. M. Hill, Direct utilization of methanol on impregnated Ni/YSZ and Ni–$Zr_{0.35}Ce_{0.65}O_2$/YSZ anodes for solid oxide fuel cells, *J. Power Sources*, 2010, **195**(13), 4002–4012.
121. P. Li *et al.*, Improved activity and stability of Ni-$Ce_{0.8}Sm_{0.2}O_{1.9}$ anode for solid oxide fuel cells fed with methanol through addition of molybdenum, *J. Power Sources*, 2016, **320**, 251–256.
122. W. Wang *et al.*, Nickel-based anode with water storage capability to mitigate carbon deposition for direct ethanol solid oxide fuel cells, *ChemSusChem*, 2014, 7(6), 1719–1728.
123. B. He, D. Ding and C. Xia, Ni–LnO_x (Ln = La, Ce, Pr, Nd, Sm, Eu, and Gd) cermet anodes for intermediate-temperature solid oxide fuel cells, *J. Power Sources*, 2010, **195**(5), 1359–1364.
124. H. Zhu *et al.*, A new nickel–ceria composite for direct-methane solid oxide fuel cells, *Int. J. Hydrogen Energy*, 2013, **38**(9), 3741–3749.
125. L. Yang *et al.*, Promotion of water-mediated carbon removal by nanostructured barium oxide/nickel interfaces in solid oxide fuel cells, *Nat. Commun.*, 2011, **2**, 357.
126. F. Wang *et al.*, Enhanced sulfur tolerance of nickel-based anodes for oxygen-ion conducting solid oxide fuel cells by incorporating a secondary water storing phase, *Environ. Sci. Technol.*, 2014, **48**(20), 12427–12434.
127. E. Fabbri, D. Pergolesi and E. Traversa, Materials challenges toward proton-conducting oxide fuel cells: A critical review, *Chem. Soc. Rev.*, 2010, **39**(11), 4355–4369.
128. L. Malavasi, C. A. Fisher and M. S. Islam, Oxide-ion and proton conducting electrolyte materials for clean energy applications: Structural and mechanistic features, *Chem. Soc. Rev.*, 2010, **39**(11), 4370–4387.
129. W. Wang *et al.*, Enhanced electrochemical performance, water storage capability and coking resistance of a Ni+$BaZr_{0.1}Ce_{0.7}Y_{0.1}Yb_{0.1}O_{3-\delta}$ anode

for solid oxide fuel cells operating on ethanol, *Chem. Eng. Sci.*, 2015, **126**, 22–31.
130. O. Costa-Nunes, R. J. Gorte and J. M. Vohs, Comparison of the performance of Cu–CeO_2–YSZ and Ni–YSZ composite SOFC anodes with H_2, CO, and syngas, *J. Power Sources*, 2005, **141**(2), 241–249.
131. R. J. Gorte *et al.*, Anodes for direct oxidation of dry hydrocarbons in a solid-oxide fuel cell, *Adv. Mater.*, 2000, **12**(19), 1465–1469.
132. H. Kim, J. M. Vohs and R. J. Gorte, Direct oxidation of sulfur-containing fuels in a solid oxide fuel cell, *Chem. Commun.*, 2001, (22), 2334–2335.
133. C. Lu *et al.*, A comparison of Cu-ceria-SDC and Au-ceria-SDC composites for SOFC anodes, *J. Electrochem. Soc.*, 2003, **150**(10), A1357–A1359.
134. E. Tsipis, V. Kharton and J. Frade, Mixed conducting components of solid oxide fuel cell anodes, *J. Eur. Ceramic Soc.*, 2005, **25**(12), 2623–2626.
135. S. An *et al.*, Characterization of Cu–CeO_2 direct hydrocarbon anodes in a solid oxide fuel cell with lanthanum gallate electrolyte. *Solid State Ionics*, 2004, **175**(1–4), 135–138.
136. S. Omar, E. D. Wachsman and J. C. Nino, Higher ionic conductive ceria-based electrolytes for solid oxide fuel cells, *Appl. Phys. Lett.*, 2007, **91**(14), 144106.
137. S. Omar, E. D. Wachsman and J. C. Nino, A co-doping approach towards enhanced ionic conductivity in fluorite-based electrolytes, *Solid State Ionics*, 2006, **177**(35–36), 3199–3203.
138. E. P. Murray, T. Tsai and S. A. Barnett, A direct-methane fuel cell with a ceria-based anode, *Nature*, 1999, **400**(6745), 649.
139. L. G. Tejuca and J. L. Fierro, *Properties and Applications of Perovskite-type Oxides*, CRC Press, Boca Raton, FL, 1992.
140. H. Kurokawa *et al.*, Ceria nanocoating for sulfur tolerant Ni-based anodes of solid oxide fuel cells, *Electrochem. Solid State Lett.*, 2007, **10**(9), B135–B138.
141. P. Blennow *et al.*, Electrochemical characterization and redox behavior of Nb-doped $SrTiO_3$, *Solid State Ionics*, 2009, **180**(1), 63–70.
142. S. P. Jiang *et al.*, GDC-impregnated $(La_{0.75}Sr_{0.25})(Cr_{0.5}Mn_{0.5})O_3$ anodes for direct utilization of methane in solid oxide fuel cells, *J. Electrochem. Soc.*, 2006, **153**(5), A850–A856.
143. X. Chen *et al.*, High-performance cathode-supported SOFC with perovskite anode operating in weakly humidified hydrogen and methane, *Fuel Cells Bull.*, 2007, **2007**(6), 12–16.
144. C. Wang *et al.*, Fabrication and performance of thin-film YSZ solid oxide fuel cells, *J. Electrochem. Soc.*, 2001, **148**(8), A864–A868.
145. M. Colomer *et al.*, Microstructure, electrical properties and phase equilibria relationships in the $ZrO_2Y_2O_3TiO_2$ system: The subsolidus isothermal section at 1500 °C, *Mater. Sci. Eng. A*, 1997, **229**(1–2), 114–122.
146. A. Sauvet and J. Irvine, Catalytic activity for steam methane reforming and physical characterisation of $La_{1-x}Sr_xCr_{1-y}Ni_yO_{3-\delta}$, *Solid State Ionics*, 2004, **167**(1–2), 1–8.

147. A.-L. Sauvet *et al.*, Surface properties and physicochemical characterizations of a new type of anode material, $La_{1-x}Sr_xCr_{1-y}Ru_yO_{3-\delta}$, for a solid oxide fuel cell under methane at intermediate temperature, *J. Catal.*, 2002, **209**(1), 25–34.
148. M. Weston and I. Metcalfe, $La_{0.6}Sr_{0.4}Co_{0.2}Fe_{0.8}O_3$ as an anode for direct methane activation in SOFCS, *Solid State Ionics*, 1998, **113**, 247–251.
149. D. Skarmoutsos *et al.*, Physical characterization of $Y_{0.25}Zr_{0.60}Ti_{0.15}O_{2-x}$ and its performance as a Ni/$Y_{0.25}Zr_{0.60}Ti_{0.15}O_{2-x}$ anode cermet in an SOFC, *Solid State Ionics*, 2004, **170**(3–4), 153–158.
150. T. Tsai and S. A. Barnett, Effect of mixed-conducting interfacial layers on solid oxide fuel cell anode performance, *J. Electrochem. Soc.*, 1998, **145**(5), 1696–1701.
151. M. Mogensen and S. Skaarup, Kinetic and geometric aspects of solid oxide fuel cell electrodes, *Solid State Ionics*, 1996, **86**, 1151–1160.
152. M. Brown, S. Primdahl and M. Mogensen, Structure/performance relations for Ni/yttria-stabilized zirconia anodes for solid oxide fuel cells, *J. Electrochem. Soc.*, 2000, **147**(2), 475–485.
153. S. Primdahl and M. Mogensen, Gas conversion impedance: A test geometry effect in characterization of solid oxide fuel cell anodes, *J. Electrochem. Soc.*, 1998, **145**(7), 2431–2438.
154. Y. Wu *et al.*, Morphology and catalytic performance of flake-shaped NiO-yttria-stabilized zirconia (YSZ) particles with nanocrystalline YSZ grains, *Ind. Eng. Chem. Res.*, 2012, **51**(18), 6387–6394.
155. Y. Wu *et al.*, Synthesis of flake-shaped NiO–YSZ particles for high-porosity anode of solid oxide fuel cell, *J. Am. Ceramic Soc.*, 2011, **94**(11), 3666–3670.
156. A. Atkinson *et al.*, Advanced anodes for high-temperature fuel cells, in *Materials for Sustainable Energy: A Collection of Peer-Reviewed Research and Review Articles from Nature Publishing Group*, World Scientific, Singapore, 2011, pp. 213–223.
157. A. S. Arico *et al.*, Nanostructured materials for advanced energy conversion and storage devices, in *Materials for Sustainable Energy: A Collection of Peer-Reviewed Research and Review Articles from Nature Publishing Group*, World Scientific, Singapore, 2011, pp. 148–159.
158. J. Canales-Vázquez, S. Tao and J. T. Irvine, Electrical properties in $La_2Sr_4Ti_6O_{19-\delta}$: A potential anode for high temperature fuel cells, *Solid State Ionics*, 2003, **159**(1–2), 159–165.
159. O. A. Marina and L. R. Pederson, Novel ceramic anodes for SOFCs tolerant to oxygen, carbon and sulfur, in *The Fifth European Solid Oxide Fuel Cell Forum*, ed. J. Huijismans, European Fuel Cell Forum, Lucerne, 2002, pp. 481–489.
160. S. Mekhilef, R. Saidur and A. Safari, Comparative study of different fuel cell technologies, *Renewable Sustainable Energy Rev.*, 2012, **16**(1), 981–989.

Subject Index